3rd International Conference on Disaster Management

IOP Conference Series: Earth and Environmental Science
Volume 1173

Online
29 September – 1 October 2022

ISBN: 978-1-7138-7290-0
ISSN: 1755-1307

Printed from e-media with permission by:

Curran Associates, Inc.
57 Morehouse Lane
Red Hook, NY 12571

Some format issues inherent in the e-media version may also appear in this print version.

Institute of Physics
Dirac House, Temple Back
Bristol BS1 6BE UK

Phone: 44 1 17 929 7481
Fax: 44 1 17 920 0979

techtracking@iop.org

Additional copies of this publication are available from:

Curran Associates, Inc.
57 Morehouse Lane
Red Hook, NY 12571 USA
Phone: 845-758-0400
Fax: 845-758-2633
Email: curran@proceedings.com
Web: www.proceedings.com

TABLE OF CONTENTS

STRUCTURES AND RETROFITTING

TSUNAMI AND SEISMIC ENGINEERING

GEOTECHNICAL ENGINEERING

WATER AND RESOURCES ENGINEERING

TRANSPORTATION AND LOGISTICS

UNDERSTANDING DISASTER MANAGEMENT

STRENGTHENING SUSTAINABILITY DEVELOPMENT

ENHANCING FRAMEWORK FOR SUSTAINABILITY

IMPROVING LESSON LEARNT IN DISASTER MANAGEMENT

Preface: The 3rd International Conference on Disaster Management (ICDM 2022)

On behalf of the conference committee, I would like to extend my warmest welcome to all of the participants to our International Conference on Disaster Management (ICDM 2022). The theme for the 3rd International Conference on Disaster Management (ICDM 2022) is Disaster Management During and After Pandemic. The aim of the conference is to delegates from around the world can share excellent opportunities and new ideas and findings for both developed and developing countries to implement policies to build disaster-resilient societies and to create new directions through integrating various fields of disaster mitigation and management.

The 3rd International Conference on Disaster Management – ICDM 2022 has been hosted by Civil Engineering Department and Disaster Study Centre, Andalas University. This conference is attended by presenters from the United Kingdom, India, Indonesia, Malaysia, Turkiye, and Palestine.

The conference was held over two days, from 30 September to 1 October 2022 by online method conference and was hosted from Convention Hall of Andalas University. The conference included two plenary sessions and four parallel sessions in which there were invaluable presentations by both national and international presenters.

I would like to thank our co-host, the Universitas Indonesia, Universitas Pertahanan Republik Indonesia, Universitas Kristen Maranatha, Universitas Indo Global Mandiri, Universitas Jenderal Soedirman, Universitas Teknokrat Indonesia, Institut Teknologi Sumatera and Universitas Metamedia for their support that have allowed us to

IOP Conf. Series: Earth and Environmental Science 1173 (2023) 011001 doi:10.1088/1755-1315/1173/1/011001

organize this event. We received a number of 137 extended abstracts for this event, and unfortunately, we could only accept a number of 79 selected ones to be published on our upcoming IOP conference series in Earth and Environmental Sciences. Therefore, I would like to congratulate all authors that have made it this far.

I would like to take this opportunity to convey my appreciation to the organizing committee for their untiring efforts to manage this conference, to the reviewers for reviewing paper and making valuable suggestion to ensure the published paper have a very best quality. I also thank all the authors for their time and effort in contributing to ICDM 2022. Thanks to all these efforts, we firmly believe that the final conference program consisted of contributions of the highest quality.

Prof. Elsa Eka Putri, S.T., M.Sc (Eng), Ph.D,
Chairman of 3rd ICDM 2022

IOP Conf. Series: Earth and Environmental Science 1173 (2023) 011002 doi:10.1088/1755-1315/1173/1/011002

Peer Review Statement

All papers published in this volume have been reviewed through processes administered by the Editors. Reviews were conducted by expert referees to the professional and scientific standards expected of a proceedings journal published by IOP Publishing.

- **Type of peer review:** Double Anonymous
- **Conference submission management system:** Morressier
- **Number of submissions received:** 158
- **Number of submissions sent for review:** 158
- **Number of submissions accepted:** 137
- **Acceptance Rate (Submissions Accepted / Submissions Received × 100):** 86.7
- **Average number of reviews per paper:** 1.92
- **Total number of reviewers involved:** 22
- **Contact person for queries:**
  Name: Elsa Eka Putri
  Email: elsaeka@eng.unand.ac.id
  Affiliation: Andalas University

IOP Conf. Series: Earth and Environmental Science 1173 (2023) 012001 doi:10.1088/1755-1315/1173/1/012001

Flexural capacity improvement of the bamboo reinforced concrete beam with perpendicular pivot hooks

A M Ummati[1,3], Michael[1], I P E Sarassantika[2,3], G T Fanna[1], S Syuhada[1] and A P Nasution[1]

[1] Department of Civil Engineering, Institut Teknologi Sumatera, Lampung Selatan, Lampung 35365, Indonesia
[2] Department of Civil Engineering, Universitas Warmadewa, Denpasar, Bali, 80235, Indonesia
[2] Department of Civil Engineering, National Central University, Zhongli, Taoyuan, 32001, Taiwan

E-mail: michael@si.itera.ac.id

Abstract. Bamboo as a natural resource is very common used in constructio, such as columns and beams, also can be used as architectural elements. The form of bamboo as a reinforcement is still under identification by many scholars to find an optimum shape and form of bamboo, thus it may be applied as concrete reinforcement since bamboo can improve concrete's flexural strength. An experimental test conduct in order to identify the role of perpendicular pivot end in bamboo reinforced concrete beams by observing the flexural capacity improvement. Four concrete beams with the dimension of 15 cm x 25 cm x 170 cm were made with 2 cm of square section of the lateral bamboo reinforcement and $\emptyset 8 - 100$ steel rebar as the shear reinforcement to prohibit the shear failure. The beams were loaded under three-point bending with the point loading in the middle of the span. Based on the experimental result, the beams which equipped by the perpendicular pivot end could improve the load capacity up to 33% higher than the beam without the pivot end. This is proofed that the pivot can improve the flexural capacity of the bamboo reinforced beam and considerable to develop in the bamboo reinforcement beam.

1. Introduction

The study of bamboo to replace steels as reinforcement in concrete structure is getting popular based on several aspects, such as low cost (compared to steel) [1] and availability of bamboo as one of the natural resources in Indonesia [2]. The use of bamboo is not literally a new innovation in human life. It has been used traditionally, especially in Indonesia, as a part of structure of traditional houses, such Sundanese traditional houses (built using both wood and bamboo). Nowadays, bamboo is a common material to construct benches, tables, architectural components of buildings, and structural components (column and beams). As structural components, bamboo must fulfill the requirements of flexural and shear capacity. The use of bamboo as reinforcement in concrete can improve the strength of the structure [3], which can be considered to be one of replacement of steel. However, as an organic natural resource, bamboo has weakness compared to the properties of steel, particularly in a perspective of concrete structure [3].

Many studies have been conducted to identify the influence of bamboo reinforcement in concrete structure. The diversity of bamboo species causes various influence, in term of its use as concrete structure reinforcement. The natural characteristic of a plant (in this case bamboo) has specific morphology in every trunk of bamboo that grow from one single seed. It means engineers need to adjust the form and dimension of bamboo before using it, particularly as structural component. The properties of the bamboo itself must be crucial to predict the whole structure's strength and characteristics. Thus, in order to achieve a uniform property of bamboo as the reinforcement component, which at last to get an optimum concrete, the shape of bamboo must be justified. Bamboo can be cut and formed to adjust its utilization in construction component. Particularly for enforcement in concrete, bamboo has been identified in many shapes to get the optimum strength, which finally the use of bamboo as an enforcement for concrete can be approved in construction sector. A study showed an influence of using different cutting and type bamboo enforcement in concrete. The specimens were evaluated under three-point bending test and compared to steel reinforced concrete. Based on the test of four shapes of bamboo reinforced concrete, the different shape of bamboo resulted various stress, strain and modulus of elasticity. However, the pattern of stress-strain for all type of cutting is not significantly different. Furthermore, the presence of bark on several type of cutting can result higher tensile strength as well as the presence of node can improve 50% of strength [4].

Flexural characteristic of bamboo reinforced concrete can be, somehow, not significantly improved compared to steel reinforced concrete. The study analyzed Petung Bamboo with several variation of notches and cross-sectional area for the reinforcement. It showed that the variation resulted un-significantly difference of its flexural strength, however it is quite poor compared to steel reinforced one [5]. In addition to notches applied on bamboo reinforcement, another study showed that U-Shaped notch strip can increase the flexural capacity compared to bamboo reinforcement concrete without notch [6]. The variation of forms as the reinforcement can also be applied through its hose-clamp distance. It impacted an increasing bond-stress and flexural capacity with 4%-36% area ratio of bamboo reinforcement compared to steel reinforced concrete with 0.89% steel area ratio. The presence of clams on bamboo reinforced concrete increased 35.71% of capacity [7]. Through several studies on the impact of variation of bamboo as concrete reinforcement, this paper will discuss the flexural capacity on bamboo reinforced concrete with perpendicular pivot end.

2. Methodology

An experimental test conduct in order to identify the flexural capacity improvement of the concrete beam which reinforced by the bamboo reinforcement with adding the perpendicular pivot in both ends. This pivot has the same role as hook in the rebar reinforcement which commonly used in construction field, with the aim of hook is to increase the bonding of the rebar and concrete. Such as hook, these perpendicular pivots also improve the bond of the bamboo and concrete with increasing the capability in grip. Considering the main issue of involving the organic material such as bamboo, as the reinforcement material combining with concrete as the inorganic material, the slip due to material shrinkage in reducing its volume due to the water loss become the main issue. This pivot end provides the transverse reinforcement to prohibit the material slip and keep the concrete and bamboo remain composited. The beam specimens and the details of the perpendicular pivot ends in the bamboo bar inside the beam described as the figures 1 and 2.

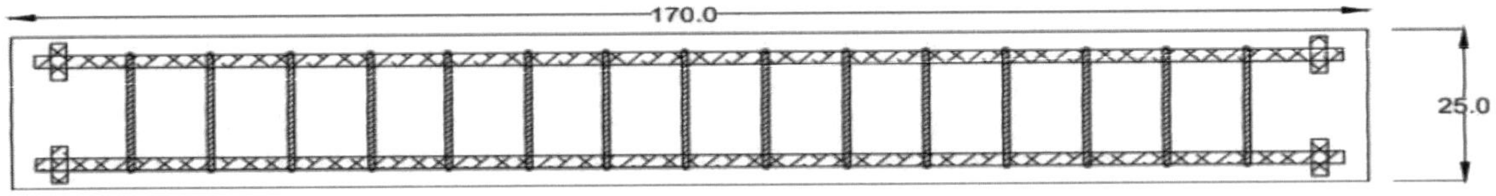

Figure 1. Bamboo reinforced beam with the perpendicular pivot ends.

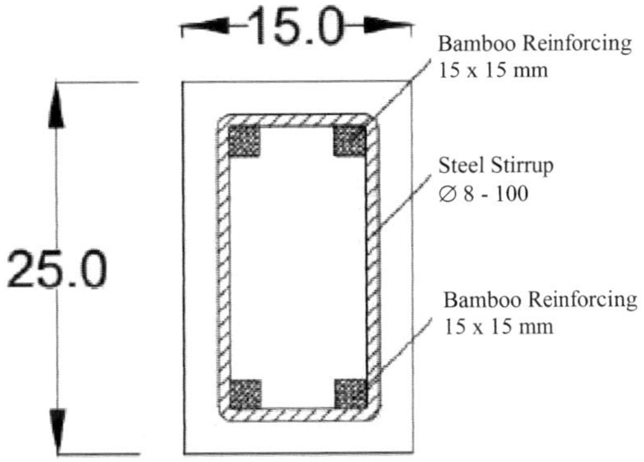

Figure 2. Cross sectional view.

The beams dimension is 15 cm x 25 cm x 170 cm made by concrete which reinforced by 4 bars of 15 mm square section of bamboo bars as longitudinal reinforcement and $\emptyset 8 - 100$ steel bar as the shear reinforcement. The concern of this experimental test is focusing on the flexural failure, thus, the rebar used as the transverse reinforcement due to the shear failure in beam is not permissible. There are 4 beams in total, 2 beams are the beam without the pivot and 2 beams with the pivot ends on their reinforcement. This beam variations arranged to observe the effect of the pivot in improving the flexural capacity of the beams. The perpendicular pivot both ends of the bamboo bar in details shown as the Figure **3**.

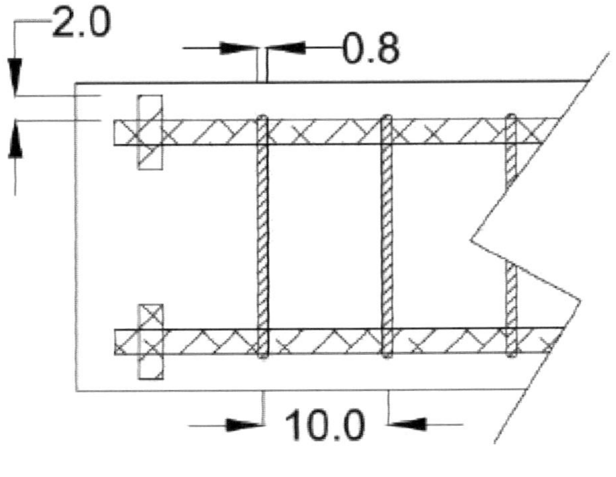

Figure 3. Perpendicular pivot details.

3. Analysis and result

The compressive strength of the concrete tested by the cylinder testing to get the $f'c = 25$ MPa. The reinforcement consists of 2 materials: bamboo as the longitudinal reinforcement with the $fy(b) = 193$ MPa, and steel as the transverse reinforcement with $fy(s) = 400$ MPa, it is known that the yield strength of bamboo is almost half of the yield strength of the steel rebar.

The beams loaded by the concentric loading in the middle span, and supported by the roller and hinge, Figure 4 described the experimental testing by three-point bending test.

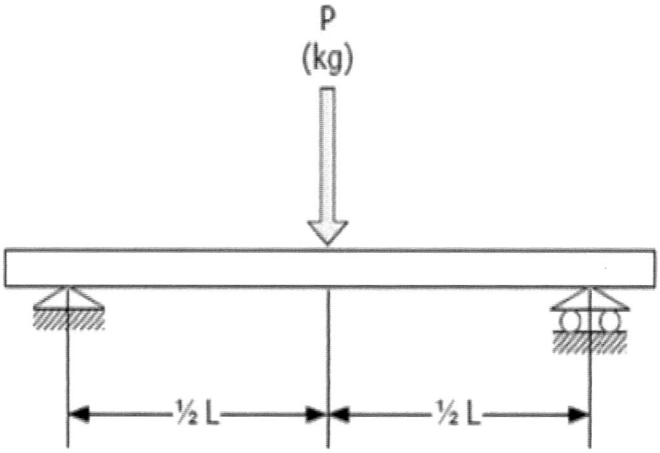

Figure 4. Static loading in the middle of the span

The beams loaded by three-point bending scheme as in Figure 4 to observe the behavior of the beam under the load control with increase periodically. The load-displacement relation identified by this experiment and represent the behavior of the bamboo reinforcement which equipped by the perpendicular pivot end. The three-point bending test result that was done by 4 beams specimens shown on table Table 1. Specimens undernamed control 1 and control 2 are 2 beams without the perpendicular pivot ends in their bamboo reinforcements, and Pivot End 1 and Pivot End 2 are 2 beams with the perpendicular pivot ends in their bamboo reinforcements.

Table 1. Loading result.

Specimens	Load (kg)	Average (kg)	Difference (%)
Control 1	1778	2105.5	33%
Control 2	2433		
Pivot End 1	2981	2806	
Pivot End 2	2631		

The load which recorded as in table above mentioned that the specimen control 1 and control 2 can bear the loading until their fracture up to 1778 kg and 2433 kg, respectively. Specimen with the pivot end 1 and pivot end 2 are able to bear the loading until they failure up to 2981 kg and 2631 kg, respectively. Furthermore, comparing the specimen of beams without and with the pivot end in their longitudinal reinforcement shows that the beams which equipped by the pivot end can sustain more loading than the beam without the pivot end, it was proofed by the average loading which can bear of the specimens' control is 2105.5 kg and 2806 kg for the beams with the pivot end, it was 33% larger than the beams without the pivot end. This result shows that the pivot end affecting the beam especially for the longitudinal bamboo reinforcement to improve the capacity of loading which can they bear.

The second observation was about the displacement. The same identity as in Table 1, the beams which labelled as the Control 1 and Control 2 are the beams without the pivot end, in the other hand, the beams which labelled by the Pivot End 1 and Pivot End 2 are the beams which equipped by the pivot end for the lateral bamboo reinforcement. The displacement record listed on

Table *2*.

Table 2. Displacement result.

Specimens	Displacement (mm)	Average (mm)	Difference (%)
Control 1	32	46	29%
Control 2	60		
Pivot End 1	40	32.5	
Pivot End 2	25		

Table *2* shows the displacement record based on the experimental test, which shown that the beams without the pivot end of control 1 and control 2 are able to displaced 32 mm and 60 mm, respectively until the beam were failure. Meanwhile, the beams which equipped by the pivot end were able to displaced 40 mm and 25 mm for the specimens' pivot end 1 and 2, respectively. Overall, for the beams without the pivot end averagely could displaced 46 mm and 32.5 mm for the beam with the pivot ends, that was 29% lower than the beams without the pivot ends.

4. Conclusion

The experimental test that was done for these four beams reflect the behavior of the bamboo reinforced concrete beams if there are the pivot end equipped the lateral reinforcement to stop the over slipped which is the main issue of the concrete beams with the bamboo material as the reinforcements. Based on the experimental test that was done by three-point bending loading, shows that the pivot end which added in the end of the reinforcement may increase the capacity of the beams to sustain the loading to improving their flexural capacity. Regarding with the ductility issue, the beam which equipped by the pivot end is more ductile than the beam without the pivot end, this is due to the pivot will suddenly break right after they reach the capacity limit. Overall, the perpendicular pivot end in the lateral reinforcement may help the bamboo reinforced beams to improve their flexural capacity.

Acknowledgement

The acknowledgements toward LPPM Institut Teknologi Sumatera which provide the research grants year 2022 No.: B/763ag/IT9.C1/PT.01.03/2022 in order to support this experimental research, Department of Technology Infrastructure and Regional Planning for the publication. This scientific research is the collaboration with Department of Civil Engineering National Central University, Taiwan, and Universitas Warmadewa as a research partner.

References

[1] Putra D, Sedana I W and Santika K B 2007 Flexural capacity of bamboo reinforced concrete panel *J. Ilm. Tek. Sip.* **11** 45-54

[2] Putra I N R D, Sinarta I N and Bagiarta I K Y 2020 Analysis of bamboo structure on entry building green school ubud *U KaRsT : J. of Civ. Eng. Res. at the Univ. of Kadiri* **4** 39-53

[3] Mali P R and Datta D 2020 Experimental evaluation of bamboo reinforced concrete beams *J. of Build. Eng.* **28** 1-25

[4] Al-Fasih M Y, Hamzah S, Ahmad Y, Ibrahim I S and Ariffin M A M 2021 Tensile properties of bamboo strips and flexural behavior of the bamboo reinforced concrete beams *Europ. J. of Env. and Civ. Eng.* **26** 6444-6460

[5] Nuraeni R, Widyarti M and Sapei A 2018 Flexural capacity assessment of Bamboo Petung reinforced concrete beams *J. Civ. Eng. And Env.* **3** 49-58

[6] Budi A S and Rahmadi A P 2019 Flexural behavior of petung bamboo strip notched reinforced concrete beams *IOP Conf. Series: J. of Phys.* **1153** 012127

[7] Muhtar 2021 Experimental study of the hose clamp distance on flexural behaviour of bamboo reinforced concrete beams *Jurnal Penelitian Ipteks* **6** 103-113

The flexural strength of green foamed concrete with polypropylene fiber

K Al Qubro[1], A Saggaff[2] and Saloma[2]

[1]Universitas Indo Global Mandiri, Jalan Jendral Sudirman No. 629, Palembang, South Sumatera, Indonesia
[2] Universitas Sriwijaya, Jl. Raya Palembang-Prabumulih Km. 32. Indralaya, OI. South Sumatera, Indonesia

E-mail: khodijah@uigm.ac.id

Abstract. Green concrete is made from eco-friendly waste materials. It reduces CO2 emissions and reduces water use. Normal concrete contributes about 40-50% of carbon in the air due to its production, impacting climate change. The waste material from coal combustion, like fly ash used as an additive to replace some cement in the concrete mixture. Foamed concrete is considered concrete with an eco-friendly material because it uses less aggregate and reduces water use. Concrete is weak to tensile forces; adding Fly ash can reduce pores allowing acidic substances to enter. Adding polypropylene fiber to prevent cracking makes foamed concrete more ductile. Green expanded concrete contains 0.8% foaming agent, with polypropylene fiber (PPF) variants of 0.15%, 0.2%, 0.3%, and 25% fly ash as a cement alternative. The findings of the flexural strength test demonstrate that fibers added to green foam concrete boost its flexural strength. Green foam containing polypropylene fibers produces a maximum flexural strength of 0.3% higher than green foam concrete without fibers.

1. Introduction

A Green building is an environment-friendly building designed, built, and operated to minimize its overall environmental impact. Carbon dioxide (CO_2) is the principal greenhouse fuel online emitted using human activities. It is claimed that 5% of global carbon dioxide emissions are attributable to the cement industry, an essential concrete component. Due to the significant contribution of CO_2 to environmental pollution, besides the moving civil construction needs, it is necessary to find an optimal solution. Cellular concrete is a new innovative technology for sustainable construction and civil engineering that meets the criteria of becoming a green material [1]. Aerated concrete is a concrete technology applied to buildings. The advantage of aerated concrete is that it is light and quick to construct. Using foam in aerated concrete reduces the density of aerated concrete [2]. Coal-fired power plants are made from the leftovers of FA combustion (fly ash). Due to growing environmental awareness, using FA has emerged as an appealing replacement for disposal. FA can strengthen concrete by replacing essential elements like sand. By developing new chances for additional usage, the negative effects of AF disposal on the environment and the economy will be lessened [3]. The addition of fly ash to partially replace the cement in concrete grade K. 350 resulted in a substantial compressive strength value of 30.20 MPa, which was beyond the target with the change of the addition of 20% FA when the results of concrete compressive strength Test on concrete samples with the addition of 30% partial replacement of cement showed a decrease [4].

Concrete has a strong compressive strength but a compressive strength that is about ten times lower. Additionally, following cracking, it exhibits brittle qualities and hinders stress transmission. In order to enhance mechanical qualities and prevent brittle fracture, fibers can be added to the concrete mixture [5]. Concrete with reinforcement scattered throughout the mixture in the form of fibers, such as glass, carbon, steel, polymer, polypropylene, and others (FRC). High-performance fiber-reinforced concrete is one of the more advanced varieties of concrete that have been created over the past 30 years as a result of developments in FRC research (HPRC), self-compacting fiber-reinforced concrete (SCFRC) (HPFRC), ultra-high-performance fiber reinforced concrete (UHPFRC), hybrid fiber reinforced concrete (HFRC), and lightweight fiber reinforced concrete (LWFRC) [5]. Fly ash can be applied to help block holes where acids can enter. Because of the high elastic modulus of the PF blend, which prevents cracking, aerated concrete is more malleable than fiberless aerated concrete. It might make concrete stronger [6]. A study was conducted on newly aerated concrete using fly ash as a partial replacement for cement and including polypropylene fibers to boost the material's flexural strength. This study compares green foam concrete that is not reinforced to green foam concrete with polypropylene fiber added to determine the effects of the fibers on the flexural strength characteristics of green foam concrete.

2. Materials and methodology

2.1. Foamed concrete

Aerated concrete, or foam concrete, is a lightweight form of concrete. Foam is inserted inside of concrete to generate low-density concrete. A hydrolyzed protein-based substance, or soapy resin, known as a foaming agent, creates air bubbles added to concrete [6]. Building materials made of aerated concrete are adaptable and light. It has various uses and may be produced in a wide range of compressive strengths.

Over conventional building materials, aerated concrete can frequently be more affordable and effective. The production process, benefits, and specifications of aerated concrete are all covered in this article. Three case studies will be discussed where aerated concrete has been used to show the material's potential. These case studies demonstrate the advantages of aerated concrete over conventional backfill materials when utilized for soil stabilization, trench rehabilitation, and arch bridge reinforcement [7].

Figure 1. Foamed concrete (geopolymerhouses.wordpress.com [8]).

One way to characterize foamed concrete is as a flowable, lightweight material that works well in various situations. It is capable of a wide variety of compressive strengths, ranging from 1 MPa to 15 MPa, as well as a large range of dry densities, typically between 400 kg/m3 and 1600 kg/m3 [7].

2.2. Foaming agent

A concentrated foaming agent is used to create aerated concrete foam. Foaming agents based on synthetic surfactants, protein hydrolyzates, or both. The foam is produced using a foaming machine. The component is diluted with water in a foaming machine to produce a pre-foaming solution, which is then expanded with air to produce foam. Shaving foam is a stiff foam with a density of 45 g/litre. The foam concrete's stable bubbles can survive the chemical and physical pressures used in the mixing, setting, and curing processes. The majority of the bubbles, 75 to 85%, range in size from 0.3 to 1.5 mm [9].

Figure 2. Foam from the foaming agent.

A gas that disperses in a liquid or solid is called foam. Foam is created when a foaming ingredient, such as soap, oil, acid, or a wetting agent, causes changes in the liquid's gas distribution. A membrane isolates the liquid from the minute bubbles that form during the production process. There are a variety of foam types with different functions. [10].

2.3. Fly ash

One of the byproducts of burning coal is fly ash. Fly ash contains fine particles resembling cement. The composition of fly ash varies, but all fly ash contains silica dioxide (SiO2) and calcium oxide (CaO), which can replace cement.

In general, fly ash consists of solid spherical particles, and some are hollow spherical particles and spherical particles containing other smaller spherical particles. The surface size of fly ash varies from 1 µm to 100 m with an area from 300 to 500 m2/kg.

Cement may be partially substituted with fly ash following ACI 1993 Concrete Practice Manual Part 1 226.3R-3 and ASTM C 618. [11,12], divided into three grades.

2.3.1. Fly ash class F. Anthracite or bituminous coal burning produces fly ash with a CaO content of less than 10%. Low calcium fly ash, non-binding, and particularly pozzolanic are additional names for class F fly ash.

2.3.2. Fly ash class C. Burning lignite or bituminous coal produces fly ash containing>10% CaO. Due to its high CaO content, and cohesive and pozzolanic qualities, Class C fly ash is often referred to as high calcium fly ash. Class C fly ash hydrates and hardens after 45 minutes when exposed to moisture or water due to its cohesive nature and somewhat high CaO content.

2.3.3. Fly ash class N. Diatomic, Schertz opaline, shale, tuff, and volcanic ash are naturally occurring pozzolans or combustion products that may or may not ignite. It also possesses advantageous pozzolanic characteristics.

Figure 3. Fly ash class F.

2.4. Polypropylene fiber

Polypropylene fibers (PPF) are extruded, orientated, and sheared polymeric materials that can be straight or distorted, according to EN 14889-2. (Figure 4). Two varieties of PPF can be recognized, according to EN 14889-2: microfiber and microfibre [13]. They vary in length, but more significantly, they serve different purposes in concrete. Because they can take the place of conventional bar reinforcement and transmit loads to the structure, macro fibers are also structural fibers. Consequently, the reinforcement construction took longer than expected, saving money. Typically, they are between 30 and 50 mm long.

However, because it is shorter than 30 mm, the microfiber cannot fulfill its carrying purpose. It primarily combats plastic shrinkage and prevents concrete cracks from forming. As a result, they make the element stronger and make it last longer. Alternatives to crack control mesh include micro PPF. They can be categorized as fibrillated or monofilament.

Figure 4. Polypropylene microfibers

2.5. Materials and mixing testing

The test specimen for this investigation, conducted at the Concrete and Materials Laboratory, is a beam mold measuring 100 x 100 x 350 mm. Fly ash is a substitute for cement in the polypropylene fiber variations of 0.15, 0.2, and 0.3%, and the binder ratio of fine aggregate to water is 1:1.5:0.45. The equipment employed includes a beam mold, a concrete mixer, an Abrams cone, a Vicat apparatus, an LVDT, and a Universal Machine Test for flexural strength testing of the test object and fresh concrete.

Table 1. The composition of the mixture of green concrete and green foamed concrete.

Sample	Foaming Agent (%)	Polypropylene Fiber (%)	Binder		Sand (%)	Water Ratio (%)
			Portland Cement (%)	*Fly Ash* 25 % (%)		
GC	0	0	25.42	8.48	50.85	15.25
GFC	0.8	0	25.42	8.48	50.85	15.25
GFCP - 0.15	0.8	0.15	25.42	8.48	50.85	15.25
GFCP – 0.2	0.8	0.2	25.42	8.48	50.85	15.25
GFCP – 0.3	0.8	0.3	25.42	8.48	50.85	15.25

The period known as "setting time" is needed for the cement mixture to bind with the fresh water and harden. Using a Vicat needle with a diameter of 1 mm and piercing the cement paste to a depth of 25 mm for 30 s after needle release, the setting time of the cement was measured. The first setting time should be at least 45 minutes, and the final setting duration should be at most 375 minutes, per ASTM C-191 [14].

Analyzing the chemical composition of the fly ash and testing the fine aggregate's physical characteristics are both parts of the materials testing process. Each component was weighed appropriately before being combined. Sand, fly ash, and cement should be added to the mixer before water. The polypropylene yarn should be added to the mixer once the mixture has been homogenized. After finishing its first mixing with the mixture, the foam is swirled in the mixing tank. If the mixture is consistent, the phase is finished and prepared to be put into the mold. The recent grout test was a settlement test. As quickly as possible, up to three layers of the same height of fresh concrete mix are gradually added to the Abrams cone. Lift the Abrams cone and the droplet slowly after that [15].

The flexural strength of concrete is the ability of a concrete beam to withstand a force perpendicular to its axis until the concrete beam fails. The test object used was 100 mm x 100 mm x 350 mm. The flexural strength test refers to the ASTM C 293 - 09 [16].

$$R = \frac{3PL}{2b.d^2} \quad (1)$$

Where L is the length of the span, b is the width of the l specimens (mm), d is the depth of the sample, P is the maximum stress (tensile load) applied to the specimen, and R is the flexural strength or modulus of failure (MPa) (mm).

Using a Universal Testing Machine with LVDT, the flexural strength of foamed concrete was tested, and the flexural and deflection values were read using the data recorder provided in the lab.

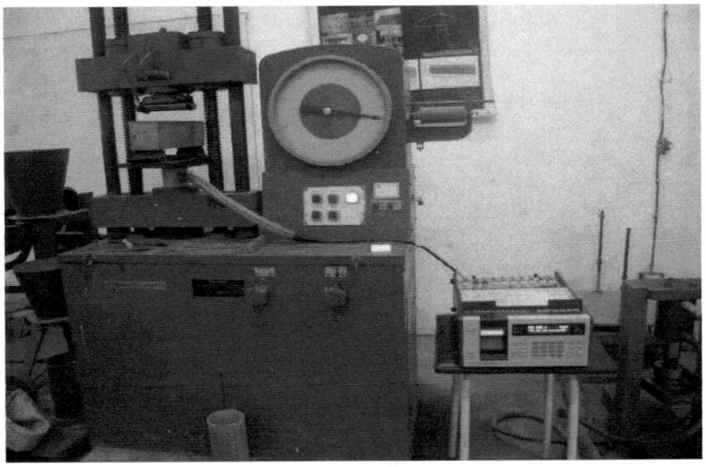

Figure 5. Flexural strength testing with LVDT and data logger.

3. Result

Results of tests on the chemical composition of fly ash as a mortar ingredient. The findings of the FA test indicate that the FA is type F, with CaO = 3.75% and SiO2 + Al2O3 + Fe2O3 = 87.54%. Results of tests on FA's chemical constituents used as mortar. The fly ash test results show that the FA is type F, with SiO2 + Al2O3 + Fe2O3 = 87.54% and CaO = 3.75%.

Table 2. The findings of the slump test change depending on the preparation.

Samples	Slump Test (cm)
GC	8.20
GFC	12.65
GFCP – 0.15	11.40
GFCP – 0.20	10.70
GFCP - 0.30	10.05

The findings of the slump test demonstrate that the foaming agent's contribution to the concrete mix impacts the slump value. A larger polypropylene fiber content might also lower the blend's slump value.

Table 3. Results of setting time tests using various test object modifications.

Samples	Initial Set. (minute)	Final Set. (minute)
GC	75	120
GFC	105	180

The setting time tests show that adding a foaming agent to the mix prolongs the setting time of the concrete.

Table 4. Flexural strength and concrete are green foamed concrete's specific gravity in the relationship.

Sample	Specific gravity (gr/cm^3)	Flexural strength (MPa)
GC	2.05	4.81
GFC	1.63	2.77
GFCP – 0.15	1.64	3.06
GFCP – 0.20	1.66	3.15
GFCP - 0.30	1.70	3.72

Test results show that polypropylene fibers affect the flexural strength of concrete. The higher the fiber content of the concrete, the higher the flexural strength. In addition, adding a foaming agent affects the density of the concrete. Concrete mix has 0.8% foaming agent; the density is lighter than concrete without a foaming agent.

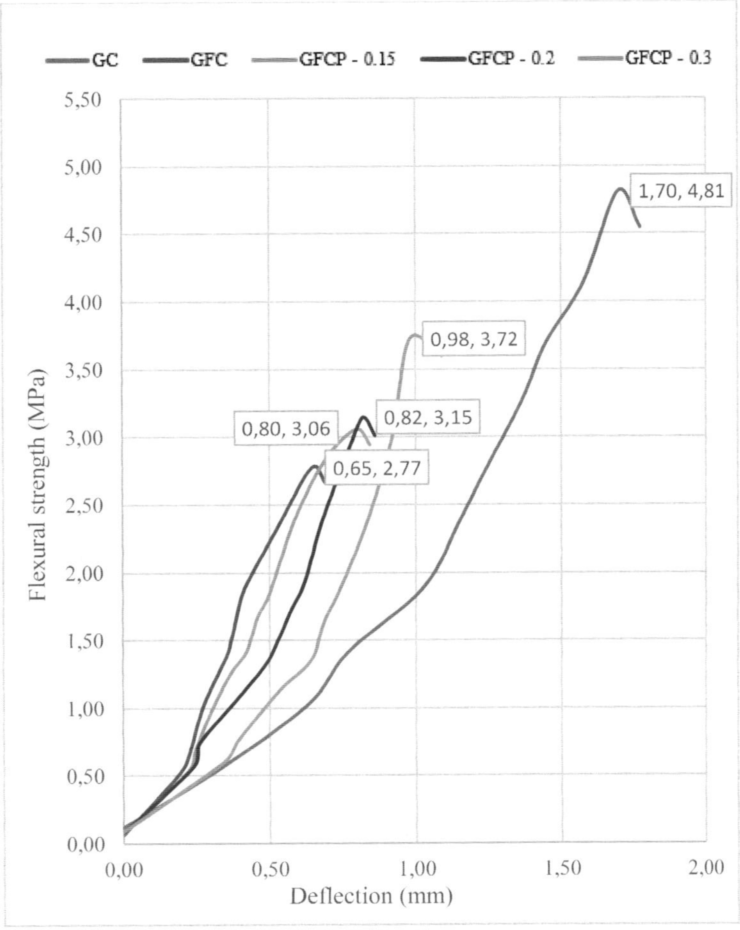

Figure 6. Relationship of flexural strength with the deflection test.

The graph of the relationship between deflection reduction and flexural strength with 1 kN load range to flexural strength with maximum load explains that the deflection decreases more as the flexural strength increases.

4. Conclusion

The following conclusions are drawn from laboratory tests of green foam concrete containing polypropylene fibers:

1. The density of green foam concrete with polypropylene fibers and without fibers produces 17% - 20% lighter than green mortar mix.
2. Testing of fresh concrete with the slump test showed that the green foam concrete has better workability than the green mortar mixture. Adding fiber to the green concrete affects the mixture's viscosity, increasing with adding fiber.
3. The flexural strength of green foam concrete is 42% smaller than that of green mortar, but with the addition of polypropylene fiber, they will increase by 10 - 25% of the flexural strength of green foam concrete as the fiber increases.
4. Adding a foaming agent to the green mortar mixture significantly affects the deflection value, reaching a difference of 61%. Suppose green foam concrete is added with polypropylene fiber. In that case, the difference in deflection will decrease to 18% as the polypropylene fiber increases with a 0.1% variation interval from green foam concrete without fiber.

Acknowledgments

I respectfully thank my Professor, Prof. Dr Ir. H. Anis Saggaff, MSCE., IPU., ASEAN.Eng. for his guidance and research funding so that I can complete this research, and I also express my gratitude to Dr Saloma, ST, MT, for his guidance in conducting the research from start to finish.

I also thank my colleagues, especially the Head of Undergraduate Program of Civil Engineering Program Study of Universitas Indo Global Mandiri, Mrs Sartika Nisumanti, S.T., MT who has supported me in completing this paper and advised me to take part in the International Conference on Disaster Management (ICDM 2022).

References

[1] Moon A S, Varghese V and Waghmare S S 2015 *Int. J. For Research In Emerging Sci. And Tech.* **2** 25–32
[2] Febryandi and Setiawan R 2020 *Jurnal Tekno Global* **9** 72–77
[3] Kumar K, Arora R, Khan S and Dixit S 2021 *Proc. Int. Conf. on Applied Research and Engineering (Cape Town)* vol 56 (Amsterdam: North-Holland/American Elsevier) p 1886
[4] Hadiyana D and Nisumanti S 2016 *Bearing : Jurnal Penelitian dan Kajian Teknik Sipil* **4** 107–113
[5] Blazy J and Blazy R 2021 *Case Studies in Construction Materials* **14** e00549
[6] Al Qubro K and Saggaff A 2021 *Int. J. of Progressive Sciences and Tech. (IJPSAT)* **29** 447–453
[7] Basiurski J W D and Wells D 2001 *The use of foamed concrete in construction and civil engineering* (England: EAB Associates) pp 65–73
[8] Sarmin S N 2015 *Lightweight Building Materials of Geopolymer Reinforced Wood Particles Aggregate* vol 802 pp 220–224
[9] British Cement Association 1994 *Foamed concrete: Compositions and properties* (UK: Camberley)
[10] Thakrele M H 2014 Experimental study on foam concrete *Int. J. of Civ. Struc. Env. and Infrastructure Eng. Res.and Dev. (IJCSEIERD)* **4** 145–158
[11] ACI 226.3R-3 Part 1 1993 *Standard Practice for Selecting Proportions for Normal, Heavy, Weight and Mass Concrete* (Washington, DC: ACI)
[12] ASTM C 618 2019 *Standard Specification for Coal Fly Ash and Raw or Calcined Natural Pozzolan for Use in Concrete* (West Conshohocken: ASTM International) vol 04 chapter 02 pp 1–3
[13] EN 14889-2 2018 *Fibres for concrete - Part 2: Polymer fibres - Definition, specifications and conformity*
[14] ASTM C 191 2008 *Standard Test Method for Time of Setting of Hydraulic Cement by Vicat*

Needle (West Conshohocken: ASTM International) vol 04 chapter 01 pp 1–6

[15] ASTM C 143M 2003 *Standard Test Method for Slump of Hydraulic-Cement* (West Conshohocken: ASTM International) vol 04 chapter 02 pp 1–4

[16] ASTM C 293M 2002 *Standard Test Method for Flexural Strength of Concrete (Using Simple Beam With Center-Point Loading)* (West Conshohocken: ASTM International) vol 04 chapter 02 pp 1–3

IOP Conf. Series: Earth and Environmental Science 1173 (2023) 012003 doi:10.1088/1755-1315/1173/1/012003

Shear capacities of reinforced concrete interior beam-column joints subjected to cyclic loading

H Maulana[1], I Amal[1], W Iqbal[1], R Thamrin[1], Zaidir[1] and R Kurniawan[1]

[1] Andalas University, Civil Engineering Department, Padang, Indonesia

E-mail: bighafiz45@gmail.com

Abstract. The primary focus of this research is the experimental investigation of internal RC beam-column joints with cyclic loads. There were two interior RC beam-column joints test specimens made without shear reinforcement. The longitudinal reinforcement ratio of the beams served as the study's test variable. The specimen test results were compared to the theoretical capacity given in SNI 2847-2019 after that. The test revealed the following: (a) shear and flexural beam capacities; (b) shear and flexural column capacities; and (c) shear capacities of joints. The results showed that there was a significant impact on the shear capacity of the beam-column joint as a result of the variation in the reinforcement bars of the beam. After the longitudinal reinforcement bars of the beam yielded in both instances, the degradation of the beam-column joint occurred. A sharp decline in the load value that the specimen can support serves as evidence..

1. Introduction

Beam-column elements are crucial in the multi-story reinforced concrete truss constructions that utilize special moment-resisting frame structures or sustain strong seismic loads. Because the seismic moments in the beams and columns operate in different directions across the joint area, beam-column joints are vulnerable to vertical and horizontal shear stresses that are typically larger than those encountered in conventional beam and column sections. Before the 1960s, structural damage to reinforced concrete buildings often occurred in the walls, columns, and beams, with the beam-column connection section being an exception. After the 1960s, earthquakes started causing serious damage to the reinforced concrete beam-column joints, which made the connection strength weaker than the capacity of the surrounding cross-section [1].

This is also corroborated by the fact that most reinforced concrete buildings built before 1970 did not include shear reinforcement in the joints section due to the lack of criteria governing the requirement for it in the design code. Some cases of failure of reinforced concrete structures at that time were caused by weak beam-column connections in the absence of shear reinforcement. It caused the capacity of the connection caused the longitudinal reinforcement of the column to be unable to withstand the acting force and did not produce plastic hinges in the beam section following the failure pattern. It should be due to earthquake loads. The concept of strong beams – weak columns that became the structural system in the 1960s and 1970s also contributed to the failure of beam-column joints [2].

Figure 1. Failure due to the absence of column reinforcement in the connection of reinforced concrete columns in the Samsat Building of West Sumatra (source: http://www.wm-site.com/civil-eng/short-course-haki-komda-bali-2010)

In reinforced concrete structures that experience lateral deformation due to earthquake loads, the beam-column connection section experiences an enormous shear force. Figure 1 shows damages or stiffness deterioration of the building structures. A good reinforced concrete structure has seismic performance, with good damage tolerance and deformation capacity. Previous studies examining reinforced concrete joints' seismic performance [3] have produced design guidelines or standards used in the ACI 352-02 standard. Three different connection failure modes exist [4], notably J-failure with the condition of joint failure occurring at the end of the beam adjacent to the joint prior to the appearance of the plastic hinge, BJ-failure with the condition of joint failure occurring after the appearance of the plastic hinge on both ends of the connection located close to the beam, and B-failure with the condition of beam failure occurring in the inelastic deformation area of the beam adjacent to the junction but the joint condition not occurring (figure 2).

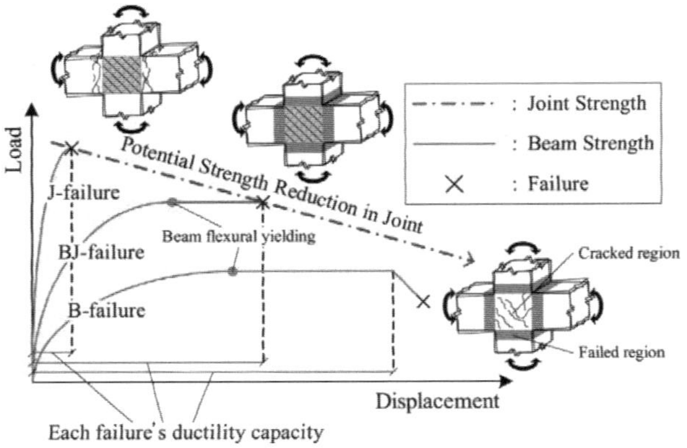

Figure 2. The three joint failure modes [4].

The experimental investigation of internal beam-column reinforced concrete joints without shear reinforcement under cyclic loads is the main topic of this research. The longitudinal reinforcement ratio of the beam served as the study's test variable. After that, the experimental findings are contrasted with the capacity predicted by SNI 2847-2019. The flexural and shear capabilities of the beams, the flexural and shear capacities of the columns, and the shear capacities of the RC beam-column connection were all noted during the testing.

2. Experimental program

Two inside RC beam-column specimen were used to simulate the interior beam-column joint section between 2 storeys of a RC building. The specimens were then tested. The interior RC beam-column joints of the test specimens experienced cyclic loads throughout the testing process.

2.1. Design and shape of specimens

Two RC beam-column joints were made in accordance with the SNI code 2847-2019 and the weak beams - strong columns design concept (figure 3). The cross-sectional dimensions of the columns for both specimens are 250 x 250 mm with 8 D25 longitudinal reinforcement. Dimensions of the beam's cross-section are 150 x 300 mm for both specimens with upper and lower beam reinforcement, 3 D13 for the BCJ-01 and 5 D13 for the BCJ-02 specimen. Transverse reinforcement bars of beams and columns is in the form of D10 steel reinforcement hoops, except in the connecting area, where stirrups are not utilized, set equally at 100 mm.

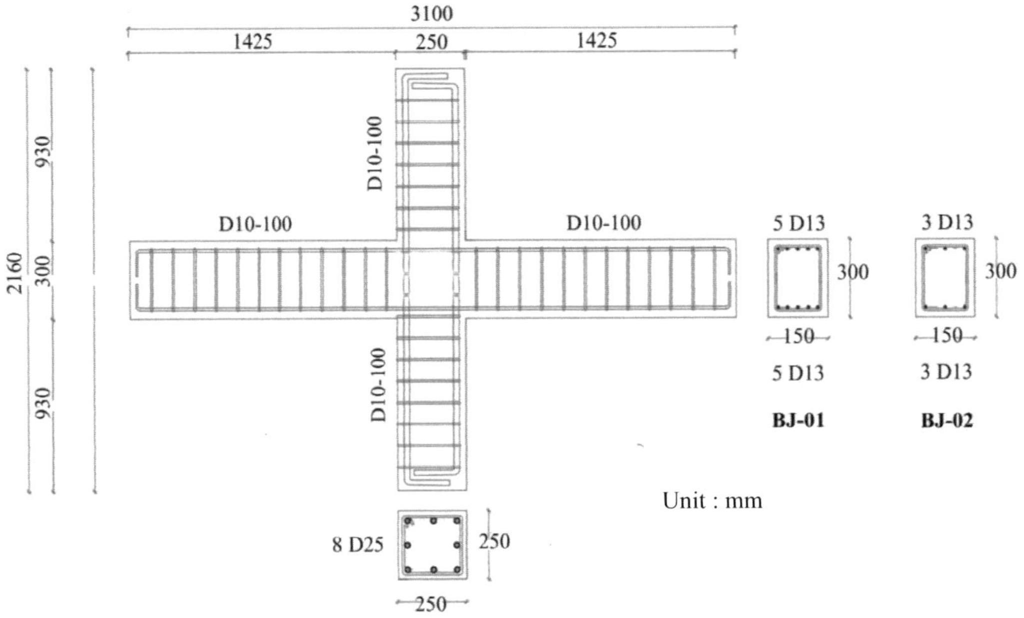

Figure 3. Geometry and reinforcing details of specimens.

2.2. Material properties

For both specimens, concrete had an average compressive strength of 27.05 MPa after 28 days. The tensile characteristics of all the steel reinforcement used in this experiment were tested by ASTM A370. table 1 displays the yield stress of each reinforcing steel.

2.3. Test setup

Figure 4 depicts the test configuration for the beam-column connection specimen used in this experiment. To replicate seismic loading, quasi-static cyclic loading on the column was used to evaluate both interior beam-column joint components. A hydraulic actuator and a load cell are used to replicate the cyclic load on the specimen. The specimen's deformation was captured using an LVDT installed at the end of the column. The specimen was supported by three supports that were joined to the loading frame, and the top of the column was fastened to a steel plate that held the load cell. The load cell and LVDT were connected to a data logger to record the applied loads and deformations.

Table 1. Properties of specimens.

Specimens	Beam							f_c' (MPa)
	Reinforcing bar (Upper and lower each)			Stirrup				
	f_{by} (MPa)	ρ_{bu}, ρ_{bl}	η_b	f_{sy} (MPa)	ρ_s	s_b (mm)	η_s	
BCJ-01	431	0.0147	5-D13	401	0.0048	100	D10	27.05
BCJ-02	431	0.0088	3-D13	401	0.0048	100	D10	27.05

Specimens	Column							f_c' (MPa)
	Reinforcing bar			Hoop				
	f_{cy} (MPa)		η_c	f_{hy} (MPa)	ρ_h	s_b (mm)	η_h	
BCJ-01	424	27.05	8-D25	401		100	D10	27.05
BCJ-02	424	27.05	8-D25	401		100	D10	27.05

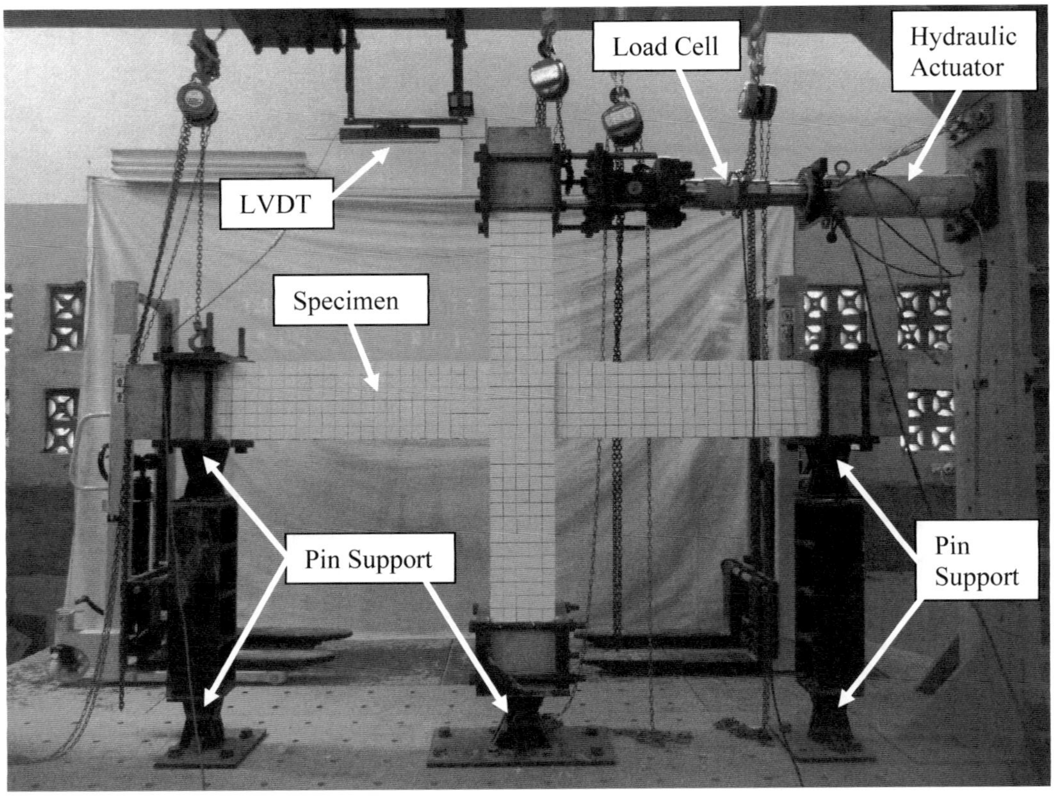

Figure 4. Specimen test setup.

2.4. Load history

Figure 5 displays the loading history of the hysteretic load applied in all tests. The tests were conducted using displacement- and load-controlled test methods. In the first three cycles, a load-controlled loaded scheme was used, and in the fourth cycle, a displacement-controlled loaded was used. The yield moment value of the beam (Mby) was used to define the load control scheme. The values of load at each incremental step were 1/3 Mby, 2/3 Mby, and Mby, respectively.

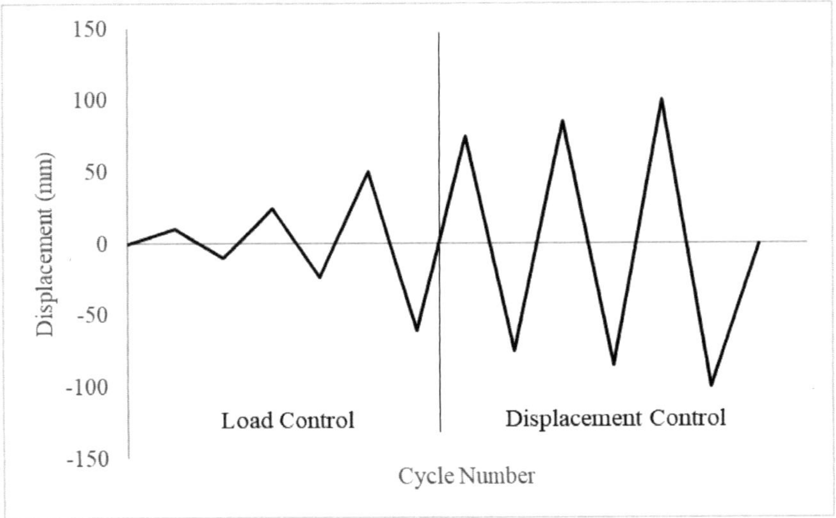

Figure 5. Typical loading history.

The deformation in the displacement control was obtained by setting the deformation values as 1.5 Mby, 1.7 Mby, and 2 Mby, respectively.

3. Cross-section capacity

The shear and flexural capacities of the beam and column sections in table 2 are obtained based on the code SNI 2847-2019 and flexural analysis using theoretical moment-curvature determination.

Table 2. Cross-section capacity of specimens.

Specimens	Beam		Column	
	Mnb (kN.m)	Vnb (kN)	Mnc (kN.m)	Vnc (kN)
BCJ-01	67.07	135.67	124.99	186.25
BCJ-02	42.62	0.285	124.99	186.25

4. The joint shear strengths

Equation (1), based on codes SNI 2847-2019 and ACI 318-19, multiplies the effective cross-sectional area of the connections by the concrete compressive strength (fc') to produce the joint shear strength (Vj). The ACI 318-19 recommendation for a non-confined connection uses the following equation to calculate the joint shear strength (Vj).

$$V_j = 1.0\,\lambda\,\sqrt{f_c'}\,A_j h_c \quad (1)$$

$$A_j = b_j \times h_c \quad (2)$$

Where λ for normal concrete is 1.0 and b_j is the effective width of the joint and h_c is the height of the joint

Joint shear strength when beam yields (V_{jby}) can be calculated by equation (3).

$$V_{jby} = \left(\frac{l_c}{z} - 1\right) V_{by} \frac{l_b}{l_c} - \frac{h_c}{z} V_{by} \quad (3)$$

Where z would be the distances between both the centers of gravity of the lower and upper beams, l_c would be the column height, h_c would be the joint height, and V_{by} would be the shear strength from the beam when the beam yields.

Table 3. Joint capacity.

Specimens	V_j (kN)	V_{by} (kN)	V_{jby} (kN)	V_j/V_{jby}
BCJ-01	325.06	52.02	480.82	0.68
BCJ-02	325.06	33.20	306.89	1.06

The connection failure mode can be predicted through the value of V_j/V_{jby} according to the failure mode differentiation method by [4] and [5], for specimens that meet $V_j/V_{jby} < 1$ it is a joint failure mode, while for a value of $V_j/V_{jby} > 1$ it is a mode of beam joint failure and beam failure.

5. Test results

5.1. Cracks patterns

The crack patterns in the specimens BCJ-01 and BCJ-02 are shown in figure 6. The type of failure in the BCJ-01 connection is different from the BCJ-02 due to the varying amount of reinforcement in the beam. The shear failure occurred on specimen BCJ-01 prior to flexural yielding, and partial spalling of the concrete and cracks were seen at the joints. On either side, the failure happened in specimen BCJ-02 after the beam had given way. They were visually obtained with a reduced number of beam longitudinal reinforcements, resulting in reduced damage to the joint. The crack in the plastic hinge region of the beam closes and widens on each side. The joint damage on the BCJ-02 specimen was less severe than it was on the BCJ-01 specimen because the failure was brought about by the crushed concrete in the plastic hinge part of the beam close to the joint. The failure mode that manifested in both specimens conformed to the forecast made by [4] and [5] utilizing the failure mode.

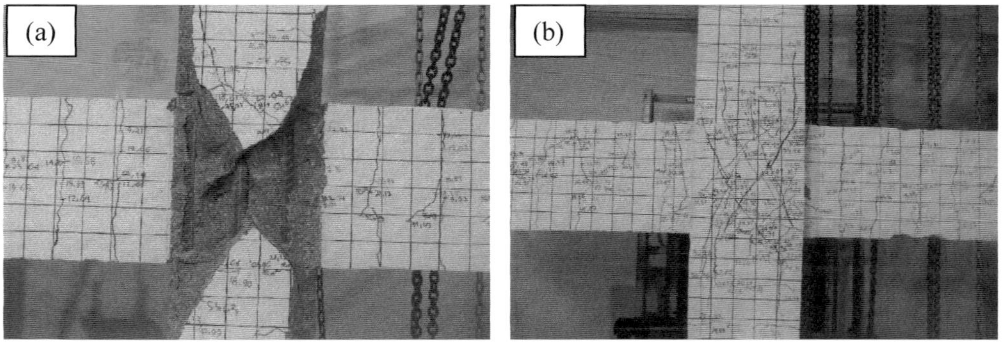

Figure 6. BCJ of specimens at peak horizontal load: (a) Specimen BCJ-01; (b) Specimen BCJ-02.

5.2. Load-deformation relation

When the displacement and load curves of the two specimens are compared, the shear strength of the joint in figure 7 improved. Increased joint shear strength due to variations in beam longitudinal reinforcement through hysteresis of envelope curve load transfer.

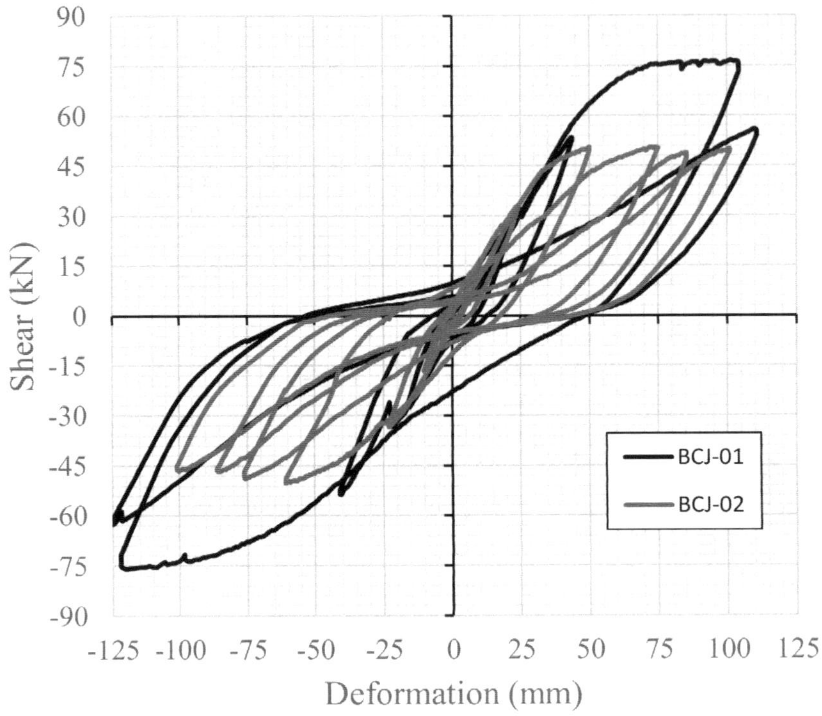

Figure 7. Load-displacement envelopes of specimens.

Table 4. Test results

Specimens	Max. Storey Shear (kN)		Max. Storey Drift Ratio (%)		Story Drift Ratio at yielding of beam bars (%)		*Story Drift Ratio at 80% of max storey shear (%)*		*Ductility Ratio* (μ_Δ)	
BCJ-01	76.56	-75.87	6.51	-7.37	6.12	-7.18	2.85	-3.40	1.75	1.87
BCJ-02	50.52	-50.22	5.94	-5.93	2.65	-3.21	1.78	-2.12	3.11	3.30

5.3. Joint ductility

The improved ductility of the two joint specimens with different longitudinal beam reinforcement is explained by comparing the ductility of the two groups. The ratio [6] is used to calculate the displacement ductility capacity in relation to the idealized yield displacement. The ductility ratio (μ_Δ) was calculated using equation (4).

$$\mu_\Delta = \frac{\Delta_c}{\Delta_y} \tag{4}$$

Where Δ_y would be the displacement at yields and Δ_c would be the displacement capacity just at the top. Table 4 displays a comparison of ductility capacity for both various specimens.

6. Conclusions

The experiment results of the two RC beam-column joint specimens revealed that the increase mostly in deformability within the beam-column joints, which was also represented by an increase in the ductility ratio, was inversely related to the decrease in the beam longitudinal reinforcement. For beam-column relationships with a V_j/V_{jby} ratio less than 1.0, cracks or damage are most commonly observed in the joint area. For beam-column joints with a V_j/V_{jby} ratio greater than 1.0, there were evident cracks in the plastic hinge area of the beam near the joint. The goal of the remaining stages of this experiment will be to strengthen the failure mode at the internal reinforced concrete beam-column connection.

References

[1] Paulay T and Priestley M J N 1992 Seismic design of reinforced concrete and masonry buildings (New York: A Wiley Interscience Publication) p. 744.

[2] ACI Committee 318 1963 *Building code requirements for structural concrete* (Farmington Hills: American Concrete Institute)

[3] Ong K C G, Lin Z S, Chandra L R, Tam C T and Pang S D 2013 Experimental investigation of a DfD moment-resisting beam-column connection *Eng. Struct.* **56** 1676–1683

[4] Lee J Y, Kim J Y and Oh G J 2009 Strength deterioration of reinforced concrete beam-column joints subjected to cyclic loading *Eng. Struct.* **31** 2070–2085

[5] Teraoka M, Kanoh Y, Sasaki S and Hayashi K 1996 An estimation of ductility in interior beam-column sub-assemblages of reinforced concrete frames *J. Soc. Mater. Sci.* **45**(9) 103341

[6] Allam K, Mosallam A S and Salama M A 2019 Experimental evaluation of the seismic performance of interior RC beam-column joints strengthened with FRP composites," *Eng. Struct.* **96** 109308

[7] ACI Committee 318 2019 *Building code requirements for structural concrete and commentary, ACI Committee 318*, ACI318-19 (Farmington Hills: American Concrete Institute)

[8] National Standardization Agency 2019 *Structural Concrete Requirements for Buildings, SNI 2847:2019* (Jakarta: BSN)

The Numerical Study of The Need for Transverse Reinforcement of a Spun Pile

Orientilize M[1], Prakoso W A[1], Lase Y[1], Dessy N W[2], Willie S[2]

[1]Civil Engineering Department, Universitas Indonesia, Depok, Indonesia
[2]Undergraduate Student, Civil Engineering Department, Universitas Indonesia, Depok, Indonesia

[1] Corresponding author: mulia@ui.ac.id

Abstract. A numerical study based on FEA was conducted to seek the optimal need for transverse reinforcement of a spun pile with 400mm-diameter and 75mm of wall thickness. Two equations from ACI 318-19 and ASCE 7-16 were compared. The parametric study was limited to a spun pile with 450mm diameter and 75mm of wall thickness. The amount of transverse reinforcement was presented as the ratio to the ACI requirement. The ratio was varied and its effect on strength and ductility was observed. Three different axial load ratios, 0.1, 0.2 and 0.3 Fc'Ag, that are usually applied on the foundation were considered. The needs for spun pile with and without concrete infill are different since it is a function of the ratio of gross area (Ag) to the confined area (Ach) of concrete. An empty spun pile require a higher amount than the filled pile. The axial load ratio is another factors that affect the need for that reinforcement. However, ACI does not recognize it. The study found that the amount of transverse reinforcement required by ACI and ASCE can be reduced to get an economical amount without significantly reducing the strength and the ductility of the spun pile.

1. Introduction

Indonesia still adopts a conventional elastic design concept for a foundation. Damage is restricted and prevented by limiting the lateral displacement of the foundation to 25 mm. However, few details should fulfil the requirement for inelastic stage performance, such as the amount of transverse reinforcement. The need for that reinforcement is mentioned in article 25.7.3.3 of SNI 2847-19 referring to ACI 318-19 [1] and article 7.14.2.2.5 of SNI 1726:2012 referring to ASCE 7-16 [2]. Both codes are different to some extent resulting in different quantity demands. The amount is a function of the ratio of concrete gross section (A_g) to core concrete section (A_{ch}). Since a precast prestressed pile is produced with a thicker concrete cover, hence it needs a higher quantity of transverse reinforcement. To cope with the requirement results uneconomical design pile.

Several studies reviewed the need for transverse reinforcements of the precast pile to sustain in severe earthquake but still consider the economic values [3], [4]. The study [3]criticized the equation in ACI 318-05 where the required amount was irrational to fulfil. An example of a precast pile with 610-mm-diameter with 76mm cover, 41.3 MPa concrete, and 413 MPa spiral reinforcement would require a spiral reinforcement ratio of 3.5%. This would call for 10M spirals pitched at 18 mm. Aside from being uneconomical to build, this spacing would violate ACI 318-05 minimum pitch. A rational design procedure to determine the required transverse reinforcement based on ACI 318-05 equation was proposed. The need for transverse reinforcement was correlated with the targeted ductility, curvature ductility demand, and the axial load. The length of the dowel that connected the pile and the pile cap was also considered in that procedure. The new procedure resulted in 20% to 40% lighter transverse reinforcement with constructible spiral pitch. Another parametric study was conducted to determine whether piles with light transverse reinforcement could act as ductile structural elements [5]. The amount of reinforcement was 20% lower than the requirements. The pile was modelled as a non-linier Winkler beam in soil. The connection types of the pile to pile caps were varied since based on the previous study [3] affected the inelastic performance of the pile-soil system. The parametric study found that the pile with lighter reinforcement performed ductile with limited ductility of 2.5 to 4.

A study conducted by [4] modified and combined two equations from ACI 318-05 and ATC 32-1996 and proposed the new equation. The changing of gross section A_g to core section A_{ch} was proposed since the transverse reinforcement was to confine the core area. The axial load and the curvature ductility capacity were other two factors that include in the revised equation. The equation results more rational need for transverse reinforcement and the values lay between the ATC and ACI's equation. However, it was limited to piles with curvature ductility greater than 18.

The spun pile is a precast prestressed pile that is intensively used in bridges and wharves in Indonesia. The piles are produced with a limited amount of transverse reinforcement which is less than 20% of than the minimum requirement set by ACI 318-19. The curvature ductility os this pile is less than 18. The experimental study conducted by [6] on a spun pile with 0.1% volumetric ratio where the minimum requirement of 2.85% found that the spiral reinforcement poorly contributed to the confinement of spun pile section. The numerical study of the typical spun pile connection in Indonesia was conducted by [7]. It was found that the spun pile with low confinement could reach limited ductility of 3. The higher amount of confinement affects the post peak behaviour of the connection and slightly improves the ductility.

Experimental and numerical studies of the spun pile connection in China have been carried out by [8], [9]. Though the volumetric ratios were not mentioned but based on the data, the spun pile had transverse reinforcement slightly lower than the minimum standard. The amount was 0.43% and the minimum requirement set by ACI was 0.623%. There were six specimens were tested with lateral cyclic loading and axial load of 0.1Fc'Ag. All of the specimens showed the flexural damage. The study found that the ductility of the connection was in the range of 2.5 to 3.00. The connection type affected the ductile performance of the specimens.

According to Article C4.7.1 [10], inelastic deformation in the piles should be limited to prevent severe damage and therefore the ductility should be restricted to four. ATC 96 and JRA 2002 limit the ductility to three and the damage is allowed near the ground surface for accessible repairing [11]. Similar regulation is also adopted in the New Zealand code for highway bridges, where design ductility is limited to four and the plastic hinges are expected at a depth less than two meters below the ground level. For deeper plastic hinges, it should be limited to three [11]. The preceding research found that the precast prestressed piles reinforced by lower transverse reinforcement still performed ductile with a limited ductility from 2 to 3.

The numerical study was conducted to investigate the need for transverse reinforcement of a spun pile. The equation proposed by [4] does not cover the spun pile produced in Indonesia since it has low curvature ductility. The study aims to find out the minimum volumetric ratio of the reinforcement where at that amount the strength and the ductility are close to those with full confinement. Since in the real case the spun pile on the connection area is strenghthened by filling the hollow with reinforced concrete, hence, the study was conducted in two conditions, with and without concrete infill. A case study was conducted on a 400mm-diameter spun pile with wall thickness of 75mm that produced in Indonesia.

2. Methodology

2.1. Validation of The Finite Element Model

A study the need foof r transverse reinforcement was performed based on the FE analysis using Abaqus software. Prior to the study, validation of the FE model was conducted based on the experiment by [12] [13]. It was a four-points loading bending test referring to JIS 5335-1987 on 400mm-diameter of a spun pile with 75mm wall thickness. It was reinforced by ten prestressing bars of 7.1mm-diameter. The yield and ultimate strength of the bars were 1275MPa and 1420Mpa, respectively with stressing level 75% of Fpu. Meanwhile, the concrete strength fc' was 52Mpa. The first experiment is a hollow spun pile and the second one is a spun pile with concrete infill.

Solid element C3D8R was chosen to model concrete of the spun pile and pile cap and truss element T3D2 with an embedded region for steel reinforcement. Spiral was modelled as hoops with similar spacing as 100mm. The constitutive law of concrete followed Meander theory meanwhile all the rebar adopted true stress-strain to represent the necking phenomenon. For spun pile with concrete infill, the interaction between precast concrete and concrete core was assumed as hard contact with friction of 0.5 whereas the reinforcement was assumed as an embedded region with concrete without slip. Push-over analysis was conducted to get the performance of the spun pile until failure. For specimens with axial load, the load was applied gradually, then the analysis was carried out.

Figure 1 shows the test setup [12] [13] and the validation results. As can be seen, the FE model could represent the experiment very well in the elastic until the initial inelastic stage. Beyond that stage, the load-displacement curve of the FE was above the experiment. The ultimate of the FE model is slightly higher than the experiment. Hence, the FE model can be considered valid and the numerical study could be carried out.

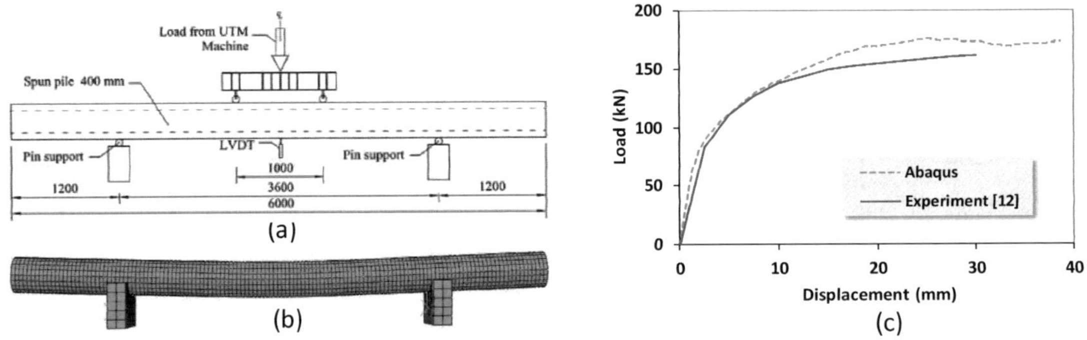

Figure 1. (a) Experiemental setup; (b) FEM Model; (c) Validasi Model

2.2. The Needs of Transverse Reinforcement Based on ACI 318-9 and ASCE 7-16

There are two equations to determine the needs for transverse reinforcement adopted in the Indonesia design code, which are article 10.9.3 of ACI 318-19 (article 25.7.3.3 of SNI 2847-19) and article 14.2.3.2.6 of ASCE 7-16 (article 7.14.2.2.5 of SNI 1726:2012) as presented in equations 1 and 2, respectively. The former equation covers the detailing of the concrete structure in the building. The equation of ASCE 7-16 is aimed to prestressed precast pile at higher seismic risk which is adopted from PCI Recommended Practice 1993 [[4]. The difference between the two equations is the axial forces. ACI does not considere the axial forces in determining the need for reinforcement meanwhile the ASCE does.

ACI 318 - 19 $$\rho = 0.45\left(\frac{A_g}{A_{ch}} - 1\right)\frac{f_{c\prime}}{f_{yt}} > 0.12\frac{f_{c\prime}}{f_{yh}} \quad (1)$$

ASCE 7 – 16 $$\rho = 0.25\frac{f_{c'}}{f_{yh}}\left(\frac{A_g}{A_{ch}} - 1\right)\left(0.5 + \frac{1.4P}{A_g f_{c'}}\right) > 0.12\frac{f_{c'}}{f_{yh}}\left(0.5 + \frac{1.4P}{A_g f_{c'}}\right) \quad (2)$$

The comparison of the equations is presented in Figure 2. This figure shows the need for transverse reinforcement of a spun pile that was used in the experiment [12]. It was a 400mm-diameter of the spun pile with a wall thickness of 75mm, and a concrete cover of 37.5mm. The pile was made of concrete with Fc' as 52Mpa and spiral wire with Fyh as 440MPa. The pile have 32mm-diameter of spiral with a pitch of 100mm. Since the ratio of A_g/A_{ch} of the spun pile with concrete infill is 1.52 and hence, based on equation 2, the minimum volumetric ratio according to ASCE 7-16 is almost similar to the requirements. The need of spiral reinforcement of empty spun pile according to ACI is 2.45 times higher than pile with concrete infill because the ratio of Ag/Ach of this pile escalate to 2.26. This is because equation 1 was derived to ensure that the capacity of the column is sufficient to resist compression load when the concrete cover spalling. The graph shows that the volumetric ratio based on ACI equation is higher than ASCE. However, The minimum requirement set by ACI and ASCE is almost similar for axial load ratio of 0.1 to 0.2.

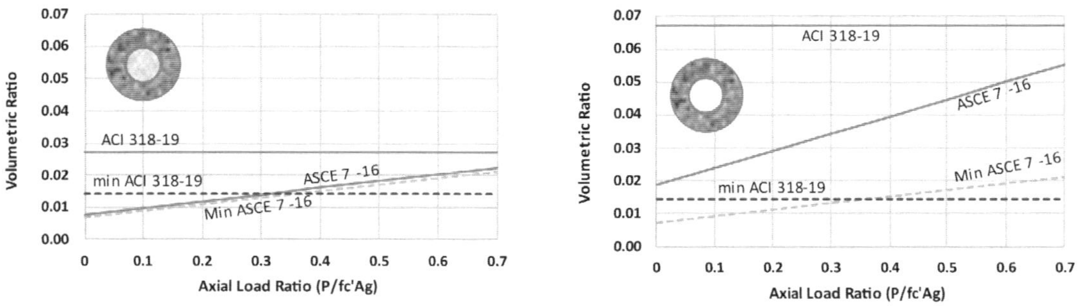

Figure 2. Volumetric ratio of a spun pile of 400mm-diameter (a) with; (b) without concrete infill

2.3. *Parametric Study*

A parametric study was based on four point load bending test in Figure 1a on the typical spun pile produced by one of the promenade precaster in Indonesia. The pile has 400mm-diameter and 75mm of wall thickness. The FE model followed the prior validation stage. The similar element type and material constitutive law were adopted. Push-over analysis was conducted to get the performance of the beam in terms of strength and ductility.

Table 1. The variation of volumetric ratio (ρ) of 400mm-diameter of empty spun pile

	Axial Load Ratio	d/pitch								
		3.2/100	6/110	8/100	10/100	10/75	10/60	12/45	14/45	14/30
ρ_{used} (%)	0	0.1	0.32	0.62	0.97	1.29	1.61	3.09	4.21	6.31
ρ_{used}/ρ_{ACI} (%)	0	1.5	4.7	9.2	14.4	19.2	24.0	46.2	62.9	94.3
ρ_{used}/ρ_{ASCE} (%)	0	5.3	17	33.2	51.9	69.2	86.5	166.3	-	-
	0.1	4.1	13.3	25.9	40.6	54.1	67.6	130	-	-
	0.2	3.4	10.9	21.3	33.3	44.4	55.5	106.6	-	-
	0.3	2.9	9.2	18.1	28. 2	37.6	47.0	90.3	123.0	-

The ratio of A_g/A_{ch} of the filled spun pile is in the range of 1.4 to 1.5 and thus, the need for transverse reinforcement according to ASCE for these products are almost similar as shown in Figure 2, where the minimum requirement is equal to the needs. Different transverse reinforcement with and without axial load was investigated. The amount of the reinforcement of the empty spun pile was varied as the ratio to the requirement of ACI or ASCE as shown in Table 1. Not all the variations are presented in this Table. The second and the third rows show the comparison of both codes. For pure bending, the requirement of ACI and ASCE is 6,7% and 1.86%, respectively. To cope with ACI, 14mm-diameter of spiral with a pitch of 30mm is needed. Meanwhile, to fulfil the requirement of ASCE depends on the axial load ratio. Higher loads need more reinforcements. A diameter of 10mm anda pitch of 50mm can fulfil the ASCE's code when the axial load is zero. Referring to the real condition, the axial load ratio investigated was varied from 0.1, 0.2 and 0.3.

Table 2 shows the variation of transverse reinforcement on the spun pile with concrete infill at similar diameter. Based on equation 1 and 2, since the needs are based on the ratio of Ag/Ach, then the spun pile with concrete infill need a smaller quantity referring to ACI and ASCE. A higher axial load need a larger confinement ratio.

Table 2. The variation of volumetric ratio (ρ) of 400mm-diameter of spun pile with concrete infill

	Axial Load Ratio	**d/pitch**						
		3.2/100	6/110	8/100	10/100	10/75	100/60	12/45
ρ_{used} (%)	0	0.10%	0.32%	0.62%	0.97%	1.29%	1.61%	1.93%
ρ_{used}/ρ_{ACI} (%)	0	4%	12%	23%	35%	47%	59%	71%
ρ_{used}/ρ_{ASCE} (%)	0	13%	42%	81%	127%	-	-	-
	0.1	10%	32%	64%	99%	132%	-	-
	0. 2	8%	27%	52%	81%	109%	-	-
	0.3	7%	23%	44%	69%	92%	115%	-

3. Result and Discussion

3.1. Spun Pile without Concrete Infill

The push-over results of pure bending FE analysis are presented in Figure 3 at different amounts of transverse reinforcement. In general, the load-displacement curves are similar until they reach the maximum condition. A slight difference was obvious at the post-peak load. At this stage, the behaviour is affected by the amount of spiral reinforcement. For those with the ratio to ACI above 50%, the strength decreases faintly as can be seen when the curve is zoomed in.

The effect of transverse reinforcement amount on the bending capacity and the ductility is presented in Figure 4. The axis is the ratio of the used reinforcement to the ACI requirement. The ACI is chosen since it is the current equation adopted in Indonesia, SNI 2847-19. The revision of SNI 1729 in year 2019 dismissed equation 2. As can be seen from Figure 3, the amount of transverse reinforcement affect the strength and ductility. To show a comparison between equation 1 and 2, the requirement according to ASCE is plotted with a red star. To obtain an almost similar bending capacity to that reinforced by 100% according to the ACI, spun piles without axial load should have at least 20% amount to the ACI need or 64% than ASCE code. At this condition, the bending capacity of the spun pile could reach 96.5% of the maximum value. For spun pile with an axial load from 0.1, 0.2, 0.3; the amount of reinforcement is 25% of ACI, which are equal to 90%, 70% and 49% of ASCE. The bending capacity of the spun pile could reach 96-97% of the maximum moment and the ductility have a difference of less than 0.2. To cope with the 100% ASCE requirement, the ratio of reinforcement to the ACI is equal to 28%, 36%, 43% dan 51% for the axial load ratio as 0, 0.1, 0.2 and 0.3, respectively.

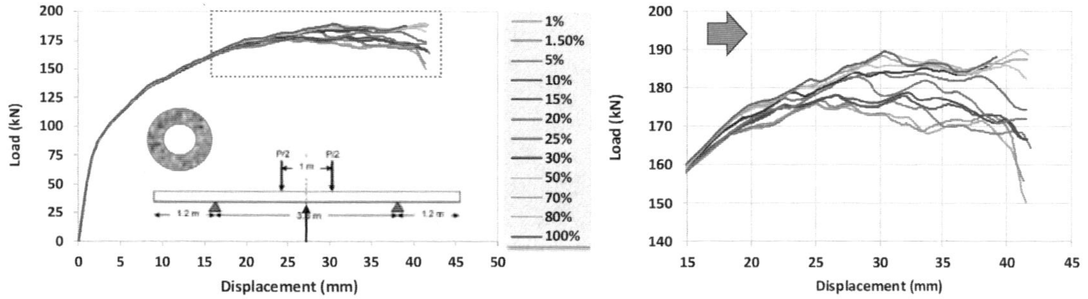

Figure 3. Vertical Load P (kN) vs maximum vertical at different ratios of ρ/ ρ_{ACI}

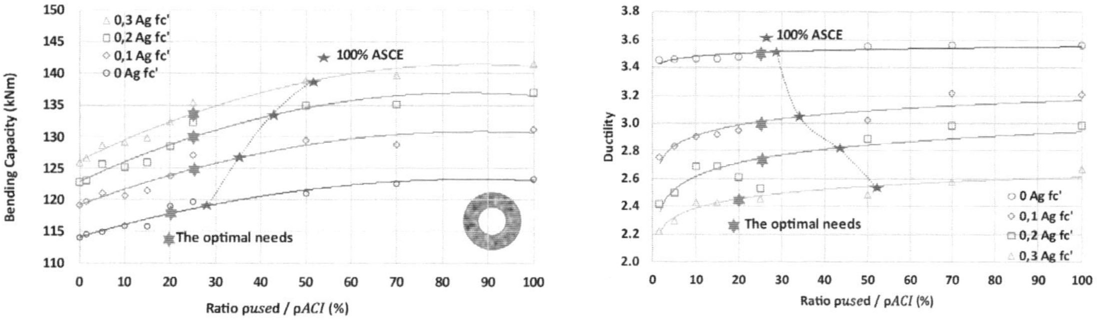

Figure 4. The effect of ρ to (a) Bending Capacity and (b) Ductility of Spun Pile at different axial load

3.2. Spun Pile with Concrete Infill

The same pile diameter was used for the parametric study of the spun pile with concrete infill referring to the experimental study conducted by [13]. The difference is the concrete strength and the yield strength of the spiral, which were 58.4Mpa and 596Mpa, respectively. A similar trend of the push-over graph as shown in Figure 3 was obtained where the amount of transverse reinforcement affects the post-peak stage. Figure 5 presents the push-over curves of a different axial load of a spun pile confined by 3.2mm-diameter and pitch of 100mm. The graph can be approached as a tri-linear line. The first line presents the elastic stage until the first yielding, the second is the inelastic phase until the maximum capacity and the last one is the post-peak stage. The axial load does nor affect the first line but it increase the capacity of the spun pile. However, the amount of axial load accelerate the failure of the pile. Srength degradation is more drastic when the axial load ratio is higher. Therefore, it affect the ductility of the spun pile. The spun pile loaded by 0.3 Fc'Ag is less ductile than that with 0.1fc'Ag.

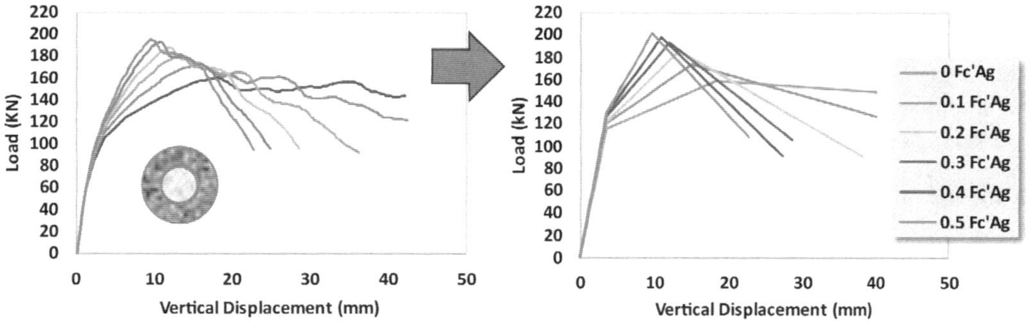

Figure 5. The Load-Displacement curves of different axial load

The strength and ductility with different transverse reinforcements are shown in Figure 6. Learning from the earlier parametric study, the maximum ratio to ACI is 70% since at this condition there is no significant change in the bending capacity and the ductility. To obtain the optimal transverse reinforcement needed, strength degradation was observed. The spun pile with the ratio of 21% to ACI requirement could reach the maximum bending where the bending capacity could reach 97% to the maximum bending when the axial load ratio is 0.1. The ductility value is 0.13 points lower than spun pile with a ratio of 70%. Meanwhile, in terms of its strength, the ratio of 28% to the ACI was considered optimal for those with the axial load ratios 0.2 and 0.3. At this condition, the bending capacity was about 94% to 95% of the maximum. However, the ductility drops significantly, which is 0.5 lower than the maximal value. Hence, to get a ductile spun pile, the reinforcement of 45% is optimal. The findings are close to the ASCE demand. The 100% reinforcement according to ASCE is equivalent to the ACI ratio of 28%, 36% and 51% for the axial load ratio of 0.1, 0.2 And 0.3, respectively.

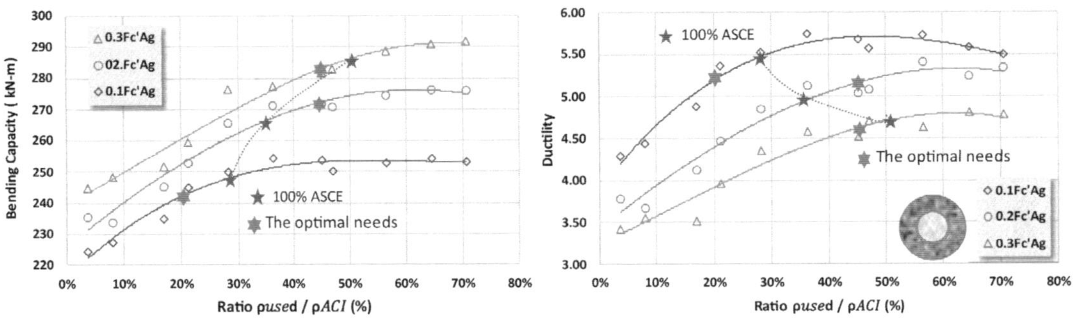

Figure 6. The effect of ρ to (a) Bending Capacity and (b) Ductility of Spun Pile at different axial loads

4. Conclusion

The volumetric ratio of transverse reinforcement based on ACI 315-19 does not suitable for the foundation design. The requirement is uneconomic to apply. The numerical study found that an empty spun pile with 400mm-diameter and 75mm of wall thickness needs only 20% to 25% of ACI demand. Referring to ASCE code, the needs depend on the axial load ratio. Based on ASCE calculation, the spun pile only have to be reinforced by 10%, 30% and 50% to the ASCE demand for the axial load ratio of 0.1, 0.2 And 0.3 Fc'Ag, respectively. With this amount, the bending capacity could reach 96-97% of the maximum value and the ductility is 0.2 points less than the pile reinforced by 100% of ACI requirement. The spun pile with concrete infill need lower transverse reinforcement than the empty pile. A case study of a 400mm spun pile needs about 40% reinforcement than an empty pile. The parametric study, found that the spun pile that was loaded with a 0.1 axial load ratio need reinforcement of only 21% of ACI requirements to obtain the maximum bending and ductility that is equal to those with 70% of ACI reinforcement. Meanwhile, 0.2 and 0.3 of axial load ratios needs volumetric ratio of 45% to the ACI quantity. The study found a similar amount with the ASCE equation.

To conclude, further study should be carried out to reformulate the need for transverse reinforcement for the empty spun piles. The study found that the rational amount is considerably lower than the requirement set by ACI 318-19 and ASCE 7-16. However, the need for transverse reinforcement of the spun pile with concrete infill could be determined based on the equation set by ASCE.

Acknowledgments

The research is supported by Indonesia Ministry of Research and Universitas Indonesia through Doctoral Research Grant, contract number NKB-101/UN2.RST/HKP.05.00/2022.

References

[1] A. C. 318, *ACI 318-19: Building Code Requirements for Structural Concrete and Commentary*. 2019.
[2] ASCE 7 - 16, *Minimum design loads and associated criteria for buildings and other structures*. 2017. doi: 10.1061/9780784414248.
[3] A. Budek-Schmeisser and G. Benzoni, "Rational seismic design of precast, prestressed concrete piles," *PCI Journal*, vol. 53, no. 5, pp. 40–53, 2008, doi: 10.15554/pcij.09012008.40.53.
[4] A. Fanous, S. Sritharan, M. Suleiman, J.-W. Huang, and A. K. Arulmoli, "Minimum Spiral Reinforcement Requirements and Lateral Displacement Limits for Prestressed Concrete Piles in High Seismic Regions," *Reports and White Papers*, pp. 1–164, 2010.
[5] A. Budek and G. Benzoni, "Obtaining ductile performance from precast, prestressed concrete piles," *PCI Journal*, vol. 54, no. 3, pp. 64–80, 2009, doi: 10.15554/pcij.06012009.64.80.
[6] C. Irawan, R. Djamaluddin, I. G. P. Raka, Faimun, P. Suprobo, and Gambiro, "Confinement behavior of spun pile using low amount of spiral reinforcement - An experimental study," *International Journal on Advanced Science, Engineering and Information Technology*, vol. 8, no. 2, pp. 501–507, 2018, doi: 10.18517/ijaseit.8.2.4343.
[7] O. Mulia, P. Widjojo Adi, and M. Nadila, "Numerical study of low confinement spun pile to pile cap connection," *E3S Web of Conferences*, vol. 347, p. 01002, 2022, doi: 10.1051/e3sconf/202234701002.
[8] T. Wang, Z. Yang, H. Zhao, and W. Wang, "Seismic performance of prestressed high strength concrete pile to pile cap connections," *Advances in Structural Engineering*, vol. 17, no. 9, pp. 1329–1342, 2014, doi: 10.1260/1369-4332.17.9.1329.
[9] Z. Yang and W. Wang, "Experimental and numerical investigation on the behaviour of prestressed high strength concrete pile-to-pile cap connections," *KSCE Journal of Civil Engineering*, vol. 20, no. 5, pp. 1903–1912, 2016, doi: 10.1007/s12205-015-0658-8.
[10] AASHTO, *AASHTO guide specifications for LRFD seismic bridge design*, 2nd ed., no. May. 2011. [Online]. Available: https://searchworks.stanford.edu/view/11842518%0Ahttps://bookstore.transportation.org/item_details.aspx?ID=1915
[11] Y. H. Chai and T. C. Hutchinson, "Flexural Strength and Ductility of Extended Pile-Shafts. II: Experimental Study," *Journal of Structural Engineering*, vol. 128, no. 5, pp. 595–602, 2002, doi: 10.1061/(asce)0733-9445(2002)128:5(595).
[12] P. S. Candra Irawan, Rudy Djamaluddin, I Gusti Putu Raka, "THE EXPERIMENTAL INVESTIGATION OF FAILURE MECHANISM OF SPUN PILE DUE TO MONOTONIC LOADING USING NEHRP 2000," in *International Conference on Protective Structures (ICPS4), Beijing, China*, 2016, no. October, pp. 313–322.
[13] C. Irawan, R. Djamaluddin, I. G. P. Raka, Faimun, P. Suprobo, and Gambiro, "The effect of the presence of infilling concrete on flexural performance of spun pile – An experimental study," *Jurnal Teknologi*, vol. 82, no. 1, pp. 85–94, 2020, doi: 10.11113/jt.v82.11974.

Cracks evaluation at a pier of a curved box girder bridge during its balanced cantilever construction sequence

N H Hidayat[1] **and H Purnomo**[2]

[1]Master Program of Civil Engineering, Department of Civil Engineering, Universitas Indonesia, Depok 16424, Indonesia
[2]Department of Civil Engineering, Universitas Indonesia, Depok 16424, Indonesia

E-mail: nur.hafidz01@ui.ac.id

Abstract. In urban area, modern role of bridges is to connect two or more locations. In that area, it is often to construct long span road crossing curved integral bridges consisting of several spans. In this study, the same type of bridge which is located at the intersection of an existing road and an elevated toll road in Jakarta showed cracks at one of the piers during construction of bridge segments. Construction time started from the beginning of 2016 until the end of 2018. Observations were done after the occurrence of cracks by installing strain gauge sensor at the lower part of the pier. Strain from the strain gauge is not only influenced by the construction activities but it is also influenced by the temperature of the surrounding environment. Construction staging analysis using MIDAS Civil software was conducted to evaluate the cracks. Numerical strain evaluation was done by considering the temperature effect taken from the measurement. Both stress which is calculated from the strain obtained by the measurement and construction staging analysis are in good concordance. Using the allowable stress limit for both obtained stress, the actual condition at the pier were verified.

1. Introduction

Horizontally Curved bridges are often found in urban crossroads. Most curved box girders are continuous spans. At the design stage, bridges at complex intersection locations often require a main span of more than 100 m. This is because it is difficult to find a location to put the position of the pier. The curvilinear nature of box girder bridges, along with their complex deformation pattem and stress filed, have led designer to adopt approximate and conservative methods for their analysis and design [1]. The global behavior of curve bridges can be very different from that of straight bridges. The curvature results in the placement of the load off-center and then the load induces torsion to the supporting elements such as bridge piers. The curve geometry of the bridge will result development of transverse moment, which can increase the normal stress on the outside edges of the bridge and can result in higher tension or and or compression stresses [2]. The magnitudes of such effects depend on the radius of curvature, span configuration, cross-sectional geometry, and load patterns among other parameters [2]. In this study, the type of box girder balanced cantilever curve horizontally bridge is continuous span. The balanced cantilever method is a construction method that utilizes the balance effect of the cantilever resting on the pier. Piers that support segment by segment arranged as cantilevers on both sides with the help of strands which are then distressed to balance with the vertical coordinates of the deck along the cantilever must be within the accepted tolerance value [3]. The total length of the bridge is 494 m consisting of 5 spans (93+145+103+97+55 m) with a radius of curvature of 283m. Construction time starts from the beginning of 2016 until the end of 2018. This bridge is part of the elevated toll road. The bridge has been carried out loading tests with static and dynamic tests carried out which are required for operational permits. The bridge in this study has been operated and can be seen in figure 1. During the construction period, this bridge underwent a process in such a way that it became a complete bridge. The process itself is referred to as the construction phase. The construction stage consists of the stages of work and the stages of work time. The construction stage is influenced by self-load factors, tool loads, environmental loads. Data at the construction stage is very important to be obtained in evaluating the behavior of the post-construction bridge. In this study, we want to know the behavior of the pier during the construction phase. Modeling analysis is done by FEM on the element pier. the monitoring process during the construction of the behavior of the pillars began to show fine cracks in the concrete cover area on the outer side of the bridge radius at segment 13 of 16 segments. Monitoring in the field during the construction phase is carried out by installing a VSWG (Vibration Wire Strain Gauge) sensor at the foot of the pier. Sensor readings produce strain and temperature during the construction phase. Later the results of the sensor readings will be compared with the results of the FEM model analysis. So that in this study, we can study the causes of cracks that occur in the box girder curved horizontally bridge pillars during the bridge construction stage.

2. Methodology

2.1. Description bridge

Type of bridge in this research is prestressed reinforced concrete box girder curve horizontally. This bridge crosses the existing road and the existing elevated toll road which can be seen in figure 1. The span configuration of the bridge consists of a continuous box girder cast in site have 5 spans (93+145+103+97+55 m) and supported on reinforced concrete piers. The bridge deck has a radius of curvature of 283 m with a total length of 464 m and a bridge width of 14 m. Plan and profile of bridge can show in figure 2. The bridge deck consists of a main span of 145 m. The piers consist of 4 monolith piers and 2 EJ piers. The highest of the pier is 16.66 m. The construction method used is segmental balance cantilever at 3 pier locations and 3 other piers using shoring assistance. The box girder cross section can be seen in figure 3.

Figure 1. Ns-link bridge

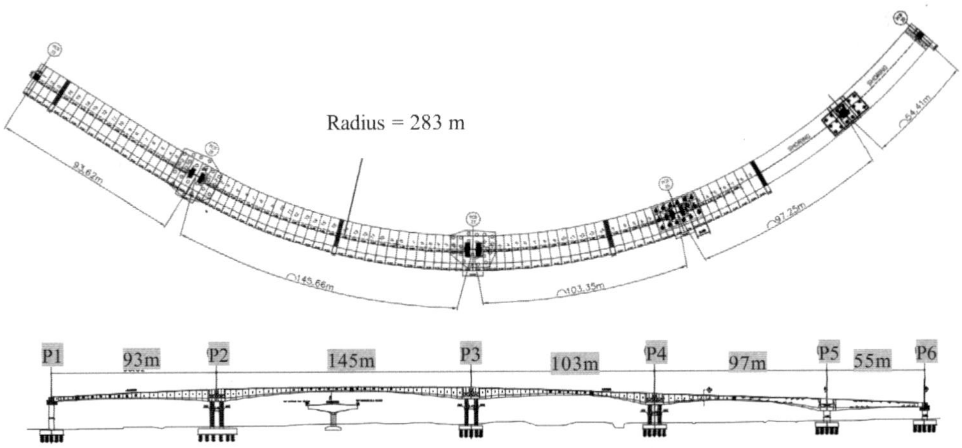

Figure 2. Plans and profile ns-link bridge

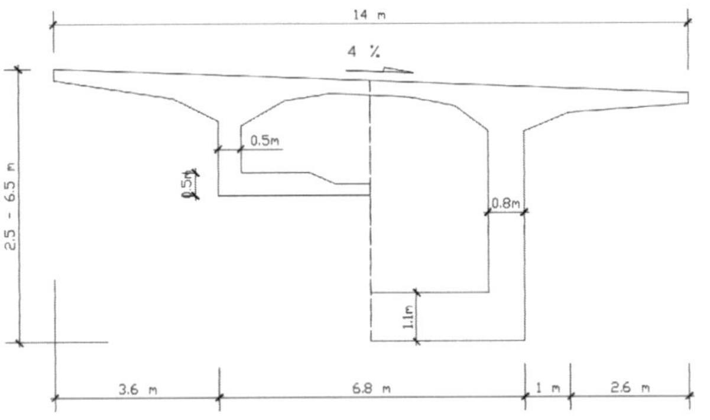

Figure 3. Cross section of deck

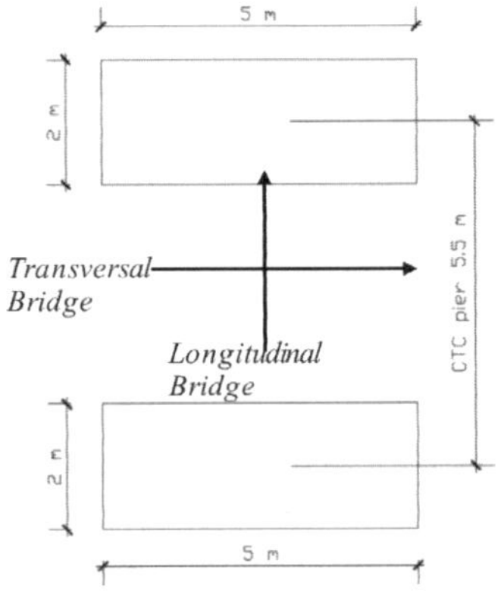

Figure 4. Cross section of pier

The materials used this bridge are concrete, reinforcement, and tendons prestressed. It is very important to obtain information related to the quality of the design materials and test materials. This is because there is often a difference between the quality of the plan and the quality used in the field based on the results of material tests. Detailed information on the quality of the design concrete characteristics and the test concrete characteristics can be seen in Table 1. Tendon characteristics can be seen in Table 2. In the numerical modeling analysis, the quality of the test results will be used, this is based on the quality of the test results that have been confirmed by the contractor and supervisor, that the quality of the material used during the construction of the bridge.

Table 1. Concrete characteristics

Element	Value (MPa)	
	Design	Tested
Box Girder Concrete Grade (fc')	49.8	59.52 ± 1.19
Hammer Head Concrete Grade (fc')	49.8	55.40 ± 2.32
Pier Concrete Grade (fc')	49.8	52.91 ± 4.88
Pile Cap Concrete Grade (fc')	29	34.95 ± 1.20
Foundation Bore Pile Concrete Grade (fc')	29	35.33 ± 0.79

Table 2. Tendon characteristics

Tendon	Value		Unit
	Design	Tested	
Ultimate Tensile Strength (UTS)	1860	1970	MPa
Diameter	15.2	15.2	mm
Area	140	140	mm^2
Friction factor (μ)	0.3	0.2	1/rad
Wobble factor (k)	0.007	0.008	1/m
Jacking Force (75% UTS)	1395	1425	MPa
Grade	ASTM 270, low relaxation		

Information regarding the number of strands for each tendon can be seen in Table 3. The location of each type of tendon has been determined by the bridge planner which can be seen in the tendon scheme of the bridge in figure5.

Table 3. Tendon properties

Type Tendon	n	Diameter (inch)
T1 - T16	19	0.6
TE1 - TE2	12	0.6
B1 - B12	19	0.6
BE1 - BE6	19	0.6

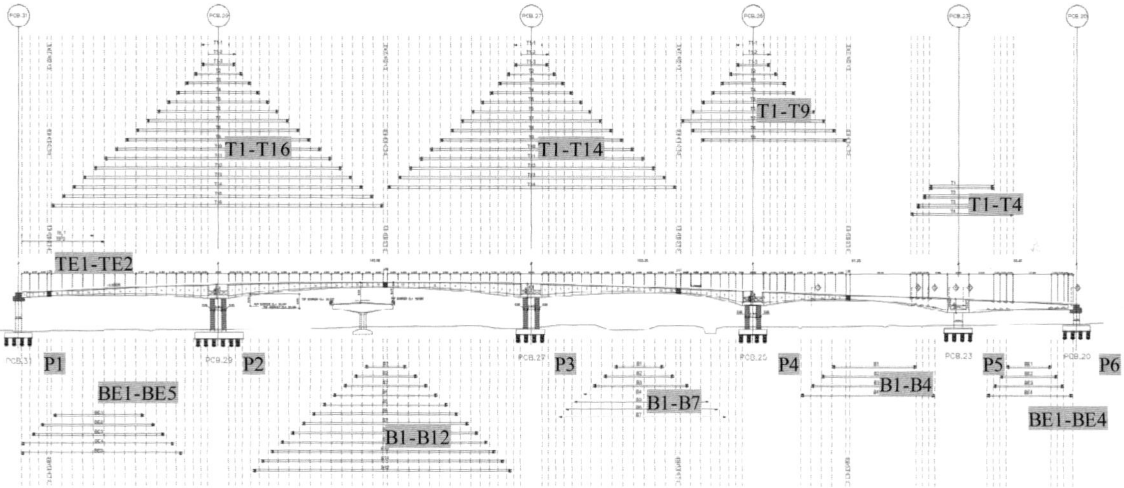

Figure 5. Tendon scheme

2.2. Construction stage

Bridge construction method with balanced cantilever. The balance cantilever method is a construction method that utilizes the balance effect of a cantilever resting on a pier, pier that support segment by segment which are arranged as cantilevers on both sides with the help of tendons which are then distressed to balance with the vertical girder coordinates along the cantilever must be within the accepted tolerance value [3]. The initial stage begins with the construction of the pier for each location. The next stage is hammer head work on each pier. The bridge erection starts from segment 1 as a support segment to the end segment and will be connected to the closure segment. The stages of construction from the beginning to connecting to a bridge structure can be seen in the figure. The duration of the processing time in 1 cycle segment is between 7 to 14 days. Load of during construction include traveller load, workers load, wet concrete, and casting support erection [4]. In this research, the position of the observation pier is located on pier P2.

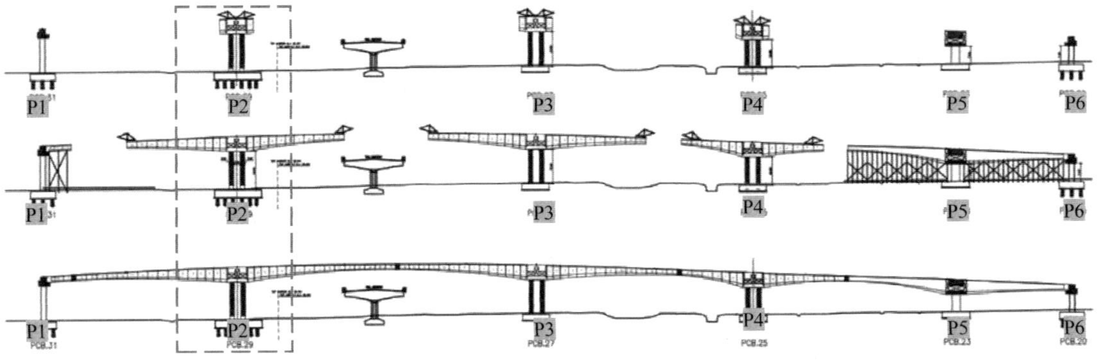

Figure 6. Construction stage ns-link bridge

Figure 7. Ns-link bridge during construction

2.3. Monitoring of concrete pier strains during construction

In this research, the method used to determine the change in strain on the concrete surface of the curved bridge pier is to install a strain gauge sensor in areas of the concrete surface that experience the greatest tension and stress. The type of strain gauge sensor installed is the VSWG (Vibration Wire Strain Gauge) type from Geokon Model 4000 [5]. The location and process of installing the sensor can be seen in figure 8 and figure 9. The number of vibration wire strain gauges installed is 4 pieces. This tool measures the strain in pier P2 during the construction phase of segments 15 – 16 until the closure and bridge work are perfectly connected. The results of the tool strain readings will be compared with numerical analysis. The value of the strain reading from the sensor still has additional factors due to the influence of temperature, it needs a follow-up analysis related to how big the temperature influence factor is on the strain reading in the construction stage conditions. The reading of the strain value is done every 5 minutes ($\Delta t = 5$ minutes).

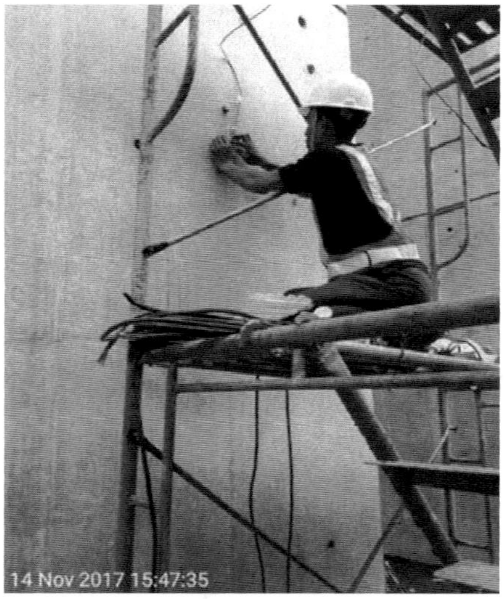

Figure 8. Process installation sensor P2

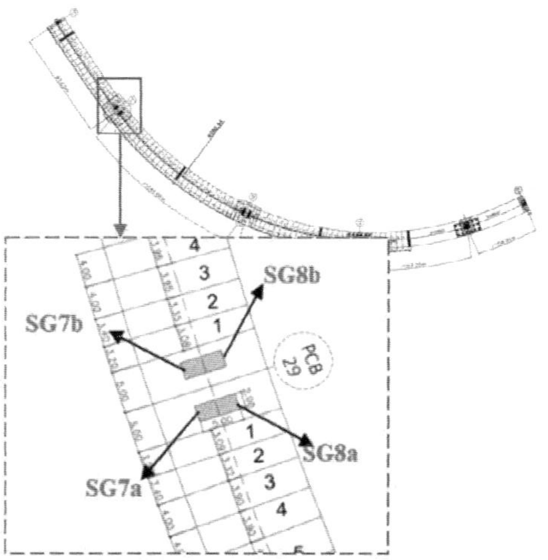

Figure 9. Plan location sensor in P2

2.4. Modelling

FEM analysis using plate and shell elements is currently accepted as the most accurate level of analysis available for box girder bridges [2]. Solid elements are used in 3D element pier modelling. In this research, FEM modelling focuses on the structural elements of the P2 pillars of the bridge. This is because it is the location of the element structure which is fitted with a strain gauge sensor. The modelling is assisted by Midas civil software. The modelling analysis considers the construction stage according to the stage that has been carried out in the field, the 3D model at the cantilever full segment stage can be seen in figure 10. The loads that work on the construction stage modelling include self-weight, equipment, stressing tendons, workers.

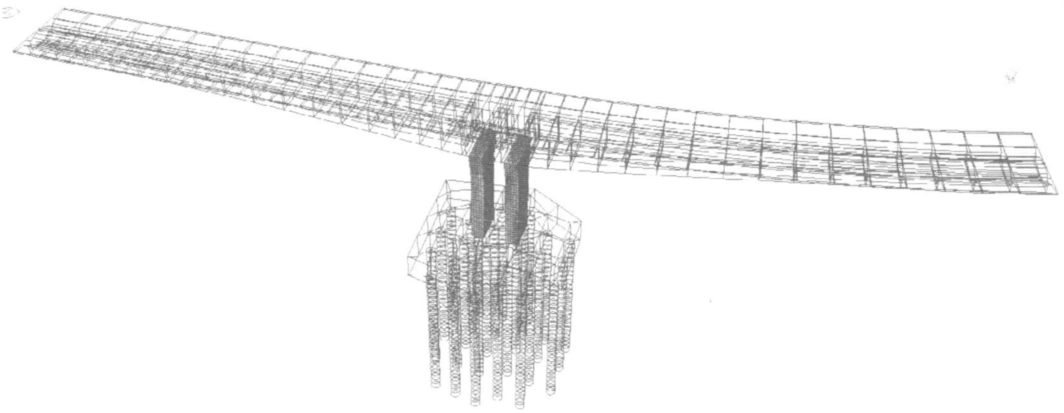

Figure 10. Model 3D P2 during construction

The modelling is continued in the full segment closure staging process until all lower tendon stressing has been completed. The final stages of the construction stage modelling can be seen in figure 11. From the results of this modelling will get the principal stress at the sensor locations for each construction. The principal stress values analysis from the modelling are only influenced by the static construction stage. There is no additional influence due to environmental factors, namely

IOP Conf. Series: Earth and Environmental Science 1173 (2023) 012005 doi:10.1088/1755-1315/1173/1/012005

changes due to temperature. Later this principal stress value will be multiplied by the modulus of elasticity of the test results to obtain the strain analysis.

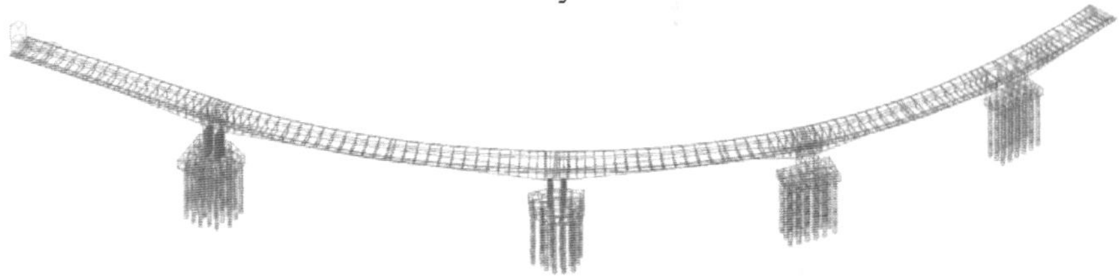

Figure 11. 3D model done construction stage

3. Result and discussion

3.1. Result strain monitoring

The result of the readings in the first 24 hours with no construction stage activity taking place [6], in figure 12 until figure 15 show measured strains and temperatures in 24 hours. Strain and temperature same time are read every 5 minutes by the strain gauge sensor. It shows that the amount strain is affected by temperature $\Delta\varepsilon = 90\ \mu\varepsilon$, So, it is carried out an analysis to separate the strain that is affected by temperature.

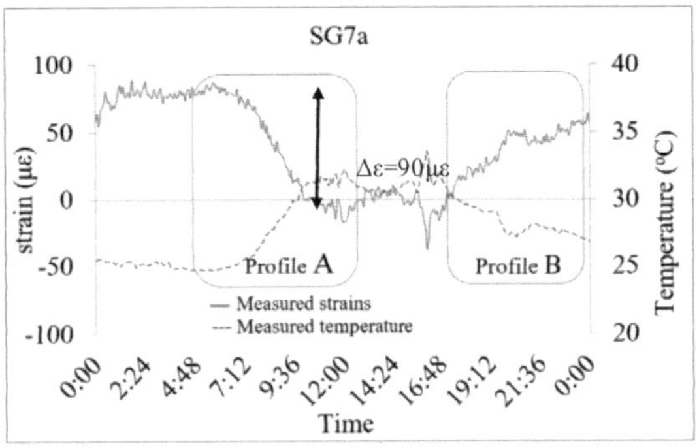

Figure 12. Measured strains & temperature SG7a

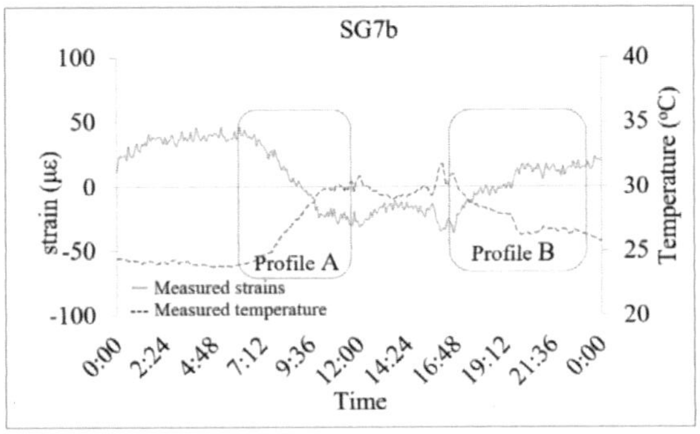

Figure 13. Measured strains & temperature SG7b

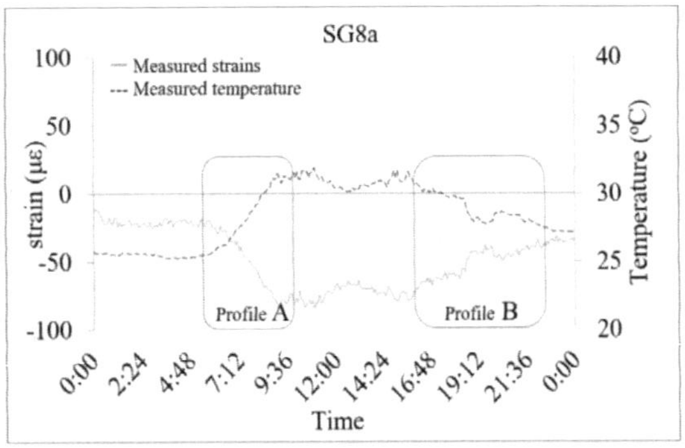

Figure 14. Measured strains & temperature SG8a

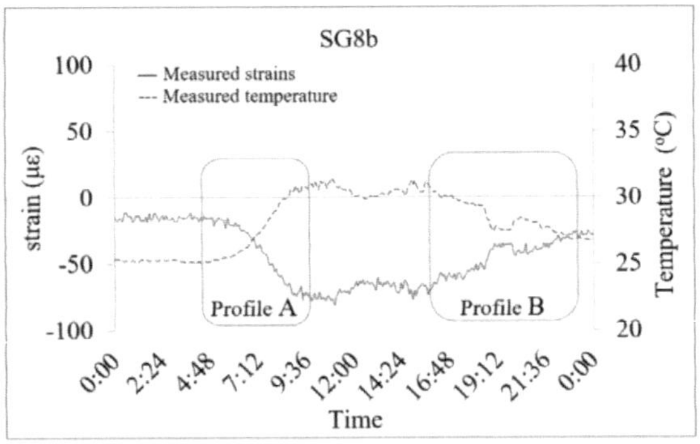

Figure 15. Measured strains & temperature SG8b

3.1.1. Analysis factor of temperature. The effect of temperature in the study can be seen from the condition that there is no erection stage activity, but the strain gauge sensor measurements produce strain that rises and falls constantly according to the amount of environmental temperature. We need to separate the effect of temperature in the process of reading the strain data so that we can get the results of strain measurements only due to the static forces of the construction stage. To determine the change in strain due to temperature alone, we need to take a strain reading of at least 24-hours [6] with the condition that there are no ongoing construction stages. Then the process continues after getting a 24-hour reading looking for a linear increase or decrease. We need to find the equation for the effect of temperature by making a linear regression equation in the region of increasing temperature and decreasing temperature. So, we will get a feeling of the influence of temperature. This can be seen in profiles A and B in figure 12 to figure 15. Profile A is a condition where temperature increases while the strain decreases. Profile B is condition where temperature decreases while the strain increases. The results of the regression analysis of each profile data at each sensor location can be seen in figure 16 to figure 19. Resume of the regression equation can be seen in Table 4.

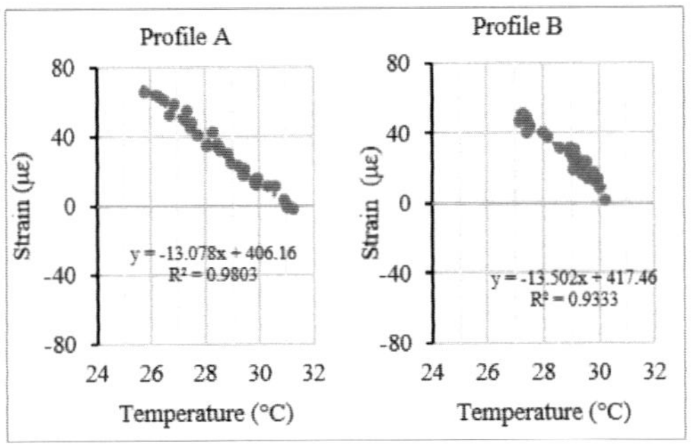

Figure 16. Regression analysis SG7a

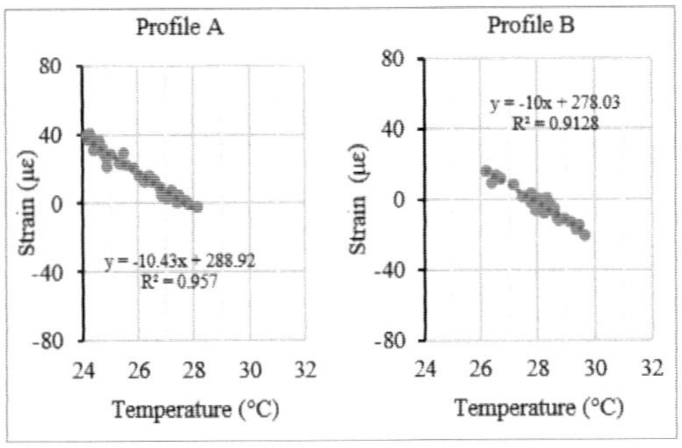

Figure 17. Regression analysis SG7b

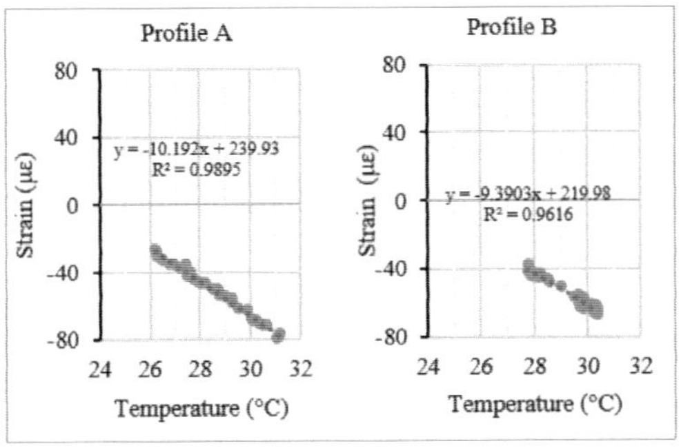

Figure 18. Regression analysis SG8a

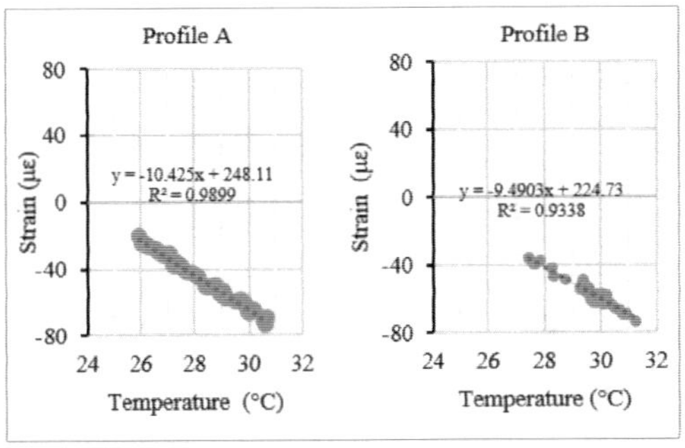

Figure 19. Regression analysis SG8b

Table 4. Resume analysis regression equation

No. Sensor	Profile	regression equation	R^2	X[1]	Y[2]	Delta
SG7a	A	y = -13.078x + 406.16	0.9803	25	79.21	0.9%
	B	y = -13.502x + 417.46	0.9333	25	79.91	
SG7b	A	y = -10.430x + 288.92	0.9570	25	28.17	0.5%
	B	y = -10.000x + 278.03	0.9128	25	28.03	
SG8a	A	y = -10.192x + 239.93	0.9895	25	-14.87	0.6%
	B	y = -9.3903x + 219.98	0.9616	25	-14.78	
SG8b	A	y = -10.425x + 248.11	0.9899	25	-12.52	0.1%
	B	y = -9.4903x + 224.73	0.9338	25	-12.53	

[1] Temperature (°C)
[2] Strain (με)

The regression equation obtained by the average R^2 value is above 0.9, which is close to 1. The researcher believes that this equation can be a way to get the strain that is affected by temperature. The result of the strain value calculated by the obtained equation will be a deduction value from the result of the sensor strain reading. So, at the end will get the strain value that has been factorized by the influence of temperature. Then the strain measurement will be compared with the strain value from the numerical analysis in figure 29 to figure 32. The corrected strains on each sensor can be seen in figure 20 to figure 23.

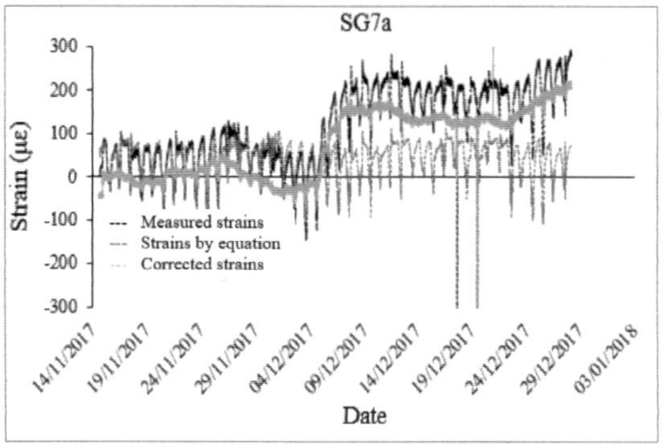

Figure 20. Corrected strains SG7a

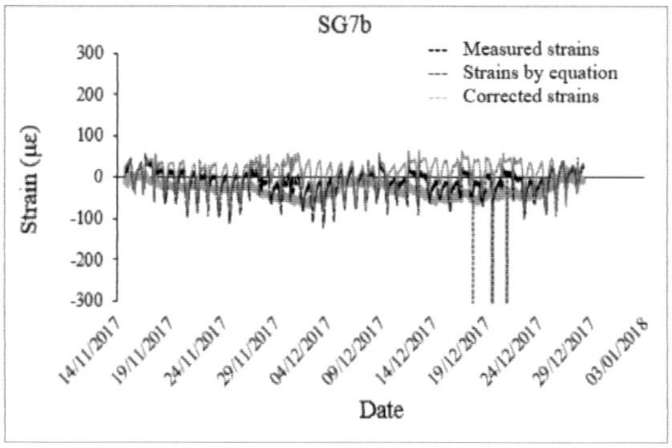

Figure 21. Corrected strains SG7b

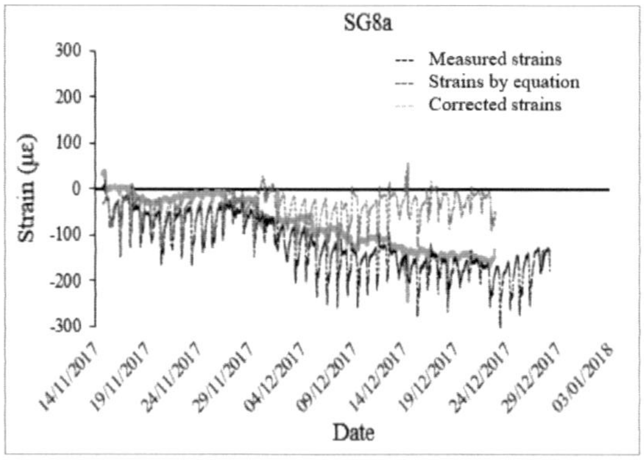

Figure 22. Corrected strains SG8a

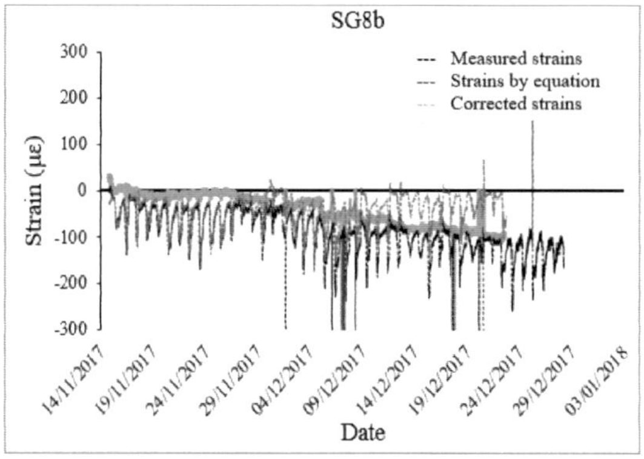

Figure 23. Corrected strains SG8b

3.2. *Result strain analysis.*

Based on the model, we can predict the cracks on concrete surface by using allowable tensile stress by Indonesia standard. From the analysis at the segment 13, the stress at the lower part of pier P2 has passed already the allowable tensile stress. The contractor has found the crack during the initial construction of segment 14. It is confirmed from the analysis and cracks finding on the field that crack appears during construction of segment 13. The value generated from numerical modeling is the stress value at each sensor position. Meanwhile, to convert it to the strain value is by using the equation of the relationship between strain and stress [7]. Where the value of strain is the stress which is influenced by the modulus of elasticity of the material used with the equation $E=4700\sqrt{fc'}$. So based on these equations we can find the strain value. Strain is analysis per point location of sensor can be seen in figure 25 until figure 28. We need to remove dead load before segment 14 stage to be able to compare with corrected strains, its because install sensor after construction stage segment 14. Model 3D in element pier of bridge used finite element method. We can get surface stress in FEM analysis where location max experience tensile stress. Surface FEM sample in max condition construction stage segment 16 can be see in figure 24.

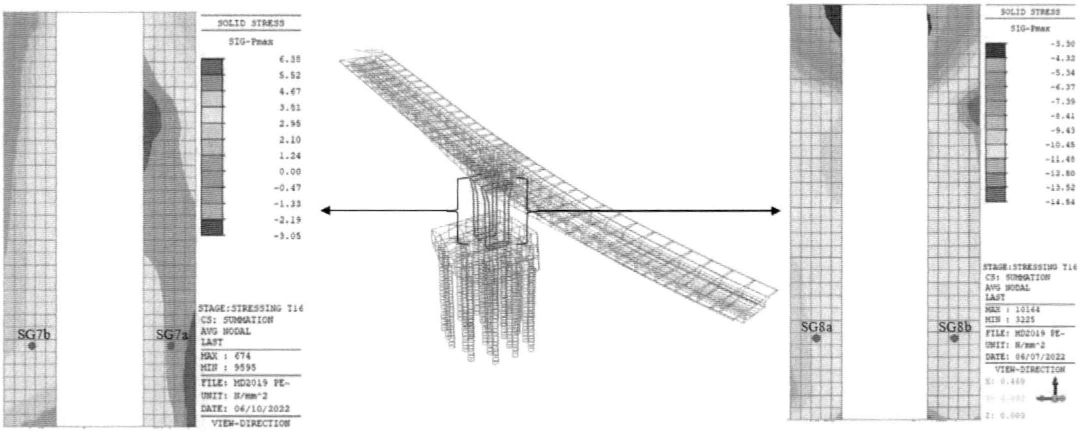

Figure 24. Model 3D FEM pier P2 during construction

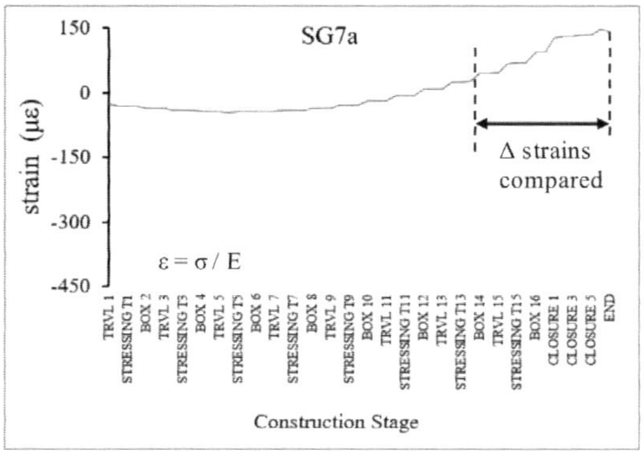

Figure 25. Strains analysis SG7a

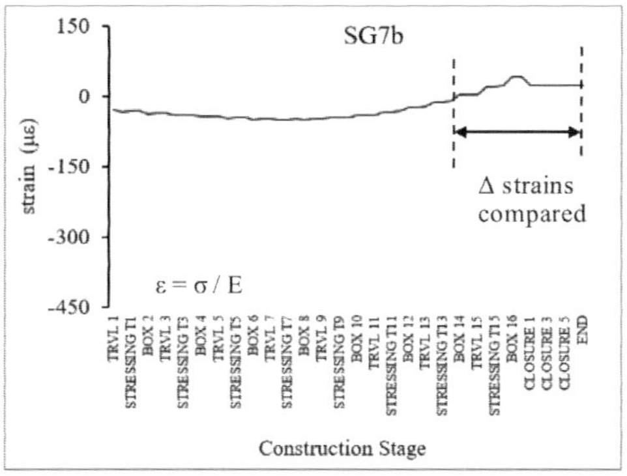

Figure 26. Strains analysis SG7b

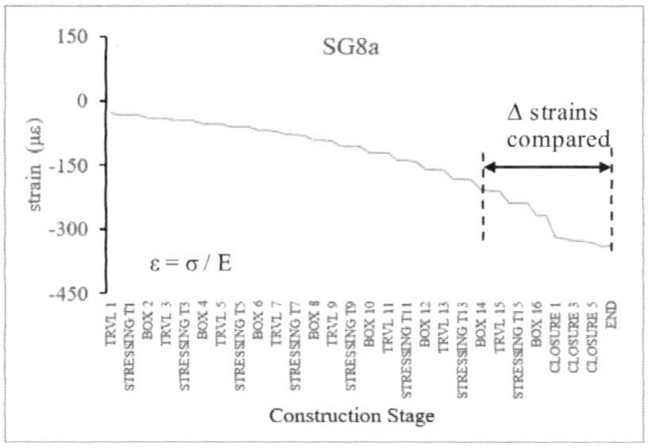

Figure 27. Strains analysis SG8a

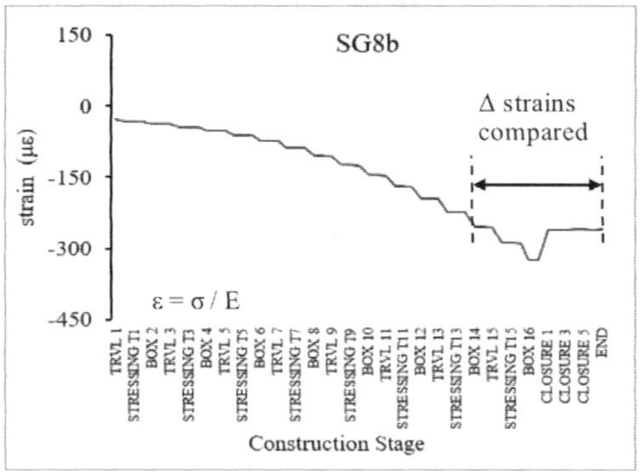

Figure 28. Strains analysis SG8b

3.3. Strain monitoring vs strain analysis

The result of this research is the strain obtained from the process of monitoring the concrete strain on the pillars during the construction period will be compared with the strain from numerical analysis. As well as observing the cracks that occur on the surface of the concrete with sensors installed. When the strain has exceeded the limit of the concrete tensile permit [7]. Then the condition of the concrete will experience cracks. The following comparison results between strains measurement and strains analysis can be seen in figure 29 to figure 32.

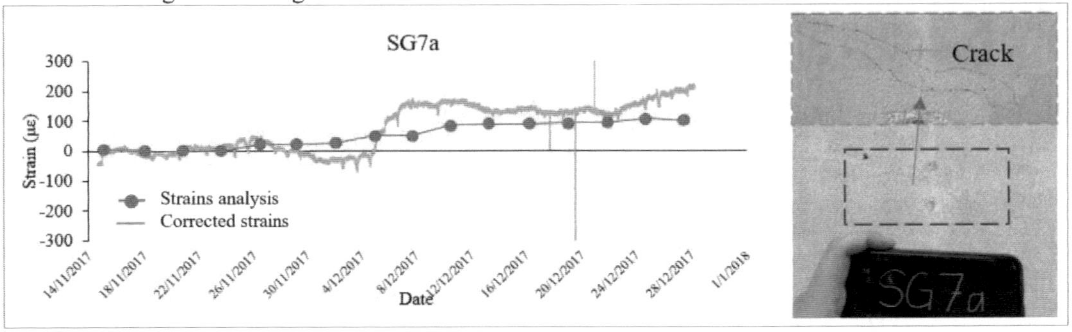

Figure 29. Compered strains analysis with corrected & visual surface concrete SG7a

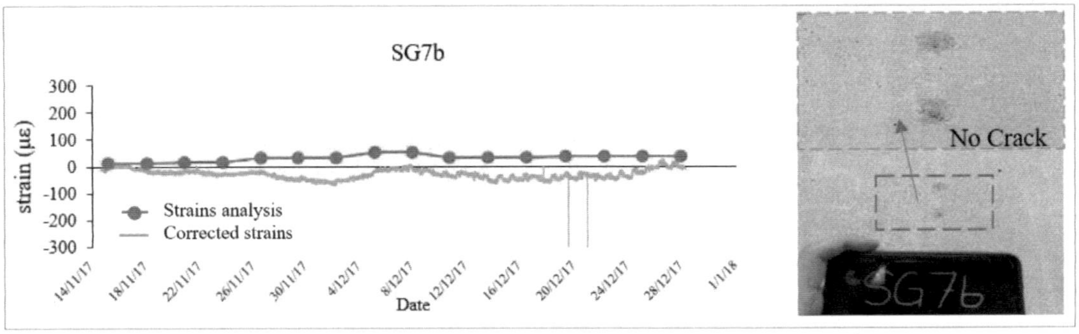

Figure 30. Compered strains analysis with corrected & visual surface concrete SG7b

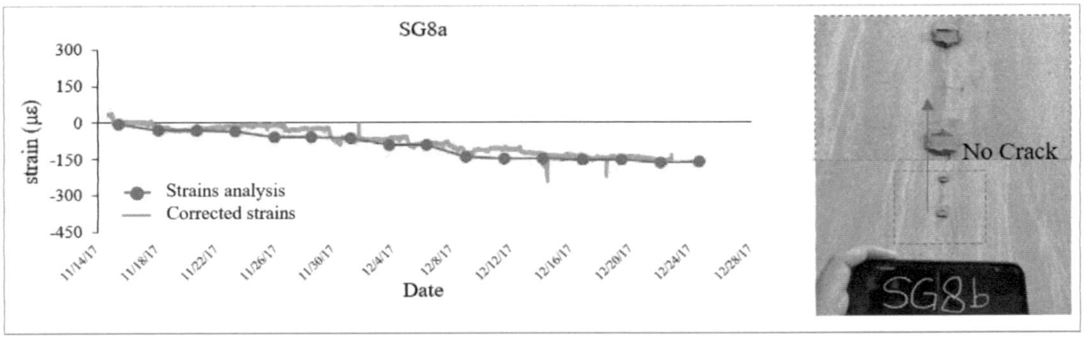

Figure 31. Compered strains analysis with corrected & visual surface concrete SG8a

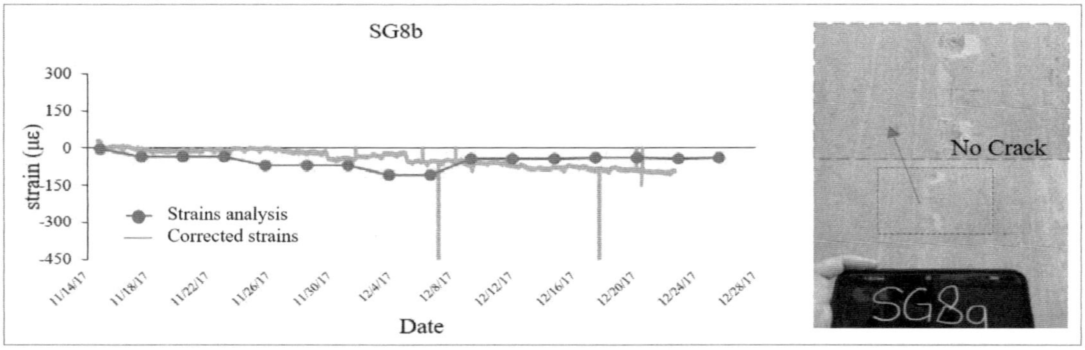

Figure 32. Compered strains analysis with corrected & visual surface concrete SG8

4. Conclusion

The result show that:

1. From the measurement, each sensor shows that the amount of strain is affected by temperature $\Delta\varepsilon = \pm 90$ με. So, it is necessary to carry out an analysis to separate the strain that is affected by temperature.
2. From the comparison of the corrected strains and strains obtained by analysis, the maximum tensile strain at position SG7a is more 100 με and it is confirmed by visual field observations that on the surface of the concrete area under the sensor is cracked.
3. From the comparison of the corrected strains and strains obtained by analysis, the maximum compression strain at position SG8a is more 150 με and it is confirmed by visual field observations that on the surface of the concrete area under the sensor is no cracked.
4. During construction of horizontally curved balanced cantilever bridge construction the pier will always experience tension on the outer side of the bend radius of bridge.

References

[1] Senna K M and Kennedy J B 2002 Literature review in analysis of curve box girder bridge *J. of Bridge Eng. ASCE* **7** 134-143
[2] NCHRP R620 2008 *Development of Design Specification and Commentaryfor Horizontally Curved Concrete Box Girder Bridge* (National Cooperative Higway Research Program)
[3] Hernandez S, Pablo O, Luis E R 2014 Optimization of PT in cantilever construction of prestressed concrete bridge *Madrid: Symposium Report IABSE* **102** 1536-1543
[4] Delta G S 2017 *Report Analysis and Traveller NS-Link PCB Bridge Becakayu. (In Bahasa)*
[5] Geokon M 4000 S VWSG 2017 Instruction Manual
[6] Risen J I R 2018 *Daily Report Instrument Bridge Tol Becakayu PCB27-PCB29 Ref.No 70-32/re c/i/2018. (In Bahasa)*
[7] RSNI T-12 2004 *Design of concrete structures for bridges*. (*In Bahasa*)

Evaluation of seismic vulnerability assessment building using rapid visual screening method: case study at the Faculty of Islamic Economics and Business of Islamic State Prof. KH. Saifuddin Zuhri University

M S H Pramesti[1], **A Widyaningrum**[1] **and N I S Hermanto**[1]

[1]Civil Engineering Departement, Jenderal Soedirman University, Indonesia

Email: arnie.widyaningrum@unsoed.ac.id

Abstract. Indonesia is one of the nations that is more prone to earthquakes due to its location at the intersection of the Indo-Australian, Eurasian, and Pacific plates. This study aims to find out the vulnerability and final score of the Faculty of Islamic Economics and Business of Islamic State Prof. KH. Saifuddin Zuhri University used the RVS method based on FEMA P-154, where this method is one of the earliest methods used to evaluate the vulnerability of a building, and has a final value limit of 2. If the final score is more than 2, then the building under review has a low vulnerability to earthquakes and vice versa. This building is one of the mid-rise buildings in Purwokerto which was established in 2019. Based on RVS analysis, the final score was 3.6 for the category of concrete moment-resisting frames (C1) and 1.4 for the category of concrete frames with unreinforced masonry infill walls (C3). Thus, the results showed this 6-story building has a vulnerability to earthquakes with a high category of a high probability of damage to level 3 and a very high probability of level 2 damage.

1. Introduction

Indonesia is a country located at the junction of three major tectonic plates, namely the Indo-Australian plate, the Eurasian plate, and the Pacific Plate. Thus, Indonesia is included in the category of countries prone to earthquakes. An earthquakes itself is a vibration that originates from the earth's surface resulting from the sudden release of energy. Earthquakes also generate seismic waves that can cause damage to infrastructure and buildings.

Two factors constitute the level of earthquake risk on buildings, the first factor is a hazard, this factor cannot be prevented or reduced because this factor is natural. Then, the second factor or vulnerability can be minimized by reducing the level of vulnerability to existing buildings or infrastructure.

One of the efforts in mitigation is the vulnerability of buildings to earthquakes. Building vulnerability is one of the various factors that can cause a building to be damaged or unable to meet the desired performance in the event of an earthquake. The desired performance is that building structure keeps the building from collapsing when an earthquake occurs. For this reason, efforts to reduce earthquake risk (earthquake risk reduction) are needed, one of which is by evaluating buildings against earthquake susceptibility with the Rapid Visual Screening method.

Rapid Visual Screening (RVS) is a method of researching a building against the potential of earthquake hazards based on visual observations of the building's exterior, and interior if possible so that its implementation is relatively fast [1]. This method is an easy first step and can determine the vulnerability of buildings built based on the Federal Emergency Management Agency P-154 (FEMA-154) guidelines [2]. By using these guidelines, if the final score is above 2, then the building being reviewed has a low seismic vulnerability but otherwise, the building has a high seismic vulnerability.

Based on the explanation above, the author will describe the research regarding the evaluation of earthquake vulnerability to the building of the Faculty of Economics and Islamic business (FEBI) UIN Prof. KH. Saifuddin Zuhri is located in Purwokerto, Central Java. This building was built in 2019 and functions as a school that has a large capacity. This building has six floors with a building height of approximately 33.5 m. The function of this building has a crucial type of function because it has a large capacity and the structure of the building is high enough that it is felt necessary to evaluate and analyse quickly using the RVS method. This study aims to determine the level of vulnerability of a 6-story building based on the final score obtained and the factors that affect it based on the RVS method.

2. Literature review

An earthquake is a seismic vibration that originates from the plates of the earth's surface. Indonesia's territory is geographically located at 6°N - 11°S and 95°E - 141°E, in which there are three active plates (triple junction) [3].

Rapid Visual Screening (RVS) is a method that was first introduced by FEMA in 1998. In 2002, due to the earthquake disaster in the 1990s, this method was modified to keep up with existing technological developments. Some countries also modify this method to suit the region of their respective countries.

The RVS method used for hazard vulnerability has been discussed in several journal series Techniques for Rapid Assessment of Seismic Vulnerability [4]. FEMA-154 was first issued in 1988, and in 2002 FEMA issued a second edition to improve existing procedures. In January 2015, FEMA issued a third edition of the procedures for the FEMA P-154. This RVS is the initial research method to evaluate the vulnerability assessment is performed by filling out a questionnaire with direct observations of a building with components that must be considered as the location of seismicity, types of dwelling, types of soil, falling hazard structural elements, building type, number of floors in the building, the proximity between building (adjacency), vertical irregularity, plan irregularity, seismicity regulations, and scoring.

The scoring in evaluating buildings is by circling the score under the type of building (building type) according to the building being evaluated. After completion, all the scores that have been circled are added up and the final score (S) is obtained. If the final score obtained is more than or equal to 2, then the building under review has a low vulnerability to earthquakes. However, if the building has an S value of less than 2, then the building has a high vulnerability to earthquakes and it is recommended to carry out further evaluation using the FEMA-310 guidelines [5].

Research using RVS based on FEMA P-154 has been carried out in several areas such as in India with the title "Earthquake Safety Assessment of Buildings through Rapid Visual Screening" [6], in Manila written by Clemente, et al with the title "Risk Assessment of Seismic Vulnerability of All Hospitals in Manila Using Rapid Visual Screening (RVS)", and in Purwokerto, Indonesia with the title "Seismic Evaluation in Engineering Faculty Building of Jenderal Soedirman University Using Rapid Visual Screening" written by Yumna Zakira A [7].

The research entitled "Study of Rapid Visual Screening of Buildings for High Potential Seismic Hazard According to Indonesian Standard" [8] to find out how important RVS from FEMA 154 can be applied in Indonesia with a case study of ATK Yogyakarta Building. The RVS FEMA 154 method with SNI 1726 can be said to have met so that RVS can be used in Indonesia by adjusting the earthquake acceleration map so that there are no errors in determining the earthquake zone. In addition, another study entitled "A Rapid Visual Screening Methodology for the Seismic Vulnerability Assessment of Historic Brick Mansory Buildings in Vienna" [9] explains that this RVS method can be used to classify the behavior of building structures without going through a long computational process. The results obtained are mapping the vulnerability of buildings almost throughout the Vienna city area so that buildings with a high level of vulnerability to earthquakes can be known.

3. Methods

The first thing to do in this study is to collect secondary data such as building construction data (as built), soil testing data, and seismicity zoning maps. Seismicity zoning maps are used to obtain Ss and S1 values. The map was obtained from the Indonesian National Standard (SNI 1726:2019) [10] seismic map as well as the application of response spectra provided by the Ministry of Public Works and Housing. Then after obtaining S_s and S_1 values based on the location of the study, the researcher can determine the type of RVS form according to the seismic region as in Table 1.

Table 1. Seismicity level of regional division (FEMA P-154) [2].

Region Seismic	Parameter Response Spectra, S_s	Parameter Response Spectra, S_1
Low	Less than 0.25g	Less than 0.1g
Moderate	Greater than or equal to 0.25g but less than 0.5g	Greater than or equal to 0.1g but less than 0.2g
Moderately High	Greater than or equal to 0.5g but less than 1.0g	Greater than or equal to 0.2g but less than 0.4g
High	Greater than or equal to 1.0g but less than 1.5g	Greater than or equal to 0.4g but less than 0.6g
Very High	Greater than or equal to 1.5g	Greater than or equal to 0.6g

Notes: g = acceleration of gravity in a horizontal direction

In the guidelines of SNI 1726:2019[10], there is each type for the irregularity of a building both vertically and horizontally. To get more accurate results, researchers conduct an irregularity analysis with the help of ETABS software.

After conducting a structural analysis with the help of ETABS software, the researcher conducted a field survey of the research site. This field survey takes 15-30 minutes [6]. In this survey, researchers conducted a visual review of buildings to determine the function of rooms and buildings, number of floors, building area, geological hazards, and exterior falling hazards. It's a good idea when conducting a field survey, researchers can be accompanied by people who understand the construction of the building so that they can get more detailed building data.

3.1. Information and building characteristics

The initial stage is to identify information and also the characteristics of the buildings according to the RVS form. Contains the address, postal code, type of dwelling, building name, coordinates, screener identity, survey implementation time, and the number of floors to the floor area of the building.

3.2. Photo, sketch, elevation, and building occupancy class

In the RVS form, there is a section to display photos and sketches along with the elevation of the building under review. For residential classes, it is a building function such as emergency services such as hospitals, commercial, or schools.

3.3. Soil types and geological hazards

Soil type is determined based on the guidelines of SNI 1726:2019[10] and FEMA P-154 (2015)[2]. The soil data used in this study was to use SPT data. Geological hazards in this RVS form, such as liquefaction, potential landslides, or cracks in the ground surface. During an earthquake, any of these three factors might enhance the probability of a building being damaged and collapsing.

3.4. Proximity and exterior falling hazards

The interaction between surrounding buildings can cause various types of damage during an earthquake. If there is no separation between buildings, then the buildings can hit each other when reacting to ground shaking. Another potential hazard is non-structural hazards falling from adjacent buildings. This can be in the form of chimneys, fences, complementary walls, tanks, cladding, or other building components.

3.5. Irregularity

Irregularity is divided into two types, those are vertical irregularity and plan irregularity. For the analysis of irregularities, this research uses ETABS software to get more accurate results. This irregularity analysis refers to SNI 1726:2019 [10] to comply with existing regulations in Indonesia.

3.6. FEMA building type and basic score

The two main characteristics of earthquake performance are the material of construction (e.g. wood, concrete) and the type of seismic resisting system (moment frame, steel frame, and shear wall). In FEMA P-154 [2], there are 17 types of buildings developed by FEMA and basic values for each seismic area, to measure the expected performance of each type of FEMA development in each region.

3.7. Modified score and final score

After determining the type of building and the basic score, the RVS form contains a modified score. The modification score consists of several determining elements, namely the number of building floors, vertical irregularities, and floor plans. The number of floors of buildings based on FEMA P-154 [2], they are classified into two types, namely mid-rise-building (4-7 floors) and high-rise-building (8 floors or more). For buildings that have several floors below 4 then the value is not taken into account.

3.8. Final scoring (S)

The sum of all scores becomes a final score (S). This final score is compared with a predetermined cut-off score. The specified cut-off-score value is two. If the final S value is more than or equal to two, then the building being reviewed is considered to have a low earthquake vulnerability. Likewise, vice versa, if the value of S obtained is less than two, the building that is evaluated to have earthquake vulnerability with a certain probability of damage. Figures 1 (a) and (b) are the five levels of classification of the final score of the RVS along with the damage that will be obtained to the building.

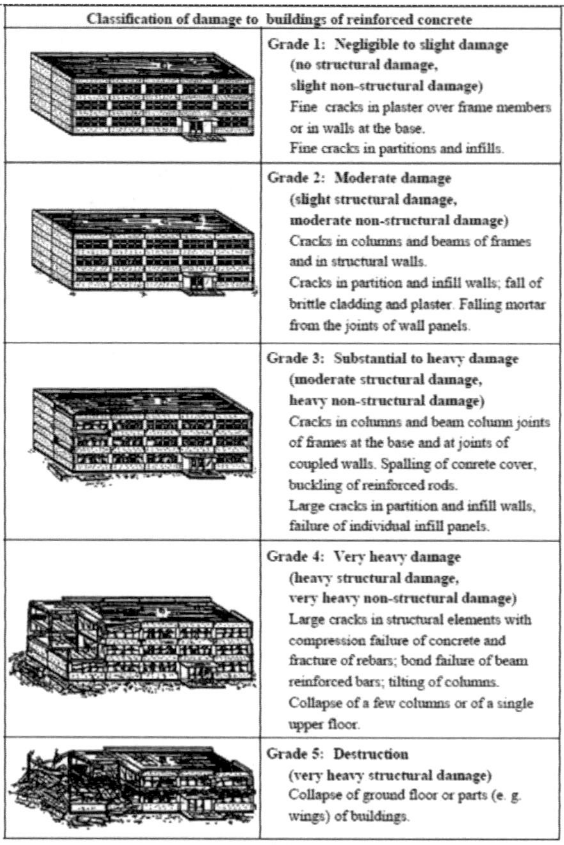

Classification of damage to buildings of reinforced concrete	
	Grade 1: Negligible to slight damage **(no structural damage, slight non-structural damage)** Fine cracks in plaster over frame members or in walls at the base. Fine cracks in partitions and infills.
	Grade 2: Moderate damage **(slight structural damage, moderate non-structural damage)** Cracks in columns and beams of frames and in structural walls. Cracks in partition and infill walls; fall of brittle cladding and plaster. Falling mortar from the joints of wall panels.
	Grade 3: Substantial to heavy damage **(moderate structural damage, heavy non-structural damage)** Cracks in columns and beam column joints of frames at the base and at joints of coupled walls. Spalling of conrete cover, buckling of reinforced rods. Large cracks in partition and infill walls, failure of individual infill panels.
	Grade 4: Very heavy damage **(heavy structural damage, very heavy non-structural damage)** Large cracks in structural elements with compression failure of concrete and fracture of rebars; bond failure of beam reinforced bars; tilting of columns. Collapse of a few columns or of a single upper floor.
	Grade 5: Destruction **(very heavy structural damage)** Collapse of ground floor or parts (e. g. wings) of buildings.

(a)

RVS Final Score (S)	Damage Potential
S<0.3	High probability of Grade 5 damage; Very high probability of Grade 4 Damage
0.3<S<0.7	High probability of Grade 4 damage; Very high probability of Grade 3 Damage
0.7<S<2.0	High probability of Grade 3 damage; Very high probability of Grade 2 Damage
2.0<S<2.5	High probability of Grade 2 damage; Very high probability of Grade 1 Damage
S>2.5	Probability of Grade 1 damage

(b)

Figure 1. (a) Five classes of damage to buildings and their explanation, (b) RVS final score and their explanation.

4. Results and discussions

This research was conducted at the Faculty of Economics and Islamic Business (FEBI) UIN Prof. KH. Saifuddin Zuhri (SAIZU) which is located at Ahmad Yani Street 57A Purwokerto, Central Java, Indonesia. This building was built in 2019 with 6 floors. The height of this building is approximately 33.5 m with a building area of approximately 632 m^2/floor. FEBI UIN SAIZU building has a main structure with an earthquake resistance system that is a concrete frame or open frame so that it has high ductile property. Figure 2 is the FEBI UIN SAIZU building.

Figure 2. The FEBI UIN SAIZU in Purwokerto.

4.1. Irregularity analysis results

Irregularity analysis in this research uses the help of ETABS software based on SNI 1726:2019. Torsion irregularity and torque overload, it is defined if the deviation between floors of the maximum level is more than 1.2 times and 1.4 times the deviation between floors of the average level at both ends of the structure. Internal angular irregularities are defined to exist if both projections of the structure plan from the inner angle are greater than 15% of the dimensions of the structure plan. Third, diaphragmatic discontinuity irregularities exist if they have a truncated or open area greater than 50% of the gross diaphragm. If there is a discontinuity in the trajectory of lateral force resistance, it can be stated that there are irregular consequences that shift horizontally. Finally, if the vertical lateral force holding elements are not parallel then there is a non-parallel system irregularity. Table 2 is the result of a recap of the irregularities in the plan which is the result of an analysis of the ETABS software. The results show that the FEBI UIN SAIZU building does not have any irregular plans in both the x direction or the y direction.

Table 2. Results of plan irregularity.

Plan Irregularity	X-dir	Y-dir
Torsion irregularity and torque overload	x	x
Internal angular irregularities	x	x
Diaphragmatic discontinuity irregularities	x	x
Irregularity consequences shift horizontally	x	x
Non-parallel system irregularities	x	x

The first type of vertical irregularity, namely torsion irregularity and torque overload is defined as existing if there is a degree where the lateral stiffness is less than 70% lateral stiffness level above it or less than 80% of the average stiffness of the three levels above it and the lateral stiffness is less than 60% lateral stiffness of the level above it or less than 70% of the average stiffness of the three levels above it. For heavy irregularities, there is if the effective mass of all levels is more than 150% of the effective mass level nearby. Thirdly, diaphragmatic discontinuity irregularities exist if the horizontal dimensions of the seismic force restraining system at all levels are more than 130% horizontal dimensions of the nearby seismic force restraining system. Irregularity consequences shift horizontally defined exist if the shift in the direction of the plane of the lateral force retaining element is greater than the length of the element. The last is non-parallel system irregularities, there is if the lateral strength level is less than 80% of the level above it and for the excess less than 65% of the lateral strength level above it.

Table 3. Results of vertical irregularity

Vertical Irregularity	X-dir	Y-dir
Soft level stiffness irregularity and excess soft level	x	x
Heavy irregularity (mass)	x	x
Vertical geometric irregularities	x	x
Irregularity due to discontinuity field on the vertical element of the lateral force bearer	x	x
Weak level irregularity	x	x

In table 3, it can be seen that neither direction x nor direction y of the FEBI UIN SAIZU Building has all five types of vertical irregularities. So it can be stated that the building has no irregularities either vertically or horizontally.

4.2. FEMA P-154 results

Based on the location survey and structural drawings, the FEBI UIN SAIZU building is a building type C1 (Concrete Moment Resisting Frame) and C3 (Concrete Frame Buildings with Unreinforced Mansory Infills Walls). The basic score is owned by each type of FEMA building.

Table 4. Results of filling put RVS form score

Building Type	C1 (Concrete Moment Resisting Frame)	C3 (Concrete Frame Buildings with Unreinforced Mansory Infills Walls)
Basic Score	+1.7	+1.4
Post Benchmark	+1.9	0
Soil Type A or B	0	0
Soil Type E (>3 stories)	0	0
Final Score	+3.6	+1.4

In its planning, the FEBI UIN SAIZU building uses the seismic regulation, namely SNI 1726:2019 [10], so that this building has a modified score for the posted benchmark. For the type of soil based on the results of the calculation of the average NSPT value owned by the FEBI UIN SAIZU building, it is included in the medium soil class (stiff soil) or D both in the guidelines of SNI 1726:2019 [10] and FEMA P-154 [2]. Thus, the modified score for soil type A or B and soil type E with the condition that the number of floors is more than three floors is not taken into account. Thus, the final score for the C1 building type is 3.6 and for the C3 building type, it is 1.4.

In this research, it was continued to fill out the optional RVS level 2 form. In the RVS level 2 form, there is a modified score for pounding or proximity between buildings. As explained by FEMA, this score is taken into account when the building is separated from adjacent structures by less than 0.5% of the lowest building height. Then, the tallest building around it is a residential house with a height of approximately 5 m and the distance from the FEBI UIN SAIZU building is approximately 11.8 m. Thus, the modified score for this pounding is not taken into account.

5. Conclusion

The RVS method is very helpful to find out the vulnerability of a building quickly. Because this method can be done without the help of computational calculations and structural engineering experts. The FEBI UIN SAIZU building is one of the mid-rise buildings with a total of 6 floors in Purwokerto and is located in the middle of the city. The function of this building is also a building for educational facilities and around it, there are residential houses. Thus, there is a need for an approach regarding the vulnerability of the building to earthquakes.

Factors that affect the final score of the FEBI IN SAIZU building are the basic score and post benchmark (score for buildings that have used seismic regulations) building types C1 and C3. The FEBI UIN SAIZU building has a final score for the C1 building type of 3.6 and the C3 building type of 1.4. The C3 type building has an S value below the cut-off-score of 2. So, based on FEMA guidelines, the FEBI UIN SAIZU building has a vulnerability to earthquakes with a high probability of level 3 damage and a very high probability of level 2 damage.

References

[1] FEMA P-154 2015 *Rapid Visual Screening of Buildings for Potential Seismic Hazards: A Handbook* (Washington D C: Federal Emergency Management Agency)

[2] FEMA 154 2002 *Handbook for Rapid Visual Screening of Buildings for Potential Seismic Hazards* (Washington D C: Federal Emergency Management Agency)

[3] Indonesia's earthquake 2011 Retrieved from: https://www.usgs.gov/

[4] Clemente S J C *et al* 2020 Risk Assessment of Seismic Vulnerability of All Hospitals in Manila Using Rapid Visual Screening (RVS) *IOP Conf. Ser.: Earth Environ. Sci.* **479** 012002

[5] Haryanto Y et al 2020 FEMA 310 Tier 1 seismic evaluation of existing building: A case study of a 7-story academic RC building of Jenderal Soedirman University, Indonesia *IOP Conf. Ser.: Mater. Sci. Eng.* **982** 012031

[6] Harirchian E, et al 2020 Earthquake Safety Assessment of Buildings Through Rapid Visual Screening *Buildings* **10** 51

[7] Adhyarini Y Z 2021 *Seismic Evaluation in Engineering Faculty Building of Jenderal Soedirman University Using Rapid Visual Screening* (Banyumas: Jenderal Soedirman University)

[8] Wahyuni, Endah, et al 2016 Study of Rapid Visual Screening of Buildings for High Potential Seisic Hazard According to Indonesian Standard *The 2nd International Seminar On Science ang Technology* (Surabaya, Indonesia: Postgraduate Program Institut Teknologi Sepuluh November)

[9] Achs G and Adams C 2012 A Rapid-Visual-Screening Methodology for the Seismic Vulnerability Assessment of Historic Brick-Masonry Buildings in Vienna *15th World Conference on Earthquake Engineering (15WCEE)* **vol 154** (Lisboa)

[10] Standar Nasional Indonesia 1726:2019 2019 *Tata Cara Perencanaan Ketahanan Gempa Untuk Struktur Bangunan Gedung dan Nongedung* (Jakarta: BSN)

IOP Conf. Series: Earth and Environmental Science 1173 (2023) 012007 doi:10.1088/1755-1315/1173/1/012007

Structural assessment of the stalled building in Department of Industrial Engineering, Andalas University, Indonesia

Fauzan[1], M W Aulia[1], D A Nugraha M[1] and Z A Jauhari[2]

[1]Department of Civil Engineering, Faculty of Engineering, Andalas University, Padang, Indonesia
[2]Department of Civil Engineering, Bengkalis State Polytechnic, Bengkalis, Indonesia

E-mail: fauzan@eng.unand.ac.id

Abstract. The construction of the Industrial Engineering Department building is an effort to support the educational infrastructure at Andalas University. However, the building construction process was stopped because the contractor had terminated the contract (breach of contract) and it took a long time to continue the construction. Before that, a structural evaluation of the building structure should be conducted due to the update of the building standards. In this study, the structural evaluation was carried out by analyzing and modelling the building using ETABS v.18 software, as well as loading analysis on the building regarding current Indonesian building standards such as SNI 1726-2019 for earthquake resistance building design, SNI 2847-2019 for requirements of the structural concrete building and SNI 1727-2020 for building minimum design loads. The results of the analysis show that several structural beams did not have a strong enough capacity to withstand the working loads, where the ultimate moment and shear values due to the influence of acting loads were more than the nominal moments and shear values. Retrofitting of the weak structural beams using the CFRP wrapping method was recommended to improve the structural capacity of the building.

1. Introduction

Andalas University is a university that is engaged in education and has several duties in improving the quality of human resources. One of the goals of the University is to improve the quality of campus services, infrastructure, facilities and technology in accordance with national and international standards and to create a conducive and beneficial academic atmosphere for the community. The Industrial Engineering building is one of the administrative buildings in the University. The construction of the Industrial Engineering Department building is one example of efforts to support the educational infrastructure. However, the building construction process was stopped because the contractor had terminated the contract (breach of contract) and it took a long time to continue the construction. Figure 1 shows the existing stalled building of the Industrial Engineering Department at the Andalas University. When the construction of the building will be continued, the building structure should be evaluated due to the update of the standards.

In this study, a structural evaluation of a stalled building structure in the Department of Industrial Engineering building, Andalas University was carried out based on the current Indonesian building standards, to determine whether the structure of the building is still capable to resist the working loads. If it is not able to withstand the acting loads, retrofitting of the building structure should be carried out before continuing the construction.

Figure 1. Industrial Engineering Department building at Andalas University.

2. Material and method

Analysis of the building structure was carried out by modelling the structure of the building using ETABS structural analysis program version 2018. This analysis will determine the ability of the structure to resist the loads such as self-weight loads and the load due to activities in the building. The simulation will also obtain the capacity of the building structure to withstand the loads acting on the structure. In this study, the dynamic earthquake load (response spectrum) in Padang City based on the current Indonesian seismic standard (SNI 1726-2019) was used in the seismic load analysis. The data of the response spectrum of Padang City was obtained from the RSA PUSKIM PUPR application by entering the global coordinates of the building location [1].

2.1. Building structural data

The parameters used in the structural analysis are the technical data from the existing condition of the building obtained by measuring the existing building directly on the field, as shown in table 1.

Table 1. Building data.

Name	Data
Name of building	Industrial Engineering Dept. building of Andalas University
Location	Limau Manis Street, Padang City
Building function	Education building
Structure type	Reinforced concrete structure
Number of stories	3 story building
Building height	22.45 m
Slab thickness	12 cm
Tie beam dimension	S1 (25x60) cm
Beam dimension	B1 (40x80) cm, B2 (40x80) cm, B3 (35x80) cm, B4 (35x80) cm and B5 (35x80) cm
Column dimension	K1 (40x70) cm
Compressive strength of concrete, fc'	25 MPa
Yield strength of steel, fy	400 MPa (deformed bar) and 320 MPa (plain bar)

2.2. Modelling of the building structure

3D modelling of the building structure was carried out using ETABS v.18 software [2]. In this modelling, the columns and beams of the building structure are modelled as frame elements and slabs are modelled as slab elements. Structural modelling was carried out according to the existing conditions or the actual condition of the Industrial Engineering Department building of the Andalas University. The result of the structural modelling is shown in figure 2.

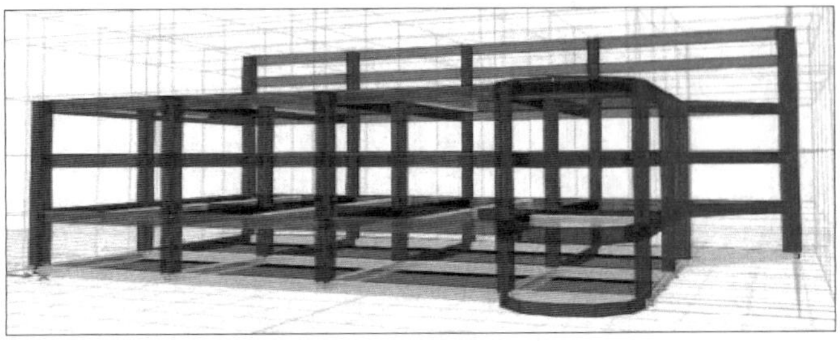

Figure 2. Modelling of the building structure.

2.3. Loading analysis

2.3.1. Dead load. Based on SNI 1727-2020 [3], the dead load is defined as the weight of all installed building construction materials including walls, floors, roofs, ceilings, stairs, fixed partition walls, finishing, building cladding and other architectural and structural components as well as other installed service equipment including crane weight and material conveying system. The values of additional dead loads are shown in table 2.

Table 2. Additional dead loads.

Load	Load Value
Reinforced concrete	2400 kg/m^3
Clay and silt (wet)	2000 kg/m^2
Wall of brick masonry	250 kg/m^2
Ceiling	20 kg/m^2
MEP installation	30 kg/m^2
Waterproofing (asphalt layer)	5 kg/m^2
Plasters (per cm thick)	21 g/m^2

The dead load acting on the floor of the Industrial Engineering Department building consists of:

- Self-weight of structural elements which are calculated directly by the structural analysis program ETABS v.18.
- Floor cover (1 cm thick) is 48 kg/m^2.
- Plaster (2 cm thick) is 42 kg/m^2.
- Weight of the ceiling is 20 kg/m^2.

2.3.2. Live load. Based on SNI 1727-2020 [3], the live load used in the design of buildings and other structures must be the maximum load that is expected to occur to the occupancy and use of the building but it must not be less than the specified minimum homogeneous load. Live loads on the Industrial Engineering Department building at the Andalas University are 1.92 kN/m^2 for classroom/study, 3.83 kN/m^2 for the corridor above the 1st floor and 4.79 kN/m^2 for the 1st corridor.

2.3.3. Wall load. The wall load is the acting load on the building other than the dead load on the floor of the building, so it needs to be taken into account in the loading analysis. According to the Indonesian loading standards for buildings (PPIUG 1983) [4], the load on the brick wall in the building is 250 kg/m^2 and the wall height is 4 m, so the homogeneous load of the wall acting on the beam structure is 4 m x 250 kg/m^2 which is equal to 1000 kg/m.

2.3.4. Earthquake load. The dynamic response seismic analysis approach method of response spectrum was used in this structural analysis of the building. Earthquake load refers to SNI 1726-2019 for procedures of planning earthquake resistance for building and non-building structures [3]. The earthquake load data for the Industrial Engineering Department building at Andalas University is shown in table 3 and figure 3.

Table 3. Earthquake parameters.

No	Data	Variable	Value
1	Risk category of educational building	K	Category IV
2	Earthquake priority factor	I	1.5
3	Ground acceleration parameters (Ss, S1)		
	(a) Accelerated response of MCE spectra from earthquake map in short period	SS	1.5
	(b) Acceleration of MCE spectral response from earthquake map in 1 second period	S1	0.6
4	Site classification	KS	SC
5	Site coefficient factor (Fa, Fv)		
	(a) Site coefficient for short period	Fa	1.200
	(b) Site coefficient for long period	Fv	1.400
6	Parameter spectrum response (SMS and SM1)		
	(a) Accelerated short-period MCE spectral response adjusted for site class (SMS = Fa. Ss)	SMS	1.800
	(b) 1 second period MCE spectral response acceleration adjusted for site class (SM1 = Fv. S1)	SM1	0.840
7	Design spectrum acceleration parameters (SDS and SD1)		
	(a) Acceleration of spectral response in a short period (SDS = 2/3 SMS)	SDS	1.200
	(b) Acceleration of spectral response over 1 second period (SD1 = 2/3 SM1)	SD1	0.560
8	Design response spectrum time		
	(a) To = 0.2 SD1/SDS	T_0	0.093
	(b) Ts = SD1/SDS	T_s	0.467
9	Seismic design category	KDS	D

Based on the earthquake spectra data in table 3, the response spectrum value was calculated and inputted into ETABS v.18 programs which results in the response spectrum, as shown in figure 3.

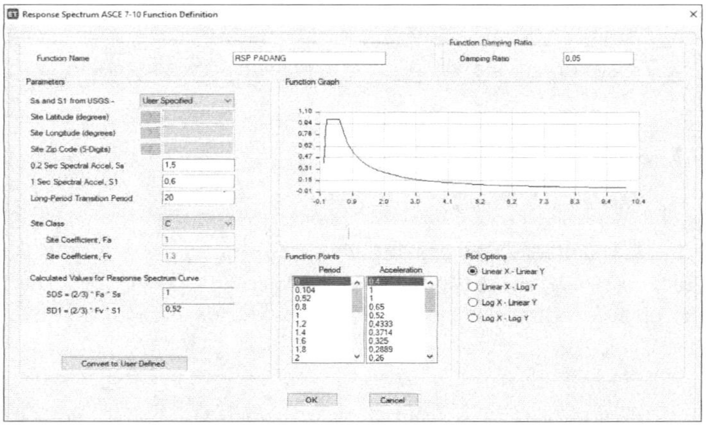

Figure 3. Earthquake response spectrum design of Padang City.

2.4. Loading combination

In this structural analysis, the load combination refers to the SNI 1726-2019 [3] as follows:

1. 1.4 DL
2. 1.2 DL + 1.6 LL
3. 1.440 DL + 1.0 LL + 1.3 EQX + 0.39 EQY
4. 1.440 DL + 1.0 LL + 1.3 EQX – 0.39 EQY
5. 1.440 DL + 1.0 LL – 1.3 EQX + 0.39 EQY
6. 1.440 DL + 1.0 LL – 1.3 EQX – 0.39 EQY
7. 1.440 DL + 1.0 LL + 0.39 EQX + 1.3 EQY
8. 1.440 DL + 1.0 LL + 0.39 EQX – 1.3 EQY
9. 1.440 DL + 1.0 LL – 0.39 EQX + 1.3 EQY
10. 1.440 DL + 1.0 LL – 0.39 EQX – 1.3 EQY
11. 0.660 DL + 1.3 EQX + 0.39 EQY
12. 0.660 DL + 1.3 EQX – 0.39 EQY
13. 0.660 DL – 1.3 EQX + 0.39 EQY
14. 0.660 DL – 1.3 EQX – 0.39 EQY
15. 0.660 DL + 0.39 EQX + 1.3 EQY
16. 0.660 DL + 0.39 EQX – 1.3 EQY
17. 0.660 DL – 0.39 EQX + 1.3 EQY
18. 0.660 DL – 0.39 EQX – 1.3 EQ

Where DL is dead load, LL is live load, EQX is X-directional earthquake load and EQY is Y-direction earthquake load.

2.5. Building priority factor (Ie) and response modification coefficient (R)

According to the Indonesian seismic standard (SNI 1726-2019), the building priority factor was determined based on the building and non-building risk categories [3], in which the building for educational facilities was categorized in level IV. From this category, the value of the building priority factor was obtained, where the building with the risk category level IV has the priority factor Ie = 1.5. For the planned earthquake response modification coefficient (R), the value was obtained in table 12-point C SNI 1726-2019 which is R = 8 for the SRPMK (special moment resistant frame system) building. Furthermore, the earthquake factor scale must be determined according to equation (1).

$$\frac{G\, I_e}{R} = 1839.38\, mm/s^2 \qquad (1)$$

Where G is the acceleration of gravity, I is the building priority factor and R is the response modification coefficient.

3. Results and discussion

3.1. Inter story drift

Based on article 7.8.6 of SNI 1726-2019 [3], the determination of the building deviation between the design levels (Δ_x) must be calculated as the difference in the deviation at the center of mass above and below the level under consideration. The center of mass deviation (mm) must be determined according to equations (2) and (3).

$$\Delta_x = \frac{Cd\ _{xe}}{Ie} \leq \delta_{permit} \qquad (2)$$

$$\delta_{xe} = \delta_1 - \delta_2 \qquad (3)$$

Where δ is the deviation between levels, δ_1 is floor drift at level x, δ_2 is floor drift below level x, Cd is lateral deviation enlargement factor in, δ_{xe} is deviation at x-level and Ie is the priority factors for earthquakes.

Table 4. Inter story drift limit.

Structure	Risk category		
	I or II	III	IV
Structures, other than brick shear wall structures, 4 stories or less with interior walls, partitions, ceilings and exterior wall systems designed to accommodate deviations between stories.	$0.025h_{xx}^{e}$	$0.020h_{xx}$	$0.015h_{xx}$
Brick cantilever shear wall structure.	$0.010h_{xx}$	$0.010h_{xx}$	$0.010h_{xx}$
Another brick shear wall structure.	$0.007h_{xx}$	$0.007h_{xx}$	$0.007h_{xx}$
All other structures.	$0.020h_{xx}$	$0.015h_{xx}$	$0.010h_{xx}$

The risk category of the Industrial Engineering Department building at Andalas University is level IV. Based on the risk category and type structure of the building (table 4), the permitted limit of the inter-story drift formula of the building was obtained in equation (4).

$$\delta_{permit} = 0.010h_{xx} \tag{4}$$

Where δ_{permit} is the deviation between permission levels and h_{xx} is the height of each story. Based on SNI 1726-2019, the calculation results of the inter-story drift of the Industrial Engineering Department building at Andalas University are shown in tables 5 and 6.

Table 5. Inter-story drift of building in X-direction.

No	Load Case/ Combo	Direction	δ_e (mm)	C_d	δ_{xe} (mm)	Δ_x (mm)	Height (mm)	δ_{permit} (mm)	Note
3	EQD- X Max	X	18.991	5.5	69.634	30.580	4000	40	Ok
2	EQD- X Max	X	10.651	5.5	39.054	39.054	4000	40	Ok
1	EQD- X Max	X	0.0	5.5	0.0	0.0	4000	40	Ok

Table 6. Inter-story drift of building in Y-direction.

No	Load Case/ Combo	Direction	δ_e (mm)	C_d	δ_{xe} (mm)	Δ_x (mm)	Height (mm)	δ_{permit} (mm)	Note
3	EQD- Y Max	Y	20.687	5.5	75.852	38.372	4000	40	Ok
2	EQD- Y Max	Y	10.222	5.5	37.481	37.481	4000	40	Ok
1	EQD- Y Max	Y	0.0	5.5	0.0	0.0	4000	40	Ok

From tables 5 and 6, the maximum story drift for X and Y-directions is 39.05 mm and 38.37 mm, respectively. These values are less than the allowable inter-story drift of 40 mm so that the inter-story building meets the requirements in Indonesian seismic standards.

IOP Conf. Series: Earth and Environmental Science 1173 (2023) 012007 doi:10.1088/1755-1315/1173/1/012007

3.2. The capacity of structural elements

3.2.1. Column cross-section capacity [5]. Columns in buildings are generally made as a monolith vertically upwards which causes all column elements to withstand combined moments between flexural and axial loads [6]. Analysis of structural elements should be carried out to determine the capacity of structural elements, such as flexural (axial-moment) and shear capacities of the columns. From this data, it can be determined the capability of the cross-section capacity to resist the acting loads on the building structure. In this study, the calculation of the P-M interaction diagram and shear capacity of columns was carried out according to Indonesian building standard, SNI 2847-2019.

- Column P-M interaction diagram. The cross-sectional capacity of columns in reinforced concrete structures to withstand a combination of axial forces and bending moments can be described in the form of an interaction diagram, which is generally called a P-M interaction diagram [7]. Figure 4 shows the column P-M interaction diagram of the building. From figure 4, it shows that the internal forces acting on the column are still within the nominal moment and axial reduction limit so that the columns are still able to resist the working loads on the building.

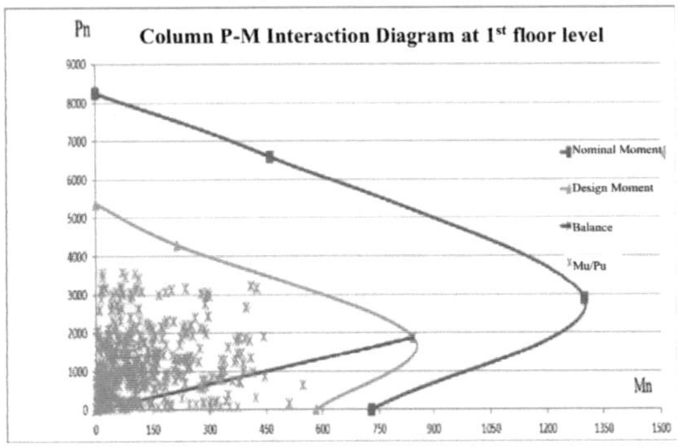

(a) P-M interaction diagram on 1st floor.

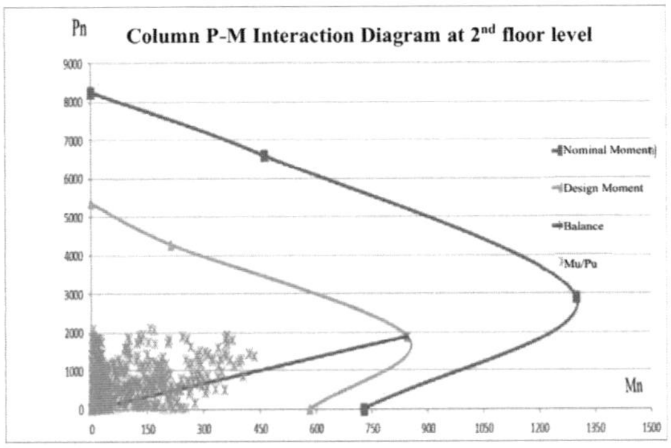

(b) P-M interaction diagram on 2nd floor.

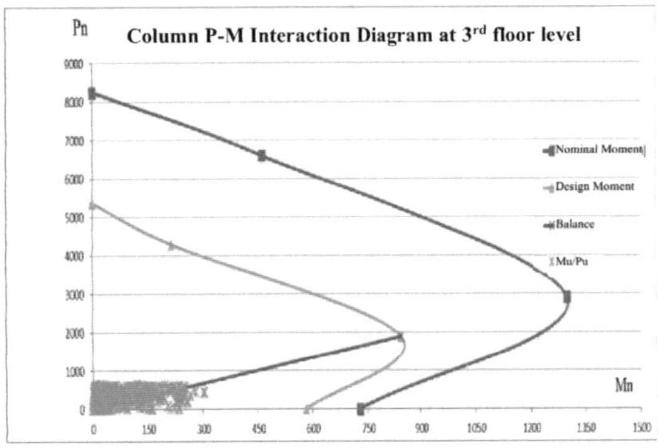

(c) P-M interaction diagram on 3rd floor.

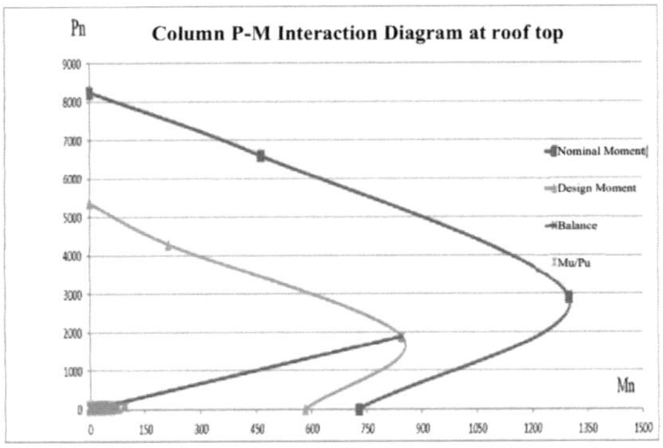

(a) P-M interaction diagram on roof top.

Figure 4. P-M interaction diagram of columns.

- Column shear capacity. The ability of the column to resist the shear forces can be evaluated from the analysis of the column's shear capacity. Table 7 shows the shear capacity of columns for every building story. As can be seen from the table, the shear capacity is strong and still sufficient to withstand the applied loads.

Table 7. Column shear capacity.

Floor Level	Code	Cross-section Dimension (BxH) mm	Bar Diameter (mm)	Space (mm)	ΦVn (kN)	Vu (kN)	Note
1	K1	400 x 700	10	130	437.73	385.04	OK
2	K1	400 x 700	10	130	437.73	247.75	OK
3	K1	400 x 700	10	130	437.73	141.59	OK
Roof Top	K1	400 x 700	10	130	437.73	56.35	OK

3.3. Beam cross-section capacity

Beams are structural elements that carry the working loads that produce bending moments and shear forces on the beam and cause three types of possible collapse of the beam, namely tensile collapse (under-reinforced), compressive collapse (over-reinforced) and balanced collapse (balance reinforced) [8]. The beam flexural and shear capacity were calculated based on Indonesian building standard, SNI 2847-2019.

3.3.1. Beam flexural capacity. Tables 8 to 11 show the calculation results of beam flexural capacity on the 1st floor, 2nd floor, 3rd floor and rooftop.

Table 8. Beam flexural capacity on the 1st floor.

Floor Level	Beam Type	Cross-section Dimension (mm)	Note	ØMn (kN)		Mu (kN)		ØMn ≥ Mu	
				Positive	Negative	Positive	Negative	Positive	Negative
1	B1	80 x 40	At support	337.20	496.72	665.07	352.18	Not Ok	Ok
			At field	417.72	337.20	346.25	543.24	Ok	Ok
	B2	80 x 40	At support	446.56	752.93	696.87	906.84	Not Ok	Not Ok
			At field	752.93	446.56	581.81	724.72	Ok	Not Ok
	B3	80 x 35	At support	335.47	492.83	294.33	464.35	Ok	Ok
			At field	415.02	335.47	265.97	360.51	Ok	Not Ok
	B4	80 x 35	At support	443.45	646.53	359.30	609.54	Ok	Ok
			At field	646.53	443.45	305.99	369.47	Ok	Ok
	B5	80 x 35	At support	335.47	335.47	461.32	522.34	Not Ok	Not Ok
			At field	335.47	335.47	439.29	345.91	Not Ok	Not Ok

Table 9. Beam flexural capacity on the 2nd floor.

Floor Level	Beam Type	Cross-section Dimension (mm)	Note	ØMn (kN)		Mu (kN)		ØMn ≥ Mu	
				Positive	Negative	Positive	Negative	Positive	Negative
2	B1	80 x 40	At support	337.20	496.72	381.55	563.25	Not Ok	Not Ok
			At field	417.72	337.20	309.31	453.52	Ok	Not Ok
	B2	80 x 40	At support	446.56	752.93	746.72	809.94	Not Ok	Not Ok
			At field	752.93	446.56	608.07	637.65	Ok	Not Ok
	B3	80 x 35	At support	335.47	492.83	203.61	414.39	Ok	Ok
			At field	415.02	335.47	186.34	332.49	Ok	Ok
	B4	80 x 35	At support	443.45	646.53	395.42	599.35	Ok	Ok
			At field	646.53	443.45	322.22	349.12	Ok	Ok
	B5	80 x 35	At support	335.47	335.47	564.09	582.20	Not Ok	Not Ok
			At field	335.47	335.47	531.45	377.56	Not Ok	Not Ok

Table 10. Beam flexural capacity on the 3rd floor.

Floor Level	Beam Type	Cross-section Dimension (mm)	Note	ØMn (kN)		Mu (kN)		ØMn ≥ Mu	
				Positive	Negative	Positive	Negative	Positive	Negative
3	B1	80 x 40	At support	337.20	496.72	224.72	311.40	Ok	Ok
			At field	417.72	337.20	211.94	247.12	Ok	Ok
	B2	80 x 40	At support	446.56	752.93	217.38	374.35	Ok	Ok
			At field	752.93	446.56	204.47	245.03	Ok	Ok
	B3	80 x 35	At support	335.47	492.83	195.85	254.16	Ok	Ok
			At field	415.02	335.47	183.69	214.86	Ok	Ok
	B4	80 x 35	At support	443.45	646.53	125.43	323.17	Ok	Ok
			At field	646.53	443.45	136.02	201.10	Ok	Ok
	B5	80 x 35	At support	335.47	335.47	202.65	101.30	Ok	Ok
			At field	335.47	335.47	110.55	39.99	Ok	Ok

Table 11. Beam flexural capacity on the roof top floor.

Floor Level	Beam Type	Cross-section Dimension (mm)	Note	ØMn (kN)		Mu (kN)		ØMn ≥ Mu	
				Positive	Negative	Positive	Negative	Positive	Negative
Roof top	B1	40 x 80	At support	337.20	496.72	13.34	173.68	Ok	Ok
			At field	417.72	337.20	83.84	11.31	Ok	Ok

From these tables, it shows that the beam flexural capacity on several beams on the 1st and 2nd floors of the building is not strong enough to withstand the working loads. Meanwhile, all beams on the third floor and rooftop floor are able to resist acting loads, in which the nominal moment was higher than the ultimate moment.

3.3.2. Beam shear capacity. The calculation results of the beam shear capacity are shown in tables 12 - 15.

Table 12. Beam shear capacity on the 1st floor.

Floor Level	Beam Type	Cross-section Dimension (mm)	Note	ØVn (kN)	Vu (kN)	ØVn ≥ Vu
1	B1	80 x 40	At support	740.47	382.99	Ok
			At field	628.82	373.73	Ok
	B2	80 x 40	At support	740.47	785.80	Not Ok
			At field	628.82	774.91	Not Ok
	B3	80 x 35	At support	708.39	451.78	Ok
			At field	596.74	434.40	Ok
	B4	80 x 35	At support	708.39	390.07	Ok
			At field	596.74	373.54	Ok
	B5	80 x 35	At support	699.19	606.71	Ok
			At field	588.99	655.84	Not Ok

Table 13. Beam shear capacity on the 2nd floor.

Floor Level	Beam Type	Cross-section Dimension (mm)	Note	ØVn (kN)	Vu (kN)	ØVn ≥ Vu
2	B1	80 x 40	At support	740.47	342.61	Ok
			At field	628.82	333.01	Ok
	B2	80 x 40	At support	740.47	756.65	Not Ok
			At field	628.82	745.45	Not Ok
	B3	80 x 35	At support	708.39	459.60	Ok
			At field	596.74	442.21	Ok
	B4	80 x 35	At support	708.39	459.44	Ok
			At field	596.74	361.24	Ok
	B5	80 x 35	At support	699.19	636.31	Ok
			At field	588.99	624.25	Not Ok

Table 14. Beam shear capacity on the 3rd floor.

Floor Level	Beam Type	Cross-section Dimension (mm)	Note	ØVn (kN)	Vu (kN)	ØVn ≥ Vu
3	B1	80 x 40	At support	740.47	223.90	Ok
			At field	628.82	219.01	Ok
	B2	80 x 40	At support	740.47	286.55	Ok
			At field	628.82	271.67	Ok
	B3	80 x 35	At support	708.39	215.09	Ok
			At field	596.74	208.26	Ok
	B4	80 x 35	At support	708.39	210.07	Ok
			At field	596.74	202.52	Ok
	B5	80 x 35	At support	699.19	384.20	Ok
			At field	588.99	373.11	Ok

Table 15. Beam shear capacity on the rooftop floor.

Floor Level	Beam Type	Cross-section Dimension (mm)	Note	ØVn (kN)	Vu (kN)	ØVn ≥ Vu
Rooftop	B1	80 x 40	At support	740.47	101.21	Ok
			At field	628.82	78.09	Ok

The several beams B2 and B5 on the 1st and 2nd floors (tables 12 and 13) have not strong enough capacity to support applied loads in the building. However, the overall shear performance of the structural beams on the 3rd floor and rooftop is reliable and sufficient to withstand the acting loads, as shown in tables 14 and 15.

In this study, other evaluation criteria according to the Indonesian seismic code SNI 1726-2019 such as the examination of mass participation, P-delta and structural irregularities of the building were also conducted, in which all the criteria met the requirements in accordance with SNI 1726-2019.

From the above analysis results, it was found that the existing building structure is not able to withstand the applied loads because the beams have insufficient capacity, so the building structure should be retrofitted before continuing the construction [9–11].

4. Retrofitting

Retrofitting is a set of operations done on a part or all structure so that it can bear more loads and overheads than the initial condition and shows better behavior characteristics [9]. One of the retrofitting methods for beams is by using fiber-reinforced plastic (FRP), which is the strengthened beam with the carbon fiber-reinforced polymer (CFRP) sheets (fabric). The CFRP external reinforcement gives a contribution to retrofitting and rehabilitation of structural elements. In this study, the retrofitting of the weak beams was designed using CFRP sheets with the example shown in figure 5. Based on the result of the previous research, the retrofitting of beams using CFRP sheets increases the flexural and shear capacity of the beams [12-16].

Figure 5. CFRP retrofitted beam [16].

The material data [17] in calculating the retrofitting of the beams using CFRP sheets is as follows:

- Material Name = SIKA CarboDur S1012
- Type = CFRP laminate tape
- Width = 100 mm
- Thickness = 1.2 mm
- Area of FRP (Af) = 100 mm x 1.2 mm = 120 mm^2
- Tensile Strength (ffu*) = 2800 MPa
- Ultimate Strain (εfu*) = > 1.7%
- Modulus Elasticity (Ef) = 165,000 MPa
- Reduction Factor (Ce) = 0.85 (Table 9.1 ACI 440.2R-08)

4.1. Retrofitted beam flexural capacity

To increase beam flexural capacity using CFRP, the design calculation was carried out based on ACI 440.2R-08 [13]. The beam flexural capacity of the structure retrofitted by CFRP sheets can be calculated by using the following equations (5) and (6):

$$\emptyset \mathrm{Mn} = \emptyset(\mathrm{M_{ns}} + \Psi_f \mathrm{M_{nf}}) \tag{5}$$

$$\mathrm{M_{nf}} = \mathrm{A_f} \times \mathrm{f_{fe}}\left(\mathrm{d_f} - \frac{\beta_1 \times c}{2}\right) \tag{6}$$

Where M_{ns} is the nominal moment of reinforcement, ψ_f is the CFRP reduction factor of 0.85, M_{nf} is the nominal moment CFRP, A_f is the area of CFRP, f_{fe} is the CFRP stress level, d_f is the effective height of the beam and c is the distance of the upper outer edge to the neutral line.

Table 16. Beam flexural capacity after retrofitting.

Floor Level	Beam Type	Dimension (mm)	Location	ϕMn (kNm)		ϕMu (kNm)		ϕMn > ϕMu		Increased Ratio	
				Positive	Negative	Positive	Negative	Positive	Negative	Positive	Negative
1	B1	80 X 40	At support	794.61	954.43	665.07	352.18	Ok	Ok	135.65%	92.14%
			At field	875.43	794.61	346.25	543.24	Ok	Ok	109.57%	135.65%
	B2	80 X 40	At support	750.21	917.57	696.87	906.84	Ok	Ok	67.99%	21.87%
			At field	917.57	727.21	581.81	724.72	Ok	Ok	21.87%	62.85%
	B3	80 X 35	At support	729.44	886.8	294.33	464.35	Ok	Ok	117.83%	79.94%
			At field	805.08	725.54	265.97	360.51	Ok	Ok	93.98%	116.27%
	B5	80 X 35	At support	728.46	728.46	461.32	522.34	Ok	Ok	117.15%	117.15%
			At field	728.46	728.46	439.29	345.91	Ok	Ok	117.15%	117.15%
2	B1	80 X 40	At support	794.61	954.43	381.55	563.25	Ok	Ok	36.87%	21.87%
			At field	875.43	794.61	309.31	453.52	Ok	Ok	21.87%	62.85%
	B2	80 X 40	At support	750.21	917.57	746.72	809.94	Ok	Ok	36.87%	21.87%
			At field	917.57	727.21	608.07	637.65	Ok	Ok	21.87%	62.85%
	B5	80 X 35	At support	728.46	728.46	564.09	582.2	Ok	Ok	117.15%	117.15%
			At field	728.46	728.46	531.45	377.56	Ok	Ok	117.15%	117.15%

Table 16 shows the calculation result of the retrofitted beam flexural capacity and the percentage increase in the ratio of the beam flexural capacity. From the table, it reveals that the retrofitting of beams using CFRP sheets increases the beam flexural capacity in the range of 21.87% to 117.15%.

4.2. Retrofitted beam shear capacity

Beam shear retrofitting using CFRP sheets improves the stiffness and strength values of ones. Based on the equation refer to ACI 440 [13], the shear capacity in beam-reinforced concrete using CFRP sheets [15] can be calculated by using the following equations (7) and (8):

$$V_n = V_c + V_s + V_f \tag{7}$$

$$V_f = \frac{A_f f_{fe}(\sin\beta + \cos\beta) d_f}{s_f} \tag{8}$$

Where V_n is the shear capacity at the support area, V_c is the shear capacity of concrete, V_s is the shear capacity of reinforcement, V_f is the shear capacity CFRP, S_f is the pitch and β is the angle of inclination of the spiral.

The calculation results of the retrofitted beam shear capacity and the percentage increase in the ratio of the beam shear capacity are shown in table 17. As seen in the table, the retrofitting of beams using CFRP sheets improve the beam shear capacity ranges from 27.84% to 80.05%.

Table 17. Beam shear capacity after retrofitting.

Floor Level	Beam Type	Dimension (mm)	Location	ϕVn (kN)	ϕVu (kN)	ϕVn > ϕVu	Increased Ratio
1	B1	80 X 40	At support	946.61	785.8	Ok	27.84%
			At field	834.96	774.91	Ok	32.78%
	B2	80 X 35	At support	1170.69	606.71	Ok	67.43%
			At field	1060.49	665.84	Ok	80.05%
2	B2	80 X 40	At support	946.61	756.65	Ok	27.84%
			At field	834.96	745.45	Ok	32.78%
	B5	80 X 35	At support	1170.69	636.61	Ok	67.43%
			At field	1060.49	624.25	Ok	80.05%

5. Conclusion

Evaluation criteria of inter-story drift, the mass participation of the structure, the P-Delta, building irregularity and column capacity for the existing Industrial Engineering Department building of Andalas University have met the requirements based on Indonesian seismic codes (SNI 1726-2019). Meanwhile, several beams on the 1st and 2nd floors do not have strong enough capacity, so the building is not able to withstand the loads applied to the structure. Therefore, the building should be retrofitted before continuing the construction.

The retrofitting of the existing building was designed using carbon fiber-reinforced polymer (CFRP) sheets because it can improve the building's structural capacity significantly based on the previous study results. The retrofitting of weak beams using CFRP sheets at the 1st and 2nd floors increased the beams' flexural capacity between 21.87% and 135.65%, while the beams' shear capacity increased in the range between 27.84% and 80.05% so that the retrofitted building have strong enough capacity to resist the working loads. Therefore, the CFRP sheets method was recommended to retrofit the beam structural elements of this building.

Acknowledgments

The authors gratefully acknowledge the financial support from Andalas University.

References

[1] National Standardization Agency 2019 *Procedure for planning earthquake resistance for building structures*, SNI *1726:2019* (Jakarta: BSN)

[2] Alfianto R and Rahmat D 2018 Analysis of calculation of buildings with methods ETABS version 9.7.2 *J. of Civ. Eng. Build. and Trans.*

[3] National Standardization Agency 2020 *Minimum load for planning of buildings and other structures, SNI 1727:2020* (Jakarta: BSN)

[4] Departemen Permukiman dan Prasarana Wilayah 2002 *Indonesian loading regulations for buildings, PPIUG 1983* (Jakarta: BSN)

[5] National Standardization Agency 2019 *Structural Concrete Requirements for Buildings, SNI 2847:2019* (Jakarta: BSN)

[6] Subrianto A, Puryanto P and Firdausa F 2020 Evaluation of cross-sectional capacity of reinforced concrete columns using interaction diagrams *Pilar Jurnal Teknik Sipil* **15** 12-18

[7] Hidayat H 2016 *Feasibility evaluation and recommendations for strengthening the structure of the existing building of the industrial engineering building of Andalas University, Padang* (Padang: Andalas University)

[8] Zaidir 2015 *Reinforced concrete construction* (Padang: Andalas University Press)

[9] Fauzan, Ismail F A and Jauhari Z A 2019 Seismic retrofitting analysis using concrete jacketing and shear wall on dental hospital building of Andalas University *IOP Conf. Ser. Mater. Sci.* **602** 012108

[10] Zaidir, Fauzan, Hakam A and Ismail F A 2017 Structural evaluation of damaged BRI Branch Office Building at Jalan Khatib Sulaiman Padang due to earthquake *Int. J. of Civ. Eng. And Tech (IJCIET)* **8** 1170-1180

[11] Fauzan, Hakam A, Yuliet R and Osman J V 2019 Structural evaluation of Ikhwatun shelter building constructed on liquefaction potential area in Padang city, Indonesia *Int. Conf. on Adv. In Civ. And Env. Eng. (ICAnCEE 2018)* volume 276 (Bali: MATEC Web of Conferences) 01019

[12] Vemmy Y 2017 *Building strengthening using carbon fiber reinforced polymer (CFRP) case study using universitas jember c-dast laboratory and lecture room layout* (Surabaya: Institut Teknologi Sepuluh Nopember)

[13] ACI Committee 440 2008 *Guide for the Design and Construction of Externally Bonded FRP System for Strengthening Concrete Structures* (Farmington Hills: American Concrete Institute)

[14] Utami S R L 2019 Effect of Carbon fiber reinforced polymer (CFRP) on reinforced concrete beams *INERSIA* **15** 23-42

[15] Rada S 2018 *Modelling element methods for reinforced concrete blocks with reinforced CFRP (Carbon Fiber Reinforced Polymer) subjected to bending and shaving* (Yogyakarta: Universitas Atma Jaya Yogyakarta)

[16] Yalburgimath C, Rathod A and Bhavanishankar S 2018 Retrofitting of reinforced concrete beam using carbon fiber reinforced polymer (CFRP) fabric *Int. Research J. of Eng. and Technology (IRJET)* **5** 362-367

[17] Sika 2018 *Sika CarboDur Carbon Fiber Laminate for Structural Strengthening in Product Data Sheet Edition 12.2018/v1* (Avenue Delmar Pointe-Claire, Quebec: Sika Canada Inc.)

Seismic assessment and retrofitting of the old URM railway station building in Padang City, Indonesia

Fauzan[1]**, Z A Jauhari**[2]**, G A Agista**[1]**, N D Carol**[1] **and Y S Putra**[1]

[1] Department of Civil Engineering, Faculty of Engineering, Andalas University, Padang, Indonesia
[2] Department of Civil Engineering, Bengkalis State Polytechnic, Bengkalis, Indonesia

E-mail: fauzan@eng.unand.ac.id

Abstract. The train is one of the means of land transportation between cities that is in demand by all levels of society. As a mass transportation capable of transporting passengers and goods in large quantities and cheaply, the train can be immediately implemented in Padang City, especially from Padang to the Minangkabau International Airport (BIM) in Pariaman regency. To support the program, re-activating the old train stations that are no longer in use such as Pulau Aer Station, should be done by the directorate general of transportation, Indonesia. This station building is located in Southern part of Padang City, Indonesia. This station building is an old unreinforced masonry (URM) building, which had heavily damaged after the earthquake on September 30th, 2009. Before re-activating this station, a building assessment and retrofitting of this station building are required. The building assessment is carried out both visually and through structural analysis, while the building retrofit is designed using the ferrocement layer method. The analysis result shows that the shear stress on the walls of the retrofitted building is able to withstand the working loads, including the earthquake load.

1. Introduction

In general, the train is one of the means of land transportation between cities that is in demand by all levels of society. With the increasing number of people using land transportation services, it should be balanced with adequate facilities such as planning and designing railway lines and improving the quality of good service, both at the train station and the train itself, so that people have more confidence and they choose to use rail transportation services.

As mass transportation is able to transport passengers and goods in large quantities and cheaply, the train is alternative land transportation that can be immediately held in Padang City, especially from Padang to Pariaman (BIM Airport). The existence of the station is the most important part as a terminal for the departure and disembarkation of passengers, as well as in the process of interaction and activity for rail transportation users who are waiting for their scheduled departure.

The selection of the location of a large-scale train station is currently made by taking into account the principles of the Transit Oriented Development (TOD) system. Basically, the TOD system is intended to reduce population mobility between regions by integrating and bringing the city transportation system closer to residential areas, business and community activity centers to create an efficient city [1]. By implementing the TOD approach, travel time and transportation costs can be reduced, so productivity and public interest in using rail transportation will increase.

By looking at the increasing need for rail transportation, it is necessary to add more stations from the existing number and activate the old stations that are not used. In order to fulfil this, the Director General of transportation will reactivate several old stations that are no longer active in Padang city, one of them is Pulau Aer Station. This railway station consists of an unreinforced masonry (URM) building. The condition of this station was quite heavily damaged after the earthquake of September 30th, 2009. For this reason, repair and retrofit of this station building are needed to enable it to be reactivated.

The primary objective in the seismic retrofitting of URM buildings is to improve the performance of the structure to ensure that they perform safely during earthquakes. Subject to the mechanical behaviour of the building, which is dependent upon the properties of the materials used and loading direction, the retrofitting methodology chosen may have differing effects. Therefore, designers and engineers need to carefully study the performance of the structures before suggesting any retrofitting methods. Different strengthening/retrofitting techniques on unreinforced masonry structures/elements subjected to seismic loads have been studied by many researchers [2-8].

This study will evaluate the existing condition of the Pulau Aer Station building and provide solutions to these problems in the form of repairing and retrofitting methods for the station building, so that the structural strength after the repair becomes strong and sufficient to withstand the load in accordance with current Indonesian building standards.

2. Evaluation of existing building conditions

The visual assessment results of the Pulau Aer Station building which has not been operating for a long time show that the existing condition of the building has suffered from heavy damage, especially the walls and other structural elements. Figure 1 shows the existing condition of the front view of Pulau Aer Station before retrofitting.

Figure 1. Front view of Pulau Aer Station before retrofitting.

The existing condition of the station wall was damaged in the form of wide cracks and loose wall connections. It is generally known that the condition of the wall damage contained cracks that mainly occurred at the top of the window corners (openings). Although the walls have a thickness of 30 cm, many walls were cracked and there were no binding castings on the walls of the building as in general

IOP Conf. Series: Earth and Environmental Science 1173 (2023) 012008 doi:10.1088/1755-1315/1173/1/012008

Unreinforced Masonry (URM) buildings built by the Dutch in Indonesia. Figure 2 shows the existing condition of the damage that occurred on the walls of the Pulau Aer Station building.

(a)

(b)

(c)

(d)

(e)

(f)

(g)

(h)

Figure 2. (a) Cracks in the axle wall I, (b) cracks in the axle wall 3 (H-I), (c) cracks in the wall in the middle aisle (as J), (d) cracks in the axle wall 3 (K-L), (e) cracks in the bottom wall of the axle 3 (K-L), (f) cracked walls and loose bricks as L, (g) cracks in the axle wall 1 (K-L) and (h) the loose wall meeting crack at point K3.

In order to re-activate this station, a new wall renovation plan was made according to the crack points, as shown in figure 3(a) which is the development of the old plan of the damaged Pulau Aer Station (figure 3(b)).

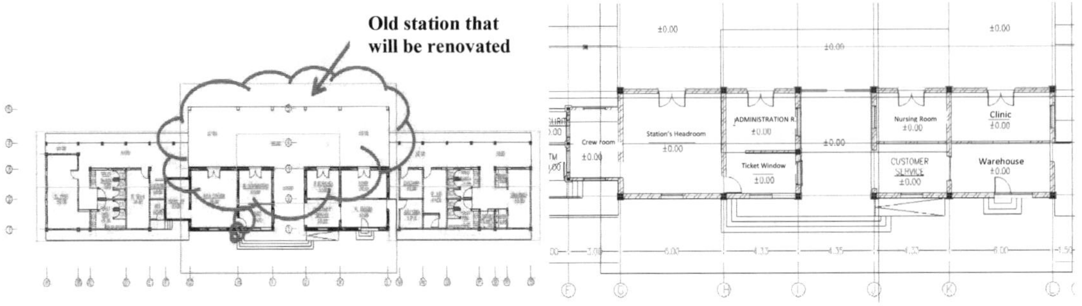

Figure 3. (a) A plan for the renovation of Pulau Aer Station and (b) the plan of the damaged old Pulau Aer Station.

3. Structural analysis of the existing station building

An analysis of the existing building structure was carried out using the ETABS v16 software. With the software simulation, the reliability of the structure can be seen against the loads received by the building structure such as the load in the form of its weight, the load due to activities in the building and the load due to the earthquake. To analyze the structure of this station, the following building standards are used: SNI 2847:2019 (requirements for structural concrete for buildings) [9], SNI 1726:2019 (procedures for planning earthquake resistance for buildings and non-buildings) [10], SNI 1727:2020 (minimum load for designing buildings and other structures) [11] and ACI 318-9 (building code requirements for concrete frame design) [12].

In this analysis, the 3D structural modelling of the Pulau Aer Station building is carried out as shown in figure 4. The technical specifications of the structure for the structural analysis are as follows:

- Building name : Pulau Aer Padang Station
- Location : Pasa Gadang, South Padang, Padang City, Indonesia
- Masonry wall thickness : 30 cm
- Type of structure : Unreinforced masonry
- Soil condition : Moderate soil (SD)
- Earthquake region : Padang City (West Sumatra Province)

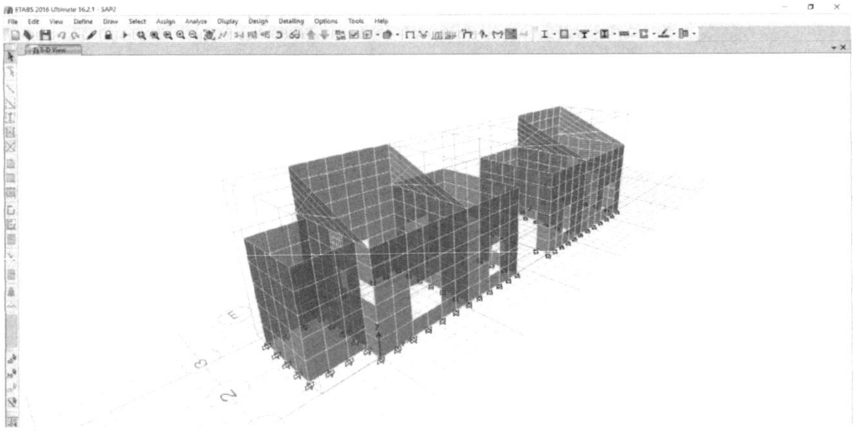

Figure 4. 3D modelling of the railway station building.

The analysis of this building structure uses an earthquake response analysis approach to the dynamic response spectrum. Response spectrum analysis is one of the few methods to determine the dynamic response of a 3D building structure to the influence of an earthquake, where the total dynamic response of the building structure is obtained as a superposition of the maximum dynamic response of each variety obtained through the design earthquake response spectrum.

For the earthquake area of West Sumatra, based on SNI 1726-2019 (procedures for planning earthquake resistance for building and non-building structures), risk category IV was obtained with a priority factor of 1.5, moderate soil type (SD) and seismic coefficient D. Figure 5 shows design earthquake response spectrum in Padang City.

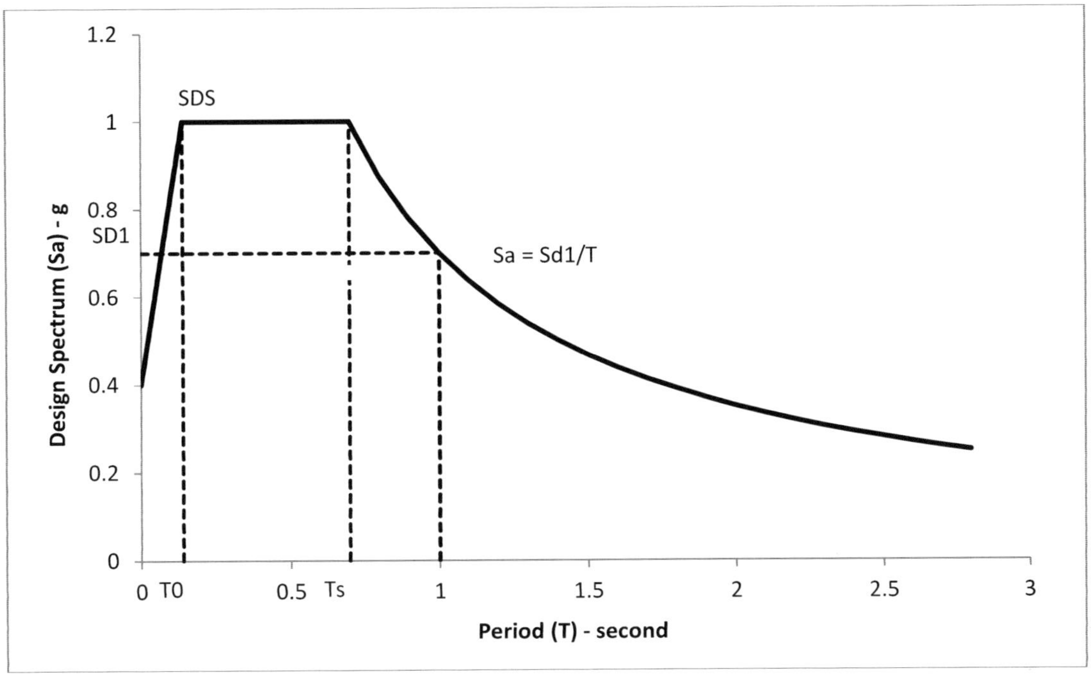

Figure 5. Design earthquake response spectrum in Padang City.

In reality, dead, live and earthquake loads will most likely work at the same time. To ensure that the building structure can withstand the working loads, the loads must be combined in such a way to the building structure, as regulated in SNI 2847-2013. Because the direction of the earthquake load is alternating, the combination of loading also considers this characteristic. Thus, the possible loading combinations will be obtained as shown in table 1.

Table 1. Earthquake load combination.

No	Combination
1.	U = 1 Ex + 0.3 Ey
2.	U = 1 Ex – 0.3 Ey
3.	U = - 1 Ex – 0.3 Ey
4.	U = - 1 Ex + 0.3 Ey
5.	U = + 0.3 Ex + 1 Ey
6.	U = - 0.3 Ex + 1 Ey

Figure 6 shows the shear stress on the walls of the Pulau Aer Station. It can be seen from the figure that the lowest shear stress is at the bottom of the wall at 0.5 kg/cm^2 and the highest shear stress due to horizontal loads is at the top of the wall which is about 7 kg/cm^2, while the allowable shear stress of the brick is 2 kg/cm^2. From figure 6, it is also seen that the shear stresses above the windows and openings are greater than the rest of the walls. This large shear stress causes cracks that occur in the walls, which matched well with the existing building condition, where there are many cracks in the window opening area and above the window.

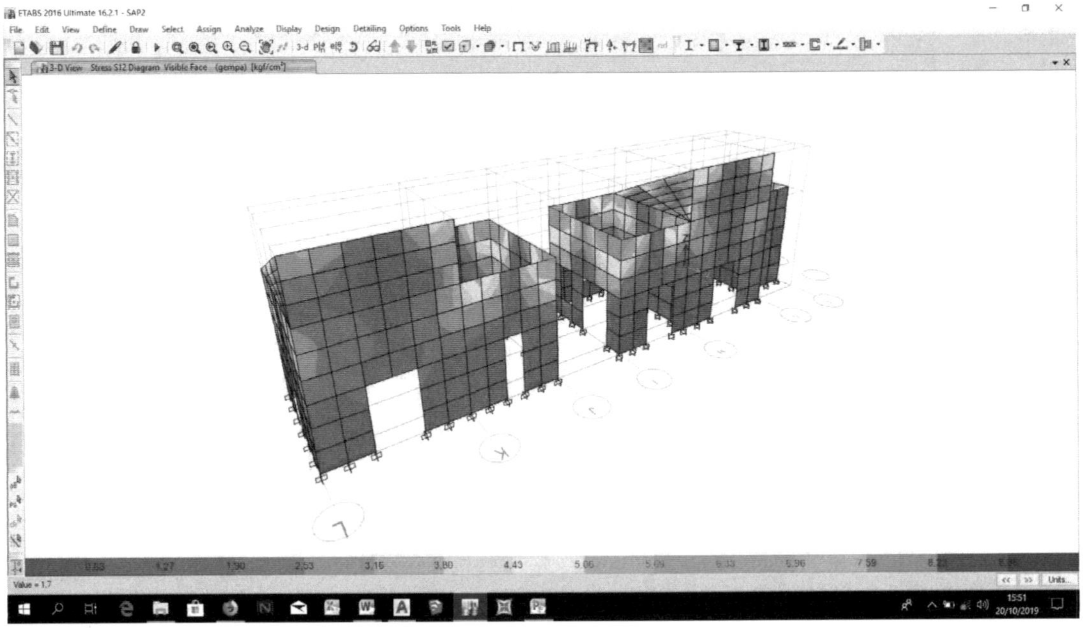

Figure 6. Shear stress due to earthquake (S12) that occurred on the walls of the station building.

4. Repair and retrofit of the building

4.1. Wall repair methods

The damage to the Pulau Aer Station building was quite heavy consisting of the damage to walls in the form of cracks with large and long gaps. As a result of the damage, the station building must be repaired so that the building can return to its original state.

Building repairs are divided into three types, namely architectural repair, restoration and strengthening or retrofitting. Architectural repair is the process of restoring the architectural form of the building so that all building components can return to their original function. The repair process on the walls of the Pulau Aer Station building that experienced crack was carried out using adhesives such as cement paste (a mixture of water and cement), mortar or additives such as Sika grout and others.

The following are the repair procedures for the cracked walls of the Pulau Aer Station building:

- First, cracked walls with a gap of less than 5 mm must be repaired by filling the crack with water cement (cement paste) in a ratio of 1 cement to 2 water, as shown in figure 7(a).
- Furthermore, the cracked gap is filled with cement water by pouring cement water into a plastic soy sauce bottle or using an injection/compressor tube for concrete grouting tools. The results of water-cement injection into a small cracked wall are, as shown in figure 7(b). Figure 7(c) shows the results of the injection of cement paste into the wall.

- For the cracked part of the wall with a gap of more than 5 mm, the gap is filled with a mortar of 1 cement: 3 sand, add enough water and mix until smooth, as shown in figure 7(d). For the implementation of filling cracks with a width of more than 5 mm, the usual way is to install plaster as shown in figure 7(e).

Figure 7. (a) Composition of the water-cement mixture, (b) injection of cement water into cracked walls, (c) injection of cement water into cracked walls < 5 mm, (d) specific mix composition of mortar and (e) filling crack gaps that are more than 5 mm with the specific mortar.

4.2. Building retrofitting method

Retrofit is a method used to repair or renovate a damaged building structure due to an earthquake. Retrofitting itself can be divided into several actions, namely repair, restoration and strengthening. The types of retrofits are global retrofit and local retrofit. Local retrofit is a method of strengthening structural elements with one example of its reinforcement using the ferrocement layer method. The ferrocement layer itself is a method of strengthening by adding woven wire/wire mesh on the wall which is coated with hydraulic cement mortar. The advantages of using this method are that the raw materials are easily available in all areas, the processing costs are relatively cheap and it is very easy for construction workers [13,14]. The procedures for retrofitting the Pulau Aer Station building are as follows:

- Installation of anchors/wires.
  The installation of this anchor/tie wire can be done using concrete wire. The concrete wire is installed to bind the woven wire inside and outside the room, so that the woven wire sticks firmly to the wall. For the installation of woven wire, holes must be made (drilled walls) with a distance of 30 cm horizontally and 30 cm vertically. Drilling of anchor installation as shown in figure 8(a).
- Installation of stucco head.
  For the thickness of the stucco to be even, before applying the plastering thoroughly, a stucco head must be made with a thickness of 1 cm and a width of 7.5 cm and the distance between the stucco heads is 25 cm, as shown in figure 8(b).

- Installation of woven wire.
  Woven wire is installed on all walls, both outer and inner walls and covers the entire wall. For the anchorage of woven wire to the wall, it is carried out as in procedure number 1, as shown in figure 8(c). The cross-section of the wall after laying woven wire and fastening woven wire with anchors is shown in figure 8(d). It is recommended to use woven wire with a diameter of 1 mm with a mesh of 3/4 inches. Figure 8(e) shows the final result of installing woven wire before plastering.
- Filling and drilling holes with cement paste.
  To further strengthen the stucco bond after installing woven wire and fastening the woven wire with anchors/wires/nails, the drill holes are first filled with cement paste figure 8(f).
- Plaster installation.
  For wall finishing, a mixture of 1:3 plaster is applied and after that it is finished as needed, as shown in figure 8(g).

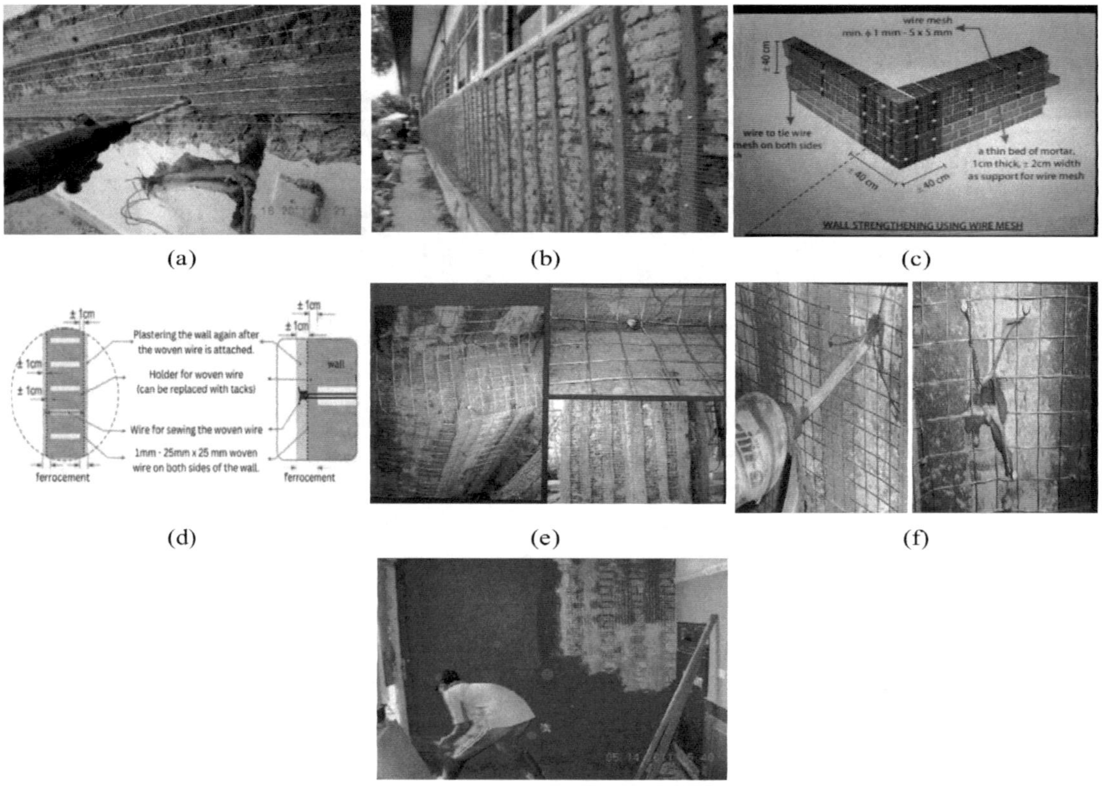

Figure 8. (a) Installation of woven wire anchors, (b) installation of stucco head, (c) installation of woven wire and anchor position, (d) cross-section of the wall with woven wire, (e) installation of woven wire, anchors and nails on the walls of woven wire installed with a diameter of 1 mm and a mesh of 3/4 inch, (f) filling of drill holes with cement paste and (g) wall plastering.

5. Structural analysis of the retrofitted building

The analysis was carried out after repairing and retrofitting the building using the ferrocement layer method. In the modelling of the specimen, there are three materials were defined, the first is a hollow brick made of a mortar mixture with a ratio of cement and sand 1: 4 with a compressive strength of 2.5 MPa, the second is plaster with a compressive strength of 9.9 MPa [6] and the third material is wire mesh with a tensile strength of 275 MPa [15]. These three combined materials are considered capable of withstanding earthquake loads that are applied to the building.

Modelling was carried out using finite element software ETABS v.16. Retrofitting of the walls with ferrocement layers was modeled with 5 layers of material consisting of inner plaster, inner wire mesh, hollow brick, outer wire mesh and outer plaster, as shown in figure 9.

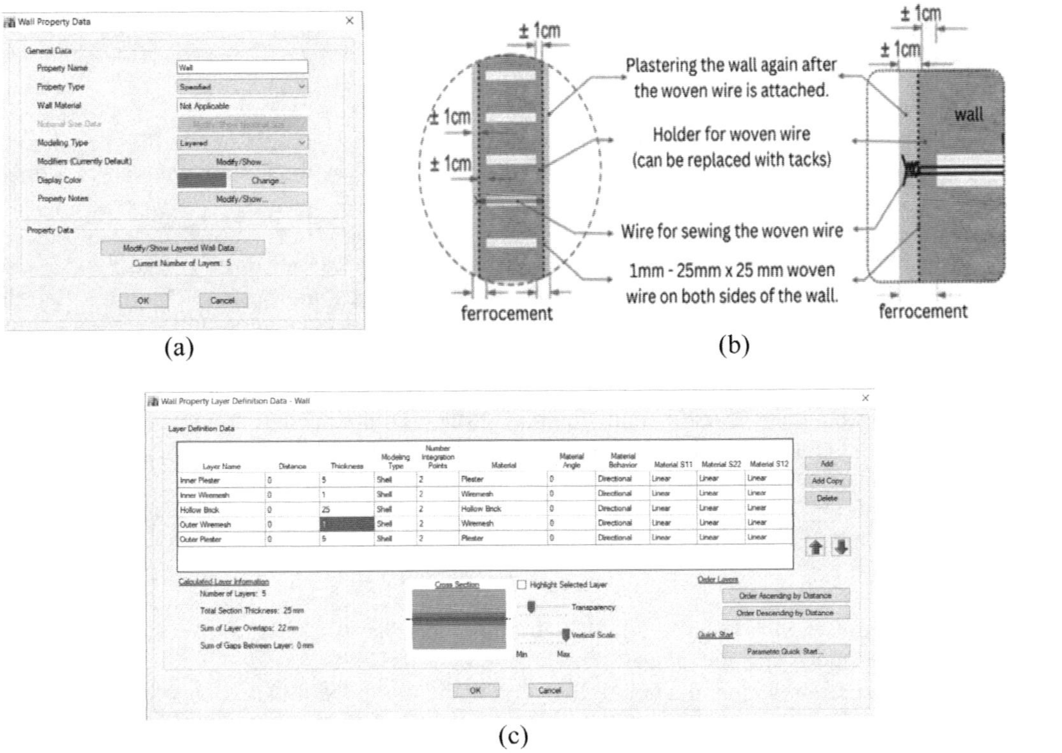

Figure 9. (a) Wall property data, (b) wall modelling and (c) building retrofitted by ferrocement layers.

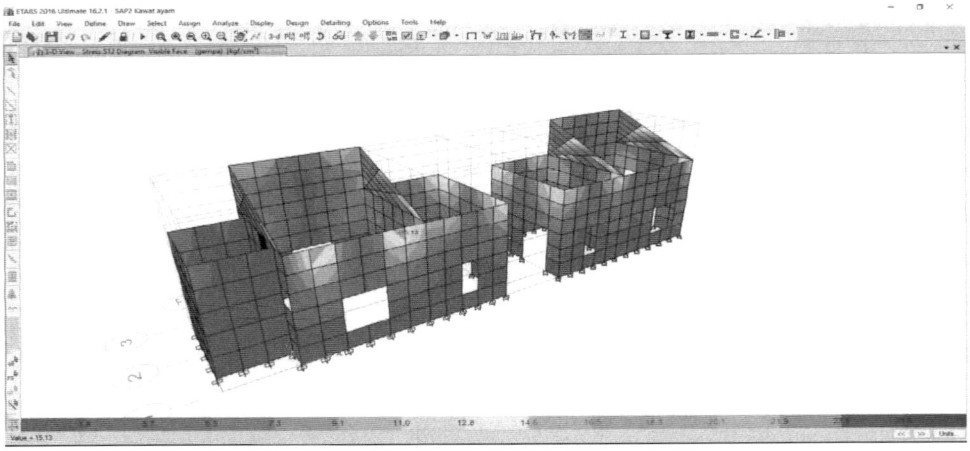

Figure 10. The shear stress due to the earthquake (S12) that occurred on the walls of Pulau Aer Station building.

The structural analysis was carried out on the station building retrofitted by using the ferrocement layers method to obtain wall stress values. Figure 10 shows the analysis results of the wall stress for the retrofitted station building. From figure 10, it can be seen that the lowest shear stress is at the bottom of the wall at 2 kg/cm^2 and the highest shear stress due to horizontal loads is at the top of the wall, which is 14 kg/cm^2, while the allowable shear stress of retrofitted wall is 140 kg/cm^2. This indicates that the retrofit of the URM old station building using ferrocement layer method improves the capacity of the station building so that the retrofitted building is strong enough to resist the working loads, including the earthquake load.

6. Conclusion

Pulau Aer Station building was heavily damaged, especially on the walls of the building which had severe cracks in the upper corner of the window (opening) walls. Although the walls have a thickness of 30 cm, many walls were loose due to the previous earthquake. In order to re-activate the station building, it is necessary to repair and retrofit the structure.

Repairs to the walls of the Pulau Aer Station building were carried out by restoring the cracked walls using adhesives such as cement paste (a mixture of water and cement), mortar or additional materials such as grout and others. Meanwhile, the retrofitting of the station building was carried out using the ferrocement layer method, a retrofitting method by adding wire mesh to walls covered with cement mortar. The result of the structural analysis shows that the retrofit of the URM old station building using ferrocement layer method improves the capacity of the station building so that the retrofitted building is able to resist all applied loads, including the earthquake load.

Acknowledgments

The authors gratefully acknowledge the financial support from Andalas University.

References

[1] Ayuningtias S H and Karmilah M 2019 Application of Transit Oriented Development (TOD) as an Effort to Realize Sustainable Transportation *Pondasi* **24** 45-66

[2] Wang C, Sarhosis V and Nikitas N 2018 Strengthening/retrofitting techniques on unreinforced masonry structure/element subjected to seismic loads: a literature review *The Open Construct. Build Technol. J.* **12** 251-268

[3] Bhattacharya S, Nayak S, Dutta S C 2014 A critical review of retrofitting methods for unreinforced masonry structures *Int. J. Disaster Risk Reduction.* **7** 51–67

[4] Meireles H, Bento R 2013 Rehabilitation and Strengthening of Old Masonry Buildings *Technical Report ICIST DTC nº02/13* (Lisbon: ICIST-IST)

[5] Ferretti E and Pascale G 2019 Some of the latest active strengthening techniques for masonry buildings: a critical analysis *Materials* **12** (7) 1151

[6] Fauzan, Ismail F A, Hakam A, Zaidir and Amailia S H 2018 Experimental Study on Masonry Building Strengthened with Ferrocement Layers *Int. J. Of Geomate* **14** 84-90

[7] Fauzan and Oswin 2016 Analysis of Simple Building Retrofitting Implementation Methods (Case Study: 43 East Rawang Elementary School, Padang) *Jurnal Rekayasa Sipil* **8** 11-20

[8] Fauzan, Ismail F A, Jauhari Z A 2016 Structural Response and Pounding of Andalas University Hospital Building Using New Indonesian Seismic Code SNI 1726-2012 *Int. J. of Applied Mechanics and Materials* **845** 274-282

[9] Plesu R, Taranu G, Covatariu D and Gradinariu I D 2011 Strengthening and rehabilitation conventional methods for masonry structures, Buletinul Institutului Politehnic din lasi *Sectia Constructii, Arhitectura* **57** (4) 165

[9] National Standardization Agency 2019 *Structural Concrete Requirements for Buildings, SNI 2847:2019* (Jakarta: BSN)

[10] National Standardization Agency 2019 SNI *Procedure for Planning Earthquake Resistance for Building Structures, 1726:2019* (Jakarta: BSN)

[11] National Standardization Agency 2013 SNI *Minimum Load for Planning of Buildings and Other Structures, SNI 1727:2013* (Jakarta: BSN)

[12] ACI Committee 318 2008 *ACI 318-9 building code requirements for concrete frame design* (Farmington Hills: American Concrete Institute)

[13] Rahman A, Bachtiar G, Daryati 2013 Study of the Flexural Strength of Ferrocement Plates with Aluminum Sheet Layers as Fixed Formwork on High-rise Building Floor Plates *J. of Menara Civil Engineering Department Faculty of Engineering* UNJ **8** 1

[14] Apriwelni S, Ismail F and Hakam A 2017 Numerical Study of Red Brick House Reinforcement Using Ferrocement Layers In: Hidayat, B and Purnawan, P (Eds.) *Prosiding 4 th Andalas Civil Engineering (ACE) Conference 2017, November 9, Andalas University, Padang, Department of Civil Engineering Unand,* 583-593

[15] Sakthivel P and Jagannathan A 2011 Ferrocement Construction Technology and its Application, – A Review *International Conference on Structural Engineering, Construction and Management* India pp.69-76

IOP Conf. Series: Earth and Environmental Science 1173 (2023) 012009 doi:10.1088/1755-1315/1173/1/012009

Numerical study of spun pile with steel jacket

J Reagen[1], Y Lase[2], W A Prakoso[2] and M Orientilize[2]

[1] Postgraduate Students, Civil Engineering Department, Universitas Indonesia, Depok, Indonesia, 16420
[2] Civil Engineering Department, Universitas Indonesia, Depok, Indonesia, 16420

E-mail: jansen.reagen@ui.ac.id

Abstract. The confinement ratio of spun piles manufactured in Indonesia has been a matter of concern since it does not meet the applicable requirements. Additional steel jacketing retrofitting was used to cope with the problem as well as improve the strength of spun pile to pile cap connection. A numerical study was performed to study the effect of steel jacketing on a spun pile with concrete infill reinforcement. Furthermore, a parametric study was conducted to study the effect of steel jacketing length to spun pile behaviour. Pushover analysis using the Opensees software will be performed with a monotonic loading approach. The results were compared in terms of load – displacement curve, shear and bending capacity, and displacement ductility. It is found that the presence of steel jacketing provides a better behaviour and significantly increases the ductility of specimens. The study also found that a minimum length 1.5D of steel jacketing retrofitting was necessary for better results and remain easily repairable for future concerns.

1. Introduction

Spun piles is a hollow circular pile made by precast and prestressing method with an outer diameter size range from 300 to 1200 mm [1]. Spun piles were commonly used as a foundation to transfer load structure to the ground. The reason for the widely used of spun piles are because of the lower cost and the high bearing capacity of spun piles. One of the most important components that need to be considered is the connection between spun piles and pile cap, where as in this region the change of area, stress, and stiffness occurred suddenly [2]. The connection region experienced a high curvature if were designed as fixed head. Therefore, a ductile connection must be achieved so that it can sustain large deformation without a significant strength reduction.

Indonesia is well-known for being prone to earthquakes due to the geographical location of Indonesia, which is located between on the tectonic zone where three plates of the earth collide (Indo-Australia, Eurasian, and Pacific). The other reason for Indonesia's vulnerability to earthquakes is because it is in the ring of fire, where a volcanic eruptions and earthquakes frequently occurred. In 2017 Pusat Studi Gempa Nasional (PuSGeN) released Indonesia earthquake map that shows an increasing earthquake acceleration compared to 2010. Due to the increasing in earthquake acceleration, seismic demand also increased. Hence, structure that have been designed using the old seismic demand potentially experiencing damage due to earthquake.

The strength of spun pile against earthquakes requires special attention. Based on the results of investigations on damage to buildings after the Kobe earthquake in Japan with a scale of 6.9 Richter, it was found that compression and shear failures in the spun pile section close to the pile cap [3]. In addition, the investigation of damage building due to Tohoku earthquake with a scale of 9.1 Richer found the same compression and shear failures at the spun pile and pile cap connection [4]. Based on the ductility analysis using NEHRP 2000, it was found that spun pile is suitable for use in areas with low to moderate earthquake risk. Thus, with the increase in earthquake acceleration and the worrying performance of spun piles in high earthquake risk areas, it is necessary to conduct further studies on spun piles [5].

Several studies have been conducted on spun pile to improve spun pile performance against earthquakes. The investigation of spun pile damage due to high earthquake suggested that the spun pile design should be ductile and provided with additional reinforcement. One of the experimental studies to increase the ductility of the spun pile is to add concrete infill to the hollow section of spun pile carried out by Irawan [6] and Bang [2]. Further research conducted by Guo where the spun pile to pile cap connection is reinforced with T-shape steel and strengthened ring, resulting in higher strength and ductility compare to unreinforced connection [7]. Increased ductility also achieved by adding steel fiber and deformed bar reinforcement as has been done by Yang [8], but unfortunately this type of reinforcement doesn't increase bearing capacity and stiffness of the connection. In addition, a parametric study was carried out by Orientilize to study the effect of steel jacketing thickness using SAP2000 and Abaqus software. It is found that the proper thickness of steel jacketing can improves the strength connection and ductility [9].

Research on spun piles in Indonesia has not been widely carried out because the design of the lower structure is still within the elastic limit, therefore has not been allowed to suffer damage, so it does not refer to performance based design (PBD). Lower structure design was guided by SNI 8640:2017 [10], where lateral displacement of pile is restricted to 25 mm for severe earthquakes. Amplification in seismic demand will lead to higher construction cost and larger size of lower structure to fulfill the requirement that damage is not allowed to happen. Since damage due to earthquake is almost impossible to avoid, hence a PBD should be allowed to optimizing lower structure to become earthquake-resistant structure.

Although the use of spun piles in Indonesia is generally given concrete infill and reinforcement in the hollow section with a span of one meter from the lower limit of the pile cap, one of the spun pile studies that has been carried out by Irawan found that the confinement spun piles made in Indonesia are not strong enough, even far below the ASCE 7-16 standard [11]. Generally, spiral reinforcement on spun piles manufactured in Indonesia has a closer distance at both ends and more far in the middle. However, the spun pile is often cut right in the middle with more insufficient confinement and connected to the pile cap. This will certainly reduce the ductility and strength of the spun pile and pile cap connections. Adding more spiral reinforcement is not an effective option since the design will become more extravagant. To cope with this problem additional steel jacketing retrofitting were added to solve confinement problem and increasing the strength and ductility of the spun pile to pile cap connection. Although Orientilize found that the thickness of steel jacketing has a significant effect on strength and ductility, a further parametric study using Opensees will be performed using a different parameter, that is the steel jacketing length. Furthermore, a numerical study that uses Opensees to perform monotonic pushover analysis was conducted to study the effect of steel jacketing on a spun pile with concrete infill reinforcement.

2. Research methodology

2.1. Validation model

In order for the model conducted able to represent the actual condition of the structure, a validation phase was carried out. An experimental study done by Tandita, et al. [12] used as a reference. In this study one specimens were used as a reference, that goes by the code SPPC 02. The specimens were a 450 mm diameter PHC pile with 80 mm thickness, 2100 mm in length, and embedded into the pile cap as deep as 100 mm. The spun pile consisted of ten 7.1 mm in diameter of prestressed bar and six rebar with 19 mm diameter as the anchorage against the pile cap. The stirrup spacing and diameter of the

IOP Conf. Series: Earth and Environmental Science 1173 (2023) 012009 doi:10.1088/1755-1315/1173/1/012009

spun pile was 4 mm and 120 mm respectively. The stirrup spacing and diameter of the concrete infill was 8 mm and 75 mm respectively. In addition to validation of specimens, an experimental study was performed on a new specimen that goes by the code SPPC 07. SPP 07 is the same spun pile as SPPC 02 but with additional steel jacketing retrofitting. SPPC 07 is retrofitted by 750 mm in length, 25 mm in thickness of grouting concrete and 0.7 mm in thickness of zincalume from the top of pile cap. The experimental setup and SPPC 02 and SPPC 07 cross section can be seen in figure 1.

The typical cross section included concrete, rebars and prestressed bar, and the steel jacket present in figure 2. Since Opensees was not able to adjust the location of prestressed bar manually, the position of the prestressed bar differs from the experiment, but this effect can be neglected due to the same input number of prestressed bar. The Concrete 02 material with linear tension softening was used for concrete infill, the Concrete 01 material was similar to Concrete 02, but without tension properties was used for grouting concrete, and Concrete 07 material was used for spun pile concrete. The Steel 02 material with isotropic strain hardening was used for concrete infill rebar and the Reinforcing Steel material was used for prestressed bar. Reinforcing Steel material was chosen because compared to Steel 02 material, Reinforcing Steel material present a better failure model, although this material was not able to include prestresses effect, the prestressed force can be neglected since it has a very small effect compared to axial load. A fully bonded connection between material was assumed by default by the Opensees program since it cannot adjust the slip between material.

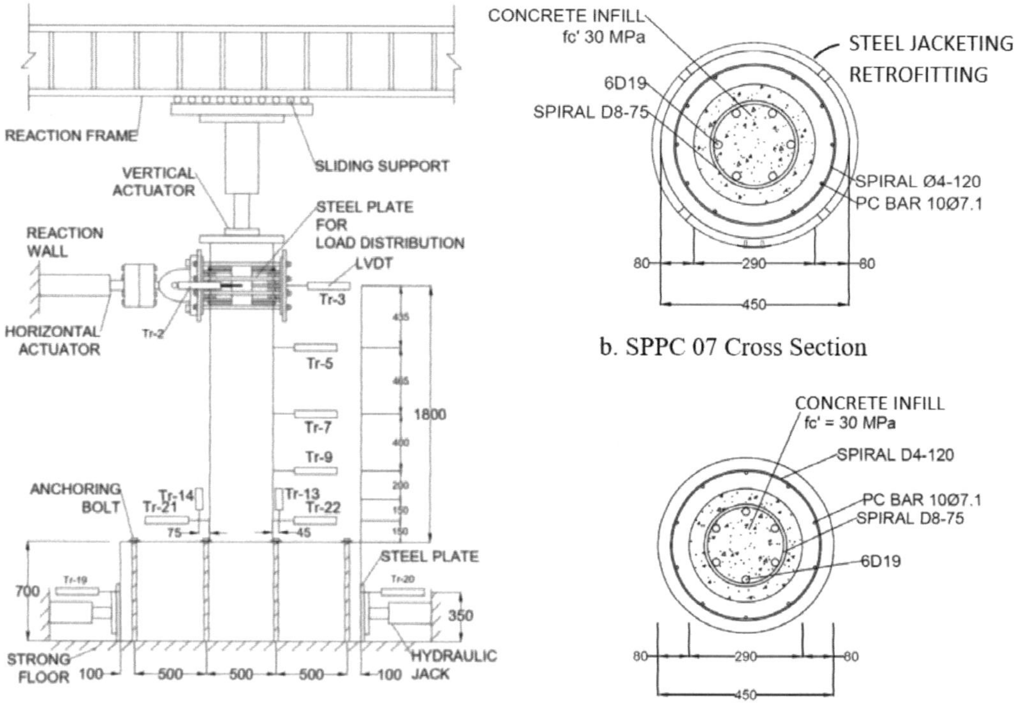

a. Experimental Setup

b. SPPC 07 Cross Section

c. SPPC 02 Cross Section

Figure 1. Experimental study.

Monotonic pushover analysis was performed to represent the experiment where in this type of analysis, the load is increased continuously and controlled by displacement until the ultimate condition achieved. The results were plotted together with the experimental results and presented in figure 3. It is shown that Opensees analysis coincided with the experimental data with a relative error 7.3% for SPPC 07 and 6.5% for SPPC 02. This means that the numerical model is valid and can represent the structure real behavior, both on the elastic stage and post yielding behavior. Therefore, further study to investigate the effect of steel jacketing on a spun pile can be carried out.

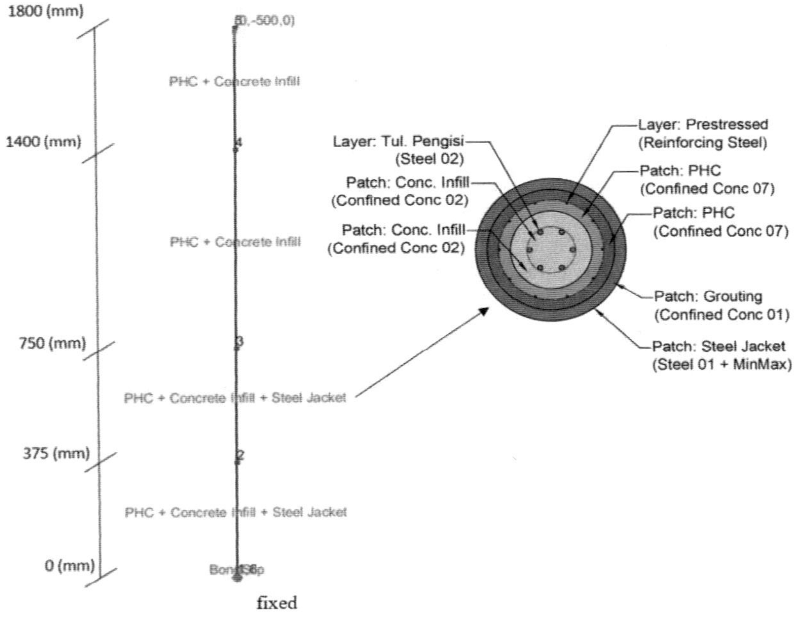

Figure 2. Modelling of spun pile with steel jacketing using opensees.

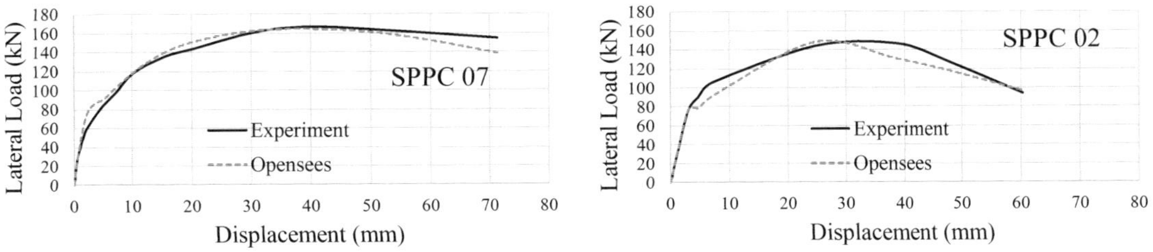

Figure 3. Validation results.

2.2. Numerical study

The main purpose of adding a steel jacketing on the spun pile is to achieve minimum confinement ratio required based on ASCE 7-16 [13], where the minimum confinement ratio needed for pile can be calculated using equation 1, as follow:

$$\rho_s = 0.25\left(\frac{f_c'}{f_{yh}}\right)\left(\frac{A_g}{A_{ch}} - 1\right)\left(0.5 + \frac{1.4P}{f_c' A_g}\right) < 0.021 \tag{1}$$

where f'_c is the compressive strength of the concrete, f_{yh} is the yield strength of spiral reinforcement, A_g is the pile cross-sectional, A_{ch} is the core area defined by spiral outside diameter, and P is the axial load on pile. Meanwhile, the pile confinement ratio can be calculated based on equation 2, as follow:

$$\rho_s actual = \frac{4A_{sp}}{D's} \tag{2}$$

where A_{sp} is the area of transverse reinforcement bar, D' is the diameter of confined concrete and s is spiral spacing. The thickness of the steel jacketing, in this case the zincalume assumed to behave as a spiral reinforcement with a spacing of 1 mm and a diameter equivalent to equation 3, as follows:

$$\frac{\pi}{4} d_{equivalent}^2 = bh \tag{3}$$

where b is the zincalume thickness and h is the 1 mm spacing. Based on the equation above, the minimum confinement ratio for a 450 mm in diameter pile with 80 mm thickness and stirrup spacing and diameter was 4 mm and 120 mm respectively is 1.02%. On the other hand, the actual ratio for pile used in this study is 0.14% which is very far below the minimum requirement. With the present of steel jacketing, the confinement ratio increases up to 4 times the previous value, that is 0.56% and satisfy 55% of the minimum confinement ratio require.

Similar to the validation stage, the pile was modelled as a single frame element with the exact fibre cross section used in validation phase. In the experiment study, the steel jacketing is applied along 750 mm from the top of pile cap. This value was obtained by considering the maximum moment location where the plastic hinge would occur. The plastic hinge length can be calculated according to AASHTO (2011) [14] by equation 4, as follow:

$$Lp = 0.1\, H' + D\ \leq 1.5D \tag{4}$$

where H' is the length of pile from ground level to contraflexure point below ground and D is the diameter of the pile. For this study, a parametric study will be performed to evaluate the effect of steel jacketing length to the pile strength and ductility. The steel jacketing length will be divided into three lengths, that is along D or the diameter of the spun pile; 1.5 D; and 2D. Since the diameter of the pile used was 450 mm, then the steel jacketing length that will be used are 450 mm, 675 mm, and 900 mm with additional length 750 mm that was used by previous study. A further analysis will be conducted to observe the process of plastic hinge or yielding.

3. Result and discussion

3.1. Effect of steel jacketing based on 2D model on Opensees

Spun pile with 750 mm steel jacketing by the code SPPC 07 and a regular spun pile without steel jacketing by the code SPPC 02 was modelled using Opensees. The results were compared in figure 4. and it shows that the steel jacketing indeed increases shear capacity of the spun pile by 10.8% from 148.1 kN to 164.1 kN. The presence of steel jacketing also increases the bending capacity of spun pile by 9% from 290.4 kNm to 316 kNm. Yielding of the specimen was calculated based on equivalent elasto-plastic energy absorption [15], and presented in Table 1. Figure 4 also shows that the presence of steel jacketing provides a better failure condition than without steel jacketing. After the ultimate condition, SPPC 02 experience a significant loss in strength that drop up to 35% from 148.134 kN to 96.660 kN in a short range of displacement compared to SPPC 07. Measured from SPPC 07 ultimate condition, the strength drop up to 7.6% from 164.119 kN to 151.601 kN. Therefore, it can be concluded the presence of steel jacketing increase the bending and shear capacity of the specimen and provide a better post ultimate condition.

Improvement in flexural strength was followed by improvement in displacement ductility. Displacement ductility was calculated using methods suggested by Park by dividing Δ_u with Δ_y. Δ_u present the ultimate displacement based on a significant load capacity after peak load and Δ_y present the yielding displacement. Table 1 shown the presence of steel jacketing increase the displacement ductility of the specimen by 49.125%. Based on FEMA 356 [16], it can be said that SPPC 02 have a moderate ductility and SPPC 07 have a high ductility. Therefore, it can be concluded that the presence of steel jacketing has a significant effect on increasing ductility of the spun pile.

Table 1. Summary of numerical study of steel jacketing effect.

Specimen	Bending Capacity Mu (kNm)	Shear Capacity V (kN)	Structure Yielding (kN)	Displacement Ductility
SPPC 02: Without Steel Jacket	290.424	164.119	131.145	3.258
SPPC 07: With Steel Jacket	316.466	148.134	143.773	4.858

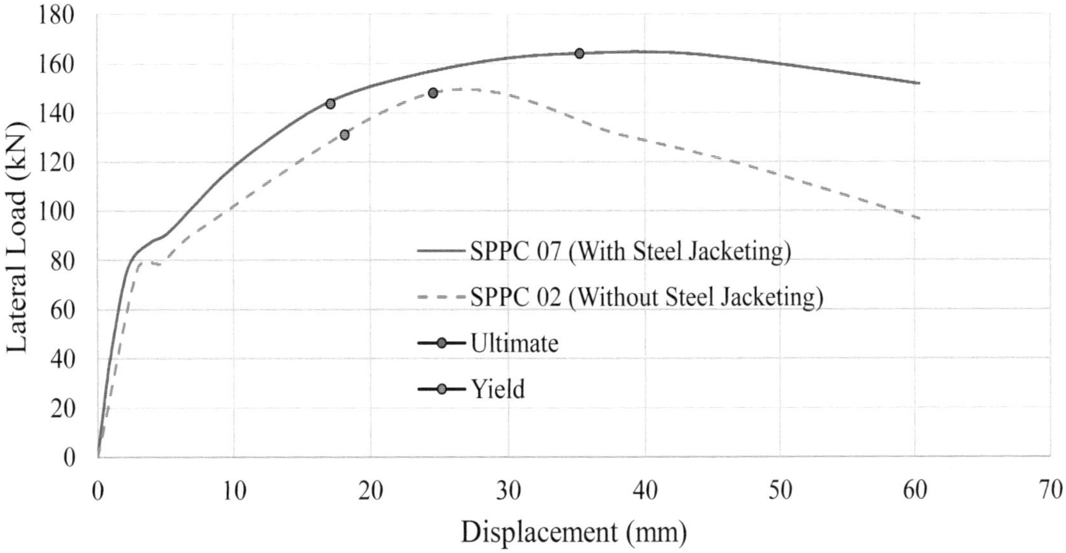

Figure 4. Monotonic Pushover Curve Comparison using Opensees between SPPC 07 and SPPC 02.

3.2. Effect of steel jacketing based on 2D model on Opensees

An experimental study specimen SPPC 07 was chosen as a benchmark for numerical analysis on the steel jacketing length and goes by the code SPPC 07 A. Three different additional specimens were modelled using Opensees, that is SPPC 07 B with D length steel jacketing or 450 mm, SPPC 07 C with 1.5D length steel jacketing or 675 mm, and SPPC 07 D with 2D length steel jacketing or 900 mm. Figure 5 shows comparison between the monotonic pushover curve from the four specimens. The specimen with a minimum length followed equation 4 shows a similar curve shape. However, specimens SPPC 07 B with D length steel jacketing shows a greater loss compared to other specimens after ultimate condition. A better failure condition was shown in the specimens with a minimum length of steel jacketing 1.5D. The length of steel jacketing has a significant effect on the strength loss after ultimate condition, therefore the minimum length of steel jacketing should be fulfilled.

Differences can be found at the value of ultimate condition between SPPC 07 A, SPPC 07 C, and SPPC 07 D, with the greatest value belong to SPPC 07 D with 2D length steel jacketing. Table 2 shows that if the length of steel jacketing fulfills the requirement governs by equation 4, adding more length to the steel jacketing provide greater shear capacity, bending capacity, and displacement ductility. An additional 0.5D length of steel jacketing increase the shear capacity of the specimen by less than 5%, therefore it doesn't significantly increase shear capacity of specimens. The same thing can be said for bending capacity and displacement ductility, with an amplification less than 5% in bending capacity and displacement ductility, an additional 0.5D length of steel jacketing doesn't increase bending capacity and displacement ductility significantly. Based on FEMA 356, all specimens were categorized to have high ductility with a value greater than four.

Table 2. Summary of Parametric Study based on Opensees.

Specimen	Length of Steel Jacketing (mm)	Bending Capacity Mu (kNm)	Shear Capacity V (kN)	Displacement Ductility
SPPC 07 A	750	316.466	164.119	4.858
SPPC 07 B	475 (D)	289.802	151.579	4.103
SPPC 07 C	675 (1.5D)	309.312	161.105	4.662
SPPC 07 D	900 (2D)	331.251	171.316	4.930

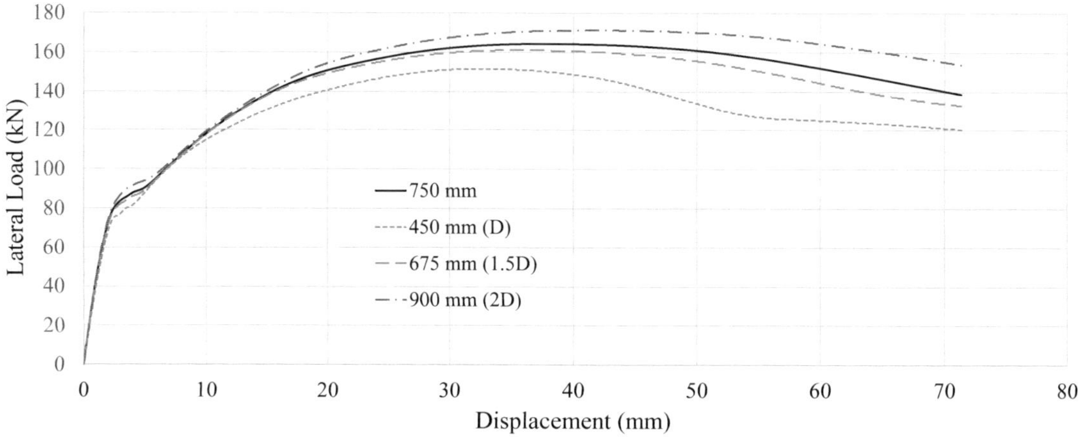

Figure 5. Effect of steel jacketing length variation.

Overstrength ratio (OSR) was calculated among four specimens. Overstrength ratio present the ratio between the ultimate strength and the strength obtained by code-based design. OSR can be calculated globally or locally. Locally, the calculation of OSR only considers the contribution of one element of the overall structural system. Globally, the calculation of OSR considers the overall contribution of the structural system. Table 3 shows the OSR value for each specimen. Vy express the value shear capacity at yielding and Mn express the value of cross section loaded with 500kN axial load. According to ASCE 41-17, for each value of OSR still under the value 1.25 (reinforcing steel tensile). This means that the OSR value is still within the expected limit value by considering α with a value of 1.25 on the concrete infill reinforcement.

Table 3. Summary of OSR Value based on Opensees.

Specimen	OSR (GLOBALLY)			OSR (LOCALLY)		
	Vmax (kN)	Vy (kN)	OSR	Mu (kNm)	Mn (kNm)	OSR
SPPC 07 A	164.319	143.773	1.143	316.466	281.575	1.124
SPPC 07 B	151.579	131.664	1.151	289.802	281.575	1.029
SPPC 07 C	161.105	141.386	1.139	309.312	281.575	1.099
SPPC 07 D	171.316	150.047	1.142	331.251	281.575	1.176

To confirm the location of plastic hinge, an additional software SAP2000 was used to obtain the value of plastic moment. Section designer feature was used, and the value obtained for plastic moment in the steel jacketing section is 309.592 kNm and 207.069 kNm for the section without steel jacketing. The moment-curvature of section with and without steel jacketing was presented in figure 7. Figure 6 shows the maximum bending capacity for each specimen (colored green) compared to the section bending capacity (colored red for section without steel jacketing and colored blue for section with steel jacketing). For SPPC 07 A, C, and D the maximum bending capacity value exceed plastic moment, hence plastic hinge formed on the section with steel jacketing retrofitting. For SPPC 07 B, the maximum bending capacity value didn't exceed plastic moment on the steel jacketing section but exceed plastic moment value of the section without steel jacketing, hence plastic hinge was formed on the section above the steel jacketing retrofitting. Therefore, it can be concluded that for specimens SPPC 07 A, C, and D the first plastic hinge was formed on the steel jacketing section and a second plastic hinge was not formed. The first plastic hinge on SPPC 07 B was formed on the section without steel jacketing and also without a second plastic hinge. A minimum length 1.5D of steel jacketing is necessary for plastic hinge to formed on the upper part of pile, so that it will remain easily repairable for future consideration.

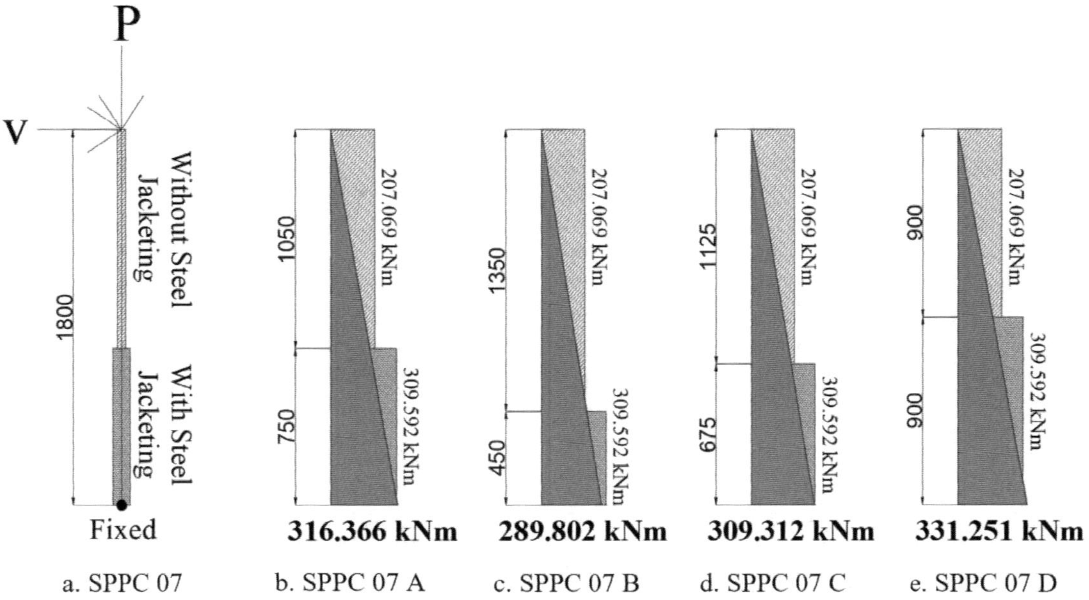

Figure 6. Bending capacity of specimens.

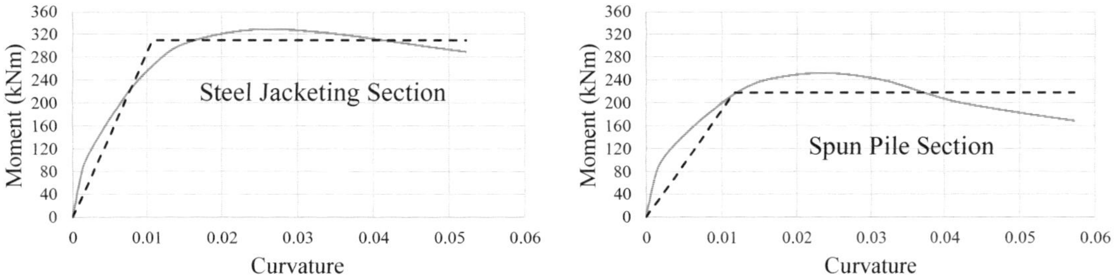

Figure 7. Moment-curvature of cross section.

4. Conclusion

Numerical results in this research shows that the application of steel jacketing retrofitting in the spun pile to pile cap connection improves shear and bending capacity and significantly improve structure ductility. The presence of steel jacketing also provides a better failure condition compared to the connection without steel jacketing. Parametric results show that a minimum 1.5D length of steel jacket is necessary for plastic hinge to formed on the upper section of the pile so that thc pile will be easily repairable and to avoid significant loss at the failure stage. If the length of steel jacketing meets the 1.5D requirement, additional 0.5D length of steel jacketing doesn't significantly improves the strength and ductility of the specimens.

References

[1] Irawan C, Suprobo P, Gusti Putu Raka I and Djamaluddin R 2014 A Review of Prestressed Concrete Pile with Circular Hollow Section (Spun pile) *J. Teknol.* **72** 115–123
[2] Bang J W, Oh S J, Lee S S and Kim Y 2016 Pile-cap Connection Behavior Dependent on the Connecting Method between PHC pile and Footing *J. Korea Inst. Struct. Maint.* 25-32
[3] Mizuno H, Iiba M, Hirade T 1996 Pile Damage During the 1995 Hyogoken-Nanbu Earthquake in Japan *11th World Conf. Earthq. Eng.* Paper No. 977
[4] Kaneko O, Kawamata S, Nakai S, Sekiguchi T, Mukai T 2018 Analytical study of the main causes of damage to pile foundations during the 2011 off the Pacific coast of Tohoku earthquake *Japan Archit.* **1** 235–244
[5] Irawan C, Djamaluddin R, Raka IGP and Suprobo P 2016 Experimental Investigation of Failure Mechanism of Spun Pile Due to Monotonic Loading Using Nehrp 2000 *Int. Conf. Prot. Struct.* (ICPS4) (Beijing, China) pp. 313–322
[6] Irawan C, Djamaluddin R, Raka IGP, Faimun, Suprobo P and Gambiro 2020 The effect of the presence of infilling concrete on flexural performance of spun pile – An experimental study *J. Teknol.* **82** 85–94
[7] Guo Z, He W, Bai X and Chen Y F 2017 Seismic Performance of Pile-Cap Connections of Prestressed High-Strength Concrete Pile with Different Details *Structural Engineering International* **27** 546-577
[8] Yang Z, Li G, and Nan B 2020 Study on Seismic Performance of Improved High-Sterngth Concrete Pipe-Pile Cap Conncetion *Adv. Mater. Sci. Eng.* **9** 264
[9] Orientilize M, Prakoso WA, Fadel M, Fernaldy RP 2021 Study of spun pile connection with steel jacket strengthening *IOP Conf. Ser.: Earth Environ. Sci.* **622** 012028
[10] SNI 2017 *Standar Nasional Indonesia (SNI) 8460 Persyaratan Perancangan Geoteknik* (Jakarta: BSN)
[11] Irawan C 2018 Confinement Behaviour of Spun Pile using Low Amount of Spiral Reinforcement *Int. on Advanced Science Engineering Information Technology* **8** 501-507
[12] Yulius, Prakoso WA, Orientilize M and Sentosa B O B 2021 Numerical Study of Spun Pile To Pile Cap Connections In Soft Soil *ICETIA Journal of Physics: Conference Series* **1858** 012087
[13] ASCE 2016 7-16 *Minimum Design Loads and Associated Criteria for Buildings and Other Structure* (Virginia: American Society of Civil Engineers)
[14] AASHTO 2011 Guide Specifications for LRFD Seismic Bridge Design. In *American Association of State Highway and Transportation Officials* (United States: American Association of State Highway)
[15] Park R 1988 Ductility evaluation from laboratory and analytical testing *Proceedings of the 9th World Conference on Earthquake Engineering* (Tokyo-Kyoto)
[16] FEMA 2000 *FEMA 356: Prestandard and Commentary for The Seismic Rehabilitation of Buildings* (Washington, D.C: Federal Emergency Management Agency)

The Ratio of Cement and Sand from Lake Bingkuang as Aggregate to Create Foam Mortar 2,000 kPa

D R Basri [1,2], E E Putri [3], B M Adji [3] and A Hakam [3]

[1] Doctoral Student, Civil Engineering Department, Faculty of Engineering, Andalas University, Padang 25163, Indonesia
[2] Civil Engineering Department, Abdurrab University, Pekanbaru 22892, Indonesia
[3] Civil Engineering Department, Andalas University, Padang 25163, Indonesia

Corresponding e-mail: elsaeka@eng.unand.ac.id

Abstract. Riau is known as the Province that has the largest peat soil where the construction of roads on peat soil will result in subsidence of the peat soil surface.The purpose of this study is to replace the top layer foundation for road construction with lighter and stronger materials according to the PUSJATAN 2013 standard with acompressive strength of more than 2000 kPa and volume weight less than 0.9 tons/m3by using foam mortar that consist of cement (PCC), sand, foam, and water. The test conducted include gradation analysis test and specific gravity test.Then proceed with making a mixture composition with a ratio of cement and sand 1C:0.4S, 1C:0.5S, 1C:0.6S, and 1C:0.7S in order to create a cylindrical sample with a diameter of 10 cm and a height of 20 cm. There are 3 samples in each ratio of cement and sand, so the total number of samples is 54. Then the volume weight test and compressive strength test were carried out using a CBR modified. From the test results, the sample with a cement amount of 525 kg and a ratio of 1C: 0.6S can float in water, with a density of 0.834 tons/m3 and compressive strength of 2,401.40 kPa.

1. Introduction

Riau province's land is mostly formed of peat soil. It covers 4,044 million ha, or 56.1% of Sumatra's total area, or about 45% of Riau's. Because of the low bearing capacity and deep hard soil, the soil reaches 16 meters deep in several locations. Peat soil is made up of organic vegetation sections that have been chemically transformed into fossils [1].

Peat soil has a low bearing capacity, the construction of the road on this soil can cause very significant subsidence on the soil surface which reaches up to 0.5 meters. Therefore, there are several efforts are needed to intensify the soil to prevent massive subsidence and permanent deformation of the soil surface itself. Treatments of light construction on peat soil are reducing subsidence on the peat soil surface.

The Research and Development Center for Roads and Bridges (PUSJATAN) 2013 created a new technique which is called Foam Mortar Light Embankment Technology. [2] This technique has a smaller density rather than water with a high bearing 800kPa. Thus, it can be used as a substitute for backfill and top layer foundation in road construction on peat soil.

The primary goal of the research is to determine the quantity of several materials like cement and sand. These materials are used to produce foam mortar regarding the standard requirements of PUSJATAN 2013. This research is using materials ratio, they are cement ratio, and also the cement and sand ratio. The sand used comes from Lake Bingkuang, Tambang Village in Kampar, Riau, and PCC cement (Portland Composite Cement) comes from Padang, West Sumatra.

2. Literature Review

Foam mortar is created by combining mortar and foam agents. It contains materials like cement, sand, and water. Meanwhile, the foam is made up of foam agents and water [3].

On the load test which has been carried out, it is seen the scheming of a dam with foam mortar material produces lower compressive value on the soil rather than in the dam with soil, stone, and sand mixing because foam mortar is more stable and lighter [4].

Previously, a study which has been done by using foam agents to create light concrete bricks, the foaming agent is submitted into the concrete mixing in producing light concrete [5]. Samples were examined by using a cube mold with cement and sand composition 1:0.5; 1:0.7 and 1:0.9. This composition produces pressure strength respectively of 17.422 kg/cm2; 14,756 kg/cm2, and 9,778 kg/cm2 with a weight of 2,510 kg, 2,748 kg and 2,808 kg [5].

In addition, there is also another research on creating foam concrete which uses a mixing ratio of cement, sand, and water 1:1.5:0.5 [6]. In this research, the addition of foam agents started from 28%, 45%, 62%, and 78%. The twenty-eight percent adding foam agent resulted in a density value of 1770 kg/m, while the seventy-eight percent adding foam agent has a density value of 500 kg/m^3 [6].

When it is compared, the alternative usage of CSS-Mortar Foam type construction helps in reducing the cost of flyover construction by around 44.7% rather than PCI-Girder type construction [7].

The results showed that the impact of polycarboxylate superplasticizer (PS) on foam concrete trait was the factor of water, cement, and polycarboxylate superplasticizer (PS) had a key effect on foam concrete performance. PS 1.35% improves the mechanical and transport properties of foam concrete [8].

According to the findings, adding ground calcium carbonate (GCC) can increase pressure strength and thermal conductivity due to the effect filling process. Furthermore, the addition of glass fiber (GF) increases the flexural strength of the concrete [9].

Meanwhile, the foaming agent concentration effect and its microwave and ultrasonic treatment on foam stability, bulk density, microstructure, and pressure strength are being researched [10]. The pore size reduction contributes to the rise of pressure strength which determines the prepared ultrasonically treated foaming agents and samples with lower microwave concentrations. There is a positive effect when the foaming agent has a lower concentration, which is, giving stability to the foam itself. However, the effect is not mentioned [10].

Furthermore, raw materials, possible modifications, and the latest structures inside the internal lining of porous materials experimental studies offer to improve the structure and operational character of autoclaved foam concrete [11].

In this assessment, the amount of cement used was 400 Kg, 425 Kg, 450 Kg, 500 Kg, 525 Kg, and 550 Kg, and the cement and sand ratio varied by 1C:0.4S, 1C:0.5S, 1C:0.6S, and 1C: 0.7S. Those variations were chosen in specifying the weight of cement and the ratio of the cement and sand mixing which produces light concrete, which has a density of less than one and indicated by the floating of the concrete sample in the water, with a compressive strength of at least 2,000 kPa.

3. Material and Methods

3.1. Material

Since its constituent materials are cement, sand, water, and foam, foam concrete is classified as mortar. Foam is created by combining foam and water in a 1:20 ratio. When there is no coarse aggregate in the foam mortar, it is classified as a mortar.

The air trapped in the foam mortar by a chemical reaction is expected to lower the mortar's specific gravity. The materials that are mixed are cement, sand, and water.

These materials were combined with the previously described compositions in the laboratory mixer. The materials must be lump-free, and the sand used has been granted by a No.10 sieve and retained by a No.200 sieve [3].

The material weight is to form one meter cubic mixed mortar, and the results are seen in table 1.

Table 1. Material weight to produce foam mortar 1 m^3

No	Materials (kg)	Cement : Sand Ratio			
		1C:0.4S	1C:0.5S	1C:0.6S	1C:0.7S
1 (C 400)	Cement		400.00	400.00	400.00
	Sand		200.07	239.85	279.96
	Water		200.00	200.00	200.00
	Foam		35.63	34.69	33.74
2 (C 425)	Cement		425.00	425.00	425.00
	Sand		212.69	254.89	297.60
	Water		212.50	212.50	212.50
	Foam		34.10	33.10	32.09
3 (C 450)	Cement		450.00	450.00	450.00
	Sand		225.09	270.14	314.87
	Water		225.00	225.00	225.00
	Foam		32.58	31.51	30.46
4 (C 500)	Cement	500.00	500.00	500.00	
	Sand	199.86	249.89	299.87	
	Water	250.00	225.00	250.00	
	Foam	30.72	31.04	28.36	
5 (C 525)	Cement	525.00	525.00	525.00	
	Sand	209.88	262.38	315.25	
	Water	262.50	236.25	262.50	
	Foam	29.26	29.59	26.77	
6 (C 550)	Cement	550.00	550.00	550.00	
	Sand	219.79	275.18	330.03	
	Water	275.00	247.50	275.00	
	Foam	27.79	28.13	25.19	

Table 2. Material weight for foam mortar composition material for 3 samples per mixture.

No	Materials (kg)	Cement : Sand Ratio			
		1C:0.4S	1C:0.5S	1C:0.6S	1C:0.7S
1 (C 400)	Cement		2.512	2.512	2.512
	Sand		1.256	1.506	1.758
	Water		1.256	1.256	1.256
	Foam		0.224	0.218	0.212
2 (C 425)	Cement		2.669	2.669	2.669
	Sand		1.336	1.601	1.869
	Water		1.335	1.335	1.335
	Foam		0.214	0.208	0.202
3 (C 450)	Cement		2.826	2.826	2.826
	Sand		1.414	1.696	1.977
	Water		1.413	1.413	1.413
	Foam		0.205	0.198	0.191
4 (C 500)	Cement	3.140	3.140	3.140	
	Sand	1.255	1.569	1.883	
	Water	1.570	1.413	1.570	
	Foam	0.193	0.195	0.178	
5 (C 525)	Cement	3.297	3.297	3.297	
	Sand	1.318	1.648	1.980	
	Water	1.649	1.484	1.649	
	Foam	0.184	0.186	0.168	
5 (C 550)	Cement	3.454	3.454	3.454	
	Sand	1.380	1.728	2.073	
	Water	1.727	1.554	1.727	
	Foam	0.175	0.177	0.158	

As seen in table 1 and table 2, water and cement weights are equal for each mixing. Further, the sand weight variations refer to the predetermined comparing the cement and sand ratio.

3.2. Methodology

The weight of each foam mortar is regarding the Circular of the Minister of Public Works and Public Housing No. 44 2015 on guidelines design for mixtures of lightweight materials with foam mortar for road construction [3].

3.3. VolumeWeight

Volume analysis was done by dividing the sample weight into tons. The density is to figure out which sample has a smaller density than the water density, so it is possible to use as lightweight material substitution dams according to The Research and Development Center for Roads and Bridges (PUSJATAN) [2]. In this research, the quested volume weight is the foam mortar volume as a substitution at the top layer foundation of road construction.

$$Volume\ Weight = \frac{Sample\ Weight\ (Ton)}{Sample\ Volume\ (m^3)} \quad (1)$$

The volume weight is the unit weight volume of oven-dried soil and is usually stated in grams/cm3 [12].

3.4. Foam Production

The mortar density is evaluated by SE No: 44/SE/M/2015 [3]. It needs liquid foam and water to produce the foam. Air cavities are created when they are mixed by using a generator and a compressor in the mortar. In the mixing process, pores appear and are lighter.

Figure 1. Foam making.

3.5. Foam Mortar Production and Flow Test

The mortar compiler materials are cement, sand, and water. These materials are stirred smoothly with a mixer and the foam is laid into the mortar. They are stirred in a mixer at low speed evenly, further, the flow test was carried out based on ASTM C230-08 (2008) [5] to examine the workability and consistency.

Figure 2 shows the measurement of mortar spread. The flow value is the diameter of the mortar after the ring flow has been removed. According to the provisions of the SE Minister of PUPR No. 44 of 2015, the discharge value must be 18 cm 2 cm. When it has done, the foam mortar is inserted into a cylindrical mold with a diameter of 10 cm and 20 cm in height. While Figure 3 is a cylindrical sample that has been removed from the mold, The sample was then wrapped for 12 days to cure. This is done to keep water from escaping during the hardening process. As a result of the cement-water reaction, there is no hydration.

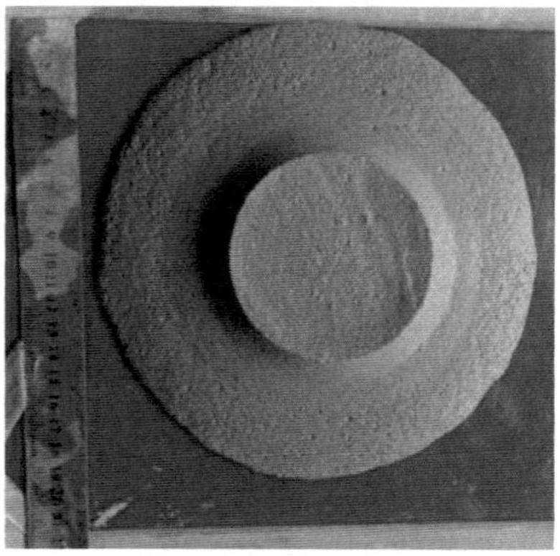

Figure 2. Flow Test

Figure 3. Curing / Sample Treatment.

Figure 4. Drying sample process

Figure 5. Weighing sample process

After that as you can see from figure 4 the cylindrical sample has been removed from a plastic wrap that has been aged for 12 days, then dried at room temperature for 2 days. Next, the samples were removed and weighed after 14 days of curing. A modified CBR tool was used to test samples at 14 days of age as can be seen in figure 5.

3.6. Compressive Strenght Test

The pressure strength test is carried out to see the strength of the made sample. It is to get information on the strength data and whether can reach 2,000 kPa referring to the specified conditions [14]. The test is conducted with a modified CBR tool following the standards recommended by PUSJATAN. the pressure strength test with the CBR tool is seen in figure 5.

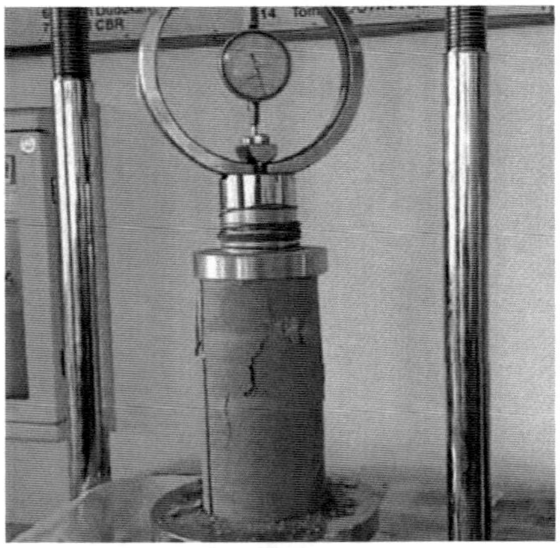

Figure 6. Compressive test of 525 kg cement sample, 1C:0.6S.

Figure 6 show a 525 kg cement sample, and the cement and sand ratio is 1C:0.6S. The pressure strength is defined by dividing the maximum loads by the sample base area of millimeter squares. Pressure Strength Analysis is conducted to determine foam mortar's ability against pressure loads. The pressure strength of this foam mortar must be equal to or bigger than the pressure strength of the road top layer (>2,000 kPa). The equation-two is based on SNI 1974:2011 and it is the determination of the pressure strength of concrete cylinders [14,15].

$$\textit{Compressive Strenght} = \frac{\textit{Max Load (Newton)}}{\textit{Area ((mm}^2)} \quad (2)$$

4. Result and Discussion

Table 3 presents the outcome of various cement weights and the ratio of cement and sand after 14-days of treatment. Two main parameters have been examined to figure out a suitable lightweight foam mortar that is going to be applied on peat soil with low bearing capacity. The parameters themselves are density and pressure strength

Table 3. volume weight and strength for all samples.

Cement	Mixture	Weight (gr)	volume weight (T/M3)	Strength (kPa)
400 Kg	1C:0.5S (1 Cement : 0.5 Sand)	965	0.615	485.68
		915	0.583	599.00
		945	0.602	518.05
		Average	**0.600**	**534.24**
	1C:0.6S (1 Cement : 0.6 Sand)	920	0.586	582.81
		960	0.611	518.05
		950	0.605	599.00
		Average	**0.601**	**566.62**
	1C:0.7S (1 Cement :	**950**	**0.605**	**599.00**
		910	0.580	161.89

	0.7 Sand)	1005	0.640	356.16
		Average	**0.608**	**372.35**
425 Kg	1C:0.5S (1 Cement : 0.5 Sand)	1050	0.669	688.04
		1025	0.653	882.31
		1045	0.666	858.03
		Average	**0.663**	**809.46**
	1C:0.6S (1 Cement : 0.6 Sand)	1170	0.745	1,133.24
		1210	0.771	955.16
		1195	0.761	890.40
		Average	**0.759**	**992.94**
	1C:0.7S (1 Cement : 0.7 Sand)	885	0.564	283.31
		915	0.583	275.22
		855	0.545	275.22
		Average	**0.564**	**277.91**
450 Kg	1C:0.5S (1 Cement : 0.5 Sand)	1200	0.764	1,311.32
		1180	0.752	1,327.51
		1160	0.739	1,554.16
		Average	**0.752**	**1,397.67**
	1C:0.6S (1 Cement : 0.6 Sand)	1244	0.792	1,078.20
		1229	0.783	1,100.86
		1312	0.836	1,214.19
		Average	**0.804**	**1,131.08**
	1C:0.7S (1 Cement : 0.7 Sand)	1075	0.685	615.19
		1000	0.637	518.05
		1110	0.707	696.13
		Average	**0.676**	**609.79**
500 Kg	1C:0.4S (1 Cement : 0.4 Sand)	1240	0.790	1,845.57
		1271	0.810	1,991.27
		1268	0.808	2,039.84
		Average	**0.803**	**1,958.89**
	1C:0.5S (1 Cement : 0.5 Sand)	1244	0.792	1,262.76
		1218	0.776	825.65
		1229	0.783	825.65
		Average	**0.784**	**971.35**
	1C:0.6S (1 Cement : 0.6 Sand)	1295	0.825	1,651.30
		1312	0.836	1,181.81
		1319	0.840	1,359.89
		Average	**0.834**	**1,397.67**
525 Kg	1C:0.4S (1 Cement : 0.4 Sand)	1264	0.805	2,136.97
		1273	0.811	2,104.59
		1393	0.887	2,072.22
		Average	**0.834**	**2,104.59**

	1C:0.5S (1 Cement : 0.5 Sand)	1220	0.777	874.22
		1217	0.775	890.40
		1180	0.752	906.59
		Average	**0.768**	**890.40**
	1C:0.6S (1 Cement : 0.6 Sand)	1295	0.825	2,638.84
		1312	0.836	2,217.92
		1319	0.840	2,347.43
		Average	**0.834**	**2,401.40**
550 Kg	1C:0.4S (1 Cement : 0.4 Sand)	1414	0.901	1,845.57
		1471	0.937	2,185.54
		1426	0.908	2,655.03
		Average	**0.915**	**2,228.71**
	1C:0.5S (1 Cement : 0.5 Sand)	1571	1.001	
		1555	0.990	
		1575	1.003	2,557.89
		Average	**0.998**	**2,557.89**
	1C:0.6S (1 Cement : 0.6 Sand)	1630	1.038	3,108.32
		1616	1.029	3,772.08
		1591	1.013	3,432.11
		Average	**1.027**	**3,437.50**

As seen in table 3, there are increases either in the cement amount or the bulk weight quantities. The cement amount in the mixing increased from 400 kg to 550 kg, while the bulk weight of the sample also increased from 0.6 T/m3 to 1,027 T/m3. However, only one mixing meets the criteria for light concrete following SE PUPR 2015. It is 525 kg of cement with a ratio of cement and sand 1C:0.6S. It is because the amount of cement and the comparing ratio results in a volume weight being more than one. It also means that the sample cannot float in the water. The compressive strength value obtained is 2,401.40 kPa with a volume weight of 0.834 T/m3. The average value of volume weight, as well as the compressive strength, is presented in Fig. 6 and 7.

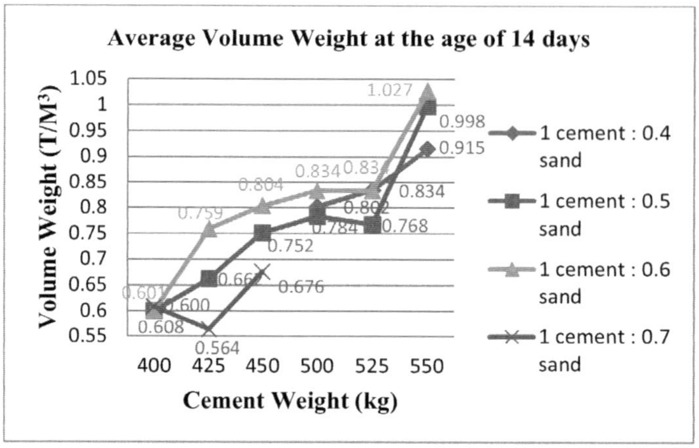

Figure 7. shows an increase in the density

The use of cement from 400 kg to 525 kg increased volume weight from 0.6 T/m3 to 0.834 T/m3. In the use of 550 kg of cement, the value of volume weight is > 1 T/m3 which is 1,027 T/m3. In a ratio of 1 cement:0.7 sand, the volume weight value is low namely 0.608 T/M3 to 0.676 T/m3, while in a ratio of 1 cement: 0.6 sand, the volume weight value does not decrease, starting from 0.601 T/m3 to 1.027 T/m3. The volume weight of the sample following the SE Minister of PUPR No. 44 of 2015 can be found in the use of cement 525 kg with a ratio of one cement: 0.6 sand and a volume weight value of 0.834 T/m3.

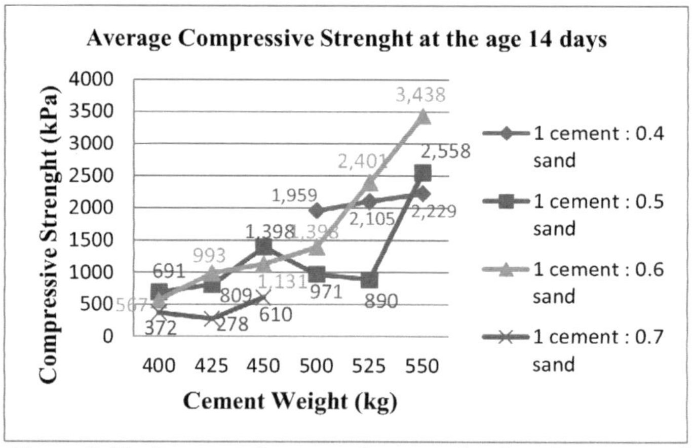

Figure 8. The increase of compressive strenght when the amount of cement used increases

In the use of cement from 400 kg to 500 kg the compressive strength value is less than 2000 kPa. The compressive strength value of more than 2000 kPa is found in the use of 525 kg and 550 kg of cement. The ratio of cement and sand with a compressive strength value of more than 2000 kPa is found in the ratio of 1-cement: 0.4 sand, 1-cement: 0.5 sand, and 1-cement: 0.6 sand.

The figures below show a sample with a cement amount of 525 kg with a ratio of cement and sand 1C: 0.6S that floats when drowned in the water. Therefore, this ratio is chosen as the foundation layer for road construction on peat soil, it is expected that this layer will not overlay the subgrade, and massive degradation due to the load it carries is avoided.

Figure 9. Sample floats in water with 225 Kg cement with cement and sand ratio of 1C:0.6S.

5. Conclusion

Based on the examination, it can be concluded that

- Lake Bingkuang sand in Kampar Riau as a material for creating foam mortar has a pressure strength value of more than 2000 kPA by using 525 kg and 550 kg of cement.
- An appropriate composition as a foundation layer for road construction on peat soil is the use of 525 kg of cement and a ratio of cement and sand of 1C:0.6S resulting in a volume weight of 0.843 T/m3 and a compressive strength value of 2.401 kPa.

Acknowledgments

The authors would like to thank the Faculty of Engineering, Andalas University for their support throughout this research.

References

[1] Panjaitan *A* S R N 2013 Study on the Value of Shear Strengthin Muara Batang Toru Peat, North Sumatra After Experiencing Initial Compression *J. Ranc. Sipil* **2**

[2] Wicaksono S and Iqbal M 2020 Effectiveness Of Foamed Mortar Light Weight Fil lAs An Approach Bridge Fill On Sheet-Pile Structures *J. Jalan-Jembatan* **37** 87–101

[3] Ministry of PUPR 2015 *Guidelines for Specifications of Lightweight Materials with Foam Mortar forRoad Construction (Pedoman Spesifikasi Material Ringan Dengan Mortar Busa Untuk Konstruksi Jalan)* Retrieved from: https://binamarga.pu.go.id/uploads/files/410/pedoman-spesifikasi material-ringan-dengan-mortar-busa- untuk-konstruksi-jalan.pdf

[4] Adi M A P, Lastiasih Y, and Mochtar I B 2019 Alternative Planning for Road Filling with Sirtu Material and Lightweight Mortar Foam on Batang– Semarang (Alternatif Perencanaan Timbunan Jalan dengan Material Sirtu dan Material Ringan Mortar Busa pada Jalan Tol Batang–Semarang) Seksi III Weleri–Kendal STA 414+ 525–STA 424+ 576 *J. Transp. Sist. Mater. dan Infrastruktur* **1** 72-77

[5] Simbolon E F 2015 Use of Foam Agent in Making Concrete Brick (Penggunaan Foam Agent Dalam Pembuatan Bata Beton Ringan) (Medan: Universitas Sumatera Utara)

[6] Sulaiman S 2011 *Water permeability and carbonation on foamed concrete* (Universiti Tun Hussein On Malaysia)

[7] Santoso H T 2020 Use of CSS-Mortar Foam as an Alternative for Selection of Flyover Construction Types on Construction Costs for Road and Bridge Construction (Penggunaan CSS-Mortar Busa Sebagai Alternatif Pemilihan Tipe Konstruksi Jalan Layang terhadap Biaya Konstruksi) *BENTANG J. Teor. Dan Terap. Bid. Rekayasa Sipil* **8** 87–98

[8] Al-Shwaiter A, Awang H and Khalaf M A 2021 The influence of superplasticiser on mechanical, transport and microstructure properties of foam concrete *J. King. Saud. Univ. – Eng. Sci.*

[9] Calis G, Yildizel S A, Erzin S and Tayeh B A 2021 Evaluation and optimisation of foam concretecontaining ground calcium carbonate and glass fibre (experimental and modelling study) *Case. Stud. Constr. Mater.* **15** e00625

[10] Kuzielová E, Pach L, and Palou M 2016 Effect of activated foaming agent on the foam concrete properties *Constr. Build. Mater.* **125** 998–1004

[11] Mestnikov A, Semenov S, Strokova V, and Nelubova V 2016 Autoclave foam concrete: Structure and properties in *AIP Conference Proceedings* **1698** 070010

[12] Foth H D 1978 Fundamentals of soil science *Soil Sci.* **125** 272

[13] ASTM 2014 *Standard specification for flow table for use in tests of hydraulic cement* (ASTM Int1–6) (Pennsylvania: ASTM International)

[14] Standar Nasional Indonesia 1974 *Concrete Compression Test procedure with Cylindrical Test (Cara Uji kuat tekan beton dengan benda uji silinder)* (Jakarta: BSN)

[15] Badan Standarisasi Nasional 2002 *Portland Cement Mortar Compressive Strength Test Method for Civil Works (Metode pengujian kekuatan tekan mortar semen Portland untuk pekerjaan sipil) (SNI 036825-2002 Indonesia)* (Jakarta: BSN)

IOP Conf. Series: Earth and Environmental Science 1173 (2023) 012011 doi:10.1088/1755-1315/1173/1/012011

Comparative Analysis on Lateral Strength of Multi-Span RC Frame with Different Types of Infill Walls

F Nugroho [1,2], Zaidir [1], J Tanjung [1] and Maidiawati [2]

[1] Civil Engineering Department, Andalas University, Padang 25163, Indonesia
[2] Civil Engineering Department, Padang Institute of Technology, Padang 25143, Indonesia

Corresponding e-mail: zaidir@eng.unand.ac.id

Abstract. The seismic behaviour studies of structures with RC frame show that the wall with the infill significantly improved the structures lateral strength. However, since the numerical analysis of the structures considers the infill walls as non-structural components, the analysis rarely includes the infills. It is strongly recommended in the further research to investigate the behaviour of RC structures under seismic load for having the infill as one of the important components in their design process. This study developed two numerical models of multi-span RC frames; 1) with walls made of brick masonry, and 2) with wall made of lightweight block masonry. Furthermore, the results were compared with the analysis of lateral strength of the models. In order to verify the results, a comparison with the prior analytical study and experimental results was conducted.

1. Introduction

The typical structure of Indonesia buildings usually uses filler of the mansonry walls for its reinforced concrete portal structures due to its location in the earthquake-prone zones. The masonry infill walls are treated only as the dead load instead of as the earthquake load-bearing for the building structure. In other words, the mansory infill walls can also be functioned as the earthquake load-bearing for the building structure but many building planners fail to take that as one of the main considerations in building planning. Analytical studies, experimental studies and post-earthquake field observations have been carried out. In analytical studies, the stiffness of the structure can be improved by the presence of infill walls [1]. The mansory infill walls in reinforced concrete portals also can strengthen the portal more than the lateral loads do as observed in the experimental studies. Another experimental study in the four-floor buildings with three spans, with masonry as a filler and with several wall openings show that a building with a full infill wall has greater lateral strength than a building with wall openings when subjected to lateral loads. Testing the performance of the infill walls in frame structures with reinforced concrete toward the seismic simulation indicates that the walls contributed significantly to both the lateral and the structural strength than those which place the openings of the wall in the center of the portal [2].

The capability of portal structures of reinforced concrete with brick walls have been tested by performing the alternating static load either experimentally or analytically. They found that walls of masonry brick significantly increase portal structure lateral strength. Furthermore, the wall lateral strength can be evaluated by observing from the height of the contact among the wall and the surrounding edging structure by determining the length of the contact among the infill wall and the edge which makes it possible to the test the load of the lateral in the infill wall and the performance of buildings with the frame in the seismic simulation as reported in the analytical study. The infill walls also contribute significantly to the structure throughout the earthquake in the analytical simulations and experimental verification. A new analytical modeling in examining buildings with reinforced concrete portal strengthened with masonry walls under the concept of diagonal strut concept has successfully been developed. This modeling showed more satisfying results than that in the experiment one, particularly in determining the lateral strength. Testing the strength of lateral wall with infill in buildings that are damaged due to the earthquake in Indonesia also applied the modeling [3].

The modelling was yet in building with portal of only one span. The portal structures in buildings could be either in single span or in multi spans. Buildings which had two spans in their masonry walls showed more endurance to the seismic occurrence than buildings with only one span. This result was concluded based on the experimental research on examining how the infill wall of multi-span structures reacts to the load from the lateral which applied the model of the diagonal strut in order to able to examine the strength of masonry wall strength analytically [4].

This study focuses on analyzing a numerical design for evaluating the performance of a portal structure lateral in multi spans by using wall filler of masonry in a portal with a multi-span reinforced concrete. The numerical model was analyzed using Seismostruct software which applied the concept of diagonal struts to replace brick walls in portals with single span and it was also able to be used in portals with multi-span reinforced concrete. Verifying the numerical results was confirmed from the results of the experiment on a portal with two-span reinforced concrete with uniform span spacing of masonry brick walls as infill that was equipped with alternated cyclic lateral loading.

2. Analytical study of RC frame with multi-span

The developed model of the portal structure with multi-span reinforced concrete was designed in accordance with the design principle of portal structure with single-span. The portal structure model with multi-span reinforced concrete was installed with two columns in the outer part and one column in the inner part as shown in Figure 1.

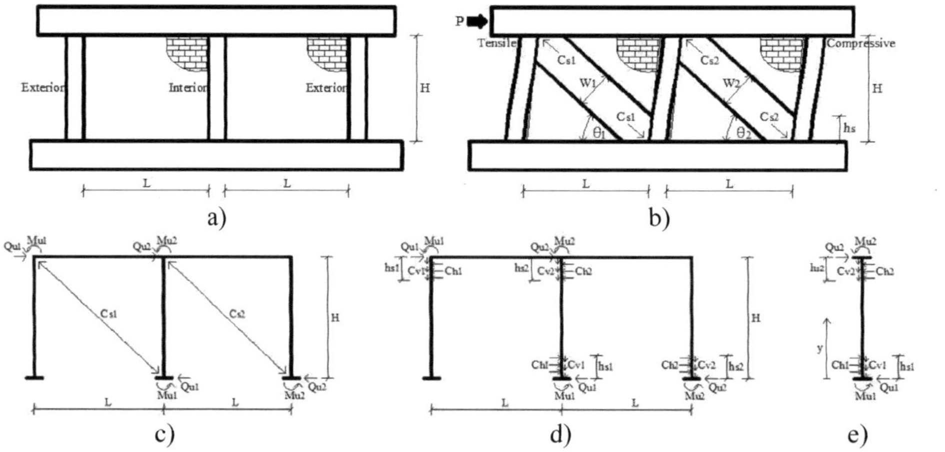

Figure 1. The portal structure model with multi-span reinforced concrete is presented as follows: a) a multi-span portal with column in the outer and inner part; b) load *P* causes the lateral deformation; c) the wall with the diagonal compression strength; d) the column with diagonal compressive forces distributed; e) interior columns with distributed forces.

The developed design can determine the shear strength and optimum moments on the outer part along with the force coming laterally from the wall [3]. Similarly, the same strength but in the inner part of column with the lateral force on the wall with uniform span are also able to be determined with the same concept as shown in Figure 1e. After the moment distribution, angular rotation and displacement were calculated and the sum is derived in order to then identify the column shear strength Q_u on the inner- part column as in the following equation (1).

$$Q_u = \frac{2M_{u2}}{H} + C_{h1} \cdot h_{s1} - \frac{C_{h1} \cdot h_{s1}^2}{H} + \frac{C_{h1} \cdot h_{s1}^3}{3H^2} \qquad (1)$$

In which h_s is calculated [3], and the optimum moment of the column M_u can be derived with equation (2) [5].

$$M_u = 0.8a_t \cdot \sigma_y \cdot D + 0.5N \cdot D \cdot (1 - \frac{N}{b \cdot D \cdot F_c}) \qquad (2)$$

3. Experimental study of multi-span RC frame

This study was carried out a series of experiments of portal structure with a reinforced concrete with two spans and with infill of masonry walls. Two testing prototypes were built; a building model with two spans using masonry wall of red brick (IF-2B-BM) and a building with two spans using lightweight masonry walls (IF-2B-LM) [6]. Both of them showed a similar distance of span. Both models were designed with a scale of 1:4 which was considered small as it was commonly found in Indonesia. The data of the reinforcement and dimensions of the building model can be found in Figure 2.

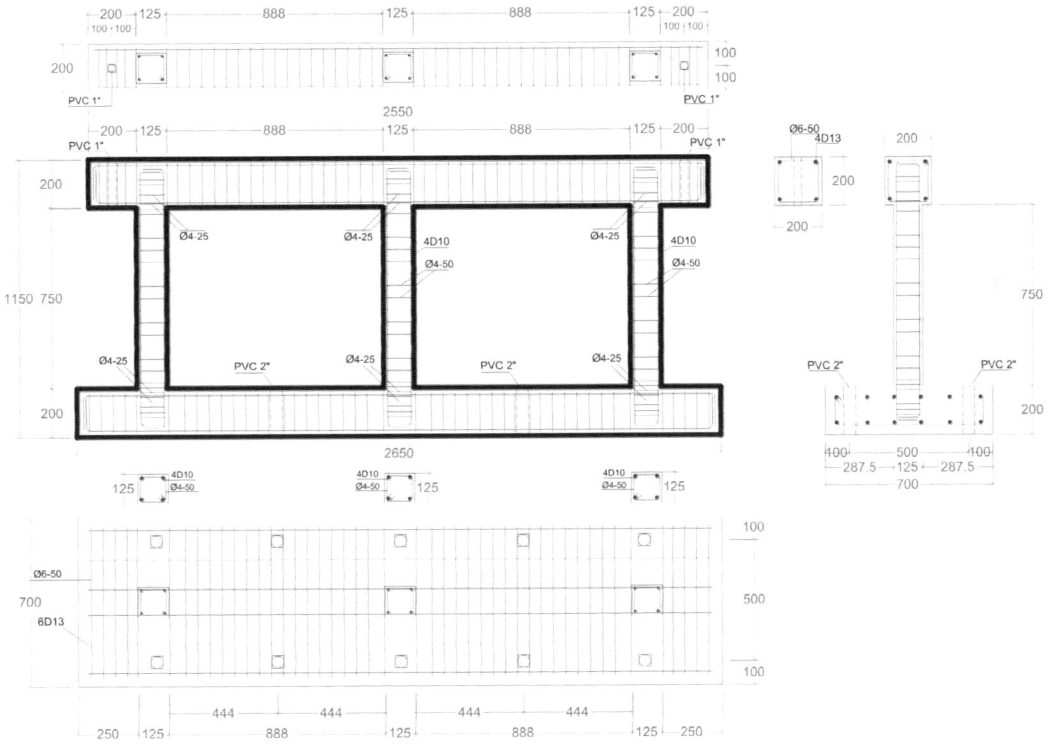

a)

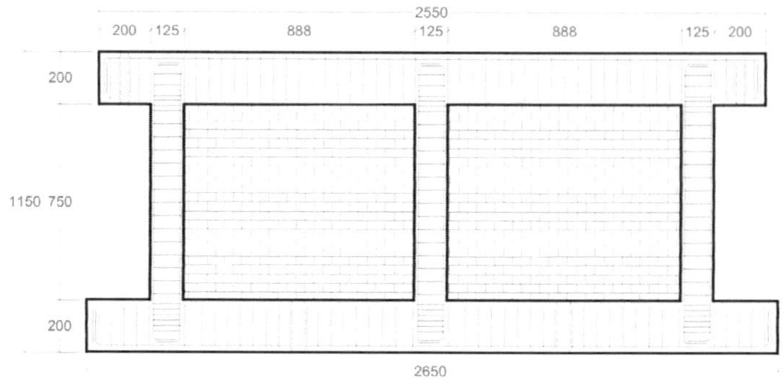

b)

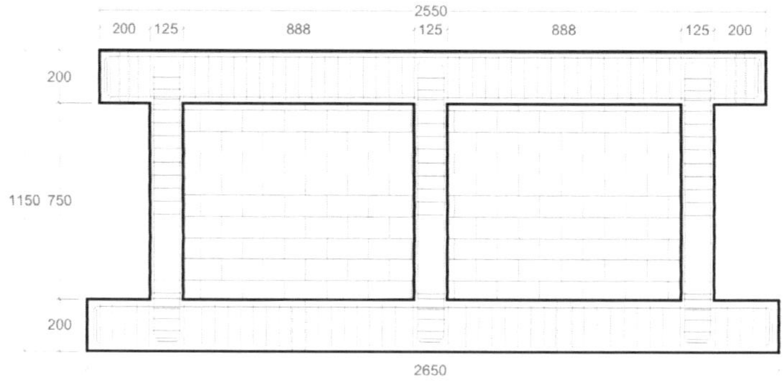

c)

Figure 2. This figure presents the dimensions of the model of the portal structures with multi-span in the experiment consisting of: a) the building model data of its reinforcement and the dimensions; b) The building with wall of brick (IF-2B-BM); c) The building with wall of lightweight block (IF-2B-LM).

Table 1 shows the compressive strength of the wall with concrete, with brick and with the lightweight block. The material data of the reinforcement are presented in table 2.

Table 1. The strength properties of concrete and masonry

The building model	F_c (N/mm^2)	E_c (N/mm^2)	$f_{m'}$ (N/mm^2)
IF-2B-BM	28.4	25045.0	7.6
IF-2B-LM	28.4	25045.0	1.7

Table 2. The properties of reinforcement strength

The diameter of reinforcement (mm)	The strength of yield (N/mm^2)	The strength of tensile (N/mm^2)	The Young modulus (N/mm^2)
Ø4	390.2	574.9	1.7x10^5
Ø6	346.8	446.3	1.9x10^5
D10	449.5	612.2	2.6x10^5
D13	443.3	603.9	2.1x10^5

The test apparatus schematic in Figure 3a indicates that the cyclic static load is used to control the lateral displacement in the column on the top part. The structure models are set on the rigid floor. There is lateral force which is considered as a load on the surface of the upper beam on the left. The upper beam only can be moved in lateral directed to alternating static loading. Figure 3b shows that the device of LVDT device is arranged to be placed on a frame mount with the basic steel in few areas in order to observe the occurrence of the displacement.

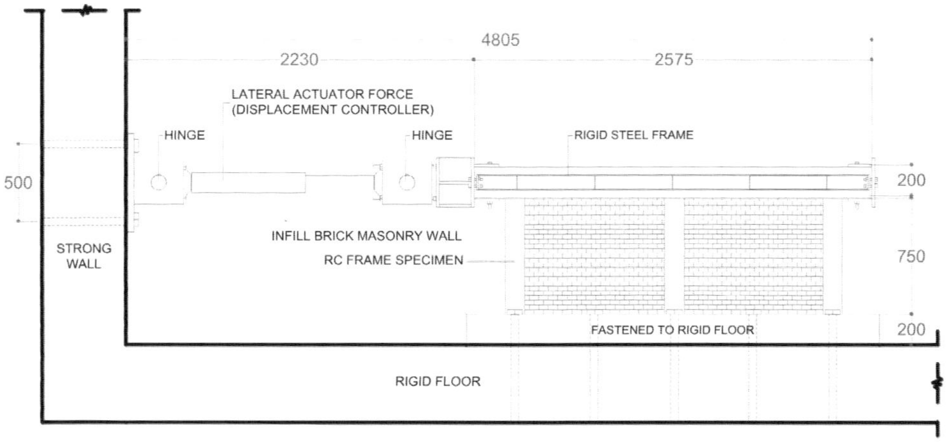

a)

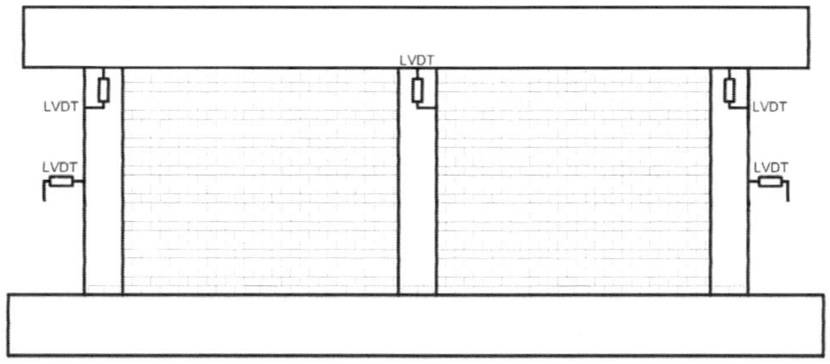

b)

Figure 3. The testing Schematic of the structure model consists of : a) The testing equipment schematic; b) the LVDT placement schematic.

The angel of the drift at R = 1/800, 1/400, 1/200, 1/100, 1/50, 1/25, and 1/12.5 controlled the column lateral displacement. Its loading data can be seen in Figure 4. If the structure had been collapsed before the last cycle, the test would have been stopped. The cracks appearing during the testing were identified in order to find out its structural failure mechanism.

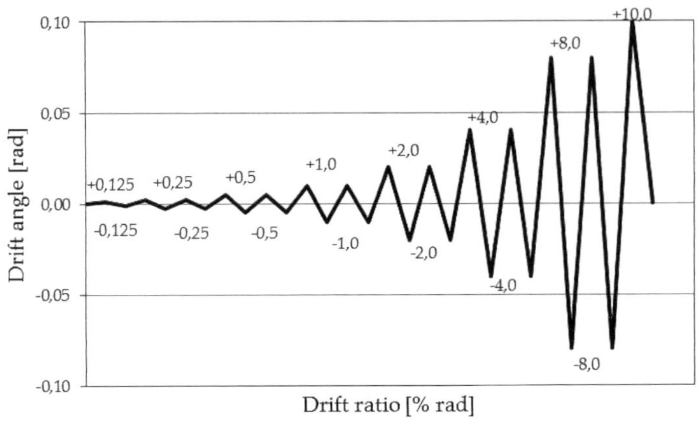

Figure 4. The lateral loads of the drift.

The seismic capacity of structures is presented in the correlation between lateral strength and drift ratio [7]. Figure 5 displays lateral strength and drift ratio correlation in both testing buildings. The lateral load in building 1 (IF-2B-BM) gets its ultimate number is at 159.7 kN with the drift ratio of 0.9 % as seen in Figure 5a, and the second one (IF-2B-LM) is at 72.4 kN with the drift ratio of 2.0 % as it is presented in Figure 5b.

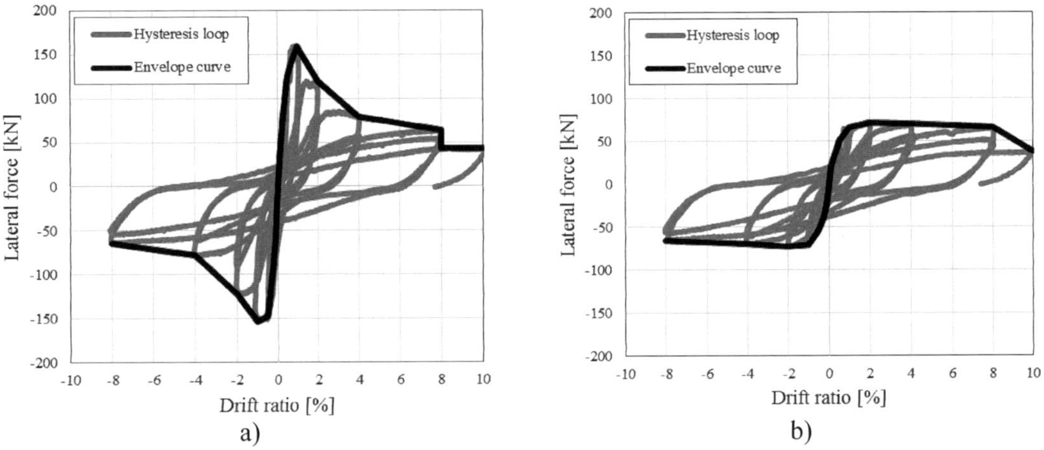

Figure 5. The lateral load and drift ratio correlation in both buildings consists of: a) IF-2B-BM test building; b) IF-2B-LM test building.

4. Numerical model for RC frame with multi spans

Anticipating the behavior of 3D frames on the large displacements in static or dynamic loading condition along with geometric non-linearities and non-linear material property consideration can be done by applying the framework of the SeismoStruct [8]. Therefore, this framework is ideal to be applied in developing the numerical model. In this case, providing the specimens detailing, material properties and loading conditions is a must. Numerical modeling with SeismoStruct is arranged based on Figure 2. The dimension and the compressive force of wall with concrete, with brick and with lightweight block were derived based on the properties in Table 1. The data of the reinforcement material are based on Table 2.

Since the model con_ma is based on the constitutive relationship principle proposed by Mander, the numerical models are applied it along with the uniaxial non-linear constant confinement model.

The infill model inf_strut is also used in the numerical models. Moreover, the steel model stl_mp is used in the numerical model where this steel model proposed by Menegotto and Pinto, which works under the principle of the stress-strain relationship and it suits with the aim of this study. Regarding other model parameters, this study applies the default principle of SeismoStruct. The numerical modelling with SeismoStruct is shown in Figure 6.

Figure 6. The numerical modeling with SeismoStruct

5. Results and discussion

Obtaining the lateral force of the entire buildings was analysed by increasing the lateral force in both parts of the columns and the wall. The comparison of the analysis, experiment and numerical test results of the buildings with wall made of brick are displayed in figure 7. The greatest lateral strength in of model 1 (IF-2B-BM) reached 159.7 kN in the experimental study, 129.9 kN in the analytical study and 109.5 kN in the numerical study. Meanwhile in model 2 (IF-2B-LM) the greatest lateral strength was at 95.2 kN in the analytical study, at 76.4 kN in the numerical study and 72.4 kN in the experimental study.

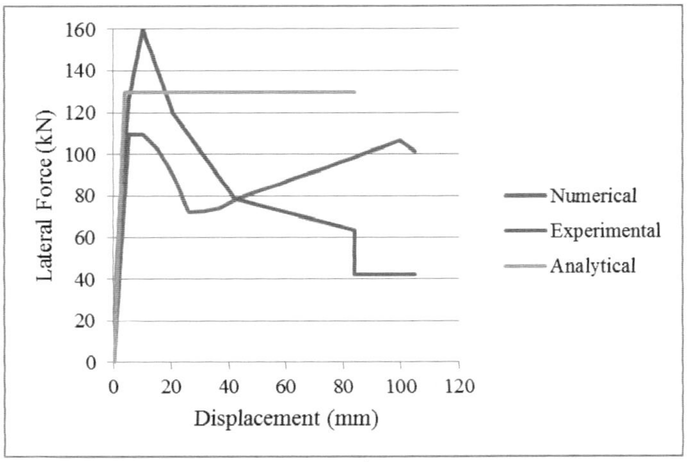

a)

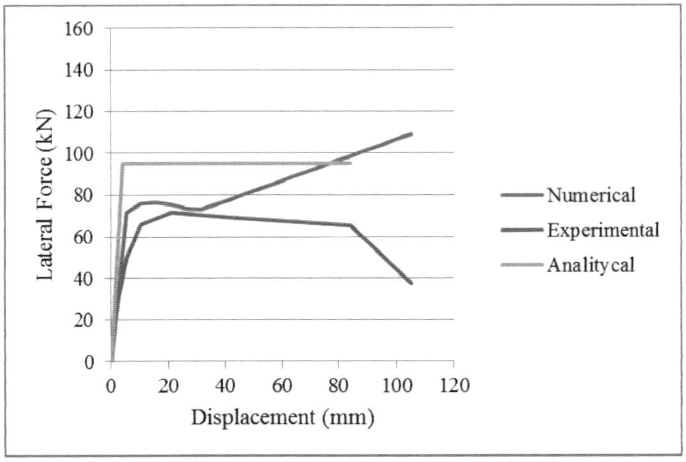

b)

Figure 7. The lateral force and displacement correlation of both models consists of: a) building 1 (IF-2B-BM); b) building 2 (IF-2B-LM).

6. Conclusions

After the developed design is tested to evaluate the seismic performance of buildings with reinforced concrete in multi spans and with filler of masonry walls, then it is verified in the experimental test with the test objects, and finally the numerical models with SeismoStruct are developed. Therefore, it comes to the conclusion that:

a. The buildings with reinforced concrete in multi spans using column in the outer and inner part can be developed based on the buildings with reinforced concrete in one span under the principle of the diagonal strut;
b. The analysis, experiment and numerical test result of the lateral strength show a slight difference. That is to say that this analytical model can be used to examine the performance of buildings with multi spans and with filler of the masonry wall under the earthquake simulation.

Acknowledgements

The authors would express his greatest and sincere gratitude toward the Ministry of Education, Culture, Research and Technology, Indonesia for their contribution in funding this research.

References

[1] Agrawal N, Kulkarni PB, Raut P. 2013. Analysis of Masonry Infilled R.C. Frame with & without Opening Including Soft Storey by using Equivalent Diagonal Strut Method. *International Journal of Scientific and Research Publications*, Vol. 3, Issue 9.

[2] Zhai C, Kong J, Wang X, Chen Z. 2016. Experimental and Finite Element Analytical Investigation of Seismic Behavior of Full-Scale Masonry Infilled RC Frames. *Journal of Earthquake Engineering*, 20(7), p.1171-1198.

[3] Maidiawati, Sanada Y. 2017. R/C Frame Infill Interaction Model and its Application to Indonesian Buildings. *Earthquake Engineering and Structural Dynamics*, 46 (2), pp. 221–241.

[4] Suzuki T, Choi H, Sanada Y, Nakano Y, Matsukawa K, Paul D, Gulkan P, Binici B. 2017. Experimental Evaluation of the in-Plane Behaviour of Masonry Wall Infilled RC Frames. *Bulletin of Earthquake Engineering*.

[5] The Japan Building Disaster Prevention Association (JBDPA), English Version, 1st, Standard for Seismic Evaluation of Existing Reinforced Concrete Buildings. JBDPA, Japan, 2005,

http://128.46.154.21/ayhan/Rabab/BRI_Japan_SeismicEvaluationofExistingRCBuildings_2001.pdf

[6] Maidiawati, Tanjung J, Nugroho F, Hayati Y. 2020. Seismic Performance Analysis of Multi-Span Masonry Infilled RC Frame Based on Analytical Strut Model. *The 2nd Aceh International Symposium on Civil Engineering (AISCE) IOP Conf. Series: Materials Science and Engineering* 933, 012042.

[7] Nugroho F, Zaidir, Tanjung J, Maidiawati. 2022. Analytical Study on Lateral Strength of Multi-Span RC Frame with Masonry Infill. *Civil and Environmental Engineering*. https://doi.org/10.2478/cee-2022-0042.

[8] Seismosoft (2022). Seismostruct. Available online at : https://seismosoft.com (accessed September, 2022).

Numerical Study of Castellated Beam with Stiffener – Case on Cantilever Structure

S Haris[1,3], P K Sari[2,4] and Masrilayanti[1,5]
[1]Civil Engineering Department, Andalas University, Indonesia
[2]Postgraduate Student, Civil Engineering Department, Andalas University, Indonesia

Email: [3]sabril.haris@eng.unand.ac.id
[4]putrikumalasari1998@gmail.com
[5]masrilayanti@eng.unand.ac.id

Abstract. A castellated beam is designed to increase the depth of the beam by slicing the beam web longitudinally in a specified form and re-unite both pieces using welding to obtain a higher flexural capacity of the beam. Due to the increase in cross-sectional height and the presence of openingsin the web plate, some potential failures may occur, for example buckling of the web and stress concentration on the opening. Addition of stiffener on the web is useful to prevent those potential failures. This paper presented a numerical study of castellated beam with additional stiffeners which is applied on a cantilever structure using MSC Nastran software. Thecastellated beam was added by stiffeners placed on three different zones. There were stiffeners on the perimeter of opening, on the web between openings, and on the web of opening. The cantilever beam was subjected by a point load increased monotonically until ultimate conditionreached. It was found that applying of the stiffeners on perimeter of opening will increase ultimate load up to 7.48 % comparing to the original castellated beam. On contrary, stiffeners onthe web between openings and the web of opening will decrease the ultimate load within the range of 0.40% - 4.86 %. The beam stiffness will slightly increase because of these three types of additional stiffeners. Regarding ductility of the beams, adding of stiffeners on the castellatedbeam showed vary numbers. Overall, the stiffeners on perimeter of opening become an alternative to increase the performance of the castellated beam in term of the ultimate load, beamstiffness and ductility.

1. Introduction

Various innovations are being developed in the field of construction to obtain an efficient and effective structure. One of them is the WF steel beam material which is modified into castellated beams which aims to increase the stiffness and load carrying capacity of steel structural elements without increasing the weight of the steel. Castellated beams are becoming increasingly popular and are widely used for long-span structures and in high-rise buildings [1].

Castellated beams are made from I-section or H steel profile which is manufactured by cutting the web in a zigzag manner to a certain size and reconnected by welding. This causes the cross-sectional height of the steel profile increased as well as the moment of inertia and the cross-sectional modulus. Several types of openings in castellated beams, including circle, square, rectangle and hexagonal [2].

In some cases of castellated beam structures with short spans, the dominant failure occurs is the shear mechanism which may reduce the load capacity. The openings in the web cause the castellated beam to be weak against shear [3]. In addition, the failure of the castellated beam can be caused by several things, including the special geometric shape, slenderness of the web plate, geometry and size of the opening, the type of loading given and the provision of lateral support [4].

The increase in cross-sectional height of the castellated can cause several failure modes due to loading, such as the formation of flexural mechanisms, shear failure, lateral torsional buckling, Vierendeel mechanism, buckling or yielding of the beams, local buckling and broken welded joints [5]. Therefore, stiffeners is given to increase the buckling strength of the web, stiffness and strengthen the moment resistance of the steel plate along the transverse and longitudinal directions as well as along the edges of the opening [6].

In this study, a numerical study was conducted using the MSC Nastran & Patran software to identify the effect of adding stiffeners to the body plate which was varied based on the type of stiffener used, the number of stiffeners, and the thickness of the stiffeners. As well as a study of the comparison of structural capacities such as load-displacement graphs, stiffness, ductility and stress distribution.

2. Research Metodology

2.1. Work Procedures

This research uses MSC Nastran & Patran Software to analysis the castellated beam with stiffener. The process carried out using this program is divided into three parts.

a. The pre-processing: modelling using MSC Patran software by inputting the data of castellated beam model.
b. Running the model using MSC Nastran.
c. The post-processing: present the results using MSC Patran to see outputs such as deformations and contour stress.

2.2. Description of Specimens

The specimens analyzed in this study were original castellated beam (without stiffener) and with additional stiffeners (stiffened castellated beam). The original castellated beam was varied based on the opening angles of 55° and 65°. The original castellated beam was added by stiffeners placed on the three different zones. There were stiffeners on the perimeter of opening, on the web between openings, and on the web of opening. Then, each of stiffeners is varied based on the number of stiffeners and the thickness. In Figure 1, the parameter of the original castellated beam structure geometry is illustrated.

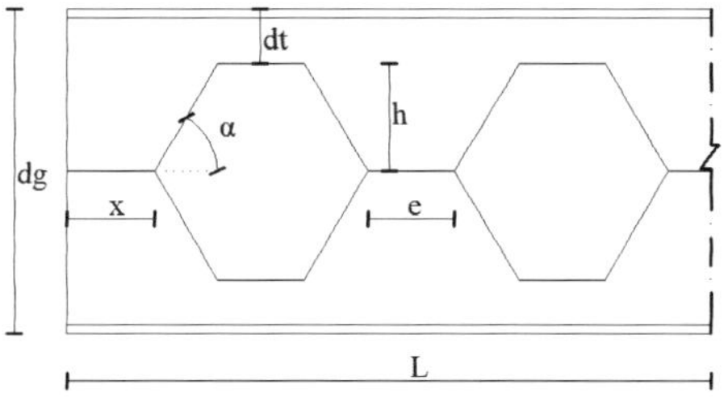

Figure 1. Geometry of Castellated Beam.

Castellated Structure Data:

a. Web Opening Type	: Honeycomb
b. Structure Type	: Cantilever
c. Angles (α)	: 55° (Profile A) and 65° (Profile B)
d. Span Length (L)	: 2000 mm
e. Castellated Height (dg)	: 280 mm
f. Opening Spacing	: 120 mm
g. Steel material	: BJ-37
h. Load	: Static Monotonic with Displacement Control

Figure 2 shows the detail of castellated beams with three types of stiffeners for profile A and B, respectively.

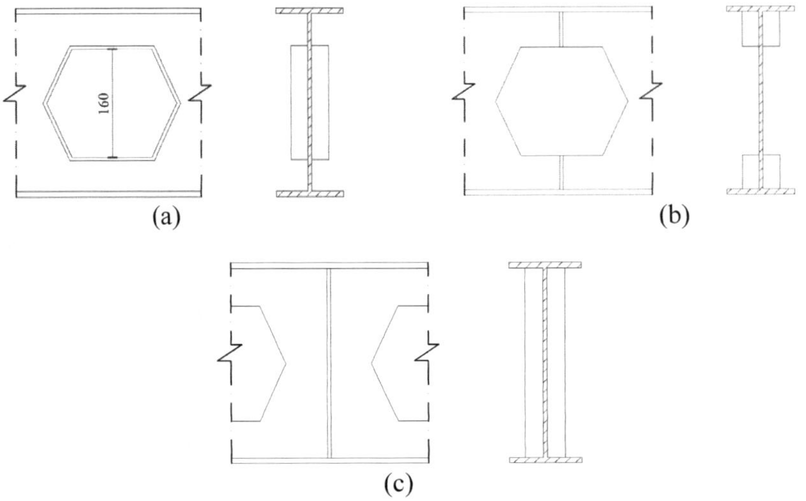

Figure 2. Types of Castellated Beam Stiffeners, (a) Type-1: Stiffeners on the Perimeter of Opening, (b) Type-2: Stiffeners on the Web Between Openings and (c) Type-3: Stiffeners On the Web of Opening.

Table 1. Castellated Model.

A Profile (55°)				B Profile (65°)			
Type	ID Model	n	t_s (mm)	**Type**	ID Model	n	t_s (mm)
1	A1-1-4	1	4	**1**	B1-1-4	1	4
	A1-1-5	1	5		B1-1-5	1	5
	A1-1-6	1	6		B1-1-6	1	6
	A1-2-5	2	5		B1-2-5	2	5
	A1-3-5	3	5		B1-3-5	3	5
	A1-4-5	4	5		B1-4-5	4	5
	A1-5-5	5	5		B1-5-5	5	5
					B1-6-5	6	5
2	A2-1-4	1	4	**2**	B2-1-4	1	4
	A2-1-5	1	5		B2-1-5	1	5
	A2-1-6	1	6		B2-1-6	1	6
	A2-2-5	2	5		B2-2-5	2	5
	A2-3-5	3	5		B2-3-5	3	5
	A2-4-5	4	5		B2-4-5	4	5
					B2-5-5	5	5
3	A3-1-4	1	4	**3**	B3-1-4	1	4
	A3-1-5	1	5		B3-1-5	1	5
	A3-1-6	1	6		B3-1-6	1	6
	A3-2-5	2	5		B3-2-5	2	5
	A3-3-5	3	5		B3-3-5	3	5
	A3-4-5	4	5		B3-4-5	4	5
	A3-5-5	5	5		B3-5-5	5	5
					B3-6-5	6	5

n = number of stiffeners
t_s = thickness of stiffener

3. Result and Discussion

3.1. Original Castellated Beam (Without Stiffener)

3.1.1. Stress Contour. The result obtained in this study are presented in the form of load-displacement curves. After observing, from both profiles A and B that have been tested, it is found that the load-displacement graph pattern is almost the same. So, the stress contours shown in Figure 3 may represent the condition of the two profiles.

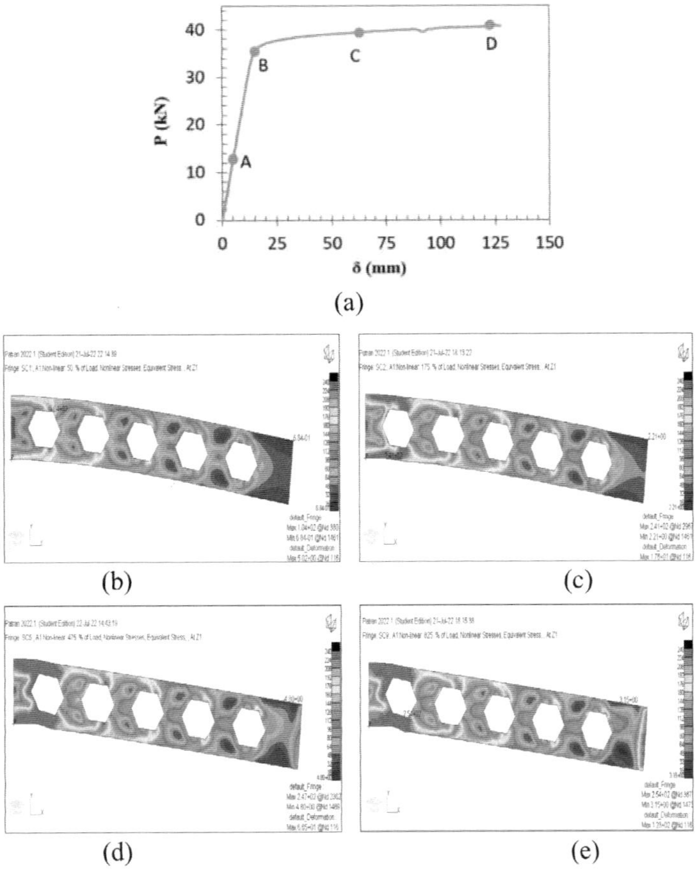

Figure 3. (a) Stress Contour Point to be Noticed, (b) at Load Step 50%, δ=5 mm, (c) Load Step 175%, δ=17.5 mm, (d) Load Step 475%, δ=66.25 mm, (e) at Load Step 825%, δ=125.274 mm.

The color on the contour describes the amount of stress that occurs in the specimen under subjected load, i.e.: the red color indicates that the structure has reached a yield stress of 240 MPa. It can be observed that in the first load step the test object has not reached the yield stress condition, the maximum stress that occurs is 104 MPa. In the second load step, the specimen hasexperienced yield stress which is marked by a red color in the area around the first opening, the maximum stress that occurs is 241 MPa. Then in the next load step, the red color on the contour of the specimen has spread around the first opening towards the support and the maximum stress value is 247MPa. in the fourth step of the load, in addition to the widening yield stress condition, it can also be observed that the specimen bends at the flange near the support, the maximum stress value is 254 MPa. From the overall contour above, it can be concluded that the yield stress is most dominant around the first opening area and the web plate near the support.

IOP Conf. Series: Earth and Environmental Science 1173 (2023) 012012 doi:10.1088/1755-1315/1173/1/012012

3.1.2. Ultimate Load, Stiffness and Ductility. The three capacity values of the original castellated beam profile are shown in the table below.

Table 2. The Value of the Structural Capacity of the Original Castellated Beam.

Profile	P-ult (kN)	k (kN/mm)	Ductility
A (55°)	40.625	2.552	7.865
B (65°)	40.730	2.519	7.900

From the results above, it can be concluded that the value of ultimate load and ductility is obtained in a larger value in profile B, while the larger stiffness value is in profile A. The stiffness value is obtained from the load-displacement trendline graph. While, the ductility value is obtained from the ratio of the displacement value of ultimate condition with the displacement value in the first yield condition.

3.2. Castellated Beam with Stiffener

3.2.1. Ultimate Load. Three types of stiffeners used in castellated beams with an opening angle of 55° and 65° are shown in Figure 4 and 5.

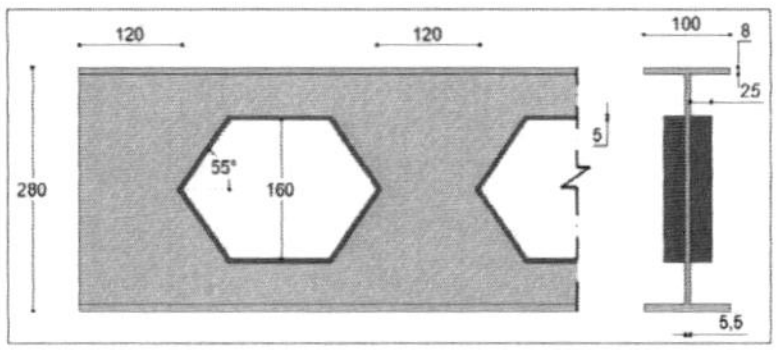

(a)

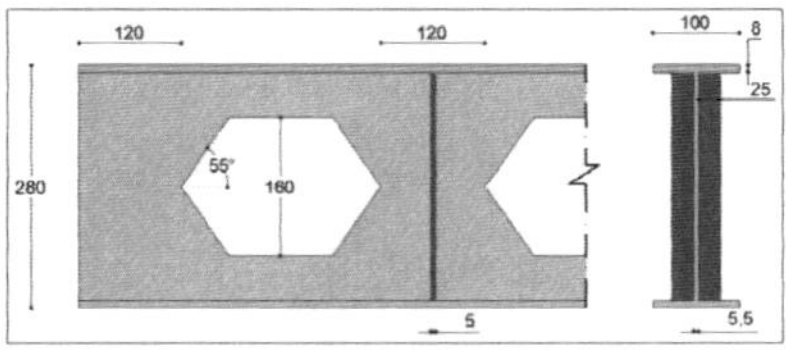

(b)

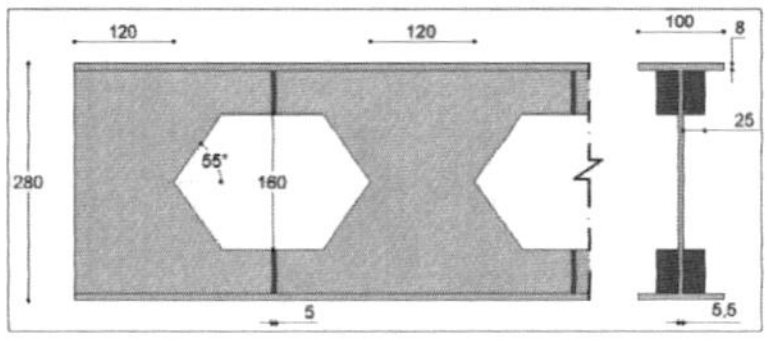

(c)

Figure 4. Castellated Beam with Opening Angle of 55°, (a) Stiffeners on the Perimeter of Opening, (b) on the Web Between Openings, and (c) On the Web of Opening.

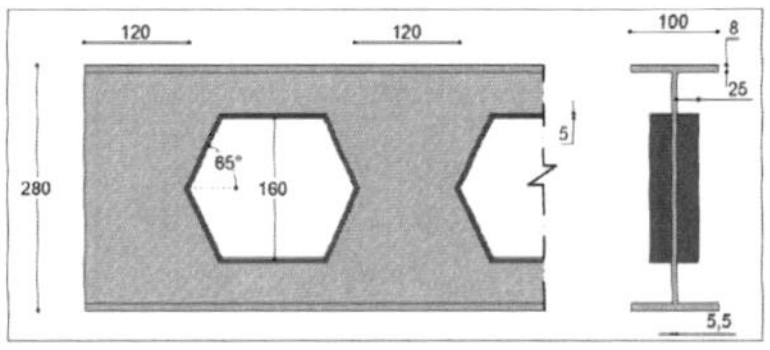

(a)

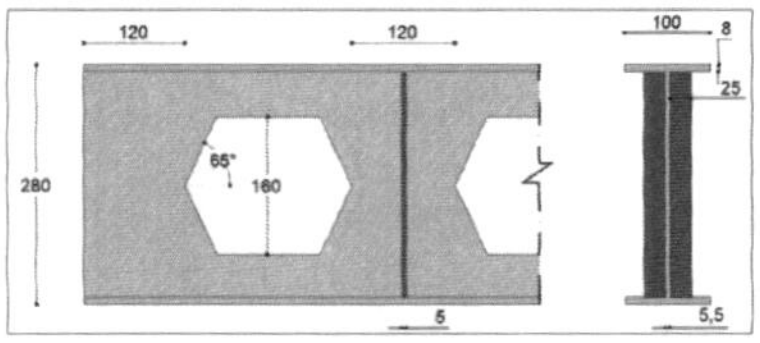

(b)

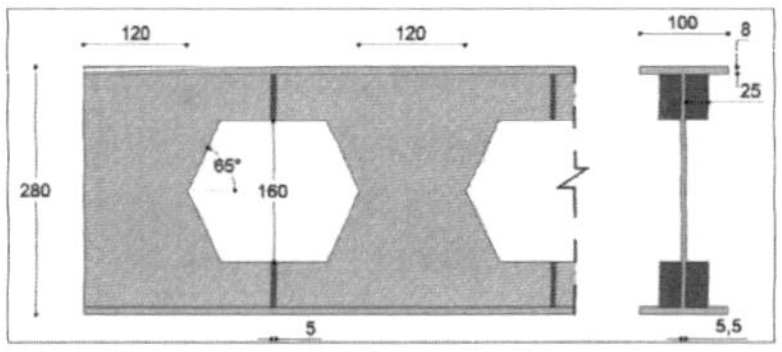

(c)

Figure 5. Castellated Beam with Opening Angle of 65°, (a) Stiffeners on the Perimeter of Opening, (b) on the Web Between Openings and (c) On the Web of Opening.

Figure 6 and 7 show the results for each type of stiffeners in Profile A and B.

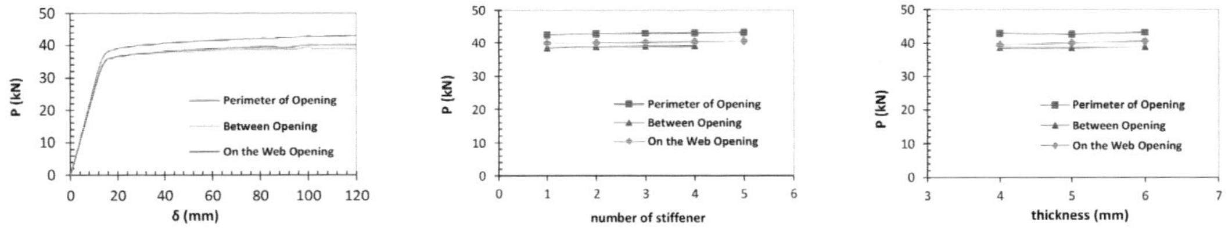

(a) (b) (c)

Figure 6. P-ult in Profile A based on (a) for each type of stiffener, (b) variations in the number of stiffeners, (c) variations in thickness.

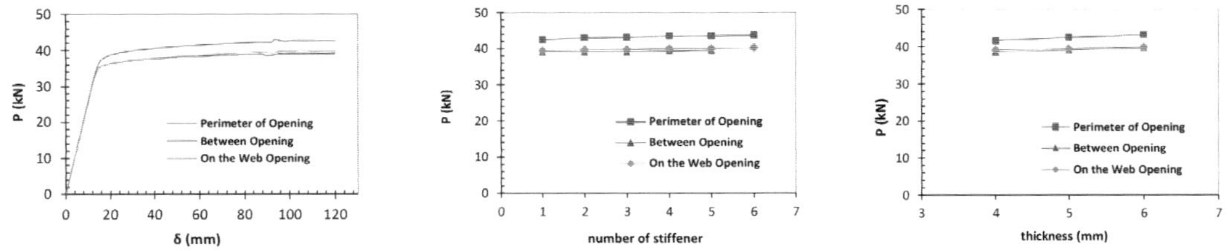

(a) (b) (c)

Figure 7. P-ult in profile B based on (a) for each type of stiffener, (b) variations in the number of stiffeners, (c) variations in thickness.

From Figure 6 and 7, it can be concluded that in the castellated profile with an angle of 55° and 65° after being given stiffeners, the largest load value is on the perimeter of opening and the smallest value is on the between opening stiffener type. In all types of stiffeners, the highest ultimate load value is obtained from variations in the maximum number of stiffeners. At the profile A, for perimeter of opening type stiffener, the P-ult value obtained is 43.269 kN with a displacement of 113.52 mm. While the between opening is 39.971 kN at a displacement of 122.50 mm. And then, the ultimate load value on the web opening is 40.634 kN with a displacement of 101.581 mm. Meanwhile, at the profile B of castellated, for perimeter of opening type stiffener, the P-ult value obtained is 43.694 kN with a displacement of 112.687 mm. While the between opening is 39.677 kN at a displacement of 122.538 mm. And then, the ultimate load value on the web opening is 40.227 kN with a displacement of 103.594 mm. So, it can be concluded that perimeter of opening type stiffeners are more effectively used to increase the value of the load capacity on castellated beam. Comparing to the original castellated beam, the stiffeners on perimeter of opening will increase ultimate load up to 6.51% and 7.28% for Profile-A and Profile-B respectively. On contrary, stiffeners on the web between openings and the web of opening will decrease the ultimate load within the range of 0.02% - 5.19%.

3.2.2. Stiffness. The stiffness of the structure can be seen from the load displacement graph for each given variation by determining the slope of the curve which is carried out by linear regression to the elastic region limit to obtain the approximate value of the stiffness of each castellated profile.

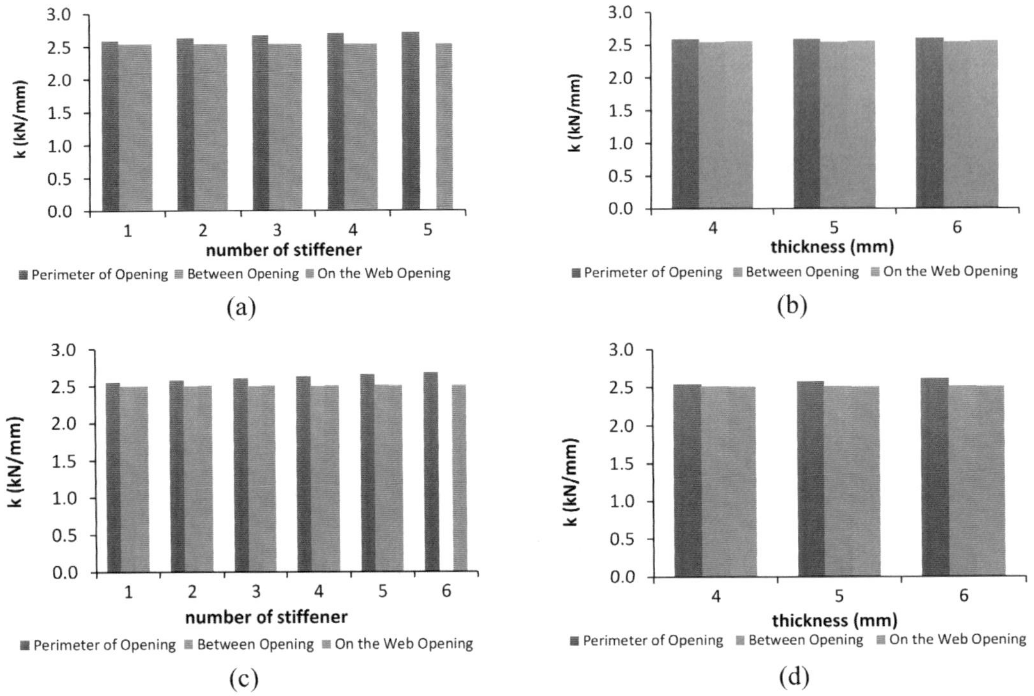

Figure 8. The stiffness value based on (a) the number of stiffeners in the profile A, (b) the thickness of stiffeners in the profile A, (c) the number of stiffeners in the profile B, (d) the thickness of stiffeners in the profile B.

From Figure 8, it can be observed that the variation does not have a significant effect on the stiffness value. However, from the variation in each types of stiffener, it is found that if the number of stiffeners increases, the stiffness also increases where the maximum stiffness value obtained is 2.716 kN/mm (Profile A) and 2.671 kN/mm (Profile B). As well as variations in thickness, with the largest stiffness value is 2.594 kN/mm (Profile A) and 2.606 kN/mm (Profile B). Comparing to the original castellated beam, the increment of ductility for all variations is not significant.

3.2.3. Ductility. In general, ductility consists of several types. Ductility can be calculated in terms of stress (strain), curvature, and displacement. The value of ductility in this structure is viewed in terms of displacement ductility. The following are the ductility values for castellated beams with stiffeners in profile A and B, see Figure 9.
Regarding ductility of the beams, adding of stiffeners on the castellated beam showed vary numbers.

- For variations in the number and thickness of stiffeners in Profile A, the highest ductility values were in stiffener between opening with values of 7.980 (n=4) and 7.998 (t_s=6 mm) respectively. While the average value of ductility obtained in all types of stiffeners with variations in the number of stiffeners is 6.998, and the average ductility stiffener thickness variation is 6.923.
- For variations in the number and thickness of stiffeners in Profile B, the highest ductility values were in stiffener on the web opening with values of 7.754 (n=4) and on the between opening 7.998 (t_s=6 mm). While the average value of ductility obtained in all types of stiffeners with variations in the number of stiffeners is 6.78, and the average ductility stiffener thickness variation is 6.72

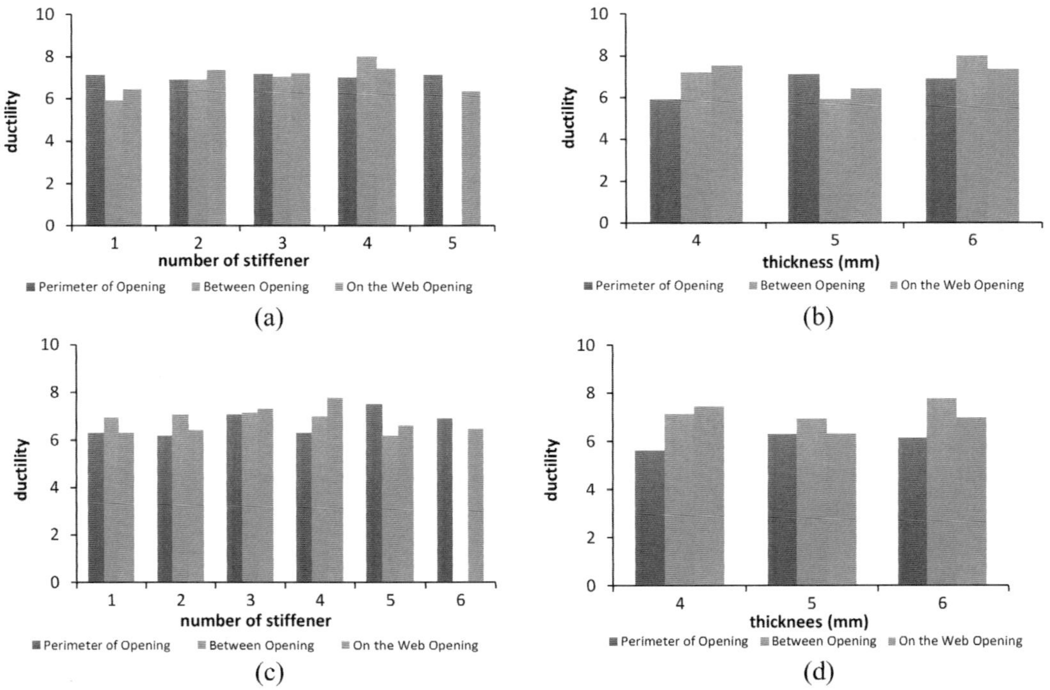

Figure 9. The ductility value based on (a) the number of stiffeners in the profile A, (b) the thickness of stiffeners in the profile A, (c) the number of stiffeners in the profile B, (d) the thickness of stiffeners in the profile B.

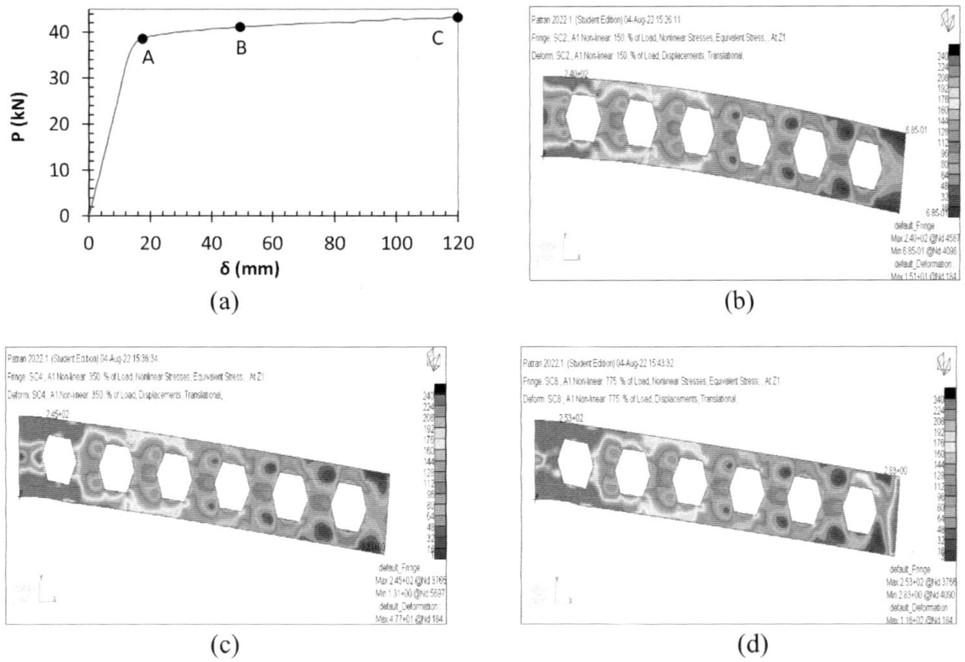

Figure 10. Typical Stress Contour for Castellated Beam Structure with Stiffener.

3.2.4. Stress Contour. For two types of profiles, similar deformation pattern and stress distribution are observed. Figure 10 shows typical stress contour for 4 stages of loading, which are represented by Stage A, B, C, and D. Stress contour at the beginning of inelastic regime (Stage A) is denoted by yielding on the outer fiber on the support zone. Yielding propagated along web direction as well as the beam axis (Stage B). More yielding are found on the web and the flange around the first opening (Stage C). At the Stage D, yielding covers almost entirely part of the castellated beam until the first opening.

4. Conclusion

Based on the results of the research that has been carried out on castella beams that were given additional stiffeners and varied based on the number and thickness of the stiffeners due to monotonic static loading, the following conclusions were obtained:

- Overall, the stiffeners on perimeter of opening become an alternative to increase the performance of the castellated beam in term of the ultimate load, beam stiffness and ductility
- For all variations, if the number of stiffeners and the thickness of the stiffeners increases, the ultimate load value and stiffness also increase
- Regarding the ductility of the beam, the addition of stiffeners to the castellated beam shows varying numbers

References

[1] G. S. NAIR and P. R. S. PILLAI, "Castellated Beam with Diagonal Stiffeners Along Hexagonal Cuts," *Int. Res. J. Eng. Technol.*, vol. 05, no. 06, pp. 2879–2883, 2018.

[2] M. * Suprapto, S. Pd, and P. T. Bangunan, "Optimalisasi Balok Castella Dengan Pergeseran Lubang Ke Bawah Yang Dibebani Pada Bagian Yang Utuh," *Rekayasa Tek. Sipil*, vol. 1, pp. 33–39, 2015.

[3] A. MURTOPO, D. FIRMANSYAH, and R. M. JANNAH, "Perilaku Geser Balok Castellated Bukaan Segi Empat Penuh dengan Pengaku Diagonal," *Rev. Civ. Eng.*, vol. 3, no. 1, pp. 7–11, 2019, doi: 10.31002/rice.v3i1.1205.

[4] R. Frans, H. Parung, D. Sandy, and S. Tonapa, "Numerical Modelling of Hexagonal Castellated Beam under Monotonic Loading," *Procedia Eng.*, vol. 171, pp. 781–788, 2017, doi: 10.1016/j.proeng.2017.01.449.

[5] F. Menkulasi, C. D. Moen, M. R. Eatherton, and D. Kuruppuarachchi, "Investigation of stiffener requirements in castellated beams," *Struct. Stab. Res. Counc. Annu. Stab. Conf. 2015, SSRC 2015*, no. March, pp. 473–490, 2015.

[6] H. W. A. Al-Thabhawee and M. A.-A. Al-Kannoon, "Improving Behavior of Castellated Beam by Adding Spacer Plat and Steel Rings," *J. Univ. Babylon Eng. Sci.*, vol. 26, no. 4, pp. 331–344, 2018, doi: 10.29196/jub.v26i4.810.

IOP Conf. Series: Earth and Environmental Science 1173 (2023) 012013 doi:10.1088/1755-1315/1173/1/012013

Spatial modelling of tsunami hazards and their exposure to settlements in Ambon City

N Sugandhi[1], Supriatna[1], E Kusratmoko[1] and H Rakuasa[1]

[1]Department of Geography, Faculty Mathematics and Natural Sciences, Universitas Indonesia, 16424 Depok, Indonesia

E-mail: nadhi.sugandhi@ui.ac.id

Abstract. Ambon City is an area that has the highest disaster risk index in Maluku Province. The direction of development of residential areas in Ambon City tends to be coastal areas which of course will have an impact on the risk of tsunami hazard. This study aims to spatially model the tsunami hazard and its exposure to settlements in Ambon City. The research variables consisted of land height, slope, distance from the river, and distance from the shoreline. This study uses an overlay technique on the variables that have been assessed and analysed using a Geographic Information System approach. The results of the study show that the tsunami hazard in Ambon City is dominated by a low level of tsunami susceptibility which has a wider area than the medium and high tsunami vulnerability classes, this is because 77.32% of the Ambon City area is at an altitude of > 64 m while the tsunami exposure level in settlements in Ambon City are nominated with a high level of vulnerability with an area of 2,835.33 ha.

1. Introduction

In her article entitled How Women Were Affected by the Tsunami: A Perspective from Oxfam MacDonald [1], explained that coastal areas are the areas where human activities are most carried out, about 70% of the world's population lives in coastal areas, thus making this area concentrated in various centres of economic activity such as fisheries, tourism, transportation, industry, settlements, defence and security [2,3]. This condition occurs in Ambon where the direction of development of residential areas in Ambon City is more inclined to coastal areas. The pattern of growth and distribution of the population living in coastal areas will certainly have an impact on the risk of tsunami hazards and ecosystem damage in the area [4]. This must be a serious concern from the AmbonCity government regarding coastal spatial planning based on disaster mitigation, which aims to maintain a sustainable coastal ecosystem in Ambon City and overcome the development of undeniable settlements in the coastal areas of Ambon City.

Population growth that continues to increase along with development in all fields results in the emergence of new problems, namely the increasing need for population settlements that are increasingly high as demands for living needs in addition to the needs of clothing and food [5]. The increasing need for settlements automatically demands the expansion of the area to be used as a residential area, so that coastal areas become a priority for settlement development in Ambon City [6]. According to Meilianda et al [7], land use change and land cover in coastal areas are particularly vulnerable to tsunamis and sea level rise [2].

According to Pranantyo & Cummins [8], Ambon City has limitations in the development of residential areas because it is in an area prone to earthquake and tsunami disasters [9], added that Ambon City is in a very active and complex tectonic area. The National Disaster Management Agency explained that Ambon City is an area that has the highest disaster risk index in Maluku Province [10], this was previously explained in the history of the earthquake and tsunami in Ambon recorded by Georg Everhard Rumphius which is mentioned as the tragedy of February 17, 1674 about 346 years ago, the earthquake that shook Ambon and its surroundings which resulted in damage to residents' homes and cost lives estimated at 2,500 people who died [11, 12]

Based on the Catalogue of Tsunamis on The Western Shore of the Pacific Ocean compiled by Solov′ev et al [13], it is explained that from the period 1600-2015 there have been 85 tsunamievents in the Maluku islands which are 40% of the 210 tsunami events that hit Indonesia and in general the tsunami events that occurred in Maluku Province were located on the island of Ambon and its surroundings. Demographically, the coastal area of Ambon city is an area that is prone to tsunami disasters because the flat and sloping topography and densely populated city inhabits the coastal areas. Based on the geographical conditions of Ambon City, which is a tsunami-prone area, with growing population growth, the built-up land will also develop, this can increase the risk of earthquake and tsunami disasters as disasters that are difficult to predict their arrival. Therefore, it is necessary to predict the spatial danger of the tsunami and its exposure to settlements in Ambon City using the Geographic Information System (GIS).

According to Fuad et al [14] the use of the Geographic Information System (GIS) is very important, effective and efficient in an effort to map and predict vulnerable and dangerous areas exposed to tsunami disasters, as a first step in disaster mitigation efforts to minimize losses that will occur in the future. Based on this background, this study aims to spatially model the dangers of tsunamis and their exposure to settlements in the city.

2. Materials and method/experimental

This research was conducted in Ambon City, Maluku Province which is geographically located at 3°34'4.80" - 3°47'38.4" South Latitude and 128°1'33.6" - 128°18'7.20 East Longitude and administratively Ambon City consists of Sirimau District, Nusaniwe, South Leitimur, Ambon Bay, Ambon Baguala Bay, with an area of Ambon City which is 32,573.68 Ha. Spatially the location of the study can be seen in figure 1. The data used in the full study include the Indonesian Earth Map (RBI) of Ambon City and National DEM data obtained from the Geospatial Information Agency (BIG). The data is then processed and analysed using ArcMap 10.8 software. Based on the data obtained, it is then processed into variables that affect the level of tsunami hazard and the level of tsunami exposure to residential areas in Ambon City which are modified from previous studies [15,16]. More of these variables can be seen in table 1 and figure 2.

Table 1. Variables affecting tsunami hazard levels.

No	Variable	Classification	Score	Weight
		< 10 m	5	
		11-25 m	4	
1	Elevation	51-100 m	3	30
		>100 m	2	
		100 m	1	
		< 2 %	5	
		3-5 %	4	
2	Slope	6-15 %	3	25
		16-40%	2	
		>40%	1	
		< 500 m	5	
		501-1000 m	4	
3	Distance from Coastline Coast	1001-1500 m	3	30
		1501-3000 m	2	
		>3000 m	1	
		< 100 m	5	
		101-200 m	4	
4	Distance from River	201-300 m	3	15
		301-500 m	2	
		>500 m	1	

These parameters were analysed using the weighted overlay method with the tsunami hazard formula = [(Distance from Coastline Coast *0.30) + (Elevation *0.30) + (slope*0.25) + (distance from river*0.15)].

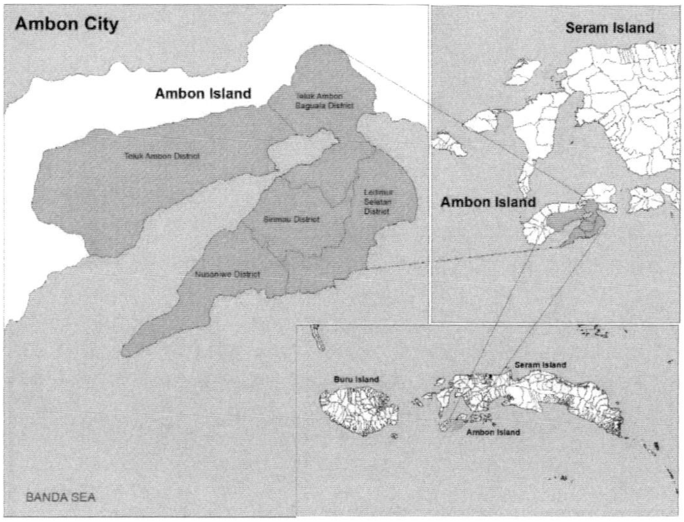

Figure 1. Research location.

Variables that affect the level of tsunami hazard include slope, land height, distance from the coastline and distance from the river which will then be given weighting before the overlay process is carried out to produce a tsunami hazard map in Ambon City (figure 2). A map of the level of tsunami exposure to residential areas in Ambon City was obtained from the results of the overlay of Ambon city settlement land in 2022 with the results of tsunami hazard modelling.

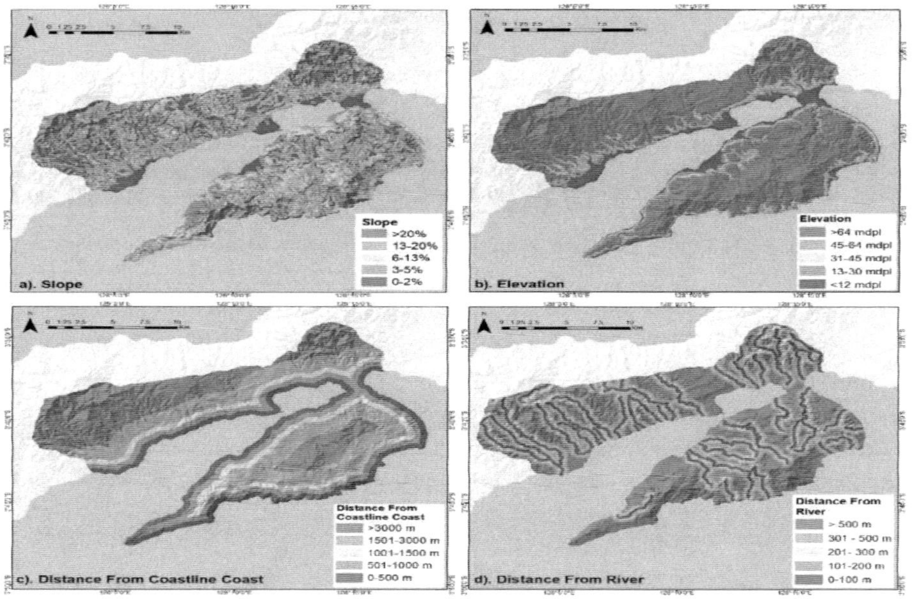

Figure 2. Research variables; a) slope, b) elevation, c) distance from coastline coast, d) distance from river.

Variables that affect the level of tsunami hazard include slope, land height, distance from the coastline and distance from the river which will then be given weighting before the overlay process is carried out to produce a tsunami hazard map in Ambon City (figure 2). A map of the level of tsunami exposure to residential areas in Ambon City was obtained from the results of the overlay of Ambon city settlement land in 2022 with the results of tsunami hazard modelling.

3. Results and discussion

3.1. Tsunami hazard level in Ambon City

Based on the results of tsunami hazard modelling in Ambon City using the four variables of physical condition can be seen spatially in figure 2 below. Based on figure 2 above, it can be seen that the low tsunami hazard level has a wider area compared to the medium and high tsunami hazard classes, this is due to the fact that 77.32% of the Ambon City area is at an elevation of land >64 meters above sea level (table 1) Based on figure 2 and the results of the area analysis, it is known that tsunamis with low classes have an area presentation of 61.12% oran area of 19,903.72 ha, tsunamis with medium classes have an area presentation of 22.91% or an areaof 7,460. 19 ha, and tsunamis with high classes have an area presentation of 15.97% or an area of 5198.66 ha.

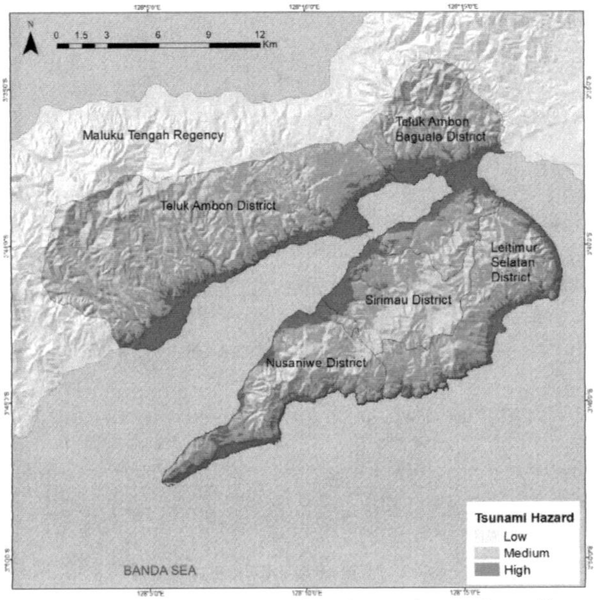

Figure 3. Tsunami hazard level map in Ambon City.

Table 2. Classification of Tsunami Hazard Levels in Ambon City Per-Subdistrict

No	District	Tsunami Hazard Level in Ambon City		
		High (Ha)	Medium (Ha)	Low (Ha)
1	Sirimau	471.13	769.73	2,461.61
2	Nusaniwe	982.11	1,740.20	1,950.58
3	Teluk Ambon	1,725.51	2,319.10	9,316.52
4	Teluk Ambon Baguala	1,268.84	1,193.59	3,612.67
5	Leitimur Selatan	751.07	1,437.57	2,562.34
	Total	19,903.72	7,460.19	5,198.66

3.2. Tsunami exposure rate in residential areas in Ambon City

Clearly and in detail the classification of tsunami hazard levels in Ambon City is explained per-sub-district in table 2 above. Based on table 2, it can be concluded that Ambon Bay District has the highest area in the Tsunami Hazard class in Ambon City, this is because Ambon Bay District is the largest district in Ambon City while Sirimau District has a relatively small area.

Based on figure 3 and table 3 above, the level of tsunami exposure in residential areas in Ambon City is dominated by a high level of danger with an area of 2,835.33 ha this is because almost a large number of residents in Ambon City live in coastal areas or in other words the pattern of population settlements in Ambon City follows the coastline / in coastal areas. Based on table 3 it is known that tsunamis with low exposure have an area of 419.95 ha, tsunamis with moderate exposure of 1,005.16 ha, tsunamis with high exposure of 2,835.33 ha.

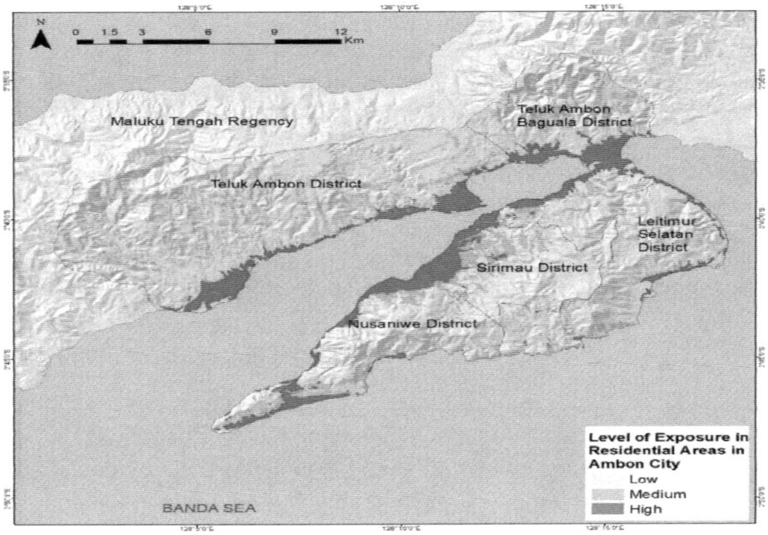

Figure 4. Map of Exposure Level in Residential Areas in Ambon City

Based on figure 3 and table 3 above, the level of tsunami exposure in residential areas in Ambon City is dominated by a high level of danger with an area of 2,835.33 ha this is because almost a large number of residents in Ambon City live in coastal areas or in other words the pattern of population settlements in Ambon City follows the coastline / in coastal areas. Based on table 3 it is known that tsunamis with low exposure have an area of 419.95 ha, tsunamis with moderate exposure of 1,005.16 ha, tsunamis with high exposure of 2,835.33 ha.

Table 3. Classification of Tsunami Hazard Levels in Ambon City Per-Subdistrict

No	District	Tsunami Exposure Level in Ambon City		
		High (ha)	Medium (ha)	Low (ha)
1	Sirimau	432.88	432.35	281.16
2	Nusaniwe	687.31	277.45	88.43
3	Teluk Ambon	939.85	80.24	14.51
4	Teluk Ambon Baguala	775.29	181.28	16.06
5	Leitimur Selatan	176.34	33.84	19.79
	Total Area	2,835.33	1,005.16	419.95

Based on table 3, the district that has the widest level of low exposure is Sirimau district, which is 281.16 ha, this is because the settlements are near the coast, and have a gentle topography where an area of 2,646.62 ha Ambon city area is at an altitude of >20 meters above sea level. On the other hand, the Ambon Bay District which only has an area of 14.51. The districts that have the widest level of moderate exposure are the 432.35 ha district and the South Leitimur district only has an area of 33.84, ha this is because the settlements in the area are relatively small and are in the mountainous area with an altitude of >64 meters above sea level. Ambon Bay District is the area with the widest level of exposure, which is an area of 939.85 ha while the South Leitimur district is the area with the smallest exposure level with an area of only 176.34 ha.

4. Conclusion

Based on the results of the analysis of tsunami hazard areas in Ambon City, it is in the low and medium class. This can be seen from the variables that affect it. Variables that greatly affect the level of tsunami hazard in Ambon City are the variables of land height and distance from the coastline. Tsunami hazards in Ambon City are dominated by low tsunami hazard levels have a wider area compared to medium and high tsunami hazard classes, this is due to the fact that 77.32% of the Ambon City area is at an elevation of land >64 meters above sea level, while the level of tsunami exposure to settlements in Ambon City is dominated by a high level of danger with an area of 2,835.33 ha, this is because almost most of the population in Ambon City lives in coastal areas, this can increase the risk of tsunami disasters as disasters that are difficult to predict their arrival. Ambon Bay Subdistrict has the highest area in each tsunami hazard class and its exposure in Ambon City is because Ambon Bay District is the largest district in Ambon City. Based on the condition of Ambon City, which is prone to tsunami, so this needs special attention from the Ambon City government in terms of disaster mitigation. The Ambon City Government plays a very important role in efforts to reduce disaster risk by continuing to improve its adaptive capacity indicators and the preparation of spatial planning basedon disaster mitigation in the future.

References

[1] MacDonald R 2005 How Women Were Affected by the Tsunami: A Perspective from Oxfam *PLoS Med.* **2**, 6 p. e178

[2] Rakuasa H, Febermen Halawa D A S 2022 Pemodelan Spasial Ketersedian Lahan Dengan Kawasan Rawan Tsunami di Kota Ambon *Geo Image* **11** 1–12

[3] Luo J, Xing X, Wu Y, Zhang W, and Chen R S 2018 Spatio-temporal analysis on built-up land expansion and population growth in the Yangtze River Delta Region, China: From a coordination perspective *Appl. Geogr.* **96** 98–108

[4] Susiloningtyas D & Lesy S A A 2021 Dynamic System Model of Land Use Affected by Sea Level Rise in the Coastal Area of Bengkulu City, Indonesia *J. Hunan Univ. Nat. Sci.* **48** 4

[5] AlFanatseh A 2021 Land suitability analysis of urban development in the Aqaba area, Jordan, using a GIS-based analytic hierarchy process *GeoJournal* **87** 4143-4159

[6] Salakory M and Rakuasa H 2022 Modeling of Cellular Automata Markov Chain for predicting the carrying capacity of Ambon City *J. Pengelolaan Sumberd. Alam dan Lingkung.* **12** 372–387

[7] Meilianda E *et al.* 2019 Assessment of post-tsunami disaster land use/land cover change and potential impact of future sea-level rise to low-lying coastal areas: A case study of Banda Aceh coast of Indonesia *Int. J. Disaster Risk Reduct* **41** p.101292

[8] Pranantyo I R and Cummins P R 2020 The 1674 Ambon Tsunami: Extreme Run-Up Caused by an Earthquake-Triggered Landslide *Pure Appl. Geophys.* **177** 1639–1657

[9] Rohadi S *et al.* 2021 The M 6.5 Ambon earthquake 26 September 2019: the source mechanism and the aftershock sequence characteristics *IOP Conf. Ser. Earth Environ. Sci.* **873** 12013

[10] BNPB 2021 *IRBI (Indeks Resiko Bencana Indonesia) Tahun 2021 Direktorat Pengurangan*

Risiko Bencana (Jakarta: BNPB) p. 115

[11] Rumphius GE1675 *History of the Terrible Earthquake that Took Place Recently, and Some Time Before, but Principally on February 17, 1674 in and around the Island of Amboina* Batavia.

[12] Horspool N *et al.* 2014 A probabilistic tsunami hazard assessment for Indonesia *Nat. Hazards Earth Syst. Sci.* **14** 3105–3122

[13] Solov'ev S L 1974 *Catalogue of Tsunamis on the Western Shore of the Pacific Ocean* (Canada: Institute of Oceanic Sciences)

[14] Fuad M A Z, Effendi V R & Dewi C S U 2022 Application Of Sar Remote Sensing And Geographic Information Systems (Gis) In Coastline Change Studies–A Case Study Of Coastline Change Due To The 2018 Tsunami In Tanjung Lesung, Banten *Asian J. Aquat. Sci.* **5** 87–98

[15] Dewi P U Oktaviana Wahdini M Prasiamratri N Alghifarry M B and Utami N A 2020 Aplikasi SIG Untuk Pemetaan Zona Tingkat Bahaya Dan Keterpaparan Pemukiman Terhadap Tsunami Kota Denpasar *J. Geosains dan Remote Sens.* **1** 80–88

[16] Akbar F S Vira B A Doni L R Putra H E and Efriyanti A 2020 Aplikasi Metode Weighted Overlay untuk Pemetaan Zona Keterpaparan Permukiman Akibat Tsunami (Studi Kasus: Kota Bengkulu dan Kabupaten Bengkulu Tengah) *J. Geosains dan Remote Sens.* **1** 43–51

IOP Conf. Series: Earth and Environmental Science 1173 (2023) 012014 doi:10.1088/1755-1315/1173/1/012014

Index active tectonic relative of Karangsambung geopark area, Central Java, Indonesia: assessment for developing geological disaster-based areas

E Hidayat[1,2], D Muslim[1], Z Zakaria[1], H Permana[3] and D A Wibowo[3]
[1] Faculty of Geological Engineering, Universitas Padjadjaran, Jatinangor, Sumedang 45363, Indonesia.
[2] Research Center for Geological Disaster, National Research and Innovation Agency, Jln. Sangkuriang, Bandung 40135, Indonesia.
[3] Research Center for Geological Resources, National Research and Innovation Agency, Jln. Sangkuriang, Bandung 40135, Indonesia.

E-mail: edih002@gmail.com

Abstract. Karangsambung as a "subduction fossil" of the Cretaceous-Paleocene on the island of Java, has a complex geological structure. The current subduction process in the south of Java with a north-south compression force has contributed to the reactivation of the geological structure in Karangsambung. The index active tectonic relative (IAT) of the Karangsambung area reflects the current level of tectonic activity. IAT is obtained from quantitative calculations of geomorphic indices, including streams length - gradients index (Sl), the ratio of valley floor width to valley height (Vf), basin asymmetry (Af), basin shape (Bs). mountain front sinuosity (Smf), the hypsometric integral (HI), and drainage density (Dd). The index active tectonic relative (IAT) calculation with an average IAT value of 2.37 indicates that tectonic activity is in the medium/moderate category. In addition, an attempt was made to compare the regional neotectonic activity with the relative tectonic activity of the study area. Based on this geomorphic analysis, the study area may have a lower seismic risk but needs to be evaluated more carefully for regional seismic hazards. The IAT value can be used as an initial reference for geological disaster assessment, especially earthquakes hazard in the development of the Karangsambung National Geopark Area.

1. Introduction

Research on neotectonic activity can be carried out using a tectonic geomorphology approach. The active tectonic response can be determined by studying geomorphic indices, such as active fault movements [1-3]. Earth movements in the past that have continued to the present are categorized as neotectonic activities [4,5]. Detailed spatial data is needed to study tectonic activity, especially active faults [6]. In addition, seismicity parameters in an area can be determined by analyzing tectonic geomorphology [7]. The interaction between uplifting and degradation processes, which results in erosion or sedimentation, is part of the characteristics of the deformed area [8]. Morphology of pointed hills, deep and narrow valleys, to plains (denudational) is evidence of tectonic activity recorded in the study area. This research focuses on studying the geological evidence recorded in the Quaternary period, and it is estimated that tectonic activity greatly influences the morphological character of the study area. The geomorphic index approach that will be used in this study includes stream length - gradient index (SI), the ratio of valley bottom width to valley height (Vf), basin asymmetry (Af), basin shape (Bs), mountain front sinuosity (Smf), integral hypsometric (HI), and drainage density (Dd). DEM is used to calculate the geomorphic index parameter with the GIS program. The geomorphic index parameter is used to see the effect of uplifting tectonic activity, which impacts the study area's morphology. The research results are expected to be the basis for developing disaster-based areas, especially earthquakes in the Karangsambung Geopark area.

2. Regional tectonic and geomorphological setting

During the Cretaceous-Paleocene, the subduction pathway crossed the study area [9-13]. In the Upper Cretaceous, the movement of the Gondwana microcontinent collided with the Sundaland edge, which resulted in a locked subduction pathway in the Northeast-Southwest direction [14]. In the Middle Eocene, a new subduction zone in southern Java was formed and was followed by the deposition of rocks of the Eocene age, namely the Karangsambung Formation. It continued to be deposited with younger rocks, namely the Totogan Formation and the Waturanda Formation [15]. The physiography of the Karangsambung area and its surroundings is part of the South Serayu Mountains [9,15-22]. The main fold formed is the Karangsambung anticline with an East-West axis and is a cap anticline that opens to the west. This anticline folds Paleogene and Neogene rocks. Structures also influence the study area in the E-W and N-S directions that cut through pre-Tertiary and Tertiary rocks [23]. The uplift of pre-Tertiary rocks in the study area is influenced by the NE-SW trending regional structure through a bouncing mechanism [20] and is depicted on the geological map of the study area [23] (figure 1).

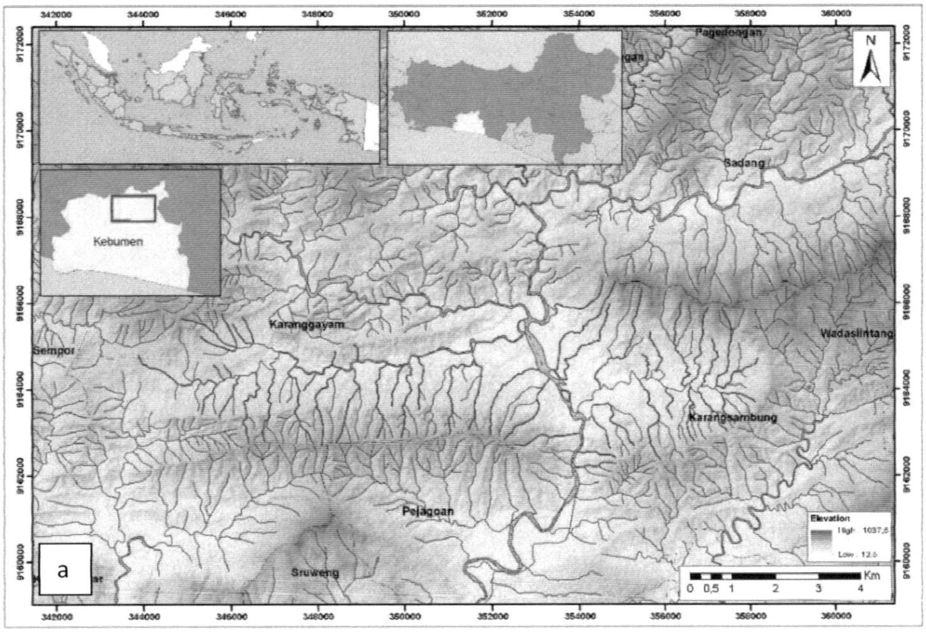

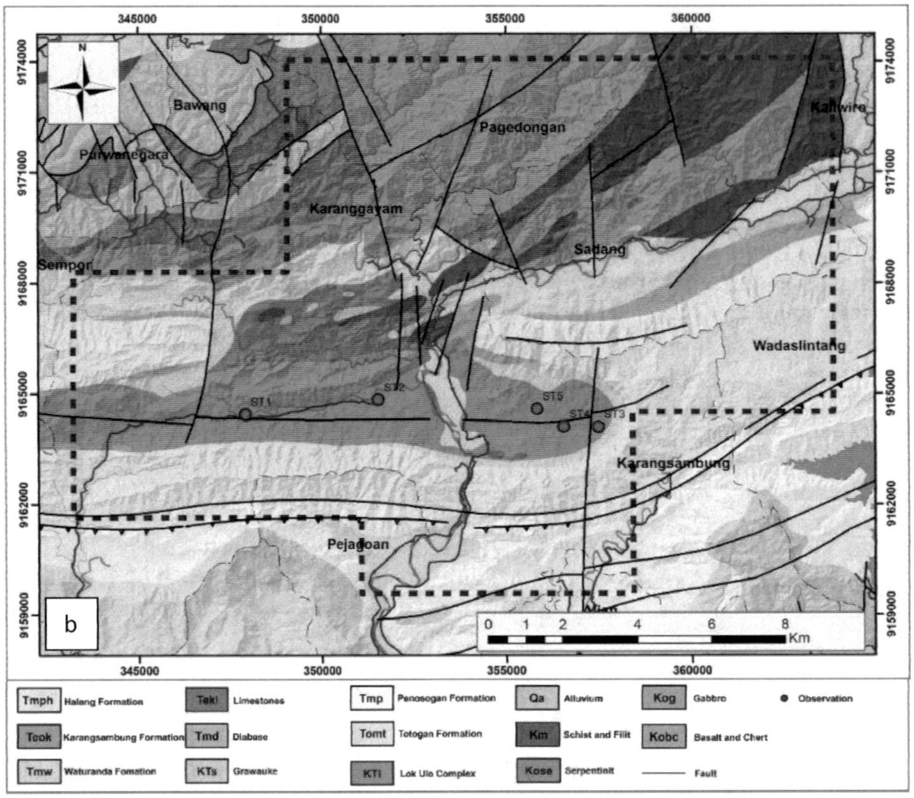

Figure 1. (a) Geographical position of the research area is the Karangsambung Central Java and (b) Geological map of the research area

3. Methods

The index active tectonic relative (IAT) value can be determined by calculating the geomorphic index parameters [26,27,1]. The geomorphic index parameters used in this study are: 1) stream length – gradient index (Sl), 2) the ratio of the width of the valley floor to the height of the valley (Vf), 3) the asymmetry of the drainage basin (Af), 4) basin shape (Bs), 5) mountain front sinuosity (Smf), 6) correlated hypsometric integral (HI), and drainage density (Dd). The base plane used in calculating the geomorphic index is the Digital Elevation Model (DEM) with an accuracy of 8 meters (DEMNAS).

3.1. Streams length - gradients index (Sl)

The geomorphic index parameter can identify the evaluation of the relationship between active tectonics, bedrock, and topography in the form of Sl value [2]. The Sl formula is:

$$Sl = (\Delta H/\Delta L) \times L \tag{1}$$

Where the elevation (ΔH) is different from the point of being counted, the ΔL is the length of the river to the point of being counted and L is the total length of the river up to the point of being counted. The level of tectonic uplift is reflected in the SL value [2].

3.2. The ratio of the width of the valley floor to the height of the valley (Vf)

Alluvial deformation, the formation of asymmetrical folds and the shape of river valleys can be affected by fault movement activity [28]. The low uplifting rate is evidenced by the high Vf value and forms a wide valley floor. A narrow valley bottom is usually represented by a low Vf value associated with a high uplift. The Vf formula is:

$$Vf = 2\, V_{fw} \,/\, (E_{ld} - E_{sc}) + (E_{rd} - E_{sc}) \tag{2}$$

Where V_{fw} is the width of the valley, E_{ld} and E_{rd} is the left-right elevation of the valley, and E_{sc} is the valley base elevation.

3.3. the asymmetry of the drainage basin (Af)

Tectonic activity influences the shape of the watershed. The slope of the watershed can affect changes in river flow in the watershed due to active tectonics. The Af value reflects the correlation between uplift tectonics and the shape of the watershed. The Af formula is:

$$Af = 100\left(\frac{A_r}{A_t}\right) \tag{3}$$

The relatively stable tectonic activity is indicated by the Af 50 value.

3.4. Basin shape (Bs)

The BS formula is [30,31,1]:

$$Bs = Bl \,/\, Bw \tag{4}$$

Where the Bl is the length of the basin (watershed) measured from the highest point, and the Bw is the width of the sub-watershed is measured the widest. Tectonics affect the sub-watersheds relatively elongated shape, while the sub-watersheds more circular shape is not affected by tectonic activity [26].

3.5. Mountain front sinuosity (Smf)

Low Smf values indicate indications of active tectonics (Smf<1.4) [3]. Smf is a geomorphic index calculated from the shape of the curved face of the mountains as a result of erosion and tectonic activity as an indication of active faults. The Smf formulas is:

$$Smf = \frac{L_{mf}}{L_s} \tag{5}$$

Where the Lmf – length of the mountain front along the fault (km), Ls – the straight line length along the front of the mountain (km). Low Smf values will indicate tectonic uplift activity. If the Smf value is significant, the erosion rate is high, and the uplifting activity is low.

3.6. Integral hypsometric (HI)

The Integral Hypometry (HI) formula is [32,27] :

$$HI = (h_{mean} - h_{min})/(h_{max} - h_{min}) \tag{6}$$

Where h_{max}, h_{min} and h_{mean} are the maximum, minimum and average elevation. The equation describes the topographical relief reflected by a hypsometric curve. The shape of the hypsometric curve shows the young, medium, and old stages, which reflect the topographical relief [2]. Tectonic activity does not correlate directly with SL values but is also influenced by incision processes [3].

3.7. Drainage density (Dd)

Stable areas will have more massive rocks than strongly deformed areas. Strongly deformed areas produce dense river drainage [33]. Dd is calculated by comparing the entire length of the river segment outside the sub-catchment. The less the drainage density, the lower the Dd value. The formula Dd is [34] :

$$Dd = \frac{\sum L}{A} \tag{7}$$

Where $\sum L$ is the total length of the river and A is the total area of the watershed.

3.8. Index active tectonic relative (IAT)

The IAT value is the sum of the geomorphic index parameters, and the IAT analysis correlates with the level of tectonic activity. The IAT formula is:

$$IAT = S / n \tag{8}$$

IAT – relatively active tectonic index, S – the number of class values of each geomorphic index, n – the number of geomorphic indices. Relative tectonic activity index (IAT) is grouped into 4 classes based on S/n values. Class 1 S/n values between 1-1.5 (very high); class 2 values S/n 1.5-2 (high); class 3 values S/n 2-2.5 (medium); and class 4 values S/n> 2.5 (low) [3].

4. Results and discussion

The topography of the amphitheater is the focus of calculating the geomorphic index. The morphology of the Karangsambung amphitheatre is an anticline valley trending west-east and disclosing Pre-Tertiary rocks [9,19,15]. Regional fault reactivation in the Cretaceous-Paleocene (NE-WS) suture resulted in the uplifting and exposure of pre-Tertiary rocks in the study area. Uplifting activity is also supported by gravity data [35].

The value of Sl is divided into 3 classes; Sl > 500 class 1, 500>Sl>300 class 2, and Sl<300 class 3 [3] (Table 1). The value of Sl in the study area is between 66.5-734.5. These values show that the study area has a variety of tectonic activity classes. Tectonic activity or erosion can affect the valley's shape, which is reflected in the value of Vf. The classification of Vf is as follows: Vf<0.5 class 1, 0.5<Vf<1 class 2, and Vf>1 class 3. Vf in the study area ranges from 0.25 to 12.75 (Table 1). V-shaped valleys are reflected by low Vf values, while high Vf values reflect U-shaped valleys. Sub-DAS 1-20 has high Vf values related to the rock hardness composed by breccias of the Waturanda Formation. Meanwhile, Sub-DAS 1-25 has a low Vf value related to the presence of a fault and joint structures and the type of rock in the form of melange rock blocks (figure 2).

Af index has 3 classes, namely class 1 (Af<45 / high), class 2 (45–55 / medium) and class 3 (Af>55 / low). The Af score in the study area from 22.29-79.49. The shape of the sub-watershed in the study area is 2/3 asymmetrical. The shape of the sub-watershed asymmetry in the study area is closely related to the tectonic uplift activity. The Bs value in the study area was between 0.81-4.30. Classification of BS grades: BS>4 class 1, 3<BS<4 class 2, and BS<3 class 3 (Table 1). Nearly 60% of the sub-watersheds studied belong into classes 2 and 3, meaning that the shape of the sub-watershed tends to be circular. Meanwhile, high Bs values are indicated by the shape of the sub-watershed, which tends to be elongated and is in the class 1 category. The Bs values relate to the shape of the valley and river profile which the Vf and SL values influence. The elongated sub-watershed is associated with steep river valleys because the significant vertical erosion rate is correlated with large flow velocities due to the difference in distance and elevation between upstream and downstream within the sub-watershed.

The Smf is divided into 3 classes based on the level of tectonic activity, namely 1.0<Smf<1.1 class 1, 1.1<Smf<1.5 class 2, and Smf>1.5 class 3 (Table 1). A smaller Smf value is an indication of active tectonics [2,3]. Higher Smf values (>3) are associated with tectonic inactivity [26]. The Smf as a fault line representation will affect the long and short distance from upstream to downstream of the sub-watershed so that it will affect the sub-watershed.

Rock type and geological structure affect the shape of the hypsometric curve [36]. The convex and concave shape of the hypsometric curve determines the value of the integral hypsometric (HI). HI is divided into three classes; class 1 HI>0.5 has a convex curve, class 2 0.4<HI<0.5 has a concave-convex curve, and class 3 HI<0.4 has a concave curve (Table 1). The shape of the "S" curve and concave dominates the study area, indicating moderate erosion of the sub-watershed (Table 1). Dd values range from 2.231 to 7.040, and most sub-watersheds are in class 2. Class 2 shows the active tectonic influence on the morphology of the study area (Table 1).

IAT (relative tectonic active index) assesses an area's distribution of active tectonic relativity [3]. IAT is grouped into four classes, as follows: 1.0<IAT<1.5 class 1 very high, 1.5<IAT<2 class 2 high, 2<IAT<2.5 class 3 medium, and 2.5<IAT<3 class 4 low (Table 1). The results showed that the S/n value was 2–2.5, with an average of 2.37 (medium). The results of IAT calculations show that most of the study areas are in class 3 (middle), this indicates that the formation of the morphology of the study area is more influenced by tectonic activity than erosion processes. The existence of intensive geological structures, especially faults and joints, especially in the western part, is a reflection of the IAT value of the study area (figure 2).

Table 1. Calculation index active tectonic relative (IAT) with its tectonic activity

No.	Sl	Vf	Af	Bs	Smf	HI	Dd	Sn	IAT	Tectonic activity
1	2	3	3	3	3	1	2	2,43	3	Medium
2	1	3	2	1	3	2	2	2,00	2	High
3	3	3	1	2	3	2	2	2,29	3	Medium
4	1	3	3	2	3	3	2	2,43	3	Medium
5	3	3	3	3	3	2	2	2,71	4	Low
6	3	3	1	2	3	2	3	2,43	3	Medium
7	1	3	3	3	3	2	2	2,43	3	Medium
8	1	3	2	2	3	2	2	2,14	2	High
9	3	3	2	3	3	2	2	2,57	4	Low
10	1	3	1	3	3	2	2	2,14	3	Medium
11	3	3	2	3	2	2	3	2,57	4	Low
12	3	3	1	3	3	2	2	2,43	3	Medium
13	3	3	3	3	3	3	1	2,71	4	Low
14	3	3	1	3	3	2	2	2,43	3	Medium
15	3	3	2	3	2	2	1	2,29	3	Medium
16	3	3	2	3	3	3	2	2,71	4	Low
17	2	3	3	3	3	2	2	2,57	4	Low

18	2	3	3	3	3	2	2	2,57	4	Low
19	3	2	3	3	3	3	2	2,71	4	Low
20	3	2	2	3	3	2	2	2,43	3	Medium
21	3	2	1	3	3	2	1	2,14	3	Medium
22	2	2	2	3	3	2	2	2,29	3	Medium
23	2	2	1	3	3	2	2	2,14	3	Medium
24	1	3	2	3	3	2	2	2,29	3	Medium
25	1	3	2	3	3	2	2	2,29	3	Medium
26	2	3	1	3	3	2	2	2,29	3	Medium
27	2	3	3	1	3	3	1	2,29	3	Medium
28	2	2	3	3	3	2	2	2,43	3	Medium
29	3	3	1	3	3	3	2	2,57	4	Low
30	2	2	2	3	3	2	2	2,29	3	Medium
31	2	1	3	3	3	2	2	2,29	3	Medium
32	2	1	3	3	3	2	2	2,29	3	Medium
33	3	3	3	2	3	3	2	2,71	4	Low
34	1	1	1	3	3	2	2	1,86	2	High
35	3	1	2	2	3	2	2	2,14	3	Medium
36	2	3	3	3	3	2	2	2,57	4	Low
37	2	3	3	3	3	3	2	2,71	4	Low
38	3	3	3	3	3	3	2	2,86	4	Low
39	3	1	1	3	3	2	3	2,29	3	Medium
40	1	2	2	3	3	2	1	2,00	3	Medium
41	3	3	1	3	3	3	3	2,71	4	Low
42	2	3	3	3	2	3	2	2,57	4	Low
43	3	1	1	3	3	3	2	2,29	3	Medium
44	2	3	3	2	3	3	2	2,57	4	Low
45	2	3	1	2	3	3	2	2,29	3	Medium
46	2	2	1	3	3	2	2	2,14	3	Medium
47	2	1	1	3	3	2	2	2,00	2	High
48	1	1	2	3	3	2	2	2,00	2	High
49	2	1	1	3	3	2	2	2,00	2	High
50	3	3	1	3	3	3	3	2,71	4	Low
51	3	2	1	3	3	3	3	2,57	4	Low
52	3	1	1	3	3	2	2	2,14	3	Medium

IOP Conf. Series: Earth and Environmental Science 1173 (2023) 012014 doi:10.1088/1755-1315/1173/1/012014

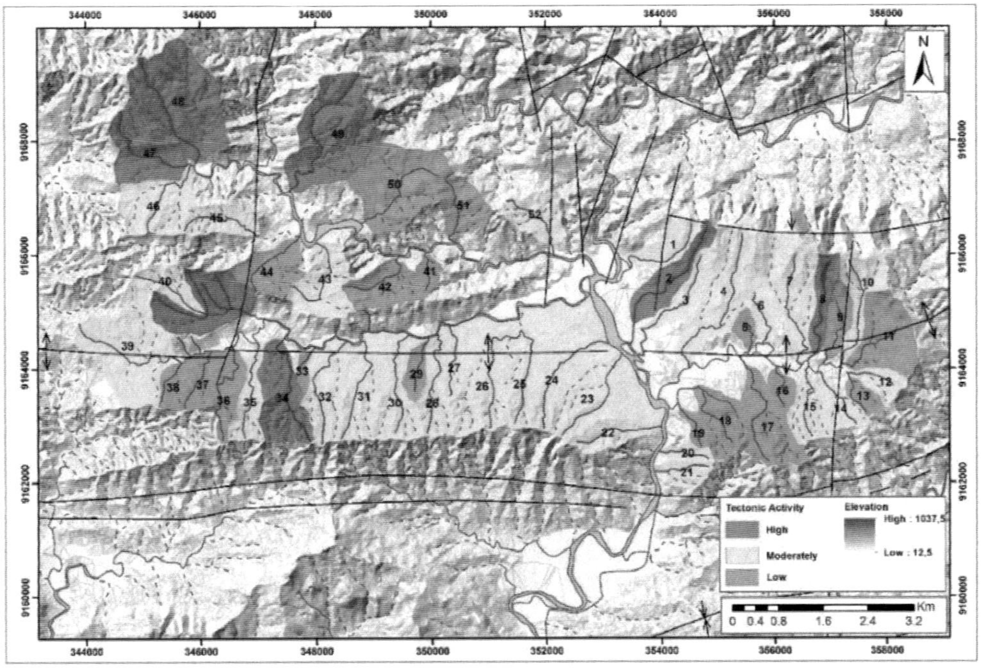

Figure 2. The study area's index active tectonic relative (IAT) map is part of the Karangsambung geopark

The Vf index reflects the valley geometry correlated with tectonic or erosional activity. Low Vf values indicate uplifting activity, resulting in high vertical erosion and V-shaped valleys. In contrast to Vf values, low Af values are associated with low tectonic activity, and the shape of the sub-catchment is symmetrical with U-shaped valleys. Bs values are influenced by uplift tectonics. The research area shows that the average sub-watershed has an elongated shape, meaning that there is an uplifting effect on the sub-watershed. Apart from being reflected in the Bs value, the effect of uplifting is also evident in the Sl value. The high flow gradient index (Sl) and the hypsometric integral curve indicate a more significant tectonic influence than the erosion process. The rough topographic relief indicates the high value of Sl in the sub-watershed. The topography that forms a significant difference in elevation from upstream to downstream on the sub-watershed also shows a high Sl value. The difference in high Sl values is influenced by active tectonics in the form of active faults. The rock type and hardness also affect the HI and Sl values. Uplift is the most influential tectonic activity on the research area's morphology, especially the Karangsambung amphitheater's topography.

The compressive force trending North-South due to the compression activity of the subduction of the Indo-Australian oceanic plate under the Eurasian plate affects the uplifting activity of the study area. This compressive force can reactivate faults that develop in the study area, such as the regional fault of the Karangsambung Fault in the NE-SW direction with a horizontal movement. Several earthquake epicentres with shallow depths (> 60 km) were recorded in the study area [37] (figure 3). The shallow depth of the seismic epicenter indicates neotectonic activity due to fault movement.

Active tectonic indications resulting from IAT values and seismic data with shallow depths should be the basis for developing the Karangsambung Geopark area based on geological disasters. Geological disasters that may occur are earthquakes due to active faults. Secondary hazards from earthquakes that must be considered are landslides, and rock falls because the study area has steep slopes and intensive geological structures. The development of geological disaster-based areas is crucial because it is one of the roles of geoparks in disaster education and sustainable development.

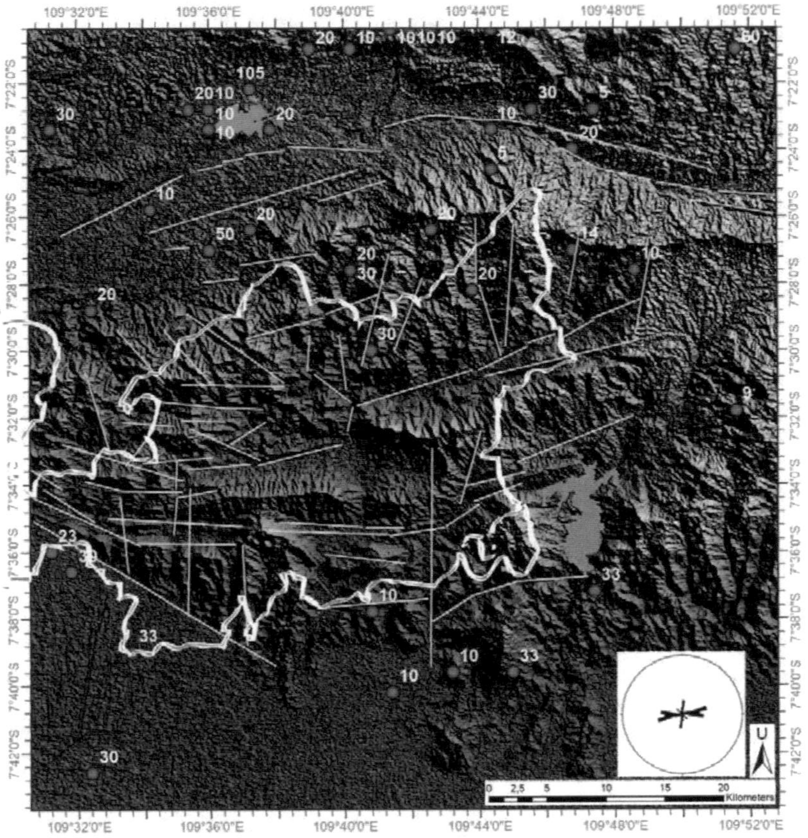

Figure 3. Map of seismic points (epicenters) with earthquake depth in kilometers (red dots) and lineament patterns (yellow lines) (source: BMKG Banjarnegara Earthquake Station, 2021). The rosette diagram shows the direction of the main structure, Southwest – Northeast is part. The white line is the boundary of the Karangsambung geopark area.

5. Conclusion

Tectonic geomorphology is an important initial study to determine active tectonic activity in an area. The study results show that the relatively active tectonic index (IAT) is in class 3 (medium) or the category of moderate tectonic activity, meaning that tectonic activity is more influential than erosion in the formation of morphology. The analysis of seismic activity shows that there are shallow seismic points (<60 km) with a magnitude (Mw) of < 5, indicating neotectonic activity in the study area. The results of this study should be a concern in regional development in the Karangsambung Geopark area, especially regarding the potential for geological hazards, especially earthquakes and their associated hazards, such as landslides, rock fall.

Acknowledgements
Authors would like to thank the head of the *National Reseacrh and Innovation Agency* for his permission to continue his research. The author also thanks to the supervisors and other parties who helped research until the research paper.

References

[1] A Faghih, A Esmaeilzadeh Soudejani, A Nourbakhsh, S Rokni 2015 *Environmental Earth Sciences*, **74** (4) 3007–3017
[2] E A Keller and N Pinter 1996 *Prentise Hall Upper Saddle River* New Jersey 07458
[3] R El Hamdouni, C Irigaray, T Fernández, J Chacón, E A Keller 2008 *Geomorphology* **96** (1–2) 150–173
[4] L S Stewart and P L Hancock 1994 *First Edition Pegamon Press London* pp 370-409
[5] R S Yeats, K Sieh, C R Allen 1997 *Oxford University*
[6] J R Arrowsmith and O Zielke 2009 *Geomorphology* **113** (1–2) 70–81
[7] F Galadini, P Galli 2002 *Nat Hazards* **22** (3) 225–268
[8] D W Burbank and R S Anderson 2001 *Blackweel Publishing* ISBN 978-0-632-04386-6
[9] S Asikin Doctoral Dissertation 1994 *Bandung Institute of Technology* unpublished 103 pages
[10] B Clements, R Hall, H R Smyth, M A Cottam 2009 *Petroleum Geoscience* **15** (2) 159–174
[11] R Hall 2012 *Tectonophysics* **570–571** pp 1–41
[12] C D Parkinson, K Miyazaki, K Wakita, A J Barber, and D A Carswell 1998 *Island Arc* **7** (1–2) 184–200
[13] K Wakita 2000 *Journal of Asian Earth Sciences* **18** 739-749
[14] H R Smyth, P J Hamilton, R Hall, and P D Kinny 2007 *Earth and Planetary Science Letters* **258** 269–282
[15] C Prasetyadi, E R Suparka, A H Harsolumakso, B Sapiie 2006 *Proceedings International Sedimentological* Congress Fukuoka
[16] H D Tjia 1966 *Ph.D Dissertation Contribut From the Dept. of Geol. ITB* No 63
[17] M E Suparka 1986 *Dissertation of Geology Engineering Department ITB* unpublished 181 pages
[18] K Wakita, Munasri, W Bambang 1994 *Journal SE Asian Sciences* **9** 29-4
[19] A H Harsolumakso 1996 *National Seminar Collection Papers*
[20] A H Satyana and M E M Purwaningsih 2002 *Indonesian Association of Ggeologists (IAGI) Yogyakarta – Central Java*
[21] C Prasetyadi 2007 *Dissertation of Geology Engineering Department ITB* unpublished 325 pages
[22] P J Cannon 1976 *Oklahoma Geology Notes* **36** (1) 3–16
[23] S Asikin, A H Harsolumakso, H Busono, and Gafoer S 1992 *Research and Development Center Bandung*
[24] Sribudiyani, Nanang Muchsin, Rudy Ryacudu, Triwidiyo Kunto, Puji Astono, Indra Prasetya, Benyamin Sapiie, Sukendar Asikin, Agus H Harsolumakso, Ivan Yulianto 2003 *Proceedings Indonesian Petroleum Association*, Twenty-Ninth Annual Convention & Exhibition
[25] W H Condon, L Pardyanto, K B Ketner, T C Amin, S Gafoer, and H Samodra 1996 *Geol. Res. And the Development Center (P3G) Bandung*
[26] W Bull, L McFadden 1997 *Proceedings Vol. of 8th Annual Geomorph. Symp.* (Edited by Doering, D.O.) State University of New York at Binghamton Binghamton NY. 116-138
[27] E A Keller, N Pinter 2002 Prentice Hall New Jersey

[28] L Khayingshing, S B Surendra, C K Girish, J 2015 *Geomorphology* **239** 58-72
[29] B B Singh, A Afzal, A R Khan, Ahmad 2016 *Arab J Geosci* **9** 390 13-17
[30] P J Cannon 1976 *Oklahoma Geology Notes* **36** (1) 3–16
[31] M T Ramírez-Herrera 1998 *Earth Surface Processes and Landforms* **23** 317–332
[32] A N Strahler 1952 *Geological Society of America Bulletin* **63** 1117–1142
[33] E Sukiyah 2009 *Dissertation, Padjadjaran University Postgraduate Program, Bandung*
[34] R E Horton 1945 *Geological Society of America Bulletin* **56**
[35] Subagio 2008 *Journal of Geological Resources* **18** No 6
[36] Flavio Giaconia, Guillermo Booth-Rea, José Miguel Martínez-Martínez, José Miguel Azañón, José Vicente Pérez-Peña, Joaquín Pérez-Romero, Irene Villegas 2012 *Geomorphology* **145–146** Pages 90-106
[37] BMKG 2021 *Banjarnegara Earthquake Station*

IOP Conf. Series: Earth and Environmental Science 1173 (2023) 012015 doi:10.1088/1755-1315/1173/1/012015

Post-disaster rapid assessment of Sunda Strait tsunami on 24th–25th December 2018 in the Regencies of Serang and Pandeglang, Province of Banten, Indonesia

D A Tiwi[1], N Sudiana[2], Prihartanto[2], I Turyana[1], F Prawiradisastra[1], Q Zahro[1] and Astisiasari[1]
[1]The National Research and Innovation Agency, Geological Disaster Research Center, South Tangerang, Indonesia
[2]The National Research and Innovation Agency – Environmental and Clean Technology Research Center, South Tangerang, Indonesia

E-mail: dwia001@brin.go.id

Abstract. Post-disaster rapid assessment of the Sunda Strait Tsunami was conducted on 24th–25th December 2018 in five selected sample sites. These study sites were located on the western coast of the Regencies of Serang and Pandeglang, and represented a range of damages, losses, and needs. The aims of this assessment are: (1) measuring the tsunami wave dimensions, (2) observing the landscape conditions, and (3) investigating the affected land cover. The physical damages were investigated based on the tsunami wave dimension analysis, by using a direct-field measurement, drone-based survey, and local interviews. The results show that (1) the greatest trace of tsunami dimensions in Tanjung Jaya Village, Panimbang District, Pandeglang Regency was subjected to the highest fatality, (2) the open landscape aggravated the tsunami impact, compared to the landscape with perennial trees or hilly topography, and (3) the severely destructed buildings were usually on the nearest coastline, where Sambolo residential area in Sukarame Village had the worst impact. These conditions were exacerbated by: (1) the absence of coastal retaining structures as physical defense, and (2) built-on facilities on reclaimed land that violated government regulation. Additionally, this paper proposes several recommendations to support the next phase of mitigation strategies and sustainable planning.

1. Introduction

The Sunda Strait tsunami occurred on December 22nd, 2018 at 20:56 local time (13.56 UTC). It caused devastating impacts of 437 fatalities, 16 missing, 14,059 injuries, and 33,719 displaced. Moreover, there were also 2,752 houses, 510 boats, 147 vehicles, and 92 lodgings damaged [1]. Figure 1 shows the tsunami impacts on five coastal regencies in two Provinces of: Banten (Serang and Pandeglang Regencies) and Lampung (Pesawaran, Tanggamus, and South Lampung Regencies) [2].

Accordingly, in order to empower a disaster risk management (prevent the new, reduce the existing, and manage the residual), one of the measures is to conduct a risk assessment that provides understanding from the diverse dimensions (hazards, vulnerabilities, and capacities). Subsequently, risk assessment can be implemented for various measures, including the disaster recovery, risk reduction, development plans, and underlying risk-drivers (e.g., poverty and weak governance) [3].

There are three stages to enable a comprehensive risk assessment: (1) preparing and scoping, (2) risk analysis, and (3) sustainable development. These steps are bound to the flexibility and timing [3]. While on a rapid assessment, a set of immediate, intensive, and effective efforts are conducted to deliver a preliminary understanding upon the disaster impacts. Generally, the post-disaster rapid assessment promptly identifies several baseline information such as: damages (infrastructure, assets), losses (lives, economic), and needs (services, reconstruction, cost). Subsequently, the result of a rapid assessment can support the quick and appropriate recovery strategies [4,5].

Corresponding to several constraints of the brief time, institutional responsibilities, and reliable results, this work concerns limitation upon the deliverable scoping on tsunami's physical impact. Correspondingly, this paper aims to deliver the fieldwork investigation on: (1) tsunami wave dimensions (inundation distance and run-up elevation), (2) landscape conditions (pre- and post-tsunami), and (3) affected land cover. The differing gap elements to previous studies related to Sunda Strait Tsunami on December 22nd, 2018 was that this paper prioritizes a focus on the post-tsunami rapid assessment on physical damages (landscapes and land covers), through a field survey (drone flights and ground observations).

The post-disaster rapid assessment of Sunda Strait tsunami was conducted on 24th–25th December 2018 in five sample sites: (1) one (1) site at Bulakan Village, Cinangka District, Serang Regency; (2) three (3) sites at Sukarame Village, Carita District, Pandeglang Regency; and (3) one (1) site at Tanjung Jaya Village (Tanjung Lesung), Panimbang District, Pandeglang Regency. These sample sites were selected to represent a range of damage, and losses, based on the field investigation. Figure 2 shows the locations of post-disaster rapid assessment, i.e., Figure 2a for the five (5) sample sites of drone flying-track for tsunami dimension and landscape damage analysis, and Figure 2b for the seven (7) points of impacted landuse and local interviews.

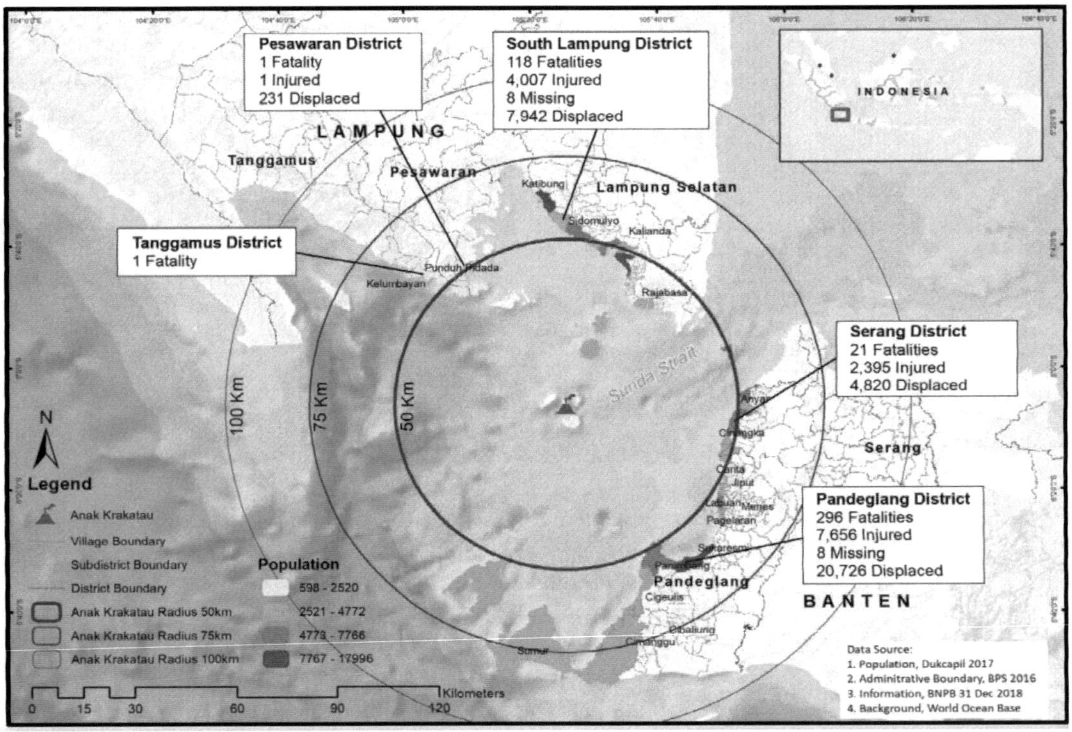

Figure 1. Impact of Sunda Strait tsunami per-January 1, 2019 [2].

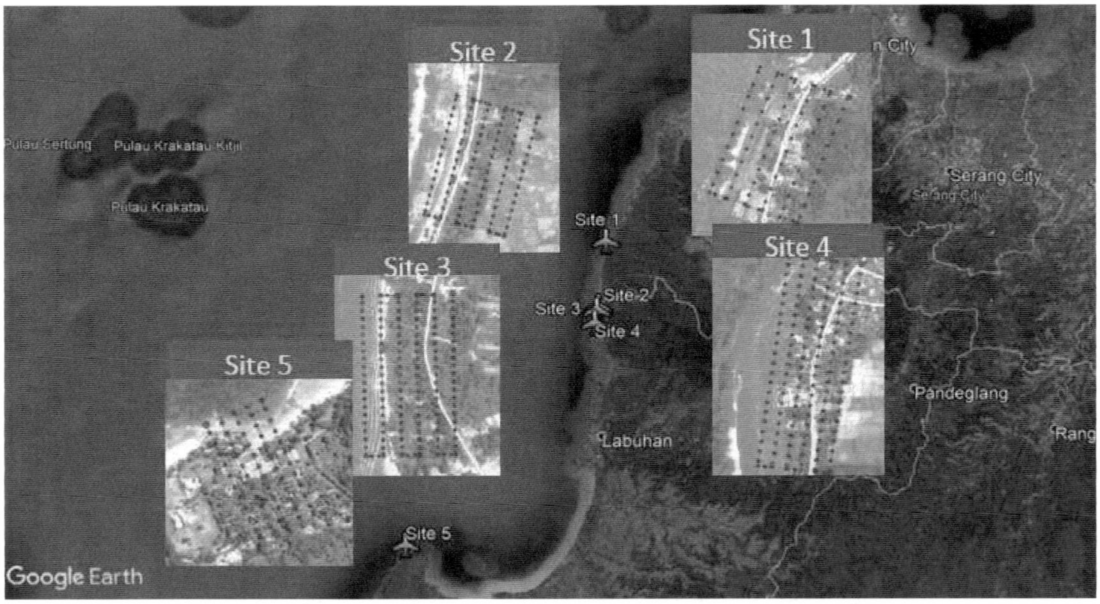

a

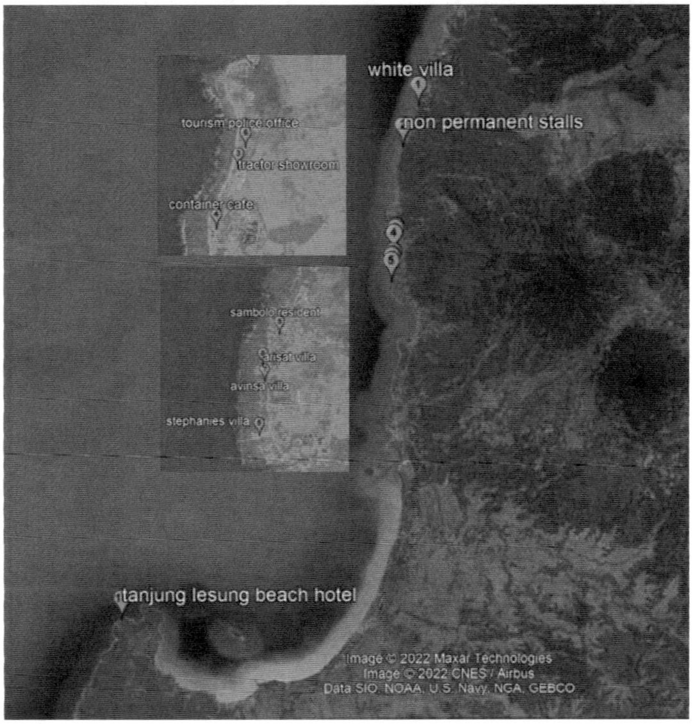

b

Figure 2. Location of the post-disaster rapid assessment: (a) five (5) sample sites of drone flying-track for tsunami dimension and landscape damage analysis, (b) seven (7) points of impacted landuse and local interviews.

1.1. Studies of Sunda Strait tsunami on December 22nd, 2018

This tsunami was considered as a "silent tsunami" for the absence of signs, and thus, no early warning was sent to the communities [6]. It was not preceded by a tectonic activity, such as a recorded earthquake. However, the flank collapse occurred during a normal-but-heightened eruption episode [7]. The tsunami itself was a cascading event, volcanically-triggered by a lateral collapse of the Mount Anak Krakatau (GAK) during the eruption. The intense phreatomagmatic explosions occurred after the failure was still not clear whether it was a response to the collapse or not [8].

Prior to the flank collapse, the height of the GAK cone initially reached up to 338 meters above sea level. Figure 3 shows the before- and after-failure of discharged volcaniclastic materials on the southwest of volcano. After losing its volume for 150–180 million cubic meters, the GAK now reaches 110 meters above sea level, remaining as a horseshoe morphology [9].

Table 1 showcases the previous studies of Sunda Strait tsunami on December 22nd, 2018. The studies are varied from the methods and point of views. Studies from [8,10 and 11] suggest the need of tsunami detection and warning system that is coupled with the alert of neglected non-seismic events (e.g., landslides and eruptions)

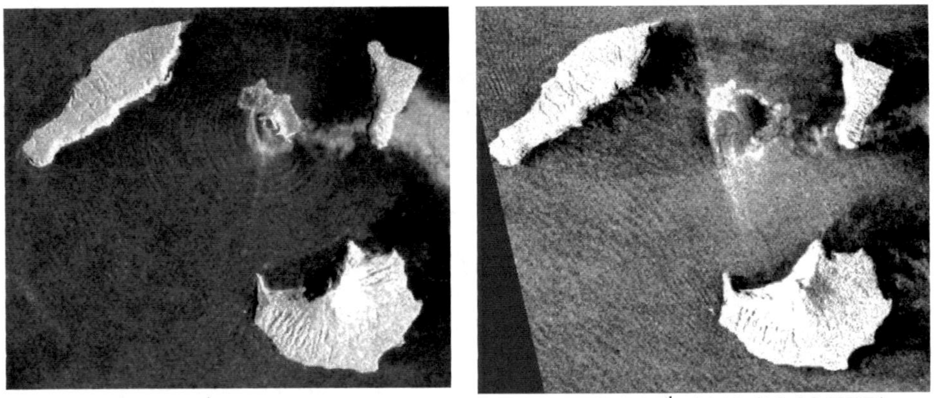

December 23rd, 2018 (5.33 WIB) December 25th, 2018 (18.22 WIB)

Figure 3. Sentinel-1A image of the GAK area on the pre- (left) and post-failure (right) [12].

Table 1. Previous studies of December 22nd, 2018 Sunda Strait tsunami

Tsunami Dimensions	Other Remarks
	[7]
Lost volume of lank failure was ~0.1 km^3.	• GAK's internal structure controls the flank failure, and the regrowth restabilizes and controls the failure interval • Failure was likely caused by the combined over-steepening flank and internal-weakening alteration
	[8]
Numerical simulation models the geometry and magnitude of the landslide source for 0.22–0.3 km^3 Initial wave elevation in 50 m height traveled to southwest direction	• Suggested landslide mechanism was as a single, rapid, en-masse failure, rather than a piecemeal process • It still remains unclear on what role the eruption had to the lateral collapse, as there was no anomalous activity preceding the collapse • It is plausible that the collapse was associated with flank instability from the GAK's long-term growth and geological characteristic
	[10]
Partial flank collapse of 0.28 Km^3 may trigger an initial wave for a height of 43 meters, towards south-westward direction	• Wave of 15–30 m height can expose GAK's adjacent coastlines (Sertung, Panjang, Rakata) in <1 minute. • On Java western coasts, the wave may reach in several tens of minutes for 1.2 (Sumur)–3.4 (Labuhan) m amplitude. • And on the Sumatera southern coast, the wave amplitude may reach 0.3 (Bandar Lampung)–2.7 (Kalianda) m after >1 hour

	• Rapid detection observatory on collapsed volcano and coastal alert system is needed
[11]	
Generally, the flow depths were found higher in Banten than Lampung. Maximum (surveyed) flow depth was 6.6 m, located in Cipenyu Beach, Tanjung Lesung, Banten	• Inferred maximum velocity was around 4.37 ms^{-1}, based on transported boulder • Tsunami arrival times, analysed from four tidal-gauge stations were considered short notice, and are slightly faster in Banten • Building damages were found more severe in Banten Province with vast majority in confined masonry brick construction
[13]	
Maximum (surveyed) runup for 13.5 m and maximum inundation distance for 330 m	• Seafloor bathymetry change for 50 m thick, from the collapsed volcaniclastic materials • Tsunami dominant period is considered short nature, ranging for 6.6–7.4 minutes
[14]	
Source (initial) model of wave dimensions: the length of 1.5–2.5 km and amplitude of 100–150 m	Potential energy ≡ 6.0–6.1 EQ magnitude (Richter scale).

2. Data and methods

This post-disaster rapid assessment prioritizes the investigation on specific elements, such as: (1) tsunami wave dimensions (inundation distance and run-up elevation), (2) landscape conditions (pre- and post-tsunami), and (3) affected land use. Field survey was conducted to observe those elements on 24th–25th December 2018, and was equipped with laser distometers, drones, and interview forms.

Firstly, the tsunami wave dimensions were directly measured on the field using laser distometers. Secondly, the aerial survey with drones processed the ortho-mosaic photos for analyzing the tsunami dimensions, landscape damages and landcover changes. These procedures aim to examine the physical damages (coastal vegetation, buildings, and public facilities). Lastly, the local interview was conducted to strengthen the collected data. Taking everything into consideration, all the information gathered was used to support the next phase of recommendations for mitigation strategies and sustainable planning (e.g., shelter location, evacuation route, structural mitigation, social empowerment, and spatial planning)

2.1. Ground investigation for tsunami wave dimensions

Field investigation was firstly conducted to rapidly observe the post-tsunami conditions. This ground observation aimed to measure tsunami dimensions and examine tsunami impact on landscape and land use. Field investigation on tsunami wave dimensions was conducted by observing the traces of tsunami, based on several parameters: flow depth, tsunami elevation, run-up elevation, and inundation distance. Figure 4 shows that: (1) flow depth is measured from the ground elevation (in meter), and observed through the watermarks traces (debris on walls, branches, damages), (2) tsunami elevation (wave height or water level) is measured vertically from the mean sea level to tsunami's maximum height, (3) inundation distance is measured horizontally from the shoreline (in meter) towards the inundated area limit, and (4) run-up elevation is measured at the ground level of the inundated area limit [15].

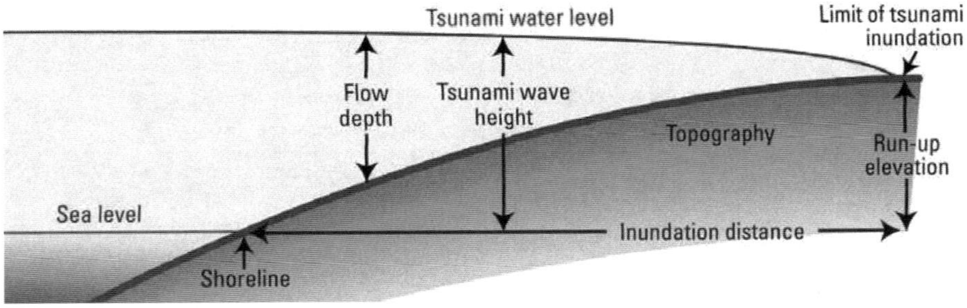

Figure 4. Terminology of tsunami [15].

2.2. Aerial investigation for post-tsunami landscape conditions

The aerial photogrammetry survey used a DJI Phantom4 pro UAV, equipped with a SONY FC-6310 8.8 camera. The flight path was as high as 80 m. Figure 3 shows the drone's flying tracks for investigating the post-tsunami landscape conditions. The drone-captured mosaic photos were processed to capture the overviews of tsunami dimensions and impacts on landscape and land use (e.g., the destruction of vegetation, buildings, and infrastructures of road and communication networks). The post-disaster rapid assessment on landscape damage was also investigated through its land system type.

2.3. Field investigation for the impacted land uses

The post-tsunami condition on impacted land use was investigated in terms of their damage condition, building construction, and neighbouring land cover. While the pre-tsunami condition was retrieved from the Google Maps Satellite Imageries. Referring to the regulated coastal land use, based on the Presidential Regulation 51/2016, the coastal buffer zone has a width of 100 m at the minimum, from the highest tide point towards the land. From this benchmark, seven (7) sample points within the five (5) sample sites were investigated further, that represent the tsunami-impacted land use (see Figure 2a and 2b).

In order to investigate the tsunami impact on land uses, the seven (7) investigation points were divided into two buffer zones; (1) first coastal buffer (0–50 m wide, from the coastline), and (2) second coastal buffer (50–100 m wide towards the land). It aims to discern the relation between land use location and its damage condition, where on the first zone, land use was directly affected by the tsunami, while on the second zone, damaged land use may also be affected by the debris overflow. In this coastal buffering, four (4) samples were located on the first buffer and three (3) samples were located on the second buffer.

3. Results and discussions

3.1. Tsunami dimensions (flow depth and inundation distance)

Table 2 shows the result of field measurement on five (5) sample sites for tsunami wave dimensions. And Figure 5 illustrates the investigation analysis on five sample sites. The site-5 in Tanjung Jaya Village, Panimbang District, Pandeglang Regency had the highest fatality, as subjected to the greatest trace of tsunami dimensions. It had the highest tsunami wave (12 m height), the furthest inundated areas (>300 m), and the deepest flow depth (6 m) (see Figure 5). Here, the watermarks and debris were found on building's top pillars and walls.

Table 2. Field measurement on tsunami wave dimensions (fow depth, inundation distance, & tsunami elevation)

Site	Locations	Coordinates	Date	Local Time	Weather	Watermarks	Dimensions
1	Karangsuraga – Cinangka, Banten	6.1761127°S 105.84667°E	Dec 24	16:10	Raining	Broken plants, wall watermarks, hanging debris	FD[a]: 1 m ID[b]: 127.8 m TE[c]: 3.7 m
2	Sukarame – Carita, Pandelang	6.262292°S 105.829239°E	Dec 24	16:10	Cloudy & raining	Debris branches, hanging on plants	FD: 4 m ID: 79.97 m TE: 4.9 m
3	Sukarame – Carita, Pandelang	6.2774615°S 105.82609°E	Dec 25	10.00	Cloudy & raining	Debris of broken branches on electric poles	FD: 1.3 m ID: +250 m TE: 6.7 m
4	Sukarame – Carita, Pandelang	6.2774615°S 105.82609°E	Dec 25	10.00	Cloudy & raining	Debris of broken branches on electric poles	FD: 1.9 m ID: +250 m TE: 6.8 m
5	Tanjung Jaya – Panimbang, Pandeglang	6.479841°S 105.653786°E	Dec 25	15.00	Heavy rain	Debris on walls, dangling plants on building's top pillars	FD: 6 m ID: +300 m TE: 12 m

[a] Flow depth; [b] Inundation distance; [c] Tsunami elevation

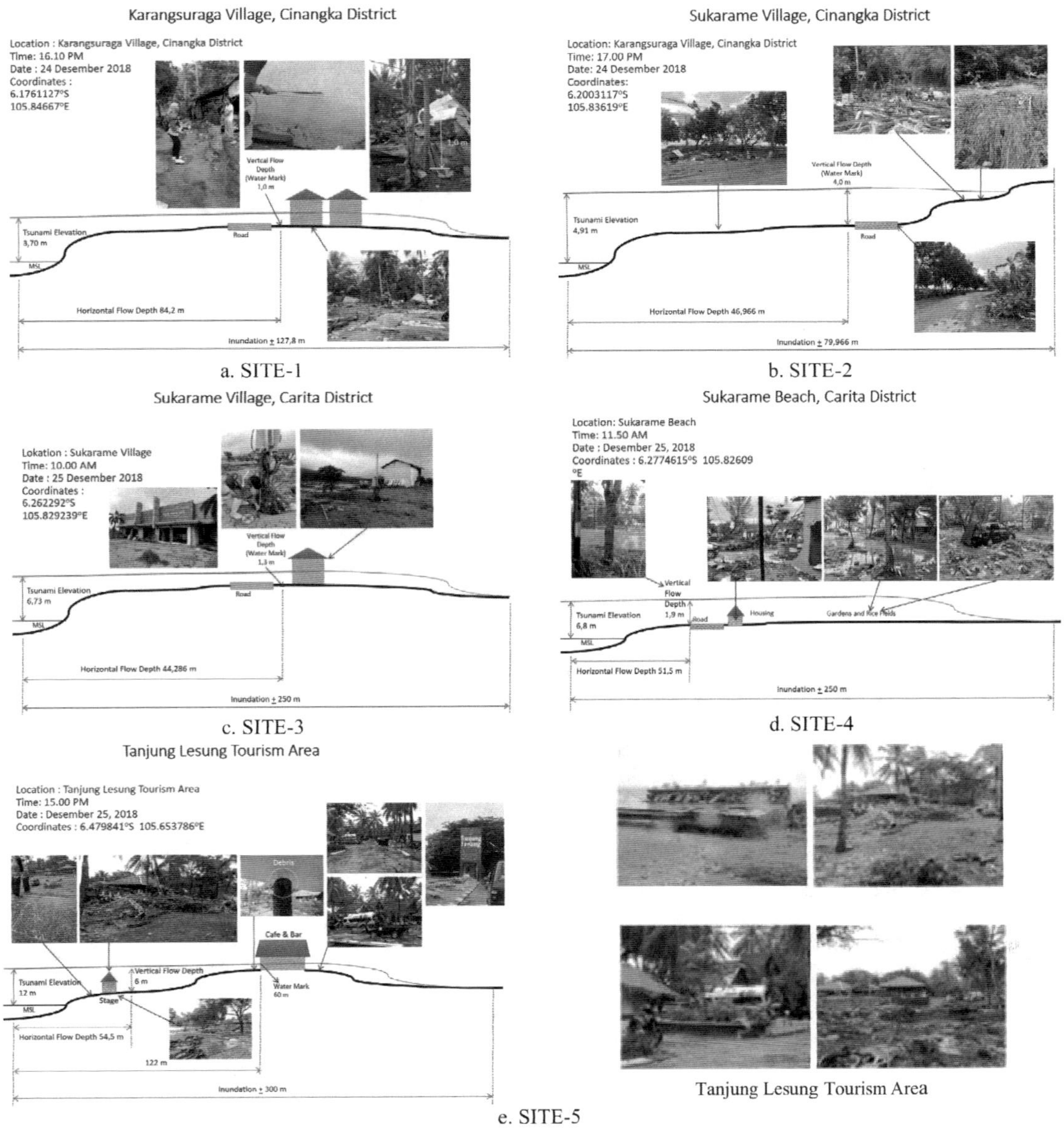

Figure 5. Field measurement for tsunami wave dimensions on 5 sample sites.

3.2. The post-tsunami landscape conditions

3.2.1. Results of photogrammetry survey. Table 3 features the selected trajectories from the ortho-mosaic photos on five (5) sample sites. The trajectories feature a more detailed landcover attribute from the common land use classes, to appropriately adapt on the ground's real situation. The land cover classes are used with as much compatibility as possible with the common land use classification system.

On the five (5) sample sites, there are nine (9) land cover types that can generally be grouped as barren land and built-up land. The open landscapes found on sample sites are the beach, shrub, and barren land. And the built-up landscapes are the scattered permanent settlement, dense settlement, lodging or villa, recreational site, public installation, and commercial service.

Table 3. Photogrammetry track lines for tsunami trace and its attributed landcover on 5 sample sites

SITE / TRACK	SITE-1	SITE-2	SITE-3	SITE-4	SITE-5
TRACK LINES	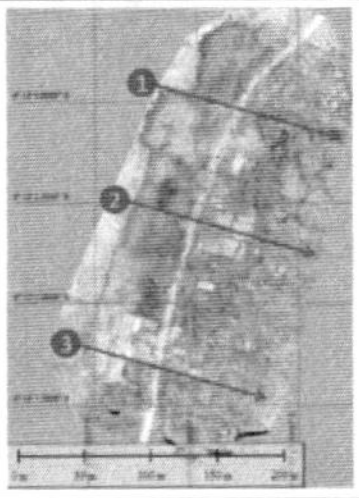	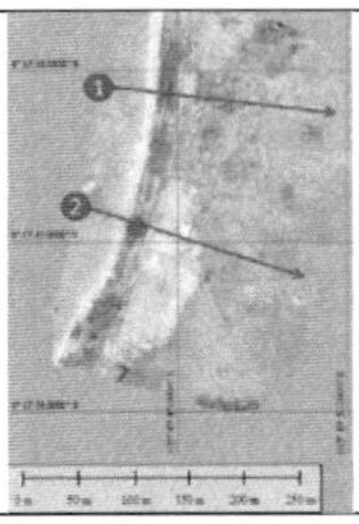	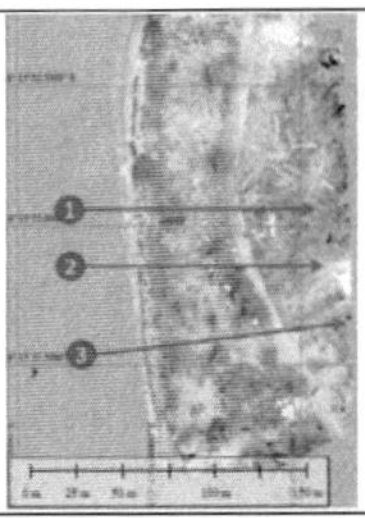	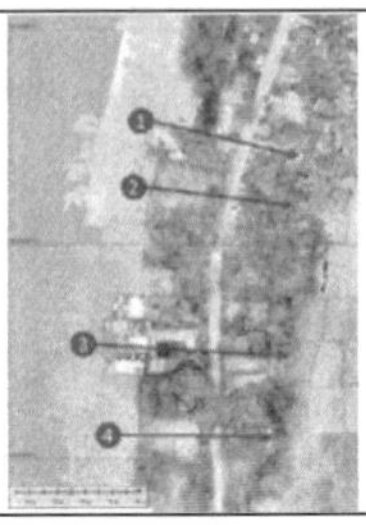	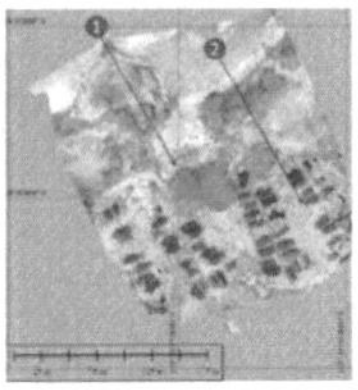
TRACK-1	B, S	B, S, C (tractor warehouses)	B, RS (containers café)	B, S, DS, RS	B, RS (swimming pool, restaurant)
TRACK-2	B, SPS	B, PI (communication tower)	B, S	B, S	B, L (Tanjung Lesung Beach Hotel)
TRACK-3	B, SPS, L (White Villa)	-	B, SPS	L, B, RS, DS	-
TRACK-4	-	-	-	B, S, BL, L (Villa Avinsa & Arista)	-

B = Beach, S = Shrubs and bushes; BL = Barren land; SPS = Scattered Permanent Settlement, DS = Dense Settlement, L = Lodging (Villa), RS = Recreational Site, PI = Public Installation, C = Commercials & Services

Table 4 shows the transect analysis of drone flying-track. It aims to analyze the relationship between the landscape condition and its tsunami dimension.

Table 4. Transect of tsunami trace on five (5) sample sites on 5 sample sites.

SITE-1

TRACK-1

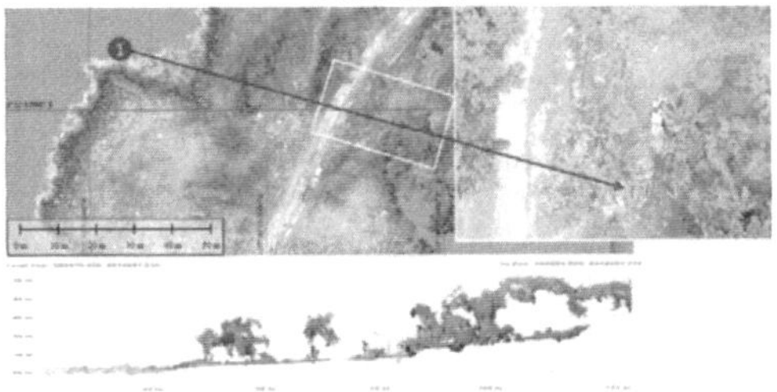

Tsunami trace of wood debris reached up to 60 m from the shoreline. No damage reported

TRACK-3

Tsunami washed up to 120 m towards the foothill. Villa Putih Lodgings and settlements were damaged

SITE-2

TRACK-1

Tsunami trace reached more than 100 m from the shoreline.

TRACK-2

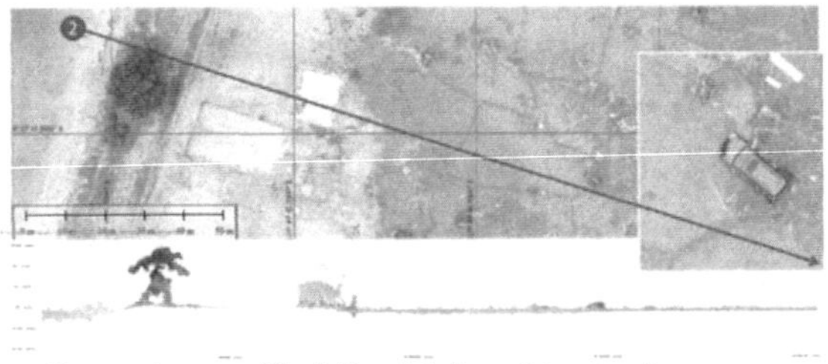

Tsunami trace of building debris and tractor showroom reached 100 m away from its warehouse

SITE-3

TRACK-1

Tsunami washed out the containers café eastward to the ricefield for 80 m away

TRACK-2

Coconut and tiliaceus trees held the containers café debris

SITE-4

TRACK-1

Dense settlements suffered heavy physical damage.

TRACK-3

Heavily damaged lodgings (Villa Avinsa & Arista)

SITE-5

TRACK-1

TRACK-2

Tanjung Lesung Beach Hotel had the highest fatality,
that along with its topographical condition, a musical event was held when the tsunami hit.

3.2.2. Landscape susceptibility. Landscape constitutes several features such as vegetation, topography, and land system. To look further on landscape susceptibility, Figure 6 briefly conveys the landscape vulnerability within the 100 m coastline-buffer, along with the its land system type [16]. Here, the susceptibility of the land system within five (5) sample sites were categorized into three classes (middle, middle-high, or high vulnerability), based on its destructed condition and land system type. Figure 6 shows that landscapes with BTG land system (an undulated beach with hilly plains on acid volcanic tuffs) have the lowest susceptibility compared to the three other landsystems (Kuranji and Pulau Rotan).

3.3. Tsunami impacts on land use

The post-disaster rapid assessment investigated seven (7) sample points that represent the tsunami-impacted land use. These sample points are located within the 100 m coastline buffer, based on the Presidential Regulation 51/2016. Table 5 and Table 6 below are focused on the pre- and post-disaster rapid assessment on the impacted land use. These tables cover several aspects on the types of building construction, surrounding land covers, and damage conditions, based on its buffered location from the coastline.

On the first coastline buffer zone (0–50 m from the coastline), the land uses are related to the economic activities, e.g., lodgings, showroom, food stalls, and cafes. Also, the buildings were mainly constructed permanently. There were also the small-scale businesses, where the locals build non-permanent building for their food enterprises (fish grill and café). This coastal zone experienced the worst destruction. The non-permanent structures were destroyed and washed away towards the land. While the permanent buildings with concrete construction suffered damage to their windows and doors that were directly facing directly to the sea.

On the second coastline buffer zone (>50 m from the coastline), the land uses are related to the local's settlement, bare land, and rice field. The settlement buildings were mainly constructed from bricks and woods. Here, houses that were located closest to the coastline and main road suffered worse destruction. The permanent settlement in Sambolo, Carita District, Pandeglang Regency suffered the worst damage. The highest fatality for 207 lives loss occurred at Tanjung Lesung Resort, Tanjung Jaya Village, Panimbang District, Pandeglang Regency. When tsunami hit, there was a music performance at the resort that highly attracted tourists. The second highest fatality for 50 lives loss occurred at Villa Stephanie, Sukarame Village, Carita District, Pandeglang Regency. This villa had an attractive tourist facility, built as projected structure on reclaimed land (used for swimming pool, meeting room, and jetty or pier). This structure was allegedly eluded the Presidential Regulation No. 51 of 2016, regarding the coastal buffer boundary.

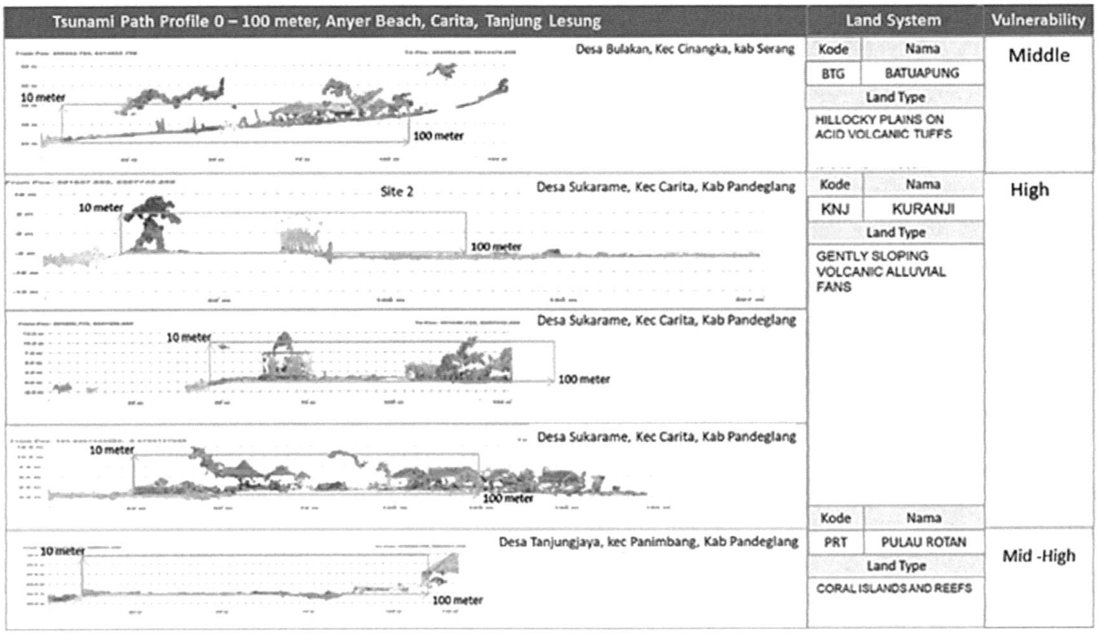

Figure 6. Land system susceptibility to tsunami hazard within 100 m coastal buffer.

3.4. Local interviews for rapid assessment on capacity

Local interview was conducted with 15 people, to identify the existing coping capacity. The result finds that there were inadequate evacuation route signs and tsunami warning systems witin the five (5) sampling sites at the time of the site visit.

In Karang Suraga Village, Cinangka District, Serang Regency, there were no evacuation route signs and shelters. So that people had to use the commonly used path by the local people to go back to their homes, up in the hill.

There was also the absence of a tsunami warning system. In Tanjung Jaya Village, Panimbang District, Pandeglang Regency, this absent warning entails to the highest fatality of 207 lives loss at Tanjung Lesung Resort. There, a live music performance was held when the tsunami hit the area.

In the tsunami event, the community did not know that the incoming water was a tsunami wave until they were informed by volunteers and outside informants. The concept of a typical tsunami that most people know was preceded by a strong shake (an earthquake) and the receding water. Meanwhile, with the Sunda Strait tsunami on 24th–25th December 2018, neither earthquake nor receding water had been experienced by the locals.

Related to coastal defenses, hard structural defence (embankment and tetrapod) was not available for the coastal and inland protection. However, the pre-existing soft structural defence was found in several areas. The natural barriers (vegetations and tsunami-ready landsystem) could retain the building debris. For examples, in site-1, where Batuapung Landsystem (BTG) could retain for its undulating landscape with a hilly plain structure, and also in site-3, where vegetations were able to retain the debris of container cafés.

Table 5. Pre- and post-damage conditions on the first coastal buffer (0 – 50 m wide from coastline)

PRE-TSUNAMI	**POST-TSUNAMI**
White Villa & non-permanent stalls (Karangsuraga Village, Cinangka District, Serang Regency)	
• White Villa was located on the beach, facing the open sea • It has a permanent, 2-storey building • Surrounded by the beach and grasses (shade trees remained rarely)	• White Villa suffered damages of broken doors and windows (as well as its supported concrete) that were facing the open sea • Two fatalities at the villa • A group of stalls, especially ones that faced the sea, was destroyed and pushed for >50 m towards the land
 The White Villa	 Non-Permanent Stall Buildings
Non-permanent stalls (Karangsuraga Village, Cinangka District, Serang Regency)	
• 35 non-permanent stalls of grilled-fish and the open space of car park area • Across the road was a hill of local settlements that was also the location of local knowledge on tsunami evacuation zone	• The stalls were destroyed, and its debris hit the hill on the eastern-side of the road • People on the stalls got injured
 The open beach, non-permanent stalls ruins	 Rubble piles of the stalls across the road
Tractor Showroom & Container Café (Sukarame Village, Carita District, Pandeglang Regency)	
• Both were located on the beach, facing directly to the open sea • Tractor showroom was consisted of two buildings in 20 m apart, built from steel construction and aluminium walls, respectively • Container Cafés were made of 2-stacked containers. There were about 5 container cafes • Surrounded by the beach and grasses	• Steel-constructed tractor showroom was destroyed and the aluminium showroom was damaged (but still standing). The displayed tractors and cars on the steel building were also destroyed and crashed the aluminium building • Container cafés were scattered towards the rice field, across the road • No casualty was reported

<table>
<tr><td>

Former locations of Tractor Showroom and Container Cafes</td><td>
Wreckage of the Container Cafés

Wreckage of destroyed showroom, tractors and cars</td></tr>
<tr><td colspan="2">Villa Stephanie (Sukarame Village, Carita District, Pandeglang Regency)</td></tr>
<tr><td>

- The villa area was located between the coastline and national road
- It was consisted of two-rows permanent lodgings complex
- It also has extended structures on the reclaimed land for swimming pool, meeting room, and pier or jetty.

The lodgings complex of Villa Stephanie

Projected structures on reclaimed land (pool, meeting room, and jetty)</td><td>

- The closest lodgings to the coastline suffered the worse damaged for its walls, windows, and doors.
- The additional structures that projected to the sea (swimming pool, meeting room, and jetty) were destroyed
- There was the second highest fatality of 55 lives loss from the visitors

Post-tsunami condition of Villa Stephanie's main lodgings complex</td></tr>
</table>

Table 6. Pre- and post-damage conditions on the second coastal buffer (> 50 m wide)

<table>
<tr><th>PRE-TSUNAMI</th><th>POST-TSUNAMI</th></tr>
<tr><td colspan="2">Tourism Police Office (Sukarame Village, Carita District, Pandeglang Regency)</td></tr>
<tr><td>

- Police station faced the open beach that was covered with grasses
- It was also surrounded by a bareland

Tourism Police Office</td><td>

- Concrete fence was damaged and hit the front part of the building.
- Main damage to the office building occurred on the doors and glass windows that were facing the sea.

</td></tr>
</table>

Open beach in front of Police Office

Damage condition of the police station

Villa Arista & Villa Avinsa (Sukarame Village, Carita District, Pandeglang Regency)

- Both vilas were permanent buildings facing the open sea.
- Material constructions of bricks and woods (Villa Arista), and steel frame (Villa Avinsa).
- Villa Arista was surrounded by bareland
- It also had the additional buildings built on the reclaimed land

- Villa Arista was damaged at the front and the back of the building, directly hit by the tsunami force and its runoff.
- Villa Avinsa suffered destruction, and its steel debris was pushed for 50 meters away towards the land.

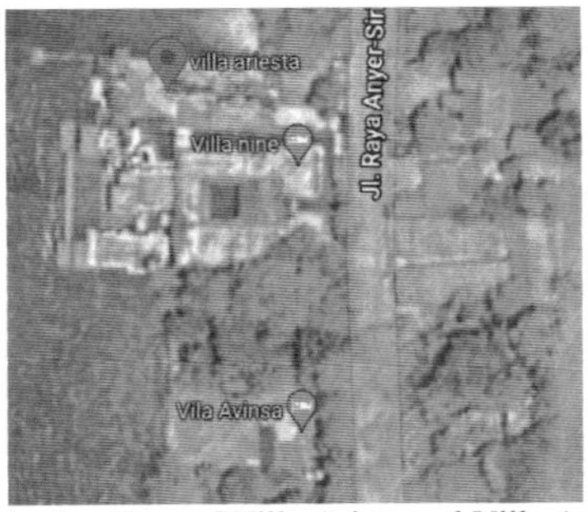

Former locations of Villa Arista and Villa Avinsa

Post-tsunami Villa Arista

Former location of Villa Avinsa

Sambolo Residential Area (Sukarame Village, Carita District, Pandeglang Regency)

- Sambolo residentials were located ± 75 m from the coastline, on the west side of the national road
- Densely occupied by permanent buildings, mainly constructed of bricks and woods.

- This was the worst damaged area in Carita District
- Buildings that were located closer to the coastline and main road suffered the worst damage.

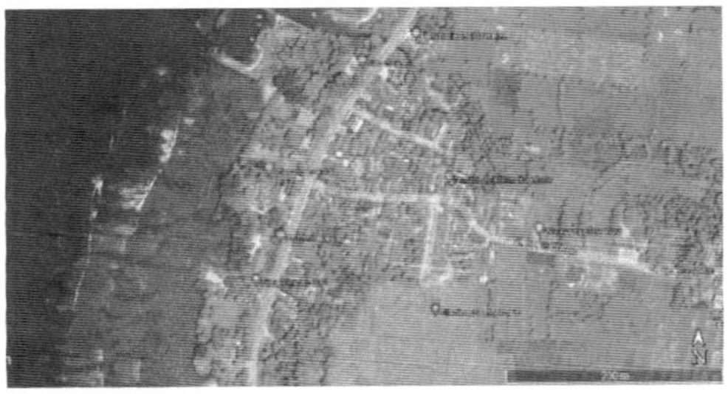

Post-tsunami condition of Sambolo Residential Area

4. Conclusion and recommendation

4.1. Conclusion

The main findings are concluded as follows:

- The largest tsunami wave dimension (12 m height, inundated areas >300 m, and 6 m flow depth) was occurred on the sample site-5 (in Tanjung Jaya Village, Panimbang District, Pandeglang Regency).
- Photogrammetry survey helped to investigate the tsunami exposures on landscape. The open landscape (e.g., open beach, shrub, and barren land) aggravated the tsunami impact. While the

coastal landscape that has either perennial vegetation or hilly topography, appeared having a reduced landscape damage.

- Damaged constructions were mostly located on the first coastal buffer (10–50 m from the coastline). Non-permanent constructions were destroyed and its debris was pushed towards inland, while the permanent buildings had damages of broken windows and doors. On the second buffer (50–100 m), permanent settlements in Sambolo (Carita – Pandeglang) suffered the worst damage, especially on the buildings that were located closer to the coastline, main road, and situated on open landscape.
- Several oversights on coastal buffer regulation and tsunami warning systems conclude that:
  - Some lodging's facilities (e.g., meeting room, pools, jetty) were allegedly violated the coastal buffer regulation, as they were built on a reclaimed land without any coastal defense.
  - Coastal areas without any vegetations also had no coastal-retaining structures (e.g., embankments and tripod) as the physical defense
  - Inadequate evacuation route signs and tsunami warning systems.

4.2. Recommendation

These following recommendations showcase several measures pertaining to the reduction of tsunami risk within the sample locations:

- Reinforcement of coastal buffer into the spatial planning, based on the Presidential Regulation No. 51 of 2016. Accordingly, several measures shall be conducted, i.e.: coastal risk mapping, spatial planning review, coastal-zoning and land use planning,
- Evaluation on the existing landuse locations (settlements, tourist facilities, and vital infrastructures),
- Enforcement on tsunami-prone buildings, e.g., feasibility on construction technology and its retrofication plans,
- Development of coastal defense structures (e.g., embankments, tripot, revetment), as well as local tree-planting buffer along the shoreline,
- Review on the existing tsunami evacuation routes and tsunami Temporary Evacuation Shelter (TES),
- Strengthened the community preparedness on hazards knowledge and its emergency measures,
- Intervention measures on tsunami-ready landscape, such as: landscape engineering, landscape management and utilization, landscape audit and certification, as well as social engineering.
- Monitoring and law enforcement succeedingly

Acknowledgments

The authors are all affiliated to The National Research and Innovation Agency (BRIN). This research was funded by the then Agency for the Assessment and Application of Technology (BPPT), and administratively supported by BRIN.

References

[1] BNPB 2019 Tsunami Selat Sunda Retrieved from: https://bnpb.go.id/berita/tsunami-selat-sunda
[2] AHA Centre 2018 AHA Centre Flash Update-Sunda Strait Tsunami Retrieved from: https://ahacentre.org/flash-updates/
[3] UNISDR 2017 National Disaster Risk Assessment: Governance System, Methodologies, and Use of Results
[4] GFDRR 2013 Post-disaster needs assessments vol. A Retrieved from: https://www.gfdrr.org/en/publication/post-disaster-needs-assessments-guidelines-volume-2013

[5] Gunasekera R *et al* 2018 Methodology Note on the Global Rapid post-disaster Damage Estimation (GRADE) approach Retrieved from: https://www.gfdrr.org/sites/default/files/publication/DRAS_web_04172018.pdf
[6] BNPB 2021 Tsunami Selat Sunda *Public Service Announcement* (https://tv.bnpb.go.id/front/article/detail/video.php/read/article10483123122021),
[7] Williams R, Rowley P, and Garthwaite M C 2019 Reconstructing the Anak Krakatau flank collapse that caused the December 2018 Indonesian tsunami *Geology* **47** 973–976
[8] Grilli S T *et al* 2019 Modelling of the tsunami from the December 22nd, 2018 lateral collapse of Anak Krakatau volcano in the Sunda Straits, Indonesia *Scientific Reports* **9**(1) pp 1–13
[9] Kurnio, H 2020 Tsunami Anak Krakatau 22 Desember 2018 sebagai Lesson Learned bagi Indonesia maupun Mancanegara Kebencanaan Geologi Kelautan dari Aktivitas Gunung Api Bawah Laut Retrived from: https://litbang.esdm.go.id/news-center/arsip-berita/tsunami-anak-krakatau-22-desember-2018-sebagai-lesson-learned-bagi-indonesia-maupun-mancanegara-kebencanaan-geologi-kelautan-dari-aktivitas-gunung-api-bawah-laut)
[10] Giachetti T et al 2012 Tsunami hazard related to a flank collapse of Anak Krakatau Volcano, Sunda Strait, Indonesia *Geological Society, Special Publication* **361** 79–90
[11] Syamsidik *et al* 2020 The 22nd December 2018 Mount Anak Krakatau volcanogenic tsunami on Sunda Strait coasts, Indonesia: Tsunami and damage characteristics. Natural Hazards and Earth System Sciences **20** 549–565
[12] PTPSW 2018 Analisis Deformasi Gunung Anak Krakatau dengan Satelit Radar Sentinel Retrieved from: https://twitter.com/PTPSW_TISDA/status/1077812975027646464
[13] Muhari A *et al* 2019 The December 2018 Anak Krakatau Volcano Tsunami as Inferred from Post-Tsunami Field Surveys and Spectral Analysis *Pure and Applied Geophysics* **176** 5219–5233
[14] Heidarzadeh M *et al* 2019 Numerical modeling of the subaerial landslide source of the 22 December 2018 Anak Krakatoa volcanic tsunami, Indonesia *Ocean Engineering* **195**
[15] USGS 2008 Tsunami Terms (https://www.usgs.gov/media/images/tsunami-terms)
[16] RePPProT (Regional Physical Planning Program for Transmigration) 1980 Indonesia Land System Map Scale 1:250,000

Study of Ground Motion Models Selection to the Concrete Frame Building Damage Probability using the 2006 Yogyakarta Earthquake Scenario

Y Muntafi[1,2], N Nojima[3] and I Ikramullah[1]
[1]Department of Civil Engineering, Faculty of Civil Engineering and Planning, Universitas Islam Indonesia, Yogyakarta, Indonesia
[2]Disaster Risk Reduction Center (DiRReC), Universitas Islam Indonesia, Indonesia
[3]Department of Civil Engineering, Faculty of Engineering, Gifu University, Japan

Email: yunalia@uii.ac.id

Abstract. Damage probability assessment using the ground motion prediction equation (GMPE) models is paramount in seismic risk analysis. Several GMPE models have been proposed and developed worldwide. However, it has its parameters and gives different spectral acceleration results. The study aimed to investigate the impact of GMPE model selection on the estimation of building damage probability due to the scenario of Yogyakarta earthquake. In the beginning, we modeled the 3D-concrete moment frame building, followed by conducting pushover analysis to determine the building responses and calculating the building damage probability based on HAZUS method to generate fragility curve and damage probability matrix. Five GMPE models were employed in this study. The results revealed that for the modeled structure, the GMPE of Ambraseys et al. (1996) gave the highest value of the peak building response compared to the GMPE of Sadigh et al. (1997), Boore et al. (1997), Spudich et al. (1999) and Idriss (2002). The probability of moderate, extensive, and complete damage levels was also dominated by the GMPE of Ambraseys et al. (1996), which indicated that in this case, the use of this GMPE would produce the most severe damage probability value.

1. Introduction

Yogyakarta experienced a devastating earthquake with 6.3 Mw or 5.9 Mb in 2006, which caused many fatalities. Elnashai et al. [1] provided data that this earthquake killed more than 5,700 people and injured over 37,000 people. The disaster also destroyed more than 156,000 residential buildings and other structures, and cost over $3 billion in financial losses. Low quality of material, improper construction installation, and inadequate construction detailing have a very high effect on building damage. This earthquake-affected area is along the fault line of the Opak River, which is a densely populated area. The damaged area extended from the Parangtritis area stretching northeast to the Bantul area and continuing northward to the Klaten region. [2]. The Yogyakarta's tectonic setting is quite similar to that of Semarang, Indonesia, which has an active fault in the south side of the province. The focal mechanism data from USGS shows the 2006 Yogyakarta earthquake was triggered by a strike-slip fault earthquake with the position N 231° E, Dip 87°, and slip 3°. The seismotectonic analysis showed that this fault is a horizontal fault to the left (sinistral strike-slip fault) [3]. Based on this experience, the vulnerability and damage probability assessment are essential to be further studied.

Perrault et al. [4] stated that the building vulnerability, both residential and special buildings can be investigated by performing fragility curves. These curves obtained by corresponding damage probability to a peak ground acceleration (PGA) or spectral displacement (Sd) as a measure of ground motion Intensity (IM). The essential parameters involved in generating these curves are model building types, seismic design levels, the IMs, and the damage parameters. All of them are considered as being the parameters of engineering demand, accommodating the best representation of specific ground motion applied. The building response in relationship with the ground motion intensity can be associated to the estimation of building damage when correlated with the thresholds of damage level. Dewobroto [5] illustrated that besides the strength level of a building, the deformation state and structural performance also have a strong influence on the building`s safety. Besides, the structure performance will be significantly affected by the probability value of structural damage. It is paramount to take into account the effect of ground motion in building structures designed in earthquake-prone areas.

There are currently many attenuation relationships or ground motion prediction equations (GMPEs) models developed and proposed by researchers around the world. It is principally applicable for the region to measure the Spectral acceleration (Sa) and Peak Ground Acceleration (PGA) values. However, it has its parameters and gives different estimation results. Some researchers discussed the ground motion selection related to the local condition [6,7]. The previous study [8] used the GMPE of Boore et al. (1997) for assessing the damage probability of Hospital as critical facilities in Yogyakarta, which was calculated based on the magnitude level, the closest distance to the rupture plane surface projection, shear wave velocity. It is necessary to study the effects of various GMPEs formulations on building damage probability values. This study aimed to find out the effect of selected GMPE models to the building damage probability by determining the peak building response, generating the fragility curve, and estimating the damage probability matrix by using a case study of a concrete moment frame building in Yogyakarta.

2. Earthquake and building structure

A variety of potential signals have been investigated around the world, but it is still difficult to reliably predict where and when an earthquake will occur. Earthquake waves have unpredictable magnitude, direction, and intensity according to time (time-varying). The vibrations are random so that the structure not only withstands the static load (dead load and live load) but also the dynamic response to the structure as a function of time.

2.1. Earthquake ground motion

Strong and moderate ground motions during an earthquake can potentially damage the structure above. In many cases, ground motions, including ground displacement, ground velocity, and ground acceleration, are amplified when transferred through a structure. All of them may trigger forces and displacements that becomes a threat to a structure. The earthquake ground motion can be combined with a targeted performance level in order to generate a performance objective. ATC-40 [9] stated the simple representation of the earthquake ground motion is defining a shaking level correlated to a given probability of occurrence (an approachment of probabilistic method). Besides, it also can be expressed by using a deterministic approach that used the highest level of shaking expected from a specified magnitude single event on a particular fault source. It is essential to express the ground motion level concerning engineering characteristics for building design applications. To make a simple illustration in building structural analysis, an external lateral force applied to the structure with an incrementally increasing pattern can represent the ground motion, as presented in Figure 1.

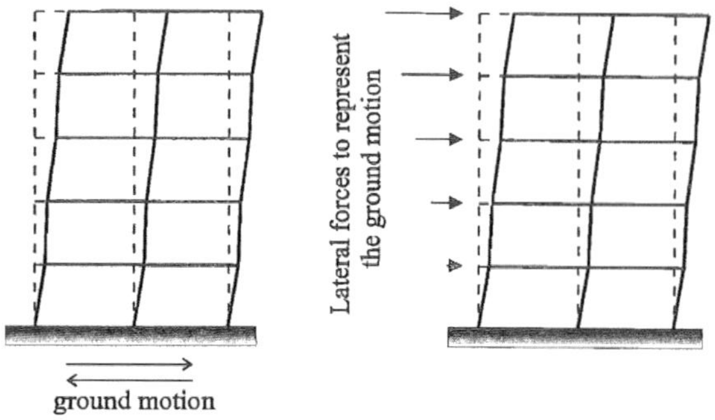

Figure 1. The application of increasing external lateral forces to represent the ground motion excitations [10]

2.2. Structural response to earthquake

The earthquake as a dynamic load in the form of wave radiation from its source is then spread in all directions with the soil and rocks properties to reach the land surface causing vibrations. The foundation as the primary support of a building will receive earthquake vibrations then proceed it into the upper structure resulting in a building sway mechanism as the inertia forces. The earthquake vibrations will be represented as the cyclic lateral loads (F_i), which may produce the roof displacement (Δroof). A simple illustration of the structural response due to an earthquake is presented in Figure 2.

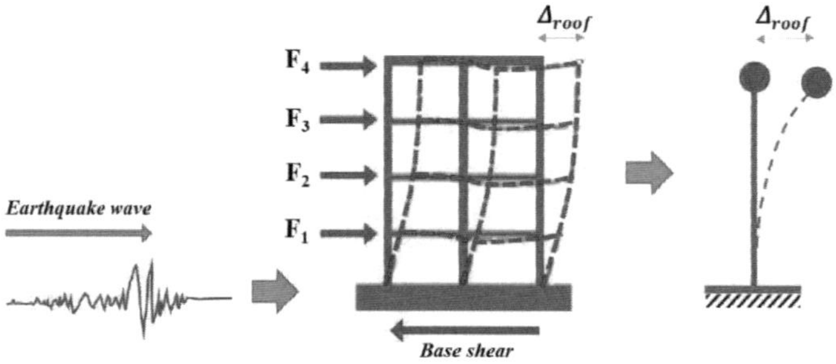

Figure 2. Structural response due to earthquake and its conversion into an equivalent SDF system

Figure 2 shows an open frame structure that idealizes the corresponding single degree of freedom (SDF) system and exhibits a nonlinear reaction to an earthquake ground motion. According to [11], a detailed nonlinear structural model that is a representation of an equivalent SDF system can be used to map the real global behavior of a detailed structural model.

2.3. Ground motion prediction models

The ground maximum acceleration value is an important parameter both in earthquake-resistant building design and building damage assessment. Thus, to get the value, we can use the ground motion attenuation, which is represented and modeled using the empirical ground-motion prediction equation (GMPE). Several GMPE models or attenuation relationships have previously been developed for many countries and regions. These equations are related to the site/source parameters, in which the majority of them are created using magnitude distributions, distances, and source mechanisms. This study used the scenario of the 2006 Yogyakarta earthquake. Some attenuation relationships which are related to the characteristics of the May 27, 2006, earthquake in Yogyakarta are the GMPE models proposed by Ambraseys et al. [12], Sadigh et al. [13], Boore et al. [14], Spudich et al. [15], and Idriss [16]. The equations in detail are as follows:

1) Ambraseys et al. (1996)

Ambraseys et al. (1996) developed an equation based on shallow earthquake data (h < 30km) that occurred in Europe and adjacent regions. The equation is recommended to use in the magnitude range of 4-7.9Ms and source distances up to 212 km. The attenuation equation provided by Ambraseys et al. (1996) is presented in the following formula:

$$Log\ (Y) = C_1 + C_2M + C_4 log\ (r) + C_A S_A + C_S S_S \quad (1)$$

$$r = \sqrt{d^2 + h^2}$$

where Y is spectral acceleration (g); M is magnitude (Ms); d is the shortest distance (km); $S_{A,S}$ is the site class dummy variable; h is constants which is determined together with C_1, C_2, C_4, C_A, and C_S as coefficients for spectral ordinates.

2) Sadigh et al. (1997)

The formulation of ground motion presented by Sadigh et al. (1997) is for spectral response accelerations and peak acceleration from shallow crustal earthquakes based on California earthquakes strong motion data. The relationship is provided for reverse-faulting and strike-slip earthquakes, with M4 to 8-moment magnitude range, and the distance is not more than 100 km. The attenuation equation is as follows:

$$Ln\ Y = C_1 + C_2\ M + C_3 \ln\left[R_{rup} + C_4^{eC_5M}\right] + C_6 + C_7\ (8.5\text{-}M)^{2.5} \quad (2)$$

where Y is spectral acceleration (g); C_1 = -2.17 for strike-slip, -1.92 for reverse and thrust earthquakes; C_2= 1.0; C_3 = 1.70; C_4 = 2.1863 for M<6.5 and C_4 = 3825 for M>6.5; R_{rup} = shortest distance to rupture surface (km), C_6 is determined together with C_7 based on earthquake source mechanism.

3) Boore et al. (1997)

The ground motion model proposed by Boore et al. (1997) is based on the shallow earthquake data that occurred in Western North America with a type of shallow earthquake (the fault rupture lies mainly more than 20km depth). The empirical attenuation formula is shown in equation (3).

$$Ln\ Y = b_1 + b_2(M\text{-}6) + b_3(M\text{-}6)^2 + b_5 \ln(r) + b_v \ln\frac{V_s}{V_A} \quad (3)$$

$$r = \sqrt{r_{jb}^2 + h^2}$$

$$b1 = \begin{cases} b_{1SS} \rightarrow \text{for strike slip earthquakes} \\ b_{1RV} \rightarrow \text{for reverse slip earthquakes} \\ b_{1ALL} \rightarrow \text{if mechanism is not specified} \end{cases}$$

where r_{jb} is the horizontal distance closest to the rupture plane surface projection (km), Y is the peak horizontal acceleration (g), M is the moment magnitude (M_w), and the pseudo acceleration response spectra are measured using the coefficients b2, b3, and b5, where Vs is the average shear-wave velocity to 30m of top soil layer (m/sec). This relationship uses the Joyner-Boore distance effect (r_{jb}), the moment magnitude (M_w), epicentral distance (r), and several types of faulting parameters (strike-slip, reverse-slip, and unspecified). The equation is to be used magnitude 5.5-7.5 Mw and distance $\leq$ 80km.

4) Spudich et al. (1999)

The GMPE developed by Spudich et al. (1999) is based on worldwide earthquake data. The equation may be used in the moment magnitude range of the 5.0 – 7.7 and the distance range of 0-100km for extensional regime ground motions. The GMPE of Spudich is presented below:

$$Log_{10}\ (Z) = b_1 + b_2(M\text{-}6) + b_3(M\text{-}6)^2 + b_5 log_{10} D + b_6 r \quad (4)$$

$$D = \sqrt{r_{jb}^2 + h^2}$$

where Z is pseudo acceleration response or peak horizontal acceleration (g); M represents moment magnitude (Mw); r_{jb} is closest horizontal distance to the rupture plane surface projection (km); r is dummy variable for the site class (0: rock site, 1: soil site); b_1, b_2, b_3, b_5, and b_6 are coefficients for velocity response spectra and peak ground acceleration.

5) Idriss (2002)

Idriss (2002) presented a relatively simple attenuation equation by taking into account the parameters of magnitude (Mw), distance (R), and earthquake mechanism (reverse, oblique, and other fault types). The equation of ground motion is as follows:

$$Ln\ Y = (\alpha_1 + \alpha_2 M) - (\beta_1 + \beta_2 M)\ ln\ (R+10) + \phi\ F \quad (5)$$

where Y is peak horizontal acceleration or pseudo acceleration response (g); R is distance (km); α_1, α_2, β_1, β_2, and ϕ are coefficients influenced by the period T ; F = 1 for reverse and oblique faults and F = 0 for other types of faults.

3. Research method

This study used pushover analysis to generate the capacity curve of the existing building and then determine the value of peak building response obtained from the building capacity curve and response curve intersection based on each attenuation relationships or ground motion prediction equations. This study uses five GMPEs that are suitable for the characteristics of the Yogyakarta earthquake. The earthquake scenario used was the Yogyakarta May 27, 2006 earthquake, which was a shallow crustal earthquake with a strike-slip mechanism. After the peak building response is obtained, the next step is to create the fragility curve using the HAZUS parameter, which represents the building vulnerability in each damage state. The estimation of the building damage probability is also generated based on the HAZUS method [17] in each GMPE model. From the analysis results, a comparison of the use of each GMPE against the estimated damage probability of building can be obtained.

3.1. Non-linear static analysis

A simplified analytical procedure that is widely used to determine the pattern of building damage due to an earthquake is the nonlinear static analysis, which also common as pushover analysis. In this method, the structure subjected to incrementally increasing lateral forces until it reached the target displacement [18]. Pushover analysis is highly recommended to be used to determine the capacity of base shear and the lateral displacement or a building structure ductility (force-displacement relationship) [19-22]. Najam [11] described that the main concept of pushover analysis is giving the gradually escalating lateral forces pattern in each floor of an inelastic building structural model as a simplified illustration of earthquake-induced forces distribution. Base shear (V) and displacement of roof (Δ_{roof}) in respect to each other is presented as a capacity curve generated using pushover analysis. A full structure capacity curves can be illustrated to express the relationship of a nonlinear force-deformation as provided in Figure 3.

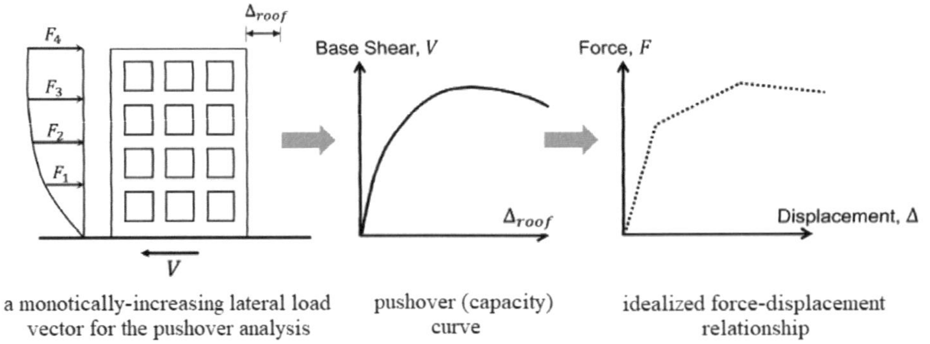

Figure 3. Illustration of the lateral load vector for a structural model in generating capacity curve [11]

3.2. Building capacity spectrum

The characteristics of a structure can be seen from the shape of its capacity curve, which shows a building's lateral resistance as a function of its typical lateral displacement [23]. In pushover analysis procedure of ATC-40, the capacity spectrum can be performed by doing step-by-step conversion of base shear (V_i) and roof displacement (Δ_{roof}) from the capacity curve to the first mode spectral coordinates. The conversion can be generated from the capacity spectrum in any point V_i, Δ_{roof} to the corresponding point S_{ai}, S_{di} using equations as follows [9]:

$$S_{ai} = \frac{V_i/W}{\alpha_1} \tag{6}$$

$$S_{di} = \frac{\Delta_{roof}}{(PF_1 x\, \emptyset_{1,roof})} \tag{7}$$

where α_1 and PF_1 stand for the first natural mode of the structure's coefficient of the modal mass and participation factors, respectively, and $\Phi_{1,roof}$ is the amplitude of the first mode's roof level. Figure 4 depicts the capacity curve that is formed into the spectrum of capacity in the Acceleration-Displacement Response Spectrum (ADRS) format.

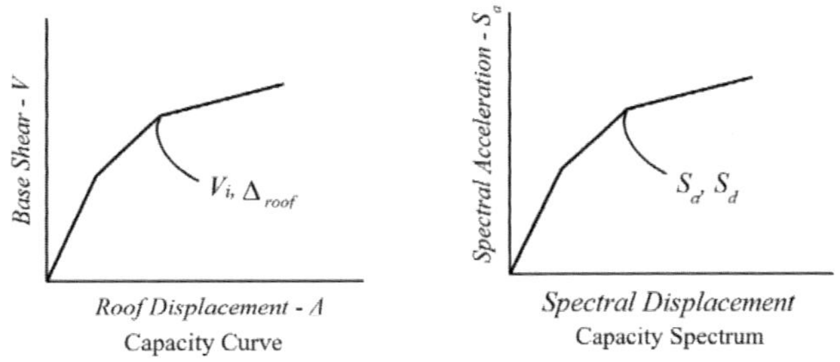

Figure 4. Capacity curve to capacity spectrum conversion in ADRS format (ATC-40, 1996)

3.3. Peak building response

A significant thing in measuring the damage probability of a building structure is determining the peak building response value, which is performed by a meeting point of capacity spectrum and response spectrum. The response spectrum of a building can be created based on the ground motion formulations related to the scenario of a particular earthquake. The peak building response will be used to generate the fragility curve for estimating the building vulnerability. The spectral acceleration result in each period obtained from ground motion analysis can be converted to the inelastic spectral displacement to create the response spectrum. The inelastic spectral displacement can be determined by using the equation as follows (NIBS, 2002) [24]:

$$S_d = 9.8 \cdot S_a \cdot T^2 \tag{8}$$

where S_a is the amplified spectral acceleration in g, T is the time period (seconds), and S_d is the spectral displacement (in).

3.4. Building damage probability

The Hazard-US (HAZUS) method is one of the procedures for analysing the probability of building damage due to earthquake disaster that produces fragility curves in four levels of damage (or performance) in detail. Fragility curve depicts the relationship between spectral displacement and the probability of reaching or exceeding the various level of building damage given peak building response due to earthquake (slight, moderate, extensive, and complete). Based on HAZUS-99 [24], in addition to the variability related to the damage levels, the median value of potential earth science hazards (PESH) is very influential in determining the shape of the building's fragility curve. The median value taken is that which corresponds to the damage level threshold. An example of typical fragility curves for each damage state used in HAZUS methodology is provided in Figure 5.

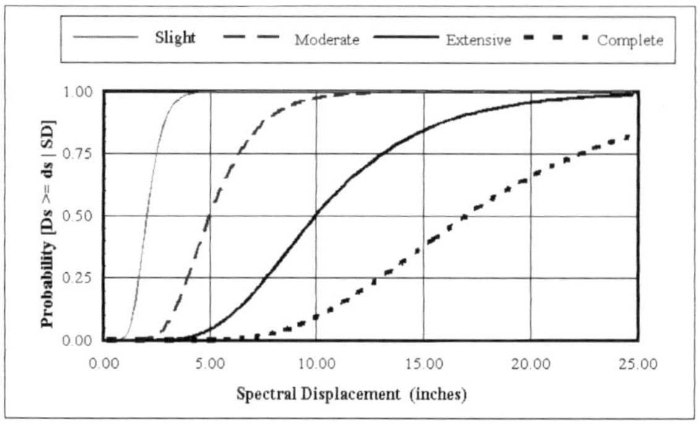

Figure 5. Example of typical fragility curves for each damage level based on HAZUS, 1999 [24]

Given the spectral displacement, S_d, the conditional probability of being in or exceeding a particular damage state, ds, is expressed by the equation below:

$$P[ds/S_d]=\Phi\left[\frac{1}{\beta_{ds}}\ln\left(\frac{S_d}{\bar{S}_{d.ds}}\right)\right] \quad (9)$$

$P[ds/S_d]$ represents the value of damage probability; S_d and $\bar{S}_{d.ds}$ is inelastic spectral displacement (inches) and the spectral displacement median value, respectively, at which the building reaches the damage state threshold; and Φ represents the function of standard normal cumulative distribution. Building resilience against earthquakes can be predicted from the form of a fragility curve as an effort to earthquakes mitigation. As part of the endeavour to mitigate earthquakes, building resilience against earthquakes can be predicted from the shape of a fragility curve.

4. Result and discussion

4.1. Structural modelling and earthquake load

This research object is a four-story reinforced concrete building that functions as a music and dance laboratory located in Yogyakarta, Indonesia, and is modeled as a 3-dimensional open frame building, as shown in Figure 6. Gravity loads, including dead load and live load were applied to the structure. Besides, the horizontal force (earthquake load) induced upon the structure based on the Indonesian Code on Seismic Resistant Design for Buildings (SNI 1726:2012) [25]. Table 1 displays the outcome of the distribution of the lateral seismic load in each floor.

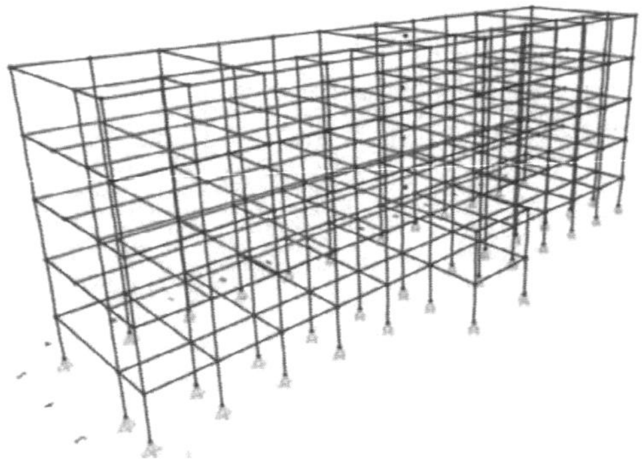

Figure 6. Open frame building in 3D-view

Table 1. Lateral earthquake loads distribution

Storey	Weight (kN)	C_{vx}	V (kN)	$F = C_{vx}.V$ (kN)
4	2381.5277	0.19	13103.68	2473.97
3	4830.1274	0.33		4375.94
2	4830.1274	0.24		3146.44
1	7492.9364	0.24		3107.33
Total				13103.68

C_{vx} is vertical distribution factor; F is lateral earthquake load (kN); and V is base shear force (kN).

4.2. Capacity and response curve

The capacity curve for x and y-direction of the building structure can be performed from pushover analysis results. The value of base shear (kN) and displacement (m) as a result of pushover analysis in both directions are shown in Figure 7. The discrete points depicted in Figure 7 illustrates the occurrence of main events in the lateral response history of building structure. The initial step shows there is a linear curve before reaching the yielding condition, followed by the nonlinear behavior of the structure. The maximum shear force that occurs in the x-direction is 2426.62kN with the displacement value of 0.235m while the y-direction is 2216.67kN of the maximum shear force with displacement value of 0.245m. The x-direction of pushover curve is more upright compared to the y-direction, which means that the x-direction portal is stronger in resisting the lateral force than the y-direction. Besides, the number of columns in the x-direction portal (long axis) is higher than the y-direction (short axis), thus affecting the structural stiffness and resulting in the displacement of the y-direction is higher than the x-direction.

In order to get the response curve, this study employed five ground motion prediction equations (GMPEs) modeled by Ambraseys et al (1996), Sadigh et al (1997), Boore et al (1997), Spudich et al (1997) and Idriss (2002) to determine the earthquake's motion's severity and create iso-acceleration graphs. The five GMPE were selected based on the similar characteristics of the 2006 Yogyakarta earthquake related to the shallow crustal earthquake (strike-slip) mechanisms, earthquake distance, and magnitude range. The response curve of each equation models is provided in Figure 8.

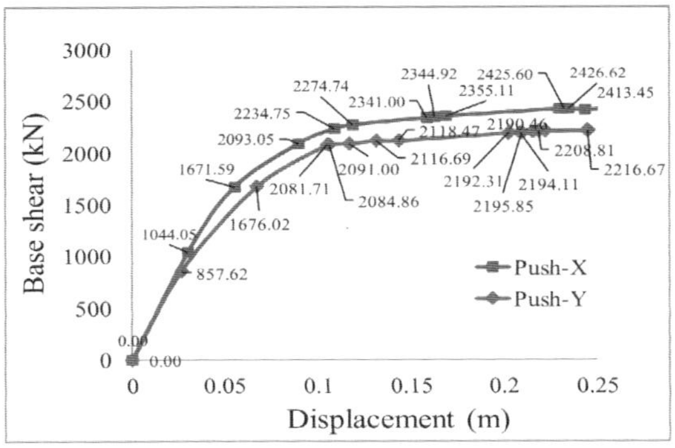

Figure 7. The x- and y-direction capacity curve

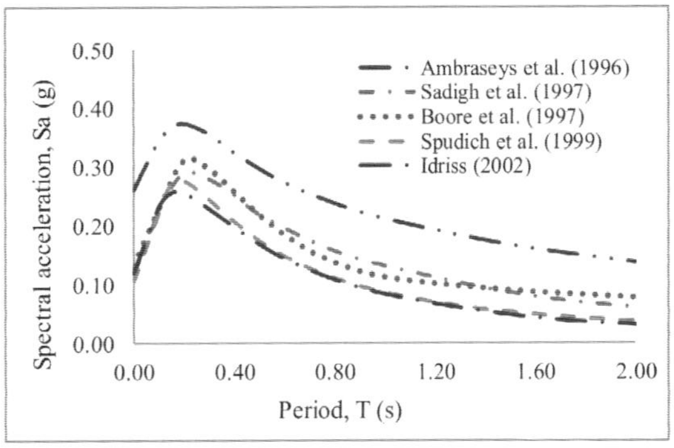

Figure 8. Response curve of each GMPE models

4.3. Peak building response value

Pushover analysis gives the result that the push-Y has the most contribution to the structural plastification and displacement; hence the capacity curve of push-Y is used to define the value of peak building response. The relationship between the response spectrums created based on each ground motion equation models and the capacity spectrum as a pushover analysis result, which has been converted to ADRS format is presented in Figure 9, while the peak building response values of each selected GMPEs are provided in Table 2.

Based on the result, the GMPE of Ambraseys et al. (1996) gives the highest value of the peak building response (3.7 in) which is about three times higher than the GMPE of Boore et al. (1997) and about four times higher than the GMPE of Idriss (2002). It can be concluded that the newer GMPE is used, the smaller the peak building response value. The S_d-value obtained then is used to determine the parameter of fragility curve.

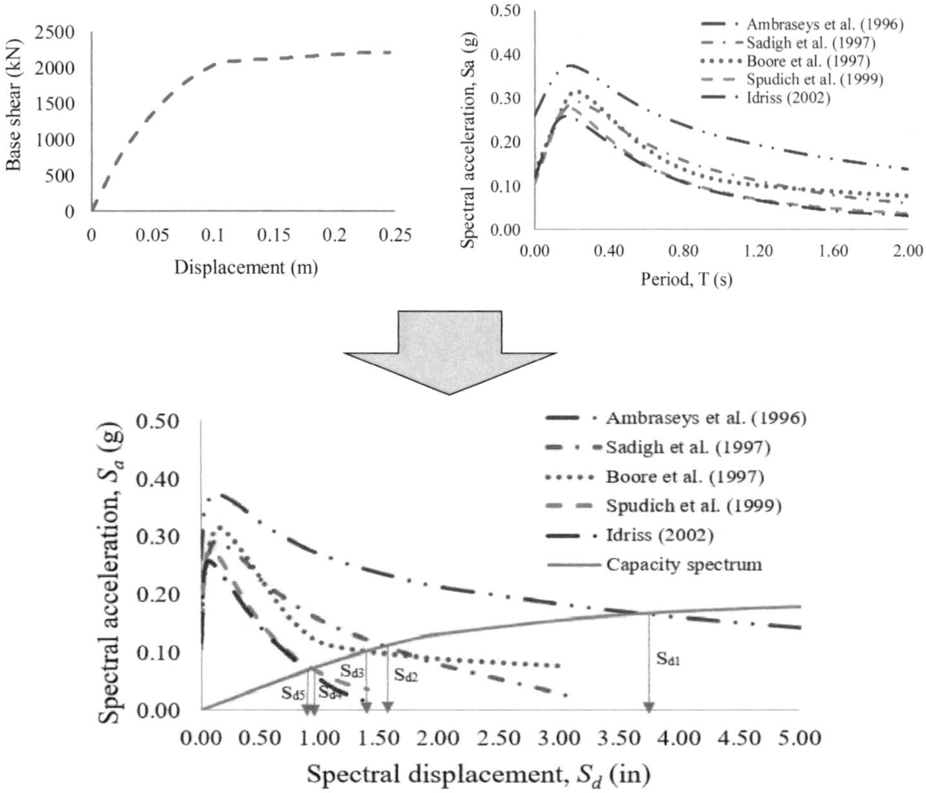

Figure 9. Relationship between response spectrum and capacity spectrum

Table 2. The result of peak building response value, (S_a, S_d)

GMPE	Sa (g)	S_d (in)
Ambraseys et al. 1996	0.188	3.7
Sadigh et al. 1997	0.116	1.58
Boore et al. 1997	0.105	1.4
Spudich et al. 1999	0.087	0.97
Idriss 2002	0.083	0.9

4.4. Fragility curve

The fragility curve parameters were obtained based on the building model type and the building seismic design level for each damage state. The assessed building is included on the C1M model (reinforced concrete buildings with a medium total number of stories) and moderate code seismic design level based on HAZUS building categories. The fragility curve parameter in each damage state is provided in Table 3. Results of cumulative probability used equation (9) are shown in Table 4, while the fragility curves generated from the calculation of cumulative probability are presented in Figure 10.

Table 3. The parameter of structural fragility curve for moderate-code seismic design level based on HAZUS, 1999

Structural Fragility Curve Parameters of C1M Model								
Type	Slight		Moderate		Extensive		Complete	
	$\hat{S}_{d.S/S}$	β_S	$\hat{S}_{d.S/M}$	β_M	$\hat{S}_{d.S/E}$	β_E	$\hat{S}_{d.S/C}$	β_C
C1M	1.5	0.7	2.6	0.7	7	0.7	18	0.89

Table 4. Building damage cumulative probability due to selected GMPE models

Ambraseys et al. 1996							
Damage stage	**Sd**	**$\hat{S}$dS**	**βds**	**X = Sd/$\hat{S}$dS**	**Ln (X)**	**Y = [Ln(X)]/βds**	**φ [Y]**
Slight	3.7	1.5	0.7	2.467	0.903	1.290	0.901
Moderate	3.7	2.6	0.7	1.423	0.353	0.504	0.693
Extensive	3.7	7	0.7	0.529	-0.638	-0.911	0.181
Complete	3.7	18	0.89	0.206	-1.582	-1.778	0.038
Sadigh et al. 1997							
Damage stage	**Sd**	**$\hat{S}$dS**	**βds**	**X = Sd/$\hat{S}$dS**	**Ln (X)**	**Y = [Ln(X)]/βds**	**φ [Y]**
Slight	1.58	1.5	0.7	1.053	0.052	0.074	0.530
Moderate	1.58	2.6	0.7	0.608	-0.498	-0.712	0.238
Extensive	1.58	7	0.7	0.226	-1.488	-2.126	0.017
Complete	1.58	18	0.89	0.088	-2.433	-2.734	0.003
Boore et al. 1997							
Damage stage	**Sd**	**$\hat{S}$dS**	**βds**	**X = Sd/$\hat{S}$dS**	**Ln (X)**	**Y = [Ln(X)]/βds**	**φ [Y]**
Slight	1.4	1.5	0.7	0.933	-0.069	-0.099	0.461
Moderate	1.4	2.6	0.7	0.538	-0.619	-0.884	0.188
Extensive	1.4	7	0.7	0.200	-1.609	-2.299	0.011
Complete	1.4	18	0.89	0.078	-2.554	-2.870	0.002
Spudich et al. 1999							
Damage stage	**Sd**	**$\hat{S}$dS**	**βds**	**X = Sd/$\hat{S}$dS**	**Ln (X)**	**Y = [Ln(X)]/βds**	**φ [Y]**
Slight	0.97	1.5	0.7	0.647	-0.436	-0.623	0.267
Moderate	0.97	2.6	0.7	0.373	-0.986	-1.409	0.079
Extensive	0.97	7	0.7	0.139	-1.976	-2.823	0.002
Complete	0.97	18	0.89	0.054	-2.921	-3.282	0.001
Idriss 2002							
Damage stage	**Sd**	**$\hat{S}$dS**	**βds**	**X = Sd/$\hat{S}$dS**	**Ln (X)**	**Y = [Ln(X)]/βds**	**φ [Y]**
Slight	0.9	1.5	0.7	0.600	-0.511	-0.730	0.233
Moderate	0.9	2.6	0.7	0.346	-1.061	-1.516	0.065
Extensive	0.9	7	0.7	0.129	-2.051	-2.930	0.002
Complete	0.9	18	0.89	0.050	-2.996	-3.366	0.000

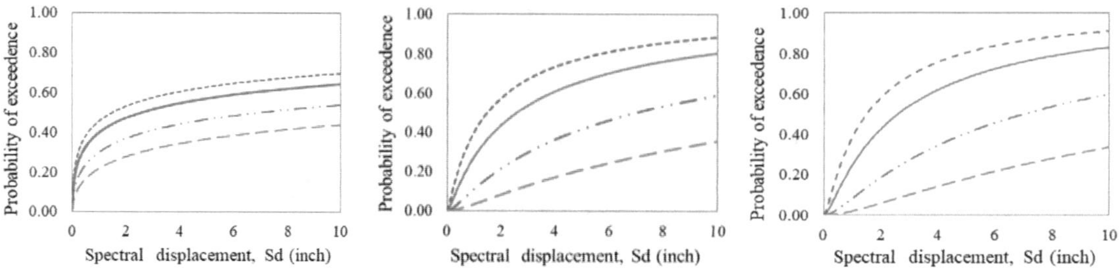

a) Ambraseys et al. (1996) b) Sadigh et al. (1997) c) Boore et al. (1997)

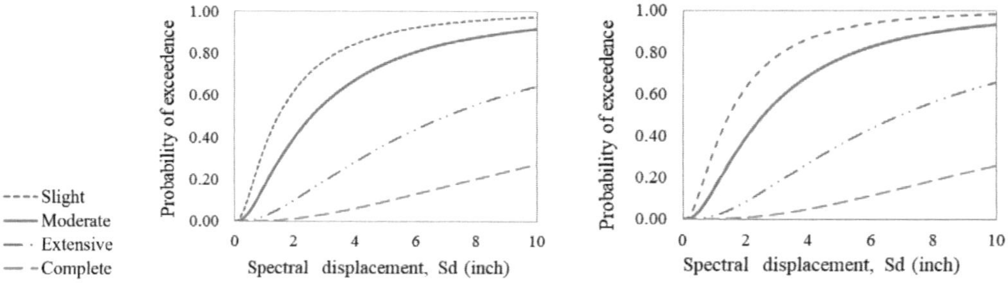

d) Spudich et al. (1999) e) Idriss (2002)

Figure 10. The fragility curves of building in each selected GMPEs models

Figure 10 depicts the building fragility curve due to the ground motion equation proposed by Idriss (2002) has the most gentle curve for the probability of a complete damage level. The value of damage appears after 0.2-inch spectral displacement and increases slowly, which also happens to the equation provided by Spudich et al. (1999). In contrast, the GMPE proposed by Ambraseys et al. (1996) shows the significant increase of complete state damage probability, which reached about 27% at 2.0-inch spectral displacement and around 40% at 10-inch spectral displacement. The values of building damage probability at a moderate level for all ground motions proposed are more than 60% at 4.0-inch spectral displacement except the equation proposed by Ambraseys et al. (1996). However, all five selected GMPE models have almost the same effect for the extensive damage probability value.

4.5. Damage probability matrix

The probability of building damage in each damage state based on GMPE models is presented in Table 5, while the matrix of building damage probabilities are described in Figure 11.

Table 5. The building damage probability value based on GMPE models

Probability of Damage	***Level of Damage***	***Damage Probability based on GMPE models***				
		Ambraseys et al. 1996	**Sadigh et al. 1997**	**Boore et al. 1997**	**Spudich et al. 1999**	**Idriss 2002**
P [C] = P [C | SD]	*Complete*	20.86%	29.12%	27.25%	18.72%	16.80%
P [E] = P [E | SD] - P [C | SD]	*Extensive*	51.17%	22.16%	17.75%	7.71%	6.31%
P [M] = P [M | SD] - P [E | SD]	*Moderate*	14.35%	1.36%	0.87%	0.19%	0.13%
P [S] = P [S | SD] - P [M | SD]	*Slight*	3.77%	0.31%	0.21%	0.05%	0.04%
Total vulnerability		90.14%	52.96%	46.07%	26.67%	23.28%
P [None] = 1 - P [S | SD]	*No damage*	9.86%	47.04%	53.93%	73.33%	76.72%

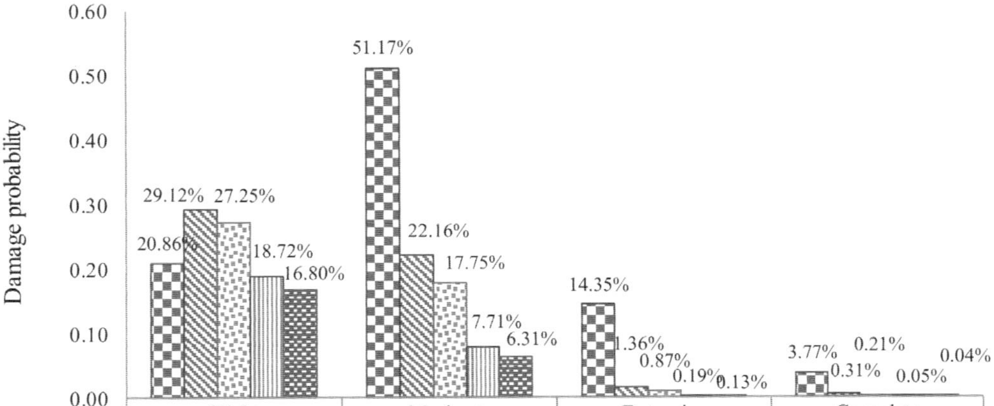

	Slight	Moderate	Extensive	Complete
Ambraseys et al. 1996	20.86%	51.17%	14.35%	3.77%
Sadigh et al. 1997	29.12%	22.16%	1.36%	0.31%
Boore et al. 1997	27.25%	17.75%	0.87%	0.21%
Spudich et al. 1999	18.72%	7.71%	0.19%	0.05%
Idriss et al. 2002	16.80%	6.31%	0.13%	0.04%

Figure 11. Damage probability matrix of each level of damage based on the selected GMPEs models

Figure 11 reveals the building damage probability values due to the scenario of the 2006 Yogyakarta earthquake. The probability value of moderate, extensive, and complete damage level is dominated by GMPE of Ambraseys et al. (1996) with values of 51.17%, 14,35%, and 3,77%, respectively. On the contrary, the use of GMPE of Idriss (2002) produces the lowest probability of damage in all damage states.

5. Conclusion

Based on pushover analysis, the x-direction portal is stronger in resisting the lateral force compared to the y-direction portal. The response curve of Ambraseys et al. (1996) has the highest value of spectral acceleration, while the lowest spectral acceleration value is the response curve of Idriss (2002). In this case, by using the 2006 Yogyakarta earthquake scenario, the GMPE of Ambraseys et al. (1996) gives the highest value of the peak building response (3.7 in) which is about three times higher than the GMPE of Boore et al. (1997) and about four times higher than the GMPE of Idriss (2002). It is showed that the newer GMPE is used, the smaller the peak building response value is obtained.

The ground motion equation which gives the most vulnerable results to building damage probability is the GMPE proposed by Ambraseys et al. (1996), while the smallest vulnerability is obtained from calculations using the GMPE of Idriss (2002). The ground motion equation proposed by Idriss (2002) has the gentlest curve for the probability of complete damage level. In contrast, the GMPE model proposed by Ambraseys et al. (1996) shows the significant increase of complete damage state probability. This research highlights that the probability value of moderate, extensive, and complete damage level is dominated by GMPE of Ambraseys et al. (1996). Discrete damage probability indicates that the use of GMPE of Ambraseys et al. (1996) will produce the most severe damage probability value due to the Yogyakarta earthquake scenario, while the GMPE of Idriss (2002) will produce the lowest damage probability value.

References

[1] Elnashai A S, Kim S J, Yun G J and Sidarta D 2006 *The Yogyakarta Earthquake of May 27th, 2006*, Mid-America Earthquake Center Report no 07-2 University of Illinois at Urbana–Champaign

[2] Tsuji T, Yamamoto K, Matsuoka T, Yamada Y, Onishi K, Bahar A, Meilano I and Abidin H Z 2009 Earthquake fault of the 26 May 2006 Yogyakarta Earthquake observed by SAR Interferometry *Journal of Earth Planets Space* vol 61 pp e29-e32

[3] Geomagz 2016 *10 Years Yogyakarta Earthquake* vol 6 no 2

[4] Perrault M and Gueguen P 2015 Correlation between ground motion and building response using Californian earthquake records *Earthquake Spectra* Earthquake Engineering Research Institute vol 31 no 4 pp 2027–2046

[5] Dewobroto W 2006 Evaluation of the performance of earthquake-resistant steel buildings with SAP2000 *Journal of Teknik Sipil* vol 3 no 1 pp 7-24

[6] Bommer J J and Stafford P J 2020 Selecting ground-motion models for site-specific PSHA: adaptability versus applicability *Bulletin of the Seismological Society of America* vol 110 no 6 pp 2801–2815

[7] Karaca H 2020 Selection of the most appropriate ground motion prediction equation for local seismic hazard analysis *Eskişehir Technical University Journal of Science and Technology A-Applied Sciences and Engineering* vol 21 no 1 pp 21-38

[8] Muntafi Y, Nojima N and Jamal A U 2020 Damage probability assessment of hospital buildings in Yogyakarta, Indonesia as Essential Facility due to an Earthquake Scenario *Journal of the Civil Engineering Forum* vol 6 no 3 pp 225-236

[9] Applied Technology Council (ATC-40) 1996 *Seismic Evaluation and Retrofit of Concrete Buildings* vol 1 (California: Redwood City)

[10] Satyarno I 2000 A Thesis of Doctor of Philosophy in Civil Engineering: *Adaptive Pushover Analysis for The Seismic Assesment of Older Reinforced Concrete Buidings* (New Zealand: University of Canterbury)

[11] Najam F A 2018 *Nonlinear Static Analysis Procedures for Seismic Performance Evaluation of Existing Buildings* – Evolution and Issues. In: Rodrigues H., Elnashai A., Calvi G. (eds) Facing the Challenges in Structural Engineering. GeoMEast 2017. Sustainable Civil Infrastructures. Springer, Cham

[12] Ambraseys N N Simpson K A and Bommer J J 1996 Prediction of horizontal response spectra in Europe *Earthquake Engineering and Structural Dynamics* vol 25 pp 371-400

[13] Sadigh K, Chang C Y, Egan J A, Makdisi F and Young R R 1997 Attenuation relationships for shallow crustal earthquake based on California strong motion data *Seismological Research Letters* Seismological Society of America vol 68 pp 180-189

[14] Boore D M, Joyner W B and Fumal T E 1997 Equations for estimating horizontal response spectra and peak acceleration from western north American Earthquakes: A Summary of recent Work, Seismological Research Letters vol 68 no 1 pp 128-153

[15] Spudich P, Joyner W B, Lindh A G, Boore D M, Margaris B M and Fletcher J B 1999 SEA99: A Revised ground motion prediction relation for use in extensional tectonic regimes *Bulletin of the Seismological Society of America* vol 89 no 5 pp 1156-1170

[16] Idriss I M 2007 An NGA empirical model for estimating horizontal spectral values generated by shallow crustal earthquakes *Earthquake Spectra* vol 24 no 1 pp 217-242

[17] FEMA: HAZUS'99 2001 *Estimated Annualized Earthquake Loss for the United States Report FEMA 366* (Washington DC USA: Federal Emergency Management Agency)

[18] Chopra A K and Goel R K 2001 *Capacity-Demand Diagram Methods for estimating seismic deformation of inelastic structure SDOF Systems* PEER Report 1999/02 Pacific earthquake Engineering Research Centre, University of California, Berkeley

[19] Priestly M J N and Calvi G M 1991 Toward a capacity design assessment procedure for reinforced concrete frames *Earthquake Spectra* vol 77 no 3 pp 413-437

[20] Priestly M J N 1995 Displacement based seismic assessment of existing reinforced concrete buildings *Proceedings of Pacific Conference on Earthquake Engineering* vol 2 Melbourne pp 225-244

[21] Park R 1996 A Static force-based procedure for the seismic assessment of existing reinforced concrete moment resisting frames *Proceedings of New Zealand National Society for Earthquake Engineering Annual Conference* New Plymouth pp 54-67

[22] Miranda E and Bertero V V 2010 Seismic performance of an instrumented ten-storey reinforced concrete building *Earthquake Engineering and Structural Dynamics* vol 25 pp 1041-1059

[23] Molina S, Lang D H and Lindholm C D 2010 SELENA-an opensource tool for seismic risk and loss assessment using a logic tree computation procedure *Comput. Geoscience* vol 36 pp 257–269

[24] National Institute of Building Sciences (NIBS) 2002 *HAZUS'99: Earthquake Loss Estimation Methodology: technical manual* In: FEMA (Editor) Technical Manual (Washington DC: Federal Emergency Management Agency) p 325

[25] National Standardization Agency of Indonesia 2012 *Earthquake Resistance Design Procedures for Building and Non-Building Structures (in Indonesian)* SNI 1726–2012 (Jakarta: National Standardization Agency of Indonesia)

IOP Conf. Series: Earth and Environmental Science 1173 (2023) 012017 doi:10.1088/1755-1315/1173/1/012017

The threat of tsunami debris due to non-permanent buildings in Cilacap City

W Widiyanto[1], **S N Purnomo**[1] **and P B Santoso**[1]
[1]Civil Engineering Department, Universitas Jenderal Soedirman, Jl Mayjen Sungkono Km 5, Purbalingga, Indonesia

E-mail: sanidhyanika.purnomo@unsoed.ac.id

Abstract. Run-up tsunamis have the ability to pick up debris along their path. More severe damage may result from the debris and their force of impact. Therefore, it is important to analyse the potential for debris in a tsunami-prone area. This study examines the threat of tsunami debris in Cilacap City, especially the source of debris from temporary or non-permanent buildings along the coast of Cilacap City. A site assessment was conducted to observe the existence of non-permanent buildings along with their number and characteristic. The results of the study show that the infrastructure of Cilacap City faces a serious threat from tsunami debris which is source from non-permanent buildings. The material transported by the tsunami run-up to cause debris was maximum 465 tons found in area around Pasir Emas Street cause potential debris for one-meter perpendicular to tsunami run-up direction 503 kg and generate tsunami impact force 28,284.3 N. In addition, the fishing port, Pendem fort and Pertamina oil refinery are also potentially affected by tsunami debris because each location has a significant number of non-permanent buildings.

1. Introduction

Along the ring of fire on the outer arc of Sumatra, Java to Nusa Tenggara are earthquake and tsunami prone areas. Cilacap City is one of the cities in the area which has a dense population and intensive industrial and economic activities. The city was hit by a tsunami in 2006 due to the Mw 7.5 earthquake [1] [2]. This magnitude is still below the potential that could occur [3], which is Mw 8.7, especially due to the West-Central Java megathrust segment [4]. The potential for a tsunami due to this segment has been simulated by Widiyanto [5]. Against this background, mitigation is an absolute must. One thing that needs to be studied is the threat of debris flow due to the run-up of the tsunami which carries large amounts of material and objects towards residential areas and areas where other human activities are carried out.

It can be seen in previous events, objects in the coastal area can be carried away from the shoreline by tsunami run-ups [6][7]. Compared to a tsunami run-up with clear water without the addition of solids, the destructive power of a tsunami will increase if the run-up flow mass is larger or the drifting objects have high stiffness. This can be explained by the tsunami impact force formula [8]:

$$F = v\sqrt{km_d} \quad (1)$$

where v = impact velocity; k = relevant debris stiffness; and m_d = debris mass. Based on previous studies, the tsunami impact velocity is quite reasonable if taken 4 m/s. As for buildings made of light weight, [9] suggested a stiffness of 0.10 kN/mm.

IOP Conf. Series: Earth and Environmental Science 1173 (2023) 012017 doi:10.1088/1755-1315/1173/1/012017

Based on equation (1), the parameters of velocity, stiffness and mass depend on the coastal situation. In this study, the discussion is limited to the mass variable which is influenced by the presence of objects on the beach. More specifically, these objects are limited to temporary or non-permanent building types. The term a non-permanent building means a building that is a temporary in nature and constructed of materials which do not last long or a movable building with a useful life of not more than 10 (ten) years; for examples barracks and dormitories made of wood and used by employees. These buildings are made of wood and bamboo which are commonly found in coastal areas in Indonesia, including Cilacap City. These buildings can function as shops and houses. For buildings that function as stalls, they generally try to be placed as close to the shoreline as possible so that consumers can enjoy selling goods and food while watching the waves breaking on the beach (Figure 1).

Figure 1. More stalls are made in places closest to the shoreline so that consumers can enjoy food while watching the incoming waves.

2. Study area

This study overview covers the coastal area of Cilacap City from point A to point B in Figure 2 (a) where there are various buildings and infrastructures. The western boundary of the study is the Donan River Estuary which is also a strait located between Nusa Kambangan Island and Java Island. While the eastern boundary is the Estuary of the Serayu River. The western part is more densely with land use including ports, oil refineries, tourist attractions, historical relics, and housing. The eastern part is more space between buildings where the important infrastructure on the eastern boundary is the steam power plant. There are two large-capacity power plants, namely PLTU Karangkandri and PLTU Adipala. Along this boundary, a field survey is conducted to determine the existence of non-permanent buildings, their situation, structural details and materials.

The study area is divided into 10 segments as shown by number 1-10 in Figure 2 (a). Each segment has a different length, which is chosen based on an important feature or location. The ten segments are: 1) Fort Pendem (340 m), 2) Pertamina Refinery (597 m), 3) Pasir Mas Street (925 m), 4) Basarnas (716 m), 5) Fish Port (965 m), 6) South Ring Road (984 m), 7) Kemiren Beach (1400 m), 8) Menganti Beach (1730 m), 9) PLTU Karangkandri (998 m), and 10) PLTU Adipala (1115 m). The management of disaster preparedness can be more easily coordinated with this division based on famous site names.

(a)

(b)

Figure 2. (a) The study area covers Cilacap City coast from point A to B. (b) Coastal area bounded by the red dotted line is a source of debris in the event of a tsunami. A runup tsunami will carry debris to areas that are densely packed with buildings which have the potential to increase the destructive power of buildings. The aerial photographs were from ©Google Earth Pro.

3. Method

Field and laboratory work is carried out to obtain data to be analysed. Field work includes assessment of non-permanent structures throughout the study area. While the main laboratory work is aimed at knowing the density of each material used for buildings. Site assessment activities include counting the number of buildings, taking pictures, identifying building materials, investigating the layout of the situation, and measuring the dimensions of the building. From the data obtained in the field, the potential mass for debris flow can be calculated.

3.1. Site assessment

Observations were made in detail on non-permanent buildings along the study area. This observation is to determine the number of buildings, sizes, shapes, functions and materials. The results are presented and discussed in the Results and Discussion section. Figure 3 illustrates the situation of the outermost road in Cilacap where on the right and left are filled with non-permanent buildings. The Pasir Emas Street as shown in Figure 3 is 85 m from the shoreline. This is a critical distance to tsunami run-up in the sense that run-ups reaching a distance of 85 m are quite common in tsunami events.

Figure 3. One of the streets in Cilacap city that is closest to the coast is Pasir Emas Street, which has a lot of temporary structures on its right and left. This area is a coastal border where buildings should not be allowed to build. The direction of the tsunami is depicted by a blue line. The photograph was from ©Google Street View.

3.2. Laboratory work

The type of building material whose data has been obtained from the field is then looked for samples but the sample is not taken directly from the existing building in the field because it will damage the structure. Each material has at least 3 samples to be tested for density. Furthermore, the mass density can be used to calculate the mass per unit material. The unit of material depends on the practice habits in the field, for example, for bamboo, the dimensions are mass per unit length with units of kg/m'. While the roof made of zinc, the dimensions are mass per unit area with units of kg/m^2. An alternative to knowing the unit weight is to read a list of unit weights, for example in a loading book. However, for material verification purposes that are empirically used in the study area, laboratory tests are still carried out.

4. Result and discussion

In high-risk coastal areas, the significance of tsunami-resistant infrastructure should be emphasized. The extreme loading conditions associated with the tsunami have received significant attention due to the damage to critical infrastructure in the tsunami-affected areas. Debris loads were one of the crucial loads recognized by earlier studies. Debris, which can include anything from building materials to shipping boats, is referred to as solid items entrained inside the inundating flows. Recent advancements in the knowledge of debris loads and impacts, particularly in assessing the impact of a single debris piece on a structure, have been made as a result of the focus on tsunami loading [10]. The impact of extreme hydrodynamic events on a free-standing building remains has been studied [11] The results of the site assessment provide information that the number of non-permanent buildings on the coast of Cilacap City reaches 1138 units. However, the density of these buildings is uneven on the coast in the study area along the 13 km. Some of the beaches, especially on the west side of the review area, are very dense, while in the east, they are relatively rare. Non-permanent densely built areas include the Teluk Penyu (Turtle Bay) tourist area, the front side of the oil refinery, around the Pendem Fort, and the fishing port area. Some agricultural areas have relatively no non-permanent buildings as well as power plant areas.

The function of the building is mostly to sell food, souvenirs, and other merchandise. About 40% of the building is used as a place to live, the rest is only used for selling and the owner lives elsewhere when not working. Several other functions also exist such as rental of tourist equipment, workshops, etc. In general, the form of the building is simple and without design. The simplest buildings are usually for stalls near the shoreline in the form of bamboo poles embedded in the ground with a bamboo roof truss as well. The roof uses lightweight materials such as pandanus leaves, zinc, asbestos, and rooftile. Buildings intended for houses generally have a stronger structure but with dominant materials of wood and bamboo as well. The size of the building ranges from 15-25 m^2 for beachfront stalls and 25-40 m^2 for residences. The height of the building ranges from 2.5 to 4 m. The distance from the shoreline ranges from 40 – 100 m. This distance is at high risk of a tsunami run-up range.

Figure 4 shows some pictures of non-permanent buildings collected from the site survey. The type of building material is an intriguing aspect of the characteristics of non-permanent structures along the Cilacap coast, and possibly throughout Indonesia in general. The survey results show that wood and bamboo are the dominant materials. In addition to the relatively affordable price, the nature of the building that is intended to be temporary is the reason for building non-permanent buildings. The people who constructed the buildings may not have used property rights but rather use rights because land along the coast is typically owned by the government, not by specific individuals. The list of materials and their specific gravity is shown in Table 1.

Table 1. Building materials and their density in Cilacap coastal area.

No.	Building Material	Density (kg/m^3)
1	Wood	980
2	Bamboo	670
3	Brick wall	1680
4	Stone wall	2160
5	Reef wall	700
6	Concrete	2200
7	Reinforced concrete	2400
8	Zinc roof	7135
9	Rooftile	6100
10	Steel	7850

The density values in Table 1 are then used to estimate the mass of debris originating from non-permanent buildings as presented in Figure 5. The density list is a truncated list from a long list of building materials used to construct non-permanent buildings on the coast of Cilacap City. The number of non-permanent buildings was calculated using simple manual counter. A building has a volume that can be approximated from measurement of sizes. Mass is the result of volume times density. Length of site observed can be obtained from Google Earth Pro application and checked by using laser distance meter. Another important result shown by Figure 5 is mass of building materials each meter. It is computed from total mass divided by site length. The maximum mass is found in Pasir Mas Street with total mass 465 tons or 503 kg/m′. Assume that tsunami run-up velocity is 4 m/s and the stiffness of material is 0.1 kN/mm as mentioned in previous section, the tsunami impact force will be 28,284.3 N working on one-meter perpendicular to tsunami run-up direction. It is important to note that only debris from building material is considered. The force will stronger when other debris sources like boats, trees, electricity poles, road utilities, and other public facilities are also considered.

Figure 4. Characteristics of non-permanent buildings in the coastal area of Cilacap City. In other areas in Indonesia also have the same characteristics with these houses. Generally, these houses are not designed from an architectural or structural aspect so that they are less aesthetic and less strong. Wood and bamboo are the dominant materials, although concrete and steel are also used.

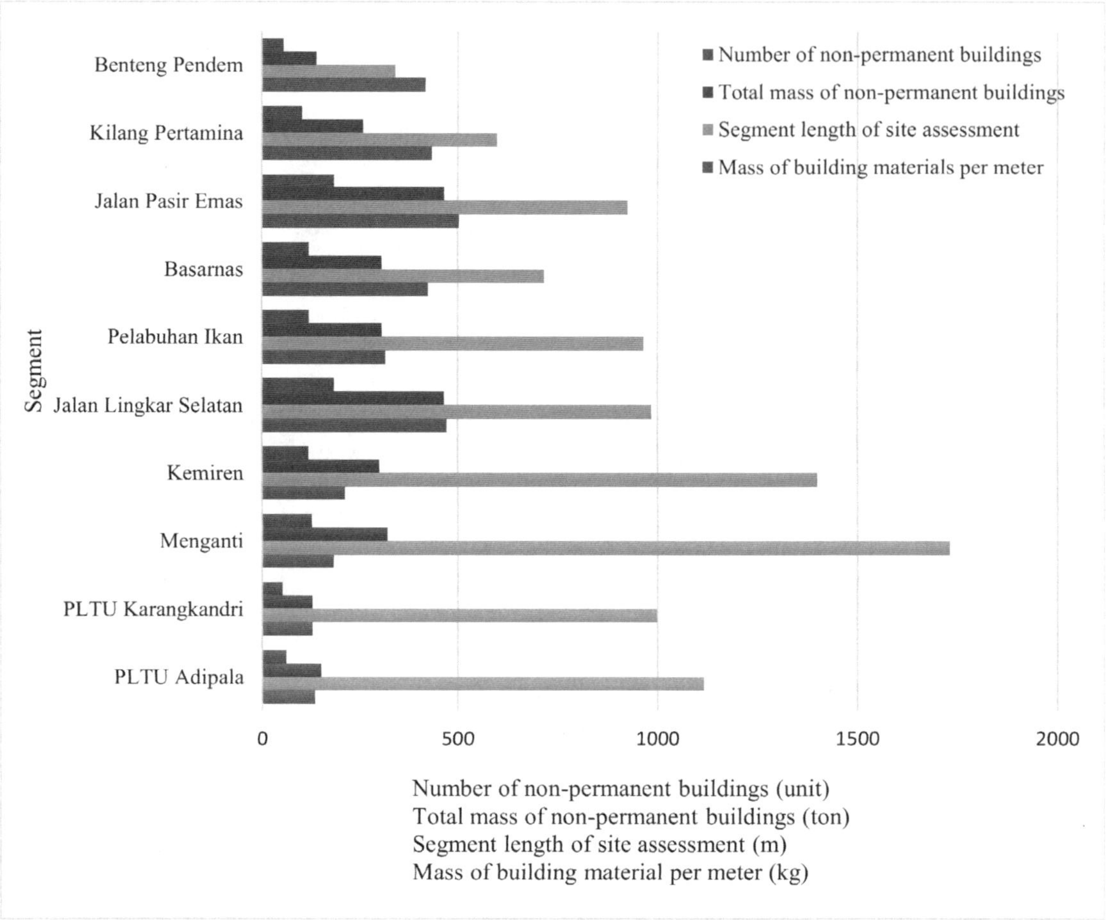

Figure 5. Number of non-permanent buildings and debris mass coming from non-permanent building in 10 segments.

In this study, the mass of debris originating from non-permanent buildings. The mass can increase or decrease due to the development and arrangement of the city.

5. Conclusion

A significant number of non-permanent buildings have been built on Cilacap Beach. The dominant building materials are bamboo and wood. The distances of the buildings to the shoreline are quite close and are within the range of the tsunami run-up. So this has the potential to be a source of debris that can be carried by the run-up to crash into more permanent buildings on land farther from the coast. The study area is divided into 10 segments, each of which has a different potential for debris mass. The highest potential is found in the Pasir Emas Street segment because of the many buildings on either side of the street. The potential for debris that also needs to be considered are segments of the fort Pendem, Pertamina refinery and Cilacap fishing Port. These facilities have a high economic value. Mitigation measures need to be implemented to reduce risks, for example by providing debris barriers to land areas that are denser and more inhabited by humans.

Acknowledgment
The author would like to thank the Institute for Research and Community Service (LPPM) Universitas Jenderal Soedirman which has provided research funds, especially the basic research scheme under the contract No. T/691/UN23.18/PT.01.03/2022. In addition, this research was also supported by research grant awarded by P.T. Maipark Indonesia under the contract No. RMI/MOU/BOD/07.20/006.

References
[1] Fritz H M 2007 Extreme runup from the 17 July 2006 Java tsunami *Geophys. Res. Lett.* **34** 1–5
[2] Lavigne F 2007 Field observations of the 17 July 2006 Tsunami in Java *Nat. Hazards Earth Syst. Sci.* **7** 177–183
[3] Widiyantoro S, Gunawan E, Muhari A, Rawlinson N, Mori J, Hanifa N R, Susilo S, Supendi P, Shiddiqi H A, Nugraha A D and Putra H E 2020 Implications for megathrust earthquakes and tsunamis from seismic gaps south of Java Indonesia. *Sci. Rep.* **10** 1–11
[4] Pusat Studi Gempa Nasional (PuSGeN) (2017). Peta Sumber dan Bahaya Gempa Indonesia Tahun 2017 (Map of Indonesia Earthquake Sources and Hazards in 2017). Pusat Penelitian dan Pengembangan Perumahan dan Permukiman, Badan Penelitian dan Pengembangan, Kementerian Pekerjaan Umum dan Perumahan Rakyat, Bandung. (in Bahasa Indonesia)
[5] Widiyanto W and Purnomo S N 2023 Tsunami Hazard in Cilacap City due to the Megathrust of West-Central Java Segment Proceedings of the 5th International Conference on Rehabilitation and Maintenance in Civil Engineering ICRMCE 2021 *Lecture Notes in Civil Eng.(Surakarta)* **225** 543-551
[6] Widiyanto W, Santoso P B, Hsiao S C and Imananta R T 2019 Post-event field survey of 28 September 2018 Sulawesi earthquake and tsunami *Nat. Earth Syst. Sci.* **19** 2781–2794
[7] Widiyanto W, Hsiao S C, Chen W B, Santoso P B, Imananta R T and Lian W C 2020 Run-up, inundation, and sediment characteristics of the 22 December 2018 Sunda Strait tsunami, Indonesia *Nat. Hazards Earth Syst Sci.* **20** 933–946
[8] Paczkowski K, Riggs H R, Naito C J and Lehmann A 2012 A one-dimensional model for impact forces resulting from high mass, low velocity debris Struct *Eng. Struct. Mech.* **42** 831–847
[9] Naito C, Cercone C, Riggs HR and Cox D 2014 Procedure for Site Assessment of the Potential for Tsunami *Debris Impact Waterway, Port, Coastal, Ocean Eng.* **140** 223-232
[10] Wuthrich D, Pfister M, Nistor I and Schleiss A J 2018 Experimental Study on the hydrodynamic Impact of Tsunami-like Waves against Impervious Free-standing *Build. Coastal Eng. J.* **60** 180-199
[11] Nistor I, Goseberg N and Stolle J 2017 Tsunami-driven Debris Motion and Loads: A Critical Review Front *Built Environ* **3** 2

Prediction of shelter needs based on simulation results of evacuees choosing evacuation facilities due to a potential tsunami earthquake in Padang, Indonesia

Purnawan[1] **and E Kharisma**[2]

[1,2]Department of Civil Engineering, University of Andalas

E-mail: purnawan@eng.unand.ac.id

Abstract. Tsunami disasters are predicted to occur in the West Sumatra region due to earthquakes. To deal with the tsunami hazard that will hit Padang City, the government has built several shelters and evacuation routes. At present, the number of evacuation facilities is still limited, this will cause community vulnerability in facing tsunami hazards. To overcome this vulnerable condition, it is necessary to determine the number of shelters needed to accommodate evacuees. In this study, the determination of the required number of shelters is done by simulating the number of evacuees who choose shelters during evacuation. The model of evacuee generation from 5 tsunami-prone zones in Padang City was created using a linear regression method. The model for selecting the type of evacuation facility in each zone uses a binary logit model. Prediction of shelter needs is carried out for minimum, normal and maximum conditions that may occur. From the simulation results, the number of evacuees who choose the shelter as a means of evacuation is obtained. The number of shelters needed in Padang City is a minimum of 6 units, normal 23 units and maximum 50 units.

1. Introduction

Indonesia is located at the confluence of three of the world's tectonic plates, namely the Indo-Australian, Eurasian and Pacific plates, besides that Indonesia is also a part of the Pacific Ring of Fire, a zone where earthquakes and volcanic eruptions are very frequent. On September 30, 2009, an earthquake measuring 7.6 M struck the west coast of Sumatra, which is close to the city of Padang, resulting in a death toll of 1,117 people, and damage to facilities and infrastructure [1]. After the tsunami disaster in Aceh in 2005 and the tsunami in Mentawai in 2010, many experts from many countries explore the future danger of the plate near Sumatra Island [2]. Aydan identify, there is an unbroken segment facing Padang city, the unbroken part is more than 400km in length, and it could result in a mega-thrust earthquake with a magnitude greater than 8.7. Padang city is located on relatively very flat ground, to reach an altitude more than 5m from the coast, one has to walk more than 3km. In the case of a tsunami with a height of more than 5m, it may be quite disastrous [3]. Most people in Padang are aware of predictions of a big earthquake with the potential to cause a tsunami due to a plate shift near the island of Sumatra. As a result of this, when a big earthquake occurs, people living in tsunami-prone zones evacuate to higher ground. This led to congestion in most areas adjacent to the tsunami-prone zone [4].

To reduce the effect of the tsunami disaster, that hit Padang city, the government already built 3 shelters in the tsunami-prone zone [5], see figure 1, but the number of shelters is still limited. The government also determine several high buildings such as hotel, mosque, and office that could be used as a shelter, but the legal aspect of this policy need more preparation because not all high building could be used as a shelter, it needs to check the design of structure and field conditions. Facing the tsunami disaster, Government also determines an evacuation route from the tsunami-prone zone to the location that has high ground level, this evacuation route is provided for evacuees who want to evacuate to a safe location on high ground. The evacuation route is equipped with signposts leading to safe locations, this route could be seen in figure 2.

Figure 1. Shelter at Ulak Karang, Padang

Figure 2. Route evacuation sign

Based on the condition above, the government should build more shelters or determine the high and strong buildings as the shelter. However, it is needed the number of refugees that will choose the shelter as the evacuation facility. This study predicts the number of refugees that will use the shelter as an evacuation facility, this data could be used as a reference to build the shelter in Padang.

2. Modeling refugee behavior

2.1. Model of refugee generation from tsunami-prone zone

This study was carried out based on modeling evacuation demand at 5 tsunami-prone zones. The number of refugees was estimated based on the evacuation generation model, this model was developed based on trip generation model [6]. The selection of evacuation facilities was based on the chosen model. Based on the number of refugees demanded and the percentage of choice the refugee evacuation facilities, the required area for shelter could be calculated. The simulation was conducted using optimistic, moderate, and pessimistic demand scenarios. The number of the refugee is determined as mature (11-65 years) and old (>65 years). The evacuation generation model was developed using regression analysis, the dependent variable (Y) is the number of refugees and the independent variable is the population in the tsunami-prone zone (X_1), the number of motorcycle (X_2), the number of car (X_3). The result of the regression analysis is shown in table 1.

Table 1. Refugee generation model for 5 tsunami-prone zone

Zone		R^2
1	$Y = 0.151+0.658\ X_1+0.411\ X_2+0.241\ X_3$	0.83
2	$Y = 0.228+0.768\ X_1$	0.73
3	$Y = 0.508+0.665\ X_1+0.0691\ X_2$	0.60
4	$Y = 0.378+0.654\ X_1+0.174\ X_2+0.187\ X_3$	0.82
5	$Y = -0.258+0.838\ X_1+0.0874\ X_2+0.060\ X_3$	0.80

Note : Y = refugee generation
X_1 = population
X_2 = the number of motorcycle
X_3 = the number of car

The proportion of refugees for each family was calculated based the equation (1). It is assumed that the person that required full space at the shelter was classified I age 11-65 years and age > 65 years, so the proportion of refugees for each family could be calculated using equation (1).

$$\text{The proportion of refugees for each family} = \frac{\text{Number of family age (11-65) + age (>65)}}{\text{Total number of family}} \quad (1)$$

The proportion of refugees for each zone could be calculated using equation (2).

$$\text{The proportion of refugees for each zone} = \frac{\text{Number proportion of regugee for each family}}{\text{Total number of the family for each zone}} \quad (2)$$

Calculation of refugees for each zone could be conducted using equation (3).

$$\text{Number of refugees for each zone} = \text{Proportion of refugees for a zone x Population for a zone} \quad (3)$$

2.2. Model of refugee choose evacuation facilities

Refugees choose the evacuation facilities based on the binary logit model [6], three attribute models were used i.e: travel time, car ownership, and accessibility-based traffic condition. The combination of attribute conditions is explained as a choice of refugees to select a shelter or evacuation route [7]. The combination of choice was developed based on refugee stated preferences, the choice of refugees based on 3 attributes could be seen in table 2.

Table 2. The alternative choice of refugees in evacuation facilities

	Shelter			Evacuation route		
Alternative Choice	Travel time	Car Ownership	Accessibility	Travel time	Car Ownership	Accessibility
1	Fast	Available	Smooth	Slow	Available	Congested
2	Fast	Not Available	Smooth	Slow	Not Available	Congested
3	Slow	Available	Congested	Fast	Available	Congested
4	Slow	Not Available	Congested	Fast	Not Available	Smooth
5	Slow	Available	Congested	Slow	Available	Congested
6	Slow	Not Available	Congested	Slow	Not Available	Congested
7	Fast	Available	Smooth	Fast	available	Smooth
8	Fast	Not Available	Smooth	Fast	Not Available	Smooth

Table 3. Value of attribute level

Atribut	Attribute level	Value
Time travel	Fast: (0-15) minutes	2
	Slow : (15-30) minutes	1
Car Ownership	Available	2
	Not Available	1
Accessibility	Smooth	2
	Congested	1

Model refugees choose the evacuation facilities were developed using binary logit ratio model, 500 respondents data from the survey at 5 tsunami-prone zones was used to produce the model. Table 4 is the result of the modeling.

Table 4. Binary logit ratio model.

Zone	Binary logit ratio model for shelter
1	$Ps = \dfrac{1}{1+3.024\dfrac{Xs}{Xj}^{(-3.647)}}$
2	$Ps = \dfrac{1}{1+3.811\dfrac{Xs}{Xj}^{(-3.879)}}$
3	$Ps = \dfrac{1}{1+3.273\dfrac{Xs}{Xj}^{(-3.507)}}$
4	$Ps = \dfrac{1}{1+3.418\dfrac{Xs}{Xj}^{(-2.855)}}$
5	$Ps = \dfrac{1}{1+3.71\dfrac{Xs}{Xj}^{(-3.614)}}$

Note : Ps = percentage of people choosing a shelter
Xs = utility model for choosing a shelter
Xj = utility model for choosing route evacuation

3. Result and discussion

3.1. Respondent characteristic

From data collection, it has been collected 500 respondents, and for each zone, it has been collected 100 respondents. Analysis of the collected data, based on their experience, 38.2% of respondents choose shelter for evacuation, and 61.8% of respondents chose route evacuation to reach a higher position. The characteristic of the respondent is shown in table 5.

Table 5. Characteristics of respondents

Characteristic	Percentage of respondent
Choose shelter	Male (49%), Female (51)
Choose evacuation route	Male (41%), Female (59.7%)
The age of the person choose shelter	<16 years (5%), 16-30 years (28%), 31-45 years (32%), 46-60 years (27%), >60 years (8%)
The age of the person choose route evacuation	<16 years (7%), 16-30 years (28%), 31-45 years (37%), 46-60 years (23%), >60 years (6%)
Number of family evacuees to shelter	single (8%), 2-4 people (39%), 5-9 people (46%), >10 people (7%)
Number of family evacuees to evacuation route	single (10%), 2-4 people (39%), 5-9 people (44%), >10 people (7%)
The reason choose to shelter is safe	Male (51%), Female (49%)
The reason choosing shelter is faster than route evacuation	Male (39%), Female (61%)
The reason choose to shelter is easy to reach	Male (43%), Female (57%)
The reason choose to shelter is high enough for safety protection	Male (41%), Female (59%)

3.2. Refugee generation simulation from tsunami-prone zone

The number of evacuees fleeing the tsunami-prone zone was calculated from the evacuee generation model in table 1. The example of calculation results is shown in table 6, in this table the number of evacuees increased by 5%, changing the choice of the evacuation route. The calculation results are shown in table 6, in this table the number of evacuees decreased by 5%, changing the choice to an evacuation route.

Table 6. Simulation of refugee evacuation from 5 zones with a 5% decrease

Zone	Refugee generation in normal codition	Simulation refugee generation (people)					
		-5%	-10%	-15%	-20%	-25%	-30%
1	19.280	18.316	17.352	16.388	15.424	14.460	13.496
2	25.493	21.315	20.193	19.071	17.949	16.827	15.705
3	30.634	28.617	27.111	25.605	24.099	22.593	21.087
4	26.919	24.820	23.514	22.207	20.901	19.595	18.288
5	48.866	45.083	42.710	40.338	37.965	35.592	33.219
Total	151.192	137.624	130.381	123.138	115.894	108.651	101.408

3.3. Refugees choose the evacuation facilities

Analysis using choosing evacuation model in table 4, the result of percentage people choose shelter and route evacuation, the percentage for normal demand refugee condition is shown in table 7.

Table 7. Prediction of the number of evacuees to shelters and evacuation routes.

Zone	Refugee generation (person)	Percentage of refugees choose shelter	Percentage of refugees choose route evacuation	Prediction of refugee use shelter (person)	Prediction of refugee use route evacuation (person)
1	19,280	31%	69%	5,895	13,385
2	25,493	27%	73%	6,947	18,546
3	30,634	29%	71%	8,857	21,777
4	26,919	27%	73%	7,204	19,715
5	48,866	27%	73%	13,230	35,636
Total	151,192			42,133	109,058

3.4. Determination of shelter area based on simulation

The shelter building area is determined based on the number of refugees that will use the shelter as an evacuation facility. The area of shelter that is needed is calculated based on the required space at the shelter. The minimum area required for shelter building space used when evacuating is 2 m^2 according to the Public Works Department, and 1 m^2 for temporary evacuation areas according to BNPB (National Disaster Management Agency). In this study, those requirements are used to obtain the space area that is needed by refugees in the city of Padang. From the data analysis, the result of the predicted shelter is shown in tables 8, 9, and 10.

Table 8. Predicted shelter for Padang city.

	Zone 1	Zone 2	Zone 3	Zone 4	Zone 5
Number of population	24,382	33,189	44,826	36,797	55,969
Number of predicted refugee	5,895	6,947	8,857	7,204	13,230
Shelter Area required when tsunami occur (stand position)	5,895	6,947	8,857	7,204	13,230
Shelter Area required when tsunami occur (lay down position)	11,790	13,894	17,714	14,408	26,461
The number of shelter (each shelter has 2 floor, area 2000m^2/floor)	3	4	5	4	7

Table 9. Predicted minimum shelter for Padang city.

	Zone 1	Zone 2	Zone 3	Zone 4	Zone 5
Number of population	24,382	33,189	44,826	36,797	55,969
Number of predicted refugee	945	892	1,470	1,723	2,004
Shelter Area required when tsunami occur (stand position)	945	892	1,470	1,723	2,004
Shelter Area required when tsunami occur (lay down position)	1,889	1,784	2,941	3,446	4,007
The number of shelters (each shelter has 2 floors, area of 2000m^2/floor)	1	1	1	1	2

Table 10. Predicted maximum shelter for Padang city.

	Zone 1	Zone 2	Zone 3	Zone 4	Zone 5
Number of population	24,382	33,189	44,826	36,797	55,969
Number of predicted refugee	13,130	16,723	19,820	14,994	30,834
Shelter Area required when tsunami occur (stand position)	13,130	16,723	19,820	14,994	30,834
Shelter Area required when tsunami occur (lay down position)	26,260	33,446	39,640	29,987	61,669
The number of shelters (each shelter has 2 floors, area of 2000m^2/floor)	7	9	10	8	16

From the result of simulation in tables 8, 9 and 10, the minimum number of shelter that is needed is 6 shelters, for normal refugee demand condition is required 23 shelters, and in the maximum demand refugees, it is needed 50 shelters. Comparison with the result of Ashar *et al*, they consider the ability of refugee to reach the shelter, the number of required shelter is 15 shelters [8], Yosritzal, *et al* also identified shelter needs, they got 36 shelters [9], and these results are still in the range of the result of simulations.

4. Conclusions

During a large earthquake that has the potential to cause a tsunami, the majority of people who live in tsunami-prone zone evacuate to evacuation routes to reach a safe place. They usually use the vehicle to reach that place, which creates traffic congestion on the main road around tsunami-prone zones. This occurs because the number of building as shelter is limited. This study has predicted the required shelter for the tsunami-prone zone. Based on the refuge simulation using the refugee generation model and refugee choose evacuation facilities model, the required number of shelters could be predicted. A normal refugee generation will require 23 shelters, however, in a minimum refugee condition scenario, it is needed 6 shelters to be built. For maximum refugee conditions, it will require 50 shelters.

Acknowledgments

The authors would like to thank the Transportation laboratory assistant who helps the data collection and the Civil Engineering Department, Engineering Faculty, University of Andalas who support this research.

References

[1] Badan Penanggulangan Bencana Daerah (BPBD) Padang. 2019. Remembering the 2009 Earthquake in Padang City www.bpbd.padang.go.id (accessed, 1 Sept 2022)

[2] McCloskey J 2008 Tsunami threat in the Indian Ocean from a future megathrust earthquake West of Sumatra *Earth and Planetary Sciences Letters* **265** 61-81

[3] Aydan O 2008 Seismic and Tsunami Hazard Potentials in Indonesia with a special emphasis on Sumatra Island *J. of The Sch. of Marine Sci. and Tech.* **6** 19-38

[4] Nurdin, W. 2016. The evacuation route in Padang is at a standstill (in Ind) www.tribunnewa.com (accessed, 1 Sept 2022)

[5] Adries 2021 Padang city has only three shelters (in Indonesia). www.hantaran.co (accessed, 1 Sept 2022)

[6] Tamin O Z 2008 *Planning, modeling and transportation engineering* (in Indonesia) (Bandung: ITB Press)

[7] Permain D and Swason J 1991 *Stated preference techniques: A guide to practice* (London: Steer Davies Gleave and Haque Consulting Group)

[8] Ashar F, Amaratungga D and Haigh R 2014 The analysis tsunami vertical shelter in Padang City *Procedia Economics and Finance* **18** 916-923

[9] Yosritzal, Kemal B M, Aulia Y B 2018 Demand versus capacity of tsunami shelters in Padang, Indonesia *Int. J. on Adv. Sci. Eng. and Inf. Tech.* **8** 1984-1990

IOP Conf. Series: Earth and Environmental Science 1173 (2023) 012019 doi:10.1088/1755-1315/1173/1/012019

Tsunami current and its impact on inundation depth in the coastal areas of West Sumatra – Indonesia

D Daoed[1], **M Syukur**[1] **and P S Ariya**[2]

[1]Civil Engineering Department, Engineering Faculty, Andalas University, Padang, West Sumatra

[2]Undergraduate Student, Civil Engineering Department, Andalas University

E-mail: darwizaldaoed@gmail.com

Abstract. The influence of cities that are in coastal areas and prone to earthquakes is very likely to be hit by tsunami waves or currents. The city of Padang is one of the cities in Indonesia which is located along the Indonesian sea coast and is also an earthquake-prone area so it has the opportunity to experience tsunami waves. Efforts to inhibit wave propagation are mostly carried out by using walls on the beach or multi-story buildings. However, in small towns or villages you rarely find multi-story buildings, there is only drainage parallel to the beach. Based on this, an experiment was carried out by imitating the existing conditions and changing the width of the drainage for the Purus and Ulak Karang Padang areas. The purpose of this study was to determine the effect of changing the drainage geometry on the tsunami current. The results showed that the drainage channel was wider, the greater the decrease in height and area of inundation or vice versa. It is recommended for stakeholders make adequate and safe drainage against tsunami flows.

1. Introduction

1.1. Background and purpose

Most of Indonesia is located in the ring of the fire area and has mountains and faults where at any time there will be an earthquake or volcanic eruption [1] The impact can cause disasters, including buildings, and bridges collapsing, landslides, or floods due to large waves or tsunamis. If the epicenter of the earthquake occurs at sea and the depth is shallow, theoretically a tsunami wave will occur [2,3]. This tsunami wave can reach the shore with the speed of propagation and the force of the waves is still large. The waves will change in the form of currents after passing the coast and into the land, as happened in the 2004 earthquake in Aceh. The speed of the current that comes is still quite large and will drown some coastal areas. The amount of energy or pushing water to land can be reduced by increasing the slope or topography of the land. Despite this, many efforts have been made to reduce energy by using wave retaining walls and indirect retaining through the arrangement of houses, high-rise buildings, or rivers that empty into the sea [4,5,6].

The morphology and direction of river flow in the coastal areas tend to bend and parallel the coast before going to the sea. This form of flow occurs due to strong waves hitting the coast and the influence of tides, so that river currents cannot directly reach the sea [7]. The existence of a drainage/river like this will be beneficial in reducing the flow of seawater entering the land [8]. Where the speed of the river water flow will be restrained and even very slow to reach the sea, and vice versa the sea water will enter the land is also restrained [9].

Based on the reasons above, the research was carried out by simulating the real conditions of the coastal area. The model is made in the form of a three-dimensional building based on actual conditions. The purpose of the study was to observe the effect of ocean currents spilling onto the land. Observations in three dimensions, namely length, width, and height of the water level inundated the area. Simulations are carried out through variations in wave height and speed of current propagation hitting the land. The location of the study was carried out in the Purus and Ulak Karang areas, Padang City, West Sumatra, Indonesia.

1.2. Inundation and run-up of water

A tsunami that reaches the coast or land consists of a series of waves that have different frequencies and amplitudes and can reinforce each other and seriously harm the coast [2,3]. The depth of air that reaches the mainland is about 15 – 30 m with a propagation speed of about 90 km/hour and deep into the ground [10,11]. Floods caused by tsunamis are categorized with two magnitudes, namely risk (inundation) and increase (run-up). Inundation is the maximum distance reached horizontally to the mainland. The run-up is the maximum height inundated by the flood compared to the normal sea level [4]. The destructive power of a tsunami when it arrives is very large, and vice versa, and can happen repeatedly within a few days [12,13].

2. Area of study and methodology

2.1. Area of study

The study was conducted in the coastal area of Padang, West Sumatra, Indonesia, namely the beaches of Ulak Karang and Purus at coordinates 0°55'24.30" S and 100°21'12.12" E. This area is directly adjacent to the West by the Indonesian Ocean, to the east by Jalan Raya Khatib Sulaiman, to the north by Ulak Karang Village, and the south by the Flood Canal. Physically, the location has neighborhood roads, houses, drainage channels, and arterial roads with relatively flat topography. The study location with an area of more than 100 ha contains many residential houses, schools, hospitals, mini-markets, and offices. At the Ulak Karang and Purus regions, there are drainage and temporary water reservoirs as well as roadside channels and environmental canals. The topography of this area is relatively flat, averaging 1.2 km from the coast to the east. The slope is 5/1000 m/m, then rises to. 3.0 km with a slope of about 8/1000 m/m. Then it is steep to very steep and ends in the Bukit Barisan mountains.

Test equipment in. Water tanks are used to generate waves and store water. In the wave generator tank, the water is regulated by varying the slope (h) of the tank. The sliding door functions as a single wave supply. The water height (H) at the sliding gate in the wave-generating tank is assumed to be the same as the wave height reaching the coast. Then the variation of the slope of the tank is to get the variation of the thrust due to waves.

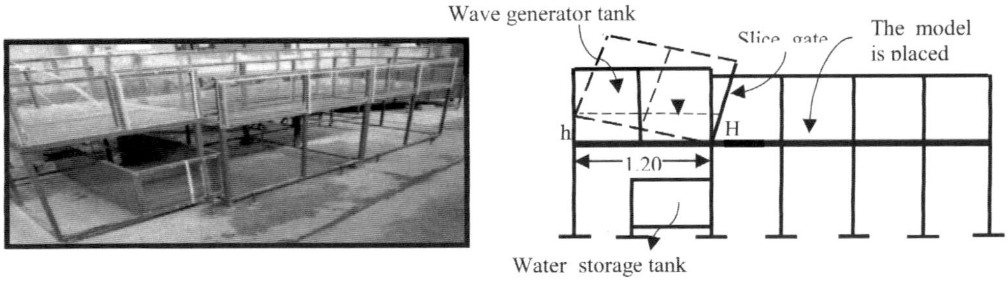

Figure 1. (a) Description of equipment and (b) wave-generating device and two water tanks.

2.2. Methodology

Simulation of the test chamber due to the tsunami was carried out by making a model. The model is imitated from the real conditions of the layout of buildings, houses, shop houses, environmental roads, and drainage. As a basis for making models of real conditions, a map of the earth from Google Earth 2021 is used. Then from the existing map, it is formed into a three-dimensional area condition. Where buildings and roads are built by making buildings using a stereo form. The topography of the Purus and Ulak Karang areas, which are relatively small with a slope of less than 0.5%, is assumed to be flat. Location maps and models are shown in figure 2 (a) maps of the study location and (b) existing model and figure 3 (a) maps of the study location at the Purus and (b) existing model below.)

The area is limited by drainage in the north and south, and from the coast to the Khatib Sulaiman highway in the east. The distance from the beach to the Khatib Sulaiman road is estimated to be approximately 1.2 km with an average slope difference of 5 m/1000 m. All buildings are designed not to deform and not shift due to a single wave blow. The assembled model can be seen in **Figure 2.** (a) maps of the study location at Ulak Karang and (b) model of existing and **Figure 3.** (a) maps of the study location at the Purus and (b) model of existing.

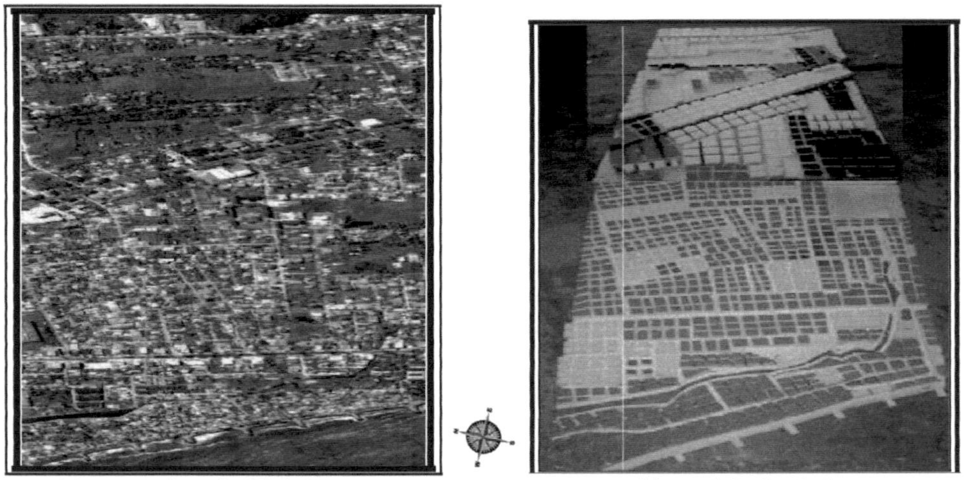

(a) maps of study Ulak Karang (b) model of existing

Figure 2. (a) maps of the study location at Ulak Karang and (b) model of existing.

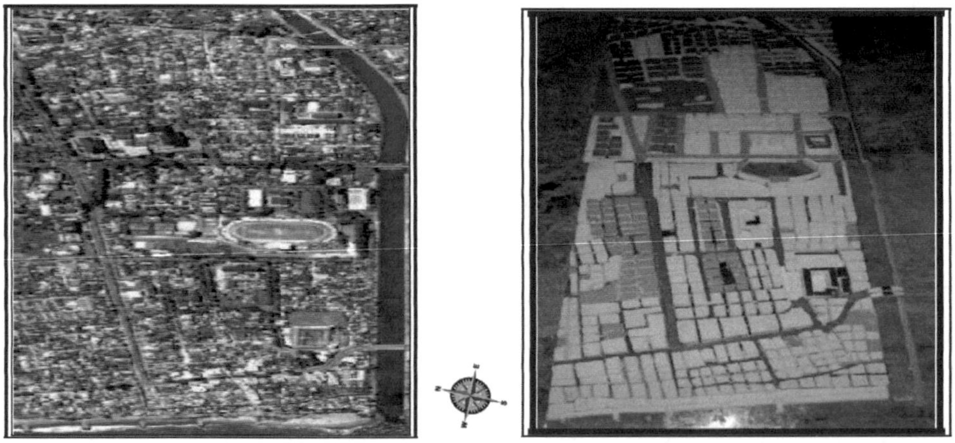

(a) maps of study Purus (b) model of existing

Figure 3. (a) maps of the study location at the Purus and (b) model of existing.

The height of the tsunami waves reaching the shore is simulated with an initial height of H-20 cm (model) equivalent to a tsunami height of 30 m. Furthermore, each treatment measured the water depth through indicator paper. Indicator paper is installed at each point in a grid, like small flags placed on the area. The area is divided into several squares or grids (2.0 x 2.0 cm). Observation of the wet part is assumed to be the depth of the water that occurs. Then to get an idea of the effect of drainage on tsunami currents, variations are made to the width of the drainage to be three as model -1 and four times the existing width model -2.

3. Result and discussion

3.1. *Result*

After testing (simulation) using a wave generator test equipment, the direction of the water flow and the water level in each grid is monitored. The height of the wet part on the indicator paper/flag is to indicate the water level that inundated the study location. Then the results of measuring the water level from each coordinate point on the grid can be contoured using AutoCAD software. The following **Error! Reference source not found.** and Figure 5. Contour lines for existing and modified models at wave height H 20 cm or tsunami height 30 m Purus Region.

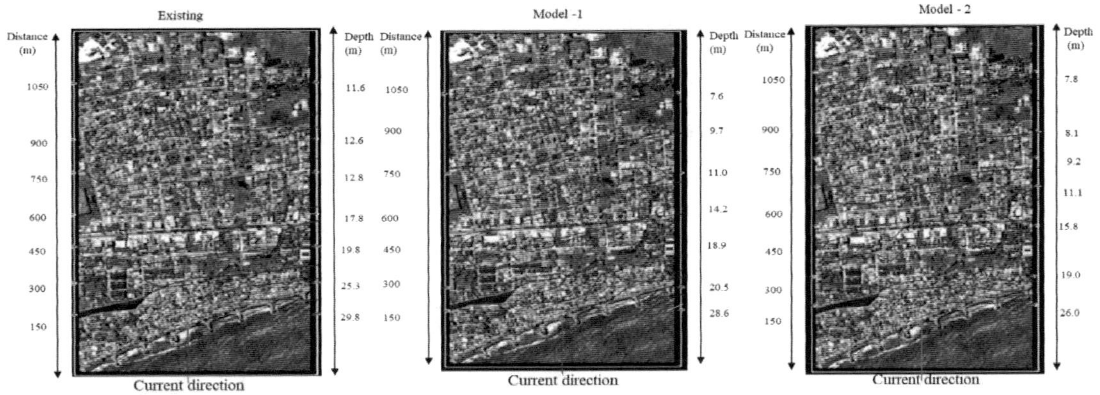

Figure 4. Contour lines for existing and modified models at wave height H 20 cm or tsunami height 30 m Ulak Karang Region.

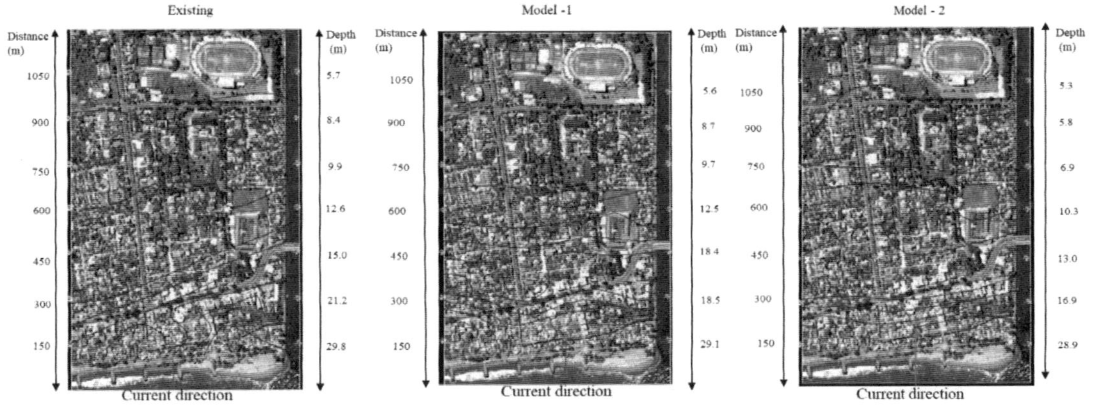

Figure 5. Contour lines for existing and modified models at wave height H 20 cm or tsunami height 30 m Purus Region.

It can be seen that the contour of the water level changes according to the magnitude of the resistance through which the water passes. At first, the water level was still high, were on the beach it was between 29 to 30 m and the depth at a distance of 1.2 m was still around 11.5 – 12.0 m, while in model -1 and model -2 the inundation height had dropped to 7.5 m. But model -3 and model - 4. are not very different. In the results of running the Purus location, there are results of the contours of the water level changing according to the magnitude of the obstacles that the water passes through. In the existing model, the water level is still high, we're on the beach it is between 29 to 30 m and the depth at a distance of 1.2 m is still around 5.7 – 6.0 m, while in model -1 and model -2 the inundation height has dropped to 5.3 m. But model -3 and model - 4. are not very different.

3.2. Discussion

When observed, it is seen that there is an effect of drainage on changes in inundation depth and inundation length measured from the coast. Figure 6. Relationship water depth to distance in Ulak Karang Region and Figure 7. Relationship water depth to distance in Purus Region shows the relationship between the length of the inundation from the beach to the water level. The water depth dropped drastically, both in the Ulak Karang and Purus study areas. In the Purus area, the water depth is up to 5.0 m at a distance of 1050 m from the beach, while in the Ulak Karang area it is around 7.0 – 8.0 m at the same distance from the beach. The depth of inundation decreases with changes in the width of the drainage. In the area of Ulak Karang model -1, the inundation depth at a distance of 600 m is still 15.0 m and the model -2 is around 11.0 m. The decrease in water level gradually decreases to a distance of 1050 m, where the low water level is almost the same between model-1 and model-2. It can be concluded that the effect of drainage widening at a distance of more than 1.0 km is relatively the same. This also occurs in the Purus area, where its behavior is similar to its neighboring area. Although the decrease in inundation depth is sharp and at a distance of 600 m from the beach model -1 is around 12.0 -13.0 m and in model -2 it is already 10.0 m. For a distance of more than 1050 m from the beach, the depth is close to the same, namely 5.0 m. this form gives the same reasons as the Ulak Karang study area with the Purus area.

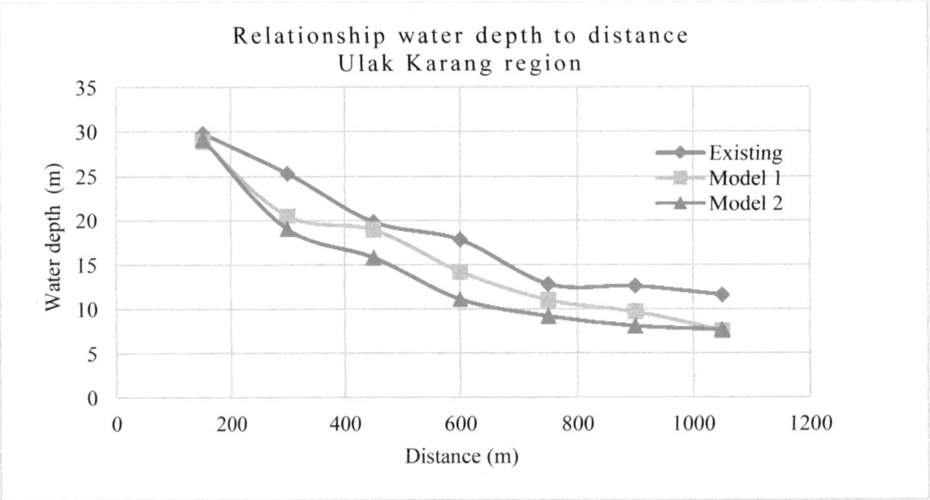

Figure 6. Relationship water depth to distance in Ulak Karang Region.

What is interesting is that at the Ulak Karang study site, the inundation height from the north to the south is rather flat. This means that the depth is almost the same, with the possibility of differences due to buildings located at that location. Buildings look orderly, so the energy loss due to obstacles is also evenly distributed. In contrast to the study site the Purus, the water level declined unevenly. Where near the canal (flood channel) the depth is different from the area far from the canal. Another difference occurs, due to the influence of the length of the drainage in the Ulak Karang area stretching from north to south, while in the Purus the drainage does not fill as wide as the area, there is also a flooded canal in the south. This canal is quite wide and perpendicular to the coast and at the same time helps in dampening the currents of seawater flowing on land. This result is also similar to the previous finding, where the flow will be faster in the area near the river to be compared to the speed of water flowing in other dense housing.

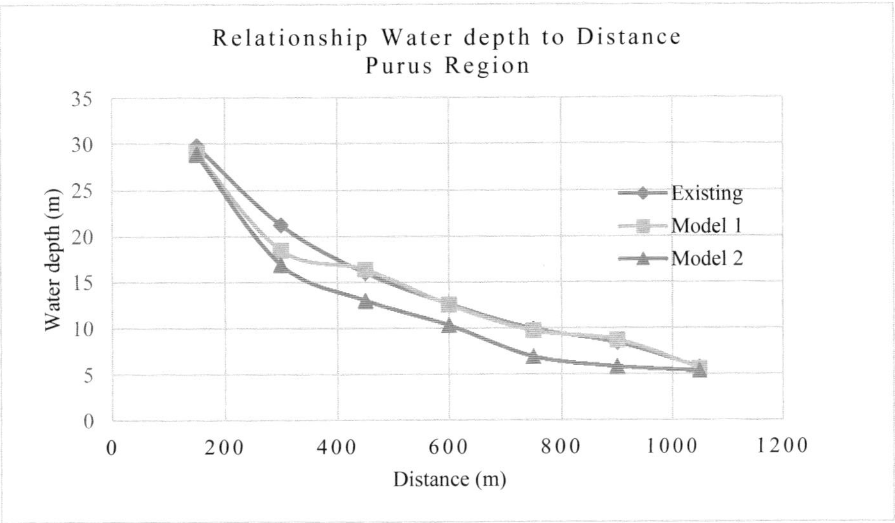

Figure 7. Relationship water depth to distance in Purus Region.

The results obtained are similar to those obtained in previous studies, namely the effect of high-rise buildings as a damper for tsunami wave currents and the effect of building density on the presence of a river estuary perpendicular to the coast [13,14,15,16]. This form also indicates that the area is safe to traverse when a disaster occurs, which is to stay away from the river and immediately look for a tsunami-safe place [5,17,18]. Based on the findings above, it can be further developed by increasing the number of drainages/channels parallel to the coast, so that the greater the resistance and the greater the energy that can be attenuated. At the same time minimizing the risk of inundation and slows the speed of water reaching a safe location against a tsunami. Deceleration of the tsunami flow that enters the land will provide opportunities for residents to move to safer places [5,19,20].

4. Conclusion

Based on the phenomenon and description of the investigation, it can be said as follows. The length and depth of water can be shortened by drains parallel to the coast. The wider the drainage, the lower the water depth or the smaller the puddle length. The larger the channel width, the greater the damped energy. This seems to further reduce the impact of the current. For the study area that is densely built and parallel to the coast, the flow depth is evenly distributed. On the other hand, for sparsely built areas, the flow depth is uneven. The length and depth of inundation can be shortened by having drainage channels parallel to the coast.

Acknowledgments

This publication was funded by Andalas University through a publication grant from the Faculty of Engineering, LPPM, Andalas University, and I would like to thank the Head of the Civil Engineering Department, Vice, Dean-1, Dean, and the Rector for their help.

References

[1] Hinga B D R 2015 R*ing of Fire: An Encyclopedia of the Pacific Rim's Earthquakes, Tsunamis, and Volcanoes: An Encyclopedia of the Pacific Rim's Earthquakes, Tsunamis, and Volcanoes* (Santa Barbara, California: ABC-CLIO) pp.1-403

[2] Gupta H K and Gahalaut V K 2013 *Three Great Tsunamis: Lisbon (1755), Sumatra-Andaman (2004) and Japan (2011)* (London: Springer Science & Business Media)

[3] Gupta H and Gahalaut V 2009 Is the northern Bay of Bengal tsunamigenic? *Bulletin of the Seismological Society of America* **99**(6) 3496-3501

[4] Nurhasanah A, et al 2008 High-rise Buildings in Coastal Areas as a Tsunami Run-up Reducer, (Indonesia language: Bangunan Bertingkat pada Kawasan Pesisir sebagai Pereduksi Run-up Tsunami) In: *Pertemuan Ilmiah Tahunan XXV Himpunan Ahli Teknik Hidraulik Indonesia, 21-23 Agustus 2008, Palembang*

[5] Daoed D, Febriansyah M D, and Syukur M 2013 Physical model of the direction of the tsunami wave flow in the Purus and Ulak Karang Padang areas, (Indonesia language: Model fisik arah aliran gelombang tsunami di daerah Purus dan Ulak Karang Padang) *Jurnal Rekayasa Sipil (JRS-Unand)* **9**(2) 20-30

[6] Bernard E N 2005 The US national tsunami hazard mitigation program: A successful state-federal partnership, in Developing tsunami-resilient communities *Natural Hazards* **35** 5-24

[7] Marsudi S and Lufira R D 2021 *River Morphology (Indonesia language: Morfologi Sungai* (CV. Ae Media Grafika)

[8] Leopold L B and Langbein W B 1966 River meanders *Scientific American* **214**(6) 60-73

[9] Hariati F, Taqwa F M L, and Qonita N S 2022 Study of Seawater Intrusion due to Backwater at the Citanduy . River Estuary, (indonesia languagae: Kajian Intrusi Air Laut akibat Backwater di Muara Sungai Citanduy) *Jurnal Komposit: Jurnal Ilmu-ilmu Teknik Sipil* **6**(2) 93-98

[10] Esteban M, et al. 2017 Overtopping of coastal structures by tsunami waves *Geosciences* **7**(4) 121

[11] Laso Bayas J C, et al. 2011 Influence of coastal vegetation on the 2004 tsunami wave impact in west Aceh in P*roceedings of the National Academy of Sciences* **108**(46) 18612-18617

[12] Ranteallo R, Nurul N and M. Nizam 2021 *Analysis of Respondents Characteristics on Understanding of Tsunami Disaster Mitigation in Maloku Village, Losari Beach, (Indonesia language: Analisa Karakteristik Responden terhadap pemahaman akan Mitigasi Bencana Tsunami di Desa Maloku Pantai Losari)* (Sensistek: Riset Sains dan Teknologi Kelautan) 92-97

[13] Darwizal D, Junaidi A, Sunaryo, Hendra S, Irawan B B 2021 The Existence of multi-story buildings to reduce tsunami effects in the city of Padang, West Sumatra *IOP Conf. Ser.: Earth and Env. Sci.* **708** 012016

[14] Diningrat L W A 2018 *Numerical Study Of Flow Depth And Tsunami Force Due To Variations Of Building Orientation, (Indonesia language: Studi numerik kedlaman aliran dana gaya tsunami akibat variasi orientasi bangunan)* (Mataram: Universitas Mataram)

[15] Triatmadja R and Nurhasanah A 2012 Tsunami force on buildings with openings and protection. *Journal of Earthquake and tsunami* **6**(04) 1250024

[16] Dean R G and Dalrymple R A 1991 *Water wave mechanics for engineers and scientists* (world scientific publishing company) Vol. 2

[17] Barrick D E 1979 A coastal radar system for tsunami warning *Remote Sensing of Environment* **8**(4) 353-358

[18] Murty T 2003 Tsunami wave height dependence on landslide volume *Pure and applied geophysics* **160**(10) 2147-2153

[19] Thusyanthan N I and Madabhushi S G 2008 Tsunami wave loading on coastal houses: a model approach in *Proceedings of the institution of civil engineers-civil engineering* Thomas Telford Ltd (London: ICE Publishing) pp.77-86 Paper 07-00037

[20] Fukutani Y, Anawat S, and Imamura F 2016 Uncertainty in tsunami wave heights and arrival times caused by the rupture velocity in the strike direction of large earthquakes *Natural Hazards* **80**(3) 1749-1782

IOP Conf. Series: Earth and Environmental Science 1173 (2023) 012020 doi:10.1088/1755-1315/1173/1/012020

The influence of helix diameter configuration on helical pile bearing capacity

G Amalia[1], M M Iqbal[2] and R Dewi[2]

[1]The Civil Engineering Departement, Engineering Faculty of Universitas Indo Global Mandiri, Indonesia
[2]The Civil Engineering Departement, Engineering Faculty of Universitas Sriwijaya, Indonesia

E-mail: ghina@uigm.ac.id

Abstract. Helical pile is an invention of pile foundation that has many advantages. Helical pile produced bearing capacity greater than other pile foundations. It because The Bearing capacity of helical pile is contributed by end bearing capacity (Qb), shaft friction bearing capacity (Qs) and bearing capacity of helix or cylindrical (Qcyc) along pile. On of the Several factors that influence bearing capacity of helical pile was the configuration of helix diameter arrangement. This research was a laboratory experiment using a box, sand as the medium and a miniature of steel pile. The diameter pile test was 1,5 cm and length was 63 cm. Each pile has three helix along the pile with a configuration of diameter from top to bottom was 10 cm – 8 cm – 6 cm and from top to bottom 6 cm – 8 cm – 10 cm. The kind of Loading test was used in this research was Quick Maintain Load (QML). From the research, It was found that the variation wich gave the most significant influence in remained the greatest bearing capacity was the pile with diameter configuration of helix from top to bottom 6 cm – 8 cm – 10 cm or pile D6810.

1. Introduction

Helical pile is a modification of pile foundation. Helical pile was invented by Alexander Mitchel in 1833. The difference between helical pile and other pile foundation is helical pile was added some steel plate named helix along the pile with certain spacing distance between helix which could make installation process easier. In general, the bearing capacity of pile foundation is a contribution of end bearing capacity and shaft friction bearing capacity along pile. But in helical pile, the bearing capacity is not only contributed by end bearing capacity, shaft friction bearing capacity but also by bearing capacity of helix. The helical pile has resistance of tensile loads, compressive loads, lateral loads and overturning momen [1,2].

The used of helical pile nowadays widespread especially on problem soils. It was caused helical pile has many advantages such as the installation process is not depended on climate and weather, neither need more effort nor cause vibration that can disturb the environment around the helical pile, do not require maintenance time after installation so that the loading test could be carried out immediately. But of the all advantages, the more important is helical pile produced bearing capacity greater than other pile foundation . Hamdy H A et all was said the similar thing that the installation process of helical pile does not damage the soil condition around helical pile because it does not remove the original soil [3].

Several previous research have been carried out to identify what factors could influenced the bearing capacity of helical pile. Hamdy H A et al in their research presented the compressive and tensile bearing capacity of helical pile on sandy soil [3]. In addition, they also research about what factors were able to increase the value of the compressive and tensile bearing capacity of helical pile. The results of the laboratory experiment concluded that the bearing capacity value was strongly influenced and directly proportional to the ratio of the depth of the foundation and the diameter of helical pile [3].

Sena bayu researched to calculated the value of bearing capacity caused by the increased of number of helix. Sand and clay was used as a test medium. The test result for the variation in the number of helix showed that the bearing capacity of the helical foundation has increased by 9.38% in clay soil and 5.58% in sandy soil due to the increase in the number of helix. While the result in variation of helix diameter increased of 50% also showed the bearing capacity increased by 19.6% on clay soil and 10.83% on sandy soil.

While likitha et al conducted an experiment to analyse the influence of the helix on the bearing capacity of helical pile. The experiment was carried out on a laboratory scale in cohesive soils [4]. So the results of the experiment conducted indicated the bearing capacity of the helical pile was significantly influenced by the helix diameter size. The value of the bearing capacity of the helical pile was directly proportional to the increase in the diameter size of the helix [4].

2. Materials and methodology

2.1. Materials

This research used several materials such as:

2.1.1. Sand. Sand was used as testing medium. Before been used, the first step was sand must been dried. Then, some dried sand was sampled for testing in the laboratory. From laboratory testing, the parameters of sand properties are obtained in table 1

Table 1. The parameters of sand properties

Soil Properties	Value
Specific Gravity, Gs	2.639
Sieve Analysis	$C_u = 3$ dan $C_c = 1,5$
Total Unit Weight, γ (KN/m^3)	13.92
Water Content, w (%)	0
Friction Angle, φ	30°

2.1.2. Testing instrument. The test was carried out on a laboratory scale using a test box measuring 210 cm of length, 100 cm wide and 95 cm high. The test box was placed under a load frame which aims to withstand the force exerted by the hydraulic jack. The hydraulic jack was used as a load on the test object had a capacity of 2 tons. To measured the settlement, two LVDT were used which placed on the left and right sides of the load-bearing plate, while to measure the load was used a 5 tons capacity of load cell. LVDT and Load cell were connected to Data logger TDS-302 series to read the value of settlement and load [5]. The installation of the testing instrument illustration could been seen in figure 1.

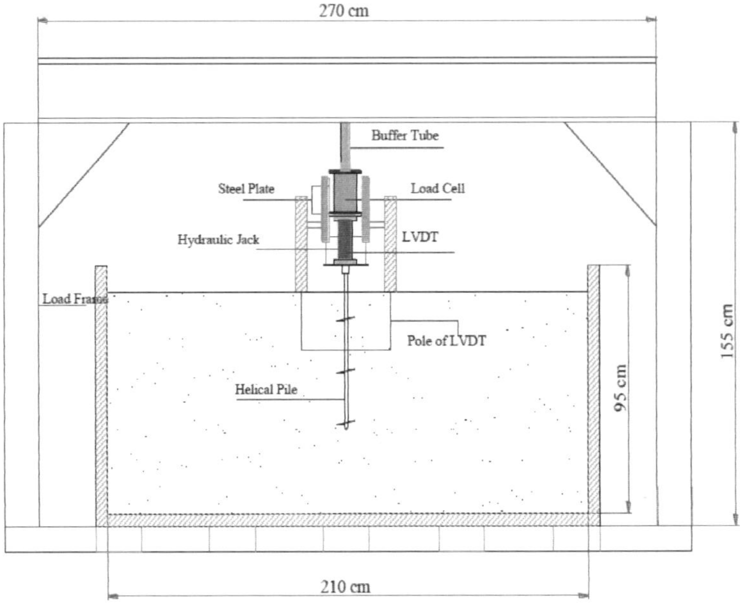

Figure 1. The illustration of testing instrument installation.

2.1.3. Helical pile. This research used a helical pile model wich made of steel with a 1.5 cm of pile diameter and 63 cm of the length. There were three helix along the pile. The helix were also made of steel plates with the diameter was 10 cm, 8 cm and 6 cm while the thickness was 1 mm. This research used 6 varies of helical pile as the test object with varying diameters and spacing were used as shown In table 2 while The illustration of helical pile could been seen in figure 2.

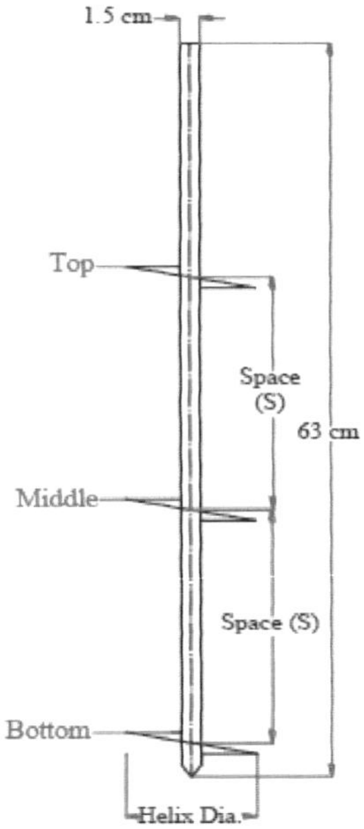

Figure 2. The illustration of helical pile

Tabel 2. Variation of the helical pile

Spacing between helix (s) (cm)	Diameter of Helix (cm)			Pile Notation
	top	middle	bottom	
10	6	8	10	D6810-10
	10	8	6	D1086-10
15	6	8	10	D6810-15
	10	8	6	D1086-15
20	6	8	10	D6810-20
	10	8	6	D1086-20

2.2. Methodology

Before loading test was started, dried sand was put into the test box which was placed under the load frame. Dried sand was added layer by layer which the thickness of each layer was approximately 30 cm and followed by a little compaction using a plate compactor. When the elevation of sand reached 90 cm, helical pile was inserted by turning it slowly to a predetermined depth of 43 cm. After that, the retaining plate was fitted on top of the helical pile as a footing for the testing instrument. On the plate successively arranged hydraulic jacks, load cells, and buffer tubes until they reach the elevation of the load frame. While LVDT was installed on the left and right sides of the plate. All instruments were arranged symmetrically so that the resultant load was parallel to the helical pile axis. Illustration of the installation of the test instrument can been seen in figure 1.

The loading test in this study was carried out using the ASTM D-1143 procedure. The type of loading test in this study includes the Quick Maintained Loading Test (QML) type of loading test. This method was relatively faster than other methods required by ASTM. In the QML test, the load increase was applied gradually every 5% of the design load which has been calculated using the cylindrical bearing method from Mitsch and Clemence until the load failure was reached. The load increase was held as fast as 4 minutes but not more than 15 minutes. The Recording was carried out at time periods of 0, 5, 10, 15, and so on at each multiple of the 5-minute time interval for each load increase. When the pile had reached failure, the loading test in the study was stopped. ASTM D 1143 stated that the test is stopped if the load reaches 1.5 to 2x the design load [5].

3. Result and discussion

The loading-settlement curve shape resulting from the loading test generally produces a curve that showed three phases of change in the behavior of the test object. There were three areas on the curve each indicating a changed phase of helical pile behavior [6,7,8].

The first phase was the linear-elastic phase whereas the load increases, the settlement that occurs was still constant so it created a linear curve. The second part was the transition phase of the object's behavior from the elastic phase to the plastic phase where in this phase an imbalance begins to occur between the given load and the settlement that occurs, in this phase the curve begins to take the form of a parabola. The third phase or linear-plastic phase is indicated by the addition of the same load or even without the addition of load, the settlement that occurs was very large, in this phase the stiffness of the test object was reduced or even lost [6,7,8].

Figures 3(a), 3(b), and 3(c) showed the load-settlement curve on the influence of helix diameter configuration represented by notated test object D06810-10 and D1086-10; D06810-15 and D1086-15; D06810-20 and D1086-20.

It could be seen in figures 3(a), 3(b), and 3(c) that helix diameter configuration on helix arrangement along the pile produced a significant difference of bearing capacity each helical pile

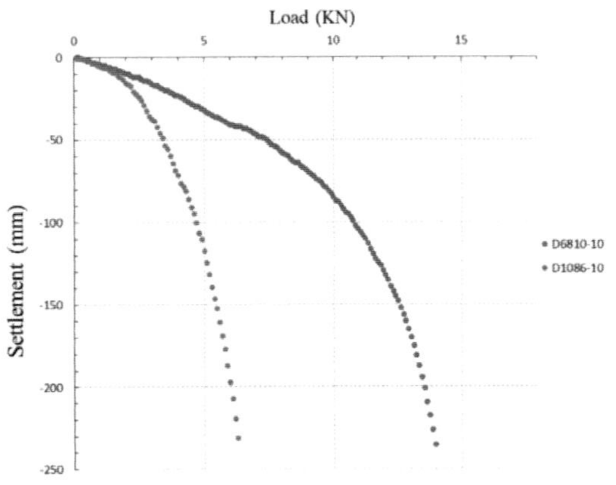

(a)

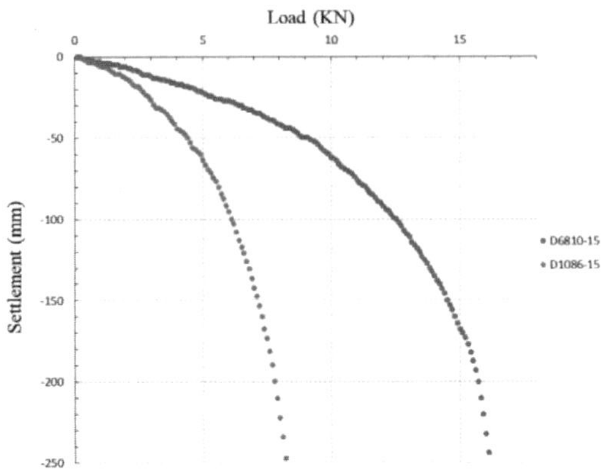

(b)

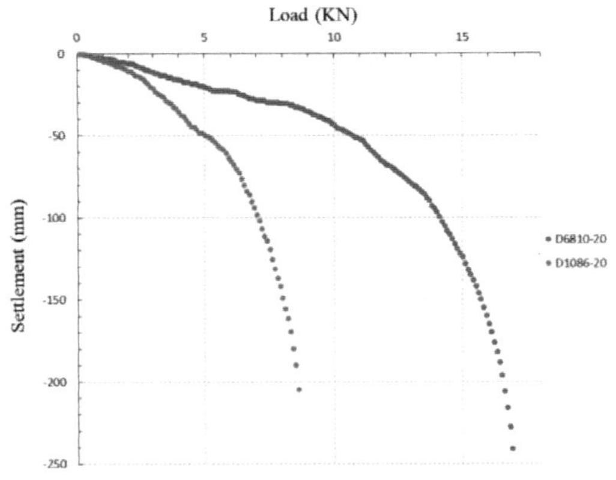

(c)

Figure 3. Load-settlement of d6810 and d1086, (a) spacing between helix 10 cm, (b) spacing between helix 15 cm, (c) spacing between helix 20 cm.

To get the ultimate bearing capacity value of each test object, in addition to using the empirical method, a graphical method was also used based on the interpretation of the loading test data. The graphic method was used in this research was the dee beer method, the chin method and the davisson method [9,4]

3.1. Bearing capacity of helical pile using empirical method

The test objects used in this research were designed to be qualified the cylindrical failure method where the specimen has a ratio of S/D ≤3 and used sand as the soil medium. So, this research used the cylindrical failure method for non-cohesive from Mitsch and Clemence as the empirical method formula to determine the bearing capacity. The formula was shown in the following [10].

$$Q_{ult} = Q_b + Q_s + Q_{cyc} \tag{1}$$

$$Q_b = N_q.\gamma'.H.\frac{\pi\,(D^2 - d^2)}{4} \tag{2}$$

$$Q_s = \pi . d . H_{eff}^{2}\left(\frac{1}{2}.\gamma'.K_s . \tan\varphi^o\right) \quad (3)$$

$$Q_{cyc} = \pi . D . H_{eff}^{2}.\left(\frac{1}{2}.\gamma'.K_s . \tan\varphi^o\right) \quad (4)$$

Where :

Q_{ult} : Ultimate bearing capacity (kN)
Q_b : End bearing capacity (kN)
Q_s : Shaft friction bearing capacity (kN)
Q_{cyc} : Cylindrical bearing capacity (kN)
N_q : Bearing capacity factor
γ' : Submerged unit weight (kN /m^3)
D : Helix diameter (m)
d : Pile diameter (m)
H : Depth of bottom helix (m)
K_s : coefficient of active earth pressure
φ^o : Angle friction
H_{eff} : Bottom helix depth – top helix depth (m)

The table below is a recapitulation of the the empirical bearing capacity for each test object calculation.

Table 3. Recapitulation of the empirical bearing capacity.

Pile	*Cylindrical Shear Failure Method* (kN)			
	Q_b	Q_s	Q_{cyc}	Q_u
D6810-10	1.20	0.03	0.20	1.43
D6810-15	1.20	0.01	0.45	1.67
D6810-20	1.20	0.00	0.81	2.01
D1086-10	0.42	0.02	0.20	0.64
D1086-15	0.42	0.01	0.45	0.88
D1086-20	0.42	0.00	0.81	1.22

3.2. Bearing capacity of helical pile using graphic method

The figure below showed several graphical methods to obtained the bearing capacity value based on field loading test data. The data interpretation to obtained the carrying capacity on the chin method was shown in figure 4. The formula below

$$Q_u = \frac{1}{c_1} \quad (5)$$

Where c_l is the slope of the linier regression line coefficient from the loading test data. Figure 4 showed c_l value was 0.0648. Therefore, the bearing capacity was 15.4 kN and after corrected by the safety factor the bearing capacity was 12.86 kN.

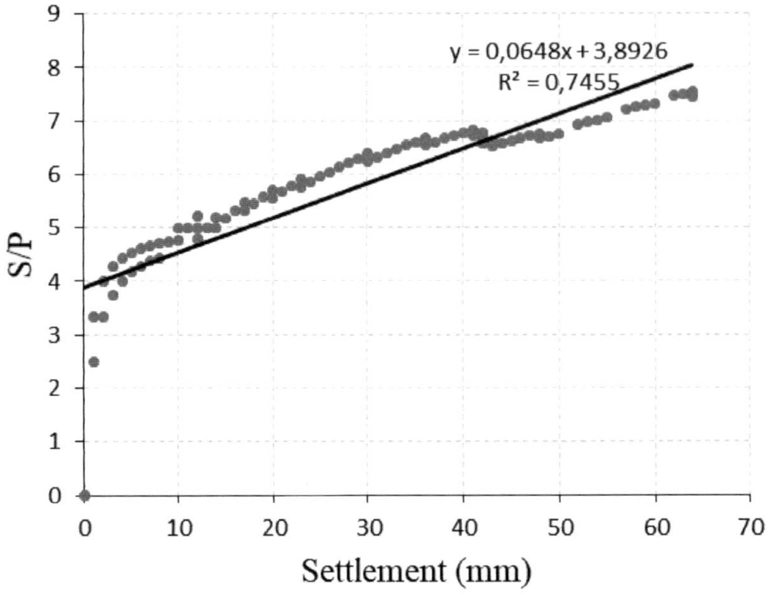

Figure 4. Bearing capacity using Chin method for d6810-10 pile.

Figure 5 showed De Beer interpretation data method. Loading-test data was plotted on double-logarithmic graph. It produced a graphic seemed to intersect each other. The ultimate bearing capacity was obtained from that intersection point. Figure 7 showed the ultimate bearing capacity value for the D6810-10 pile was 8.7 kN.

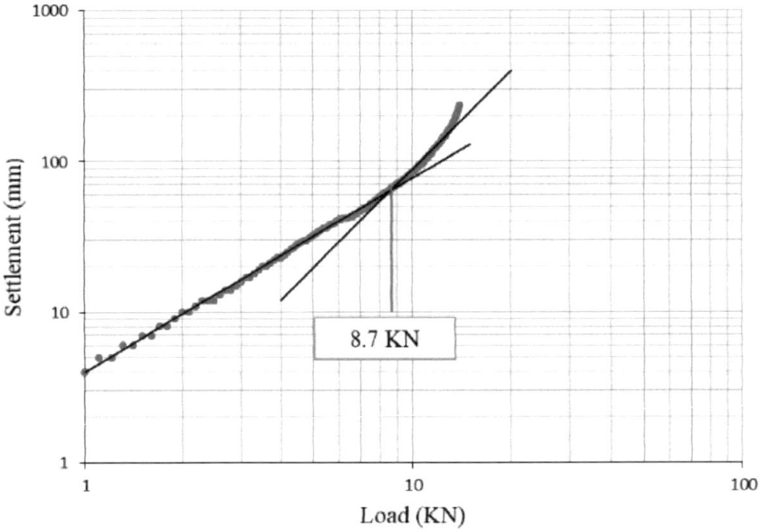

Figure 5. Helcial bearing capacity using De Beer method for d6810-10 pile.

Davisson's method shown in figure 6 was obtained the ultimate bearing capacity was the intersection of the second linier line against the loading graph. In the graph above, bearing capacity for the D6810-10 pile was 2.5 kN.

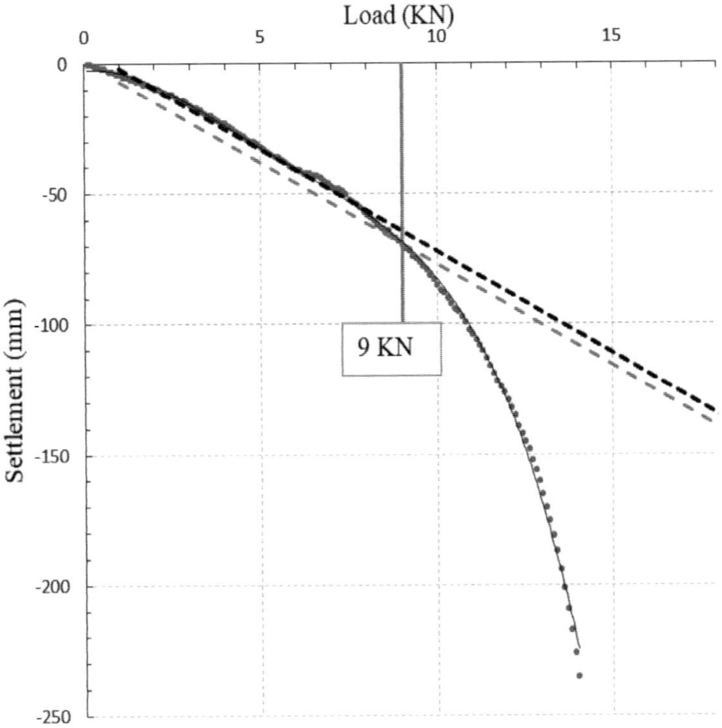

Figure 6. Bearing capacity using Davisson method for d6810-10 pile.

Table 4. Recapitulation of ultimit bearing capacity using graphic method.

Pile	Ultimit Bearing Capacity, Qult (kN)		
	De Beer	Davisson	Chin
D6810-10	8.7	9	12.86
D6810-15	9.3	9.5	14.05
D6810-20	10	9.5	18.08
D1086-10	4	2.8	5.72
D1086-15	4.8	3	8.04
D1086-20	5.7	5.1	8.43

3.3. The influence of helix diameter configuration D810, D1086 variables

The influence of helix diameter configuration were represented by D6810 and D1086 variables will be known after bearing capacity data of helical pile been analyzed. Table 5 showed the recapitulation of the ultimate bearing capacity of the helical pile based on the results of empirical calculations and data interpretation analysis also the ratio of comparison between the ultimate bearing capacity using the analysis method for interpreting the load testing data and the ultimate bearing capacity using the empirical cylindrical method.

Table 5. Recapitulation of bearing capacity based on empirical method and interpretation data analysys.

Pile	Ultimit Bearing Capacity, Qult (kN)						
	Empiris (Cylindrical)	Analysis of Load Testing Data Interpretation					
		De Beer		Davisson		Chin	
D6810-10	1.43	8.7	(6.08)	9	(6.29)	12.86	(8.99)
D6810-15	1.67	9.3	(5.57)	9.5	(5.69)	14.05	(8.41)
D6810-20	2.01	10	(4.98)	9.5	(4.73)	18.08	(8.99)
Rata-Rata Rasio			(5.54)		(5.57)		(8.80)
D1086-10	0.64	4	(6.25)	2.8	(4.38)	5.72	(8.94)
D1086-15	0.88	4.8	(5.45)	3	(3.41)	8.04	(9.13)
D1086-20	1.22	5.7	(4.67)	5.1	(4.18)	8.43	(6.91)
Rata-Rata Rasio			(5.46)		(3.99)		(8.33)

Note: ( ) = Ratio between ultimate bearing capacity method of data interpretation loading test

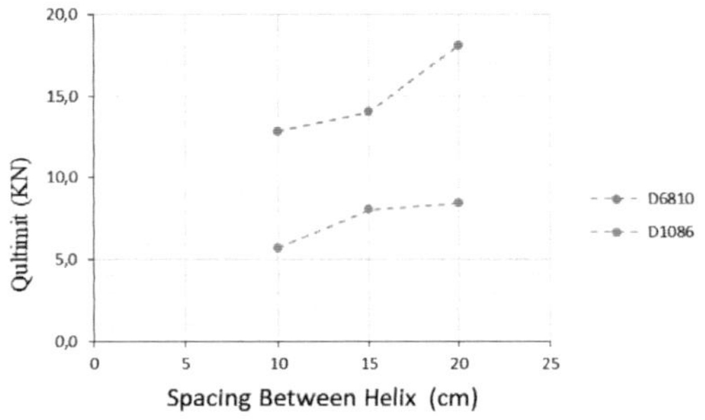

(a)

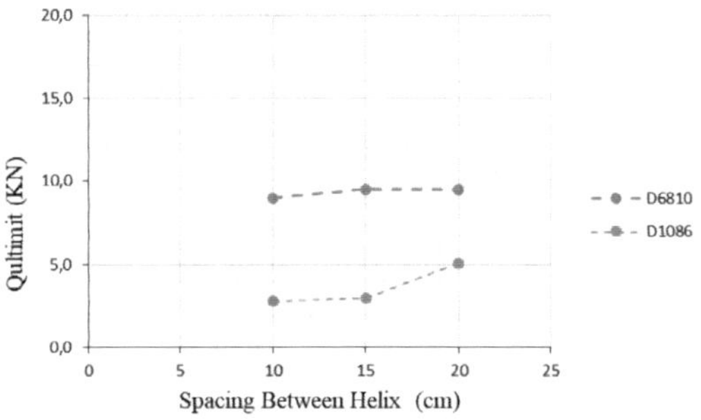

(b)

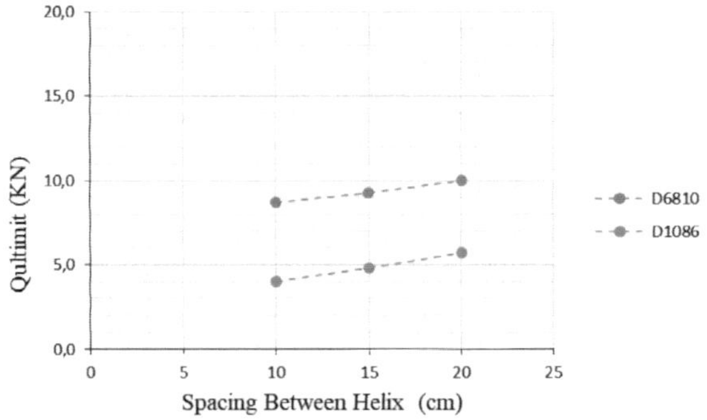

(c)

Figure 7. The comparison of ultimit bearing capacity between helix diameter configuratioan D6810 and D1086, (a) using chin method, (b) using davisson method,(c)using de beer method.

Table 6. Persentage of bearing capacity increment due to the influence of helix configuration D1086 and D6810

Analysis Method	Pile	Persentage of Bearing Capacity Increment (%)	Persentage of Avarege Increment
Chin	D1086-10 and D6810-10	125%	105%
	D1086-15 and D6810-15	75%	
	D1086-20 and D6810-20	114%	
Davisson	D1086-10 and D6810-10	221%	175%
	D1086-15 and D6810-15	217%	
	D1086-20 and D6810-20	86%	
De Beer	D1086-10 and D6810-10	118%	96%
	D1086-15 and D6810-15	94%	
	D1086-20 and D6810-20	75%	

Based on figures 7a, 7b, 7c and table 6, it can be analysed how the influence of the helix diameter configuration on the ultimate bearing capacity produced by calculating the percentage of bearing capacity Increment that occurs due to differences in helix diameter configurations. Meanwhile, the influence of the spacing between helix can be seen visually from the form of the relationship between the ultimate bearing capacity on each spacing variabel, whether it forms a steep or gentle curve. When the curve formed stepper, the influence of these variables on the ultimate bearing capacity of the pile more significant.

The pictures above showed clearly that there was a very big difference between the bearing capacity produced by the D6810 pile and the bearing capacity of the D1086 pile configuration where the ultimate bearing capacity curve produced by the D6810 pile was always above the D1086 pile curve. The average percentage of ultimate bearing capacity increment due to differences in helix configuration with the Chin, Davisson, De Beer and Empirical methods, respectively was 105%, 175%, 96% and 93%. So that the configuration that was able to gave a very significant infuence on the helical pile ultimate bearing capacity increment was the D6810 configuration. While the influence given by the spacing between helix on the helical pile bearing capacity was not significant. This can be seen from the slope of the ultimate bearing capacity curve that was produced a gentle curve.

If viewed from the total ultimate bearing capacity composition of helical pile with the Cylindrical Shear Failure method, as previously explained that the end bearing capacity or Qb gave the most dominant contribution. The end bearing capacity or Qb was strongly influenced by the diameter of the bottom helix, so that a helcial pile with a large bottom helix diameter will produce larger Qb value. While Qcyc, the diameter used was the average diameter of the three helix, so that both D6810 and D1086 will produce the same Qcyc. This was what causes the D6810 pile had a bigger bearing capacity than the the D1086.

4. Conclusion

This research get the following conclusion such as :

1. There was a very big difference between bearing capacity produced by the D6810 pile configuration and the bearing capacity of the D1086 configuration where the curve produced by the D6810 was always far above the D1086 curve. The average percentage of ultimate bearing capacity increment due to differences in helix diameter configurations using the Chin, Davisson, De Beer and Empirical methods, respectively was 105%, 175%, 96% and 93%.
2. The most significant influenced variable was the D6810 configuration, this was due to the value of the end bearing capacity which will increase as the bottom helix diameter increases [12,12]. While the value of the cylindrical bearing capacity is not affected by the configuration arrangement of the thread diameter which enlarges downwards (D6810) or shrinks to the bottom. Or in other words, the effectiveness of the configuration can be seen in the

arrangement of the thread diameters, the deeper the thread diameter, the larger the thread diameter [13].

Reference

[1] Muhajerani A, Bosnak D, mwich D B 2016 *The Japanese Geotechnical Society, Souls and Foundation* **56** 115-128 [1]
[2] Sprince A and Pakrastinsh L 2010 Helical pile behaviour and load transfer mechanism in different soils In *Proc. 10th International Conference Modern Building Materials, Structures and Techniques* pp.1174-1180 [2]
[3] Hamdy H.A. Abdel-Rahim, Yehia Kamal Taha and Walla El din El sharif Mohamed 2013 *J. of Eng. Sci.* **41** 2055-2064 [8]
[4] Likhitha H, Raghavendra H N, Rakesh K P, Uday Shrihari P 2017 *International Journal of Innovative Research in Science, Engineering and Technology* **6** (7): 12666 [11]
[5] ASTM.D1143 *Standard Test Methods for Deep Foundations Under Static Axial Compressive Load* (Pennsylvania: ASTM International) [3]
[6] Livneh B, El Naggar M H 2008 *Canadian Geotechnical Journal* **45**(8) 1142-1155 [4]
[7] Hardiyatmo H C 2017 *Mekanika Tanah I* (Bandung: Gajah Mada University Press) [9]
[8] Hardiyatmo H C 2018 *Mekanika Tanah II* (Bandung: Gajah Mada University Press) [10]
[9] Fatnantaa F, Ongko A 2018 Bearing Capacity Analysis of Helical Pile *EDP Sciences* [7]
[10] Das B M and Sobhan K 2014 *Principle of Geotechnical Engineering* (Stamford: CT Cengage Learning) [6]
[11] Nugroho S A 2011 *Jurnal Teknik Sipil ITB* **18** 31-40 [12]
[12] Yenginar Y 2017 *In Proc. 3rd Introductional Soil-Structure Interaction Syimposium-izmir* (Turkiye) pp.692-702 [13]
[13] Birid, Kedar 2017 In: Abu-Farsakh, M., Alshibli, K., Puppala, A. (eds) Advances in Analysis and Design of Deep Foundations *GeoMEast Sustainable Civil Infrastructures* (Egypt: Springer, Cham) [5]

The influence of various parameters of physical and mechanical properties on susceptibility to liquefaction of sandy soils

R Yuliet[1], A L Silvy[1], A Hakam[1], Fauzan[1], M Mera[1] and S Syuhada[2]

[1]Departement of Civil Engineering, Universitas Andalas, Padang, Indonesia, 25175
[2]Civil Engineering Study Program, Civil Engineering Department of Infrastructure and Regional Technology, Institut Teknologi Sumatera, Lampung Selatan, 35365, Indonesia

E-mail: rina@eng.unand.ac.id

Abstract. On September 30, 2009, there was liquefaction in the sandy soils around the coast of Padang city, which was triggered by an earthquake measuring 7.6 SR. Liquefaction causes land subsidence and lateral displacement of many buildings in Padang. The level of liquefaction vulnerability of Padang city is in the moderate zone based on qualitative methods; on the contrary, quantitative methods obtained a very high level of vulnerability to liquefaction. Based on this, quantitative methods by conducting geotechnical investigations in the field and laboratory tests are needed to determine the liquefaction potential. This study aims to determine the potential for liquefaction using quantitative methods based on testing the physical properties and mechanical properties of sandy soils in the laboratory. The standard for laboratory testing is the Indonesian National Standard (SNI), and soil sampling was carried out in seventeen areas. The results of the physical and mechanical properties, i.e., relative density (Dr), the particle size of the soil, and internal friction angle, showed that the sandy soils along the coast in Padang are poorly graded sand (SP). The sandy soils are dominated by fine sand with a very loose consistency, which has potential liquefaction.

1. Introduction

In this world, liquefaction disasters have occurred repeatedly, so researchers around the world have identified layers of soil that are easily liquefied. A layer of saturated soil is an area with the potential for liquefaction in an earthquake. Seed and Idriss (1971) stated that liquefaction has the potential to occur in earthquake-prone areas where the soil layer is composed of low-density sand deposits [1]. The phenomenon of soil liquefaction is primarily associated with medium to fine-grained saturated cohesionless soils [2]. The potential for liquefaction based on a quantitative approach can be estimated by analyzing the results of field and laboratory tests. Field tests include the Standard penetration test (SPT) and the Cone penetration test (CPT). Tests in the laboratory include grain size analysis and relative density of soil (Dr) [2, 3].

Sand is one of the interesting granular soils because it is hidden in the mystery of how these particles were deposited according to their environment, whether sand in an offshore, coastal, or river environment. Sand has excellent technical properties as long as the soil is relative density. Relative density (Dr), shape, and grain size are technical properties of the soil that significantly affect the shear strength and compressibility of the soil. The denser the soil, the higher the shear strength and the smaller the settlement. The larger and coarser the grain surface, the greater the shear strength. The better the gradation, the greater the shear strength. The mixture and shape of sand grains are factors that affect the technical properties of sand. That is, shear strength needs to be considered to be maximized in construction related to sand [4]. The uniform soil has a small cohesion value (c), internal friction angle (ϕ), and γ_{dmax} [5]. If the soil has a low shear strength, it can impact its load. For this reason, it is necessary to compact the soil so that the air in the soil pores can be accommodated by solid grains [6].

The Padang beach area has a liquefaction potential base on soil gradation. The depth of the oil layer that potentially liquefies in Padang beach is less than 10 m. The method of predicting the liquefaction potential of soil deposits using soil grain grading is one of the simplest and easiest prediction methods [7].

2. Methodology

Soil sampling was carried out at seventeen locations along the coast of Padang city (Table 1), and the map of the sand sampling location can be seen in Figure 1. The soil sampling was carried out using a hoe for disturbed soil samples (depth ± 30 cm), and undisturbed soil samples were taken using a cylindrical ring. Testing of the physical and mechanical properties of sands was carried out in the Soil Mechanic Laboratory of Universitas Andalas. The standard for laboratory testing is the Indonesian National Standard.

Table 1. Sampling locations.

Locations	Naming
Belakang Tangsi	1
	2
	3
Purus	4
	5
	6
Ulak Karang	7
	8
Air Tawar Barat	9
	10
	11
Parupuk Tabing	12
Pasie Nan Tigo	13
	14
	15
	16
	17

Figure 1. Sampling map.

3. Analysis and discussion

3.1 Results of index properties of soil

The results of the physical properties test of the sandy soil*s* in the coastal area of Padang city can be seen in Table 2. Natural moisture content ranges from 13% - 30%; wet volume weight between 1.1-1.9 gr/cm3; specific gravity between 2.65-2.68; dry density between 1.1-1.9; the void ratio between 0.7-1.94; the relative density (Dr) in ten locations is in the range of 0-15% which indicates that the sandy soil*s are* very loose. Seven locations have relative density values between 15% -50%, which describes loose sandy soil deposits.

Table 2. Index properties of soil.

Locations	Natural water content, w (%)	Wet density, γ (gram/cm^3)	Specific Gravity, G_s	Dry density, $\gamma_{d(lap)}$ (gram/cm^3), sand cone test	Dry density, γ_d (gram/cm^3)	Minimum dry density, $\gamma_{d(min)}$ (gram/cm^3)	Maximum dry density, $\gamma_{d(max)}$ (gram/cm^3)	Relative Density, Dr (%)	Description
1	17.20	1.53	2.66	1.88	1.30	1.84	1.99	28.09	Loose
2	20.08	1.87	2.67	1.88	1.56	1.87	1.93	12.73	Very loose
3	23.62	1.78	2.66	1.85	1.44	1.79	1.99	29.57	Loose
4	15.24	1.94	2.66	1.85	1.69	1.82	1.92	33.72	Loose
5	25.52	1.48	2.65	1.52	1.18	1.50	1.69	14.48	Very loose
6	27.81	1.17	2.67	1.45	0.92	1.42	1.66	14.02	Very loose
7	25.75	1.24	2.67	1.41	0.98	1.39	1.54	14.31	Very loose
8	17.61	1.37	2.67	1.41	1.17	1.39	1.54	14.52	Very loose
9	17.09	1.40	2.66	1.45	1.20	1.43	1.57	14.75	Very loose
10	18.36	1.40	2.67	1.44	1.18	1.42	1.56	14.83	Very loose
11	18.44	1.78	2.66	1.56	1.50	1.52	1.65	29.85	Loose
12	22.05	1.60	2.67	1.41	1.31	1.39	1.52	21.23	Loose
13	13.45	1.36	2.68	1.48	1.20	1.42	1.90	14.67	Very loose
14	30.45	1.67	2.67	1.27	1.28	1.23	1.39	30.21	Loose
15	27.10	1.34	2.65	1.37	1.05	1.34	1.56	14.45	Very loose
16	27.64	1.39	2.66	1.24	1.09	1.22	1.38	18.71	Loose
17	22.58	1.11	2.66	1.16	0.91	1.13	1.36	14.93	Very loose

Figure 2 shows the degree of saturation of the sandy soils along the coast of Padang. There are ten locations with moist soil conditions where the degree of saturation (S) is between 0.26-0.50. Seven locations with very moist soil conditions, where the value of the degree of saturation (S) is between 0.51-0.75.

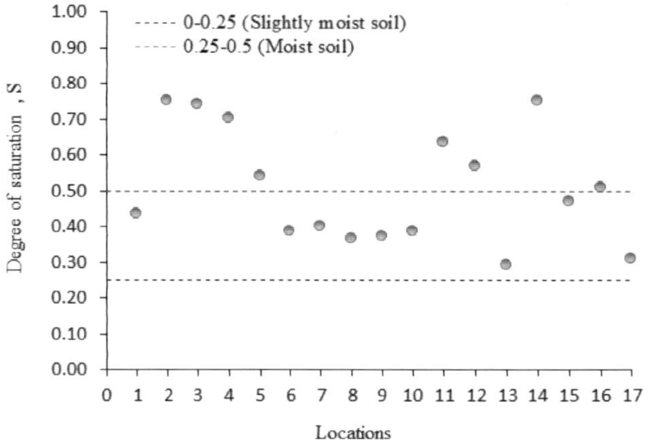

Figure 2. Result of the degree of saturation along the coast of Padang.

Figure 3 shows that there are ten locations with a relative density (Dr) between 0-15%, where the sandy soils deposited at that location have a very loose consistency. There are seven locations with relative density (Dr) between 15-50%, where the sand deposits at these locations are in a loose condition. Both of these conditions are very susceptible to liquefaction potential.

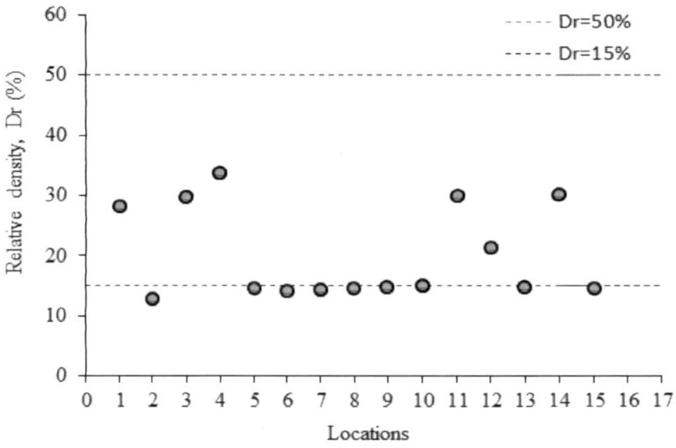

Figure 3. Result of relative density test for sand soils along the coast of Padang.

Table 3 shows that the sandy soils along the coast of Padang city are dominated by fine sand. Figure 4 shows the results of the sieve analysis test from sandy soils taken at seventeen locations along the coast of Padang. According to the Unified Soil Classification System (USCS), less than 50% pass through the No. 200 sieve, and 50% or more pass the No. 4 sieve; the group symbols are S or Sands. If the uniformity coefficient (Cu) is smaller than six, and the gradation coefficient is between 1 and 2, then the soil is classified as poorly graded sand (SP).

Table 3. Result of sieve analysis

Locations	Coarse sand, (%)	Medium sand, (%)	Fine sand, (%)	Clay, (%)	D_{10}	D_{30}	D_{50}	D_{60}	C_u	C_c	USCS Classification
1	0	9.10	89.53	1.37	0.11	0.19	0.25	0.28	2.59	1.17	SP
2	0.17	22.67	75.20	1.97	0.15	0.21	0.28	0.33	2.24	0.84	SP
3	0.07	20.77	78.43	0.73	0.16	0.21	0.27	0.32	2.04	0.88	SP
4	0.00	7.17	92.43	0.40	0.15	0.20	0.25	0.27	1.82	0.94	SP
5	0.00	0.70	96.23	3.07	0.16	0.18	0.23	0.23	1.50	0.93	SP
6	0.00	1.07	96.83	2.10	0.15	0.17	0.23	0.25	1.63	0.77	SP
7	0.43	21.63	77.60	0.33	0.16	0.18	0.28	0.23	1.50	0.93	SP
8	0.00	9.77	88.77	1.47	0.11	0.17	0.24	0.27	2.48	0.91	SP
9	0.10	14.60	82.83	2.47	0.10	0.17	0.24	0.27	2.79	1.11	SP
10	0.00	16.77	81.33	1.90	0.11	0.18	0.25	0.28	2.45	1.01	SP
11	0.10	22.67	75.93	1.30	0.11	0.18	0.27	0.31	2.71	0.96	SP
12	0.00	7.27	91.80	0.93	0.10	0.11	0.22	0.27	2.83	0.47	SP
13	0.13	36.67	62.57	0.63	0.16	0.22	0.33	0.38	2.36	0.79	SP
14	0.00	7.07	90.40	2.53	0.11	0.17	0.23	0.26	2.45	1.10	SP
15	0.00	4.23	92.77	3.00	0.09	0.16	0.23	0.26	2.85	1.07	SP
16	0.00	0.43	94.77	4.80	0.09	0.16	0.20	0.24	2.71	1.18	SP
17	0.13	2.03	97.83	2.03	0.10	0.17	0.22	0.26	2.64	1.12	SP

By plotting the grain size distribution curve of the sandy soil from the results of the sieve analysis into the Tsuchida graph (1970), it can be seen that the sample is on the inside of the curve, then the liquefaction potential.

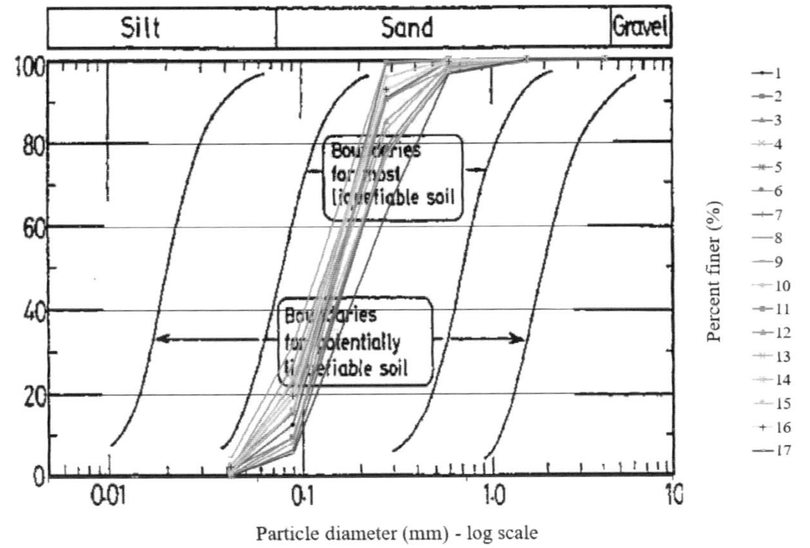

Figure 4. Grain size distribution curve and the gradation curves separating liquefiable and non-liquefiable soils.

3.2 Result of mechanical properties of soils

Figure 5 shows the results of the compacted soil are determined in the laboratory at seventeen locations. The results of the standard proctor test showed that the dry unit weight was in the range of 1.38-1.93 gram/cm3 and the optimum moisture content was in the range of 9.98%-27.65%. Based on Figure 5, it can be seen that locations 1, 2, 3, and 4, respectively, have a higher dry density value than locations 5 to locations 17. This type of soil is a poorly graded sand, where the voids between the grains are large so that air is abundant and complex to close when compacted even though it has been compacted as much as possible, so the compaction is not maximal, so necessary to improve the soil.

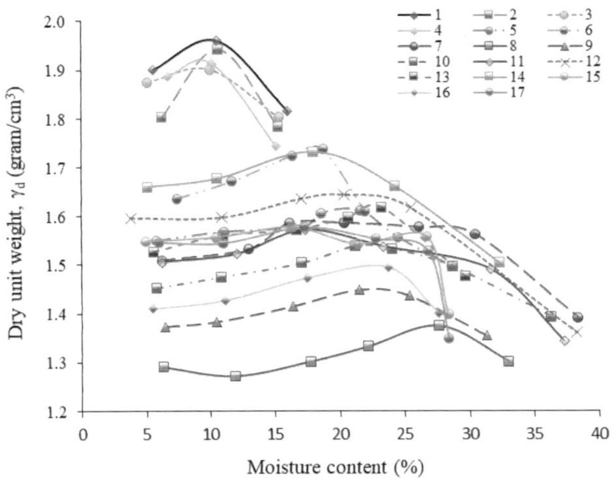

Figure 5. Standard proctor compaction test results for sandy soil.

Figure 6 shows that the cohesion (c) at locations 1 to 17 ranges from 3.6-7 kPa, meaning that the soil has a low binding capacity between particles. The cohesion value obtained from the test results shows that the soils are sandy soil with low cohesion (< 12 kPa). It can be concluded that these sandy soils have a low inter-grain bond (non-cohesive soil) because it has a sizeable inter-grain cavity.

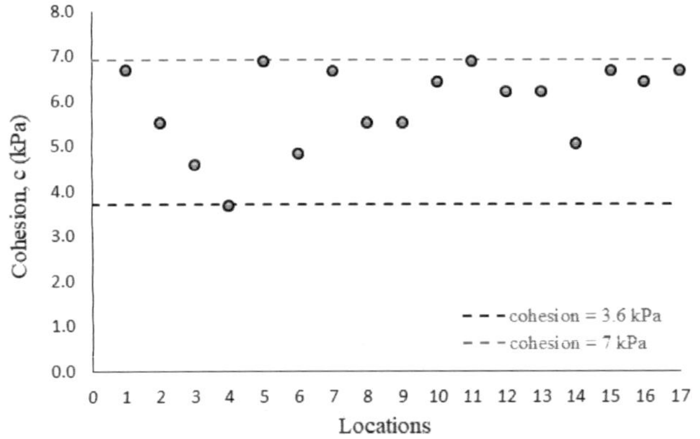

Figure 6. Cohesion value for sand along Padang coast.

Figure 7 shows that the internal friction angle (f) in the sand samples from location 1 to location 17 is 28o-30o. This means that the sandy soils are classified as uniformly graded sand in a loose state. In loose sand, the small sand grains roll easily under shear forces.

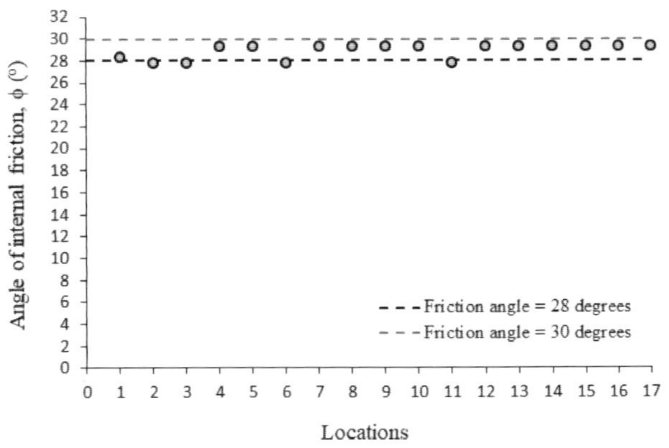

Figure 7. Internal friction angle value for sand along Padang coast.

4. Conclusions

According to the USCS method, the sandy soils in the coastal area of Padang are poorly graded or uniformly graded (SP). More than 50% is sandy soils dominated by fine sand. Based on the relative density (Dr), the sandy soils in the coastal area have consistency from very loose to loose. Based on the grain size gradation, the sandy soil has potential liquefaction. From the mechanical properties test, the soil shear strength test in the coastal area has the lowest cohesion values and internal friction angles with a very loose consistency. It has the potential for liquefaction if there is an earthquake.

5. Acknowledgments

This research was supported by the Engineering Faculty of Andalas University and conducted at the Faculty of Engineering.

6. References

[1] Seed H B and Idriss I M 1971 Simplified procedure for evaluating soil liquefaction potential *J. of the Soil Mechanics and Foundations* **97** 1249-1273

[2] Das B M and Ramana G V 2011 *Principles of Soil Dynamic Second Edition* Cengage Learning USA p 368

[3] Widyaningrum R 2012 *Penyelidikan Geologi Teknik Potensi Liquifaksi Daerah Palu, Provinsi Sulawesi Tengah* (Bandung: Kemen ESDM) p 29

[4] Kurniawan Y, Yusa M and Fatnanta F 2021 Pengaruh bentuk dan campuran butiran pasir terhadap perilaku kuat gesernya *J. APTEK* **13** 42-49

[5] Prihatin K and Wahyudi H 2009 *Korelasi antara Gradasi Material Timbunan Reklamasi dengan Kepadatan dan Kuat Geser Tanah* p E19–E28

[6] Febriana P E and Agustina D H 2021 *Effect of Compaction on Soil Shear Sterngth Sigma Teknika* **4** 262-266

[7] Hakam A and Darjanto H 2013 Penelusuran potensi likuifaksi Pantai Padang berdasarkan gradasi butiran dan tahanan penetrasi standard *J. Teknik Sipil* **20** 33-38

[8] Yuliet R, Hakam A, Mera M and Fauzan 2019 Structural evaluation of Nurul Haq Shelter Building Constructed on Liquefaction Prone Area in Padang City-Indonesia *Int. J. of GEOMATE* **17** 106-114

[9] Yuliet R, Hakam A, Mera M and Fauzan 2021 The effects of water-saturated sand-fractions on the sand-boiling phenomenon *Int. J. of GEOMATE* **21** 41-47

[10] Tsuchida H 1970 Prediction and countermeasure against liquefaction in sand deposits *Abstract of the seminar of the port and Harbour Research Institute, Ministry of transportat, Yokosuka, Japan* p 3.1-3.33

A Comparison of 1-D S-wave Velocity Profile Resulting from Microtremor Array Method and Lithology Profile from Borehole Data in Jakarta

Mohamad Ridwan[1], Asdani Soehaimi[1], Rio Alcanadre Tanjung Moechtar [2,3]

[1]Research Center for Geological Disaster, National Research and Innovation Agency

[2] Bandung Institute of Technology

[3] Center for Geological Survey, Geological Agency

*Corresponding author: [a)]: m.ridowan@gmail.com

Abstract. The subsurface investigation is very important in seismic hazard assessment for the city scale because it greatly affects the propagation of seismic waves on the surface related to the amplification phenomenon caused by the thickness and hardness of the surface sediment. Local geology and geotechnical conditions for seismic hazard analysis are represented by site class which is determined based on the value of Vs30 or NSPT30, besides that it is also necessary to know the depth of the bedrock. In general, the availability of these data in the cities in Indonesia is very limited, so it is necessary to conduct a field survey which is one of the methods using the microtremor array survey method. This paper has presented the results of subsurface investigations in Kembangan Jakarta using the microtremor array method and their comparison with borehole data that was carried out at adjacent locations. The analysis results of those methods show a positive correlation between the 1-D shear wave velocity profile and lithology where the top layer has low S-wave velocity (Vs) correlated with soft rock. The S-wave velocity increases with depth which is correlated with hard rock that is indicated by lithology as hard sandstone. The results of this study are very useful for applying the array microtremor method in other cities considering it is very simple in operation.
Keywords: Jakarta, site class, microtremor array.

1. Introduction

A lot of earthquakes event in Indonesia often give a very heavy impact on the affected area such as building damage and casualties. The level of building damage due to earthquakes can be seen in the field that is very diverse which requires a deeper understanding. There are several factors influence building damages in an area i.e: earthquake characteristics, the condition of the building, and local geology. In general, the buildings located on thick and soft sediment will have stronger seismic motion than those on hard soil, so it will have a heavy impact. Therefore, it is required a detailed study on site characterization based on the results of site investigations by using both geophysical and geotechnical methods.

The amplification phenomenon has been presented [1] where the soil response to earthquake shaking varies widely for each type of soil will cause very different characteristics of ground shaking depending on acceleration and frequency of the earthquake. The local soil condition will contribute to the seismic wave amplification factor [2,3]. This shows that a study in detail about geotechnical and local geology is required for assessing the level of seismic hazard in an area that is very useful for various purposes. However, in most of the major cities in Indonesia are very difficult to obtain local geotechnical and geological data because of the rarity of soil investigation, so a solution is required to carry out soil/rock investigation that is possible to be conducted in a major city.

The observation site is located in West Jakarta where by referring to the regional geological map the city of Jakarta is covered by Quaternary sediment [4]. Some soil parameters required for site response analysis include site class which is determined based on Vs or SPT parameters down to a depth of 30m. In addition, the depth of the bedrock is very important information because the estimation of the amplification factor must consider the thickness of the sediment layer over the bedrock. The Microtremor array survey method is one of the Geophysical methods that has been widely used for a deep-layer investigation to estimate bedrock depth [5,6]. Besides that, the subsurface survey can also be done by drilling to find out the lithology profile that can be used for data verification resulting from the microtremor array.

2. Data and Method

Microtremor array observations in this paper are part of a site characterization study in the city of Jakarta for the development of a seismic microzonation map which is the verification by borehole data required. In the location of microtremor array observation in site Kembangan, there is deep borehole data until the depth of 300 meters that describes the lithology profile [7]. It will present good information about a correlation between S-wave velocity and lithology profile.

Since the available borehole data is until 300 m in depth, then largest the array size 600 m for the microtremor array survey is sufficient to obtain a subsurface feature related to borehole depth. The others triangular array configurations are 300 and 150 m in size were used for data acquisition in the field to obtain more wide frequency range. By using this configuration, the vertical components Rayleigh wave microtremor was recorded simultaneously by using four seismographs which is each position is indicated in Figure 1.

Microtremor data processing was conducted after data series was digitized by certain frequency. By using those triangular array configuration, fundamental mode Rayleigh wave dispersion curve can be calculated by using the Spatial Autocorrelation (SPAC) method [8,9,10,11,12]. The SPAC coefficient was derived by averaging coherences between two stations, then dispersion curves estimation using the SPAC coefficient via the Bessel function of the first kind zeroth-order.

Estimation of 1-D S-wave velocity structure was carried out by Genetic Algorithm (GA) using some parameter of GA that has been widely applied for seismic data[13]. An initial model of 1-D S-wave velocity profile can be constructed that consist of some parameters i.e: velocity, thickness, and density. It is created by considering Rayleigh wave dispersion curve of microtremor and local geological condition or borehole data.

3. Result and Discussion

Microtremor array observation site were at the coordinates (longitude, latitude) 106.755, -6.179 in Kembangan area, West Jakarta, to obtain subsurface feature based on the S-wave velocity structure. It is very close to the location of a deep borehole (300 m depth) at the coordinates of 106.754, and 6.18489 degrees[7]. According to the previous study, 1-D S-wave velocity structure has been estimated until 800 m depth where the bedrock position was at a depth of 615 m[14]. Meanwhile, for correlation purpose between the Vs profile and lithology, the microtremor data was then reprocessed using an initial model by referring to the available borehole data.

Data processing was conducted by using Seisimager software developed by Geometrics where an example of microtremor data was recorded for each triangular array shown in Figure 2a. The first step of microtemor array data processing using Spatial Autocorrelation (SPAC) method is calculation of the azimuthally average of coherence for each distance between two stations. The result is shown in Figure 2b where it can be seen that the coherence is getting smaller at longer distances. Figure 3a shows the picking of the phase velocity to the frequency where this site has a fairly wide frequency range (0.5 – 3.5 Hz) with a velocity range between 170 – 550 m/sec which is sufficient to describe the subsurface profile to a depth of 300 m.

The results of inversion show the lowest misfit between the dispersion curve of the observed and calculated one (Figure 3b) while the estimated 1D S-wave velocity structure shown in Figure 4a which hasa vertical distribution of the velocity is in a range of 190 – 600 m/sec. The comparison of the 1-D profile resulting from both microtremor array method and lithology profile (Figure 4b) demonstrates a positivecorrelation where rock hardness is reflected in the value of Vs which is getting harder with depth.

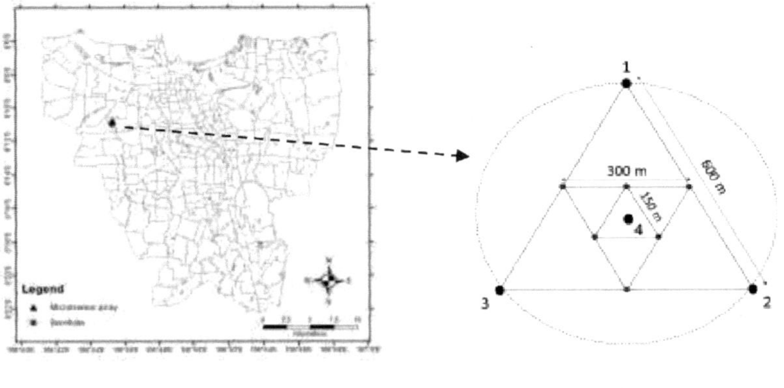

(a) (b)

Figure 1. (a) The location of the borehole and microtremor array observation site in the city of Jakarta,Indonesia, and (b) the triangular array configuration and size

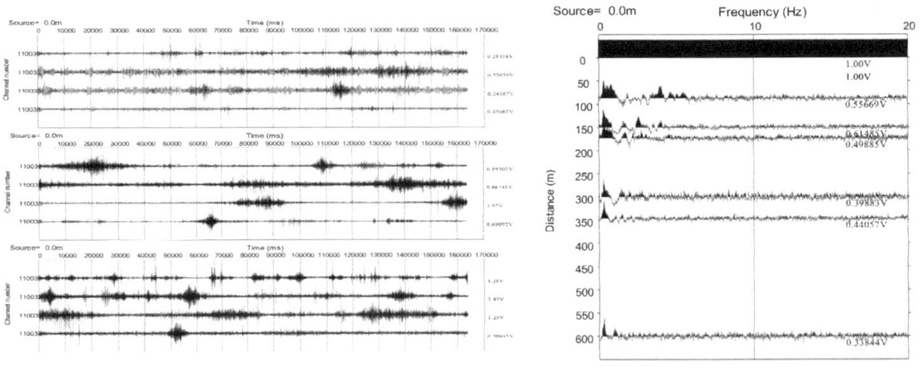

(a) (b)

Figure 2. (a) The example of microtremor array data record on the array size 600, 300, and 150 m (b) The averaged coherence curve on each distance between two stations.

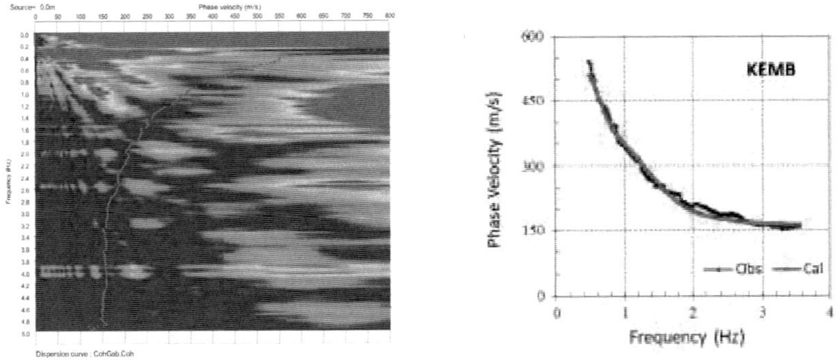

(a) (b)

Figure 3. (a) Picking of phase velocity to frequency for all array configurations, (b)Fitting of dispersion curve between calculated and observed one

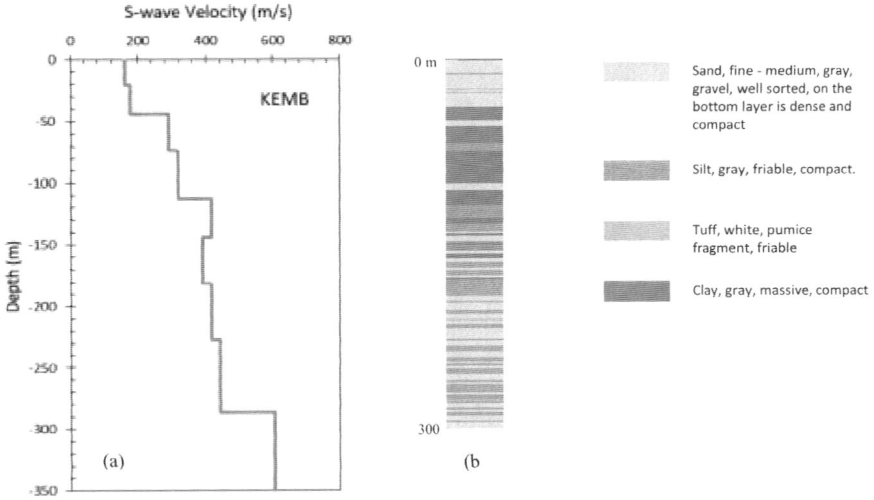

Figure 4. Comparison of subsurface profile between (a) 1-D S-wave velocity structure, and (b) Lithology.

4. Conclusions

Subsurface investigations in the observation site using microtremor array survey method were carried out to estimate 1-D S-wave velocity structure and determination of bedrock depth as one of the parameters for seismic hazard analysis on the surface related to the amplification factor of the sediments layer over the bedrock. The lithology of this area is illustrated by borehole data where the location of the borehole is very close to the microtremor observations site so that correlation of subsurface features canbe conducted.

1-D S-wave velocity structure resulting from the microtremor array which has a velocity variation between 180 – 600 m/s to a depth of 400 m with 9 layer model which is the vertical distribution increases with depth. This is reflected in the lithological profile obtained from borehole data down to a depth of 300 m consisting of alternating sandstone and claystone where the upper layer is dominated by sandstoneto a depth of 40 m, then claystone to a depth of 140 m. Lithology changes to a depth are alternation of sandstone and claystone where sandstone is more dominant. That comparison of subsurface profiles resulting from microtremor array analysis and borehole data demonstrates a strong positive correlation.

5. Acknowledgments
The authors gratefully acknowledge the Research Institute for Human Settlement (RIHS), Ministry of Public Works and the Centre of Geological Survey, Ministry of Energy and Mineral Resources, the Republic of Indonesia, for supporting data.

References

[1] Seed HB, Muraka R, Lysmer J, Idriss I (1976) Relationships of maximum acceleration, maximumvelocity, distance from the source, and local site conditions for moderately strong earthquakes. Bull Seismol Soc Am, 66, 4: 1323-1342
[2] Chimoto, K., H. Yamanaka, S. Tsuno, H. Miyake, dan N. Yamada. 2016. "Estimation of Shallow S-wave Velocity Structure using Microtremor Array Exploration at Temporary Strong Motion observation Stations for Aftershocks of the 2016 Kumamoto Earthquake." Earth, Planets, andSpace 68 (1): 1–10. https://doi.org/10.1186/s40623-016-0581-3.
[3] Irsyam M, Hutabarat D, Asrurifak M, Imran I, Widiyantoro S, Hendriyawan, Sadisun I, HutapeaB, Firmanti A, Ridwan M, Afriansyah T, Pindratno H, Haridjono SW, Pandhu R (2015) Development of seismic risk microzonation maps of Jakarta city. Geotechnics for Catastrophic Flooding Events Iai (Ed), CRC Press
[4] Turkandi T, Sidarto, Agustiyanto DA, Hadiwidjojo MMP (1992) Geological map of Jakarta andKepulauan Seribu quadrangles Jawa. Geological Research and Development Centre, Ministryof Energy and Mineral Resources, the Republic of Indonesia
[5] Asano, K., Iwata, K., Kunikazu Yoshida, K., Naoto Inoue, N., Kazuhiro Somei, K., Ken Miyakosh, K., and Ohori, M (2022) Microtremor array surveys and development of the velocity model in the Hakodate Plain, Hokkaido, Japan, Earth, Planets, and Space.
[6] Ridwan M, Widiyantoro S, Irsyam M, Afnimar, Yamanaka H (2016) Development of an engineering bedrock map beneath Jakarta based on microtremor array measurements. Geol Soc London Spec Publ, SP441.7. http://doi.org/10.1144/SP441.7
[7] Moechtar, R.A.T., Rohman, M., Yuniarni, R., Dipowiguno, Setiawan, R. (2019) Study of stratigraphy and the distribution of quaternary facies in Jakarta and its surroundings, Center for Geological Survey, Geological Agency, Ministry of Energy and Mineral Resources. (Report in Indonesian).
[8] Aki K (1957) Space and time spectra of stationary stochastic waves, with special reference to microtremors. Bull Earthq Res Inst Tokyo Univ 35: 415-456
[9] Morikawa H, Sawada S, Akamatsu J (2004) Method to estimate phase velocities of Rayleigh wave using microseisms simultaneously observed at two sites. Bull Seismol Soc Am, 94, 3: 961-976
[10] Okada H (2003) The microtremors array survey method. Geophysical Monograph Series, 12, Society of Exploration Geophysicists, 135 pp
[11] Okada H (2006) Theory of efficient array observations of microtremors with special reference tothe SPAC method, Exploration Geophysics, 37:73–85
[12] Asten, M.W., Dhu, T. & Lam, N. 2004. Optimized array design for microtremor array studies applied to site classification; comparison of results with SCPT logs. Paper 2903 presented at the 13th World Conference on Earthquake Engineering, 1 –6 August 2004, Vancouver, BC, Canada.
[13] Yamanaka H. and Ishida H (1996) Application of genetic algorithms to an inversion of surface wave dispersion data. Bull Seismol Soc Am, 86, 2: 436-444
[14] Ridwan, M., P. R. Cummins, S. Widiyantoro, dan M. Irsyam. 2019. "Site Characterization usingMicrotremor Array and Seismic Hazard Assessment for Jakarta, Indonesia." Bulletin of the Seismological Society of America 109 (6): 2644–57. https://doi.org/10.1785/0120190040.

Placement of geotextile tubes as coastal management of way hawang seashore area at Bengkulu Province

O C Pattipawaej[1]

[1]Civil Engineering Department, Universitas Kristen Maranatha, Jl. Prof. drg. Suria Sumantri, MPH No. 65, Bandung 40164, Indonesia

E-mail: olga.pattipawaej@eng.maranatha.edu

Abstract. Way Hawang Beach, one of coastal tourist attraction, located in Way Hawang Village, Maje, Kaur Regency, Bengkulu Province is suffering from erosion. The regular erosion speed is approximately 2 to 3 meters/year, taking the distance from Way Hawang coastline to West Sumatra Causeway which around 250 meters. Efforts to eradicate abrasion on Way Hawang Coast are mostly carried out using coastal protective structures. One of the breakwater structures that can reduce wave energy is geotextile tubes, which is a low-threshold breakwater structure. Geotextile tubes create a synthetic barrier along shorelines and beaches to help control erosion. Geotextile tubes placement plays an important role in the overall protection system performance. The geotextile tubes are simulated in the abrasion and accretion positions on the Way Hawang coast. The simulation results indicated that the position between abrasion and accretion are the most effective position to manage and to avoid erosion in Way Hawang coast's area at Bengkulu Province.

1. Introduction

One of the provinces in Indonesia is Bengkulu which is located on the southwest coast of Sumatra. The length of the coast is about 576 km, bordering the Indian Ocean in the western part of Bengkulu Province and has a 550 km long national road. This national road is called the West Sumatra Causeway which is the only land transportation route between districts in Bengkulu Province and is used to reach West Sumatra and West Lampung [1]. The function of the West Sumatra Causeway is very vital, heavy vehicles cannot be passed if this road is damaged and will significantly disrupt the economic acceleration of the Bengkulu population. Apart from being very important infrastructure used for traffic services, goods, agricultural products, mining, plantations and oil distribution, Bengkulu's population, the majority of whom are agrarian, trusts severely on agricultural products and requires land transportation via this national road [2].

This West Sumatra causeway is extremely defenceless to being eroded by ocean water starting from the Indian Ocean which takes very resilient hydrodynamics and sea waves to erode the road barriers. Coastal abrasion that erodes roads has a severe value, where the regular erosion rate is 2 meters/year and occasionally up to 3 meters/year [2]. Figure 1 shows the abrasion at South Kaur District, Bengkulu Province. Residents' settlements in the coastal area of South Kaur District, Kaur Regency are threatened with abrasion. The reason is that the distance from the beach to the road and residents' houses is only a few meters. This makes residents hope that the local government can immediately build the coastal protection structure. Some blisters have been managed by the Ministry of PUPR. However, abrasion remains to transfer to other parts of the coast, so a comprehensive clarification is desired to address this problem [3-7].

Figure 1. Abrasion at South Kaur District, Kau Regency, Bengkulu Province

Way Hawang Beach is approximately 250 meters away from West Sumatra Causeway. This tourist attraction is in Way Hawang, Maje, Kaur Regency, Bengkulu Province. Way Hawang coastal area is experiencing abrasions. Efforts to eliminate abrasion on the Way Hawang Coast are mostly carried out using coastal protective structures, including breakwater structures [8-10]. One of the breakwater structures that can reduce wave energy is geotextile tubes, which is a low-threshold breakwater structure. Geotextile tubes create a synthetic barrier used along shorelines and beaches to help control erosion (see Figure 2). Geotextile tubes placement shows a useful responsibility in the performance of the complete safety scheme [11]. In this study, geotextile tubes were simulated in the position of abrasion and sedimentation on the Way Hawang coast. The simulation results indicated that the position between abrasion and sedimentation was the most effective position to manage and to prevent abrasion in Way Hawang coast's area of Bengkulu Province.

Figure 2. Geotextile Tubes

2. Materials and Methods

2.1. The study area

Way Hawang Village is in Kaur Regency, Bengkulu Province. The latitude of Way Hawang is 4.8632° South and longitude is 103.4360° E. Figure 3 shows Way Hawang Village and coastal area which was taken from Google Earth platform [12] and Batu Jung is in the middle of Way Hawang Beach which is a tourist spot. The research site was in the coastal area of Way Hawang.

Figure 3. Way Hawang Village and coastal area from Google Earth Engine

2.2. The forecasting of wave height on Way Hawang Beach

Wind data used for wave forecasting is sea level data at the generation site. This data was gained from direct measurements above sea level or measurements on land near the forecasting setting which was then transformed into wind data at sea. The results of processing wind data collected from the Bengkulu wind recording station from 2011-2020, with statistical analysis data, were used to calculate the hour and direction of wind deception. The maximum wind comes from the South, directly from India Ocean.

The calculation of waves, generated by the wind, was done by hindcasting [14]. The basic principle of this method is to estimate the wave height and period calculated from wind data and effectively fetched using empirical methods. The maximum wind speed was used in the calculation, which was intended to analyse the extreme conditions of the waves. The visualisation of wave height [15] at Way Hawang Beach can be seen in Figure 4. The maximum wave height at Way Hawang Beach arises from the South directly from Indian Ocean.

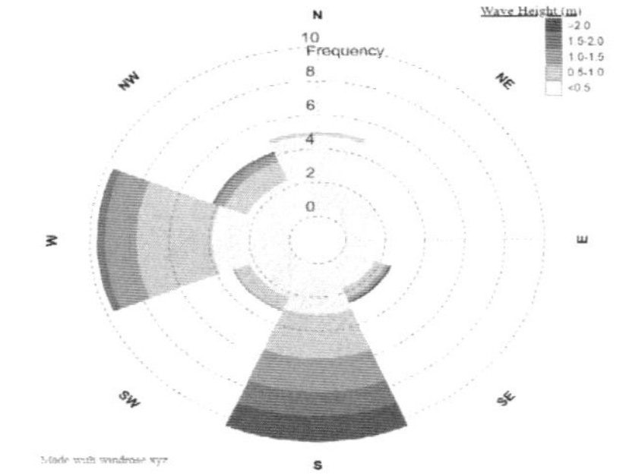

Figure 4. Wave rose of Way Hawang coastal area

The design wave height is the wave height of a certain return period at the location of the coastal protection structure, where the wave has transformed during its propagation from the deep sea. The wave data used is the period and wave height which is the result of a 10-year wind data hindcasting process. Furthermore, extreme wave analysis was carried out in the form of various return periods for each direction [16] and the 50-year return period wave height will be modelled using SMADA software [17]. The design wave height is 6.39 m from the results of this wave transformation model.

Furthermore, tides are an oceanographic physical phenomenon that needs to be studied as an effort to understand the circulation pattern of seawater masses. This tidal parameter generally determines the movement of water from midday to daily period depending on the type of tide that occurs in these waters. From the observation results, tide measurements are also used to determine the water level elevation.

The Mean High-Water Level, Mean Sea Level, and Mean Low Water Level at Way Hawang Beach are 1.08, 0.72, and 0.35 meters, respectively.

2.3. *Design of Geotextile Tubes*

The design and location of geotextile tubes was determined based on wave data. The result of forecasting wave height for the 50-year return period stands $H_{50} = 6.39$ m; therefore, the wave height (H), wave period (T), and the wavelength in the deep waters (L_0) are obtained as follows:

$$H = 33\%\ H_{50} = 2.1 \text{ m}$$
$$T = 3.88\sqrt{H} = 5.63 \text{ second}$$
$$L_0 = 1.56\,T^2 = 49.52 \text{ m.}$$

Furthermore, the height of breaking wave is 2.4777 m at the depth of the breaking wave is 2.7255 m.

The coastal protection structure using sand-filled geotextile tubes material at Way Hawang Beach has dimensions of 2 m high, a base width of 4.2426 m and a cross-sectional area of 6.2575 m². The geotextile tubes can be seen in Figure 5, where P_h is the pressure caused by waves (a combination of hydrostatic pressure and dynamic pressure) and P_p is the passive pressure caused by water pressure behind the structure. The geotextile tubes structure is stable against overturning, shear forces, and soil bearing capacity; therefore, it is in a stable and safe condition.

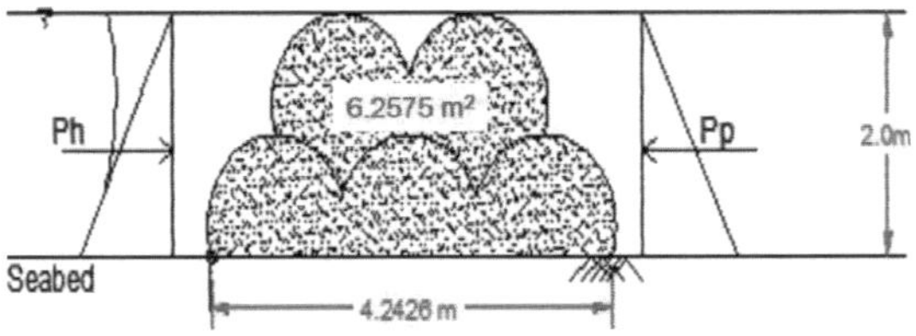

Figure 5. Geotextile tubes at Way Hawang Beach

2.4. *Numerical Modelling System of Shoreline Change*

The initial position of Way Hawang's shoreline is firstly done as an input by making the grids at a certain distance accordance to the long coastline that will be simulated. Data on the position of the Way Hawang's shoreline can be acquired from Google Earth Pro engine to AutoCAD application [18]. The distance between the grids is 66.3 m with 86 grids as shown in Figure 6. Wave data is the result of daily wind data processing in the form of wind speed and direction using a hindcasting program. The wave parameters obtained are the period, height and direction of the incoming wave. The number of wave data generated in one year is 8760 data.

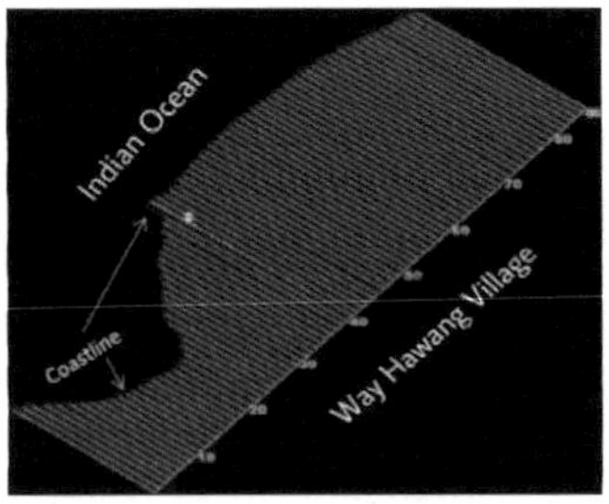

Figure 6. Grid of the Way Hawang Coastline

The prediction of shoreline change can be obtained using numerical method over a given period. A numerical model called GENESIS (GENEralized model for SImulating Shoreline change) [19], is used to calculate shoreline changes caused mostly by sea wave. This model is based on the one-line theory, which assumes that the beach profile remains unchanged, thus allowing shoreline change to be uniquely described in terms of shoreline position. The central task of this model is to simulate the shoreline response to constructions situated on or near the shore. GENESIS generally means that a simple user interface allows the system to be affected in a variety of situations involving the number, location, and combination of coastal protection structures. If all the input data needed for shoreline change prediction is obtainable, then the program will run through the START file. All comments in the START file are filled in according to the existing input and as required by GENESIS. The START input is also carried out, for conditions without or with the geotextile tubes.

3. Results and Discussion

Due to wave load, the shoreline at Way Hawang beach has changed. Figure 7 presents the shoreline change at Way Hawang coastal area. It can be seen from Figure 7 that the abrasions occur at grid 10-30 and the accretions occur at grid 30-50.

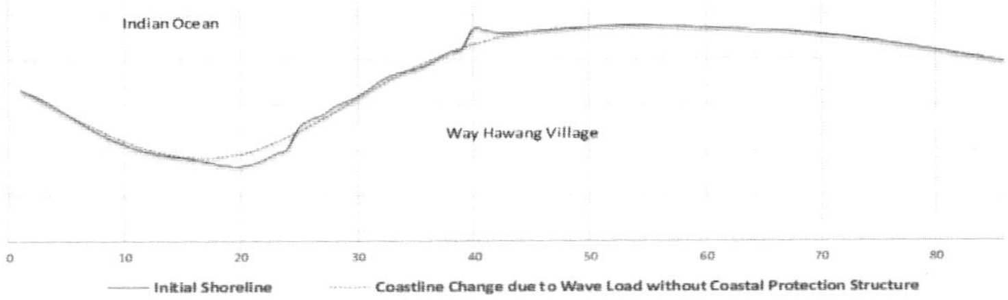

Figure 7. Coastline change at Way Hawang Beach due to wave load without coastal protection structure

The next step is to analyse the position of the most effective geotextile tube structure in order to minimise the occurrence of changes in the Way Hawang coastline. Three plans for the placement of geotextile tube structures were analysed. The placement of the geotextile tube structure at the location of the occurrence of abrasion and sedimentation was based on the results of coastline change analysis without the presence of a coastal protective structure. The first location was geotextile tube structure placement on grids 10 to 30 (the occurrence of abrasion), the second location was on grids 30 to 50 (occurrence of sedimentation), and the third location was on grids 10-50 (occurrence of abrasion and sedimentation). Figure 8 shows the coastline change due to the placement of the geotextile tube structure on the Way Hawang coast.

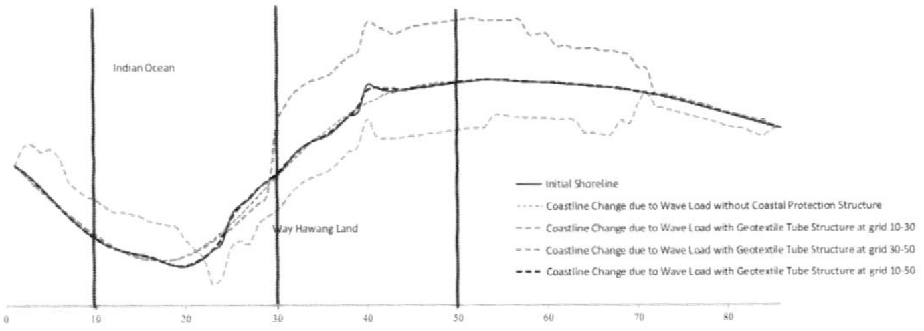

Figure 8. Shoreline change due to wave load with and without geotextile tube structure

The results of shoreline changes without the presence of a coastal protective structure and shoreline changes with the placement of geotextile tube structures at three different locations show that the average width of the occurrence of abrasion or sedimentation is as follows,

1. Abrasion width 0.2184 m where there is no coastal protection structure
2. Abrasion width of 2.7021 m for placement of geotextile tube structures on the grids 10 to 30
3. Abrasion width of 2.3252 m for placement of geotextile tube structures on the grids 30 to 50
4. Abrasion width of 0.0373 m for placement of geotextile tube structures on the grids 10 to 50.

The most effective placement of geotextile tube structures at the Way Hawang Beach is on the grids 10 to 50.

4. Conclusions

Shorelines are certainly dynamic and abrasion cycles are often a significant feature of their ecological character. Winds, waves and currents have normal forces that simply transfer sand and unconsolidated soil in shore areas, so that the position of the coastline changes rapidly. Development in coastal areas can increase erosion problems; this necessitates a great deal of effort to manage the problem of shore erosion and to return coastal function, to adapt both short- and long-term transformations caused by human activities, natural disasters and sea level rise.

Way Hawang coastal area at Kaur Regency, Bengkulu Province is experiencing abrasions. Efforts to eliminate abrasion on the Way Hawang Coast are mostly carried out using coastal protective structures. Erosion problems are exacerbated when preventive measures (i.e. selection of hard or soft structures) are inappropriate, inappropriately planned, constructed, or maintained and if effects on nearby shores are not carefully assessed. The geotextile tubes were simulated in the position of abrasion and sedimentation at Way Hawang coastal area. The simulation results indicated that the position between abrasion and sedimentation are the most effective position to manage and to prevent abrasion in Way Hawang coast's area of Bengkulu Province. Further research is how to maintain and retrofit the geotextile tubes at the coastal area.

Acknowledgments

The author is very grateful to the Institute for Research and Community Service at Universitas Kristen Maranatha, Indonesia which has provided the grant funds.

References

[1] Ministry of Public Works of Republic Indonesia 2015 Ministerial Decree No 290/KPTS/M/2015
[2] Suwarsono, Supiyati and Suwardi 2011 *Makara J. Technol.* **15**(1) 31
[3] Davidson M A, Lewis R P and Turner I L 2010 *Coast. Eng.* **57**(6) 620
[4] Suresh P L and Sundar V 2011 *J. Hidro-Environ Res.* **5**(3) 157
[5] Kermani S, Boutiba M, Guendouz M, Guettouche M S and Khelfani D 2016 *Ocean Coast. Manag.* **132** 46
[6] Cappietti L and Simonetti I 2020 *J. Coast. Res.* **95** 197
[7] Karambas T V and Samaras A G 2021 *J. Mar. Sci. Eng.* **9** 1108
[8] Choi K, Kim J and Lee J 2016 *J. Coast. Res.* **75** 1062
[9] Setyandito O, Purnama A C, Yuwono N, Juliastuti and Wijayanti Y 2020 *ComTech* **11** 19
[10] Pranzini E, Anfuso G, Cinelli I, Piccardi M and Vitale G 2018 *Water* **10** 1647
[11] Tayade B R, Mahalingaiah A V. Gokhale N V and Kudale M 2015 *Aquat. Procedia* **4** 190
[12] Warnasuriya T W, Kuddithamby G and Sudheera S G 2018 *Mar. Geodesy* **41** 1
[13] Sundari T, Samra B and Saptono A B 2020 *IOP Conf. Ser.: Earth Environ. Sci.* **469** 012017
[14] Hadadpour S, Etemad-Shahidi A, Jabbari E and Kamranzad B 2013 *J. Coast. Res.* **65** 237
[15] Marinho B, Coelho C, Larson M and Hanson H 2018 *J. Coast. Conserv.* **22** 475
[16] Chairunnisa, Atiah S N and Wahyuni R 2019 Jurnal Teknik: Media Pengembangan Ilmu dan Aplikasi Teknik **18** 08
[17] Khosravil G, Majidi A and Nohegar A 2012 *Int. J. Probab. Stat.* **1** 160
[18] Yasmin N. 2019 Introduction to AutoCAD 2019 for Civil Engineering Application (KS: SDC Publications)
[19] Hanson H 1989 *J Coast. Res.* **5**(1) 1

Analysis of clay improvement as subgrade using Palm Oil Fuel Ash (Pofa)

Andriani [1], H G Putra[1], R Yuliet[1], R Maulana[2] and S P Marel[2]

[1] Civil Engineering Departement, Andalas University, Padang, 25163, West Sumatera
[2] Student at Civil Engineering Departement, Andalas University, Padang, 25163, West Sumatera

E-mail: andriani@eng.unand.ac.id

Abstract. Road construction on clay soils often causes problems, because clay have a low bearing capacity and high swelling shrinkage values. Damage to roads on soft clay cannot be avoided; therefore, soil improvement is necessary. Soil stabilization is one of the efforts to improve soil properties, and the aim is to increase the bearing capacity of the soil. So far, the stabilizing materials commonly used for soil improvement are cement and lime. In study, oil shell waste in the form of palm oil fuel ash (pofa) will be used as a stabilizing agent. Pofa is one of the industrial wastes that contain a high percentage of silica (Si). The high silica (Si) content is beneficial for reducing the shrinkage properties of clay soils. Pofa ash content used in the study was 4%, 8%, 12%, 16%, and 20% of the total weight soil. The curing time of the sample was seven days with unsoaked and soaked (four days) conditions. The results of this study indicate that mixing pofa ash can increase CBR value and decrease the swelling value of clay.

1. Introduction

Every year road construction in Indonesia is increasing. The road is an infrastructure that supports the economy because it will increase the distribution of goods/services and form a spatial structure. The progress of the times and economic growth caused infrastructure development to accelerate, causing a reduction in stable soil that development was carried out on soft soil and peat. The main problem when building roads on soft clay and peat soils will cause damage to the road subgrade before the design life; subsidence in the road, cracks, and potholes often occur due to low bearing capacity and large settlements. It is because soft clay has low shear strength, low California Bearing Ratio (CBR) values, and considerable potential for shrinkage. Soft soils are often found on the east coast of the island of Sumatra and some parts of the west coast of Sumatra island [1,2]. Soil serves as a pedestal to accept the load from the construction of the building and forward it to the subgrade layer, so the subgrade must be able to bear the given load so that there is no construction failure on it. Geotechnical conditions are the essential thing that must be considered when building a road on soft clay; soil characteristics are analyzed to obtain the appropriate type of stabilizing material so that it can increase the shear strength of soil, CBR value, and reduce shrinkage properties [3-5]. In general, soft clay have a fine grain size of less than 0.002 mm, low permeability, high capillary water rise, is highly cohesive, a high shrinkage rate, and slow consolidation process [6-8]. The causes of soil conditions become unstable, and the low bearing capacity.

Soil improvement can be carried out using mechanical and chemical soil stabilization to increase the soil's bearing capacity. Soil stabilization is the combination and manipulation of soil with or without additives. Mechanical soil stabilization is mixing two or more different types of soil to obtain a strong material that meets the requirements of a building material or subgrade. In field applications, soil can be mixed at the project site, factory, or inventory collection point, then spread and compacted at the project site; at the same time. Chemical stabilization improves soil properties by adding additional materials such as cement, fly ash, lime, Portland cement, husk ash. and others [9]. In this research, chemical stabilization will be carried out with a mixture of materials using Palm Oil Fuel Ash (Pofa) and lime. Palm Oil Fuel Ash (Pofa) is industrial waste ash with a high silica (Si) composition. Indonesia is one of the largest producers of Crude Palm Oil (CPO) in the world and has the largest oil palm plantation in the world. Indonesia is also the world's largest palm oil producer, with a production of 43 million tonnes, ranking first [10]. The problem that arises is a large amount of palm oil shell waste, causing a large area to accommodate the waste; the waste is burned into ashes called Pofa (Palm Oil Fuel Ash). Pofa is industrial waste ash with pozzolanic content or has a high silica (Si) content, so it has the potential to be used as an alternative material for cement mixtures. This ash is a residue produced from burning oil palm shells at a temperature of 700 °C - 800 °C. Oil palm shell ash is in the form of powder material [11]. The utilization of polyethylene as a stabilizing material can reduce road construction costs and improve environmental quality by reducing waste and increasing soil-bearing capacity.

Oil palm shell ash is pozzolanic, a material containing silica and alumina, which are reactive when combined with lime and water. Oil palm shell ash is one of the industrial wastes whose chemical composition contains high levels of silica and has a high potential to be used as a cement substitute. Pozzolans are natural and rock additives, mainly composed of the elements silica (Si) and active alumina (Al). Silica is a chemical whose utilization and application range from electronics, medicine, and mechanical arts to other fields. One of the widely used uses of silica powder is as an absorber of water content in the air to extend the material's life as a mixture for making art ceramics. Table 1 shows the composition of Pofa.

Table 1. Composition of pofa

Material content	Composition (%)	Method
K_2O (Kalium)	7.40	SNI 02.2803.2000
MgO (Magnesium)	3.19	SNI 02.2804.2005
CaO (Calcium)	5.32	SNI 02.2804.2005
SiO_2	52.2	SNI 02.2804.2005

Source: Central Laboratory of Palm Oil Research.

Adding a certain amount of pofa can increase the stability of the clay, especially the CBR value in unsoaked and soaked conditions [12-16]. Some researchers also use Pofa to increase the bearing capacity of peat soils, and the results show that adding pofa can increase the CBR value, compressive strength, and shear strength of peat soils and reduce swelling [17]. Figure 1 shows some of the types of damage that occurred on the road.

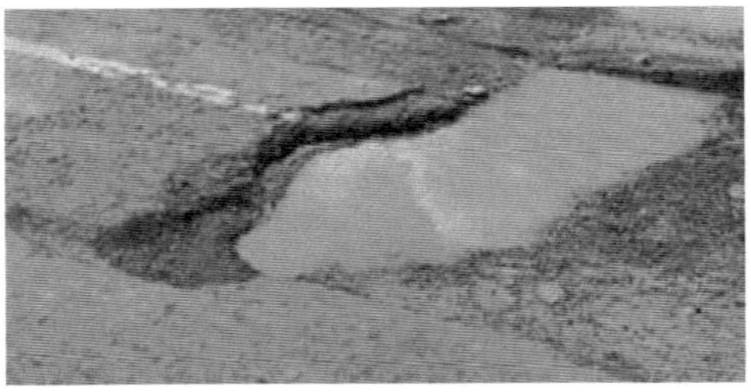

a. potholes

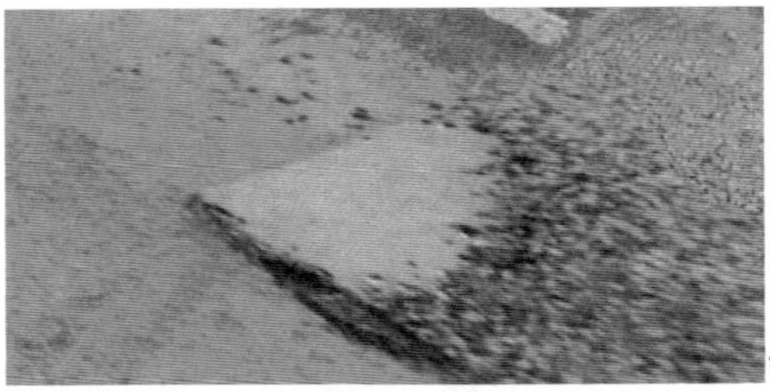

b. raveling

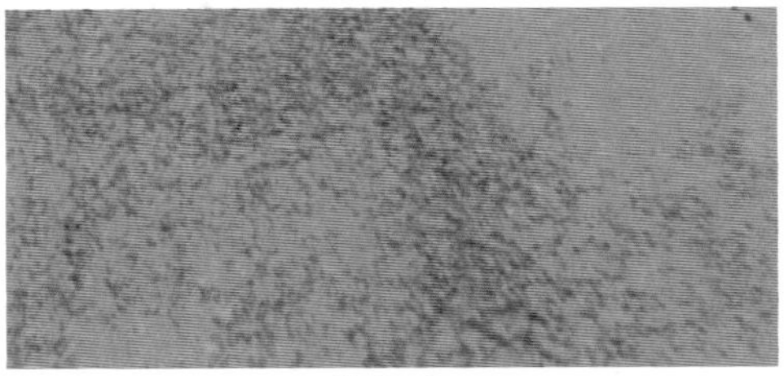

c. polished aggregate

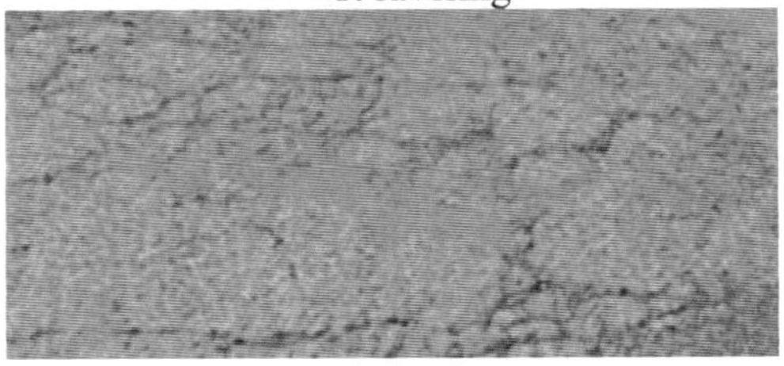

d.shoving

Figure 1. Types of damage to roads on soft soil.

Some researchers use a mixture of lime and pofa to produce high bearing capacity values in clay soils, the results show that adding lime and pofa at certain levels can increase the bearing capacity of the subgrade and reduce settlement [18-20]. Clay and lime will react to form a stable mixed structure so that it is easy to work and compact, changing the soil properties and reducing the stickiness and plasticity of the soil. In this study, lime and pofa were added to increase the clay's bearing capacity and reduce the swelling value.

2. Methodology

The soil sample used was clay with disturbed conditions located in the Limau Manis campus area, Andalas University, Padang, West Sumatra. For the stabilization material in the form of extinguished lime (Ca(OH)2), which is the result of extinguishing quicklime with water and forming a hydrate, and pofa, which pass sieved no. 200.

Figure 2. Soil sampling location

This research consists of preliminary testing, which aims to determine the physical characteristics of clay soil (specific gravity, Atterberg limits, soil type, and density of clay soil). Further tests were carried out by adding lime and pofa to the original soil. The lime content used was 5%, while the percentage of pofa was 4%, 8%, 12%, and 16% of the total weight of the mixture. The curing time of the soil sample was seven days with unsoaked and soaked conditions (four days). The tests carried out were compaction test, California Bearing Ratio (CBR), and swelling test for four days. A compaction test was carried out on all sample variations to obtain the maximum dry unit weight (γdry max) and water content optimum (Wopt) of all sample variations. The type of compaction carried out is standard Proctor, with a total of 3 layers, and each layer is given 25 blows. Furthermore, compaction results are used as the basis for making CBR samples. A swelling test was carried out on soil samples using a soaked CBR test; during a certain time, the changes in sample height that occurred for four days were recorded, then the percentage of swelling that occurred was calculated using the formula below.

$$\text{swelling} = \frac{H_t - H_0}{H_0} * 100\% \quad (1)$$

With
H_t : Final height (cm)
H_0 : Initial height (cm)

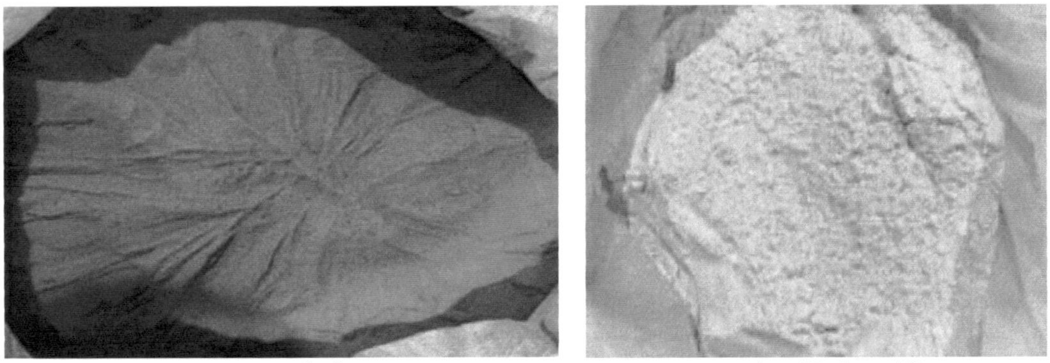

a. Pofa b. Hydrated lime

Figure 3. Additives material used in research.

The number and variation of samples tested in the study are shown in Table 2.

Tabel 2. Variation of the research sample

Sample	CBR(samples)		Swelling
	unsoaked	**soaked**	
Soil without additives matter	3	3	3
Soil 95 % + lime 5 %	3	3	3
Soil 91 % + lime 5 % + Pofa 4 %	3	3	3
Soil 87 % + lime 5 % + Pofa 8 %	3	3	3
Soil 83 % + lime 5 % + Pofa 12 %	3	3	3
Soil 79 % + lime 5 % + Pofa 16 %	3	3	3

The CBR value that will be selected in the laboratory CBR test is the CBR value at the penetration of 0.1". If there is a test result where the CBR value at 0.2" penetration is greater than the CBR value at 0.1" penetration, then the test must be carried out again. If the retest has been carried out. The results still show a greater CBR value at 0.2" penetration, the CBR value that will be selected will be the CBR value at 0.2" penetration. Mathematically, according to ASTM D, finding the CBR value is poured into the formula below.

$$\text{CBR Value at Penetration } 0{,}1" = \frac{A}{6{,}9 \text{ MPa}} \text{x } 100\% \quad (2)$$

$$\text{CBR Value at Penetration } 0{,}2" = \frac{B}{10{,}2 \text{ MPa}} \text{x } 100\% \quad (3)$$

With,
A = corrected load reading (MPa) penetrated 0.1" on the graph
B = reading of the corrected load (MPa) penetrated 0.2" on the graph

3. Results and discussion

3.1. Results of soil properties testing

The results of sieve analysis obtained the percentage value of grains passing sieve no. 200 is 85.77% (> 50%); this indicates that the soil is fine-grained. The Atterbergs limit result shows that the liquid limit value (LL) is 51.942%, the plastic limit (PL) is 30.161%, and the plasticity index (PI) is 21.781%. Soils with a plasticity index value > 17 are clay with high plasticity. Soil with high plasticity will easily change shape without changing volume; the higher the plasticity of the soil, the greater the potential for swelling and shrinkage that will cause the construction on it to be easily damaged. Based on the Unified Soil Classification System (USCS) results, it is included in the OH class of fine-grained soils, namely organic clays with medium-high plasticity. Meanwhile, according to the American Association of State Highway and Transportation Officials (AASTHO), category A-7-5 is not suitable for use as a subgrade. Based on the soil classification above, it means that the soil must be stabilized to increase its bearing capacity to be used as a road subgrade. Table 3 shows the results of preliminary testing on the soil.

Table 3. Results of soil properties testing.

No	Testing	Results
1	Water content, w (%)	8.374
2	Volume weight, γ (gr/cm^3)	1.288
3	Specific gravity, G	2.589
4	A sieve analysis (%)	
	a. Pass sieved no.10	96.97
	b. Pass sieved no.40	93.20
	c. Pass sieved no.200	85.77
5	Atterberg Limit	
	a. Liquid Limit, LL (%)	51.942
	b. Plastic Limit, PL (%)	30.161
	c. Plasticity, PI (%)	21.781
6	Maximum dry unit weight, $\gamma_{dry\ max}$ (gr/cm^3)	1.088
7	Optimum water content, w_{opt} (%)	40.150
8	CBR of soil without additive, unsoaked (%)	12.870
9	CBR of soil without additive, soaked (%)	8.980
10	Swelling soil without additive matter (%)	4.362

3.2. Result of Compaction Soil with additive matter

Soil compaction is a process of removing air from the soil, and the distance between soil particles becomes small so that the bearing capacity of the soil will increase. Parameters obtained from compaction are maximum dry unit weight and optimum water content. In soil compaction, the volume of air decreases while the volume of water remains constant. Table 4 shows the soil compaction results from adding lime and pofa.

Table 4. Results of soil compaction in the laboratory.

Sample	**Compaction**	
	$\gamma_{dry\ max}$ (gr/cm^3)	w_{opt} (%)
S1. Soil without additives matter	1.088	40.150
S2. Soil 95 % + lime 5 %	1.288	35.350
S3. Soil 91 % + lime 5 % + Pofa 4 %	1.297	33.681
S4. Soil 87 % + lime 5 % + Pofa 8 %	1.307	30.761
S5. Soil 83 % + lime 5 % + Pofa 12 %	1.322	29.990
S6. Soil 79 % + lime 5 % + Pofa 16 %	1.347	28.661

Adding lime and pofa to clay causes the maximum dry unit weight to continue to increase while the optimum water content decreases. This shows that the soil pores are filled with lime and pofa, so it forces air out of the soil. The increase in the number of solid particles in the soil impacts its dry density when compared to the original soil. This decrease in optimum water content is caused by the stabilizer material urging water out and filling the pores in the soil so that the percentage of water contained is reduced. The results showed that the highest maximum dry weight value was found in a mixture of 79% native soil + 16% POFA + 5% lime with a value of 1.347 gr/cm³ and the optimum water content was 28.661%.

3.3. Result of CBR value and swelling

The CBR value was developed to measure the bearing capacity, soil load, and road pavement load. CBR testing is used to evaluate the potential strength of subgrade, sub-base, and foundation materials, including materials that are recycled for road and airfield pavements. The CBR value of the soil strongly influences the value of the bearing capacity of the subgrade. The correlation between bearing capacity and CBR graphs was carried out to obtain the value of the bearing capacity of the subgrade. The greater the CBR value of the subgrade in road construction, the greater the value of the soil-bearing capacity of the road. A high CBR value will cause the thickness of the road pavement to be thinner, and the cost will be cheaper; on the contrary, if the CBR value is low, it will cause the pavement layer to be unstable, so a thick pavement layer is needed. Table 5 shows the increase in CBR values due to adding lime and pofa to the clay soil.

Table 5. CBR value of clay due to addition of lime + pofa.

Sample	**CBR**		**Swelling value** (%)
	Unsoaked (%)	Soaked (%)	
S1. Soil without additives matter	12.870	8.980	4.362
S2. Soil 95 % + lime 5 %	20.870	13.254	4.223
S3. Soil 91 % + lime 5 % + Pofa 4 %	23.623	19.130	1.631
S4. Soil 87 % + lime 5 % + Pofa 8 %	38.961	21.435	0.845
S5. Soil 83 % + lime 5 % + Pofa 12 %	45.624	36.568	0.566
S6. Soil 79 % + lime 5 % + Pofa 16 %	48.508	47.887	0.424

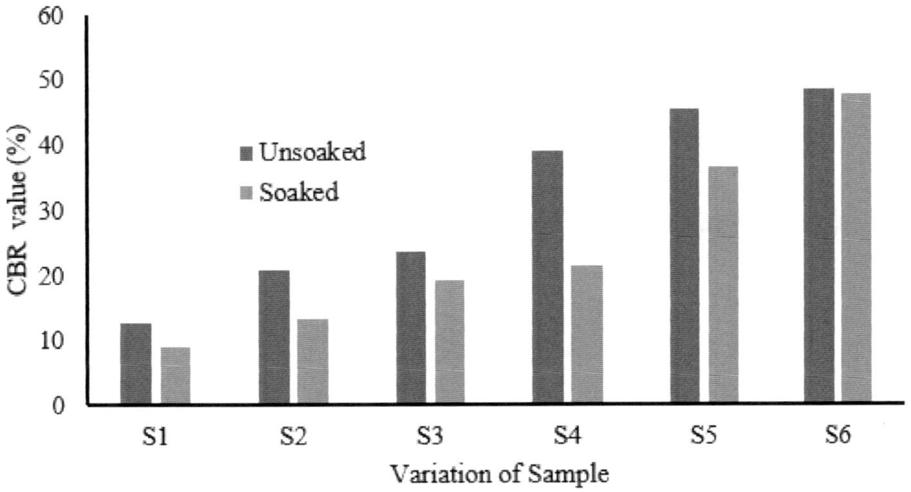

Figure 4. CBR value of clay due to the addition of lime + pofa with the unsoaked and soaked condition.

Figure 4 shows that adding lime and pofa causes an increase in the CBR value of the soaked and unsoaked conditions. In the S6 sample variation, it showed a slight difference in the unsoaked and soaked CBR values; this indicates that the S6 sample has a high density, so its difficult for water to penetrate the soil layer. The S6 variation is very well used as a subgrade because the soil layer will remain stable so that the influence of water does not easily damage the road. The presence of water in the soil will affect the bonds between the grains because water acts as a lubricant between the soil grains so that the soil grains are loose and not solid. The composition of the loose soil will be easily decomposed so that it will not be able to carry the load of the working vehicle, and the road will be damaged. Meanwhile, a dense and compact soil composition can support the working load and transmit it to the soil layer below it.

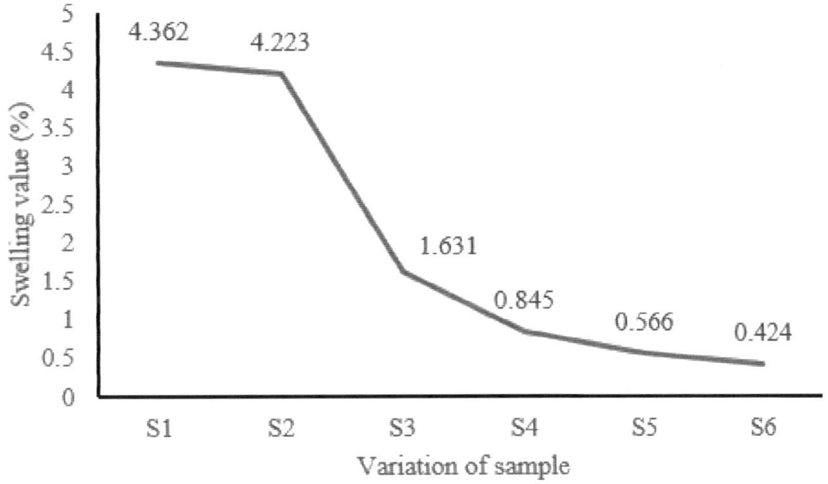

Figure 5. Swelling value of clay due to the addition of lime + pofa.

The shrinkage of clay soil is caused by the water content in the soil, where if the water content in the clay is reduced or dry, the soil will shrink and cause the soil to crack on its surface. The soil bearing capacity can increase; on the contrary, if the water content increases, the clay will swell, and the soil bearing capacity can be reduced. Its condition can trigger problems on construction, especially highway construction, such as can cause the pavement becoming bumpy, cracked, and deformation. The properties and behavior of clay soils significantly affect the composition of minerals, chemical elements, particles, clay textures, and the influence of the surrounding environment. Figure 5 shows that the highest swelling occurred in the original soil. The addition of lime and pofa causes a decrease in the swelling value of the clay soil. Samples S4, S5, and S6 showed a swelling value of <1%, meaning that the soil will become more stable and not change in volume due to the influence of water. The S6 sample is impermeable, so it is difficult for water to penetrate the soil layer, so the soil remains stable and does not change in shape and volume due to the influence of water.

4. Conclusions

a. Based on the USCS classification, the soils studied are clay soil with medium-high plasticity (OH) and A-7-5 (AASHTO), so that it susceptible to the influence of water. To increase bearing capacity, the soil must be stabilized.
b. The addition of lime and pofa on clay soil can increase soil density, CBR value and reduce swelling value in clay soil. The addition of 5% lime and 16% pofa can increase unsoaked and soaked CBR values and reduce swelling value to < 1%.

Acknowledgment
This research was funded by Andalas University in accordance with the Basic Research research contract Number T/1/UN.16.17/PT.01.03/IS-RD/2022 for the 2022 fiscal year.

References
[1] Wardoyo, Sarwondo, Farah D, Wahyudin, Wiyono, Ginda H and William PS 2019 Atlas Persebaran Tanah Lunak Indonesia **1** 10
[2] Andriani, Shubhi NH and Fuji A 2020 *Journal of Physics: Conference Series* **1500** 1-8
[3] Asgari MR, Baghebanzadeh DA, and Bayat M 2015 *Arabian Journal of Geosciences* **8**(3) 1439–1452
[4] Little D, and Nair S 2009 Texas, USA: National Cooperative Highway Research Program, Transportation Research Board of the National Academies
[5] Phetchuay C, Horpibulsuk S, Arulrajah A, Suksiripattanapong C, and Udomchai A 2016 *Applied Clay Science* 127-128
[6] Yilmaz I, and Civelekoglu B 2009 *Appl. Clay Sci.* **44** 166–172
[7] Marto A, Hassan MA, Makhtar AM, Othman BA, and Ash BC 2013 *APCBEE Procedia* **5** 116–221
[8] Bergaya F, Lagaly G 2006 General introduction: Clays, clay minerals, and clay science. Handbook of clay science **1** 1-8
[9] Fondjo AA, Theron E, and Ray RP 2021 *Civ. Eng. Archit* **9** 1295–1308
[10] Amelia U, Anisa DU, and Suharno S 2022 *J.Pen. Kelapa Sawit* **30**(2) 95-108
[11] Tangchirapat WC, Jaturapitakkul and Chindaprasirt P 2009 Use Palm Oil Fuel Ash as a Supplementary Cementitious Material for Producing High Strength Concrete. Construction and Building Materials **23**(7) 2641-2646
[12] Awal ASM, and Hussin W 1997 The effectiveness of palm oil fuel ash in preventing expansion due to Alkali-silica reaction. *Cem. Concr. Compos.* **19** 367–372
[13] Sukmak P, Sukmak G, Horpibulsuk S, Setkit M, Kassawat S, and Arulrajah A 2017 *Road Materials and Pavement Design* **20**(1) 110–131
[14] Al-Hokabi A, Hasan M, Amran M, Fediuk R, Vatin NI, and Klyuev S 2021 *Sustainability* **13** 10910
[15] Daud NNN, Muhammed AS, and Yusoff ZM 2016 *Malays. J. Civ. Eng.* **28** 1–9
[16] Yusof MBBM, Brown O, and Umar BA 2011 *J. Teknos-2k* **11** 74–86

[17] Khalid N, Mukri M, Kamarudin F, Sidek N, and Arshad MF 2014 *Strength of Soft Soil Stabilized Using Lime-POFA Mixtures*. In InCIEC 2013; Springer: Singapore 501–510
[18] Ahmad J, Abdul R, Mohd Ali M R, and Khif KF 2011 Peat soil treatment using POFA. In 2011 *IEEE Colloquium on Humanities, Science and Engineering Research Penang, Malaysia* 66–70
[19] Ouhadi VR, Yong RN, Amiri M, and Ouhadi MH 2014 *Appl. Clay Sci.* **95** 111–118
[20] Senol TB, Edil HA, Acosta CH, and Benson 2006 *Resour. Conserv. Recycl.* **46** 365–376

IOP Conf. Series: Earth and Environmental Science 1173 (2023) 012025 doi:10.1088/1755-1315/1173/1/012025

Clayshale stabilization using active natural lime to increase the shear strength of soil

M F R Hasan[1,2], I Z Utama[1], A F A Razzak[1], A Salimah[1,3] and P A M Agung[1]

[1] Department of Civil Engineering, Politeknik Negeri Jakarta, Depok, Indonesia
[2] Department of Physics, Brawijaya University, Malang, 65145, Indonesia
[3] Department of Civil Engineering, Faculty of Engineering, Universitas Indonesia, Depok, Indonesia

E-mail: aisyah.salimah@sipil.pnj.ac.id

Abstract. Clayshale not used in construction because it is easily weathered when exposed to air and water to reduce the soil shear strength. As a result, this decline can cause geotechnical engineering problems, and clayshale needs stabilization. This study aims to determine the shear strength of clayshale by adding variations of 5%, 8%, and 11% lime. Tests carried out to obtain soil parameters include soil index and engineering properties. Based on laboratory tests, the original clayshale soil has a 33.2% liquid limit, a 6,1% plasticity limit, and a 27,1% plasticity index. Hence, clayshale soils belong to the CL(clay-low plasticity) category based on the USCS. Adding lime variations can increase the specific gravity and the optimum moisture content of the clayshale soils. Still, it can reduce the dry unit weight and the plasticity index value on clayshale soils. The triaxial test on the original clayshale soil sample obtained a cohesion(c) value of 0.20kg/cm^2 and friction angle(f) of 6.38°. In contrast, the values of c and f occurred the highest increase from the original clayshale soil with the addition of 8% lime, so lime can be used as an alternative in soil stabilization because it can increase the shear strength of the soil.

1. Introduction

Generally, clayshale are layers not used in the construction world and must be disposed of (replaced). Because it often causes many geotechnical problems such as difficulty choosing heap materials, foundation bearing capacity, slope stability, underground construction, and so on [1]. In addition to causing many problems, experts also have difficulty in classifying them, and this is because clayshale has transitional properties between soil and rock that exhibit both "soil-like" and "rock-like" behaviours [2]. It is easily weathered when exposed to air and water, so it can reduce the shear strength it has [3].

Most problems in geotechnical engineering, such as low bearing capacity, landslides, and high deformation, are caused by a decrease in soil shear strength. Phenomena landslide on clayshale-layers is initiated by a condition of full saturation [4]. Thus, this study utilized the clayshale layer as the base soil layer through the stabilization process.

Soil stabilization as one of the soil improvement methods in geotechnical engineering, can be generally divided into three categories: mechanical stabilization, chemical stabilization, and physical stabilization. One of the parameters in soil shear strength, such as bearing capacity, can be increased by mechanical stabilization using additional material to the existing soil; improvement of soil gradation; and or compaction with more energy during construction. Chemical stabilization aims to add chemical material, such as lime, cement, bitumen, or special additives, to reduce or remove undesirable material properties of the existing soil. Finally, the addition of material can increase the strength and or the bearing capacity as well. Physical stabilization is more purposed to improve the shear strength without changing the soil grain structure, such as the installation of geotextiles in the existing soil layers [5,6].

In chemical stabilization, lime material has been used widely as a stabilizer or an improvement material. Lime material in soil stabilization can improve soil shear strength for road construction, railway, airport runway, and others. [7]. As a chemical process, the lime material can react with grain structure and change the existing soil's mineral properties. Calcium with positive ions can be a stimulant material in creating soil adhesion, decreasing moisture content, reducing the plasticity index, reducing swelling, and increasing soil stability [4, 5, 8, 9]. Some previous studies have shown that lime material can stabilize the existing soil during the compaction process and improve the mechanical properties of the existing soil [5, 9, 10].

Based on several studies in chemical stabilization, this paper will discuss the improvement of using CaO lime in stabilizing clayshale. Some studies were performed and resulted effectiveness assessment of CaO lime in increasing the parameter shear strength, such as angle of internal friction (ϕ) and cohesion (c). Evaluation of both parameters refers to ASTM D-2850-295 (Triaxial Test Method of Unconsolidated Undrained or UU). A case history aims to evaluate the values of c and ϕ in total or effective stress in a state condition and after adding some variations of CaO lime content in the stabilization process.

2. Research methods

2.1. Sample location

Clayshale sampling for testing was carried out in the Hambalang Citereup area, Bogor, West Java, located at coordinates 6°33'12" S and 106°54'02" E. Clayshale sampling is a disturbed sampling technique with a depth of up to 3 meters.

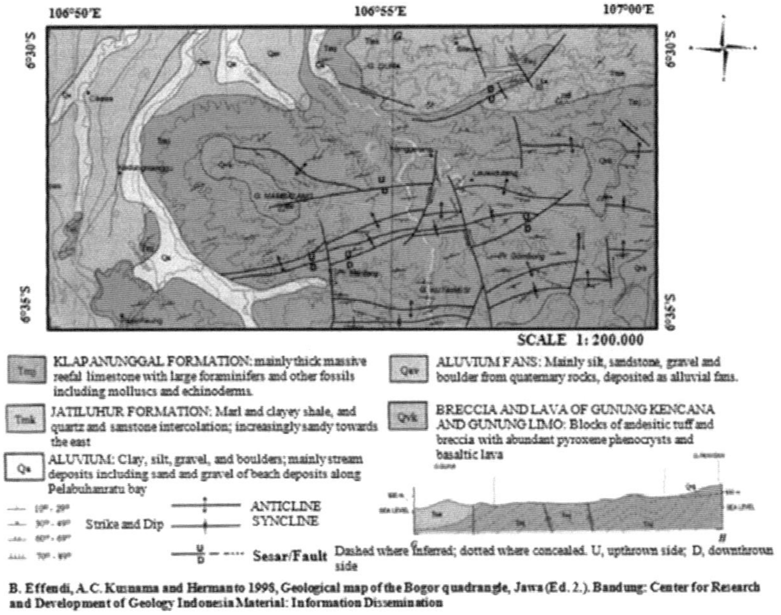

Figure 1. Geological map as a reference of Hambalang, Bogor, Indonesia.

This study examines the soil's physical changes, durability, and shear strength. Variations are made with the following differences:

Table 1. Current studies of CaO lime stabilization materials in clayshale specimens.

Sample	Lime	Description
C0L	-	Clayshale only
C5L	5%	Clayshale + 5% lime
C8L	8%	Clayshale + 8% lime
C11L	11%	Clayshale + 11% lime

The method used in this study is the research method in the laboratory. This test was carried out using one 1-day curing for clayshale soils not mixed with lime and three days of curing for clayshale samples mixed with active natural lime. The laboratory tests carried out are Physical Properties Tests which consist of Water Content, Specific Gravity, Bulk Density, and Atterberg Limits Tests. After that, Proctor Standard and Triaxial Test are carried out to find the cohesion and friction angle value.

2.2. Physical properties test

2.2.1. Moisture content test. The moisture content test measures the amount of water in the existing soil. The percentage of water is expressed by the ratio between the weight of water and the weight of dry grains of a soil specimen [11].

2.2.2. Specific gravity test. The specific gravity of soil (G_s) is expressed by the ratio between the unit weight of solid particles (γ_s) and the unit weight of water (γ_w) at the same temperature and volume [12].

2.2.3. Bulk density test. The unit weight of solids (γ_s) is a proportion of the weight of solid particles (W_s) with a volume of solid particles (V_s) [13].

2.2.4. Atterberg limits test. Atterberg limits are water content where the soil consistency and behavior changes consisting of liquid limit is the boundary of the moisture content between liquid and plastic conditions. Liquid limit is measured at the time of several blows 25 times on a straight line called the flow curve formed logarithm of several blow counts and the moisture water content. The plastic limit is defined as the moisture content at the soil thread that just crumbles when it is carefully rolled out to a diameter of 3 mm (1/8 in) [14].

2.3. Mechanical properties test

2.3.1. Standard proctor test. Several samples of the state condition after the addition of Cao lime in chemical stabilization at different water contents. The standard proctor test can measure optimum moisture content (OMC) in determining maximum dry density graphically [15]. The optimum moisture content in each percentage level of additional CaO lime obtained during the compaction test will be used as a baseline for the triaxial test.

2.3.2. Triaxial unconsolidated undrained test. The Triaxial test is the testing of a cylindrical test object wrapped in waterproof rubber. Then, it is given pressure in all directions and applied axial pressure until a collapse occurs. This Triaxial Unconsolidated Undrained (UU) test uses an unconsolidated and undrained triaxial test. This condition is a condition whose test example is not consolidated first, and friction angle water is not allowed to go outside or enter the test example during the test [16]. In this study, triaxial testing was carried out with variations in lime mixtures to determine changes in cohesion values (c) and friction angles (ϕ) after the addition of CaO lime.

3. Results and discussion

Based on several tests conducted in the soil mechanics laboratory, clayshale samples in a state condition have characteristics as shown in table 2.

Table 2. Characteristics of original clayshale soil

Parameters	Unit	Value
Physical Properties Test		
Moisture Content (ω)	%	0.214
Specific Gravity (G_S)		2.616
Liquid Limit (LL)	%	33.15
Plastic Limit (PL)	%	6.08
Plasticity Index (PI)	%	27.07
Mechanical Properties Test		
Maximum Dry Density (γ_d maks)	(g/cm^3)	1.773
Optimum Moisture Content (ω opt)	%	13.67
Cohesion (c)	(kg/cm^2)	0.201
Friction Angle (ϕ)	(°)	6.38

Based on several tests conducted in the soil mechanics laboratory, clayshale samples in a state condition have characteristics as shown in Table 2. Table 2 shows that the water content was less than 50%, so it can be assumed that all pores of clayshale were not thoroughly watered during sampling. Assessment for specific gravity value shows some characteristics were close to clayey soil. Limits of consistency identify clayshale samples were not classified as expansive soil, and it can be assumed that there were no montmorillonite contents significantly in clayshale samples since the plasticity index is less than 30%. Most of the clayshale samples were more shown creep behavior. The creep occurs as a process due to weathering, which occurs over a relatively long period in the exposed soil. Standard proctor tests identified that remoulded clayshale could strongly govern its engineering behaviour. However, cohesion and the angle of internal friction value are closer to clayey soil.

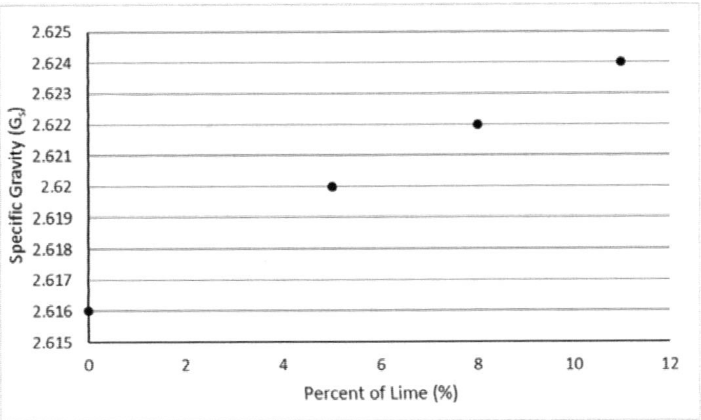

Figure 2. The relationship between specific gravity and lime addition

Figure 2 shows that the addition of lime can increase soil's specific gravity. This is because lime can fill the pores of clayshale soil by irradiating the water in the soil mixture and gluing between hard particles, making it difficult for water to penetrate. This results in a reduction in the unit weight of water and an increase in the unit weight of solids, thereby increasing the specific gravity value, which is the ratio between the unit weight of solids and the unit weight of water [17].

Table 3. Atterberg limit test results

Sample	Liquid limit (%)	Plastic limit (%)	Plasticity Index (%)
C0L	33.2	6.1	27.1
C5L	26.0	19.2	6.8
C8L	31.7	22.2	9.5
C11L	31.6	22.6	9.0

Table 3 shows that the addition of lime can lower the liquid limit value and plasticity index value and increase the plasticity limit value. A decrease in the Plasticity Index (PI) value can reduce the soil's potential for development and shrinkage. This is due to the process of hydration of lime added to the soil. This process strengthens the bonds between soil particles, resulting in more rigid and stable grains. The filling of soil pores reduces the occurrence of seepage in the soil-lime mixture, which reduces the likelihood of shrinkage and can increase the bearing capacity of the soil [18].

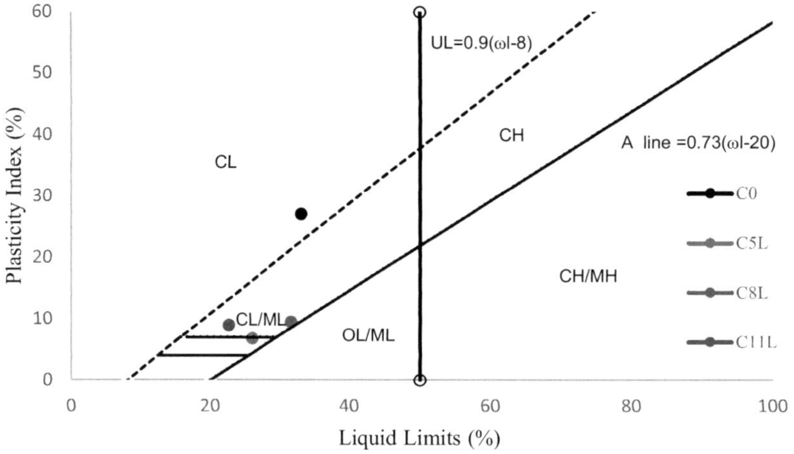

Figure 3. Classification by USCS.

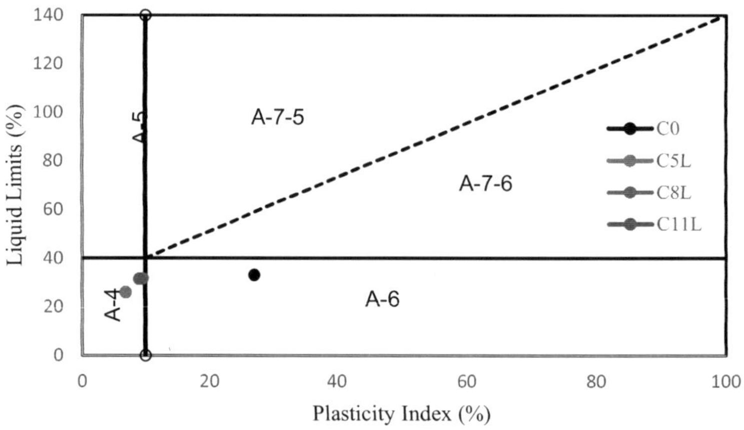

Figure 4. Classification by AASHTO.

Figures 3 & 4 show the results of atterberg limit testing, obtained clayshale classification based on USCS included in non-organic fine-grained soils belonging to the category of CL (clay-low plasticity) with low to medium plasticity [19]. And clayshale is included in sub-A-6 (clayey soil) based on AASHTO, but a change is caused by mixing active natural lime into sub-A-4 (silty soil). So, stabilization with active natural lime can affect the physical properties of clayshale.

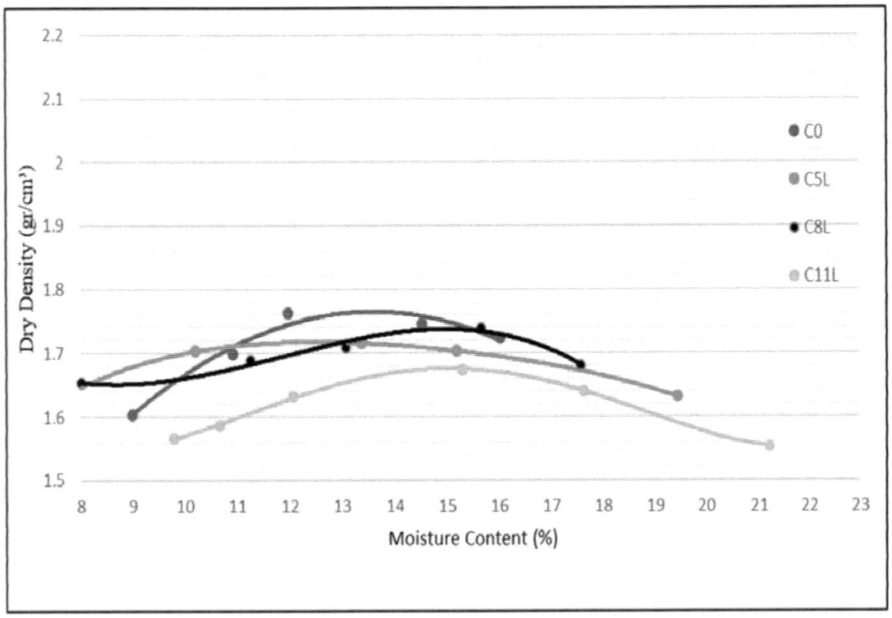

Figure 5. Comparison of water content and dry density.

Figure 5 shows the results of the standard proctor test, from which the water content and dry density values of each sample were obtained. The figure can be processed into the following table.

Table 4. Standard proctor test results.

Sample	ω optimum (%)	$\gamma r\ maks\ (gr/cm^3)$
C0L	13.7	1.77
C5L	14.1	1.73
C8L	15.7	1.74
C11L	15.5	1.68

Table 4 proves that the addition of CaO lime to clayshale soils can affect the optimum moisture content value and dry weight of the sample. The C0L example had an optimum moisture content value of 13.7%, with a dry density of 1.77 gr/cm³. The addition of CaO lime mixture can increase the optimum moisture content value, with the peak being the C8L sample by 15.7%. Meanwhile, the addition of lime can reduce the dry weight value of clayshale soil samples and make the compaction curve more flattened [20]. This change occurs because when the lime is mixed with the soil, there is a direct cation exchange reaction, and clay particles are interrelated. This process forms air cavities between particles and creates a porous medium with a lower maximum dry weight. Therefore, more water is needed to fill the hole so that the optimal water content increases [7].

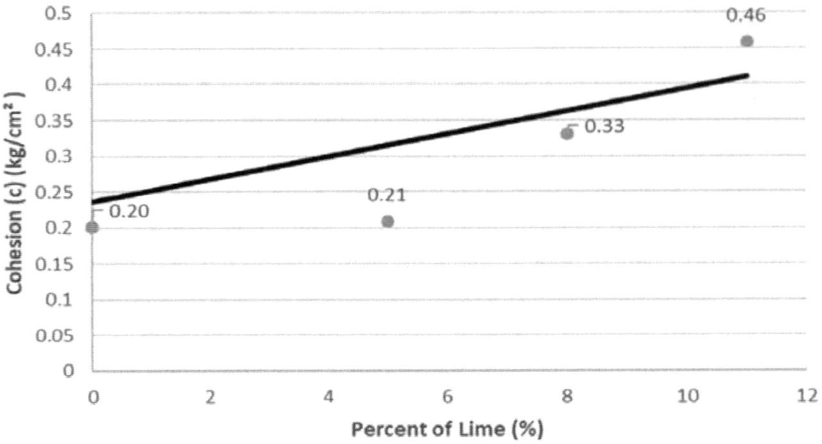

Figure 6. Relationship of cohesion analysis due to the effect of lime addition.

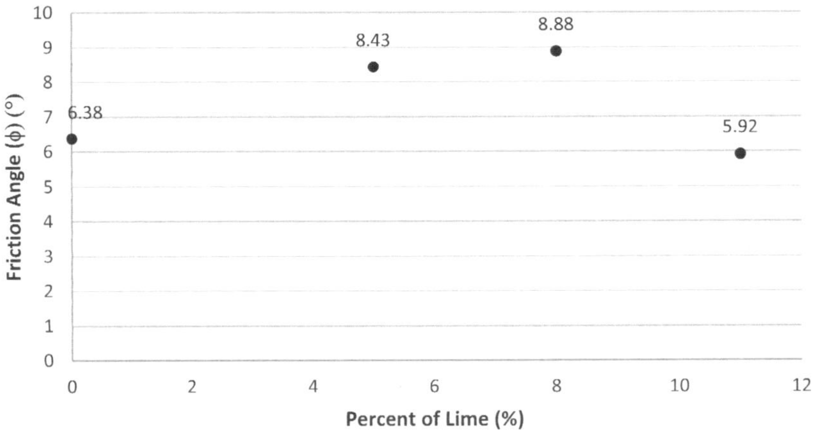

Figure 7. Relationship of friction angle analysis due to the effect of lime addition.

Previous research explained that the addition of CaO lime to soil stabilization mixtures could increase the soil's friction angle and cohesion value [21]. An increase in the friction angle and cohesion value occurs because lime can lower the value of the soil plasticity index and strengthen the particle bonds between the soils resulting in more rigid and more stable grains. The increasing cohesion value and friction angle can increase the shear strength of the soil. Figures 6 & 7 prove the addition of CaO lime can increase the cohesion value and friction angle after the Triaxial UU test. The maximum increase occurred in the C8L sample, and there was a decrease in the C11L example. This decrease is thought to occur due to the addition of lime that is too high, which makes the mixture brittle so that it breaks easily [17]. In general, it can proven that the addition of lime can increase the shear parameters on clayshale soils.

4. Conclusions

The addition of lime to clayshale soil stabilization can increase the optimum moisture content and lower the maximum dry density value so that the compaction chart is further flattened. This is because the addition of lime can reduce the liquid limit value and plasticity index value in clayshale soils so that the workability increases. The addition of lime can increase the value of cohesion (c) and friction angles (ϕ) in the soil so that the strength of the soil and soil resistance increase. The highest increase in shear strength value is found in the addition of lime by 8%, but the shear strength begins to decrease although the percentage of lime increases. Overall the test results showed that clayshale soil stabilized with 8% lime content and curing for three days produced the highest shear strength value, equal to a friction angle of 8.88° and cohesion of 0.331 kg/cm^2 between the original clayshale soil specimens and all clayshale soils stabilized with lime and Triaxial UU test. Based on the research that has been carried out regarding the influence of lime stabilizing materials on clayshale, the author advises that: it is required to conduct an SEM-XRD test to determine the properties of lime in stabilizing clayshale and also needed to test variations in the duration of curing and the content of the mixture in the lime.

References

[1] Rahayu W 2019 Stabilization of clay shale using propylene glycol and laterite on California bearing ratio *IOP Conf. Ser. Mater. Sci. Eng.* **620** 012042

[2] Agung P A M, Yuwono, Sutikno and Mursid 2017 *Landslide and Collapse Potential of Clayshale Layers through Observation by using XRD; SEM; and Petrography* (Lambert Academic Publishing)

[3] Putera A M A, Pramusandi S and Damianto B 2017 Identification and classification of clayshale characteristic and some considerations for slope stability *African J. Environ. Sci. Technol.* **11** 163–97

[4] Faray and Rahayu W 2020 Durability and strength improvement of clayshale using various stabilized materials *IOP Conf. Ser: Earth and Environmental Science* **426** 012028

[5] Jawad I T, Taha M R, Majeed Z H and Khan T A 2014 Soil stabilization using lime: Advantages, disadvantages and proposing a potential alternative *Res. J. Appl. Sci. Eng. Technol.* **8** 510–20

[6] Afrin H 2017 A Review on Different Types Soil Stabilization Techniques *Int. J. Transp. Eng. Technol.* **3** 19–24

[7] Roohbakhshan A and Kalantari B 2013 Stabilization of Clayey Soil with Lime and Waste Stone Powder *Int. J. Sci. Res. Knowl.* **1** 547–56

[8] Salimah A, Hazmi M, Fathur Rouf Hasan M, Agung P A M and Yelvi . 2021 A comparative study of red brick powder and lime as soft soil stabilizer *F1000Research* **10** 777

[9] Chemeda Y C, Deneele D, Christidis G E and Ouvrard G 2015 Influence of hydrated lime on the surface properties and interaction of kaolinite particles *Appl. Clay Sci.* **107** 1–13

[10] Andavan S and Pagadala V K 2020 A study on soil stabilization by addition of fly ash and lime *Mater. Today Proc.* **22** 1125–9

[11] ASTM-D-2216-98 1998 Standard Test Method for Laboratory Determination of Water (Moisture) Content of Soil and Rock by Mass *ASTM Int.* **04.08** 1–5

[12] ASTM-D854-10 2000 Standard Test Methods for Specific Gravity of Soil Solids by Water Pycnometer *ASTM Int.* **2458000** 1–7

[13] ASTM-C29/C29M-17a 2009 Standard Test Method for Bulk Density (" Unit Weight ") and Voids in Aggregate *ASTM Int.* **C29** 1–5

[14] ASTM-D4318-17 2017 Standard Test Methods for Liquid Limit, Plastic Limit, and Plasticity Index of Soils *ASTM Int.* **04.08** 1–20

[15] ASTM D698−7 2007 Standard Test Methods for Laboratory Compaction Characteristics of Soil Using Standard Effort (12 400 ft-lbf/ft3 (600 kN-m/m3)) *ASTM Int.* **3** 15

[16] ASTM-D2850 2015 Standard Test Method for Unconsolidated-Undrained Triaxial Compression Test on Cohesive Soils *ASTM Int.* **D2850−15** 1–6

[17] Fathurrohman A M, Syahril S and Somantri A K 2020 Repairment of clay shale soil by stabilization method using a cement binder *IOP Conf. Ser. Mater. Sci. Eng.* **830** 022032

[18] Utami G S 2014 Clay soil stabilization with lime effect the value CBR and swelling *ARPN J. Eng. Appl. Sci.* **9** 1744–8

[19] Gadouri H, Harichane K and Ghrici M 2018 Assessment of sulphates effect on the classification of soil–lime–natural pozzolana mixtures based on the Unified Soil Classification System (USCS) *Int. J. Geotech. Eng.* **12** 293–301

[20] Malathi N, Komala D ., Shabeena S and Jaya Saahithya V C 2021 Stabilization of Expansive Soil by Using Lime and Reinforcement With Geo-Textile *IOP Conf. Ser. Mater. Sci. Eng.* **1112** 012023

[21] Elmahallawy A and Rashed A 2012 Effect of Lime-Fiber Column on Cohesive Soils in Triaxial Test *Egypt. J. Eng. Sci. Technol.* **15** 112–9

Rock slope stability assessment using kinematic analysis at Gunung Batu, Lembang, West Java, Indonesia

A P Martireni[1], K Sugianti[1], K Hermawan[1], A Tohari[1], S Wibawa[1] and E Soebeowo[1]

[1]Research Center for Geological Disaster, National Research and Innovation Agency, Bandung, Indonesia Jl. Cisitu, Sangkuriang Bandung 40135

E-mail : antonina.martireni84@gmail.com

Abstract. Discontinuity plane in the Gunung Batu area associated with the geological structure of the activity of the Lembang fault has the potential to induce rock slope failure hazard to the surrounding residential areas. A series of slope stability analyses of the Gunung Batu was carried out using rock mass rating and kinematics analysis to assess the rock slope stability in the Gunung Batu and determine the rock failure type. According to the results of the rock mass rating, the outcrop of Gunung Batu is classified as a good rock. Meanwhile, kinematic analysis from three joint sets shows that the rock failure type is characterized by wedge and toppling failures. However, the analysis results of the Slope Mass Rating (SMR) indicate that the rock slope stability in Gunung Batu is considered in partially stable to stable condition.

1. Introduction

Rock engineering characteristics are essential parameters for rock slope stability evaluation. The rock engineering characteristics are influenced by the geological structure condition. The presence of intensive joint sets in a rock mass affects rock shear strength and slope stability of rock mass. Thus, rock slope stability evaluation necessitates joint sets mapping and determination of the rock mass strength.

Nowadays, many slope stability analysis methods are available, such as numerical modeling, empirical methods, and analysis of kinematic [1]. Analysis of slope stability in the discontinuity plane uses numerical modeling to analyze the movement until a failure occurs [2,3]. Various software has been developed to support slope stability analysis. However, its application depends on input data, physical and mechanical properties, as well as the analytical ability of the engineer.

Empirical methods in slope stability analysis are the most used in rock mass classification to evaluate rock strength. The fastest method in the decision-making for slope stability analysis is an empirical method. A few of the empirical methods used for the rock mass classification are Q-System [4], Rock Mass Rating (RMR) [5], Mining Rock Mass Rating (MRMR) [6], and Slope Mass Rating (SMR), that is an RMR modification [7-10]. The SMR method uses the discontinuity plane as the boundary factor. The SMR method cannot be used on soil and soft rock materials based on the characteristics of the research area.

Kinematic analysis analyzes slope stability without considering the forces that cause the slope material to slide [2,7,9,11-13]. The kinematic analysis method analyzes the effect of geological structures in the form of discontinuities plane, including joints, folds, fractures, and faults. Kinematic analysis results estimate slope stability and failure types that may happen. This analysis projects the discontinuity plane using dip and dip direction information.

The results of this study present a slope stability analysis conducted in the Gunung Batu outcrop using the empirical method. The aims of this study are to (1) specify the type of rock failure based on the joint sets and (2) obtain the slope mass rating of the Gunung Batu. The Gunung Batu is located in Lembang District, West Bandung Regency, at the coordinates S 06o49'94.4" E 107o38'06.9" (Figure 1), and is part of an active fault, well known as the Lembang Fault Zone. This active fault runs about 29 km, extending from Padalarang to Jatinangor [14]. The Gunung Batu outcrop is an andesitic igneous lava rock of the Middle–Late Quarternary age [15], formed from Mount Sunda's eruption. The outcrop is intensively jointed, probably associated with the activation of the Lembang Fault. While the Lembang Fault movement can trigger an earthquake of Mw 6.5-7.0 [14], the earthquake shaking may induce failure of the Gunung Batu outcrop and pose a risk to the surrounding inhabitants. Thus, evaluating the Gunung Batu slope stability is necessary to provide hazard assessment and mitigation options for reducing the earthquake risk.

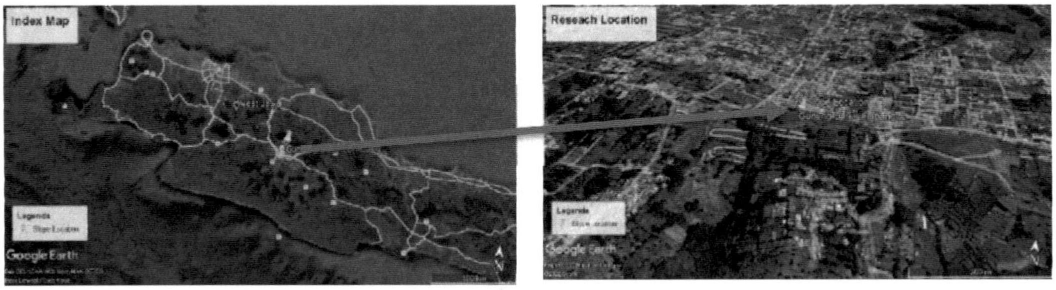

Figure 1. Study location in the Gunung Batu, Lembang District, West Bandung Regency, West Java.

2. Method

The study was conducted by joint plane mapping in the Gunung Batu outcrops. Rock mass classification was evaluated using RMR [4], and rock slope stability was calculated using the SMR method [16, 17]. The joint set data was used for kinematic analysis to determine failure types that may occur.

2.1. Rock slope stability in the Lembang Fault zone

The slope stability in the Lembang Fault zone was analyzed using the RMR method of rock mass classification. The rock mass classification method is carried out by mapping and collecting discontinuity data using the scanline method. RMR uses five parameters that are rock strength (Uniaxial Compressive Stress / UCS), Rock Quality Designation (RQD), spacing and conditions of discontinuity, also groundwater conditions [4].

The Schmidt Hammer test was used to obtain the rock strength value that was based on the value of the Schmidt hardness rebound numbers (H_r). The conversion of H_r values to UCS values has been studied previously [18-21]. Dincer's formula [21] is the most sufficient for converting H_r values (Formula 1). The average H_r value from 135 test points is the rock strength data.

$$U \quad = 2.75\, H_r - 36.83 \tag{1}$$

Rock quality designation (RQD) is the main parameter calculated in rock mass classification. RQD evaluated rock quality in boreholes based on a joint degree in a rock mass calculated as the drill core percentage in lengths of 10 cm or more [22,23]. This study calculated RQD without a drill core using the Priest equation [24] (Formula 2).

$$R \quad = 100e^{-0.1\lambda}(0.1\lambda + 1) \tag{2}$$

Where e for negative exponential distribution value and λ for the average value of discontinuities in 1 meter. Other parameters of this analysis are spacing and condition of discontinuity planes, also groundwater conditions. The discontinuity planes taken into the calculation are the strike, dip, and dip direction.

2.2. Kinematic analysis

Geological structure data obtained in the research area were analyzed kinematically using stereographic projection [25-27]. Kinematics analysis of landslides uses data on the strike angle and dip angle of discontinuity structures and slopes to produce a relative angle of a rock mass. The kinematics analysis result is the potential types of rock mass movements used as the value for calculating slope stability using the SMR method by Romana [16] (Tabel 1). The SMR method is an adjustment of the RMR method by adding the effect of excavation F_4 (Formula 3).

Table 1. Parameter correction of Slope Mass Rating [16].

Factor	Slope Condition	Failure Type	Very favorable	Favorable	Normal	Unfavorable	Very unfavorable
F_1	P $\lvert \alpha_j - \alpha_s \rvert$ T $\lvert \alpha_j - \alpha_s - 180^o \rvert$ W $\lvert \alpha_i - \alpha_s \rvert$		> 30°	30 - 20°	20 - 10°	10 - 5°	< 5°
		P / T / W	0.15	0.40	0.70	0.85	1.0
F_2	P/T $\lvert \beta_j \rvert$ W $\lvert \beta_i \rvert$		< 20°	20 - 30°	30 - 35°	35 - 45°	> 45°
		P / W	0.15	0.40	0.70	0.85	1.0
		T	1.0	1.0	1.0	1.0	1.0
F_3	P $\lvert \beta_j - \beta_s \rvert$ W $\lvert \beta_i - \beta_s \rvert$		> 10°	10 - 0°	0°	0 - (-10°)	< (-10°)
	T $\lvert \beta_j + \beta_s \rvert$		< 110°	110-120°	> 120°	-	-
		P/ T / W	0	-6	-25	-50	-60
	Excavation method						
F_4	Method		Natural slope	Pre-splitting	Smooth blasting	Normal blasting or mechanical excavation	Deficient blasting
			15	10	8	0	-8

P (planar type failure), T (toppling type failure), W (wedge type failure), α_s (slope strike), α_j (discontinuity plane strike), α_i (plunge direction of discontinuity plane), β_s (the dip angle of slope), β_j (the dip angle of discontinuity), β_i (plunge dip of discontinuity plane).

SMR calculation uses the following formula [16], and the classification of slope stability according to SMR is given in Table 2:

$$S = R + F_1F_2F_3 + F_4 \quad (3)$$

Table 2. Stability classes [16].

Class	SMR	Description	Stability	Structure
I	81 - 100	Very good	Completely stable	None
II	61 - 80	Good	Stable	A few of the block
III	41 - 60	Normal	Partially stable	Some or many of the block
IV	21 - 40	Bad	Unstable	Planar or major failure
V	0 - 20	Very bad	Completely unstable	Big planar plane or soil

3. Result and discussion

The RMR value is obtained by summing up the rating value of UCS, RQD, joint spacing, joint condition, and groundwater. Rock mass strength (UCS) is obtained without drilling from the mean value of the Schmidt Hammer test of 135 points (Table 3). The mean value of H_r is 27.56, equal to UCS 38.97 MPa. The RQD value is obtained without a drill core using the Priest method [24]. This method calculates the joint frequency (λ) per unit length. From the observation, the joint frequency is 2.11, and the RQD value is 98.06%. The field observations showed that the mean value of the discontinuity plane's spacing was 0.38 m. The joint conditions on the slope are a moderately rough surface, joint separation under 1 mm, and moderately weathered walls. Groundwater conditions on the slope are damp. It is proved that the slope has a moist condition. The results of RMR analysis obtained rocks with classification number II with good rock descriptions (Table 4).

Table 3. Result of H_r value

	Value
Number of data	135
Minimum H_r	14
Maximum H_r	52
Standard deviation H_r	7,87
Mean H_r	27.56

Table 4. Result of RMR classification

Parameter	Value	Rating
UCS (MPa)	38,97	4
RQD	98.06%	20
Groundwater	Damp	10
Joint Spacing	0.38	10
Joint characteristics	moderately rough surface, joint separation under 1 mm, and moderately weathered walls	25
Ratings		69
RMR Class		II
Description		Good

Joint stereographic projection in the study area is separated according to the possibility of rock slope failure types. Stereographic projection results obtained 3 joint sets in the slope that form 2 types of rock slope failure. Field observation obtained a slope dip of 41° with a strike of 242° (Table 5). This discontinuity plane projection is analyzed to obtain the type of rock failures that would occur. Analysis using software obtained a plunge value of N 352°E / 60°.

Based on the joint orientation analysis results, there are two possibilities for rock failures in the study area: wedge and toppling landslides. Analysis of joint set 1 and joint set 2 on the slope runs into wedge failure, and this analysis is proved by the presence of 2 discontinuous planes intersecting the slope face's position. Furthermore, the analysis of joint set 3 on slope runs into a toppling failure type. This analysis is proved by the dip value of joint set 3, that greater than the dip slope. This analysis follows the requirements of the toppling-type landslide.

The SMR value was obtained from the adjustment factor for the RMR value to the direction of the slope orientation and the excavation method. Joint orientation projections were obtained for 3 types of joint sets (Figure 2). Analysis result using kinematic analytical software shows that the rock slope failure types are wedge failure and toppling failures.

Table 5. Result of discontinuity plane orientations.

Slope Orientation	Discontinuity Plane Orientation		
	Set 1	Set 2	Set 3
N 242°E / 41°	N 30°E / 48° (wedge)	N 55°E / 24°(wedge)	N 132°E / 78°(toppling)

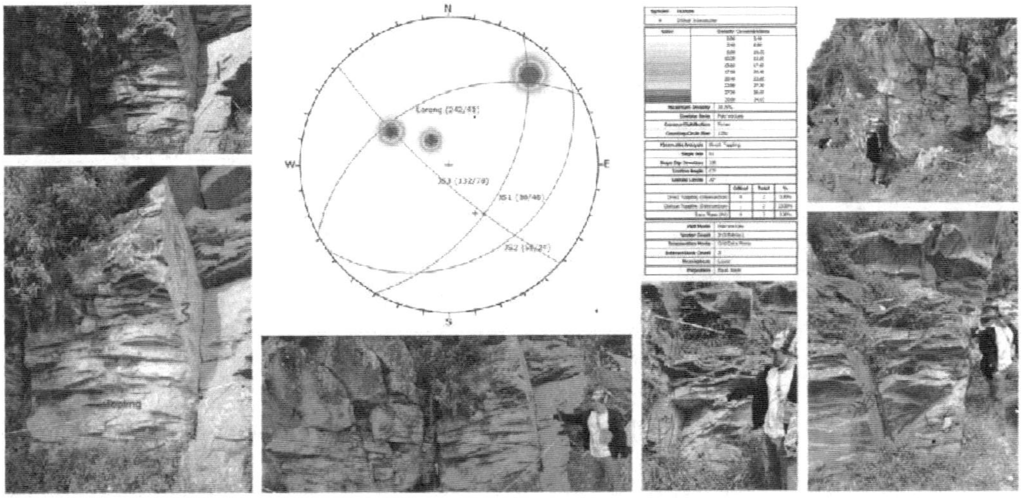

Figure 2. Stereographic analysis.

Table 6. Result of SMR calculation.

Parameter	Wedge		Toppling	
	Value	Rating	Value	Rating
RMR Score		69		69
F_1	110°	0.15	290°	0.15
F_2	60°	1	78°	1
F_3	19°	0	119°	-6
F_4	Deficient blasting	-8	Deficient blasting	-8
SMR Rating		61		60.1
Description		Good		Normal
Stability		Stable		Partially Stable

Based on the analysis results, the rock slopes in Gunung Batu are partially stable to stable. Thus, the geological structure of Gunung Batu can affect slope stability. This condition is essential to consider because the Gunung Batu area is highly populated and located on the active Lembang fault. Field observations and kinematic analysis showed that some slope sections are partially unstable and have the potential for wedge and toppling types of failures. Thus, slope monitoring is required to evaluate long-term slope stability, and prevention of rock slope instability is needed, such as installing rock nailing, ground anchors, and wire mesh. Thus, further research is required to provide mitigation options to prevent slope instability. The following research is required by testing the rock strength in the laboratory and calculating the effect of the Lembang active fault movement.

4. Conclusion

Based on research and field observations, it was found that the rock conditions in the Gunung Batu area were andesite lava rocks with normal and good rock classification. The results of the SMR adjustment show that the Gunung Batu rock slope is within a partially stable to stable condition. Field observations showed that present intensive joints are present, and it caused some areas to become to slide. This observation is validated by the software analysis results showing the potential for wedge and toppling failures. The results of the SMR adjustment indicated that mitigation action is necessary to prevent possible instability during a big earthquake in the Lembang Fault zone. Regular slope monitoring is also recommended. Further research is necessary to focus on determining the rock strength in the laboratory and adding the effect of the Lembang active fault movement on the rock slope stability analysis.

Acknowledgments

The authors wish to acknowledge to Research Center for Geological Disaster, National Research and Innovation Agency (BRIN), for providing a research grant through the Disaster Research seed grant of the National Research and Innovation Agency. We also like to express gratitude for laboratory assistance by technical staff, Mr. Wahyudin, and encouragement from colleagues.

References

[1] Basahel H and Mitri H 2017 *Journal of Rock Mechanics and Geotechnical Engineering* **9** 993–1009
[2] Ankah M L Y and Kincal C 2022 *Modeling Earth Systems and Environment* (S) 1-9
[3] Kumar S Pandey H K and Venkatesh K 2022 *Arabian Journal of Geosciences* **15** 1-13 (Springer Science and Business Media)
[4] Barton N R Lien R and Lunde J 1974 Springer Science and Business Media *Rock Mechanics and Rock Engineering* **6** 189–236
[5] Bieniawski Z T *Rock Classification Systems for Engineering Purposes* (ASTM International) 17
[6] Laubscher D H 1990 Southern African Institute of Mining and Metallurgy *Journal of the South African Institute of Mining and Metallurgy* **90** 257-273
[7] Rusydy I Al-Huda N Jamaluddin K Sundary D and Nugraha G S 2017 *Riset Geologi Dan Pertambangan* **27**
[8] Triana K and Hermawan K 2020 Springer Science and Business Media *Geoenvironmental Disasters* **7**
[9] Yanuardian A R Hermawan K Martireni A P and Tohari A 2020 *Rudarsko-geološko-naftni zbornik (The Mining-Geological-Petroleum Bulletin)* **35** 67-76 (Faculty of Mining, Geology and Petroleum Engineering)
[10] Awang H Salmanfarsi A F Misbahuddin A Z and Ali M I 2021 *IOP Conference Series: Earth and Environmental Science* **682** 012015 (IOP Publishing)
[11] Gurocak Z Alemdag S and Zaman M M 2008 *Engineering Geology* **96** 17–27
[12] Diabat A A and Rumman W A 2021 *The Iraqi Geological Journal* 11-33
[13] Yanuardian A R Indrawan I G B and Warmada I W 2018 *Riset Geologi Dan Pertambangan* **28** 101(RISET Geologi dan Pertambangan)
[14] Daryono M R Natawidjaja D H Sapiie B and Cummins P 2019 *Tectonophysics* **751** 180-191
[15] Dam M A C Suparan P Nossin J J Voskuil R P G A and Group GTL 1996 *Journal of Southeast Asian Earth Sciences* 14 101-115
[16] Romana M R 1993 *A Geomechanical Classification for Slopes: Slope Mass Rating* (Rock Testing and Site Characterization) 575–600
[17] Zheng J, Zhao Y, Lü Q, Deng J, Pan X and Li Y 2016 *Engineering Geology* **206** 42-49
[18] Armaghani D J, Mohamad E T, Momeni E, Monjezi M and Narayanasamy M S 2016 *Arabian Journal of Geosciences* **9** 1-16 (Springer Science and Business Media)
[19] Gupta V 2009 *Bulletin of Engineering Geology and the Environment* **68** 409-416 (Springer Science and Business Media)

[20] Aydin A and Basu A 2005 *Engineering Geology* **81** 1-14
[21] Dinçer I Acar A Çobanoğlua I and Uras Y 2004 *Bulletin of Engineering Geology and the Environment* **63** 141-148 (Springer Science and Business Media)
[22] Deere D U Hendron A J Patton F D and Cording E J 1967 *Proceedings of 8th US Symposium Rock Mechanics* (New York: OnePetro) 237-302
[23] Deere D U and Deere D W 1989 *Rock Quality Designation (RQD) After Twenty Years* (Vicksburg: Contract Report GL-89-1. U.S. Army Engineer Waterways Experiment Station)
[24] Priest S D 1993 *Discontinuity Analysis for Rock Engineering* (London: Chapman & Hall)
[25] Hoek E and Bray J D 1981 *Rock Slope Engineering* (London: Taylor & Francis)
[26] Hoek E 2007 *Rock Mass Classification, in Partical Rock Engineering* (North Vancouver: Evert Hoek Consulting Engineer Inc) 1–23
[27] Goodman R E 1989 *Introduction to Rock Mechanics. Second Edition* (New York: John Wiley and Sons) 562

Hydraulic gradient analysis on embankment dam foundation associated with internal erosion

R A M Putra [1], D Pratiwi [1], A D Putra [2], E P Wahono [2], D I Kusumastuti [2] and A Fitri [1]

[1]Civil Engineering Department, Faculty of Engineering and Computer Science, Universitas Teknokrat Indonesia, Z.A. Pagar Alam Street No. 9-11 Labuhan Ratu, Bandar Lampung, 35148, Indonesia.
[2]Civil Engineering Department, Faculty of Engineering, Universitas Lampung, Sumantri Brojonegoro Street No. 1, Rajabasa, Bandar Lampung, 35145, Indonesia.

Email: rioaditomo@teknokrat.ac.id

Abstract. Internal erosion is susceptible to occur in conditions where there are cracks or pre-formed cavities at the interface between the dam foundation and embankment. The initiation of internal erosion is related to the critical gradient (I_c) and hydraulic gradient (I_{xy}) at a certain position. The addition of repair materials in foundation improvement will change the total head, especially on the dam foundation, so the value of the hydraulic gradient will also change simultaneously from upstream to downstream. This study examines the value of the hydraulic gradient (I_{xy}) at certain points on the dam foundation under the condition of foundation with and without improvement. Several scenarios were simulated using Geostudio Seep/W software to get an estimated value of the hydraulic gradient. The upstream blanket foundation improvement method increases the hydraulic gradient in the upstream zone when compared to the curtain grouting method, thereby increasing the potential for internal erosion in a related area. Increasing the depth of curtain grouting from 25 m to 35 m does not significantly reduce the value of the hydraulic gradient in the dam foundation. Research results can be used for analysis to improve design and construction, and as a reference for similar dams.

1. Introduction

Dams have unique structures because, in addition to their size and variety, they also have a certain relationship to the hydrological and geological aspects of the field. One of the main causes of dam failure is internal erosion of the embankment or its foundation [1]. Internal erosion is the erosion of soil and rocks by groundwater flow and seepage pressure [2]. In certain cases, seepage flow can occur in the dam, but under certain conditions, it can cause internal erosion. Gradual internal erosion can form ducts or large holes in the dam body or dam foundation. Erosion continues and water seeps downstream through the holes created. The formation of large holes or pipes can alter the stability of the dam and lead to structural failure. The statistic shows that about 50% of the failures of water-retaining structures such as dams are caused by piping [1].

Seepage control methods in permeable dam foundations vary widely, its effectiveness depends on many factors. Grout curtains and upstream blankets are examples of seepage control methods. Those methods will change the geometry of the dam due to the addition of grout material in the curtain grouting method or impermeable core material in the upstream blanket method. The addition of repair materials will change the total energy height (total head), especially on the dam foundation, so that the value of the hydraulic gradient in the direction of the *x-axis* and *y-axis* (I_{xy}) will also change simultaneously from upstream to downstream. The initiation of internal erosion is related to critical gradient (I_c) and hydraulic gradient (I_{xy}) at a certain position [3]. If hydraulic gradient (I_{xy}) exceeds critical gradient (I_c), the soil grains will be released from their bonds so that it can initiate internal erosion on the dam foundation.

Based on these conditions, this study examines the value of the hydraulic gradient (I_{xy}) at certain points on dam foundation under the condition of foundation with and without improvement. The object of this research is Way Sekampung Dam located in Pringsewu Regency, Lampung Province. Way Sekampung Dam is a rock-fill dam with a symmetrical clay core. This rock-fill dam has 55.0 m of peak elevation and 49.0 m of designed static water level [4]. This research was carried out when the dam was still under construction. Several scenarios were simulated using Geostudio Seep/W software to get an estimated value of the hydraulic gradient.

1.1. Internal erosion initiation

As described earlier, critical gradient (I_c) and the hydraulic gradient (I_{xy}) are related to the initiation of internal erosion. Critical gradient (I_c) is defined as the gradient that initiates internal erosion [5]. In 1922, Terzaghi proposed the well-known critical hydraulic gradient formula described in equation (1), where γ' is unit weight (t/m^3); γ_w is water unit weight (t/m^3); G_s is specific gravity (dimensionless); and e is void ratio (dimensionless).

$$I_c = \frac{\gamma'}{\gamma_w} = \frac{G_{s-1}}{1+e} \quad (1)$$

Hydraulic gradient (I_{xy}) is a gradient that describes the difference in total energy height (total head) in the zone. Internal erosion begins when the hydraulic gradient (I_{xy}) exceeds the critical gradient (I_c). Based on this statement, it can be concluded that the hydraulic gradient value must be less than the critical gradient value so that internal erosion does not occur. Increasing hydraulic gradient (I_{xy}) will further increase the initiation of internal erosion

1.2. Foundation dam improvement

Seepage control of a dam can be carried out by selecting one of the foundation repair methods or a combination of several methods. Selection considerations are based on three important factors, namely (1) Condition of the dam foundation; (2) Dam Type; and (3) Dam dimensions. The existing foundation improvement method at Way Sekampung Dam utilizes curtain grouting. In this study, the curtain grouting and upstream blanket improvement methods will be analyzed to get insight into the distribution of hydraulic gradient values in dam foundation.

The curtain grouting method aims to form a kind of watertight curtain in the foundation to reduce internal water flow through the foundation by allowing the internal water flow to flow through the bottom of the curtain [6]. As the internal water flow or seepage at the bottom of the curtain, the seepage flow pattern lengthens and the internal water flow discharge from the dam is reduced. Grouting is the process of injecting a liquid under pressure into cavities or pores, rock cracks and fissures, and allowing the liquid to solidify physicochemically for a period of time. Injection materials are mixtures of cement and water in specific ratios or in the form of specific chemicals.

In conditions where the impermeable rock layers are too deep so that the application of the curtain grouting method will be more difficult and require higher construction costs, the upstream blanket method can be considered. The purpose of the dam foundation upstream blanket method is to reduce the internal water flow discharge in the foundation layer by lengthening the internal water flow path with the horizontal barrier system [6]. The material used for the upstream blanket may be the same as the material for the impermeable core zone of the dam body and is arranged at the bottom of the upstream reservoir with a specified length and thickness.

1.3. Conditions susceptible to internal erosion

Internal erosion can occur under conditions where cracks or pre-formed cavities are present at the interface between the dam foundation and embankment. These existing cracks or cavities are often caused by tensile cracks due to different subsidence during and after construction. Cracks or pre-existing cavities can be outlets for internal water flow, creating continuous tunnels or pipes. This phenomenon is called internal erosion due to concentrated leaks [7] – [9]. A concentrated leak with erosion is driven by flow in the cracks or other openings. Dam collapse statistics clearly show that concentrated leak is the most dangerous mechanism over other internal erosion mechanisms [1]. A previous study was performed using a hole erosion test (HET) apparatus that can simulate erosion phenomena in open cracks. The experimental results show that the effects of compaction energy, moisture content, degree of saturation, and the percentage of fines play a key role in erosion characteristics [10].

2. Methodology

2.1. Determination of dam cross-section

The foundation of Way Sekampung Dam lies on tuff, breccia, and quartzite. This study uses a typical cross-section of the Way Sekampung dam, which is a cross-section of BD-3 (figure 1). The cross-section of BD-3 has a base elevation of 75.00 m located in quartzite rock (figure 2).

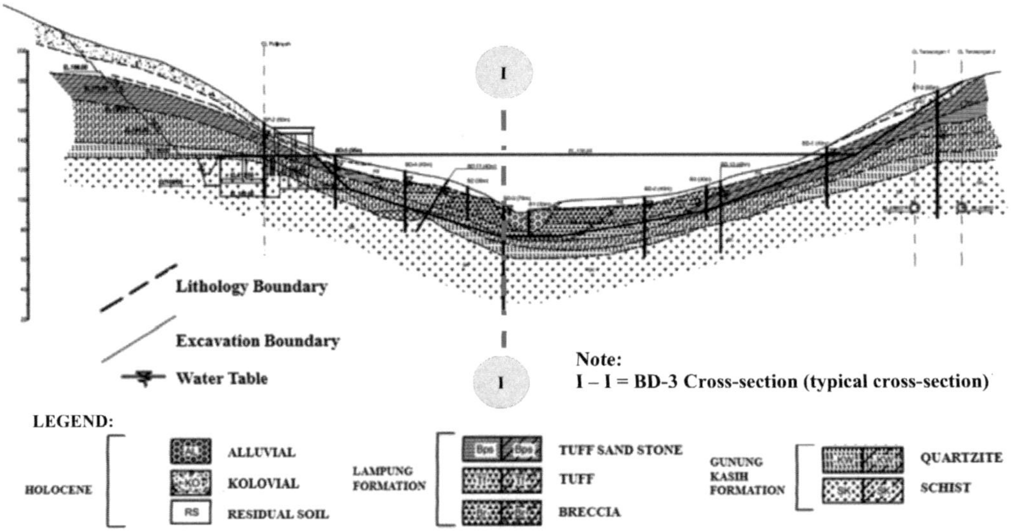

Figure 1. Longitudinal section of the dam [4].

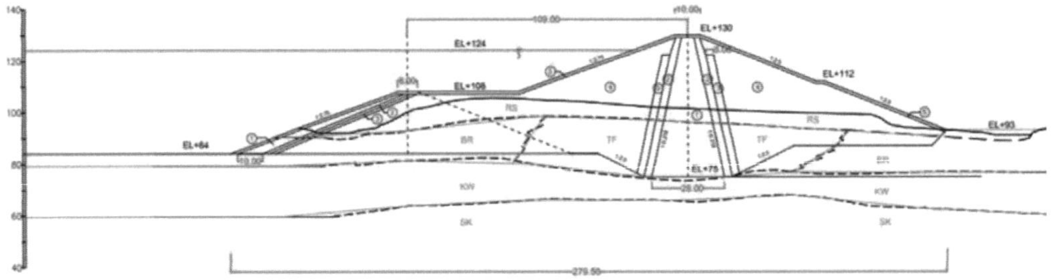

Figure 2. BD-3 cross-section as a typical cross-section [4]. Notes: 1 = Impermeable core; 2 = Fine filter; 3 = Coarse filter; 4 = Stone embankment; 5 = Rip-rap; TF = Tuff; BR = Breccia; KW = Quartzite; SK = Schist.

2.2. Basic parameters of modeling

The basic parameters used are the results of secondary data analysis in the form of permeability coefficient (k), specific gravity (G_s), saturated water content (W_s), and void ratio (e) as described in table 1.

Table 1. Parameters of the body and foundation of Way Sekampung Dam [4].

Material	k (cm/s)	w_s (%)	G_s	e
Core	1×10^{-6}	34.0	-	-
Fine Filter	5×10^{-3}	6.5	-	-
Coarse Filter	1×10^{-2}	4.0	-	-
Stone Embankment	1×10^{-1}	4.0	-	-
Rip-Rap	1×10^{-1}	5.6	-	-
Tuff	5×10^{-4}	21.28	2.555	1.17
Breccia	5×10^{-5}	29.41	2.289	1.04
Quartzite	5×10^{-4}	22.35	2.671	0.69
Schist	5×10^{-5}	6.88	2.671	0.2

Curtain grouting method has weaknesses, even the effectiveness of grouting in foundation improvement is difficult to measure with certainty [6]. Based on subjective and qualitative assessments, the effectiveness of grouting in this study was assumed to be 90% (good). Changes in permeability will be different in each layer because the initial permeability of each layer has a different value. Foundation permeability before and after grouting can be seen in table 2.

Table 2. Foundation permeability before and after grouting.

Material	Permeability before grouting (cm/s)	Permeability after grouting (cm/s)
Tuff	5×10^{-4}	5×10^{-5}
Breccia	5×10^{-5}	5×10^{-6}
Quartzite	5×10^{-4}	5×10^{-5}
Schist	5×10^{-5}	5×10^{-6}

The upstream blanket method may be advised if the employment of the curtain grout method is challenging due to the deep-seated impermeable rock layer. The material used for the upstream blanket in this research was assumed to be the same as the material in the impermeable core zone of the dam body (1×10^{-6} cm/s).

2.3. Combination modeling scenarios

The depth of the existing curtain grouting at the Way Sekampung dam varies from 30 meters to 35 meters. General approximation to determine the depth of curtain grout can be expressed by equation 2 [6], where d is curtain grout depth (m); h is static water level; and c is coefficient (8 to 20).

$$d = \frac{1}{3}h + c \tag{2}$$

By using equation 2, the estimated effective curtain grout depth in Way Sekampung Dam was from 25.0 m (c min. = 8) to 37.0 m (c max. = 20). In this study, three parameters of the curtain grout depth were determined, namely: (1) The theoretic minimum size = 25.0 m; (2) The real minimum size at Way Sekampung Dam = 30.0 m; and (3) The real maximum size at Way Sekampung Dam = 35.0 m. Based on previous studies [11]–[13], the parameter of blanket thickness used in this study was specified to be 1.0 m, with variations in length dimensions: (1) 110 m (< four times the total head); (2) 160 m (four times the total head); and (3) 210 m (> four times the total head). Based on the combination of two different alternative foundation improvement methods (curtain grouting and upstream blanket), seven modeling scenarios were created (table 3). All modeling scenarios were simulated using Geostudio/Seep W by providing certain boundary conditions. Figure 3a is a visualization of foundation improvement modeling scenario on a dam typical section with a 35.0 m depth of curtain grouting. While figure 3b is a visualization of foundation improvement modeling scenario with 210.0 m length of upstream blanket.

Table 3. Combination Modelling Scenarios.

Improvement	Parameter	Modeling
No Improvement	N/A	Scenario 1
Curtain Grouting; 4,0 m wide	25 m depth	Scenario 2
	30 m depth	Scenario 3
	35 m depth	Scenario 4
Upstream Blanket; 1,0 m thickness	110 m length	Scenario 5
	160 m length	Scenario 6
	210 m length	Scenario 7

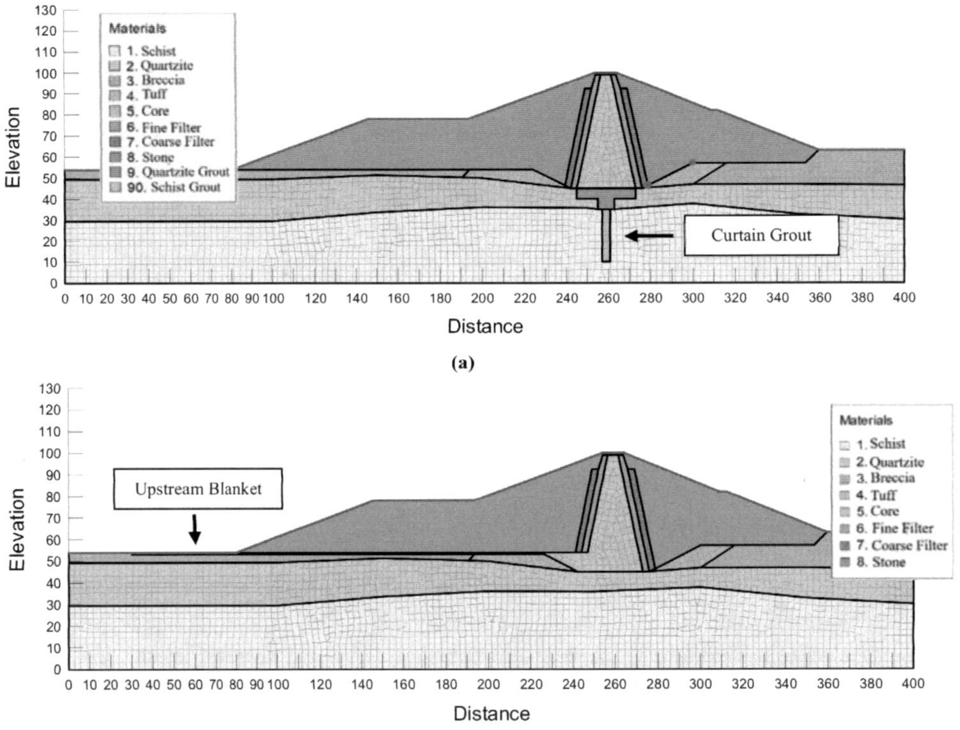

(b)

Figure 3. Modeling visualization of dam using Geostudio Seep/W; (a) Scenario 4; Curtain grouting method 35.0 m depth; and (b) Scenario 7; Upstream blanket method 210.0 m length.

3. Result and Discussion

Foundation improvement with curtain grouting and upstream blanket methods will change the total head, especially on the dam foundation, so that the value of the hydraulic gradient in the direction of the *x*-axis and *y*-axis (I_{xy}) will also change simultaneously from upstream to downstream. In this study, the changes in the value of the hydraulic gradient (I_{xy}) would be reviewed at a certain point location (n_n) on the foundation of the dam (figure 4). Changes in the hydraulic gradient (I_{xy}) in the cross-section of the dam according to the location of the specified point (n_n) can be seen in (figure 5).

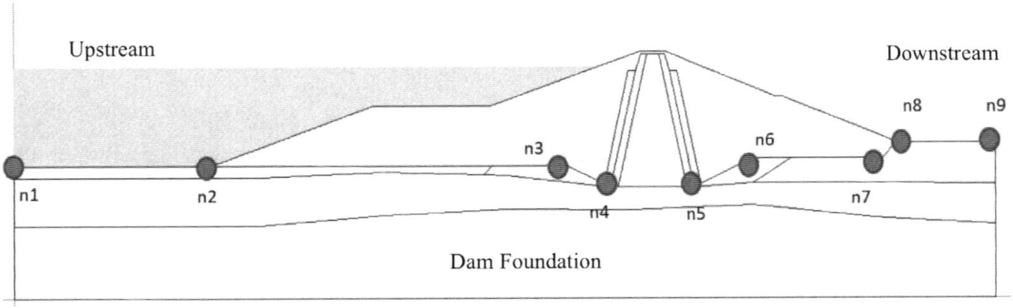

Figure 4. Location of review point (n_n).

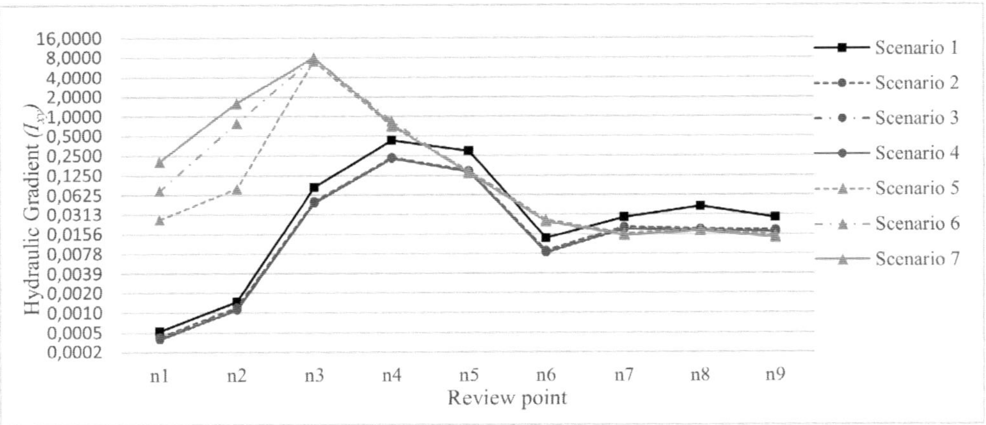

Figure 5. Hydraulic Gradient (I_{xy}) curve at review point (n_n).

In Scenarios 1, 2, 3 and 4 there is a similar pattern of changes in hydraulic gradient (I_{xy}) from upstream to downstream of the dam. However, in Scenarios 5, 6, and 7, the values of the hydraulic gradient (I_{xy}) at locations n_1, n_2, and n_3 yielded higher values up to 8.12 when compared to Scenarios 1, 2, 3, and 4. This was caused by the addition of upstream blanket material up to 210 m in length (Scenarios 5, 6, and 7). The longer the dimension of impermeable material used upstream, the greater the hydraulic gradient (I_{xy}) in the upstream area. Due to these high gradient hydraulic values, cracks in the upstream zone should be avoided to reduce the potential of internal erosion caused by the concentrated leak mechanism. Well-compacted upstream blanket material is an example of an effort to reduce the possibility of a concentrated leak mechanism. Voids must be avoided at the interface between the upstream blanket material and the foundation layer. In addition, an investigation of the upstream foundation layer must be carried out thoroughly so that if pre-existing cavities were found, the layer must be excavated and replaced. From the point of view of the hydraulic gradient analysis in the upstream area, the upstream blanket improvement method is more at risk of causing internal erosion hazards when compared to the curtain grouting method.

Increasing the depth of curtain grouting from 25 m to 35 m does not significantly reduce the value of the hydraulic gradient in the dam foundation. In terms of construction costs, grouting curtains with a theoretical minimum depth (25 m) is more recommended than the existing maximum depth in the field (35 m). Hydraulic gradient (I_{xy}) in the curtain grouting method (Scenarios 2, 3, and 4) and the upstream blanket method (Scenarios 5, 6, and 7) yielded a smaller value in the downstream when compared to the foundation without improvement (Scenario 1). This condition will reduce the possibility of internal erosion downstream of the dam. Both upstream blanket and curtain grouting methods have relatively similar values in the downstream zone. However, the smallest hydraulic gradient value in the downstream zone occurs in the upstream blanket method at review point n_9 with a value of 0.014 (Scenario 7).

4. Conclusion

In this research seven scenarios of the Way Sekampung Dam modeling have been carried out to examine the value of the hydraulic gradient (I_{xy}) at certain points on the dam foundation under the condition of the foundation with and without improvement. The largest hydraulic gradient value occurred in the upstream blanket method at review point n_3 with a value of 8.12 (Scenario 7). The smallest hydraulic gradient value in the downstream zone occurred in the upstream blanket method at review point n_9 with a value of 0.014 (Scenario 7). The upstream blanket foundation improvement method increases the hydraulic gradient in the upstream zone when compared to the curtain grouting method, thereby increasing the potential for internal erosion in a related area. The longer the dimension of impermeable material used upstream, the greater the potential for internal erosion to occur in the upstream zone.

Increasing the depth of curtain grouting from 25 m to 35 m does not significantly reduce the value of the hydraulic gradient in the dam foundation. Both upstream blanket and curtain grouting methods have relatively similar hydraulic gradient values in the downstream zone. In an improved foundation, the hydraulic gradient in the downstream zone has a smaller value when compared to the unimproved foundation. This condition will reduce the possibility of internal erosion downstream of the dam. Research results can be used as an analysis for design and construction improvements and can also be used as a reference for similar types of dams.

Acknowledgments

The authors would like to thank *Universitas Teknokrat Indonesia (UTI)* and *Balai Besar Wilayah Sungai (BBWS) Mesuji -Sekampung* in Lampung Province for their support of this research.

References

[1] J. J. Fry 2016 Lessons on internal erosion in embankment dams from failures and physical models doi: 10.1201/9781315375045-6.

[2] M. S. Fleshman and J. D. Rice, 2014 Laboratory Modeling of the Mechanisms of Piping Erosion Initiation, *Journal of Geotechnical and Geoenvironmental Engineering*, vol. 140, no. 6 doi: 10.1061/(asce)gt.1943-5606.0001106.

[3] P. Ming, J. Lu, X. Cai, M. Liu, and X. Chen 2020 Multi-Particle Model of the Critical Hydraulic Gradient for Dike Piping, *Soil Mechanics and Foundation Engineering*, vol. 57, no. 3, doi: 10.1007/s11204-020-09656-1.

[4] BBWS Mesuji Sekampung 2015 *Laporan Ringkas: Penyempurnaan Desain Regulating Dam Way Sekampung*. Jakarta: PT. Indra Karya - Consulting Engineers.

[5] D. S. Chang and L. M. Zhang, 2013 Critical Hydraulic Gradients of Internal Erosion under Complex Stress States, *Journal of Geotechnical and Geoenvironmental Engineering*, vol. 139, no. 9, doi: 10.1061/(asce)gt.1943-5606.0000871.

[6] S. Sosrodarsono and K. Takeda 2016 *Bendungan Tipe Urugan*, 6th ed. Jakarta: Jakarta: Balai Pustaka

[7] S. Bonelli and F. Nicot 2013 *Erosion in Geomechanics Applied to Dams and Levees*. doi: 10.1002/9781118577165.

[8] R. Fell and J. J. Fry 2013 State of The Art on the Likelihood of Internal Erosion of Dams and Levees by Means of Testing, in *Erosion in Geomechanics Applied to Dams and Levees*. doi: 10.1002/9781118577165.ch1.

[9] S. Savage, K. Douglas, R. Fell, W. Peirson, and R. Berndt 2019 Modeling the Erosion and Swelling of the Sides of Transverse Cracks in Embankment Dams, *Journal of Geotechnical and Geoenvironmental Engineering*, vol. 145, no. 5. doi: 10.1061/(asce)gt.1943-5606.0002040.

[10] N. Benahmed and S. Bonelli 2012 Investigating concentrated leak erosion behaviour of cohesive soils by performing hole erosion tests, *European Journal of Environmental and Civil Engineering*, vol. 16, no. 1 doi: 10.1080/19648189.2012.667667.

[11] F. Salmasi and M. Nouri 2019 Effect of upstream semi-impervious blanket of embankment dams on seepage, *ISH Journal of Hydraulic Engineering*, vol. 25, no. 2, pp. 143–152, May 2019, doi: 10.1080/09715010.2017.1381862.

[12] K. M-Gharrib Noori, H. Hashim Ibrahim, and Dr. A. Mohammed Hasan 2018 Influence of Upstream Blanket on Earth Dam Seepage, in *IEC2018 Proceedings Book* pp. 178–188. doi: 10.23918/iec2018.14.

[13] A. Ghanbari and S. Zaryabi, 2014 A simple method for calculating the seepage at the foundation of embankment dams with blanket and clay trench, *Journal of GeoEngineering*, vol. 9, no. 1, pp. 21–32. doi: 10.6310/jog.2014.9(1).3.

Comparative Analysis of Riverbanks Retrofitting Planning Using Strauss Pile and Ground Anchor

Yelvi [1], W A Prayoga [1], W M Fachmi [1], A Salimah [1*], Y Setiawan[1], M Hazmi[2]

[1] Department of Civil Engineering, Politenik Negeri Jakarta, Depok. Indonesia
[2] Directorate General of Water Resources, Ministry of Public Works and Housing, Indonesia

*Corresponding author: aisyah.salimah@sipil.pnj.ac.id

Abstract. This research was conducted to reinforce the sheet pile located on the embankment of the Ciliwung Lama river – Inspection Road (RSCM) in Central Jakarta. The road is beside the riverbanks of the collapsed river for 40 m and the position of the sheet pile that slopes towards the river. Based on field observations and analysis of the stability of the slopes carried out, the road's collapse was not due to the unstableness of the embankment but due to the inability of the sheet pile to receive lateral soil pressure. The research method used Plaxis 2D Software to analyze the reinforcement applied to the sheet pile in withstanding lateral soil pressure. The planned reinforcement of the sheet pile uses a Strauss pile with a sling and ground anchor. In retrofitting, using a Strauss pile, a diameter of 0.4 meters, a length of 9 meters, and a sling with a diameter of D32 with a distance between slings of 2.5 meters. Ground Anchor retrofitting is planned to be 27 m long, 159 mm in diameter, and 88 mm in body grout. Compared to the Safety Factor that occurs between the Strauss pile with the sling and the ground anchor, the safety factor of the Strauss pile with the sling is greater. The Strauss pile's safety factor was obtained by 1.75, while the ground anchor safety factor was 1.55 without considering the earthquake load.

Keywords: ground anchor, sheet pile, strauss pile, riverbanks

1. Introduction

The Special Capital Region of Jakarta (DKI Jakarta), which is traversed by 13 rivers, has problems if there is an overflow of the river water and floods the surrounding area. Since 2005 the DKI Jakarta Provincial Government has been trying to raise the river level by installing sheet piles along the river embankment. A sheet pile is a relatively thin vertical wall, flat and long, usually made of material, wood, steel or concrete, used to secure the ground and prevent water from entering the excavation [1]. Sheet piles are often used to build walls used as retaining walls and can be large or small in size [2]. Sheet piles are widely used as soil retainers because they are not easily damaged and have lighter weight and relatively thin dimensions, so they are suitable for use with medium-height soil piles between 2 to 10 meters [3]. If the sheet pile is not planned correctly, the ground pressure will push the retaining wall, causing construction failure and landslides [4].

In general, the area next to the embankment, the riverbank, is made a road that can function as an inspection road, access road, local road, and arterial road, which is part of the road network in the city. Population density causes an increase in the number of vehicles that pass through the road. The construction of houses around it is also getting denser. This adds to the sheet pile's pressure due to the increasing load. As a result, this sheet pile can no longer accept the lateral load it receives, so there is a movement of soil towards the river by shifting the sheet pile. The road on the riverbank also suffered damage due to the soil movement. The traffic passing through the road naturally became disrupted.

This situation occurred on the Ciliwung Lama River - Inspection Road (RSCM) in February 2022. Sheet piles already built on river embankments can no longer withstand lateral pressure. Heavy rains that fell for several days also triggered the collapse of the 40 m long road [5]. The riverbanks experienced deflection, so there were faults on the floor of the riverbanks. And several capping beams have been cracked. The sheet pile experienced displacement from its supposed position due to ground movement. After investigation, there was no sign of a slope collapse. This is due to the inability of the sheet pile to withstand the working lateral pressure. The lateral load acting at this time is impossible to reduce. Appropriate countermeasures should be selected or designed based on the results obtained from surveys and analyses [6]. The strength is planned and calculated with good analysis to ensure the stability of the building according to the working load [7]. Countermeasures are carried out to strengthen the existing Sheet pile by strengthening the Strauss ground anchor on the river embankment to increase the number of safety factors.

Condition analysis shortly after disaster and immediately after repair with Plaxis2D. The rationale for the finite element method using the Plaxis2D method is that the element-to-medium approach has been widely used in finite element methods over the past few years for landslide susceptibility modeling [8]. Plaxis is a finite element program for 2D analysis of deformation and stability in geotechnical engineering [9]. The finite element method for slope stability analysis has been widely practiced by researchers in carrying out further countermeasures [10], [11], [12], [13].

2. Methods

This research was conducted in Senen District, Central Jakarta City. The study site was on a landslide river embankment where the traffic load increased. Cladding countermeasures are carried out in several stages: survey investigations, slope stability analysis, and landslide mitigation efforts by applying slope strengthening.

2.1. *Investigation Survey Stage*

This includes surveys of slope conditions and soil investigations. A survey investigation of the condition of the slopes after the movement of the soil is necessary to determine the required tests and formulate a preliminary hypothesis of the cause of the landslide. Next, a description of the strength of the soil can be generated to prevent landslides in the future. The contours of slope measurement are also depicted in this stage.

Soil tests are carried out using cone penetration tests (CPT) and soil sampling with a hand drill. Penetration tests determine the soil's carrying capacity, while soil collection by hand drill is carried out to obtain soil parameters. Soil samples are tested in the laboratory to detect moisture content, fill weight, specific gravity, particle analysis, soil consistency limits, and soil shear strength tests. This test also refers to the ASTM standard [14][15][16].

2.2. *Slope Stability Analysis*

Properties of the model material. Mohr-Coulomb Model This simple but reliable model is based on soil parameters that have been well known in civil engineering practice. The Mohr-Coulomb model can be used to calculate a realistic supporting voltage. Table 1 shows the soil parameters used as input to the Plaxis2D software.

Table 1. Parameters used in the input of the Plaxis program

Properties	**Name**	**Layer 1**	**Layer 2**	**Layer 3**	**Unit**
Depth	**-**	**0 - 6**	**16 - 20**	**16 - 20**	**m**
Material model	Model	Mohr-Coulumb	Mohr-Coulumb	Mohr-Coulumb	-
Type of material behavior	Type	Undrained	Undrained	Undrained	-
Soil unit wight above phreatic level	γ_{unsat}	10.88	18	19	kN/m^3
Soil unit wight below phreatic level	γ_{sat}	16.57	20	20	kN/m^3
Permeability in horizontal direction	Kx	0.00864	0.00864	0.00864	m/hari
Permeability in vertical direction	Ky	0.00864	0.00864	0.00864	m/hari
Young's modulus constant	Eref	5393.66	14406.12	18958.5	kN/m^2
Poisson's ratio	ν	0.35	0.25	0.25	-
Cohesion (constant)	Cref	11.77	1	1	kN/m^2
Friction angle	Ø	11.2	35.57	35.1	°
Dilatancy angle	Ψ	0	0	0	°

Slope stability analysis using Plaxis 2D. Material property inputs in the form of graphical geometric models such as soil plating, structural elements, construction stages, loading, and boundary conditions are carried out with graphical procedures facilitated by Plaxis2D. from geometric models, elements up to 2D can be easily obtained. Plaxis will automatically generate elements up to 2D.

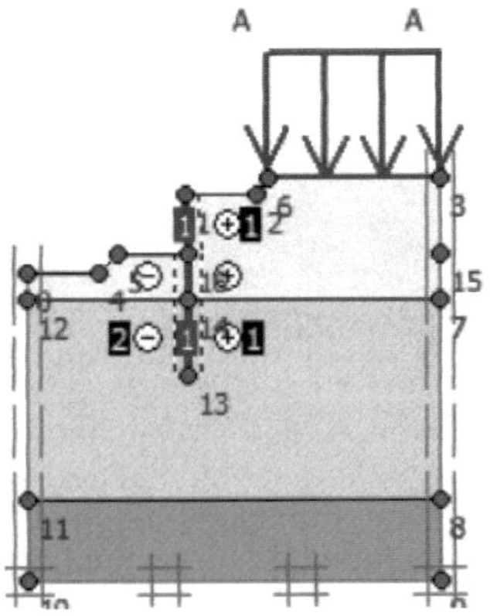

Figure 1. The initial contour design after slope failure

Figure 1 depicts the initial contour conditions after the slope collapse.

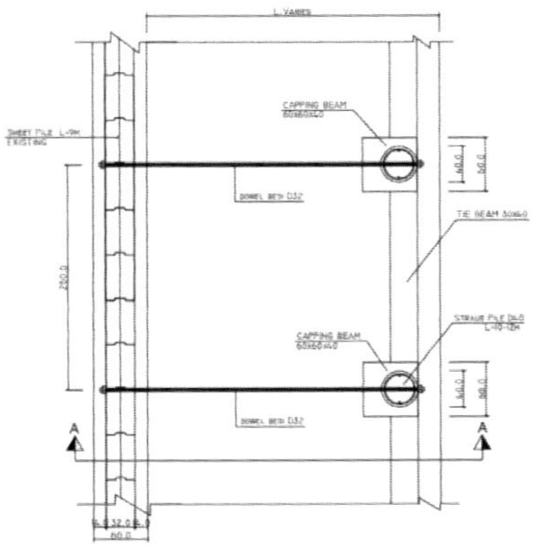

Figure 2. The refined contour design after strauss pile reinforcement

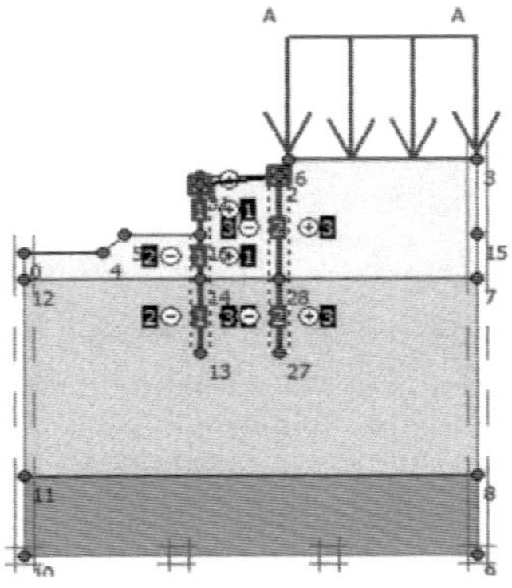

Figure 3. Design Perspective in 2D

Figure 3 shows the condition of the slope after strengthening using the Strauss pile sling.

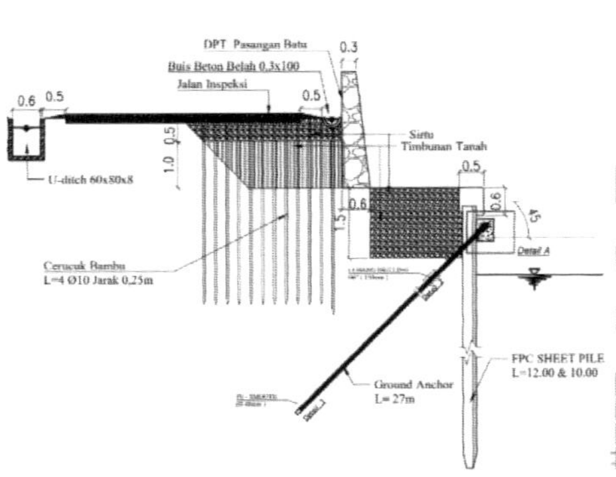

Figure 4. The refined contour design after ground anchor reinforcement

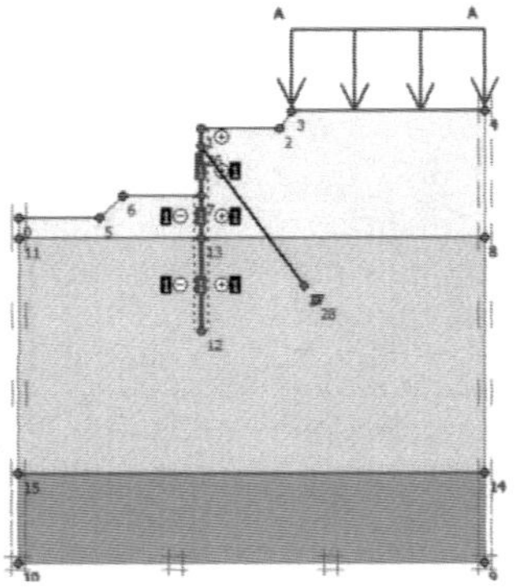

Figure 5. Design Perspective in 2D

Figure 5 shows the condition of the slope after strengthening using a ground anchor. The beige area depicts Layer 1, the brown area depicts Layer 2, and the green area shows Layer 3 in the contour design. The slope gets a traffic load, as illustrated in this picture. The input for the tower load is 27 kN/m2.

The safety factor is calculated by searching for the collapse area (FS) in the soil/rock structure. FS is calculated by gradually subtracting the cohesion value and friction angle in the soil until a collapse zone is formed [17]. Slope collapse occurs when shear strength decreases due to saturation and expansion associated with seepage conditions, resulting in shear strength equal to or less than shear stress [18]. In soil mechanics, it is important to understand the forces acting on the slope, which are the pushing force and the holding force, which prevents the collapse of the slope. The ratio of the holding force to the driving force is Factor Safety (FS). Slopes with a lower FS mean their potential for collapse is higher than slopes with a higher FS [19]. Therefore slopes with higher FS are safer.

2.3. Landslide countermeasures

The countermeasures are carried out by strengthening the river embankment to enhance the sheet pile using a Strauss pile and ground anchor. The action is to increase the number of safety factors on the river embankment.

3. Results and Discussion

2.1. Survey Investigation

The shape of the road. The inspection road is still quite accessible, but the damage to the inspection road interferes with the traffic of alternative lanes, from 2 lanes to 1 lane, as the road has been cracked due to slope avalanches, as shown in Figure 6.

Figure 6. The shape of the road after the slope

Slope conditions. Based on the initial survey results, the slope condition after the collapse showed that the floor of the embankment and capping beam was broken, and the sheet pile experienced a slope from its proper position. The analysis results show that the soil's condition is below the safe limit. The safety factor, in this case, is 1.20, while the minimum value is 1.25, where the collapse zone is just below the traffic load.

2.2. Analysis and countermeasures

Slope stability analysis using Plaxis 2D. The first analysis is carried out on the conditions after the collapse of the slope. The slope receives an outside load in the form of a traffic load. In Figure 1, it can be seen that the most prominent movement occurs in the sheet pile. This movement can develop because the location in the area is bypassed by a large traffic load, which can lead to land displacement in the area. Figures 7, 8, and 10 illustrate the numerical analysis results by applying the parameters shown in Table 1.

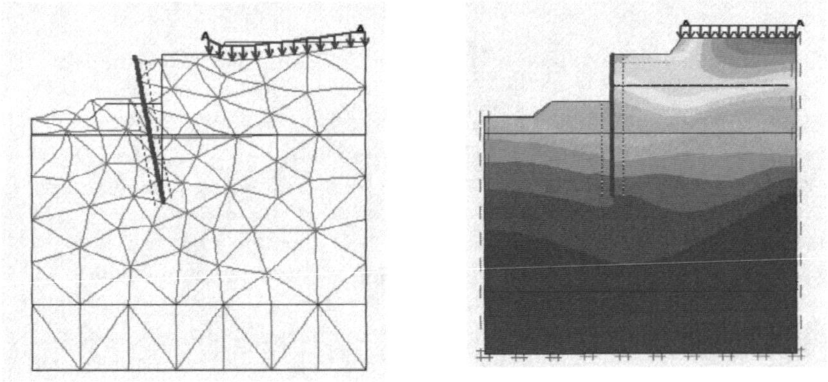

(a) (b)

Figure 7. (a) The net deformed after the landslide, (b) Field collapse before repair

The deformed mesh is shown in Figure 7a. The potential slip zone is described in Figure 7b, depicting the scale of displacement and total regional displacement. The Finite Element Method (FEM) analysis can simulate a collapse field reaching a river embankment under current conditions. The FS, in this case, is 1.20, as in Figure 9a, which shows that the slope is very likely to collapse.

The second analysis was obtained by assuming reinforcement on the floor of the embankment using a Strauss pile. The dimensions of the Straus pile are 0.4 meters in diameter, 9 meters in length, then for slings with a diameter of D32 with a distance between slings of 2.5 meters.

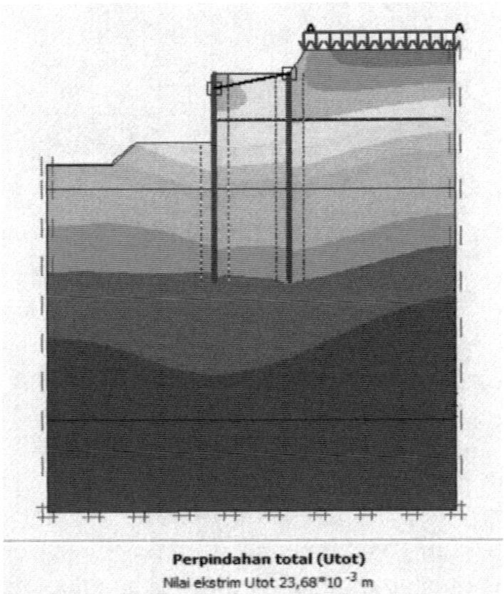

Figure 8. Condition of the collapse field after strengthening the Strauss pile

The collapse condition that occurs on the slope with Strauss pile reinforcement in Figure 8 can increase the safety factor to 1.75. It can be concluded that the use of Strauss piles on slopes can be applied. Figure 8 shows that a 24% reduction occurs on slopes given the Strauss pile system, while the direction of its descent for the land at the top of the slope is downwards so that it will shift towards the river when the slope shifts.

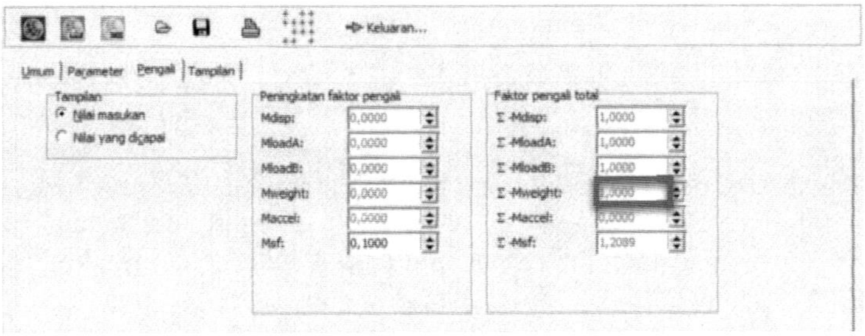

(a)

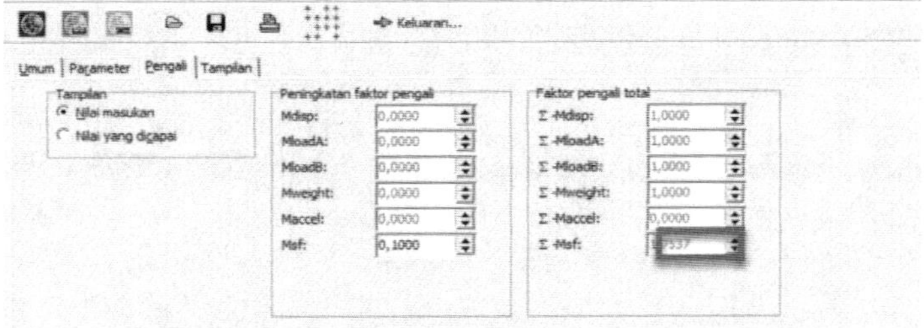

(b)

Figure 9. (a) Safety Factor of existing conditions,
(b) Safety Factor Strauss pile strengthening conditions

Then countermeasures using ground anchor construction with a length of 27 meters, without eliminating the existing sheet pile construction, calculated to produce.

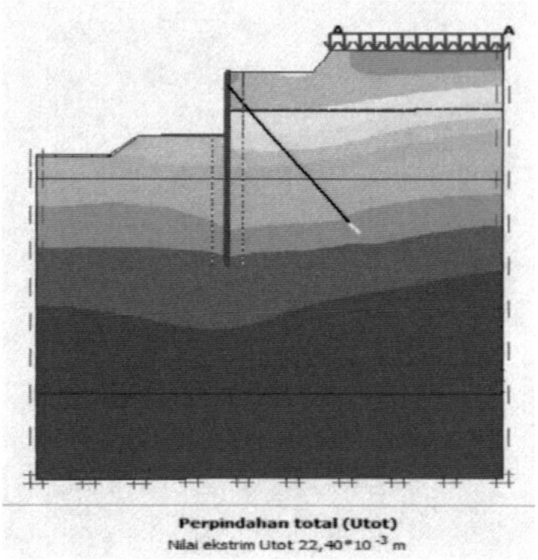

Figure 10. Conditions of the collapse field after strengthening the ground anchor

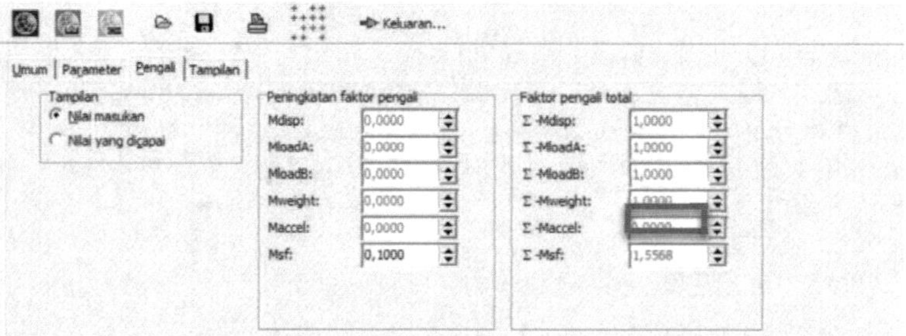

Figure 11. Safety Factors of ground anchor strengthening conditions

Figure 11 shows the results of the calculations that occurred on the slopes that the ground anchor strengthened showed that the SF value rose from the previous condition without reinforcement with a result of SF 1.55 > 1.25, smaller than the slopes that were strengthened by the Strauss pile, Figure 10 showed that the decrease that occurred on the slopes given the anchor system was 28% of the conditions before being strengthened.

4. Conclusions

Based on data from survey results, several tests, and calculations, it can be concluded that landslide mitigation on river embankments can be done in several ways, including building reinforcement using Strauss pile slings and ground anchors. Such measures can increase the value of the safety factor. In both alternatives to strengthening the Strauss pile and ground anchor, a Strauss pile strengthening is chosen, which has a more excellent safety factor value than using a ground anchor, which is 1.75. Using a Strauss pile combination with a sling is a practical application for the Inspection-RSCM road section so that safety conditions can be achieved in accordance with the needs of strengthening the river embankment.

Acknowledgments

The author would like to express his sincere gratitude to the Research and Community Service Unit (UP2M) of the Jakarta State Polytechnic for posing this research. The author also says his appreciation to all parties who have supported this research.

5. References

[1] Zainuri, G. Yanti, and S.W. Megasari, “Analisis karakterisktik tanah dan stabilitas lereng fakultas ekonomi Universitas Lancang Kuning”. Jurnal Teknik Sipil Siklus.2(2):125-1, 2016.
[2] P.T. Simatupang, Modul Jenis Turap Kantilever, Jakarta: Universitas Mercu Buana, 2008.
[3] I.R. Hertiany and A. Asyifa, “Perencanaan konstruksi sheet pile wall sebagai alternatif pengganti gravity wall”, Inersia. 10(1):53-65, 2014
[4] K. Diliaristianto, N.S. Surjandari, and R.H. Dananjaya, “Penggunaan sheet pile untuk perkuatan lereng di Desa Tambakmerang Kecamatan Girimarto Kabupaten Wonogiri”. e-Jurnal Matriks Teknik Sipil. 592-597, 2017
[5] https://www.republika.co.id/berita/r6rsfh314/jalan-ambles-di-jalan-inspeksi-kali-ciliwung-1
[6] B. R.M.S., S. M., L. Indrathilaka, and N. Amarathunga, “The manual of Landslide Monitoring, Analysis and Countermeasures,” p. 151, 2013, [Online]. Available: http://open_jicareport.jica.go.jp/pdf/12112116_05.pdf
[7] R. Katzenbach, S. Leppla, H. Ramm, M. Seip, and H. Kuttig, “Design and construction of deep foundation systems and retaining structures in urban areas in difficult soil and groundwater conditions,” *Procedia Eng.*, vol. 57, pp. 540–548, 2013, DOI: 10.1016/j.proeng.2013.04.069.
[8] A. Kainthola, D. Verma, S. S. Gupte, and T. N. Singh, “A Finite Element Approach of Stability Analysis of Internal Dump Slope in Wardha Valley Coal Field,” *Am. J. Min. Metall.*, vol. 1, no. 1, pp. 1–6, 2013, doi: 10.12691/ajmm-1-1-1.
[9] R. B. J. Brinkgreve, W. Broere, and D. Waterman, “Plaxis 2D Manual - Version 8,” no. January 2006.
[10] A. Kainthola, D. Verma, S. S. Gupte, and T. N. Singh, “A Coal Mine Dump Stability Analysis—A Case Study,” *Geomaterials*, vol. 01, no. 01, pp. 1–13, 2011, DOI: 10.4236/gm.2011.11001.
[11] D. V. Griffiths and P. A. Lane, “Slope stability analysis by finite elements,” *Géotechnique*, vol. 51, no. 7, pp. 653–654, 2001, DOI: 10.1680/geot.51.7.653.51390.
[12] J. M. Duncan, “State of the Art: Limit Equilibrium and Finite-Element Analysis of Slopes,” *J. Geotech. Eng.*, vol. 122, no. 7, pp. 577–596, 1996, doi: 10.1061/(asce)0733-9410(1996)122:7(577).
[13] T. Matsui and K. C. San, “Finite element slope stability analysis by shear strength reduction technique,” *Soils Found.*, vol. 32, no. 1, pp. 59–70, 1992, DOI: 10.3208/sandf1972.32.59.
[14] A. C. D.-18 on Soil and Rock, “Standard Practice for Classification of Soils for Engineering Purposes (Unified Soil Classification System) 1,” vol. I, 2017.
[15] ASTM C127, “Standard Test Method for Density, Relative Density (Specific Gravity), and Absorption of Coarse Aggregate,” *Annu. B. ASTM Stand.*, pp. 1–5, 2004.
[16] ASTM, “D4318 - 00: Standards,for Liquid Limit, Plastic Limit, and Plasticity Index of Soils This c of soils, ASTM D 4318-00,” *ASTM Int.*, vol. 04, pp. 1–14, 2000.
[17] H. Niroumand, K. A. Kassim, A. Ghafooripour, R. Nazir, and S. Y. Z. Far, “Investigation of slope failures in soil mechanics,” *Electron. J. Geotech. Eng.*, vol. 17 R, no. September, pp. 2703–2720, 2012.
[18] Neoh, Cheng Aik 2001 Practical Design Aspects for Slopes in Mountainous Terrains Proceedings of National Slope Seminar
[19] Hamed, H, Khairul A K, Amin G, Ramli N, Sayyed Y Z F 2012 Investigation of Slope Failures in Soil Mechanics Electronic Journal of Geotechnical Engineering Vol. 17. http://www.ejge.com/2012/Abs12.227.htm, accessed on August 13, 2019

Understanding of flow liquefaction in Lolu Village based on seismic refraction tomography method

P P Andika[1,3], A Tohari[2], T Yudistira[1], E Soebowo[2] and J Arifin[2]
[1]Faculty of Mining and Petroleum Engineering, Bandung, Institute of Technology, Bandung, West Java, Indonesia 40132
[2]Research Center for Geological Disaster, National Research and Innovation Agency, Bandung, West Java, Indonesia 40135
[3]Department of Geophysical Engineering, Sumatera Institute of Technology, Lampung, Indonesia 35131

E-mail: adrin.tohari26@gmail.com

Abstract. A big earthquake may induce flow liquefaction phenomena in lowland areas made up of sedimentary deposits. To understand the factors controlling such a hazard, it is necessary to have knowledge of the sub-surface geological conditions of the liquified area. This paper presents the results of a series of seismic refraction measurements in Lolu Village where flow liquefaction occurred during the 2018 Palu-Donggala earthquake. The results of seismic refraction tomography indicate the presence of horst-graben structures, probably associated with the localized pull-apart basin. The results of this study suggest that the sub-surface geological structure is also a governing factor for the occurrence of flow liquefaction phenomena in an area with a complex fault system.

1. Introduction

On 28 September 2018, a significant earthquake (7.4 Mw) in Central Sulawesi generated a tsunami as well as flow liquefaction in numerous places in Palu City, Donggala Regency, and Sigi Regency. The liquefaction phenomenon caused substantial fatalities and damage to residential and agricultural areas in Balaroa, Petobo, Lolu, Jono Oge, and Sibalaya Villages. Mitigation effort of such liquefaction hazard requires knowledge of factors controlling such phenomena.

Several studies have been conducted to understand the mechanisms and causative factors of flow liquefaction in Palu City and Sigi Regency by using a remote sensing approach [1, 2, 3, 4, 5]. Several researchers have tried to link the existence of irrigation canals with the phenomenon of liquefaction flow in the Petobo, Lolu, Jono Oge, and Sibalaya areas [1, 2, 4]. Using the electrical-resistivity tomography method, several studies attempted to identify shallow aquifers [6,7], and sub-surface topography that controlled the liquefaction phenomenon [8]. Based on inversion HVSR, [9] concluded that the flow liquefaction phenomenon in the Balaroa area was caused by the presence of sub-surface structures such as a subsurface ridge that blocked the groundwater flow causing the surface soil layer to saturate during the earthquake shaking.

Of many recent studies conducted to understand factors controlling the massive flow liquefaction phenomena in Balaroa, Petobo, Jono Oge, and Sibalaya, however, none have attempted to assess the factor governing the small-scaled flow liquefaction occurring in Lolu Village. This liquefaction phenomenon not only involved the flow of liquefied soil layer but also strike-slip fault movement which caused a half block of residential houses to be displaced a few hundred meters away from their original position (Figure 1). Based on field observation, [10] It was hypothesized that the higher block of soils did not liquefy but instead operated as fragmented blocks and flowed as a whole to the west. The presence of the strike-slip fault may suggest the complexity of sub-surface conditions in this flow liquefaction area. Thus, the understanding of the flow liquefaction in Lolu Village requires knowledge of sub-surface structure. Such knowledge can be gained through a sub-surface investigation using an inexpensive and non-destructive method such as geophysical surveys.

Figure 1. Arial photo of flow liquefaction impact on residential housing at Lolu Village, Sigi Regency [10].

This paper presents the results of a geophysical survey conducted in the flow liquefaction area in Lolu Village, using the seismic refraction method. The objectives of this study were to (1) obtain the seismic refraction tomography images, and (2) analyze the sub-surface structure controlling the flow liquefaction phenomenon

2. Geological and tectonic setting

The geological and stratigraphic conditions of the study area have been well organized by [11] from the oldest to the youngest rock formation. The oldest formations of the Triassic period comprise the metamorphic Gumbasa complex (TRIgg), the intrusive rocks (Tmpi), and the metamorphic Wana complex (TRw). The next rock formation overlaying the Triassic rock formations is the Latimojong Formation (Kls) of the Cretaceous period which consists of sedimentary rocks alternating with meta-claystone, and intrusive rocks (gr) comprising granite and granodiorite can be found on the western hills. Another Cretaceous rock formation is the metamorphic complex (Km) which is present in the eastern part of the high-altitude ridge. Meanwhile, the Tertiary sedimentary Tinombo Formation (Tt) is found in the western ridge at a high altitude. Overlapping unconformably the older rock formation, the Pleistocene Celebes Molasse of Sarasin and Sarasin (QTms) is present on both ridges at lower altitudes. The youngest rock formation in the area is the Holocene alluvium and coastal deposit (Qap), deposited unconformably on the oldest rocks in flat areas.

Physio-graphically, the Palu and its surrounding area consist of the east and west ridges, both of which are directed north-south and are separated by the Palu valley. The structure of this area is dominated by the Palu fault zone with a north-northwest direction, the shape of which now resembles a depression bounded by active faults. The eastward-sloping faults in the metamorphic rock complex and the Tinombo Formation show compression properties in some of the older faults. The earthquake in 1968 near Tambo revealed a normal fault zone with a northwest trend causing the ground level to drop 5 meters [11]. The most recent earthquake (7.5 Mw) on 28 September 2018 showed the major Palu-Koro fault rupture zone, visible from Sigi Regency, Palu City, and Palu Bay [12-14]. The Palu-Koro fault movement produced a maximum slip of 6 $\pm$ 0.5 m within Palu City, with an average movement of up to 4.7 m in the city's northern and southern parts [12].

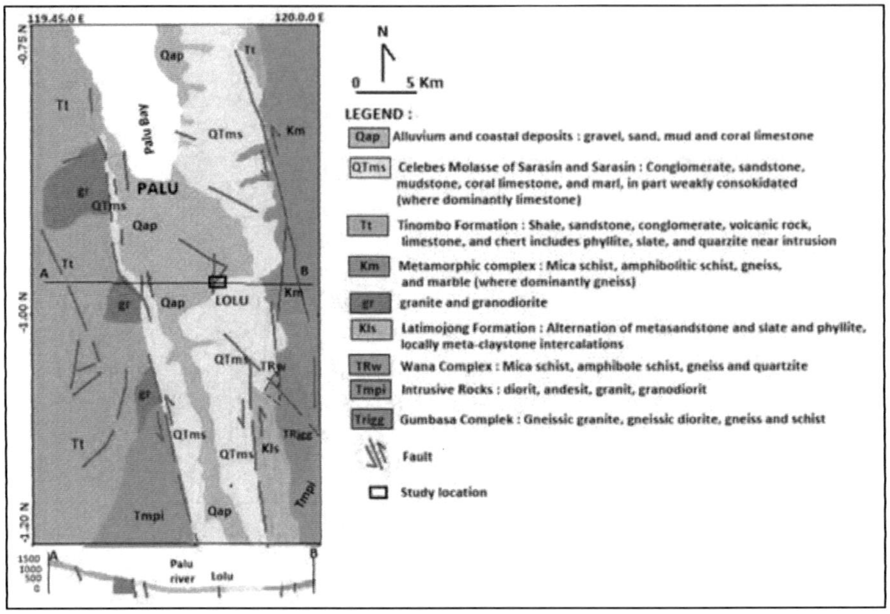

Figure 2. Geological map of Palu and its surroundings and location of seismic refraction survey.

3. Methods

3.1. Data acquisition

To achieve the objective of this study, a series of seismic refraction surveys were conducted in the flow liquefaction area using a 48-channel seismic data acquisition system McSEIS-SX48, and 28 Hz geophones. A total of four survey lines were set up crossing the liquefaction area in the North-South direction as shown in Fig. 3. The seismic data for a 240-m long profile was acquired with 5-m geophone spacing and 5 shoot points at 60-m intervals on each measurement profile: two shots were placed at the end of the profile with a 2.5-m offset, and the other three shots were performed between geophones 12 and 13, 36 and 37, and in the middle of each profile (Fig. 4). When the length of the seismic profile was extended to 460 m, the seismic cable overlapped for 3 geophones. Each shot was recorded with a 0.2-millisecond sampling rate and recording length of 820 milliseconds with a 3-kg sledgehammer as a P-wave source. The details of the geophone location and the total length of the survey line at each survey location are summarized in Table 1.

Table 1. Coordinates and length of the seismic refraction survey line.

Line code	First Geophone		Last Geophone		Length (m)
	UTM X	UTM Y	UTM X	UTM Y	
I	824280	9892730	824257	9892444	235
J	824127	9892709	824126	9892448	235
K	824405	9892756	824307	982316	460
L	924200	9892722	824167	9892267	460

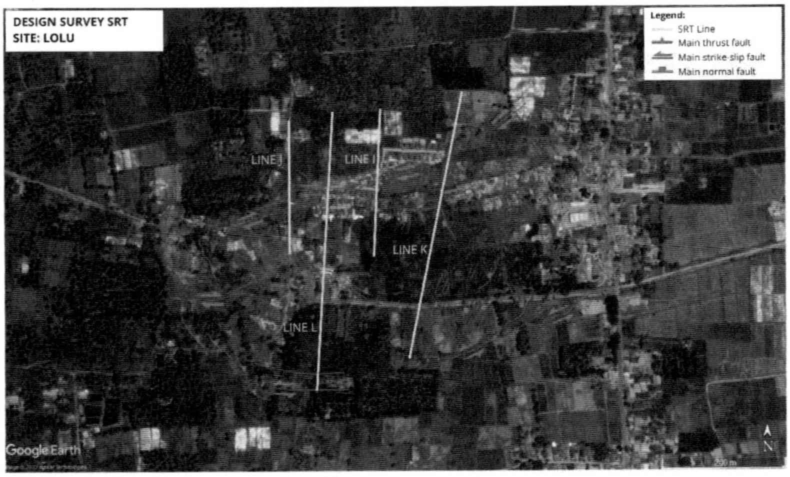

Figure 3. Seismic refraction survey lines (yellow line) at the Lolu flow liquefaction area. The red lines are fault systems that are redrawn from [2].

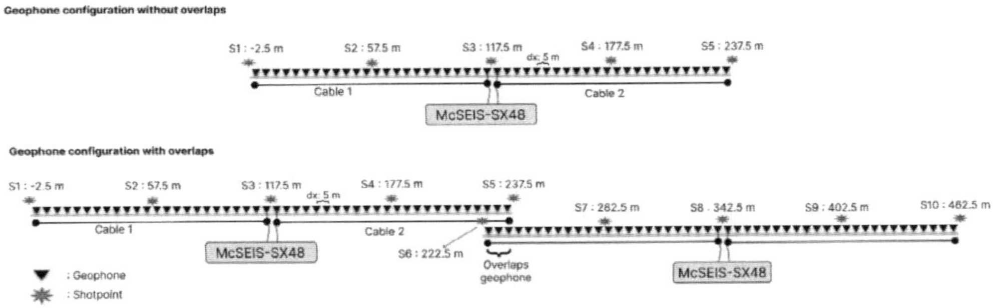

Figure 4. Configuration of geophones and location of shoot points in each survey line.

3.2. Data processing

The SRT survey data were processed using the SeisImager/2DTM software package (Pickwin and Seisrefa) by following the flowchart shown in Fig. 5. For each shot along the profiles, the software package allowed the first arrivals to be selected from the seismic data to generate the 2D velocity model based on the tomographic inverse method. We started with gradual velocity as our initial model, and by leveraging a delay between the observation and calculation data, we constructed an iterative least square inversion to update the model's parameters until the misfit criteria were fulfilled with a root-mean-square error below 2-millisecond.

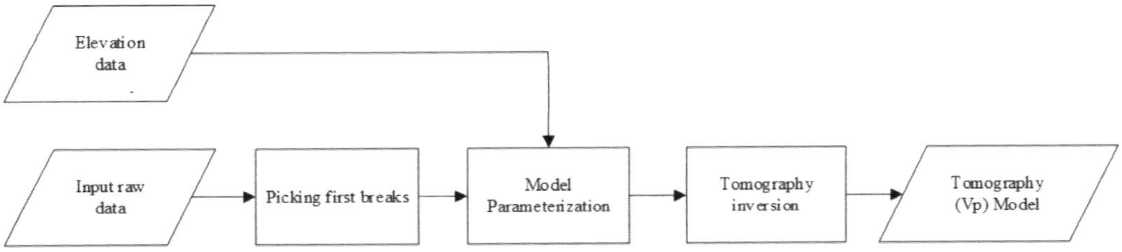

Figure 5. General data processing flow to perform SRT data analysis using SeisImager software.

4. Results and discussions

4.1. P-wave velocity

The resulting refraction tomographic images are shown in Figs. 6 to 9. The SRT images show that the P-wave velocity is relatively low in the near-surface area and relatively high at deeper depths in all profiles. The SRT image of each line shows three sub-surface layers. The upper layer has a low seismic velocity ranging from 300 to 700 m/s and a thickness of around 10 m. The middle layer has a velocity ranging from 900 to 2000 m/s and thickness varies from 20 to 50 m. The lower layer is indicated with a velocity larger than 2000 m/s.

4.2. Seismic refraction tomography (SRT) model

Line-K is entirely within the liquefaction-affected area, indicated by the presence of a basin structure in the north and getting thicker to the south part (Fig. 6a). The interpretation of this seismic tomography model is shown in Figure 6b. The presence of the horst and graben system is seen in this figure. This model clearly shows the presence of a pull-apart basin.

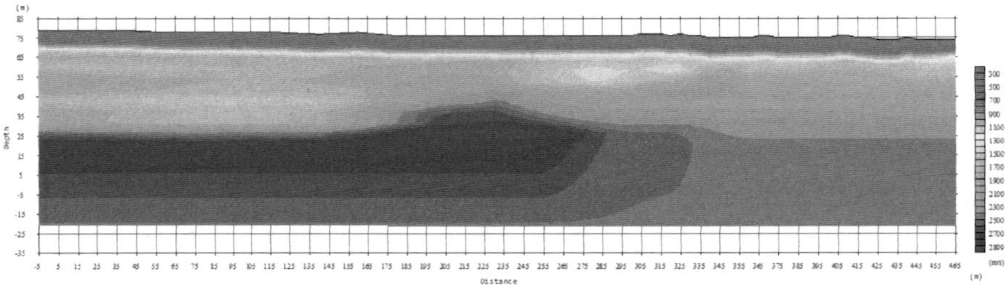

(a) Seismic refraction tomography model

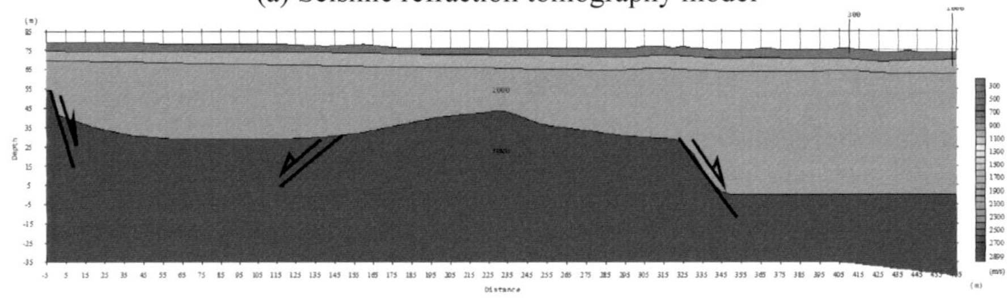

(b) Interpretation of sub-surface structure.

Figure 6. SRT model and interpretation of sub-surface structures for Line K.

As seen in Fig. 7(a), the thickness of the second rock layer thickness in Line I varies from around 20 m in the northern part then becomes thicker up to 35 m in the middle of the line and decreases to 20 m in the southern part. The area where the flow liquefaction occurred is associated with the thickest part in the middle of Line I. Thus, this seismic topography model indicates the presence of a basin structure beneath the liquefied zone, probably associated with a local graben structure within the bedrock layer. The interpretation of the graben structure is shown in Figure 7(b) as a part of localized a pull-apart basin.

Meanwhile, Line L has a relatively thicker second layer (around 60 m) compared to other lines as shown in Fig. 8a. A localized thickening of the second rock layer can be seen to exist in this seismic model to the south. This also can be interpreted as a localized shallow graben that exists in the southern part of the study area (Fig. 8b).

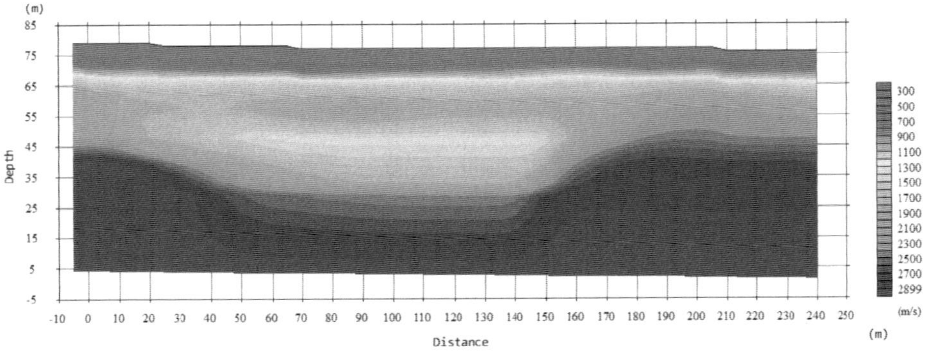

(a) Seismic refraction tomography model

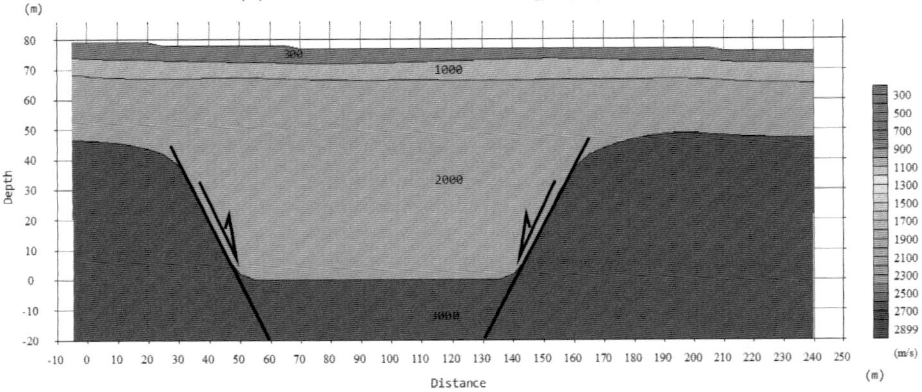

(b) Interpretation of sub-surface structure.

Figure 7. SRT model and interpretation of sub-surface structure for Line I.

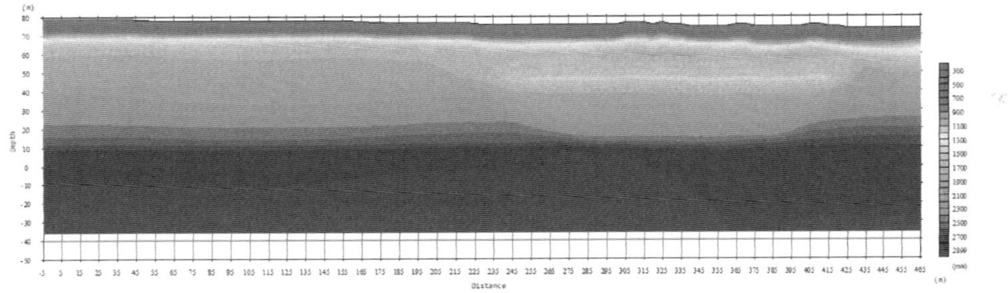

(a) Seismic refraction tomography model

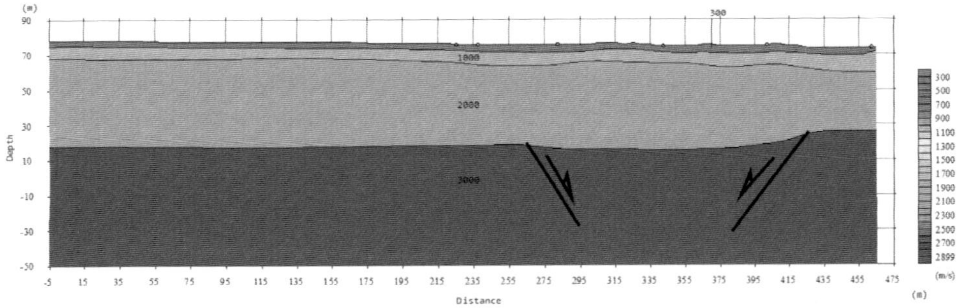

(b) Interpretation of sub-surface structure.

Figure 8. SRT model and interpretation of sub-surface structures for Line L.

The seismic tomography model for Line-J is shown in Figure 9. According to Figs. 3 and 9, the area affected by the flow liquefaction starts from the middle of the survey line towards the south. From the tomogram, the thickness of the second rock layer is more profound in the southern part. This thickening of the second rock layer implies the presence of localized graben in the high-velocity layer from the middle part of the seismic model. The presence of a strike-slip fault and graben structure may indicate a sub-surface structure associated with the localized pull-apart basin (Fig. 9b), as also exits in Line I.

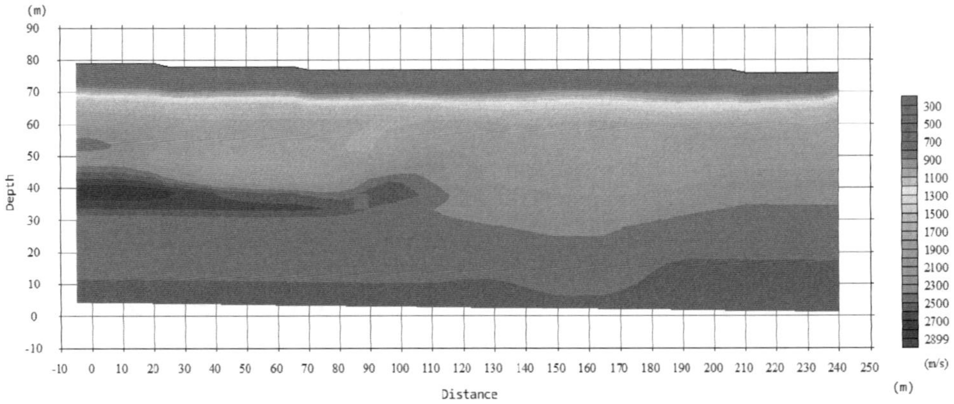

(a) Seismic refraction tomography model

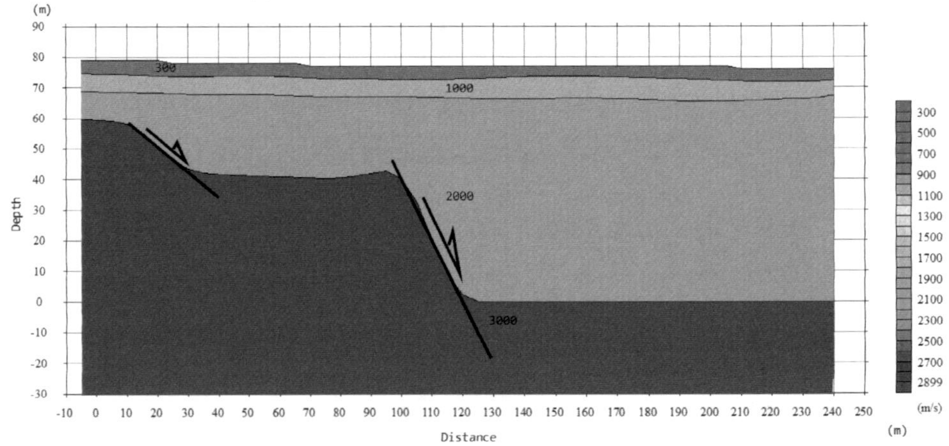

(b) Interpretation of sub-surface structure.

Figure 9. SRT model and interpretation of sub-surface structures for Line J.

5. Conclusions

The refraction seismic tomography method applied in this study effectively obtains the subsurface structure based on seismic velocity distribution. The resulting tomographic images are able to determine the thickness of the layer model and the depth of the seismic bedrock and its undulation. Based on the satellite image and pattern of layer undulation for each SRT model is fitted with the pattern of damage on the surface due to liquefaction occurring at Lolu village. From the interpretation of the seismic model, the pull-apart basin is more pronounced in the northern part of the study area. A further study aiming to have a comprehensive understanding of the sub-surface structure controlling the flow liquefaction phenomena in Lolu village is recommended.

References

[1] Bradley K, Mallick R, Andikagumi H, Hubbard J, Meilianda E, Switzer A, Du N, Brocard G, Alfian D, Benazir B, Feng G, Yun S, Majewski J, Wei S, and Hill E M 2019 *Nat. Geosci.* **12** p 935.

[2] Watkinson I M and Hall R 2019 *Nat. Geosci.* **12** p 940.

[3] Kiyota T, Furuichi H, Hidayat R F, Tada N, and Nawir H 2020 *Soils and Foundations* **60** p 722.

[4] Okamura M, Ono K, Arsyad A, Minaka U S, and Nurdin S 2020 *Soils and Foundations* **60** p 1050.

[5] Kusumawardani R, Chang M, Upomo T C, Huang R-C, Fansuri M H, and Prayitno G A 2021 *Landslides* **18**, p 3163.

[6] Arsyad A, Samang L, Harianto T, Muhiddin A B, Djamaluddin A R, and Aswad S 2019 *Proceedings of the 7th International Conference on Earthquake Geotechnical Engineering*, Rome, Italy p 1178.

[7] Nurdin S, Harianto T, Aswad S, Arsyad A, and Alexsander S 2019 *Proceeding of the 23rd Annual National Conference on Geotechnical Engineering*, Jakarta, Indonesia, pp. 208.

[8] Tohari A, Wardhana D D, Hanif M, and Koizumi K 2019 i*n E3S Web of Conferences* **331**, 03002 (EDP Sciences – France) p 1.

[9] Cipta A, Rudyanto A, Afif H, Robiana R, Solikhin A, Omang A, Supartoyo and Hidayati S 2021 *Geological Society*, London, Special Publications, **501**.

[10] Rahardjo P P 2020 In *Understanding and Reducing Landslide Disaster Risk*, edited by V. Vilímek, F. Wang, A. Strom, K. Sassa, P. T. Bobrowsky, K. Takara Springer, Cham. p 25.

[11] Soekamto R 1973 *Geological Map, Palu Triangular*, Research, and Development Center for Geology, Bandung.

[12] Bao H, Ampuero J P, Meng L, Fiedling E J, Liang C, Milliner C W D, Feng T, Huang H 2019 *Nat. Geosci.* **12**, p 200.

[13] Abdullah A I and Abdullah 2020 *National Seminar on Physics* 2019, J. Phys.: Conference Series **1434** (IOP Publishing, Bristol, UK), p 1.

[14] Natawidjaja D H, Daryono M R, Prasetya G, Udrekh, Liu P L-F, Hananto N D, Kongko W, Triyoso W, Puji A R, Meilano I, Gunawan E, Supendi P, Pamumpuni A, Irsyam M, Faizal L, Hidayati S, Sapiie B, Kusuma M A, and Tawil S 2020 *Geophys. J. Int.* **224** p 985.

Acknowledgments

This study was funded by the Disaster Research Seed Grant, and the Expedition and Exploration Seed Grant from the National Research and Innovation Agency of the Republic of Indonesia in 2022.

IOP Conf. Series: Earth and Environmental Science 1173 (2023) 012030 doi:10.1088/1755-1315/1173/1/012030

Liquefaction potential analysis for Palu City based on CPT method

A Tohari[1], E Soebowo[1], S Wibawa[1], K Hermawan[1] and O F Saputra[2]
[1]Research Center for Geological Disaster, National Research and Innovation Agency, Bandung, West Java, Indonesia 40135
[2]Research Center for Geospatial, National Research and Innovation Agency, Cibinong, Indonesia

E-mail: adrin.tohari26@gmail.com

Abstract. During the 28 September 2018 Palu-Donggala earthquake, liquefaction was also a prominent hazard causing significant damage to buildings and infrastructures in Palu City. To mitigate such a hazard in Palu City, knowledge of the depth of the liquified soil layer is necessary. This paper presents the results of CPT-based liquefaction potential analysis in some locations around the city where sand boiling and ground settlement occurred. The analysis shows that liquefaction occurs at various depths less than 15 m and may induce ground settlement up a few centimeters. In the Palu-Koro Fault zone, the liquified sand layer is likely thicker than in other locations. Consequently, the total ground settlement is higher than in other locations The results of this study suggest that the liquefaction potential should be accounted for in the development of Palu City to reduce future earthquake risk.

1. Introduction

The occurrence of big earthquakes in a coastal area can induce not only a tsunami but also a liquefaction phenomenon. The earthquake-induced liquefaction is a phenomenon of loss of strength of the loosely-packed sand layer due to an increase in pore water pressures during the earthquake shaking. Thus, the liquefaction phenomenon commonly occurs in earthquake-susceptible coastal areas composed of low-density water-saturated sand deposits, and in areas with co-seismic ground motion. The occurrences of earthquake-induced liquefaction have caused significant damage to buildings and infrastructure as well as loss of lives. For example, the 1995 Kobe earthquake (Mw 6.9), the 1999 Turkey earthquake (Mw 7.4), the 2004 Andaman earthquake (Mw 9.1), the 2005 Nias earthquake (Mw 8.6), the 2011 Christchurch earthquake (Mw 6.2), the 2011 Tohoku earthquake (Mw 9.1) and the 2018 Palu earthquake (Mw 7.5) were followed by liquefaction disasters associated with the ground settlement, lateral spreading, sand boiling, loss of bearing capacity and flow liquefaction. Thus, efforts to mitigate such a collateral hazard are also necessary to reduce the earthquake risk.

There are three common field methods to analyze the liquefaction potential, i.e. cone penetration test (CPT), standard penetration test (SPT), and shear wave velocity (V_s) methods. Of those three methods, the CPT-based method is most commonly used because of its advantages in data quality control and repeatability, as well as its ability to detect soil layer variations [1]. So, the thickness and depth of the liquefiable soil layers can be identified more accurately than the other two methods. [2] used the CPT method to map the liquefaction vulnerability in the Padang City area. The mapping results show good agreement with the liquefaction phenomenon observed during the 2009 Padang earthquake.

The 2018 Palu earthquake (Mw 7.5) caused some significant damage to some government buildings and houses in Palu City associated with the occurrences of sand boiling and ground settlement, as shown in Figure 1. From this field evidence, liquefaction potential in Palu City was not well understood. Thus, knowledge of the liquefaction potential of alluvial deposits is necessary to reduce the risk associated with earthquake-induced liquefaction phenomena in Palu City. Some previous studies have been conducted to analyze liquefaction phenomena, especially in flow liquefaction areas using SPT [3, 4] and Vs methods [5]. This paper presents the results of liquefaction potential analysis based on the CPT method in some locations around the city where sand boiling occurred during the 2018 earthquake. The objectives of this analysis were to (1) determine the depth and thickness of the liquefiable soil layers, and (2) calculate the ground settlement induced by liquefaction.

(a) Sand boiling at the tax office in Palu (b) Sand boiling in the front yard of a house

Figure 1. Examples of damaged buildings due to sand boiling.

2. Geological and tectonic setting

2.1. Geological setting

According to [6], the geology of Palu City is primarily represented by two distinct rock units i.e. coastal and alluvial deposits, and the Celebes Molasse. The former, as the youngest rock of the Holocene in age, comprises silts, sands, gravels, and coral limestones that were deposited in the river, delta, and shallow sea environments. [7] divided the alluvial deposit unit into young and old alluvial, alluvial fan, river channel, and flood deposits. The alluvial fans are observable on the east and west sides of the hilly lowlands.

The Celebes Molasse is composed of poorly consolidated conglomerates, sandstones, marls, mudstones, and coral limestones derived from older formations [6]. At low altitudes on both the west and east ridges, this rock unit is observable to overlap unconformably with the metamorphic rock complex and Tinombo Formation.

2.2. Tectonic setting

The tectonic setting of Palu City is mainly dominated by the Palu-Koro Fault, a main active strike-slip fault with a sinistral movement and an NNW–SSE direction [8-13]. The Palu-Koro Fault noticeably traverses the lower portion of old alluvial fan deposits and it is distinguished from the adjacent surfaces by distinct interbasin ridges in the moderately uplifted ground surface [13]. The big earthquake (Mw 7.5) occurring on 28 September 2018 uncovered the main Palu-Koro Fault rupture zone, which is distinctly identifiable from Sigi Regency to Palu City [14, 15]. The movement of the fault produced a typical slip of approximately 4.7 m in the proximity of Palu City [14]. According to [16], the recorded horizontal *PGA* at the bedrock for Palu City is 0.33 g.

3. Methods

Cone penetration tests were carried out using piezocone equipment, to obtain cone tip resistance (q_c), local resistance (f_s), and pore-water pressure (u) data every 0.01 m at 7 locations where sand boiling occurred during the 2018 earthquake, as summarized in Table 1. The penetration was conducted using a 2.5 tons mechanical penetrometer up to a depth of 15 m.

Evaluation of the CPT-based liquefaction potential in this current study was performed in 3 (three) stages of calculation, namely the calculations of (1) the cyclic stress ratio (*CSR*), (2) the CPT-based cyclic resistance ratio (*CRR*), and (3) the safety factor against liquefaction (FS_L).

Table 1. Location of cone penetration tests in Palu City.

CPT No.	Location/ Sub-district	Latitude	Longitude	Penetration depth (m)
CPT-01	Balaroa, Palu Barat	-0.905804°	119.841367°	6.81
CPT-02	Donggala Kodi, Palu Barat	-0.900577°	119.846260°	13.12
CPT-03	Birobuli Utara, Palu Selatan	-0.914394°	119.892759°	6.50
CPT-04	Lolu Utara, Palu Selatan	-0.898527°	119.886703°	5.86
CPT-05	Besusu Barat, Palu Timur	-0.885701°	119.866459°	6.58
CPT-06	Besusu Timur, Palu Timur	-0.895230°	119.880065°	11.63
CPT-07	Lolu Selatan, Palu Selatan	-0.903777°	119.879575°	6.14
CPT-08	Besusu Tengah, Palu Timur	-0.887892°	119.875903°	7.22

3.1. Calculation of CSR

The *CSR* was calculated by using the equation of [17] as follows:

$$CSR = 0.65(a_{max}/g) \times \left(\sigma_{v0}/\sigma'_{v0}\right) r_d \quad (1)$$

, where a_{max} is the maximum horizontal acceleration at the ground surface, g is the gravitational acceleration (9.81 m/s^2), σ_{vo} is the total vertical stresses, σ'_{vo} is the effective vertical stresses, and r_d is the stress reduction factor according to [18]. Considering the peak bedrock acceleration of 0.33 g [16], and an amplification factor of 1.4 designated for the Palu City area [5], the a_{max} value of 0.46 g was used in the calculation of *CSR*.

3.2. Calculation of CPT-based CRR

Calculation of the CPT-based *CRR* was performed using a simple procedure according to [19], as follows:

$$CRR = \begin{cases} 0.833 \cdot \left[\frac{\left(q_{c1N}\right)_{cs}}{1000}\right] + 0.05 & \text{if } \left(q_{c1N}\right)_{cs} < 50 \\ 93 \cdot \left[\frac{\left(q_{c1N}\right)_{cs}}{1000}\right]^3 + 0.08 & \text{if } 50 < \left(q_{c1N}\right)_{cs} < 160 \end{cases} \quad (2)$$

, where $(q_{c1N})_{cs}$ is the clean sand equivalent value for normalized cone tip resistance (q_{c1N}), which is affected by the soil behavior index (I_c). Based on this method, a layer of soil that has a value of $I_c > 2.6$ and $(q_{c1N})_{cs} > 160$ kg/cm^2 is included in the criteria as a non-liquefiable soil layer.

3.3. Calculation of FS_L

The FS_L was calculated using the following equation:

$$FS_L = \left(\frac{CRR}{CSR}\right) \cdot MSF \tag{3}$$

, where *MSF* is the magnitude scaling factor according to [1]. By convention, liquefaction is most likely to occur when FS_L is less than 1. On the other hand, when FS_L is higher and equal to 1, liquefaction is unlikely to occur.

3.4. Calculation of total ground settlement

Total ground settlement due to liquefaction (S_{tot}) was calculated using the following equation according to [20]:

$$S_{tot} = \sum_0^z (\varepsilon_v/100) \cdot dz \tag{4}$$

, where the value of ε_v is the volumetric strain, which was obtained from the curve of [21]

4. Results and discussion

4.1. Liquefaction potential

Figures 2 to 5 present the results of the liquefaction potential analysis for all CPT locations. Soil layers having the potential to liquefy are typified by *CRR* values smaller than the *CSR* values or a safety factor of less than 1.0. Based on these results, the liquefaction potential differs from one CPT location to another due to the variation of q_c and f_s profiles which govern the types of soil layer at each location. In general, liquefied sand layers are characterized by q_c values lower than 15 MPa and f_s values lower than 150 kPa, up to maximum depths of 15 meters. Because the a_{max} value influences the liquefaction potential, the lower the a_{max} value, the lower the q_c and f_s value criteria for sand layers to liquefy. The liquefaction potential analyses also indicate that soil layers with a q_c value less than 5 MPa and f_s less than 125 kPa are unlikely to be liquefied. According to [22] classification, the range of q_c and f_s values characterizes the clay-behaving soil layer.

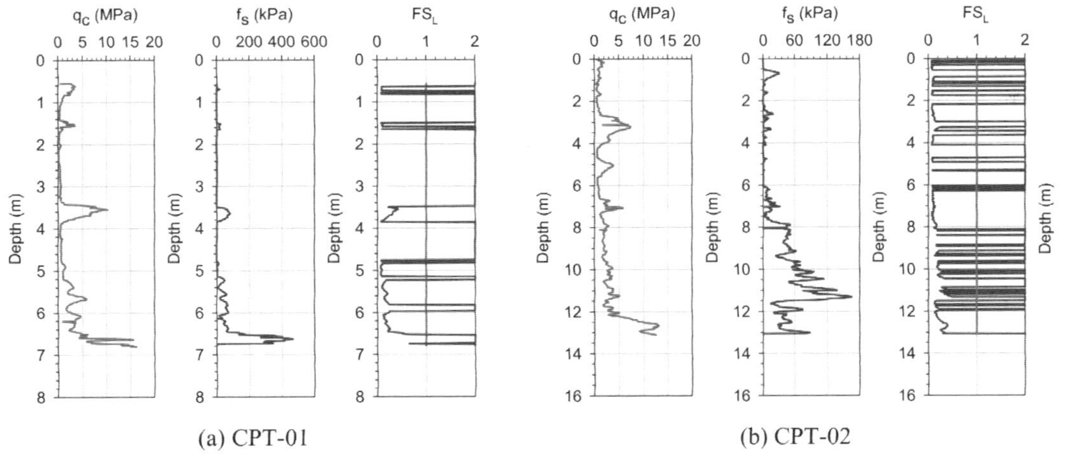

Figure 2. Graphs of q_c, f_s, and FS_L for CPT-01 to CPT-02

Figure 2 also shows that the liquefied soil layer in CPT-02 located at the Palu-Koro surface rupture is thicker than in other locations. This suggests that the soil layers in the rupture zone are mainly composed of thick layers of loosely-packed sandy soil. Furthermore, the result of the analysis of liquefaction potential in CPT-01 located in the Balaroa flow liquefaction area implies that the liquefied soil layer may still experience liquefaction again during a strong earthquake.

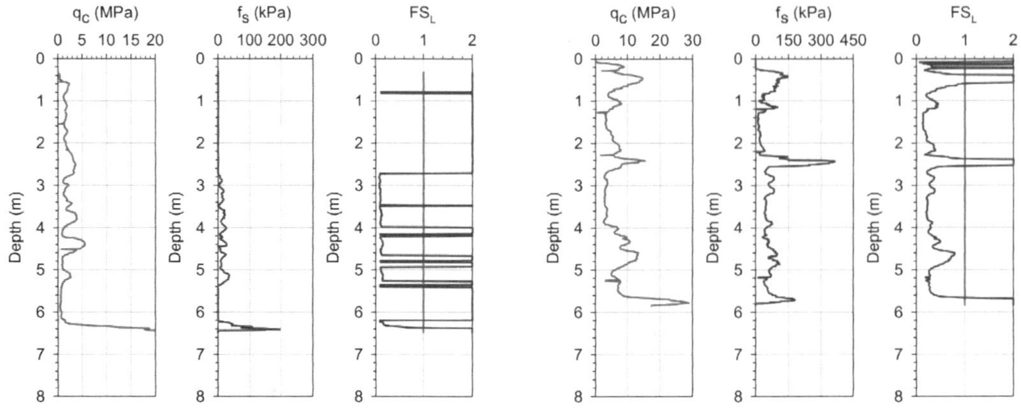

(a) CPT-03 (b) CPT-04

Figure 3. Graphs of q_c, f_s, and FS_L for CPT-05 to CPT-06.

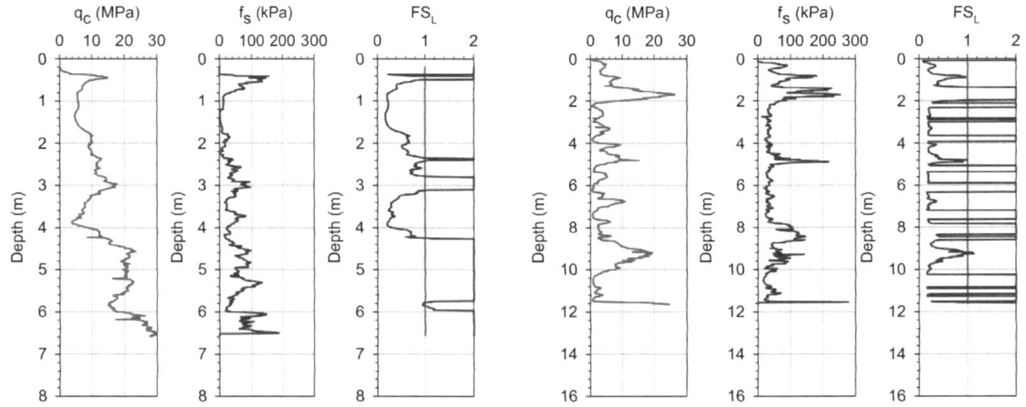

(a) CPT-05 (b) CPT-06

Figure 4. Graphs of q_c, f_s, and FS_L for CPT-05 to CPT-06.

4.2. *Total ground settlement*

The results of the total ground settlement calculation are shown in Figure 6. According to this figure, the total ground settlement varies from 5 cm to 30 cm, with the maximum ground settlement occurring in CPT-02 located at the Palu-Koro rupture zone. This variation is dependent upon the maximum depth of penetration and the thickness of liquefied soil layers. Thus, the soil type and thickness will specifically govern the total amount of liquefaction-induced ground settlement in Palu City.

The results of this current study indicate the importance of assessing liquefaction-induced ground settlement in liquefaction potential analysis because ground settlement may impose damage to building such as tilting, subsidence, and toppling, especially building constructed with shallow foundations. The results of liquefaction potential analysis may assist in the design of types of foundations of buildings to reduce the earthquake risk in Palu City. The study also implies the significance of the analysis of ground settlement due to liquefaction in the area built in young alluvial deposits like Palu City. Thus, a detailed sub-surface investigation to provide a comprehensive knowledge of the earthquake collateral hazard associated with liquefaction-induced ground settlement in Palu City is recommended.

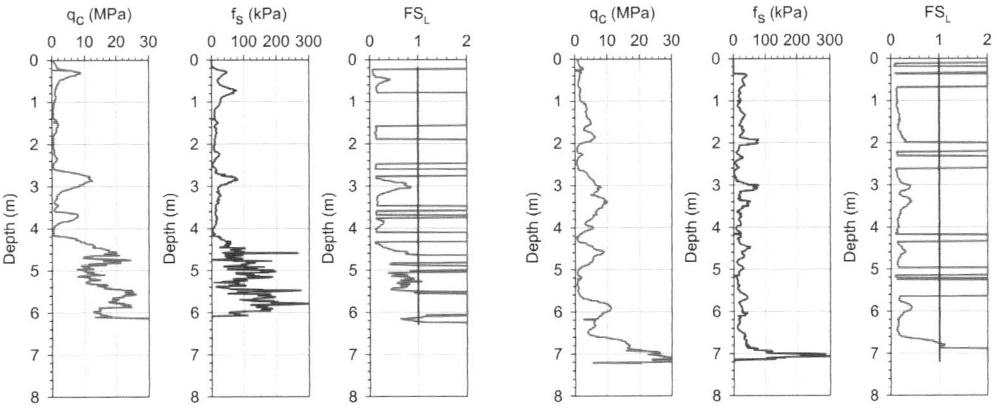

(a) CPT-07 (b) CPT-08

Figure 5. Graphs of q_c, f_s, and FS_L for CPT-07 to CPT-08.

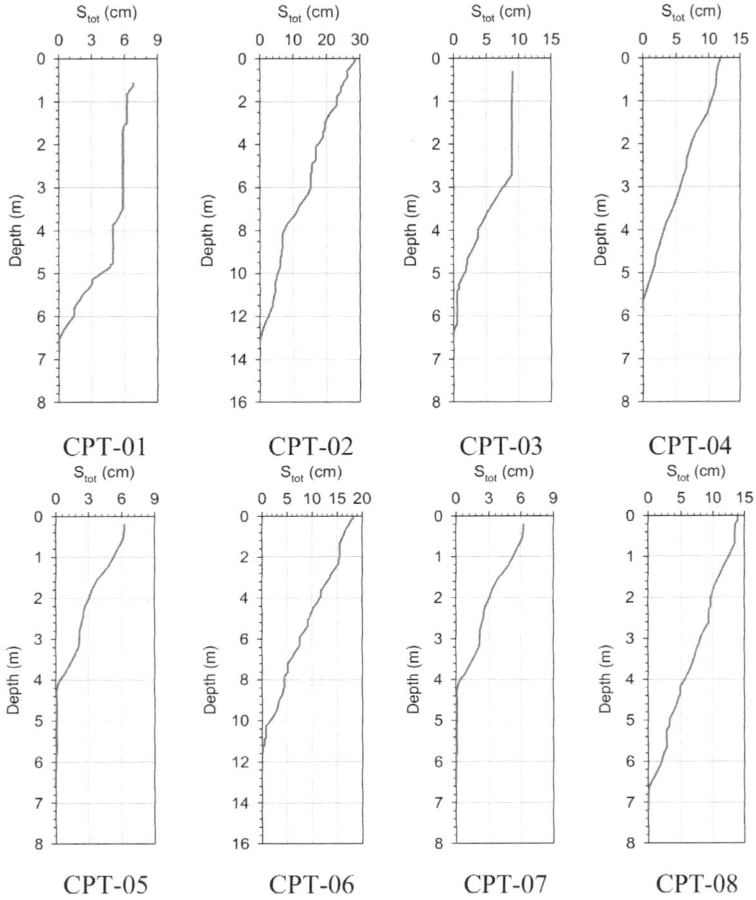

Figure 6. Graphs of S_{tot} due to liquefaction for all CPT locations.

5. Conclusions

Based on the CPT-based liquefaction potential analyses conducted in sand boiling locations in Palu City, liquefiable sand layers exist at various depths and thicknesses. The liquefiable sand layer is characterized by q_c values of 15 MPa and f_s values of 150 kPa. In the location of the Palu-Koro rupture zone, the liquified sand layer is likely thicker than in other locations. Consequently, the total ground settlement is higher than in other locations. In general, the earthquake of M_w 7.5 with a *PGA* of 0.46 g would induce ground settlement in flat areas in Palu City up to 30 cm. Further studies aiming to provide a comprehensive understanding of liquefaction potential and associated ground settlement in other areas in Palu City are recommended.

Acknowledgments
This research was partially funded by the Disaster Research Seed Grant from the National Research and Innovation Agency of the Republic of Indonesia in 2022, and the INSINAS research grant from the Ministry of Research, Technology, and Higher Education of the Republic of Indonesia in 2019.

References
[1] Youd T L et al 2001 *J. Geotech. Geoenviron. Eng.* **127**(10) p 817.
[2] Tohari A, Sugianti K and Soebowo E 2011 *J. Riset Geol. Mining.* vol **21**(1) p 7.
[3] Muhanifah H, Adi A D and Faris F 2021 *IOP Conf. Series: Earth and Environmental Science* ***861*** (IOP Publishing, Bristol, UK) p 1.
[4] Sauri S, Rifa'i A, and Hardiyatmo H C 2021 *IOP Conf. Series: Earth and Environmental Science* **930** (IOP Publishing, Bristol, UK, 2021) p 1.
[5] Tohari A, Muttaqien I and Syifa R W 2022 *E3S Web of Conferences* **340**, edited by A. Rahman *et al.* (EDP Sciences, France) p 1.
[6] Sumadirdja H, Suptandar T, Hardjoprawiro S, Sudana D *1973 Reconnaissance geological map of the Palu quadrangle, Sulawesi",* Research and Development Center for Geology, Bandung.
[7] Soehaimi A, Firdaus M, Effendi I 2000 *Earthquake Susceptibility Zonation Map of Palu and its surrounding area*, Research and Development Center for Geology, Bandung.
[8] Katili J A 1970 *Geol Rundsch.* **59** p 581.
[9] Tjia H D 1978 *Bull. Geol. Soc. Malaysia* **10** p 73.
[10] Bellier O, Sébrier M, Seward D, Beaudouin T, Villeneuve M and Putranto E 2006 *Tectonophysics* **413** p 201
[11] Daryono M R 2016 *Paleoseismologi Tropis Indonesia (Dengan Studi Kasus di Sesar Sumatra, Sesar Palukoro-Matano, dan Sesar Lembang,* Dissertation Doctoral Program. Institut Teknologi Bandung.
[12] Watkinson I M and R. Hall R 2017 *Fault systems of the eastern Indonesian triple junction: evaluation of Quaternary activity and implications for seismic hazards Geohazards in Indonesia* in *Earth Science for Disaster Risk Reduction*, edited by Cummins P and Meilano I (Geol. Soc. London, London, 2017) **441** p 71.
[13] Patria A and P. S. Putra P S 2020 *Geosci. Lett.* **7** p 1.
[14] Bao H, Ampuero J P, Meng L, Fiedling E J, Liang C, Milliner C W D, Feng T, Huang H 2000 *Nat. Geosci.* **12** p 200.
[15] Abdullah A I, and Abdullah 2020 *J. Phys.: Conference Series* **1434** (IOP Publishing, Bristol, UK) p 1.
[16] Kiyota T, Furuichi H, Hidayat R F, Tada N, and Nawir H 2020 *Soils and Foundations*, **60**(3) p 722.
[17] Seed H B and Idriss I M 1971 *J. Soil Mech. Found. Div.*, **97**(SM9) p 1249.
[18] Liao S S C and R. V. Whitman R V 1986 *J. Geotech. Engrg.* **112**(3) p 373.
[19] Robertson P K and Wride C E 1998 Canadian Geotech. J. **35**(3), 442–459
[20] Ishihara K dan Yoshimine M 1991 *Soils and Foundations* **32**(1) p 173
[21] Zhang G, Robertson P K, and Brachman R W I 1991 *Can. Geotech. J.* **39** p 1168
[22] Robertson P K 1991 *Can. Geotech. J.* **28** p 176

Spatial and Temporal Analysis Prediction of Landslide Susceptibility Using Rainfall Infiltration and Grid-based Slope Stability Methods in West Bandung area of West Java-Indonesia

Koko Hermawan[1a)], Khori Sugianti[1], Antonina Martireni [1)], Nugroho Aji Satrio[1], Yunarto[1)]

[1)]Research Center for Geological Disaster, National Research and Innovation Agency
a)Corresponding author: koko.hermawan@gmail.com

Abstract. West Bandung, West Java, is an area with a high level of landslide susceptibility. Landslides in West Bandung occurred 142 times during rainfall in the last ten years. This paper presents the results of landside susceptibility modeling in the West Bandung area of West Java Province, Indonesia, considering the spatial characteristics of the rainfall data, slope and soil properties using the TRIGRS model. This research is based on conditions in the field in the form of landslide locations, soil engineering properties, soil thickness, Digital Elevation Model, and rainfall data. The effect of one-day antecedent rainfall intensity was considered in this study, i.e., 12 hours of antecedent rainfall. The results of the TRIGRS modelling showed that the intensity of rainfall antecedent of rainfall influenced the slope stability in the study area. The TRIGRS model results indicate that the predicted landslide susceptibility distribution agrees with the historical landslide events.
Keywords: rainfall intensity, slope stability, landslide susceptibility, TRIGRS.

1. Introduction

A landslide is a geological disaster that often occurs in West Java, especially during high rainfall. Refers to the landslide susceptibility map in West Java [1], West Bandung is included in the medium to high landslide susceptibility category. A natural disaster that has frequently occurred in West Java over the last ten years is landslides, from 2010 to 2020 [https://dibi.bnpb.go.id/]. Generally, landslides in West Java occur during the rainy season [2]. One of the main factors triggering temporary landslides in landslide hazard assessment is rainfall [3; 4; 5]. Long-lasting rainfall infiltration can increase the soil's groundwater table and pore water pressure [6; 7; 8]. Previous studies have shown that rainfall triggers shallow landslides [9; 10; 11; 12]. Landslide disasters will cause environmental damage, human casualties, psychological impacts, and property loss. [13; 14]. Knowledge of slope stability analysis for the future is needed to reduce the occurrence of landslides. From these problems, it is necessary to map landslide hazards in areas with potential landslides. Susceptibility mapping is carried out to anticipate and minimize the risks posed.

Previous researchers in modeling slope stability used a Geographical Information System (GIS). To make landslide hazard zone maps, several researchers, such as [15], and [16], have applied statistical methods and [17] used the logistic regression method approach. For slope stability [18] and [19] using the AHP (Analytical Hierarchy Process) method for slope movement susceptibility. [20] applied the weighted value method for mapping landslide hazards. [21] and [16] Multiple logistic regression analysis to evaluate the probabilistic model. Statistical models depend highly on input data quality, affecting the model results. Predicting regional landslides can be focused on the landslide mechanism based on a deterministic model. Determining landslide hazard analysis was studied by [22], [23], and [24] through a deterministic approach. Development to predict the triggers of shallow ground motion vulnerability due to rainfall can use hydrogeological models and physical-based models such as the SINMAP, TiVaSS, GEOtop-FS, TRIGRS and SHALSTAB. Landslide vulnerability mapping using TRIGRS can determine the distribution of safety factors due to rainfall and estimate spatial-temporal changes [25; 26; 5; 27; 28; 10; 11; 12]. [30] In Sukabumi Regency, West Java, the TRIGRS application maps landslide susceptibility. Meanwhile, in Cililin Regency, West Java, this application is used to determine the increase in rainfall and the area of distribution of slope stability.

In this study, landslide-prone area mapping is conducted quickly and efficiently by modeling the safety factor to determine the potential for landslides in the West Bandung area using the TRIGRS model. The purpose of this study is to use the TRIGRS application for presentation in modeling erosion susceptibility in the West Bandung area and the surrounding area, considering rainfall, soil lithology and spatial characteristics of slope using the TRIGRS model. This paper aims to determine (1) the relationship between soil behavior and the occurrence of landslides and (2) the one-day antecedent rain intensity to changes in landslide probability.

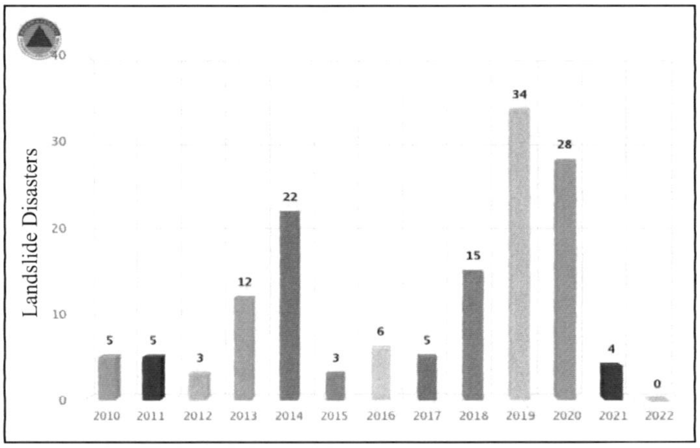

Figure 1. The number of landslides in the West Bandung area.

2. Research Location

The research was conducted in West Bandung Area, West Java Province, Indonesia. West Bandung Regency has regional boundaries, Bandung Regency and Cianjur Regency in the south; Cianjur Regency in the west; Subang Regency and Purwakarta Regency in the north; and Cimahi Regency and Bandung City in the east; and (Figure 2). Geographically, the research location is located at coordinates 107°16'35.57" - 107°34'51.05" East Longitude and 7°4'47.40" - 6°46'10.68" South Latitude. The research area is about 1,145 km2. Topographically this area has an average altitude of 110 m to 22,429 m above sea level.

IOP Conf. Series: Earth and Environmental Science 1173 (2023) 012031 doi:10.1088/1755-1315/1173/1/012031

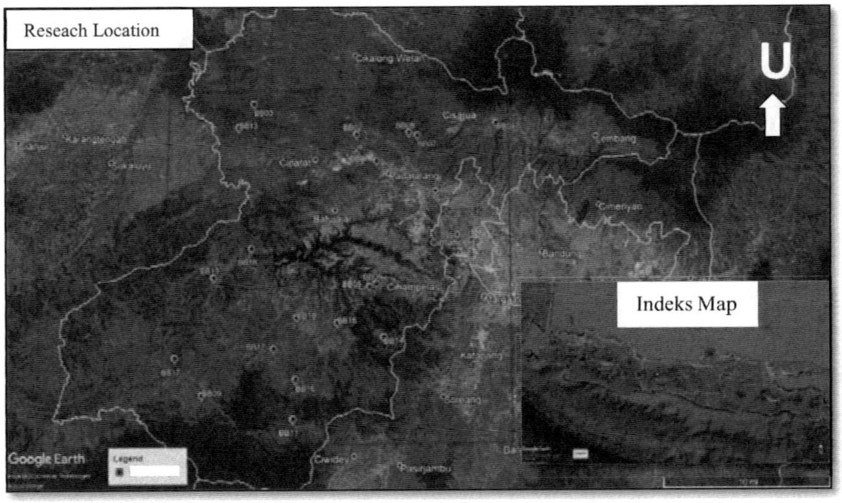

Figure 2. Research location [https://www.google.com/maps/bandungbarat].

Based on the West Java physiographic division above, the research area is in the Bandung Zone, a volcanic area. Geomorphologically, this study area dominantly consists of hilly areas with steep slopes to very steep slopes, which cause the hilly areas to be prone to landslides during a rainfall period. The study area geology consists of Holocene alluvial deposits, Quarter volcanic rocks, Quarter igneous rocks, and Tertiary sedimentary rocks (Figure 2). The volcanic rocks consist of dacite, tuff, and andesitic breccia. Meanwhile, sedimentary rocks include tuffaceous sandstone and limestone, widespread in the study area's middle part. In the research area, the developed geological structure consists of folds and faults, namely the Cipongkor Syncline, thrust faults, shear faults [31], and the Cimandiri Fault, which runs southwest-northeast direction [32]. As seen in Figure 1, landslide occurrences are mainly distributed in the hilly areas of Quaternary volcanic rocks. Landslide locations are in hilly areas and have strong geological structures. Thus, volcanic rocks are more pronounced to landslides than sedimentary and igneous rocks. In addition, the presence of geological structures contributes to the susceptibility to landslide hazards.

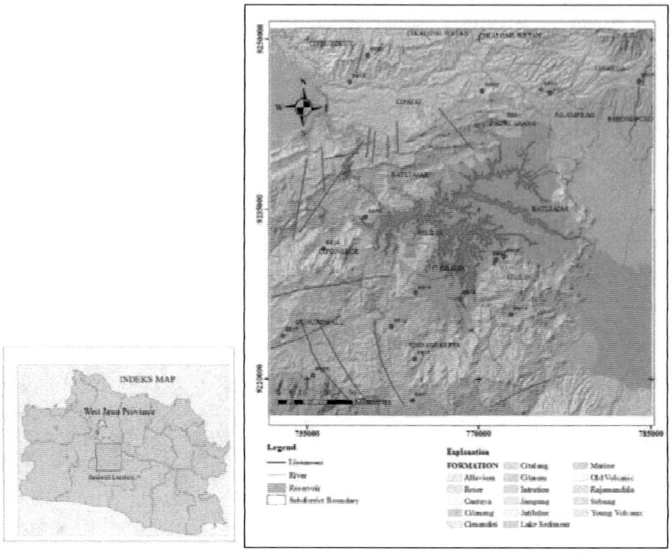

Figure 3. Geological map of the research location [modified from Sudjatmiko, 1972].

3. Method And Parameter

3.1. Trigrs

The application model is designed to model landslide safety factors by merging an infinite slope stability calculation and a one-dimensional analytical solution for pore-pressure diffusion in a soil layer of finite depth in response to time-varying rainfall [25; 26]. This application model is crucial for landslide susceptibility mapping studies and landslide hazard zoning for land-use planning on a regional scale. Rainfall infiltration modeling in TRIGRS used a linear formula [6] and an extension of Richard's equation [25]. The transient pore water pressure formula was given by the following equation 1 [25]:

$$\psi(Z,t)=[Z-d]\beta+2\sum_{n-1}^{N}\frac{I_{nz}}{K_k}H(t-t_n)^{\frac{1}{2}}\sum_{m=1}^{\infty}\left\{ierfc\left[\frac{(2m-1)d_{LZ}-Z}{2[D_1(t-t_n)]^{\frac{1}{2}}}\right]+ierfc\left[\frac{(2m-1)d_{LZ}+(d_{LZ}-Z)}{2[D_1(t-t_n)]^{\frac{1}{2}}}\right]\right\}$$
$$-2\sum_{n-1}^{N}\frac{I_{nZ}}{K_z}H(t-t_n)[D_1(t-t_n)]^{\frac{1}{2}}\sum_{m=1}^{\infty}\left\{ierfc\left[\frac{(2m-1)d_{LZ}-(d_{LZ}-Z)}{2[D_1(t-t_n)]^{\frac{1}{2}}}\right]+ierfc\left[\frac{(2m-1)d_{LZ}-(d_{LZ}-Z)}{2[D_1(t-t_n)]^{\frac{1}{2}}}\right]\right\} \quad (1)$$

where:

ψ - pore pressure of groundwater at a certain time (t) and depth (Z)

Z - $z/\cos\delta$ where Z - normal coordinate direction of the slope (positive downwards) and the depth below ground level

Δ - slope angle

d - Steady state of groundwater table measured in the vertical direction, this parameter uses a groundwater level assumption of 2.5 m based on a model approach that only predicts shallow landslide events that have a slip plane close to the slope surface.

- the depth of the impermeable basalt measured in the vertical direction,
B - $cos^2\ \delta$ - $(IZLT/Ks)$
K_s - hydraulic conductivity in the Z (vertical) direction
I_{ZLT} - steady surface flux
- surface flux with a certain intensity for an n time interval
D_0 - D_0/cos2 δ, where D_0 is the saturated hydraulic diffusivity ($D_0 = K_s/S_s$, where K_s is the saturated hydraulic conductivity and Ss is the specific storage)
N - total number of time intervals
$H(t\text{-}tn)$ - step function of t and tn was the time at n-time interval in the sequence of rainfall infiltration
$erfc\ (\eta)$ - complementary error function
$ierfc(\eta)$ - $(1/\sqrt{\pi})\ exp(-\eta 2)\text{-}\eta erfc\ (\eta)$ m is an index of an infinite series that returns an odd number in the complementary error function.

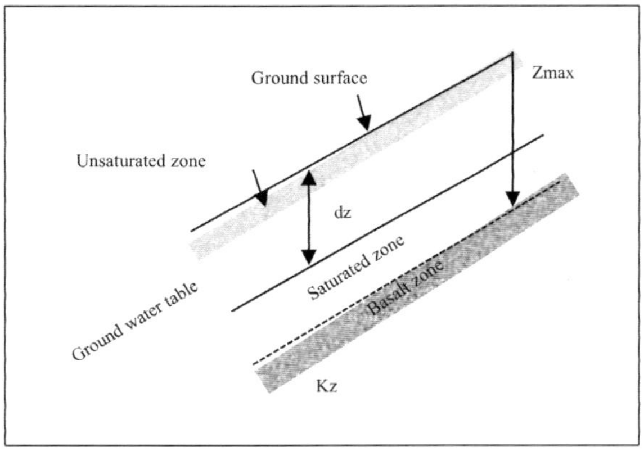

Figure 4. Conceptual diagram of the TRIGRS model [25; 26].

3.2. Slope Stability

Slope stability calculation uses analytical solutions to calculate the factor of safety value using an infinite slope model approach [6]. The method compares the total resisting basal coulomb friction against gravity and resisting basal coulomb friction. The homogeneous continuous slope factor of the safety equation used in the TRIGRS modeling [6] is as follows:

$$\mathrm{FS} = \frac{\tan\varphi'}{\tan\theta} + \frac{c' - \psi(Z,t)\gamma_w \tan\varphi'}{\gamma_s Z \sin\theta\cos\theta} \qquad (2)$$

where: φ' is the soil friction angle for effective stress, θ is the angle of the sliding plane, c' is the soil cohesion for effective stress, ψ (Z,t) is the groundwater pressure head at time t and depth Z, γ_w is the unit weight of groundwater, and γ_s is the unit weight of soil.

3.3. Parameters Model

Generally, data of slope, flow direction and digital topographic map scale as input data for modeling. For runoff-routing calculations in slope stability modeling, a topo index is needed. Topographic index, slope, friction angle and soil cohesion, permeability, porosity, soil thickness, unit weight, water table depth, diffusivity, hydraulic conductivity, infiltration rate, and rainfall intensity are input files for TRIGRS (Table 1). Compilation of all data from these Parameters, then made into a thematic map with pixels of 25 m × 25 m made with ArcGIS software, then extracted to ASCII.

Table 1. Input parameter for TRIGRS model.

No.	Parameter Data	Unit
1.	Historical landslide	Point
2.	DEM	Pixel of 25 x 25 m
3.	Flow direction	Direction
4.	Slope	Degree
5.	Effective Cohesion (c')	kN/m^2
6.	Unit Weight (γ)	kN/m^3
7.	Permeability (Ks)	m/s
8.	Difusivity (D0)	m^2/s
9.	Soils Depth (Z)	m
10.	Groundwater	m
11.	Rainfall Intensity	m/s
12.	Effective Friction Angle (phi')	Degree

3.4.Landslides Susceptibility Zone

The shear stress and shear resistance along the slip's plane will affect the slope's safety factor. The slope safety factor value is declared stable if it has a value greater than 1.0 and unstable if it is less than 1.0. Table 2 presents the zoning of landslide susceptibility based on the slope factor of safety (FS) value, which is a modification [33]. The safety factor (FS) value of less than 1.0 is a very high susceptibility zoning.

Table 2. Landslides susceptibility zone based on safety factor value [modification [33]

Factor of Safety	Landslides
FK<1,0	Very High
1,0<FK<1,2	High
1,2<FK<1,7	Medium
1,7<FK<2,0	Low
FK > 2,0	Very low

3.5.INPUT DATA DEM PARAMETERS

Digital Altitude Model (DEM) is one of the topographic parameters for input data in modeling. DEM data was obtained from SRTM satellite imagery from the DEMnas. DEM is also used to generate raster maps at a pixel size of 25 m x 25 m grids of flow direction and slope (Figure 5 and Figure 6). DEM and flow direction to calculate runoff routes in the Topoindex program. The research area has a topographical elevation of 209 m to 2,049 m with an area of 1,143 km² and is dominated by an altitude of 500-1000 m, about 64.3% of the area of 1,143 km². The distribution of the vulnerability inventory lies in the morphology of hills with a moderate slope (21%) to a steep slope (54%).

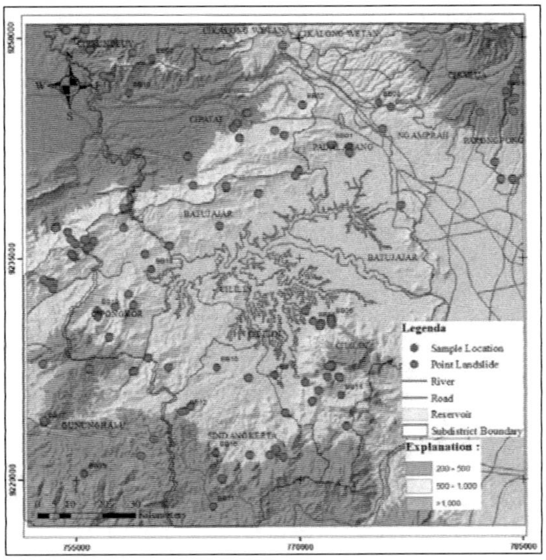

Figure 5. DEM raster map of the research area.

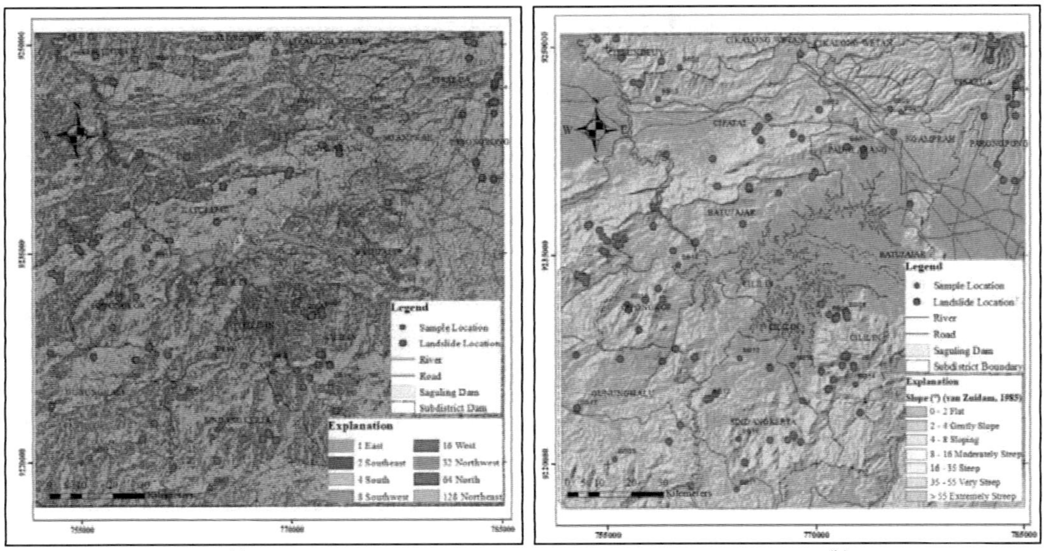

(a) (b)

Figure 6. Flow direction (a) and slope (b) raster map of research area.

3.6. Soil Mechanical Parameters

These parameters were obtained from laboratory sample testing required for calculating the slope safety factor in the TRIGRS modeling. Soil sampling represented each formation geology in the research area (Table 3). This research conducted soil testing to obtain the value of the unit weight (γs), friction angle (φ'), cohesion (c'), and the coefficient of saturation permeability (Ks). The soil hydraulic parameter, namely the steady condition infiltration rate (Iz), was obtained from the infiltration test results using the double ring infiltrometer method carried out at each soil sampling location. The diffusivity value was obtained using the relationship between the coefficient of saturated permeability (Ks), which is 100 times the value of Ks [34; 35], expressed as:

D0 = 100 x Ks
Where the D0 is hydraulic diffusivity, the Ks is the raster saturated permeability coefficient.

The study uses the assumption of the groundwater table depth parameter of a groundwater level of 2.5 m. In addition, the TRIGRS model only predicts shallow landslide events with slip planes near the sloped surface. Soil depths affect infiltration and slope stability, model. Many spatial interpolation methods are available to map non-uniform soil thickness of the soil using the angle of inclination, curvature, and elevation. In this study, the soil depths were mapped using a linearly decreasing function of the elevation (DEM).

3.7. Duration Rainfall and Rain Intensity

In this study, the rainfall intensity in modeling uses the assumption of the influence of one-day antecedent rain intensity, i.e., 12 hours of antecedent rainfall. This paper uses the rainfall intensity during the return period of one decade obtained in the research report by [37], i.e., a value of the rainfall intensity of 7.43 mm/hour for 12 hours that compares with the initial condition without involving rainfall influence in modeling.

Table 3. Input data of mechanic soils.

Formation	Effective Cohesion c' (Pa)	Effective Friction Angle phi' (θ)	Unit Weight (N/m^3)	Porosity (n)	Diffusifity (D0)	k_{sat} (m/det)
Alluvium	6905.87	29.49	16498.66	0.56	0.004	0.00004
Lake Deposite	13528.18	18.47	14479.22	0.603	0.0175	0.00018
Young Volcanic	3478.9	16.54	14105.90	0.62	0.044	0.00044
Old Volcanic	6905.87	29.49	16498.66	0.56	0.004	0.00004
Intrution	9197.57	29.9	16482.79	0.54	0.015	0.00015
Cimandiri Formation	13585.5	23.89	14474.03	0.6	0.012	0.00012
Jatiluhur Formation	6514.87	27.4	14387.42	0.59	0.024	0.00024
Beser Formation	6514.87	27.4	14387.42	0.59	0.024	0.00024
Citarum Formation	6320.37	22.49	15515.41	0.55	0.0058	0.00006
Rajamandala Formation	6320.37	22.49	15515.41	0.55	0.0058	0.000058

4. Result and Discussions

4.1. Initial Condition on Landslides Susceptibility

The initial conditions for the TRIGRS model results are shown in Figure 7. In the initial condition, the high susceptibility hillslope was mainly concentrated only in the north of the research area, which can be seen from the thickness of the soil and slopes that are upright to very upright slopes. So, slope angle and soil thickness are the main factors controlling landslide susceptibility in the research area. Many studies of landslide susceptibility show that slope angle has an active role in controlling shear forces for modeling slope stability. [38; 2]. In previous studies, shallow slides were controlled by one important factor, namely soil thickness [36; 39].

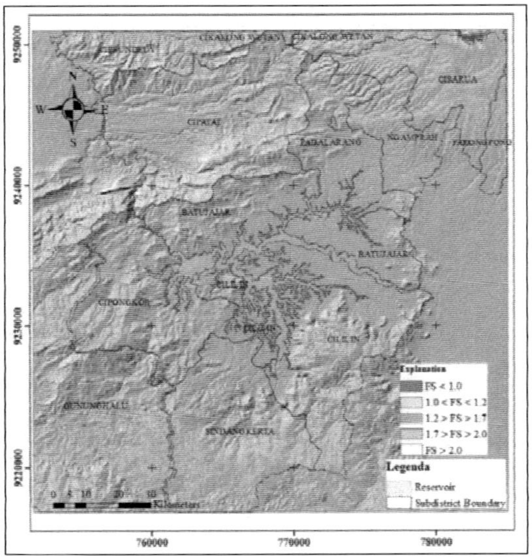

Figure 7. TRIGRS modeling results on the initial condition

4.2. Landslides Susceptibility doe to the Influence of Rainfall

The distribution of landslides in the research area during the first hour of susceptibility is very high (FS < 1.0) in the northern part. In the following duration, landslide susceptibility areas developed from very low to very high (Figure 8). Figure 8 (a) shows a very high susceptibility formed in some hilly areas in Cisarua. In this condition, landslide susceptibility is controlled by soil engineering parameters and rainfall. Furthermore, modeling results with the same intensity for 4 hours showed a very high vulnerability developing in the Cililin and Sindangkerta areas (Figure 8. b). This case was due to the formation of transient pore air pressure, which caused the slope stability, resulting in a very high susceptibility zone growth, especially in slope areas in a high susceptibility zone. Then Figure 8 (c) shows the intensity of rainfall equal to the duration of the rain for 8 hours, causing the addition of a very high susceptibility zone. Thus, rainfall duration can reduce slope stability in a shallow to medium susceptibility zone in Cipongkor, Batujajar, and Padalarang areas. The following modeling results in Figure 8 (d) show an expansion of shallow susceptibility in almost all hilly areas. It is interpreted that rainfall intensity equal to the duration of rain for 12 hours causes the addition of a very high susceptibility zone. This case is due to the long enough period for water to enter the soil, which will cause soil saturation and greater slope susceptibility. This susceptibility develops in the hilly areas of Cisarua, Cililin, Sindangkerta, Cipongkor, and Batujajar.

The results show that landslide susceptibility forms during the first hour up to 12 hours. This study proves that rainfall infiltration for 12 hours develops the soil's saturated zone, disrupting slope stability and transient pore water pressure so that slope stability decreases, especially on very steep slopes. The slope distribution in the study area has a high vulnerability value (FK < 1.0), especially in hilly areas with relatively steep slope morphology. The model results also show low susceptibility areas because rainfall infiltration can reduce slope stability in sloping hilly areas and areas with higher shear strength values.

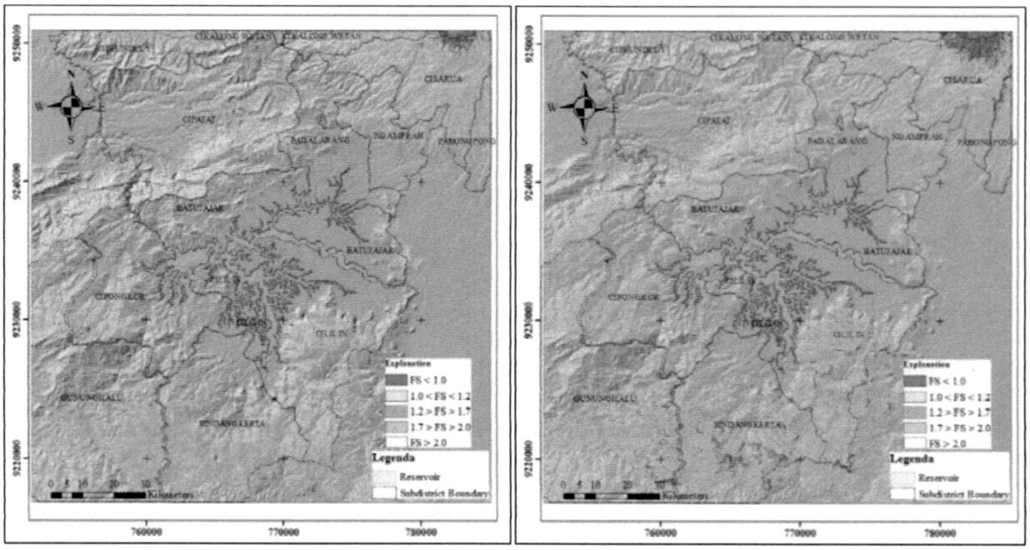

(a) **(b)**

Figure 8. TRIGRS modeling results at (a) 1 hour dan (b) 4 hours.

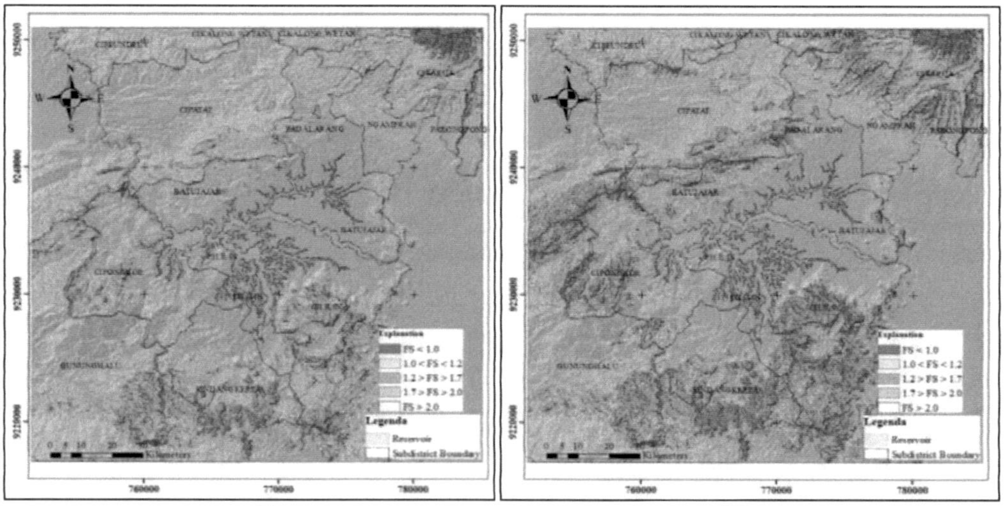

(c) **(d)**

Figure 8. TRIGRS modeling results at (c) 8 hours, and (d) 12 hours.

Figure 9 presents graph changes in landslide-prone areas in the study area. This histogram graph shows an increase in the expansion of the slope area, which is in the very high susceptibility zone (FS<1.0) from the initial to the 12-hour condition. The increase in areas prone to landslides also occurred in high and medium susceptibility zones. Thus, the influence of rainfall intensity before the landslide for 12 hours is a factor causing the increase in the area of high and very high susceptibility. The increase in high and very high susceptibility areas is associated with a decrease in low to medium susceptibility areas. So, the results of the TRIGRS modeling showed that the landslide susceptibility in the study area is affected by the rainfall intensity of antecedent rainfall.

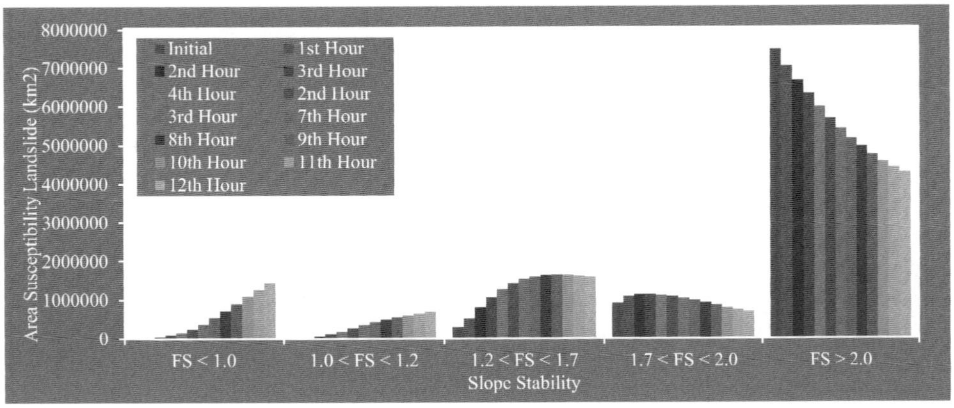

Figure 9. Histogram of TRIGRS modeling results.

4.3. Validation of Landslides Susceptibility

Validation uses landslide inventory data from the PVMBG database, field surveys, and visual identification through Google Earth. The validation process is carried out by overlapping the landslide event inventory data with the TRIGRS modeling results. Figure 10 shows the suitability of landslide occurrence data with moderate to very high landslide susceptibility. The TRIGRS model results indicate that the predicted landslide susceptibility distribution agrees with the historical landslide events in the study area. This model is very good at calculating the spatial vulnerability of landslides in the research area.

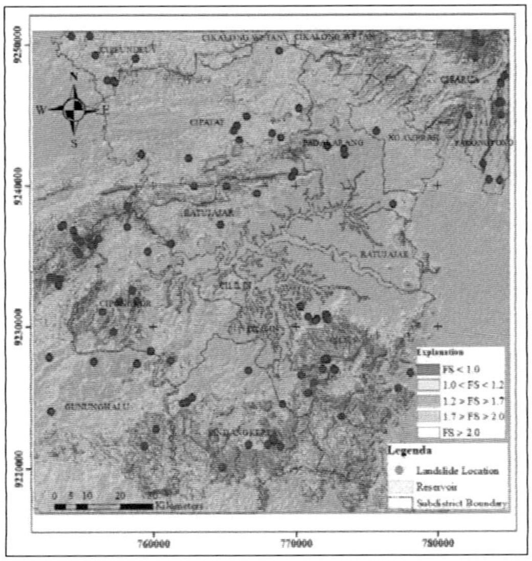

Figure 10. Histogram of TRIGRS modeling results.

5. Conclusion

This study concluded that the main factors controlling the rank of landslide susceptibility in the research area are slope and soil thickness shown in the initial condition model. Other factors are soil mechanical characteristics and rainfall intensity of antecedent rainfall shown in the transient condition model. Rainfall infiltration for 12 hours causes the soil's saturated zone, increasing pore water pressure and disturbing slope stability on steep slopes. Therefore, the effect of one-day antecedent rainfall intensity on landslide susceptibility is significant in this area. The TRIGRS model can predict landslide susceptibility distribution in the future in the study area, and TRIGRS is quite good at determining the spatial landslide susceptibility in the study area. This study benefits in assessing potential threats and landslide risks in residential areas of West Bandung.

Acknowledgments

The authors wish to acknowledge to Research Center for Geological Disaster, National Research and Innovation Agency (BRIN), for providing a research grant through the Disaster Research seed grant of the National Research and Innovation Agency. We also like to express gratitude for laboratory assistance by technical staff, Mr. Wahyudin, and encouragement from colleagues.

References

[1] PVMBG, (2009): *Landslide Susceptibility Zone Map of West Jawa Province*, Center for Volcanology and Geological Hazard Mitigation, Bandung.

[2] Sugianti, K., Yunarto, Y., Sadisun, I., & Kartiko, R., 2022. Analysis of Maximum-Rainfall-Infiltration-Induced Slope Stability Using the Transient Rainfall Infiltration and Grid-based Regional Slope-stability Model in Cililin, West Java, Indonesia. *Indonesian Journal on Geoscience*, Indonesia, 9.

[3] Guzzetti, F., Peruccacci, S., Rossi, M., & Stark, C. P. (2007). Rainfall thresholds for the initiation of landslides in central and southern Europe. *Meteorology and atmospheric physics*, *98*(3), 239-267.

[4] Tan, C. H., Ku, C. Y., Chi, S. Y., Chen, Y. H., Fei, L. Y., Lee, J. F., dan Su, T. W. (2008): Assessment of regional rainfall-induced landslides using 3S-based hydro-geological model, *Landslides and Engineered Slopes,* London: Taylor & Francis Group, 1639 – 1645.

[5] Park, D. W., Nikhil, N. V., dan Lee, S. R. (2013): Landslide and debris flow susceptibility zonation using TRIGRS for the 2011 Seoul landslide event, *Natural Hazards and Earth System Sciences*, **13**(11), 2833 – 2849.

[6] Iverson, R. M. (2000): Landslide triggering by rainfall infiltration, *Water Resources Research*, **36**(7), 1897 – 1910.

[7] Tohari, A., Sugianti, K., dan Hattori, K. (2013): Monitoring and modeling rainfall-induced landslide in volcanic soil slope, *Landslide Science and Practice: Early Warning, Instrumentation and Monitoring*, **2**, 503-510.

[8] Bai, S., Wang, J., Thiebes, B., Cheng, C., & Yang, Y. (2014). Analysis of the relationship of landslide occurrence with rainfall: a case study of Wudu County, China. *Arabian Journal of Geosciences*, *7*(4), 1277-1285.

[9] Saadatkhah, N., Kassim, A., dan Lee, L. M. (2015): Hulu Kelang, Malaysia regional mapping of rainfall-induced landslides using TRIGRS model, *Arabian Journal of Geosciences*, **8**(5), 3183 – 3194.

[10] Alvioli, M. dan Baum, R. L. (2016): Parallelization of the TRIGRS model for rainfall-induced landslides using the message passing interface, *Environmental Modelling & Software*, **81**, 122 – 135.

[12] Zhuang, J., Peng, J., Wang, G., Iqbal, J., Wang, Y., Li, W., Xu, Q., dan Zhu, X. (2017): Prediction of rainfall-induced shallow landslides in the Loess Plateau, Yan'an, China, using the T

[13] Highland, L. dan Bobrowsky, P. T. (2008): *The landslide handbook: a guide to understanding landslides*, US Geological Survey Reston.

[14] Kjekstad, O. dan Highland, L. (2009): Economic and social impacts of landslides, *Landslides–Disaster Risk Reduction*, Springer, 573 – 587.

[15] Lee, S. (2005): Application of logistic regression model and its validation for landslide susceptibility mapping using GIS and remote sensing data, *International Journal of Remote Sensing,* **26**(7), 1477 – 1491.

[16] Lee, S. dan Sambath, T. (2006): Landslide susceptibility mapping in the Damrei Romel area, Cambodia using frequency ratio and logistic regression models, *Environmental Geology.* **50**(6), 847 – 855.

[17] Chauhan, S., Sharma, M., dan Arora, M. K. (2010): Landslide susceptibility zonation of the Chamoli Region, Garhwal Himalayas, using logistic regression model, *Landslides,* **7**(4), 411 – 23.

[18] Hadji, R., Achour, Y., dan Hamed, Y. (2017): Using GIS and RS for slope movement susceptibility mapping: comparing AHP, LI and LR methods for the Oued Mellah Basin, NE Algeria, *Euro-Mediterranean Conference for Environmental Integration,* Springer, 1853 – 1856.

[19] Zhao, H., Yao, L., Mei, G., Liu, T., dan Ning, Y. (2017): A fuzzy comprehensive evaluation method based on AHP and entropy for a landslide susceptibility map, *Entropy,* **19**(8), 396.

[20] Wang, X., Zhang, L., Wang, S., dan Lari, S. (2014): Regional landslide susceptibility zoning with considering the aggregation of landslide points and the weights of factors, *Landslides,* **11**(3), 399 – 409.

[21] Kawagoe, S., Kazama, S., dan Sarukkalige, P. R. (2010): Probabilistic modeling of rainfall induced landslide hazard assessment, *Hydrology Earth System Sciences,* **14**(6), 1047 – 1061.

[22] Dietrich, W. E., de Asua, R. R., Coyle, J., Orr, B., dan Trso, M. (1998): A validation study of the shallow slope stability model SHALSTAB in forested lands of Northern California, *Stillwater Ecosystem, Watershed & Riverine Sciences, Berkeley, CA*

[23] Pack, R. T., Tarboton, D. G., Goodwin, C. N., dan Prasad, A. (1998): SINMAP, a stability index approach to terrain stability hazard mapping, *SINMAP user's manual, Terratech Consulting Ltd.*

[24] Tran, T. V., Lee, G., An, H., dan Kim, M. (2017): Comparing the performance of TRIGRS and TiVaSS in spatial and temporal prediction of rainfall-induced shallow landslides, *Environmental Earth Sciences*, **76**(8), 4.

[25] Baum, R. L., Savage, W. Z., dan Godt, J. W. (2002): TRIGRS—a Fortran program for transient rainfall infiltration and grid-based regional slope-stability analysis, *US Geological Survey Open-File Report*, **424**, 38

[26] Baum, R. L., Savage, W. Z., dan Godt, J. W. (2008): TRIGRS — a Fortran program for transient rainfall infiltration and grid-based regional slope-stability analysis, Version 2.0, *U.S. Geological Survey Open-File Report*, 75.

[27] Kim, J., Lee, K., Jeong, S., dan Kim, G. (2014): GIS-based prediction method of landslide susceptibility using a rainfall infiltration-groundwater flow model, *Engineering Geology*, **182**, 63 – 78.

[28] Cascini, L., Ciurleo, M., Di, S., dan Gullà, G. (2015): A new – old approach for shallow landslide analysis and susceptibility zoning in fine-grained weathered soils of Southern Italy, *Geomorphology*, **241**, 371 – 381.

[29] Sarah, D., Sugianti, K., Lestiana, H., 2015. Regional slope stability assessment of Tawangmangu District, Central Java, Proceeding of International Conference on Landslide and Slope Stability, N3-1 - N35, Bali, Indonesia.

[30] Sugianti, K., Sukrisyanti, Tohari, A., 2016. Model kerentanan gerakan tanah wilayah Kabupaten Sukabumi secara spasial dan temporal. Jurnal Riset Geologi dan Pertambangan 26(2), 117 - 129.

[31] Sudjatmiko. (1972): *Peta Geologi Lembar Cianjur, Jawa, Skala 1:100.000*, Direktorat Geologi, Bandung.

[32] Bronto, S., & Langi, B. B. (2017): *Geologi Gunung Padang dan Sekitarnya, Kabupaten Cianjur–Jawa Barat.* Jurnal Geologi Dan Sumberdaya Mineral, 17(1), 37–49.
[33] Ward, T. S., 1976. Factor safety approach to landslide potential delineation. Ph.D. Dissertation. Department of Civil Engineering, Colorado State University, Fort Collins, Colorado
[34] Viet, T. T., Lee, G. H., Thu, T. M., dan An, H. (2016): Effect of digital elevation model resolution on shallow landslide modeling using TRIGRS, *Natural Hazards Rev.* **18**(2), 1 – 12.
[35] Lee, G., An, H., dan Kim, M. (2017): Comparing the performance of TRIGRS and TiVaSS in spatial and temporal prediction of rainfall-induced shallow landslides, *Environmental Earth Sciences*, **76**(8), 315.
[36] Saulnier, G. M., Beven, K., dan Obled, C. (1997): Including spatially variable effective soil depths in TOPMODEL, *Journal of Hydrology*, **202**(1 – 4), 158 – 172.
[37] Sugianti, (2022) The Effect Of Increased Rainfall On Slope Stability Using Trigrs Model In West Bandung And Surrounding Area, West Jawa, *Thesis Report,* pp. 56-57.
[38] Lee S (2014) Geological application of geographic information system. Korea Inst Geosci Min Resour 9–15:109–118.
[39] Segoni, S., Rossi, G., & Catani, F. (2012). Improving basin scale shallow landslide modeling using reliable soil thickness maps. *Natural hazards*, *61*(1), 85-101.

Effect of wet – dry cycles on CBR value and grain size distribution particle on compacted residual tuff soil

S Syuhada[1*], A H Muzakky[2], R Kurniawan[1], E T Utami[1] and J Hayati[1]

[1] Study Program of Civil Engineering, Department of Infrastructure and Regional Technology, Institut Teknologi Sumatera, Jalan Terusan Ryacudu, Way Hui, Kecamatan Jati Agung, Lampung Selatan, Indonesia
[2] Study Program of Civil Engineering Alumni, Department of Infrastructure and Regional Technology, Institut Teknologi Sumatera, Jalan Terusan Ryacudu, Way Hui, Kecamatan Jati Agung, Lampung Selatan, Indonesia
Email: syahidus.syuhada@si.itera.ac.id

Abstract. Uncertain climate change is a natural phenomenon that is common nowadays. Soil is the result of weathering of rocks and climate change is one of the causes of this weathering. Residual soil is soil formed on top of the original rock layer. In addition to the parent rock, climate change also affects the residual soil, both in terms of grain size and shear strength. This climate change modeling was carried out by giving wet-dry treatment on residual tuff soil samples at the Institut Teknologi Sumatera (ITERA) which were compacted with a wet-dry cycle of 2 cycles, 4 cycles, 6 cycles, 8 cycles, and 10 cycles. From this treatment, it will be seen the change in gradation and the magnitude of the CBR value. The change in grain gradation is obtained which is calculated based on the concept of grain breakage, namely the Br value in each cycle of 15.89%; 21.51%; 25.35%; 32.01%; 36.28%. For the CBR values obtained for the initial conditions and each cycle are 50.74%, 40.34%; 29.763%; 25.16%; 22.825%; 20.580%. These results indicate that the longer the cycle occurs in the soil, the greater the change in grain gradation and the CBR value of the residual tuff soil. For this reason, the wet-dry process needs to be considered for the use of residual tuff soil as the subgrade of the road pavement.

1. Introduction

Soil is a very unique material because of its various types and properties, so that the treatment of it must be adjusted to the type and its origin. This research will examine one type of residual soil, namely tuff soil. Residual soil is soil resulting from weathering that is still in its original place or where the weathering process occurs. Meanwhile, transported soil has been transported or is no longer in its original place.

In Lampung Province there is Mount Rajabasa which is one of the active volcanoes. Specifically, Mount Rajabasa is located in the South Lampung Regency. Based on regional geological maps, Mount Rajabasa consists of andesite-basalt lava composition, breccia, and tuff. Residual soil from the weathering process in the area around volcanic deposits has characteristics and parameters with older andesite units. The unit is composed of andesitic lava which is spread from the west to the southeastern part of Mount Rajabasa as a product of tertiary volcanoes. The surface rock in the southern and northern parts of Bandar Lampung is the Lampung Formation which is composed of tuff, tuffaceous claystone and tuffaceous sandstone. In the western part is the Young Volcano Formation which is composed of lava (andesite-basalt), tuff and breccia [1]. The eruption of Mount Krakatau affects the characteristics of the soil in Lampung Province. The eruption produced sedimentary material scattered around the Gubug Gaam coastal area, South Lampung.

Extreme and erratic weather conditions are a common phenomenon today. This weather change is also a factor that causes weathering of rocks. On the scale of laboratory testing, these weather conditions can be modeled by giving wet-dry treatments to the test object. The wet-dry cycle conditions that occur in the field have attracted the attention of geotechnical experts to examine the microparticle structure of rock and soil which focuses on damage and internal mechanisms of rock and soil structures [2]. The wet-dry cycle is considered to be the most destructive factor in the environment that can cause damage to the foundation structure or subgrade of roads and road pavements [3].

This wet-dry process greatly affects the physical properties of the soil and its mechanical properties, so it is necessary to investigate how much impact this phenomenon has. The source of residual tuff soil is quite widely spread in Lampung Province so that it can be used as a basic material for road pavements. However, it is not yet known how big the effect of this wet-dry behavior is on the residual tuff soil. For this reason, it is necessary to investigate further for the effects of this phenomenon so that the utilization of tuff residual soil can be optimized.

In the laboratory, the wet-dry process is carried out by immersing and leaving the sample in an air-dry state for 1 day for soaking and 1 day for leaving, which is assumed to be 1 cycle. This wet-dry cycle was carried out with 2 cycles, 4 cycles, 6 cycles, 8 cycles, and 10 cycles to see how much effect each cycle had. Then in each cycle, sieve analysis and CBR will be tested to see changes in shear strength and index properties of residual tuff soil, which will later be used as considerations in the design of using tuff residual soil.

2. Experimental Section

2.1. Sample of residual tuff

The residual tuff soil used is from around Building F at the ITERA as shown in figure 1. Visually, the soil is white to grey in colour, with a slightly coarse texture. The soil was taken in disturbed condition (disturbed sample) and dried in the sun to obtain an air-dry state. This is done to facilitate the standard compaction test in determining the optimum water content in the soil. The soil used is that which passes sieve number 4. Sampling coordinates are -5.362211149878805, 105.31254441766205.

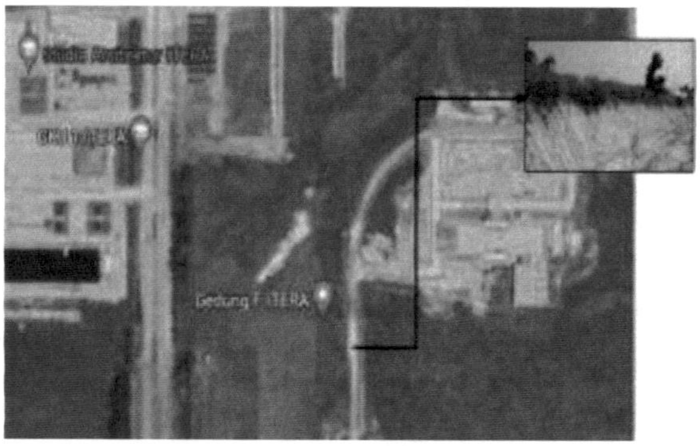

Figure 1. Sampling location.

2.2. Index properties test

This test is important to do, because it is a test that can describe the type of soil. The properties index tests carried out include the water content test in air dry conditions, specific gravity which will be used in the standard compaction test, the Atterberg limit in the form of testing the liquid limit and the plastic limit of the soil and sieve analysis testing to determine the grain size distribution in the initial conditions. For standard testing using ASTM according to each test [4-7].

2.3. *Standard proctor test*

Compaction testing is useful for determining how much optimum water content the soil can accept in order to produce maximum dry density. From this compaction test, the optimum moisture content (OMC) data will be used in mixing the manufacture of CBR specimens which will later be carried out with a wet-dry process on the soil. Compaction testing is also guided by ASTM [8].

2.4. *Wet-dry cycle schemes*

The wet-dry process is a model of climate change that occurs in the field. 1 cycle is calculated from 1 time without soaking process and 1 time immersion process. The soaking process is always the end of the cycle. The following table 1 is the schematic of the wet-dry process carried out in the laboratory.

Table 1. A slightly more complex table with a narrow caption.

Cycle	Specimen Time (Day)																				
	1	2	3	4	5	6	7	8	9	10	11	12	13	14	15	16	17	18	19	20	21
0	U																				
2	D	W	D	W	T[a]																
4	D	W	D	W	D[b]	W[c]	D	W	T												
6	D	W	D	W	D	W	D	W	D	W	D	W	T								
8	D	W	D	W	D	W	D	W	D	W	D	W	D	W	D	W	T				
10	D	W	D	W	D	W	D	W	D	W	D	W	D	W	D	W	D	W	D	W	T

Desc ;
T = Testing time
D = Dry time
W = Wet time

2.5. *California bearing ratio (CBR) test*

CBR testing is carried out at the initial conditions and after the wet - dry process is carried out according to the specified cycle. This CBR test was carried out to see how much change occurred in the shear strength of the soil due to changes in wet-dry conditions. The CBR value is commonly used for pavement planning with the relationship between the CBR value and the thickness of the pavement. The greater the CBR value, the thinner the pavement thickness and vice versa. The CBR testing carried out is also guided by ASTM [9] regarding the California Bearing Ratio test in the Laboratory. CBR testing tools in the laboratory can be seen in figure 2.

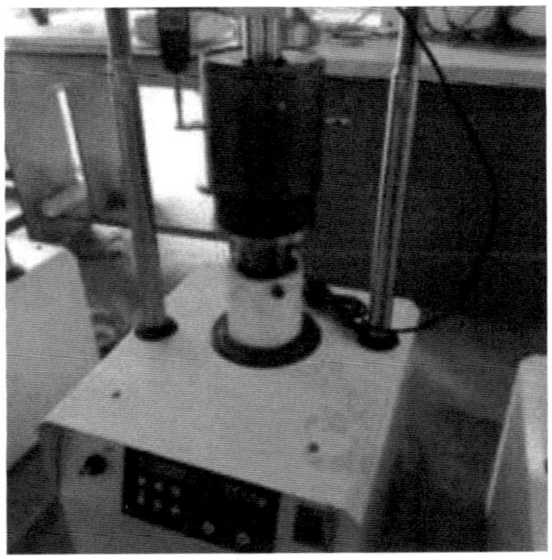

Figure 2. CBR testing machine.

2.6. Grain size distribution test and degree of change in grain size distribution curve of residual tuff

After the Laboratory CBR test is carried out, the soil sample from the CBR test will be used to see the grain size gradation so that the effects of wet dry can be known. The grain size analysis test used is a sieve analysis test. This test observes the gradation of particles that are larger than 0.075 mm or more for materials larger / retained on sieve no. 200. For this test, the ASTM guidelines regarding sieve analysis tests are used [7].

To quantify the changes that occur in the grain size distribution as a result of each applied cycle, the concept of a change in area is used which is formed from the initial grain size distribution curve with a new grain size distribution curve due to the wet-dry process that occurs. The following is an illustration of the change in the shape of the grain size distribution curve in figure 3.

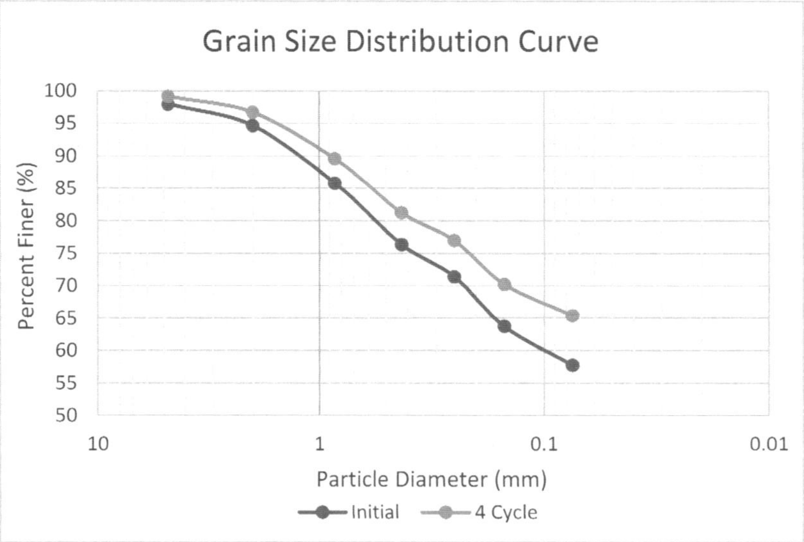

Figure 3. Change of grain size distribution curve.

Calculating the size of the area formed from the grain gradation curve (shaded area) can be used an approach to calculate the area in small areas that can be calculated using the following equation. To calculate area upper grain size curve:

$$A = A_1 + A_2 + \cdots \ldots + A_n \tag{1}$$

$$A_1 = \frac{(100 - f_1) + (100 - f_2)}{2} . \left[Log \left({}^{d_1}/_{d_2} \right) \right] \tag{2}$$

Where *An* is the piecewise area upper grain size distribution, *fn* is the percent finer and *dn* is the particle diameter. The degree of change in the grain size distribution curve uses a concept similar to particle breakage. [10] If particle breakage is the destruction of granules due to stress, in this case the granules become smaller due to the dissolution of the granules by water. The following equation calculates the degree of change in the grain size distribution curve.

$$B_r = \frac{B_t}{B_p} \tag{3}$$

Where *Br* is the degree of change in grain size distribution curve or the relative breakage, *Bt* is the area between initial grain size curve and grain size curve after cycle or total breakage and *Bp* is the area upper initial grain size curve or breakage potential.

3. Results and Discussions

The index properties test carried out on the residual soil of ITERA tuff is in the form of specific gravity (Gs), Atterberg Limit (LL, PL), and grain size distribution with the results which can be seen in table 2.

Table 2. Index properties results of ITERA residual tuff.

Gs	LL (%)	PL (%)	PI (%)	*Grain* Size (%)		
				Gravel	Sand	Silt
2,58	40,96	29,99	10,96	0,00	31,3	68,7

Based on index properties, ITERA tuff soil is classified as MH in the USCS soil classification. By using the results of index properties in the form of PL, it is possible to estimate the optimum water content for the standard compaction test. The objective of standard compaction is to obtain the optimum water content to be used in soil mixing for CBR test samples in the laboratory. The following is the standard compaction test scheme carried out on the remaining ITERA tuff soil which can be seen in table 3. The result of standard compaction test was carried out on the residual soil of ITERA tuff which can be seen in figure 4. The optimum moisture content was 20.55% and the maximum dry density was 1.32 g/cm3. The optimum water content will be used for mixing soil compaction in the CBR test.

Table 3. Standard proctor result of ITERA residual tuff.

Sample	Water content (%)	Dry unit weight (gr/cm^3)
1	16,85	1,27
2	**20,55**	**1,32**
3	26,04	1,29
4	29,07	1,24
5	32,72	1,18

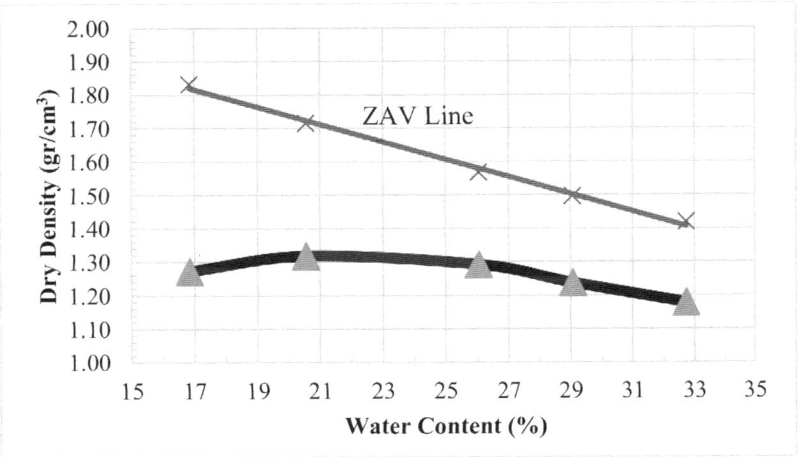

Figure 4. Standard proctor curve of residual tuff ITERA.

After obtaining the optimum water content value, compaction was carried out for the CBR sample. Then the unsoaked CBR test was carried out to determine the initial CBR value without a wet-dry cycle. The CBR value is 49.23% for the initial conditions. [11] also tested the wet-dry cycle on silty soils in Karaj, Iran. The result is that the CBR value will decrease with the number of wet-dry cycles given to the test object. In this test, the values for each 2 cycles, 4 cycles, 6 cycles, 8 cycles, and 10 cycles were 40.34%, 28.74%, 25.16%, 22.82%, 19.39%. Significant changes in the CBR value were obtained in cycle 4 with a reduction in the CBR value of 41.62% from the initial CBR value. The results obtained where the number of wet-dry cycles causes a decrease in the CBR value. This can also be affected by an increase in the water content in the soil which decrease the soil strength. The graph of the CBR test for each cycle can be seen in figure 5 and the results of the CBR test are in table 4.

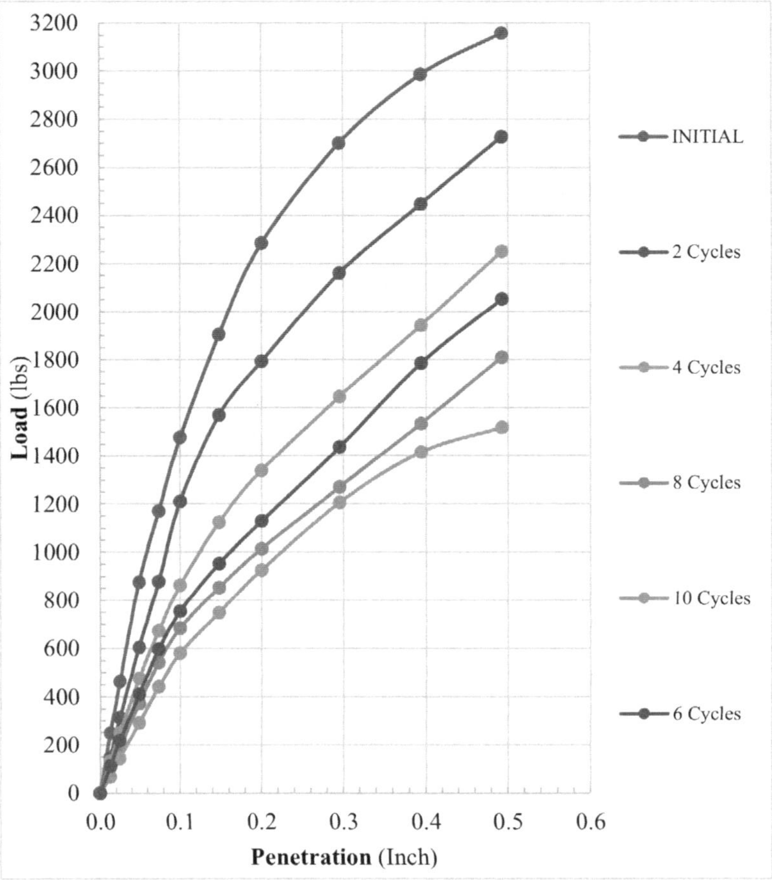

Figure 5. California bearing ratio (CBR) result of residual tuff ITERA.

For the relationship between the CBR value and the application of the wet-dry cycle, it can be seen in figure 10 which produces the equation CBR = -2.9334 x + 45,614 with the value of x being the sum wet-dry cycle.

The result of the wet-dry cycle also causes changes in the water content of the test object. Changes in water content with the CBR value are indicated by the equation CBR = -0.4971 w + 61.029 where w is the water content (%). The greater the value of the water content, the smaller the CBR value.

As a result of the wet-dry process, there is also a change in the grain size distribution curve in each cycle. The tendency that occurs is that the more cycles given to the test object, the degree of change in the curve increases. Figure 7 shows the changes that occur due to the wet-dry process on the CBR test object.

Table 4. CBR test result of ITERA residual tuff using wet – dry cycles method.

Wet – Dry Cycles	CBR Result (%)	Moisture Content (%)	CBR Difference from Initial Value	% Change of CBR from Initial Value
Initial	49.23	22.81	0	0.000
2 Siklus	40.34	52.85	8.89	18.058
4 Siklus	28.74	60.34	20.49	41.621
6 Siklus	25.16	73.21	24.07	48.893
8 Siklus	22.82	79.57	26.41	53.646
10 Siklus	19.39	87.97	29.84	60.613

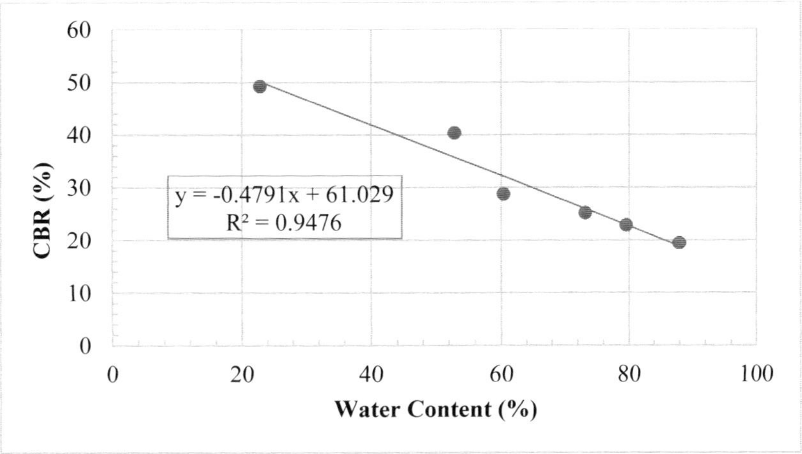

Figure 6. Effect of water content on California bearing ratio (CBR) result of residual tuff ITERA.

After obtaining the grain size distribution curve for each of the applied cycles, it will be determined how the degree of change in grain gradation or Br value occurs. In figure 8 below, there is an example of getting the calculation parameter for Br. In detail, the values of Bp, Bt, and Br can be seen in table 5. In table 5, it can be seen the relationship between the wet-dry cycle with the Br value.

Table 5. Relationship of wey – dry cycles with degree of change in grain size distribution curve.

Cycle	Parameter		
	Bp (Area)	Bt (Area)	Br (%)
2		5.682	15.89
4		7.692	21.52
6	35.74958	9.064	25.35
8		11.447	32.02
10		12.970	36.28

From the concept of particle breakage theory, the breakage of particles are mostly caused by working stresses more than the strength of the grains in receiving the load. However, in the case of the ITERA tuff residual soil, the wet-dry behavior also affects the change in the grain size distribution. [12] conducted a shear strength test on residual soil by applying a wet-dry cycle and one of them resulted in a decrease in shear strength and resulted in cracks in the tested samples. From this it can be concluded that this wet-dry behavior can lead to changes in the grain structure of the residual tuff soil.

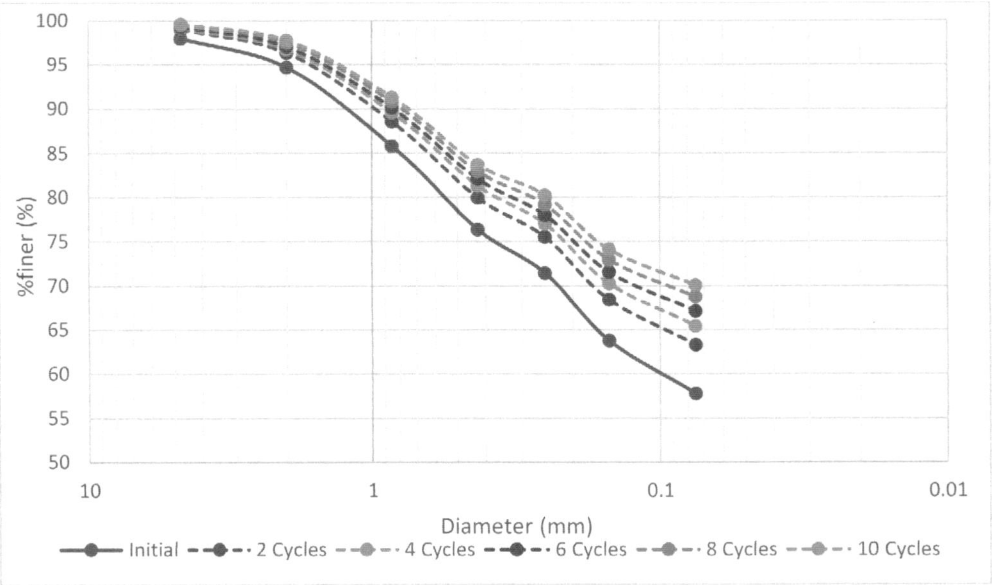

Figure 7. Grain size distribution curve on wet – dry cycles result of residual tuff ITERA.

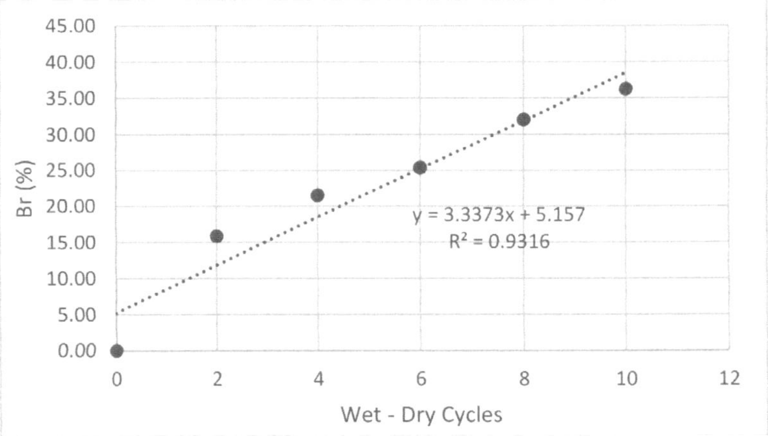

Figure 8. Influence wet – dry cycles result to Br of residual tuff ITERA.

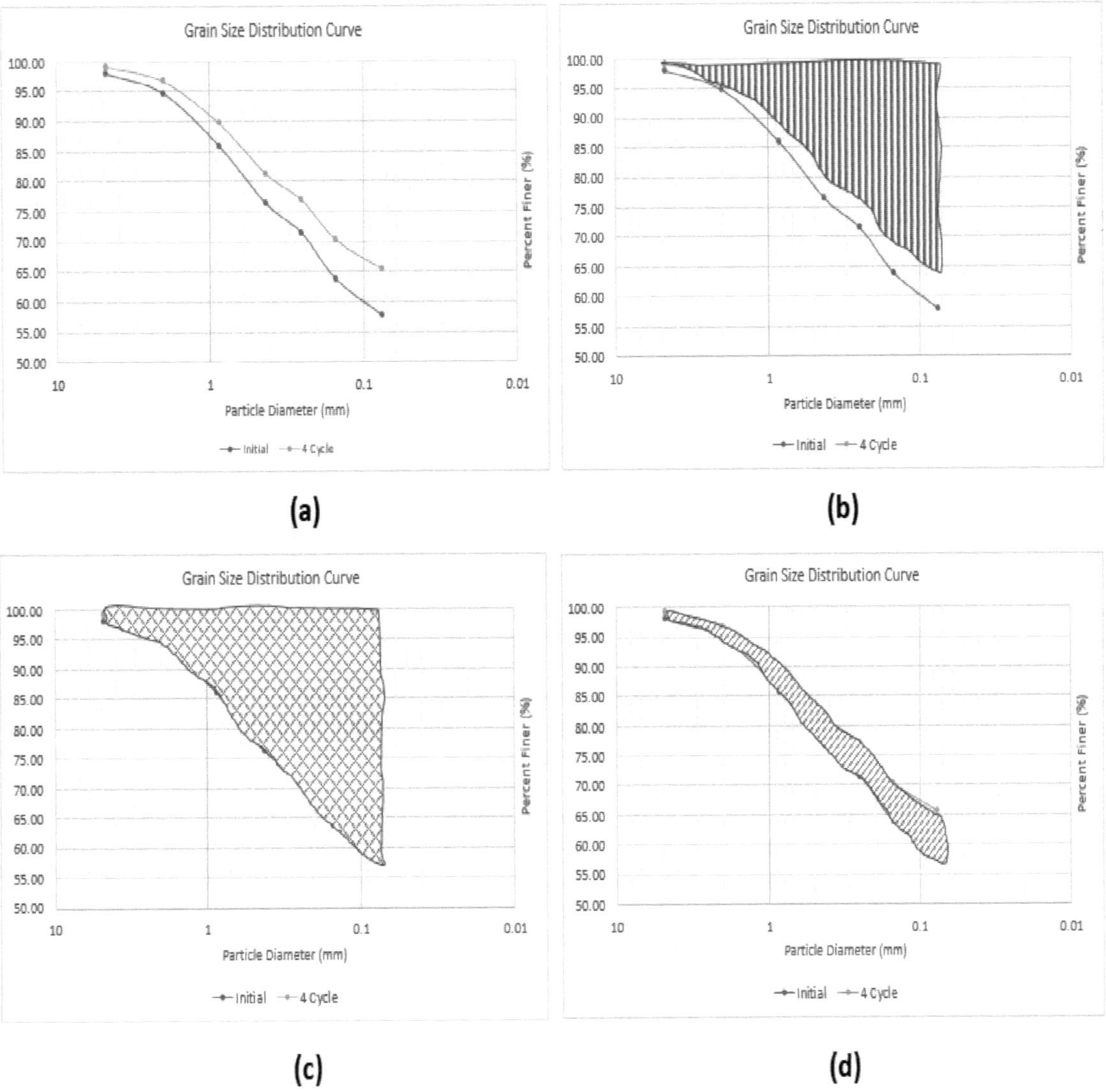

Figure 9. Calculate of area upper grain size distribution curve, a) initial grain size distribution curve and after wet – dry cycle; b) area upper grain size distribution curve after wet – dry cycle; c) area upper initial grain size distribution curve (B_p); d) area between initial grain size distribution curve and grain size distribution curve after cycle (B_t).

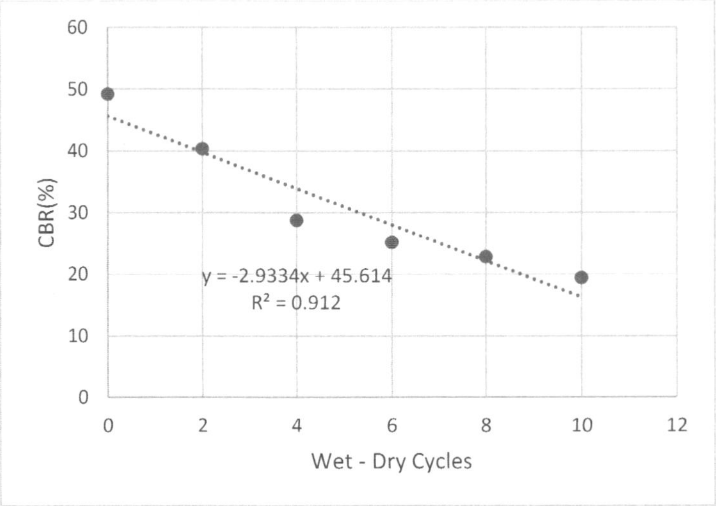

Figure 10. Influence wet – dry cycles result to CBR of residual tuff ITERA.

4. Conclusion

After conducting a series of tests on ITERA tuff soil, the following results were obtained:

1. The ITERA tuff soil is composed mainly by silt that has higher plasticity condition. Hence, the interaction of soil and water will greatly affect this soil behavior.
2. There is an increase in the value of the degree of change in the grain size distribution curve or the relative breakage (Br) value in the particle breakage theory at each additional cycle. It can be stated that the number of the wet-dry cycle is directly proportional to the weathering level.
3. There is a decrease in soil strength (CBR) from the initial conditions due to the wet-dry cycle. It was found that there was a tendency to decrease the CBR value along with the increase in the wet-dry cycle. It is also indirectly related to the Br value, the longer the cycle, the greater the value obtained. This indicates that the change in grain gradation results in changes in the shear strength of the soil. The larger the Br value, the more the grains change into smaller grains.
4. The shear strength produced in compaction of residual tuff soil is high enough so that it has the potential to be used as subgrade material or road embankment. However, it is necessary to make corrections to the value of the shear strength of the laboratory test results to be used as a design parameter, because the influence of wet - dry is very large on the CBR value of the residual tuff soil.

References

[1] S Mangga, T Amirudi, S Suwarti, Gafoer and Sidarto 1993 Peta Geologi Lembar Tanjung Karang (Bandung : Pusat Penelitian dan Pengembangan Geologi Indonesia)
[2] Rayahani M H T, et. al 2008 Physical Modeling of Desiccation Cracking In Plastic Soils *Engineering Geology 97* (Science Direct. Elsevier) p 25–31
[3] Hossain Md, Sayem, Ling-wei Kong and Song Yin 2016 Effect of Drying-Wetting Cycles on Saturated Shear Strength of Undisturbed Residual Soils *American Journal of Civil Engineering* Vol.4 no.4 (U.S) p 143 – 150
[4] American Society for Testing and Material (ASTM) International 2005 Standard Test Method for

IOP Conf. Series: Earth and Environmental Science 1173 (2023) 012032 doi:10.1088/1755-1315/1173/1/012032

Laboratory Determination of Water (Moisture) Content of Soil and Rock Mass (ASTM D 2216)
[5] American Society for Testing and Material (ASTM) International 2002 Standard Test Method for Specific Gravity of Soil by Water Pycnometer (ASTM D 854)
[6] American Society for Testing and Material (ASTM) International 2005 Standard Test Method for Liquid Limit, Plastic Limit, and Plasticity Index Soils (ASTM D 4318)
[7] American Society for Testing and Material (ASTM) International 2007 Standard Test Method for Particle-Size Analysis of Soils (ASTM D 422)
[8] American Society for Testing and Material (ASTM) International 1989 Standard Test Method for Laboratory Compaction Characteristics of Soil Using Standard Effort (ASTM D 698)
[9] American Society for Testing and Material (ASTM) International 2002 Standard Test Method for CBR (California Bearing Ratio) (ASTM D 1883)
[10] Hardin. Bobby O 1985 Crushing of Soil Particles *Journal of Geotechnical Engineering* (ASCE. U.S) Vol. III No. 10 p 1177 – 1192
[11] R Z, Moayed and B P Lahiji Y Daghigh 2013 Effect of wetting- drying cycles on CBR values of silty subgrade soil of Karaj railway *Proceeding of the 18th International Conference on Soil Mechanics and Geotechnical Engineering* (Paris. France) Volume 1. p 1321 – 1324
[12] Xu, Xu-tang, et. al 2021. Effect of Wet – Dry Cycles on Shear Strength of Residual Soil. *Soils and Foundations* (Japan)

IOP Conf. Series: Earth and Environmental Science 1173 (2023) 012033 doi:10.1088/1755-1315/1173/1/012033

Slope stability analysis at coal mining sites in Central Kalimantan

H Prabowo[1], E Barlian[1], N Syah[1] and A T Sastra[2]

[1] Environmental Science Doctoral Program, Universitas Negeri Padang, Indonesia
[2] Department of Mining Engineering, Universitas Negeri Padang, Indonesia

E-mail: heri.19782000@ft.unp.ac.id

Abstract. Slope stability is an important factor in the planning and operation of open pits and mines. Currently the company has been mining up to a height of 20 meters. In the implementation of mining activities, the potential for landslides on the side walls of the slopes of sector X occurs on one of the slopes, namely at an elevation of +70 masl. The company made two design choices, one of which was confirmed by researchers in overcoming the potential for landslides. In carrying out slope analysis, information about the general properties of rock mass consists of testing the physical properties of rocks and testing the mechanical properties of rocks. The results of the safety factor analysis of the A-A cross-section with 2.476 saturated MAT (Stable) and dry MAT 3.159 (Stable), while the B-B cross-section with saturated MAT 2.161 (Stable) and dry MAT MAT is 2.771 (Stable). Meanwhile, for the two segments, A-A' with saturated MAT of 2,459 (Stable) and dry MAT of 3,131 (Stable), while the B-B' section with saturated MAT of 1,653 (Stable) and dry MAT of 2,464 (Stable). The researcher recommends using option 1 based on the safety factor value which shows a higher value.

1. Introduction

One of the important problems in mining activities using the open pit mining method is the stability or stability of the slopes (1). The process of excavation and stockpiling is an important problem because it involves the safety of people, equipment, and buildings around the slope (2). Therefore, before mining activities, it is necessary to conduct a study on slope stability and slope design that considers the safety aspect (3). Referring to the rules in mining science, the basic principle that must be understood in determining the design of a mine opening is to determine the optimal slope angle of the mine opening, in the sense of determining a stable overall slope angle to the mining end (4).

A material in nature is in a stable state if the stress distribution on the material is in a state of equilibrium. This balance can be disturbed by various activities, such as excavation, blasting, or earthquakes, which causes the driving force to be greater than the resisting force (5). Coal mining activities in open spaces in the form of excavation and stockpiling will always face slope stability problems. Slope stability analysis is carried out to assess the level of stability of a slope (6). The term Slope Stability can be defined as the resistance of the block on an inclined surface (measured from the horizontal line) against collapsing and sliding (7).

Slope stability, both natural slopes and artificial slopes (man-made) and embankment slopes, is influenced by several factors that can be stated simply as the resisting and driving forces that are responsible for slope stability (8). When the retaining force (against landslides) is greater than the driving force, the slope will be in a stable (safe) condition (9).

In dealing with landslides, so as not to interfere with mining activities, the company makes two design options that will overcome the effects of avalanches (10). Of the two options, the author will analyze which option is more optimal to be recommended to the company. The basis for selecting recommendations is seen from the value of the safety factor for each option (11).

2. Literature review

2.1. *Types of avalanches*

Slope stability, both natural slopes and artificial slopes (man-made) and embankment slopes, is influenced by several factors that can be stated simply as the resisting and driving forces that are responsible for slope stability. When the retaining force (against landslides) is greater than the driving force, the slope will be in a stable (safe) condition. However, if the resisting force is less than the driving force, the slope is unstable and landslides will occur. (11)

A torrential slide is a characteristic interaction that occurs to obtain another oblique solidness condition (new equilibrium), where the resisting force is more prominent than the main thrust. Torrent slides can be ordered considering the accompanying torrent slide types: (12), (13). Types of landslides that will occur include: Field avalanche, Wedge avalanche, Rolling avalanche, Bow avalanche field avalanche.

2.2. *Slope safety factor*

The slope is said to be stable if the resisting force is greater than the pushing force. To express the level of stability of a slope, known as the Safety Factor. Safety factor required to determine the stability of a slope to prevent the danger of avalanches in times to come. Theoretically, the slope is said to be stable at FK conditions > 1 (stable slope) (14).

Table 1. The level of theoretical FK scores.

FK	Keterangan
> 1	Stable
1	Critical
< 1	Unstable

Table 2. Level of practice FK scores

FK	Keterangan
> 1,3	Stable
1,07 < FK < 1,3	Critical
< 1,07	Unstable

3. Methodology

Mine slope stability analysis was carried out to evaluate and recommend the two designs that have been made by the company (15). The design is made because there has been a landslide on one side wall slope. The type of avalanche that occurs is a local arc avalanche (16). Landslides are local, meaning that they only occur on a few slopes (single slope), not on the overall slope (17).

Referring to the problems that have occurred before, the landslide occurred not because of the unstable value of the safety factor during the design, but because there were several anomalies on the slope (18). Some of the causes of landslides, namely, the actual field does not match the design, the material properties are weathered, blasting vibrations are close to the slope and there is a swamp behind the slope which causes the slope to become more saturated (19).

The modeling is done by entering material property data with the help of slide v.6.0 software to calculate the value of the single slope, intermediate slope, and overall slope safety factor values of the A-A' and B-B' sections using the Bishop method (20).

3.1. Primary data

Primary data is data obtained directly from the field. Analysis of the data obtained from the field is as follows:

a. Outcrop and slope geometry measurement data
b. Slope data
c. Actual groundwater level
d. Field documentation.

3.2. Secondary data

Secondary data are data obtained from PT. Asmin Bara Bronang to support research data, such as:

a. Rainfall data.
b. Area achievement map
c. Map of the research area
d. Regional geological map
e. Topographic maps
f. Pit situation map
g. Summary bor
h. Seismic Load
i. Parameter data of rock layers in the mining area (cohesion, density, and internal shear angle).
j. Mining plan section in area X.

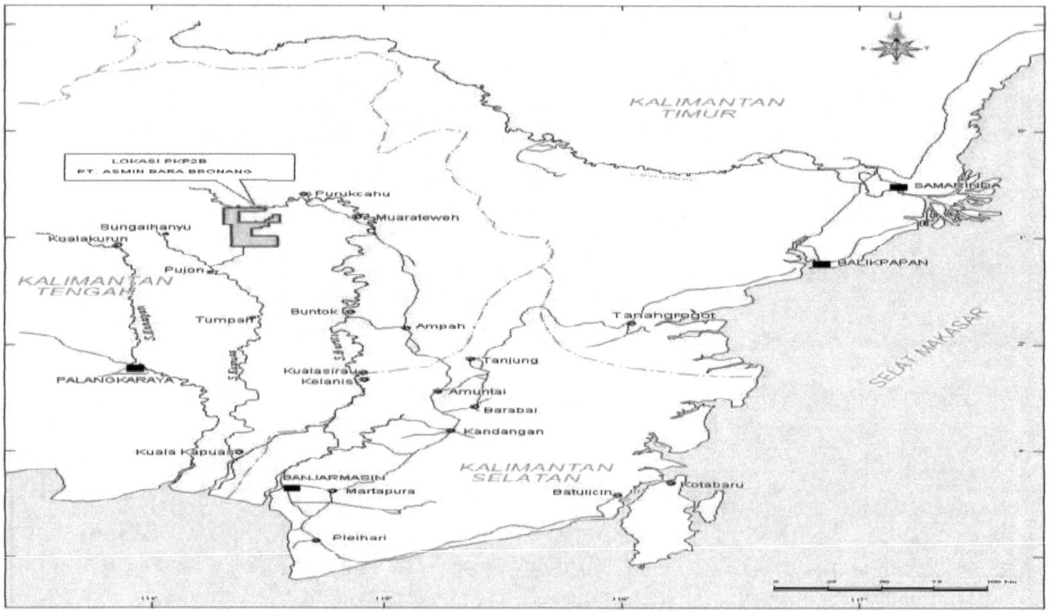

Figure 1. Research location map.

3.3. Data analysis method

For data processing in this study, the calculation of the slope stability value can be done using the boundary equilibrium method with Bishop with the help of Slide V.6.0 software (21). This program can be used to calculate the value of the slope safety factor and can simulate the desired slope design. Data processing and data analysis carried out include:

- Geotechnical Analysis
  This stage is intended to determine the physical properties and mechanical properties of the material through the material properties owned by the company (22).
- Modeling with Slide V.6.0. software
  The modeling is carried out by entering material property data with the help of Slide v.6.0 software to calculate the value of the safety factor for single slope, intermediate slope, and overall slope of the A-A' and B-B' sections using the Bishop method (23).
- To perform analysis using Slide V.6.0 Software, the required parameters are in the form of rock mechanical properties data which is divided into three criteria (24) as follows:
  a. Mohr-Coulomb strength, data required are cohesion and internal shear angle
  b. Hoek-Brown Strength, the data needed are UCS and Hoek-Brown constant (s and m)
  c. Generalized Hoek-Brown Strength, the data needed are UCS values, constants s, a and m, and GSI values.

4. Results and discussion

4.1. Results

4.1.1. Geological cross. The geological cross-section is an incision from rock layers that is used as a reference for the overall slope stability analysis, namely:

- Cross section A-A' option 1
- Cross section B-B' option 1
- Cross section A-A' option 2
- Cross section B-B' option 2

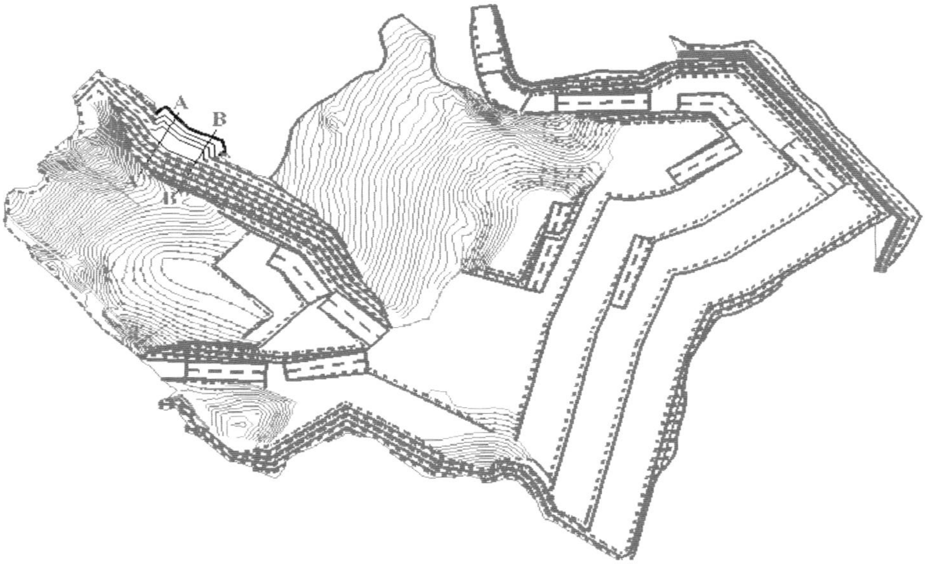

Figure 2. Design option 1 with A-A' and B-B' cross-sections.

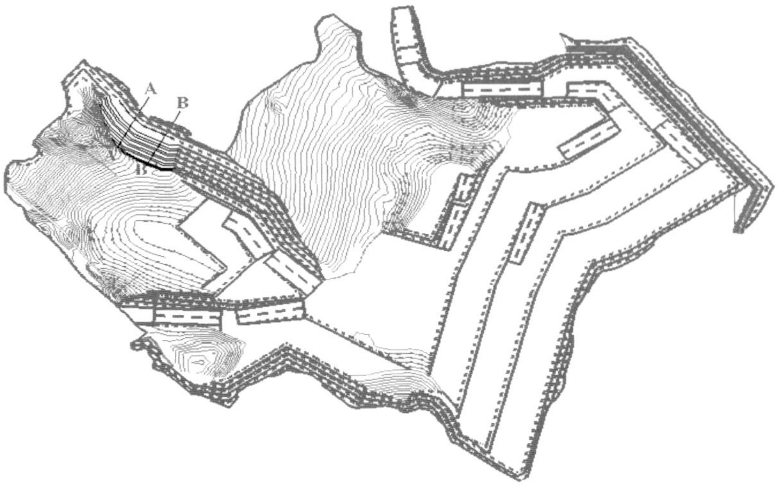

Figure 3. Design option 2 with sections A-A' and B-B'.

4.1.2. Pit X slope stability analysis. Investment field conditions actual conditions in pit X coal mining are carried out from the west to the east, and the current mining position is at an elevation of +30 masl with a total depth of 70 meters. In-pit X, investigations of several types of landslides were found namely arc slides and transverse faults along the side wall slopes.

Geotechnical Section Modeling Geotechnical cross section is made from topographic data and coal seam height obtained from exploration data. The analyzed sections are sections A-A' and BB'. Furthermore, the cross-section will be inputted into the Slide V 6.0 software. The input data in the Slide V 6.0 program is the data on the Mechanical Properties of Rocks. In this program, information is needed on the overall idea of rock mass consisting of specific gravity (unit weight), immersed weight (wet weight), pore tension, and vibration or tremor coefficient (seismic load coefficient). In addition, different information is also required, depending on what power standard is used.

4.1.3. Side wall pit X slope stability analysis. January The analysis was carried out to determine the actual slope conditions in January 2022. At the time of the research on the slope conditions, an arc-type landslide had occurred. On the side wall slope, the landslide effect only occurs on one slope, namely at an elevation of +70 masl. Figures 18 to 20 are the area for conducting research where landslides have occurred at an elevation of +70 on the side wall slope.

Figure 4. Slope side wall.

4.1.4. Slope Stability Analysis of Two Design Options in Pit X

The analysis carried out is to determine the level of the safety factor of the two design options that have been made. From this design, two cross sections were made to show the shape and direction of the coal seams to facilitate the actual description of the slopes in the field and can perform simulations with the help of slide V.6.0 software.

a. Option One

1) Section A-A' Saturated Groundwater Condition

In saturated groundwater conditions, the value of the Safety Factor (FK) cross-section A-A' overall slope (Overall) is 2,476 (stable), with an overall slope angle of 26 degrees and a height of 68.3 meters.

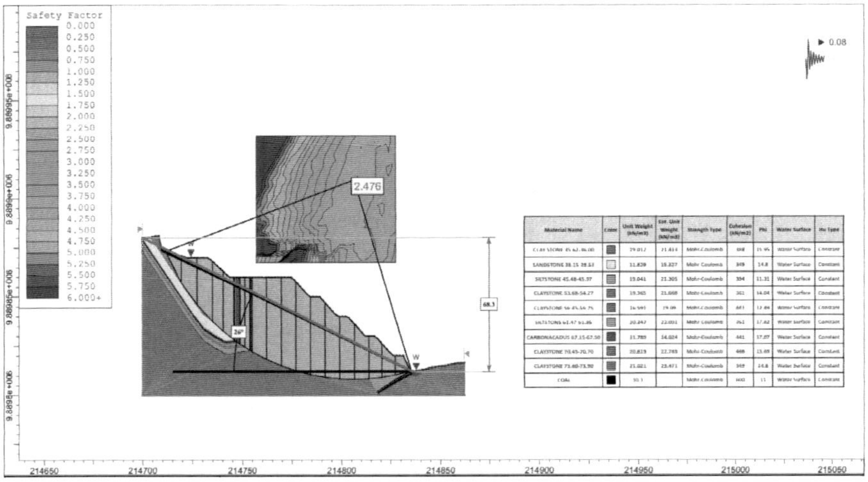

Figure 5. Option 1 segment A-A' saturated MAT.

2) Segment A-A' Dry Ground Water Level

In dry groundwater conditions, the value of the Safety Factor (FK) cross-section A-A' overall slope (Overall) is 3,159 (stable), with an overall slope angle of 26 degrees and a height of 68.3 meters.

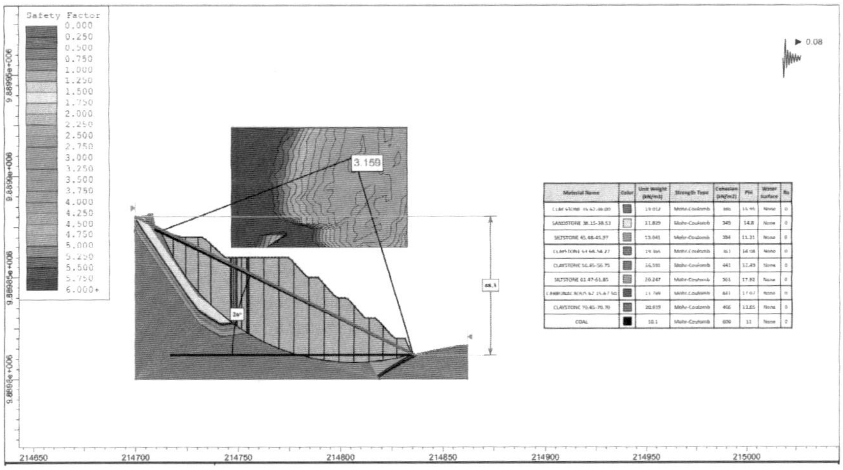

Figure 6. Option 1 segment A-A' dry MAT.

3) Segment B-B' Saturated Groundwater Condition

In a wet groundwater level, the value of the Safety Factor (FK) across segments B-B' overall slope is 2.161 (stable), with an overall slope angle of 28 degrees and a height of 78.8 meters.

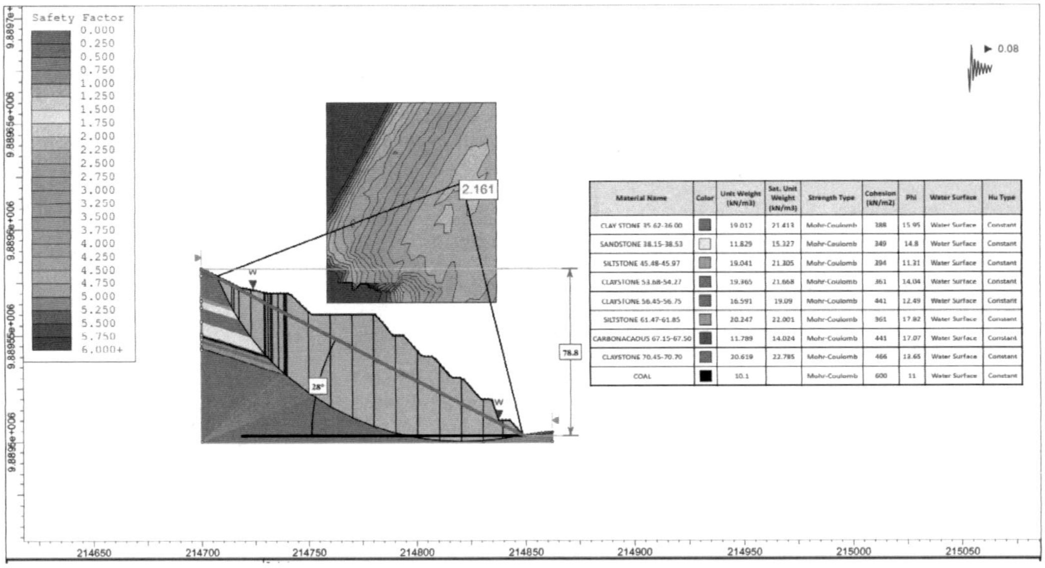

Figure 7. Option 1 segment B-B' saturated MAT.

4) Segment B-B' Dry Ground Water Level

In dry groundwater conditions, the value of the Safety Factor (FK) cross-section B-B' overall slope (Overall) is 2.771 (stable), with an overall slope angle of 28 degrees and a height of 78.8 meters.

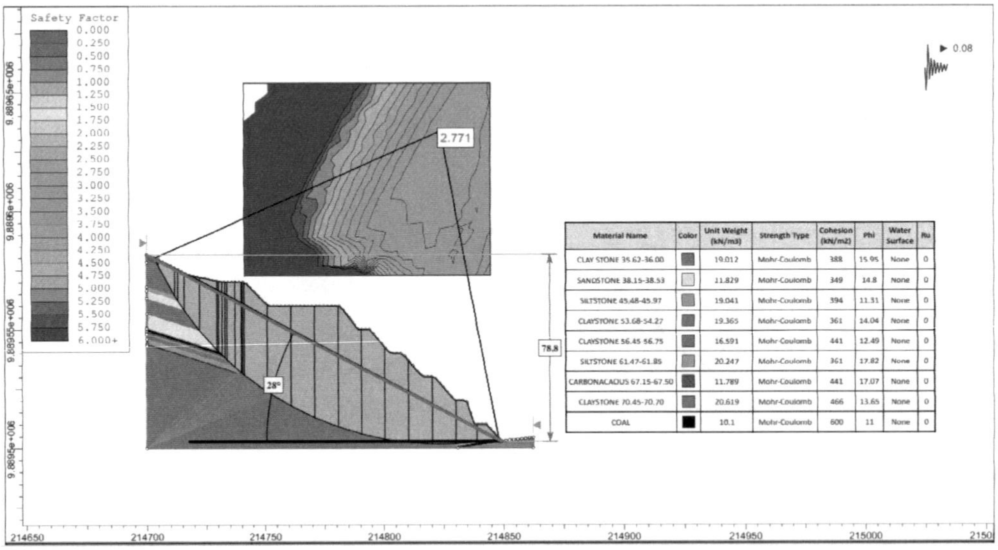

Figure 8. Option 1 segment B-B' dry MAT.

5) Segment A-A' Saturated Groundwater Condition

In a wet groundwater level, the value of the Safety Factor (FK) across segments A-A' general slope (Overall) is 2,459 (stable), with an overall slope angle of 26 degrees and a height of 68.3 meters.

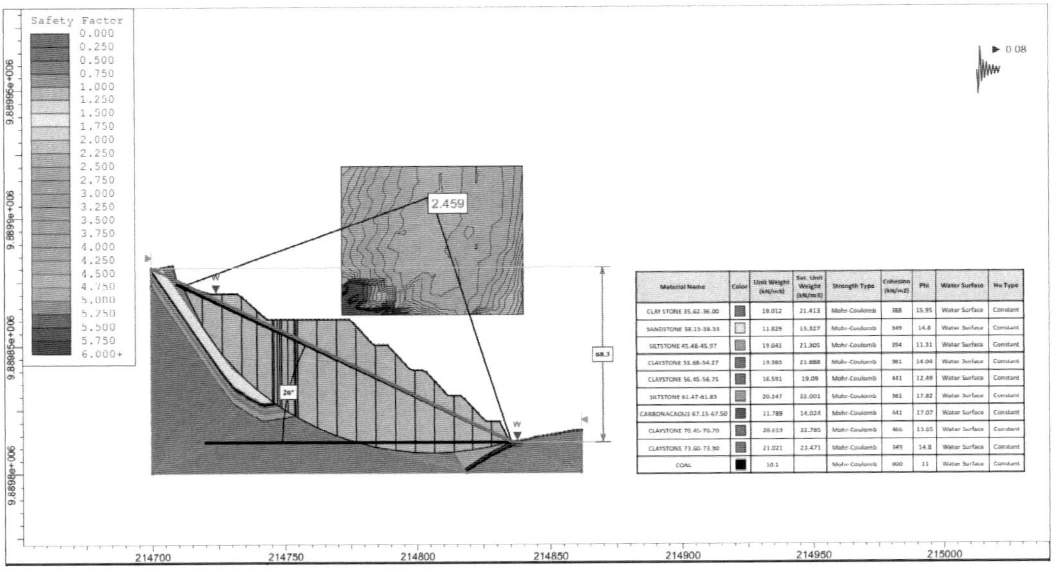

Figure 9. Option 2 segment A-A' MAT saturated.

6) Segment A-A' Dry Ground Water Level

In a wet groundwater level, the value of the Safety Factor (FK) across segments A-A' overall slope (Overall) is 3.131 (stable), with an overall slope angle of 26 degrees and a height of 68.3 meters.

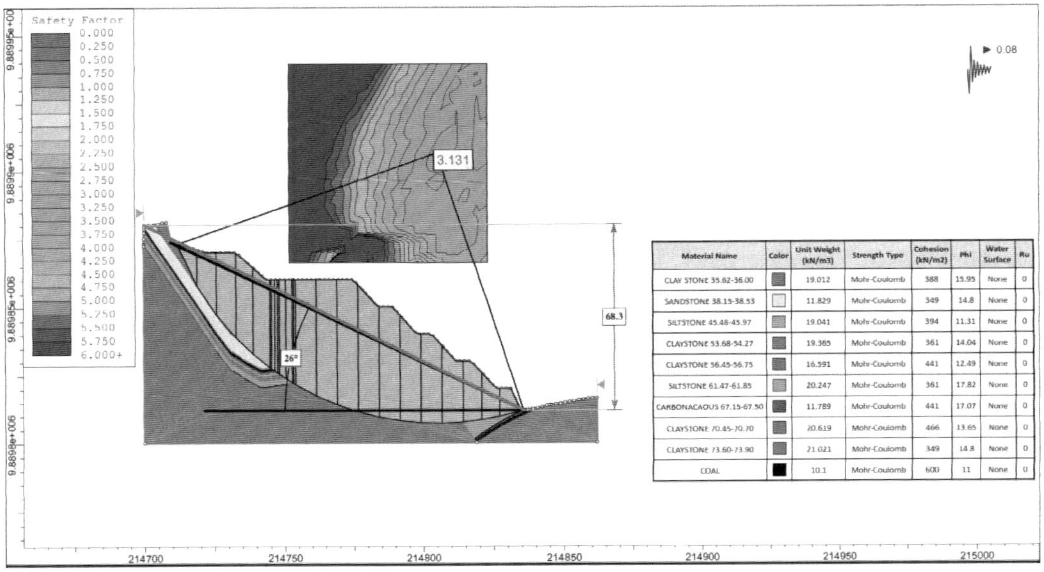

Figure 10. Option 2 section A-A' dry MAT.

7) B-B' Cross-section of Saturated Groundwater Condition

In a wet ground water table, the value of the Safety Factor (FK) across segments B-B' The overall slope (Overall) is 1.653 (stable), with an overall slope angle of 28 degrees and a height of 78.8 meters.

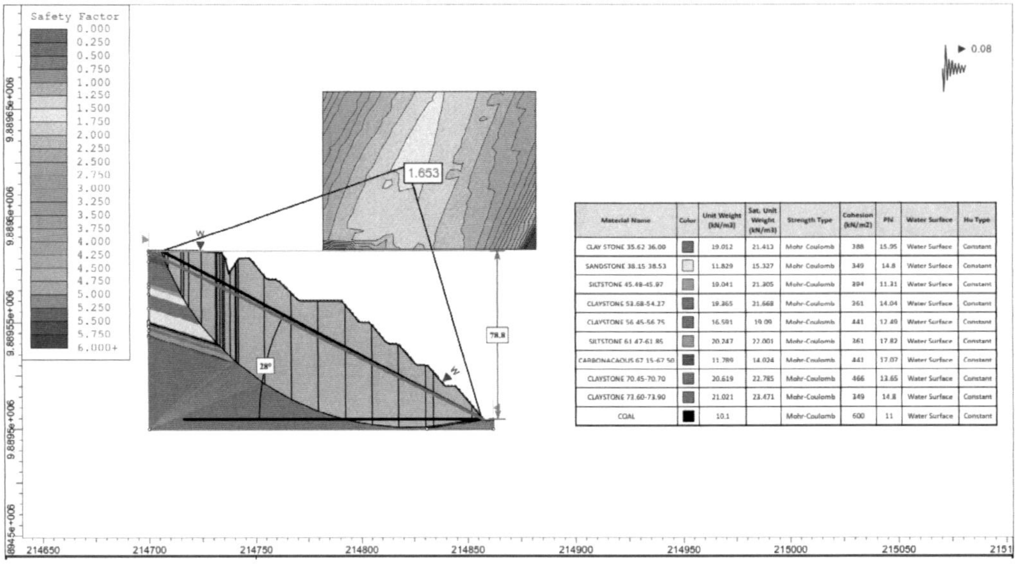

Material Name	Color	Unit Weight (kN/m3)	Sat. Unit Weight (kN/m3)	Strength Type	Cohesion (kN/m2)	Phi	Water Surface	Hu Type
CLAY STONE 35.62 36.00		19.012	21.413	Mohr Coulomb	388	15.95	Water Surface	Constant
SANDSTONE 38.15 38.53		11.829	15.327	Mohr Coulomb	349	14.8	Water Surface	Constant
SILTSTONE 45.48-45.97		19.041	21.305	Mohr-Coulomb	394	11.31	Water Surface	Constant
CLAYSTONE 53.68-54.27		19.365	21.668	Mohr-Coulomb	361	14.04	Water Surface	Constant
CLAYSTONE 56.45-56.75		16.591	19.09	Mohr-Coulomb	441	12.49	Water Surface	Constant
SILTSTONE 61.47-61.85		20.247	22.001	Mohr-Coulomb	361	17.82	Water Surface	Constant
CARBONACAOUS 67.15-67.50		11.789	14.024	Mohr-Coulomb	441	17.07	Water Surface	Constant
CLAYSTONE 70.45-70.70		20.619	22.785	Mohr-Coulomb	466	13.65	Water Surface	Constant
CLAYSTONE 73.60-73.90		21.021	23.471	Mohr-Coulomb	349	14.8	Water Surface	Constant
COAL		10.1		Mohr-Coulomb	600	11	Water Surface	Constant

Figure 11. Option 2 Segment B-B' saturated MAT

8) Segment B-B' Dry Ground Water Level

In a wet groundwater level, the value of the Safety Factor (FK) across segments B-B' overall slope (Overall) is 2.464 (stable), with an overall slope angle of 27 degrees and a height of 78.8 meters.

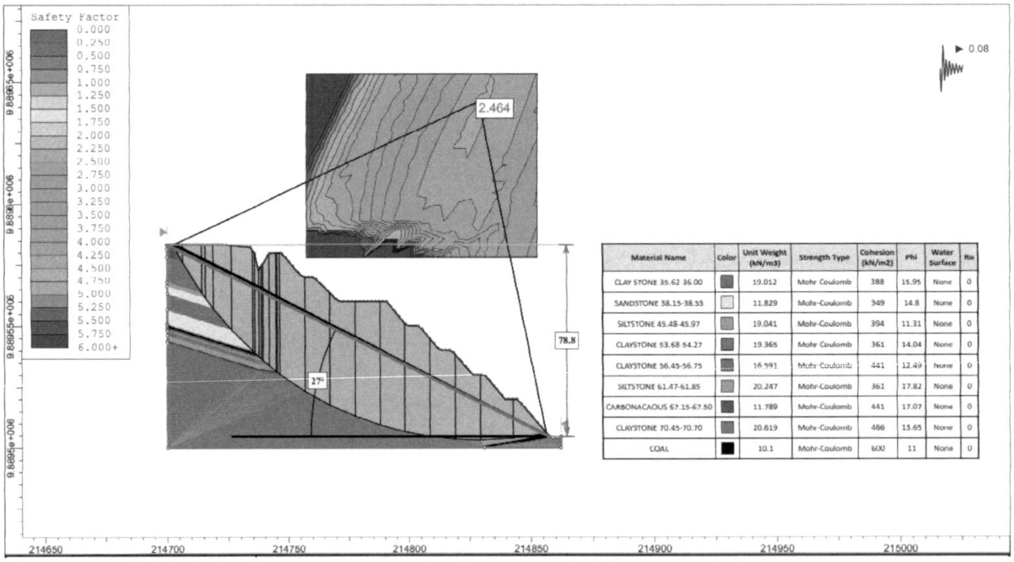

Material Name	Color	Unit Weight (kN/m3)	Strength Type	Cohesion (kN/m2)	Phi	Water Surface	Ru
CLAY STONE 35.62 36.00		19.012	Mohr Coulomb	388	15.95	None	0
SANDSTONE 38.15-38.53		11.829	Mohr-Coulomb	349	14.8	None	0
SILTSTONE 45.48-45.97		19.041	Mohr-Coulomb	394	11.31	None	0
CLAYSTONE 53.68 54.27		19.365	Mohr Coulomb	361	14.04	None	0
CLAYSTONE 56.45-56.75		16.591	Mohr-Coulomb	441	12.49	None	0
SILTSTONE 61.47-61.85		20.247	Mohr-Coulomb	361	17.82	None	0
CARBONACAOUS 67.15-67.50		11.789	Mohr-Coulomb	441	17.07	None	0
CLAYSTONE 70.45-70.70		20.619	Mohr-Coulomb	466	13.65	None	0
COAL		10.1	Mohr-Coulomb	600	11	None	0

Figure 12. Option 2 section of dry B-B' MAT.

Table 3. Groundwater Level along with Safety Factor (FK), Overall Angle, and Slope Height

Opsi	Section	Condition MAT	FK	Overall Angle	Height (m)
1	A – A'	Saturated water	2,476	26°	68,3
		Dry	3,159	26°	68,3
	B – B'	Saturated water	2,161	28°	78,8
		Dry	2,771	28°	78,8
2	A – A'	Saturated water	2,459	26°	68,3
		Dry	3,131	26°	68,3
	B – B'	Saturated water	1,653	28°	78,8
		Dry	2,464	27°	78,8

4.2. Discussion

From the results of the analysis of the two options, the value of the safety factor is categorized as stable, with the overall slope angle and slope height between option 1 and option 2, the cross-sections A-A' and B-B' are the same. From these results, we can judge if using option 1 is more profitable in terms of the value of the safety factor.

Not only that, if using option 1, the slope that has been landslide can be peeled back which will then increase the strength of the slope. This is in line with the weathering of rocks caused by climate change that has not occurred. Judging from the theory of factors that affect slope stability, namely weathering. If the weathering is higher, the strength of the rock will decrease (25).

If using option 2, we must first calculate the ability of the widened bench to withstand the event of an even larger avalanche. In option 2, we only do prevention so that landslides do not interfere with mining activities below, not dealing with existing landslides [22]. In addition, if using option 2 there is coal that cannot be mined due to being under the widened bench and that can be detrimental to the company.

Judging from the results of the analysis of option 2 cross-section B-B' with a saturated MAT the value of the safety factor shows 1.653. Meanwhile, in the initial design with a safety factor value of 1.941, landslides occurred. If option 2 is used, the risk of landslides is even higher [15].

5. Conclusion

The actual slope conditions at the time of the research were arc landslides on the side wall slopes, precisely at an elevation of +70 masl. After analyzing the condition of the landslide, it shows a safety factor value of 4.86 which in theory has reached new stability. The results of the safety factor analysis of the A-A' cross-section with saturated MAT 2,476 (Stable) and dry MAT 3,159 (Stable), while in section B-B' with saturated MAT 2,161 (Stable) and dry MAT 2,771 (Stable). Meanwhile, for the option of two sections A-A' with a saturated MAT of 2,459 (Stable) and a dry MAT of 3,131 (Stable), while a section B-B' with a saturated MAT of 1,653 (Stable) and a dry MAT of 2,464 (Stable). Of the two options with saturated and dry MAT showing the safety factor, all of them are in stable condition.

Based on the value of the safety factor, option 1 indicates a higher value and can prevent landslides. Because with the cutback, the strength of the slope can increase along with the opening of new material. Meanwhile, if using option 2, the widened slope must be ensured in advance of the ability to hold the material in the event of an even larger landslide. In addition, if option 2 is used, there is a loss of coal that can harm the company. Not only that, in option 2 cross-section B-B' with saturated MAT the slope safety factor value is 1.653 in the initial design with a safety factor value of 1.941 the slope landslide occurs. So to overcome landslides at an elevation of +70 masl on the slopes of the Pit X side wall, the author recommends using option 1.

References

[1] Zhu H, Azarafza M, Akgün H. Journal of Rock Mechanics and Geotechnical Engineering Deep learning-based key-block classification framework for discontinuous rock slopes. J Rock Mech Geotech Eng [Internet]. 2022;14(4):1131–9.

[2] Steger S, Mair V, Kofler C, Pittore M, Zebisch M, Schneiderbauer S. Correlation does not imply geomorphic causation in data-driven landslide susceptibility modelling – Benefits of exploring landslide data collection effects. Sci Total Environ. 2021;776.

[3] Moussas VC, Diamantis K. Predicting uniaxial compressive strength of serpentinites through physical, dynamic and mechanical properties using neural networks. J Rock Mech Geotech Eng [Internet]. 2021;13(1):167–75.

[4] Zhang W, Ji J, Gao Y, Li X, Zhang C. Spatial variability effect of internal friction angle on the post-failure behavior of landslides using a random and non-Newtonian fluid based SPH method. Geosci Front [Internet]. 2020;11(4):1107–21. Available from: https://doi.org/10.1016/j.gsf.2020.02.003

[5] Rechberger C, Fey C, Zangerl C. Structural characterisation, internal deformation, and kinematics of an active deep-seated rock slide in a valley glacier retreat area. Eng Geol [Internet]. 2021;286(May 2020):106048.

[6] Stanley Rungwa, Arpa G, Sakulas H, Harakuwe A, Timi D. Phytoremediation – An Eco-friendly and Sustainable Method of Heavy Metal Removal from Closed Mine Environments in Papua New Guinea. Procedia Earth Planet Sci [Internet]. 2013;6:269–77.

[7] Ali SA, Parvin F, Vojteková J, Costache R, Linh NTT, Pham QB, et al. GIS-based landslide susceptibility modeling: A comparison between fuzzy multi-criteria and machine learning algorithms. Geosci Front. 2021;12(2):857–76.

[8] Kardani N, Zhou A, Nazem M, Shen SL. Improved prediction of slope stability using a hybrid stacking ensemble method based on finite element analysis and field data. J Rock Mech Geotech Eng [Internet]. 2021;13(1):188–201.

[9] Wistuba M, Gorczyca E, Malik I. Inferring precipitation thresholds of landslide activity from long-term dendrochronological and precipitation data: Case study on the unstable slope at Karpenciny, Poland. Eng Geol. 2021;294.

[10] Malan D, Napier J. A limit equilibrium fracture zone model to investigate seismicity in coal mines. Int J Min Sci Technol [Internet]. 2018;28(5):745–53.

[11] Pradhan SP, Siddique T. Stability assessment of landslide-prone road cut rock slopes in Himalayan terrain: A finite element method based approach. J Rock Mech Geotech Eng [Internet]. 2020;12(1):59–73.

[12] Steger S, Schmaltz E, Glade T. The (f)utility to account for pre-failure topography in data-driven landslide susceptibility modelling. Geomorphology [Internet]. 2020;354:107041.

[13] González Tejada I, Monteiro-Alves R, Morán R, Toledo M. Cellular automata modeling of rockfill dam failure caused by overtopping or any other extreme throughflow. Eng Struct [Internet]. 2021;245:112933.

[14] Beni T, Casagli N, Gigli G, Lombardi L, Carl T. International Journal of Rock Mechanics and Mining Sciences A method for full three-dimensional kinematic analysis of steep rock walls based on high-resolution point cloud data. 2022;157(May).

[15] Rafiei Renani H, Martin CD. Factor of safety of strain-softening slopes. J Rock Mech Geotech Eng. 2020;12(3):473–83.

[16] Tatnell L, Dyson AP, Tolooiyan A. Coupled Eulerian-Lagrangian simulation of a modified direct shear apparatus for the measurement of residual shear strengths. J Rock Mech Geotech Eng [Internet]. 2021;13(5):1113–23.

[17] Achour Y, Pourghasemi HR. How do machine learning techniques help in increasing accuracy of landslide susceptibility maps? Geosci Front [Internet]. 2020;11(3):871–83.

[18] Zhang J, Fu M, Hassani FP, Zeng H, Geng Y, Bai Z. Land use-based landscape planning and

restoration in mine closure areas. Environ Manage. 2011;47(5):739–50.

[19] Wang Z, Cheng Y, Qi Y, Wang R, Wang L, Jiang J. Experimental study of pore structure and fractal characteristics of pulverized intact coal and tectonic coal by low temperature nitrogen adsorption. Powder Technol [Internet]. 2019;350:15–25.

[20] Li BQ, Casanova M, Einstein HH. Journal of Rock Mechanics and Geotechnical Engineering Laboratory study of fracture initiation and propagation in Barre granite under fl uid pressure at stress state close to failure. J Rock Mech Geotech Eng [Internet]. 2022;(xxxx):4–17.

[21] Ng CWW, Crous PA, Zhang M, Shakeel M. Static liquefaction mechanisms in loose sand fill slopes. Comput Geotech [Internet]. 2022;141:104525.

[22] Zhao T, Crosta GB, Liu Y. International Journal of Rock Mechanics and Mining Sciences Analysis of slope fracturing under transient earthquake loading by random discrete element method. Int J Rock Mech Min Sci [Internet]. 2022;157(December 2021):105171.

[23] Balogun AL, Rezaie F, Pham QB, Gigović L, Drobnjak S, Aina YA, et al. Spatial prediction of landslide susceptibility in western Serbia using hybrid support vector regression (SVR) with GWO, BAT and COA algorithms. Geosci Front. 2021;12(3).

[24] Butcher A, Stork AL, Verdon JP, Kendall JM, Plenkers K, Booth F, et al. Evaluating rock mass disturbance within open-pit excavations using seismic methods: A case study from the Hinkley Point C nuclear power station. J Rock Mech Geotech Eng [Internet]. 2021;13(3):500–12.

[25] Morikawa DS, Asai M. A phase-change approach to landslide simulations: Coupling finite strain elastoplastic TLSPH with non-Newtonian IISPH. Comput Geotech. 2022;148(January):104815.

IOP Conf. Series: Earth and Environmental Science 1173 (2023) 012034 doi:10.1088/1755-1315/1173/1/012034

Liquefaction Mitigation Analysis Using Vertical Drain

H Novasari [1], A Hakam [1] and Andriani [1]
[1] Civil Engineering Dep., Andalas University, Padang, West Sumatera, Indonesia

E-mail: hafizahnovasari@gmail.com

Abstract. Liquefaction in semi-permeable soils such as fine sands occurs when the required time to transfer the pore water pressure into the form of effective stress is longer than the increased time of pore water pressure. Therefore, liquefaction mitigation needs to be done to speed up the air pressure flow time during an earthquake such as using vertical drains in this study. The data used here are adopted from previous studies that have been taken from a field test in Air Tawar, Padang. The results showed there is liquefaction potential based on soil data processing. The equivalent circle diameter value for the vertical drain is 1.2 m or a radial equivalent value of 0.6 m. Based on the analysis, the 60% pore water pressure dissipation within 20 seconds was obtained at a distance of 0.291 m in the radial direction. From the data above, it can be obtained that the vertical drain is capable of dissipating 60% of water pressure within 20 seconds to avoid liquefaction within the soil mass. Then it can be concluded that the application of vertical drain at a certain distance and depth can mitigate liquefaction in soil layers.

1. Introduction

Indonesia is located astronomically at 6°N (North Latitude) - 11°S (South Latitude) and 95°E (East Longitude) - 141°E (East Longitude). Based on this astronomical location, Indonesia is on 3 main plates of the world, namely the Eurasian Plate, the Indo-Australian Plate, and the Pacific Plate or also known as the Ring of Fire. This condition causes Indonesia to have the potential for earthquake disasters in various regions of the archipelago.

The impact of this earthquake can cause other natural disasters such as tsunamis, and landslides, and what has recently caused liquefaction as happened in the 2018 Palu Earthquake. There are some cases of phenomenal liquefaction in several countries, for example, liquefaction in the Patobo sub-district, Palu in Indonesia on September 28, 2018, due to the 7.4 magnitude earthquake. Liquefaction of the Niigata Building in Japan in 1964 due to the 7.6 magnitude earthquake. Liquefaction in New Zealand on February 22, 2011, due to a 6.3 magnitude earthquake [1].

Liquefaction occurs when the water pressure in the underground layers increases significantly during an earthquake. If an earthquake occurs on soil layers that are usually fine coarse-grained and loose sands, the water pressure will increase and easily flow upward and carry soil materials. At the moment the soil will behave like a mass of liquid until it comes to the surface in terms of sand boiling. A comprehensive study has been conducted by [2] on soil liquefaction based on several past events based on standard penetration tests. The application of the liquefaction potential analysis using the cone penetration test has also been proposed recently. Considering the limitations of the data obtained in the field which complicates the calculation of the liquefaction analysis, the study of the liquefaction potential analysis was carried out again by calculating the liquefaction parameters, namely relative density and average grain size [3].

During the time of an earthquake, permeable soil such as gravelly sands or coarse sands will increase the pore water pressure. In this condition, water will be able to flow in surrounding areas that have lower pressure. However, in semi-permeable soils such as silty sands and fine sands, the time for increasing pore water pressure is shorter than the time for flow transfer [3]. The condition of increasing pore water pressure will reduce the effective stress in soil mass which causes liquefaction. For this reason, it is necessary to carry out mitigation efforts to prevent the risk of liquefaction disasters.

The liquefaction mitigation analysis that has been carried out in [4] which compares the effectiveness of Dynamic Compaction (DC) before and after being used to reduce liquefaction hazards. Analysis of the use of biogas injection to reduce the degree of sand saturation is also quite effective in reducing the potential for the liquefaction of sand [5]. The use of a Prefabricated Vertical Drain to accelerate the flow of water during an earthquake can be analyzed [6].

Efforts to mitigate liquefaction by accelerating the rate of water flow can be carried out such as the use of vertical drainage. Previously [7] stated that the use of gravel drainage can provide an efficient method to prevent the development of pore water pressures that are too high and the distance between vertical channels can be made less than the distance required for water to flow vertically to the free surface. A comparison study has been done on the effectiveness of horizontal drainage in different geometric layouts [8]. The study in [9] has analyzed the performance of vertical drainage under the building together with the dynamic response of the foundation shown by the dissipation behavior of soil excess pore pressure. Experimental measurements as well as numerical simulations were carried out to describe the increase in excess pore pressure to liquefaction [10]. The calculation of the effective radial distance from the drainage center has never been done by previous researchers, for this reason, this study will analyze the radial direction distance and the equivalent circle of diameter 'de' in terms of dissipation of excess pore water pressure to the time of liquefaction.

2. Research methodology

This study uses the data from previous studies based on documentation of disasters that have occurred. The data used in this study were taken from research that discusses the parameters needed in the calculation of liquefaction [11]. The data was obtained in the form of data for 1 point Cone Penetration Test (CPT), relative density (Dr), and average grain size (D50) in Air Tawar, Padang. The area has the potential for liquefaction with indicating evidence of sand boils during the Padang earthquake on September 30, 2009 [12].

The stages of the research to be carried out are:

- Analyzing the liquefaction potential using the Dr-D50 method [3] which compares the relative density and average grain size to the maximum soil acceleration value (a_{max}).
- Mitigate liquefaction by using a vertical drain (sand drain).
- Using the finite difference method in calculating the radial drainage pore air pressure distribution, the optimal radial distance and effective diameter will be obtained to mitigate the occurrence of liquefaction in the area.

The finite difference method for radial drainage has become the right method to solve the problem of radial drainage [13]. The formula that will be used in the excel program is: Second level heading Please type text here.

$$\bar{u}_{0,t+\Delta t} = \frac{\Delta \bar{t}}{(\Delta \bar{r})^2}\left[\bar{u}_{1,\bar{t}} + \bar{u}_{3,\bar{t}} + \frac{\bar{u}_{3,\bar{t}}-\bar{u}_{1,\bar{t}}}{2(\bar{r}/\Delta \bar{r})} - 2\bar{u}_{0,\bar{t}}\right] + \bar{u}_{0,t} \tag{1}$$

Where:

$\bar{u}_{0,t}$ = initial pore water pressure = 100
r = radial distance
Δt = time interval
Δr = distance interval

Distance data, distance intervals, and time intervals are assumed in advance for initial planning, but must be controlled with values:

$$\frac{\Delta\bar{t}}{(\Delta\bar{r}^2)} \leq \ 0{,}5 \tag{2}$$

Then, proceed with the pore water pressure distribution using equation 1 as shown in Figure 1 below.

Based on the results of the discussion, experts in Japan agree that liquefaction can occur when the pore water pressure value is 60% of the total stress which is considered 100%. So, to reduce liquefaction, in this analysis, the radial distance from the vertical drain will be calculated for 20 seconds, with a pore water pressure dissipation of 60%.

The full dissipation time from excess pore water pressure is in the range of 20-250 seconds, i.e. sand with lower permeability will tend to liquefy longer [14]. So, in this analysis, 20 seconds will be used to determine the effective distance for dissipation of 60% pore water pressure by using a vertical drain, in this case, a sand drain without taking into account the smear zone.

0	0,00375	0,0075	0,01125	0,015	0,01875	0,0225	0,02625	0,03	$= r_e$	Meter	
	0,00375	0,0075	0,01125	0,015	0,01875	0,0225	0,02625	0,03	$= \bar{r}$	time	
	1	0,0000	0,0000	0,0000	0,0000	0,0000	0,0000	0,0000		0	sec
	1	0,6239	0,2477	0,1239	0,0000	0,0000	0,0000	0,0000		0,000013	sec
	1	0,6250	0,2500	ơ,1710	0,0920	0,0460	0,0000	0,0000		1	sec
	1	0,6592	0,3184	0,2061	0,0937	0,0659	0,0380	0,0190		2	sec
	1	0,6601	0,3203	0,2315	0,1427	0,0909	0,0390	0,0383		3	sec
	1	0,6783	0,3567	0,2506	0,1445	0,1128	0,0810	0,0600		5	sec
	1	0,6792	0,3584	0,2712	0,1840	0,1335	0,0830	0,0818		6	sec
	1	0,6939	0,3878	0,2870	0,1862	0,1548	0,1234	0,1031		7	sec
	1	0,6948	0,3896	0,3059	0,2222	0,1741	0,1259	0,1245		8	sec
	1	0,7082	0,4164	0,3206	0,2248	0,1944	0,1640	0,1449		9	sec
	1	0,7093	0,4186	0,3384	0,2583	0,2127	0,1671	0,1654		10	sec
	1	0,7218	0,4435	0,3524	0,2614	0,2321	0,2027	0,1849		11	sec
	1	0,7230	0,4460	0,3694	0,2927	0,2495	0,2063	0,2044		12	sec
	1	0,7347	0,4693	0,3828	0,2962	0,2680	0,2397	0,2230		14	sec
	1	0,7361	0,4721	0,3988	0,3256	0,2846	0,2437	0,2416		15	sec
	1	0,7470	0,4940	0,4116	0,3293	0,3022	0,2750	0,2594		16	sec
	1	0,7485	0,4970	0,4269	0,3569	0,3181	0,2793	0,2770		17	sec

Figure 1. The example process of calculation

3. Result and discussion

The results of the analysis of the potential for liquefaction based on the Dr-D50 method [3] in Figure 1, it is found that in 1 point of CPT with the liquefaction potential is at a depth of 4 – 7.2 meters with sand mixture soil types (silty sand to sandy silt) and silt mixtures (clayey silt to silty clay).

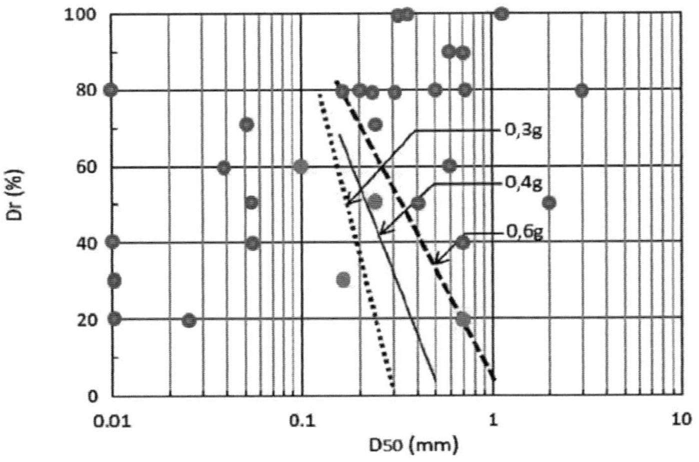

Figure 2. The result of calculation of the liquidity potential of the Dr-D50 method

Furthermore, the calculation of the pore water pressure distribution in radial drainage using the finite difference numerical method. The data needed in this calculation is the coefficient of horizontal consolidation (c_h). To obtain this c_h value, the permeability (k), modulus (m), and vertical consolidation coefficient are needed which can be calculated using the table and formula below.

Table 1. Table of estimation of soil permeability values based on CPT values [15]

SBT Zone	SBT	*Range of k (m/s)*
1	Sensitive fine-grained	3×10^{-10} to 3×10^{-8}
2	Organic soils - clay	1×10^{-10} to 1×10^{-8}
3	Clay	1×10^{-10} to 1×10^{-9}
4	Silt Mixture	3×10^{-9} to 1×10^{-7}
5	Sand Mixture	1×10^{-7} to 1×10^{-5}
6	Sand	1×10^{-5} to 1×10^{-3}
7	Dense sand to gravelly sand	1×10^{-3} to 1
8	*very dense/Stiff soil	1×10^{-8} to 1×10^{-3}
9	*very Stiff fine-grained soil	1×10^{-9} to 1×10^{-7}

According to [15], there is a relationship between CPT values and limited modulus values.

$$M = \alpha m\,(q_t - \sigma_0) \tag{3}$$

Where:
If the value Ic > 2.2, then use:

$$\alpha_M = Qt \quad \text{if } Qt < 14 \tag{4}$$

$$\alpha_M = 14 \quad \text{if } Qt > 14 \tag{5}$$

Calculating the value of c_v

$$c_v = (k\,M) / \gamma_w \tag{6}$$

According [16], the value of $c_v = (c_h/1.75)$, then:

$$c_h = 1.75 \times c_v \tag{7}$$

Where:
c_v = vertical consolidation coefficient
c_h = horizontal consolidation coefficient

After obtaining the c_h value, we can calculate the pore water pressure distribution with a finite difference [13] for an example can be seen in Figure 1.

With the acquisition of the distance value, and the decrease in pore water pressure every time interval, it can be made into a graph as below. Figures 3, 4, and 5 are graphs of the decrease in pore water pressure that occurs at each time interval and distance interval. By looking at the graph, it can be obtained at what distance 60% of the pore water pressure dissipated in 20 seconds.

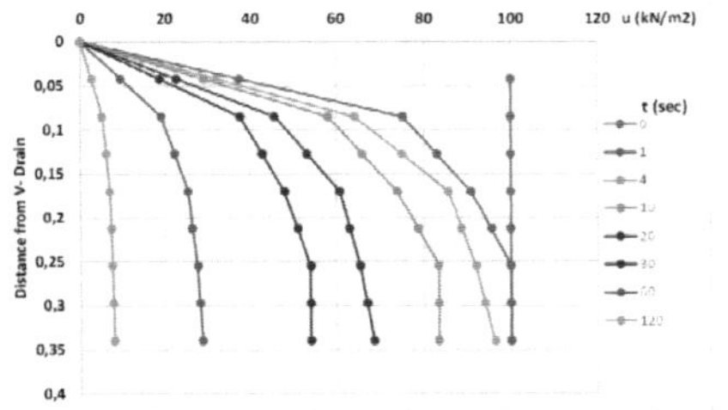

Figure 3. Pore water pressure distribution diagram over time depth 4 – 4.2 m

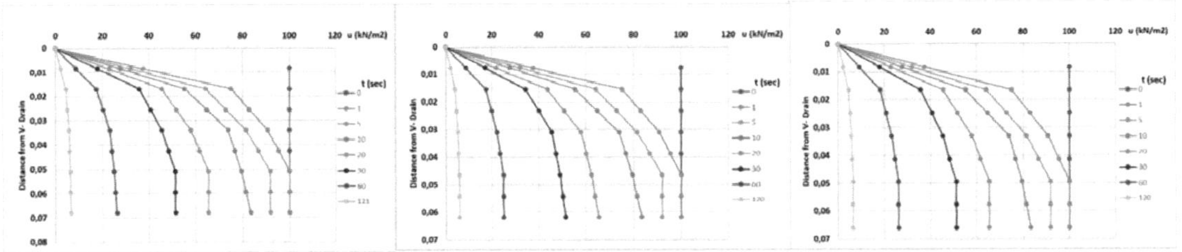

Figure 4. Pore water pressure distribution diagram over time depth 4.2–5.2 m

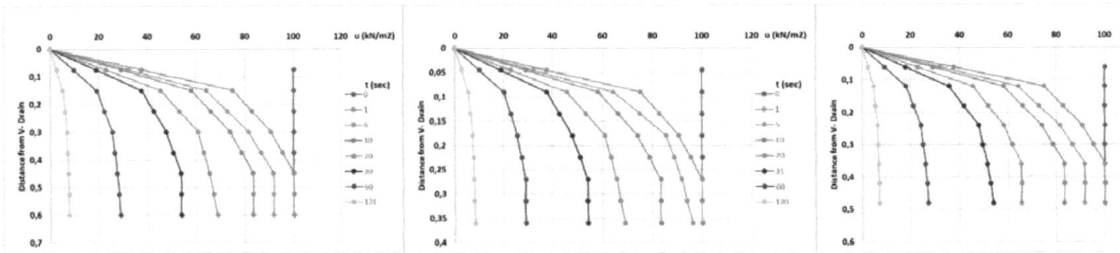

Figure 5. Pore water pressure distribution diagram over time depth 5.2 – 7.2 m

Table 2. Recapitulation of distance and equivalent diameter of liquefied layer vertical drain

No	Depth (m)	Soil Type		R 20 seconds (m)	De (m)	Potential Liquefaction
10	4 – 4.2	Sand Mixtures – Silty sand to sandy silt		0.165	0.68	Liquefaction
11	4.2 – 5.2	Silt mixture – Clayey silt to silty clay	Max	0.039	0.14	Liquefaction
			Min	0.038	0.12	
			Ave	0.037	0.13	
12	5.2 – 7.2	Sand mixture – Silty sand to sandy silt	Max	0.291	1.2	Liquefaction
			Min	0.175	0.72	
			Ave	0.272	0.96	

Form Table 2 the layer that has the potential for liquefaction based on the analysis using the Dr-D50 method is at a depth of 4 – 7.2 m with the soil type being sand mixture and silt mixture which has a permeability value of 3 x 10-8 m/sec and 1 x 10-6 m, respectively. The horizontal consolidation coefficient value obtained at the 4-4.2 m layer (sand mixture) is 0.00325 m^2/s which has a distance from the vertical drain which reaches 60% dissipation in 20 seconds (the time range for liquefaction occurs) is 0.165 m with an effective diameter of 0.68 m.

At a depth of 4.2 – 5.2 m (silt mixture) the maximum, minimum, and average horizontal consolidation coefficient values are 0.000138 m^2/s, 0.000118 m^2/s, and 0.00013 m^2/s. With this, the distances from vertical drain (VD) are 0.039 m, 0.038 m, 0.037 m, and the effective diameters of VD are 0.14 m, 0.12 m, and 0.13 m, respectively. At a depth of 5.2 – 7.2 m (sand mixture), the maximum, minimum, and average horizontal consolidation coefficient values are 0.0107 m^2/s, 0.0036 m^2/s, and 0.0067, respectively. m^2/s. With this, the distances from VD are 0.291 m, 0.272 m, and 0.175 m, and the equivalent circle of diameters of VD is 1.2 m, 0.72 m, and 0.96 m. This layer has the potential for liquefaction, but the potential is small because, in addition to having a Dr value of 60%, it also has a small permeability.

The minimum equivalent diameter value for the use of vertical drain is 0.12 m and a maximum of 1.2 m. The minimum distance to reach 60% u in 20 seconds is 0.037 m and the maximum is 0.291 m at Air Tawar, Padang.

For planning the use of vertical drains in the field, the equivalent diameter de that will be used is a maximum value of 1.2 m in Figure 6, because not only does it facilitate the field process, but the costs to be incurred will also be cheaper. The presence of dissipation of pore water pressure of 60% within 20 seconds with a fairly large radial distance, then it can be said that the vertical drain is capable of mitigating liquefaction.

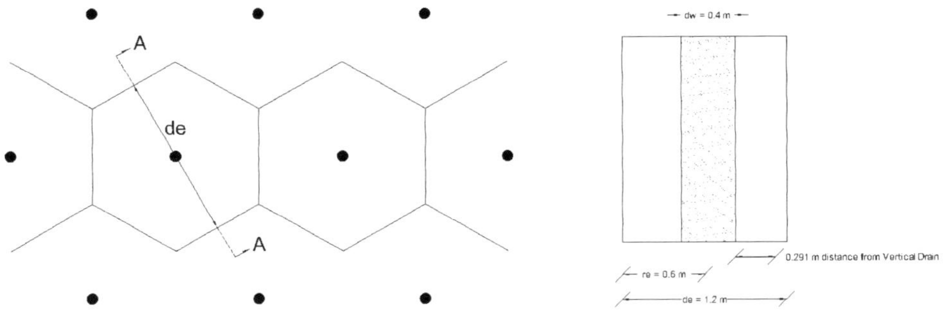

Figure 6. Pore water pressure distribution diagram over time depth 5.2 – 7.2 m

4. Conclusion

Based on the results of calculations using the finite difference method, the following conclusions are gained:

- ✓ The value of the equivalent circle diameter for the use of a vertical drain is 1.2 m. With an 're' value of 0.6 m, 60% pore pressure dissipation within 20 seconds was obtained at a distance of 0.291 m in the radial direction.
- ✓ Based on the results above, it can be concluded that the vertical drain is able to dissipate the pore water pressure by 60% in 20 seconds so it can mitigate liquefaction.

5. Acknowledgements

The authors would like to sincerely thank the Andalas University which has supported this research.

References

[1] Kurniawan P, Basuki M 2021 *Applied Geotechnics for Engineers.* Yogyakarta.
[2] Seed HB 1982 *Ground motions and soil liquefaction during earthquakes*. Earthq Eng Res insititue.
[3] Hakam A 2020 Analisis Praktis Potensi Likuifaksi. Padang, 53 .
[4] Shen M, Juang CH, Chen Q 2019 *Mitigation of liquefaction hazard by dynamic compaction—a random field perspective*. Can Geotech J.;56(12):1803–15.
[5] He J, Chu J, Ivanov V 2013 *Mitigation of liquefaction of saturated sand using biogas*. In: Bio- and Chemo-Mechanical Processes in Geotechnical Engineering: Géotechnique Symposium in Print, 116–24.
[6] Rollins KM, Goughnour RR, Anderson JKS, Wade SF 2004 *Liquefaction hazard mitigation by prefabricated vertical drains.* University of Missouri--Rolla
[7] Seed HB, Booker JR 1977 *Stabilization of potentially liquefiable sand deposits using gravel drains*. J Geotech Eng Div. 103(7):757–68.
[8] Fasano G, De Sarno D, Bilotta E, Flora A 2019 *Design of horizontal drains for the mitigation of liquefaction risk*. Soils Found.;59(5):1537–51.
[9] García-Torres S, Madabhushi GSP 2019 *Performance of vertical drains in liquefaction mitigation under structures*. Bull Earthq Eng. 17(11):5849–66.
[10] Bouckovalas GD, Papadimitriou AG, Niarchos DG, Tsiapas YZ 2011 *Sand fabric evolution effects on drain design for liquefaction mitigation*. Soil Dyn Earthq Eng. 31(10):1426–39.
[11] Hakam A, Febriansyah D, Adji BM 2020 *Liquefaction Mapping Procedure Development: Density and Mean Grain Size Formulations*. Geomate J. 18(70):155–61.
[12] Hakam A 2012 *Soil Liquefaction in Padang due to Padang Earthquake 30 September 2009*. Civ Eng Dimens.;14(2):64–8.
[13] Das BM 2019 *Advanced Soil Mechanics*. 712 p.
[14] Wang B, Zen K, Chen GQ, Zhang YB, Kasama K 2013 *Excess pore pressure dissipation and solidification after liquefaction of saturated sand deposits*. Soil Dyn Earthq Eng.49:157–64.
[15] Robertson PK, Cabal KL 2010 *Estimating soil unit weight from CPT. In: 2nd International Symposium on Cone Penetration Testing*.. p. 2–40.
[16] Poon B, Chan and K 2015 *Assessment of the coefficient of consolidation for staged preloading operations*.

Application of X-Band rainfall radar data-based on soil water index for sediment Disaster monitoring in Boyong River (BO-D6)

H R D[1], A Syarifudin[2]
[1]Civil Engineering Department, Engineering Faculty, University of Indo Global Mandiri, Palembang, Indonesia
[2]Civil Engineering Department, Engineering Faculty, University of Bina Darma, Palembang, Indonesia

E-mail: henggarrisa@uigm.ac.id

Abstract. Disasters those related to rainfall, will always come unexpectedly which can make huge potential of risk to human and facilities. Rainfall contribute a benefit and disaster to us, so we have to provide some research about the relationsip of rainfall and disaster. This research produce the information related to application of x-band radar data which have combination from soil water index. It shows the calculation of water content in soil layer by using x-band radar data. The rainfall dataset is extracted from June 2019 until June 2020. SWI can be used as preliminary information and works as a database for monitoring and observation tools for Boyong drainage area. The record of occurence sediment movement on 20th October 2019 proves in that day, the condition of soil water index reach maximum values with 80,28 mm and there is a small occurence of sediment movement around that month. The subsequent conctancy of SWI needs to be paid attention for possibility of disaster. As rainfall intensity reach the highest values, soil water index will also increase. A higher value of SWI show a higher risk of sediment disaster initiation.

1. Introduction

Sediment disaster monitoring is kind of works that provide an information for utilizing disaster risk reduction. This kind of disaster can be impacted to people and facilities so that we can build such an information to monitoring and predict from rainfall and response of current drainage area. When disaster and weather are expected, this information shows possible incident based on rainfall conducted from x-band radar and some indices [11]. The data from X-band MP (Multi Parameter) rainfall radar will provide information of rainfall intensity in the form of a more proportional grid so the area of rain whose value will be different from that will not exposed to rain [10]. Rain data that derived from X-band MP rainfall radar estimates in graphical formats with color differences for any higher intensity areas. In this way, the rainfall over the drainage area can be better represented. X-band MP rainfall radar also has higher spatial resolution and suitable for mountainous area due to being small in size and weight so mean regional rainfall can also be better predicted at any time because the movement and development of rain can be continuously monitored by scanning.The combination of setting rainfall indices and X-band MP (Multi Parameter) rainfall radar data which is more scientific precision of timing and location, sensitive to rain attenuation phenomena and monitoring spatial distribution of rainfall can be useful to identify current stages of drainage area and rainfall characteristics to make sediment-related disaster information for monitoring [5]. The aim of this research is to analyze the fluctuation of time series and spatial distribution using X-band MP (Multi Parameter) rainfall radar and to understand current stage response information of drainage area against rainfall by using short and long term rainfall on a snakeline based on several rainfall indices (2-min rainfall intensity and soil-water index) and the application of X-band MP (Multi Parameter) rainfall radar data on soil water index in the upstream of Boyong river (BO-D6) at Mount Merapi.

Table 1. Identity name of Boyong drainage area

River Name	ID	Outflow Control Point	Coordinate Position	
			Latitude	Longitude
Boyong	1-0-0-0-0	Sabo BO-D6	-7.609552	110.417386

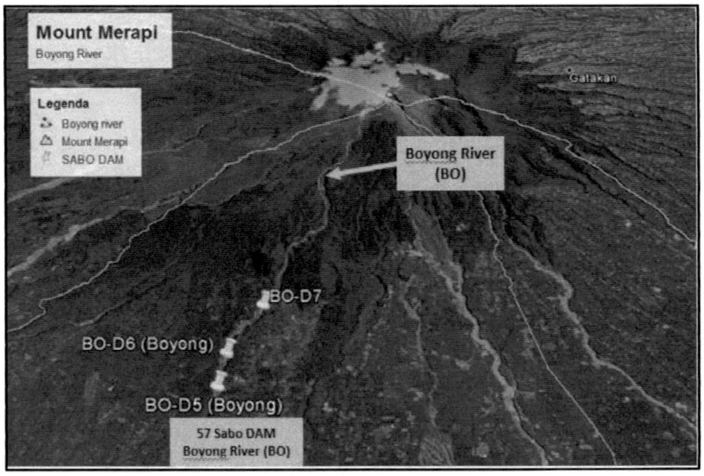

Figure 1. River flow of Boyong river with SABO DAM from google earth

Figure 2. Current status of Boyong River at BO-D6

2. Material and method

2.1. Global digital surface model (DSM)

The Japan Aerospace Exploration Agency (JAXA) generated the global digital elevation/surface model (DEM/DSM) and orthorectified image (ORI) using the archived data of the Panchromatic Remote-sensing Instrument for Stereo Mapping (PRISM) onboard the Advanced Land Observing Satellite (ALOS, nicknamed "*Daichi*"), which was operated from 2006 to 2011 [6]. This project is named "*ALOS World 3D*" and the dataset created consists of fine resolution DSM (0,15 arcsec, approx. 5 m). The ALOS has three remote-sensing instruments: the Panchromatic Remote-sensing Instrument for Stereo Mapping (PRISM) for digital elevation mapping, the Advanced Visible and Near Infrared Radiometer type 2 (AVNIR-2) for precise land coverage observation, and the Phased Array type L-band Synthetic Aperture Radar (PALSAR) for day-and-night and all-weather land observation. In order to utilize fully the data obtained by these sensors, the ALOS was designed with two advanced technologies: the former is the high speed and large capacity mission data handling technology and the latter is the precision spacecraft position and attitude determination capability. They will be essential to high-resolution remote sensing satellites in the next decade. ALOS have been successfully launched on an H-IIA launch vehicle from the Tanegashima Space Center, Japan.

2.2. X-Band MP rainfall radar

This research is using X-band MP rainfall radar that has been installed in Merapi Museum to monitor spatial distribution of volcanic ash and rainfall distribution around Mount Merapi This radar system were conducted to extract some data and apply the data to get realtime estimation of discharge rate, monitoring teh risk of sediment disaster in several catchment, analysze the dispersion of volcanic ash, forecasting multimodal sedment disaster as an effect of volcanic ash and rainfall. Multi-Parameter radar uses two polarization waves, while conventional meteorological radars use a single-polarization wave. The principle of rainfall estimation using multi-parameter radar is based on the fact that raindrop shapes change from round to flat with increase in rainfall intensity.

Table 2. Integrated observation data of X-band MP rainfall radar

Type	Data interval	File transfer interval	File size	File type	Remarks
X-band MP rainfall radar	2 min/data	2 min	3~150 kb/file	Csv (gz compressed)	Surface only 150 m mesh

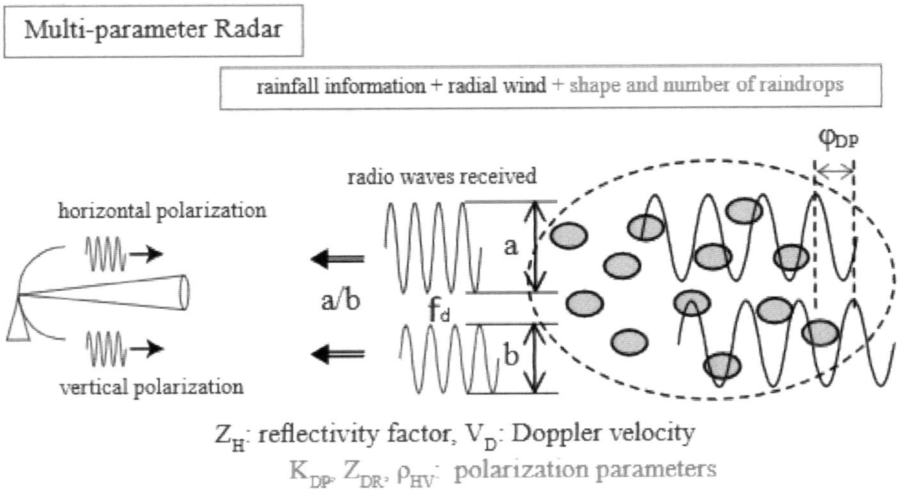

Figure 3. Polarization parameters of multi parameters radar

2.3. Soil water index (SWI)

Soil Water Index (SWI) represents conceptual soil water contents in the near-surface soil layers that can be easily calculated using the tank model based on rainfall analysis.[3] Unlike many other hydrological models that require a significant amount of data, such a surface materials, soil thickness, and sliding surface features of each slope, the SWI provides a simplified model using only rainfall with fixed parameters to allow application to wide areas. SWI using uniform nationwide parameters and does not consider the characteristic (degree of vegetation, geology, and weathering). It modeled the ground of the surface layer relatively. It is not intended for deep underground situations. Precipitation of rainfall will flow into first tank immediately regardless of snow. So falling snow that is accumulated on the ground as snow cover or melting process and immersing on the ground does not considered. The concept of water contents in using soil water index was conducted by using three-tank model using fixed parameters. Although both Indonesia and Japan has different characteristics in geological condition, but its expected that SWI only represent the conceptual of water contents and time series with current fixed parameters [8]. Tank model also used to analyzing the relationship between rainfall and runoff and modelling hydrologic response of wide range of drainage area.

Table 3. Soil water index parameters

Tank Layer	**Outflow Height (mm)**	**Outflow coefficient (1/h)**	**Coefficient of Permeability (1/h)**
First tank	$L_1 = 15$	$\alpha_1 = 0.1$	$\beta_1 = 0.12$
	$L_2 = 60$	$\alpha_2 = 0.15$	
Second tank	$L_3 = 15$	$\alpha_3 = 0.05$	$\beta_2 = 0.05$
Third tank	$L_4 = 15$	$\alpha_4 = 0.01$	$\beta_3 = 0.01$

There are some criteria often used for the lifting of sediment disaster information for monitoring, when the rainfall has decreased below the critical limit for publishing warning information and is estimated not to exceed the limit again in a short time and if the rainfall does not decrease below the critical limit regardless of the continued state of no rain the decision should be made based on the condition of the second tank storage decrease of the soil water index, etc. (second tank of the soil water index : amount of water contained in the surface layer) [9].

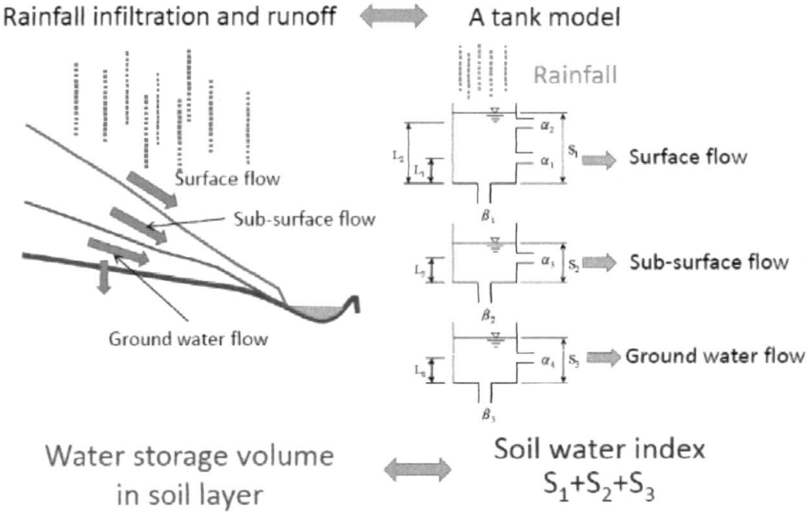

Figure 4. The concept of soil water index in using tank model

2.4. Identifying Boyong drainage area

Delineation of drainage area was conducted by using command to execute and get drainage area as ESRI shape and geotiff files with outlet point's coordinate which already defined using terminal window.[1]

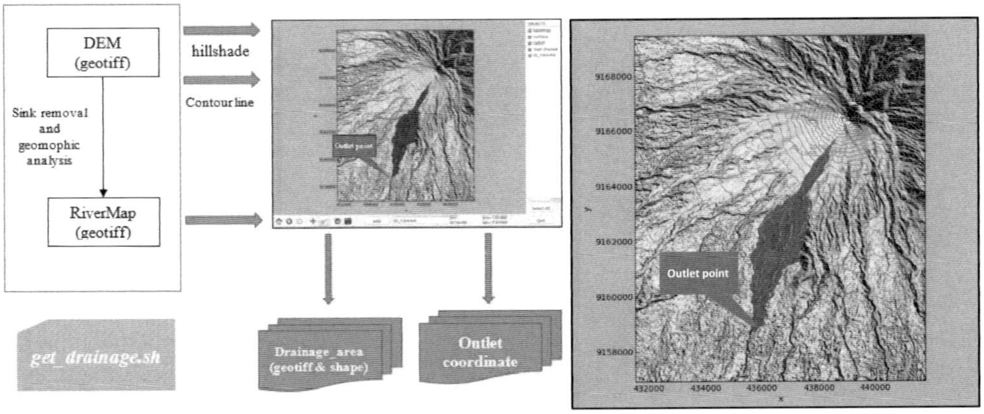

Figure 5. The concept of identifying Boyong drainage area

3. Result and discussion

Structure component for extracting event chain in this research analysis were conducted to make it simple to understand and analyze rainfall pattern and disaster chain in response of drainage area.

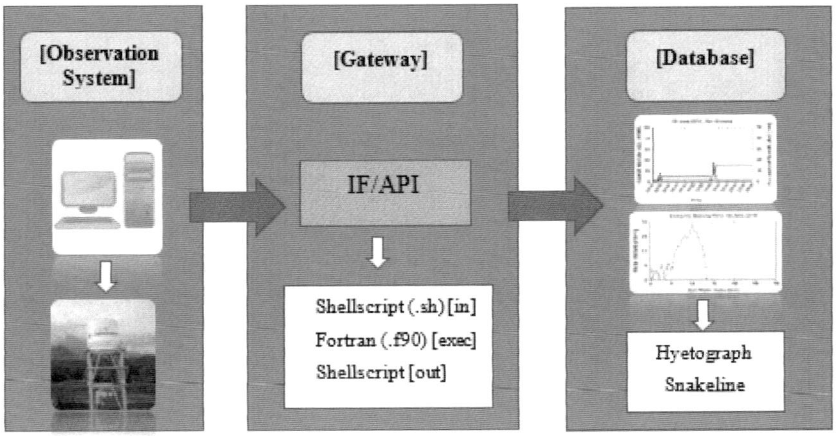

Figure 6. System for extracting data event chain from radar

This research use rainfall intensity with 2 minute time interval of observation by X-band MP rainfall radar as short-term rainfall index. Rainfall intensity were conducted by calculation of average rainfall intensity in each pixel in certain drainage area and certain time. The values of average rainfall intensity is the sum value of drainage pixels (mm/h) divided by number of masked pixels. The value of average rainfall intensity is an input time series correspond to short-term rainfall index. Figure 6 show the time series that were collected from the result of extracting process and identify the sum of rainfall data, average of rainfall data and accumulation of rainfall data every 2 minute.

The time series consist of sum value of drainage pixel (mm/hr) , mean value of drainage pixels (mm/hr) and accumulated rainfall (mm). Mean value of drainage pixels correspond to average rainfall intensity, which is used as rainfall intensity per 2 minutes. The peak of rainfall intensity was 53.5102 mm/hour occurring on October 20th at 15.20 as shown in hyetograph in Figure 7.

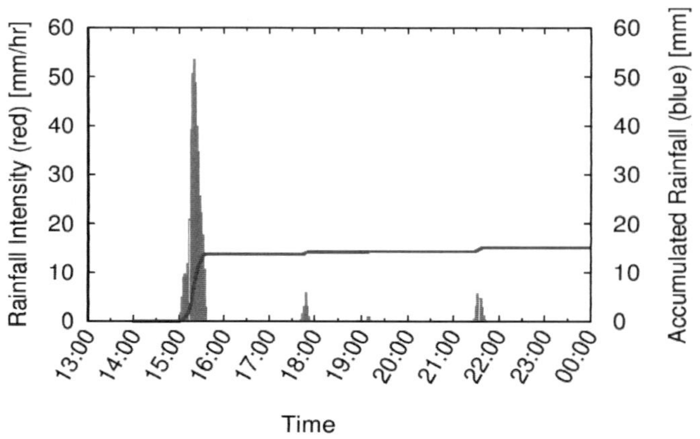

Figure 7. Hyetograph of rainfall on October 20th 2019 in Boyong drainage area (BO-D6)

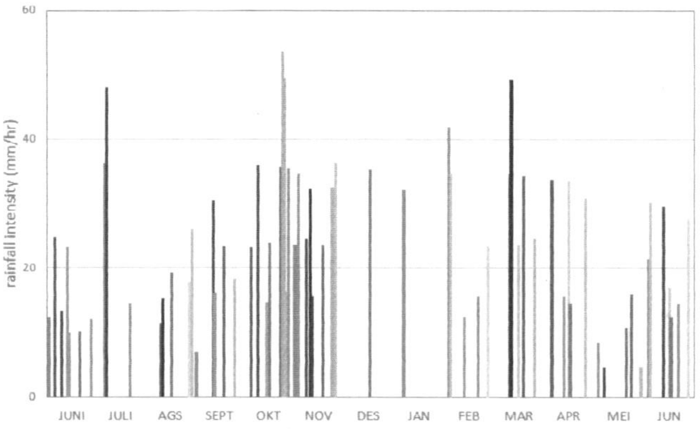

Figure 8. Rainfall intensity database from extracted x-band radar

Figure 8 shows the average frequency on rainfall intensity from dataset (June 2019 – June 2020) in Boyong drainage area which is divided into several interval of rainfall intensities. By using X-band MP rainfall radar, Boyong drainage area had the highest amount of rainfall event that were recorded on October 2019 followed by March 2020 and July 2019.

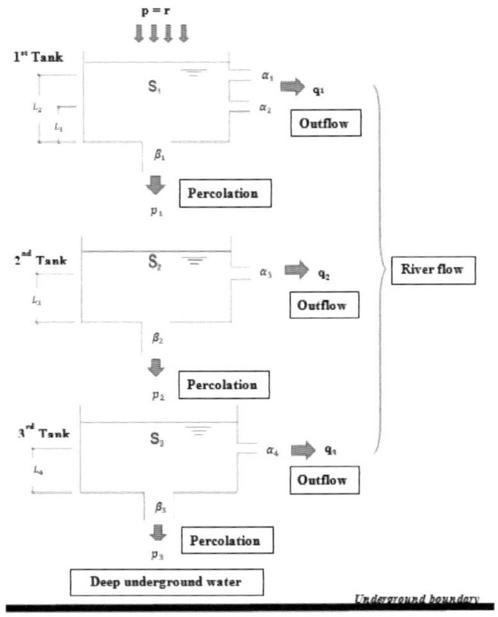

Figure 9. Schematic diagram of three-layer tank model

The three-layer tank model program was coded and debugged by using Fortran program that supported numerical analysis and scientific computation, structured programming, array programming, high performance computing on supercomputers and reasonable degree of portability between computer systems. The fixed parameter were identified by the statistical analysis of the relationship between rainfall and discharge that were conducted by Ishihara and Kobatake. Therefore, this research applied the SWI in Boyong river (BO-D6) using fixed parameters in granite region.

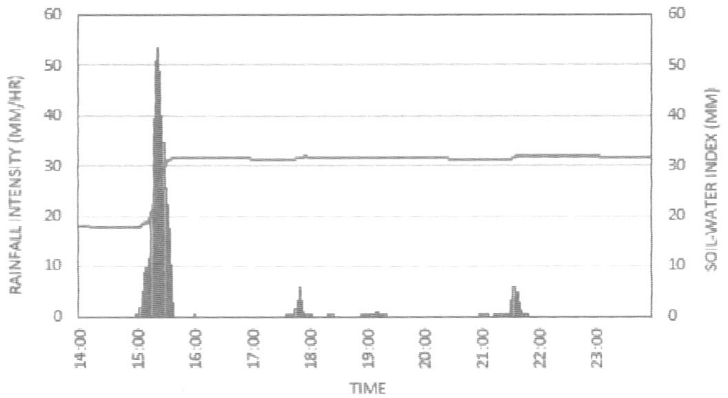

Figure 10. The relationship of rainfall intensity and soil water index on October 20th 2019 as the highest intensity

Figure 10 show the relationship between time series of rainfall intensity (mm/hr) and SWI. Figure 10 indicates the behavior of snakeline during October 20th 2019, there were occurrence of rain with the highest intensity within several minutes and the relation with SWI. SWI is represent the concept model of soil water content and elucidate soil water content during the rainfall event. As rainfall intensity reach the highest value, soil water index will also increase. A higher value of SWI show a higher risk of sediment disaster initiation.

Soil water index using tank model is a conceptual model that simulates the behavior of soil layers such as infiltration, percolation and run off. When it rains, rainwater will flow on the surface of the soil, flow into the river and soak into the ground. The water depth of three layer tanks representing the surface for first tank, middle layers for second tank and deeper layer for third tank. The example of relationship between rainfall intensity and S1, S2, S3 as the water storage from each tank that correspond to surface flow, sub-surface flow and groundwater flow with initial condition is 0 (zero) in June 2019.

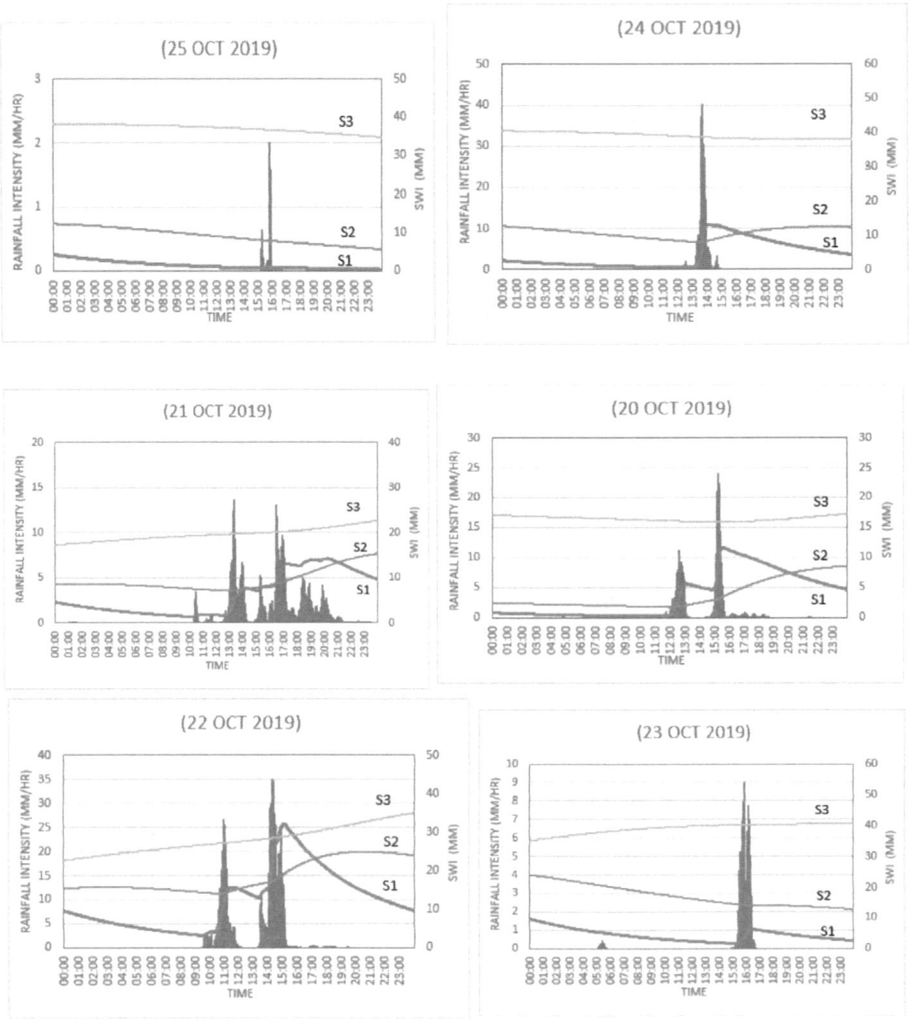

Figure 11. The relationship of rainfall intensity and water flow (S1, S2, S3) from 20th – 25th October 2019

According to the result from Figure 5.29 until Figure 5.33, it shows that surface flow (S1) has a linear relationship and responsive to rain due to the first layer that contains direct rainwater. The responsive character of surface flow is from the behaviour of rainwater inside tank. When the rain falls to the surface layer then it has direct effect to the rising of water level in the layer. Sub-surface flow (S2) as the second layer is not too affected on the movement of surface layer correspond to the smaller coefficient of permeability from surface layer. The level of water in surface layer is depend on rise of the outflow for surface flow and precipitation process to sub-surface layer then flows out to the surrounding soil. Amount of rainwater in sub-surface layer is from the precipitation process of surface layer.Groundwater flow (S3) as the third layer has time lag in infiltration process due to the small coefficient of permeability from sub-surface layer. Amount of rainwater in ground water layer is from the precipitation process of sub-surface layer.

The section of this research shows the contribution of rainfall to soil water index (SWI) so the response of water storage from each of three-soil layer (S1, S2 and S3) to rain can be seen. This simulation was carried out to show the soil water response to a single rainfall event of 1 mm/hr.

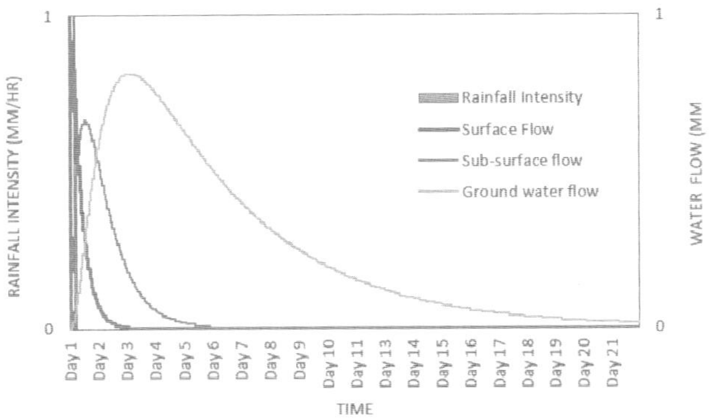

Figure 12. Hydrograph of water flow in soil layer in 21 days

Figure 12 show the movement of water storage in each three-soil layer (S1, S2, S3). Hydrograph above shows the inclusion of rainwater into the soil layer. Rising limb line shows the responsive line due to rain falling on the soil layer, then reach the peak. After reach the peak, the flow will go down until approach 0 (zero) value).

According to the result above, the duration of water retention in surface layer, au-surface layer, and groundwater layer is four days, six days and three weeks, respectively. It can be also regarded that any rainfall event within three weeks after previous rainfall corresponds to accumulated SWI of the previous rainfall.

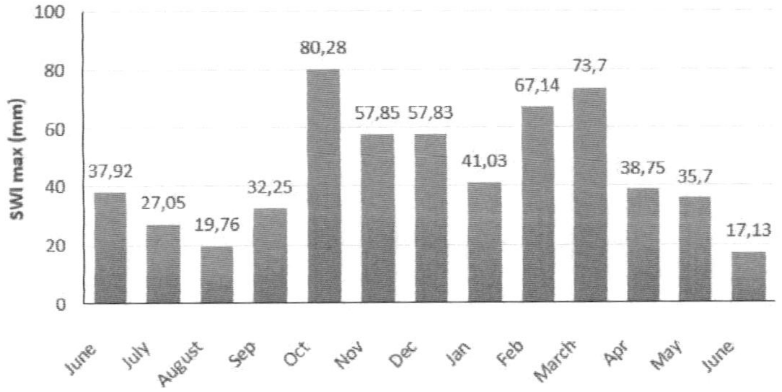

Figure 13. SWI max changes in Boyong drainage area (BO-D6) from June 2019 – June 2020

October 2019 until March 2020 shows the rapid increase in SWI changes correspond to rainfall conditions tend to be higher and continuously, due to rainy season, than other month but still not exceeding the standard reference value that was set by JMA. The research is using criteria that was set by JMA due to lack of sediment disaster occurrence data within dataset range and of course, the criteria for warning should be set based on rainfall data recorded as triggering disaster and the occurrence of disaster itself in Boyong drainage area.

4. Conclusion

X-band MP (Multi parameter) rainfall radar can be used to estimate rainfall amounst and precipitation particle sze distribution like raindrops. It is effective to establish precipitation estimation methodology and produce real-time public disclosure of rainfall information using the internet for preliminary sediment disaster prediction. By combinating database from x-band rainfall data and topographical data from ALOS3WD can be used to reflect the current state of Boyong drainase area especially BO-D6. The application of rainfall data from x-band radar, we can provide the highest amount of soil water index which reflect the water content layer from current state of drainage area. As rainfall intensity reach the highest value, soil water index will also increase. A higher value of SWI show a higher risk of sediment disaster initiation.

5. Acknowledgements

We acknowledge the support received from University of Indo Global Mandiri and also from my professor Mr. Kuniaki Miyamoto from University of Tsukuba who have given me fully support mentally amd knowledge during the research.

References

[1] Aqua Terra Consultants 2017 Watershed Delineation (Utah State University: EPA.gov)
[2] Chen, C.-W and Saito H Oguchi T 2017 Analyzing rainfall-induced mass movements in Taiwan using the soil water index *Landslides* **14** pp 14:1031-1041
[3] Destania H R 2017 Fieldtrip to Mt. Sakurajima and Mt. Unzen Tsukuba Japan
[4] Fujita M 2016 Sediment Disasters and a Warning System using Multi-hazards Simulator (Kyoto: DPRI Kyoto University)
[5] Fukao S and Hamazu K 2014 Radar for Meteorological and Atmospheric Observations (Kyoto : Springer Japan)
[6] Huaxing B 2006 Digital Terrain Analysis Based on DEM (Higher Education Press and Springer-Verlag)
[7] Hunt E D, Hubbard K G, Wilhite D A and Arkebauer T J 2009 The development and evaluation of a soil moisture index. *International Journal of Climatology* pp. 747-759
[8] Ishihara Y and Kobatake S 1979 Runoff Model for Flood Forecasting *Disaster Prevention Research Institute* pp 27-43
[9] MLIT 2007 Sediment-related Disaster Warning and Evacuation Guidelines (Japan: MLIT)
[10] Destania H R Pengaruh Intensitas Hujan Terhadap Soil Water Index (SWI) Dengan Menggunakan Data Radar *Jurnal Tekno Global* **8** no 2 2020
[11] Destania H R Radar-Based Rainfall Estimation in Boyong River (BO-D5) *Indonesian Journal of Engineering and Science* **3** no 1 2022
[12] Osanai T, Shimizu K, Kuramoto S, Kojima and T Noro Japanese Early-Warning for Debris Flows and Slope Failures Using Rainfall Indices with Radial Basis Function Network *Landslides* **7** no 3 pp 325–338 2010

An investigation of the changing patterns of rainfall in the Indravathi subbasin utilizing the Mann–Kendall and Sen's slope methods

Istijono Bambang[1], Subbarayan Saravanan[2], Nagireddy Masthan Reddy[2] and Devanantham Abijith[2]

[1] Department of Civil Engineering, Universitas Andalas, West Sumatra, Indonesia, 25175

[2] Department of Civil Engineering, National Institute of Technology Tiruchirappalli, Tamil Nadu, India, 620015

E-mail: bistijono@gmail.com

Abstract. In hydro-meteorological studies, precipitation is an important parameter that is utilized in irrigation system design and management, as well as agricultural planning. Under the effects of climate change, precipitation is predicted to alter, which will have an impact on sustainable development. Using rainfall data for the Indravathi subbasin, Godavari basin, from 1998 to 2016, the spatial variability and temporal trend of precipitation were examined over the region, which has a humid tropical climate. Using non-parametric tests like Mann-Kendall (MK) and Sen's slope approach, analysis for trend detection was conducted. The test data were loaded into ArcGIS software, which then performed monthly, seasonal, and annual analyses of the spatial and temporal trends of rainfall. According to an analysis of the monthly rainfall trends, February had the largest spatiotemporal declining trend in rainfall, while September had the highest spatiotemporal increasing trend. An analysis of seasonal rainfall data revealed a considerable increase in the tendency for summer rainfall over nearly 80% of the basin. Data on annual precipitation revealed an increase in annual precipitation for the basin's central eastern sections. The findings of this research may help the decision makers and stakeholders make the most efficient use of hydrological resources by providing insight on the effects of climate change and climatic changes on precipitation patterns in the Indravathi subbasin.

1. Introduction

Climate change is one of the most serious issues the world is now experiencing. Any notable long-term shift in the expected pattern of the planet's average weather over an extended period of time is referred to as climate change. It relates to unusual climate fluctuations and their aftereffects on Earth. These climate changes could occur in ten thousand, or perhaps millions, of years. However, in recent years, the rate of climate change has accelerated due to significant greenhouse gas emissions brought on by anthropogenic activities such as industrialization, urbanization, deforestation, agriculture, and changes in land use patterns. The quantity and frequency of yearly precipitation could rise or decrease, which would be one of the most important effects of the identified global warming [1,2]. Numerous studies undertaken at local, regional, and/or global sizes have found variations in yearly precipitation [3,4].

The amount of precipitation is a crucial climatic factor that controls how meteorological drought develops and progresses. The monsoon season, which makes up more than one-third of the yearly rainfall in the Indian subcontinent, is chiefly responsible for its distinctive characteristics. Rainfall and cloudy days heavily influence agriculture in the country's rain-fed regions, which accounts for 44% of the nation's total crop production. A minor change in precipitation and spread will have a significant influence on crop productivity, runoff, reservoir capacity, hydropower generation, and groundwater levels. Thus, rainfall and rainy-day trends are of major interest in the water resources sector, and they also aid in understanding the variance of underlying patterns linked to climate change [5].

As per IPCC, there was a 0.74 C rise in average temperatures between 1906 and 2005, along with significant spatial and temporal precipitation variance. According to IPCC 2021, "global average temperatures climbed by 1.09 °C between 2011 and 2020, compared to 1850-1900." Extreme heat, extreme precipitation, and drought will all increase with each additional 0.5 °C of warming. As a result of increasing warming, the Earth's carbon sinks, including vegetation, soils, and the ocean, will be reduced" [6]. Drought, flood, and high-intensity rainfall have been documented to occur at a rising rate all over the world [7–9]. The Asian continent has a wide range of climatic precipitation, including monsoon rains. Much research on monsoon rainfall has been conducted to demonstrate the impact of climate change [10]. Precipitation is increasing in eastern North Europe, North and South America, and Central Asia while decreasing in the Mediterranean, southern Africa, and South Asia [11]. Australia's rainfall is likewise increasing [12,13]. Reduced mean annual precipitation trends have been observed in north and north-east China [14,15], Russia [16], Pakistan's arid plains and coastal areas [17], Italy [18], and eastern Brazil [19].

Trend analysis using statistical methods has grown in popularity over the years, particularly for evaluating spatial and temporal variations in precipitation. There are several similar studies in the literature; however, they are generally limited to the use of traditional trend tests, such as linear regression, Mann-Kendall (MK), Sen Slope (SS), and modified Mann-Kendall (MMK, an enhanced version of MK) [20]. The direction of the trend (i.e., increasing/decreasing) is identified by the value of the slopes in the SS. Many researchers applied Mann-Kendall and Sen's slope method for various regions in India, Northeast India [21], Karnataka [22], Telangana [23], Mahi River basin [24], and Haryana [25].

More information on rainfall dynamics is required in order to understand better how the region's weather pattern has changed and to identify areas with increasingly frequent rainfall that is either above or below average. Government decision-makers might use this data to manage existing resources more efficiently while making timely economic choices across the basin by pinpointing micro-level locations at risk of drought or flooding. Using statistical approaches, such as the Mann-Kendall method, Sen's slope method, and ArcGIS software, the current study tried to assess the trend of rainfall across the Indravathi Subbasin, consisting of 56 points of IMD gridded precipitation based on rainfall data collected from 1998 to 2016.

2. Methods

2.1. Study area

The Godavari is the third biggest river basin in India, covering an area of 3,14,835 square kilometers and accounting for approximately 9.5 percent of the country's total land area. From latitudes of 16.50° north to longitudes of 73.50° east, it is located. Located on the Deccan Plateau, the basin encompasses significant swaths of Andhra Pradesh, Madhya Pradesh, Chhattisgarh, and Maharashtra, as well as a few minor parts in Karnataka and Orissa. There are two major tributaries running through it: one from west to east. The Manjira/Manjra, the Pranhita, the Indravati, and the Sabari are four of the most prominent tributaries in this river. The Indravati river begins at the height of 914 meters on the eastern ghats of the Dandakaranya hills in the Kalahandi district of Odisha. It then reaches the city of Jagadalpur in the Indian state of Chhattisgarh by traveling west. Near a height of 82.3 m, the river flows southward until joining the Godavari at the borders of Maharashtra, Chhattisgarh, and Andhra Pradesh. Between 80° 16' 19" East and 83° 07' 10" East, and 18° 43' 25" North and 19° 26' 46" North, is the Indravati river basin. The annual precipitation average for the basin is 1.58 meters. It has a catchment area of approximately 38,306 km2, and the length of the Indravati river basin is 535 kilometers [26]. The significant right bank tributaries of the river Indravati include the Bhaskal River, the Narangi River, the Boarding River, the Nibra River, the Kotri River, and the Bandia River, whereas the left bank tributary is the Nandiraj River. One of India's largest dams is the Upper Indravati hydroelectric project (Fig. 1).

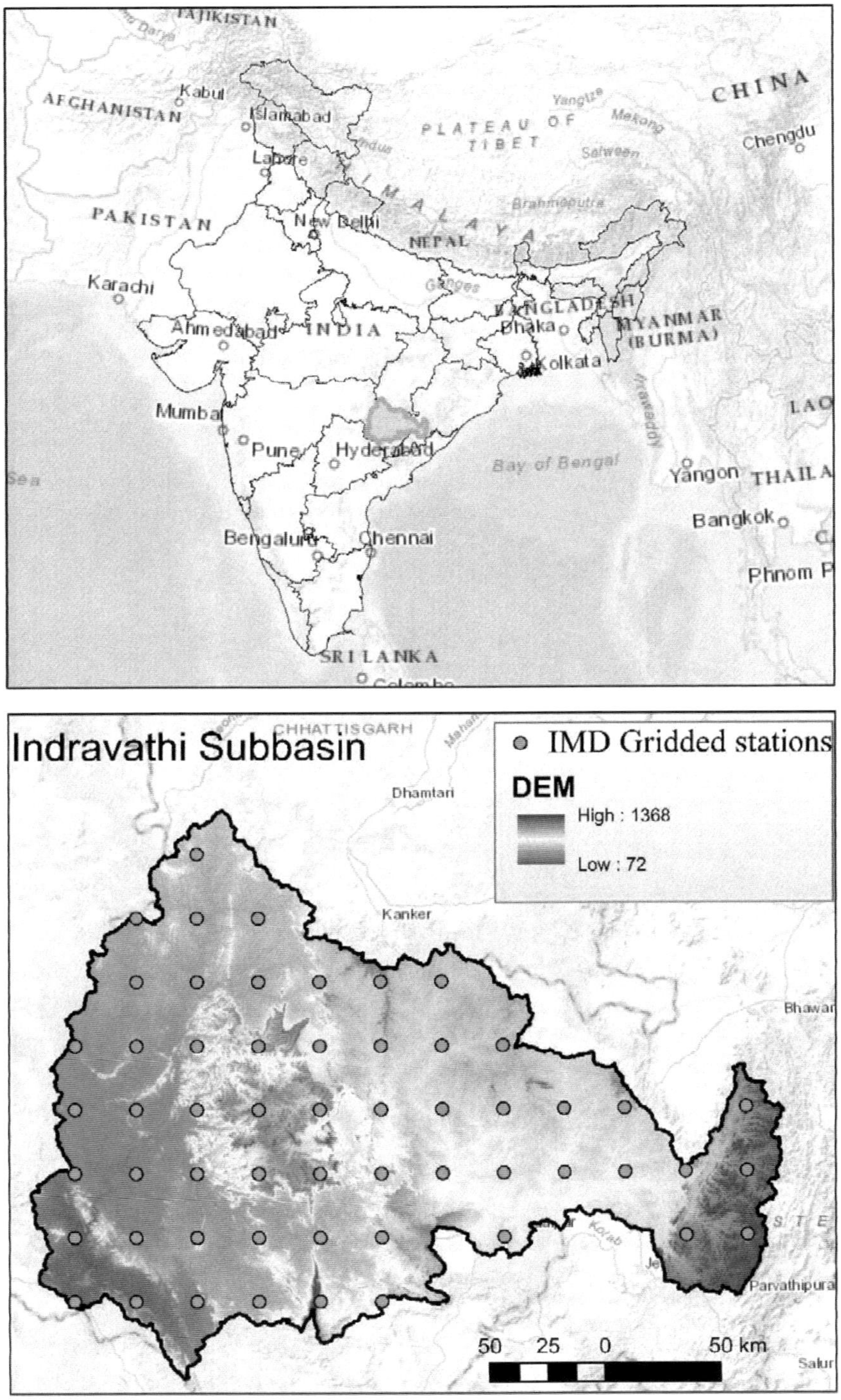

Figure 1. Study area

IOP Conf. Series: Earth and Environmental Science 1173 (2023) 012036 doi:10.1088/1755-1315/1173/1/012036

2.2. Data

The IMD gathered daily rainfall data for 56 gridded stations in the research region between 1998 and 2016, which was then converted to monthly rainfall data at a resolution of 0.25° [33]. The HydroSHEDS (Hydrological data and maps based on SHuttle Elevation Derivatives at different Scales) database provided the geo-referenced watershed map layer necessary for the research. The data sources and their details are shown in Table 1.

Table 1. Data Sources

Dataset	Time period	Source webpage
IMD Gridded	1998-2016	https://www.imdpune.gov.in/
Basin boundary	-	https://www.hydrosheds.org/

2.3. Methodology

Following figure 2 represents the flow process of methodology. Data extraction has been carried out using Climate Data Operators (CDO) [27]. Further procedures have been carried out using ArcGIS 10.3.

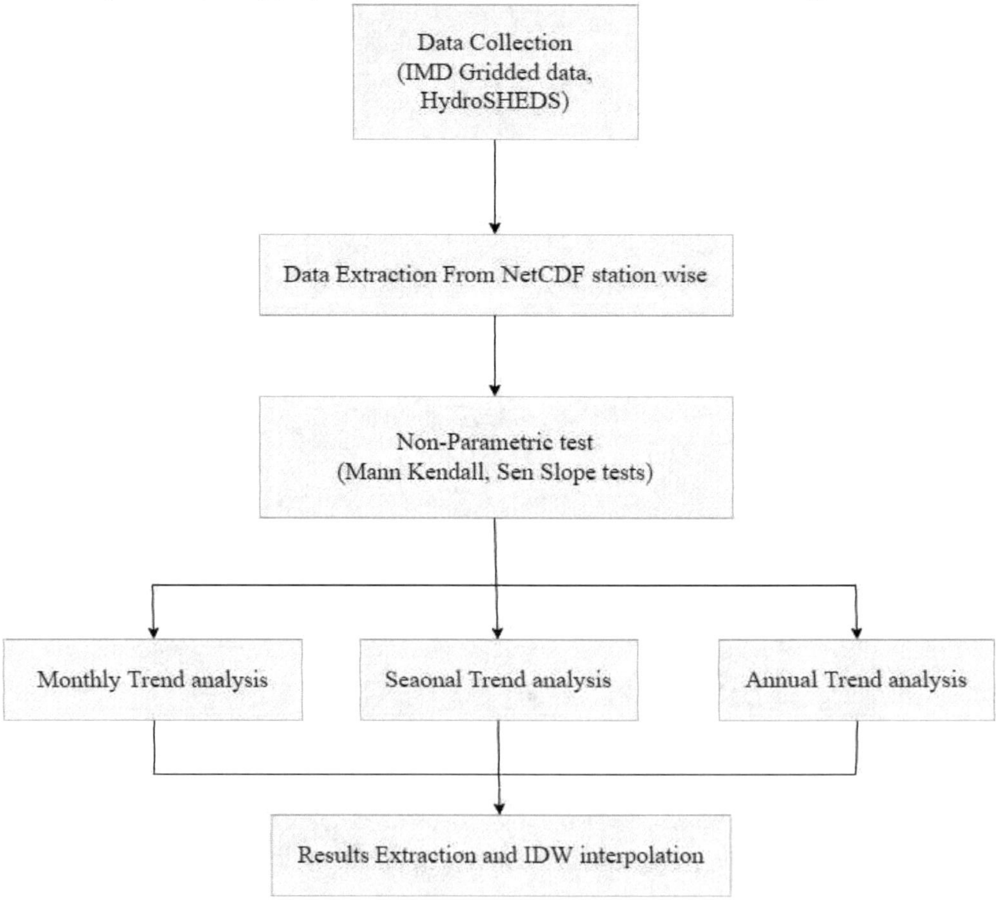

Figure 2. Methodology flow chart

IOP Conf. Series: Earth and Environmental Science 1173 (2023) 012036 doi:10.1088/1755-1315/1173/1/012036

2.3.1. Mann Kendall (MK) Test

The purpose of this research is to analyze changes in the pattern of precipitation across the chosen region by Mann Kendall. Mann (1945) [28], Kendall (1975) [29] are the researchers responsible for discovering patterns in the data pertaining to rainfall and precipitation. This is a non-parametric test for trends. It was determined that the monthly, seasonal, and yearly rainfall data had a progressive tendency. Missing data may be handled easily with this strategy, which avoids the need to make assumptions about normalcy and is resistant to bias. In comparison to parametric tests such as the t-test, it has more power. For a given time series X (x_1; x_2; x_3;... x_n), the null hypothesis (H_0) indicates no trend, but the alternative hypothesis (H_a) indicates either a steady growing or decreasing trend. The approach for analyzing MK trends is detailed below. Using the following equation, we can get the test statistic for the series, which we will refer to as "S" is shown in equation no. 1.

$$S=\sum_{j=1}^{n-1}\sum_{j=i+1}^{n}\operatorname{sgn}(x_j-x_i) \qquad (1)$$

Where x_i and x_j are the values of rainfall that correspond to the years i and j (j=i+1), respectively. Where "sgn" is shown in below equation no. 2.

$$\operatorname{sgn}(x_j-x_i)=\begin{cases}+1 & x_j>x_i\\ 0 \ \ if & x_j=x_i\\ -1 & x_j<x_i\end{cases} \qquad (2)$$

If the number of observations is more than eight, then the variable S follows what may be thought of as an approximation normal distribution. This distribution has a mean of zero and a variance that can be calculated using the following equation no. 3.

$$Var(S)=\{n(n-1)(2n+5)-\sum_{i=1}^{n}ti(i-1)(2i+5)\}/18 \qquad (3)$$

Where ti represents the total number of ties that were found in the sample analysis.
The statistics for the standardized test, which are denoted by the letter Z, are computed as follows equation no.4.

$$Z=\begin{cases}S-1/\sqrt{Var(S)} & S>0\\ 0 & S=0\\ S+1/\sqrt{Var(S)} & S<0\end{cases} \qquad (4)$$

A rising trend is indicated by a positive Z value, whereas a declining trend is shown by a negative Z value. The significance threshold for the test was set at 5% throughout its execution.

2.3.2. Sen's slope estimator

Sen's slope, a non-parametric estimator approach, was used to estimate the strength of a trend in the time series data. This was done in order to determine the overall size of the trend [29,30]. It is a test that employs a linear procedure that is not parametric. In the time series, a rising trend is shown by the slope of the trend line pointing upwards, whilst a declining trend is indicated by the slope of the trend line pointing downwards. Equation no. 5 is given as,

$$\beta=Median\left(\frac{x_j-x_i}{j-i}\right) for\ j>i \qquad (5)$$

where is the slope of the line that connects the sample points x_i and x_j on the graph. When the value is positive, it indicates a "growing trend in the time series," but when the value is negative, it indicates a "decreasing trend in the data series". The slope values for MK and Sen's slope were computed for each month from January to December, each season (winter, pre-monsoon (summer), monsoon, and post-monsoon), and for yearly precipitation data.

3. Results and Discussion

Each station's seasonal and annual precipitation totals for each year were computed using the stations' monthly rainfall data. Pre-monsoon (summer) is April–June, the monsoon (rainy season) is July–September, post-monsoon (fall) is October–November, and winter is December–February. Data were subjected to the Mann-Kendall and Sen's slope tests, yielding the Z-statistic, Sen's slope, and p-value, respectively. Using ArcGIS, these findings were utilized to determine if a statistically significant ($p < 0.05$) or non-significant rising or falling trend existed over the research region. The remaining results consist of the output acquired in ArcGIS following data import, interpolation, extraction, and relevant observations.

3.1. Monthly trend

There was no evidence of a statistically significant increase in rainfall trends anywhere in the region throughout the months of January and February; nevertheless, a statistically significant decrease in rainfall trends was seen over 85 percent of the research area during the month of February (Figures 3). The southwestern section of the Dhantewada district had a substantial rise in rainfall during the month of March, while the north-eastern parts of Bastar, Kankar, and neighboring regions noticed a considerably increasing trend in rainfall during the month of April (Figures 3). There was not a statistically significant increase in the amount of rainfall that was seen in the region during the months of May, June, or July. However, throughout the months of May and June, the northwestern area of the basin recorded a large reduction in the amount of precipitation that fell there. During the months of July and August, a markedly downward trend in rainfall was seen in the southern areas of Dantewada and the eastern territories. (Figures 4).

On the other hand, the month of September reported a substantial rise in rainfall trend throughout about 78% of the basin in the inner basin excluding southwestern region with the highest Z-value (2.917) in the whole dataset. This was the case since September had the highest Z-value in the entire dataset (Figure 5). During the months of October, November, and December, there was no discernible rise in the amount of precipitation that fell. On the other hand, a considerably declining trend was found for the North and eastern basin for the month of October, and a tendency quite similar to this was observed for the eastern areas of the basin during the month of November (Figures 5).

3.2. Seasonal trend

In the case of the summer (pre-monsoon) season, it was noticed that there was a considerable increase in rainfall in the region south of Bastar and in the north-eastern basin with a Z value of 2.311. (Figure 6). Still, fewer sections of the basin in the south-eastern half exhibited a large rise in rainfall during the monsoon, while the south-west showed a decreasing tendency. This was due to the fact that the monsoon occurs in the south-eastern part of the basin. However, most of the basin tended toward increased monsoon rainfall that was not statistically significant. (Figure 6). During the post-monsoon and winter seasons, no portion of the basin had a discernible upward trend in the amount of precipitation. During the post-monsoon time frame, over 80 percent of the basin had a declining tendency. During the winter months, practically the whole southern basin and the entire eastern area exhibited a notable tendency toward lowering temperatures, with the exception of the northwestern inner region. (Figures 6).

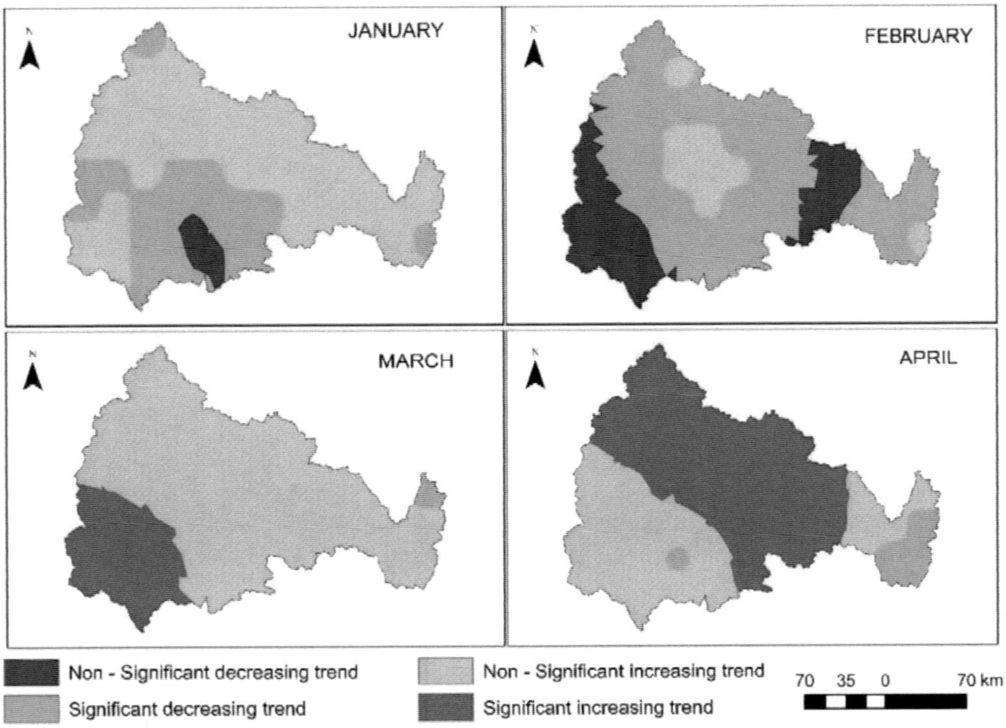

Figure 3. The monthly trend of precipitation (January, February, March, April)

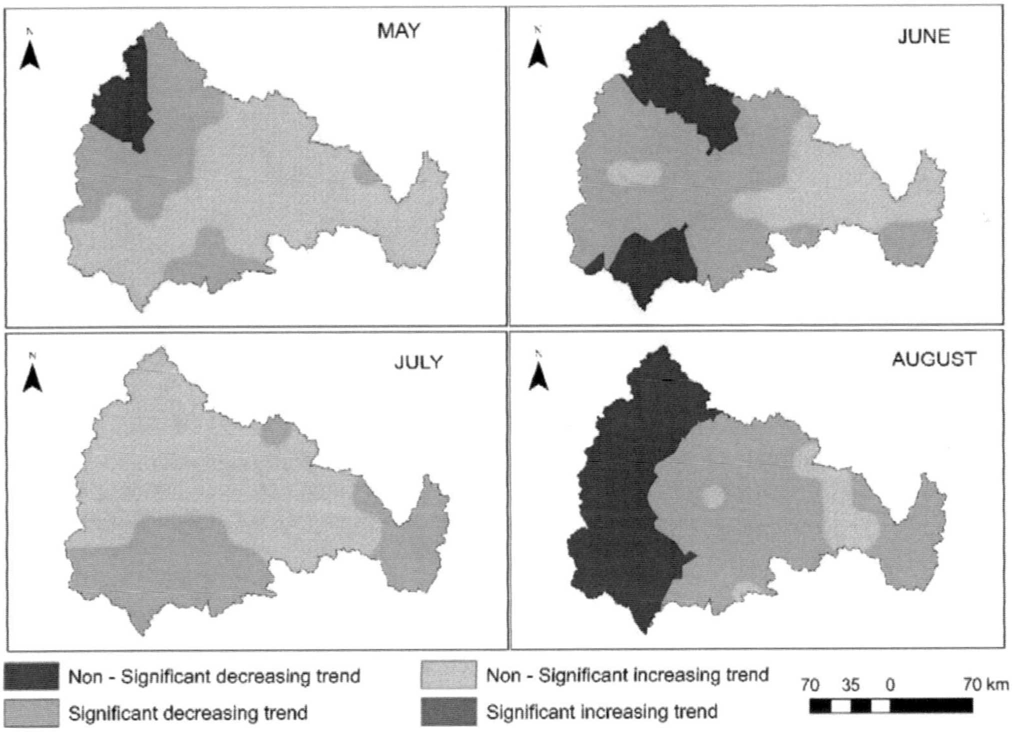

Figure 4. Monthly trend of precipitation (May, June, July, August

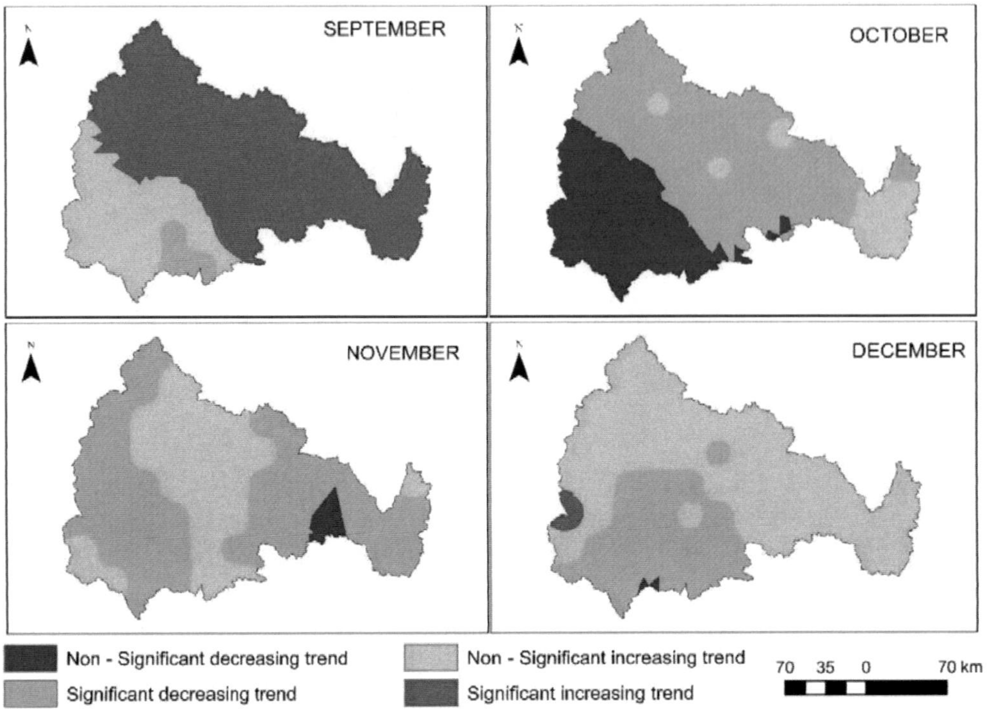

Figure 5. The monthly trend of precipitation (September, October, November, December)

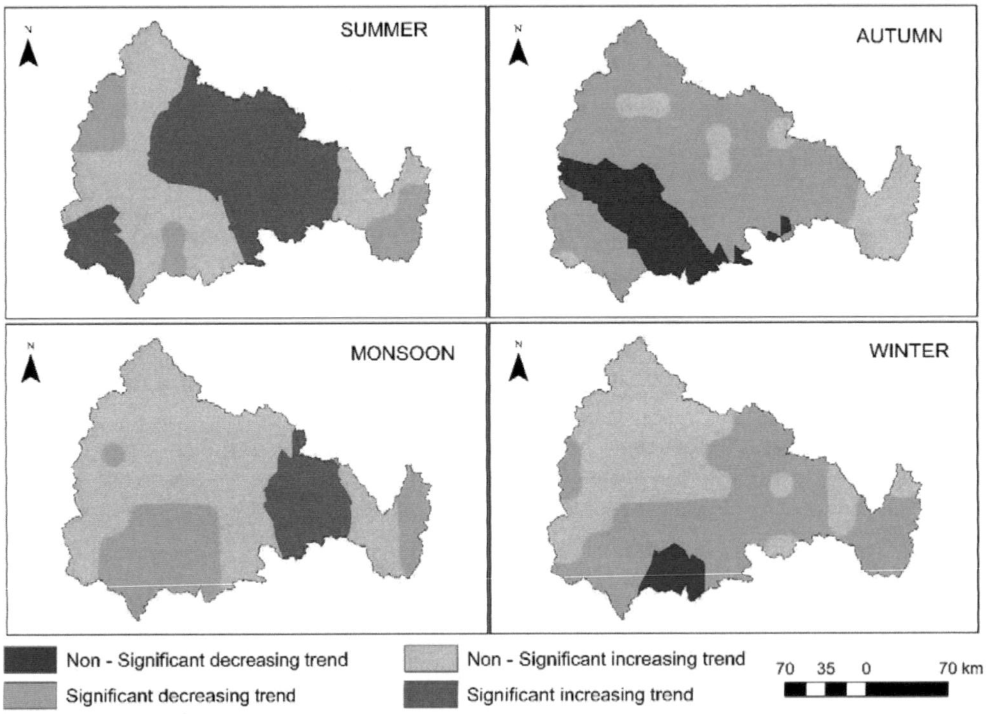

Figure 6. Seasonal trend of precipitation (Summer, Monsoon, Autumn, Winter)

3.3. Annual trend

The yearly precipitation trend over the subbasin is represented here in figure 7, which you may follow. According to the statistics on the yearly rainfall trend, the eastern central section of the basin and a few central areas of the basin had an increase in the rainfall trend on an annual scale that was significant. The lower southwestern basin, on the other hand, was a significant declining trend. The amount of precipitation that fell in the Indravathi subbasin as a whole demonstrated a clear increasing trend.

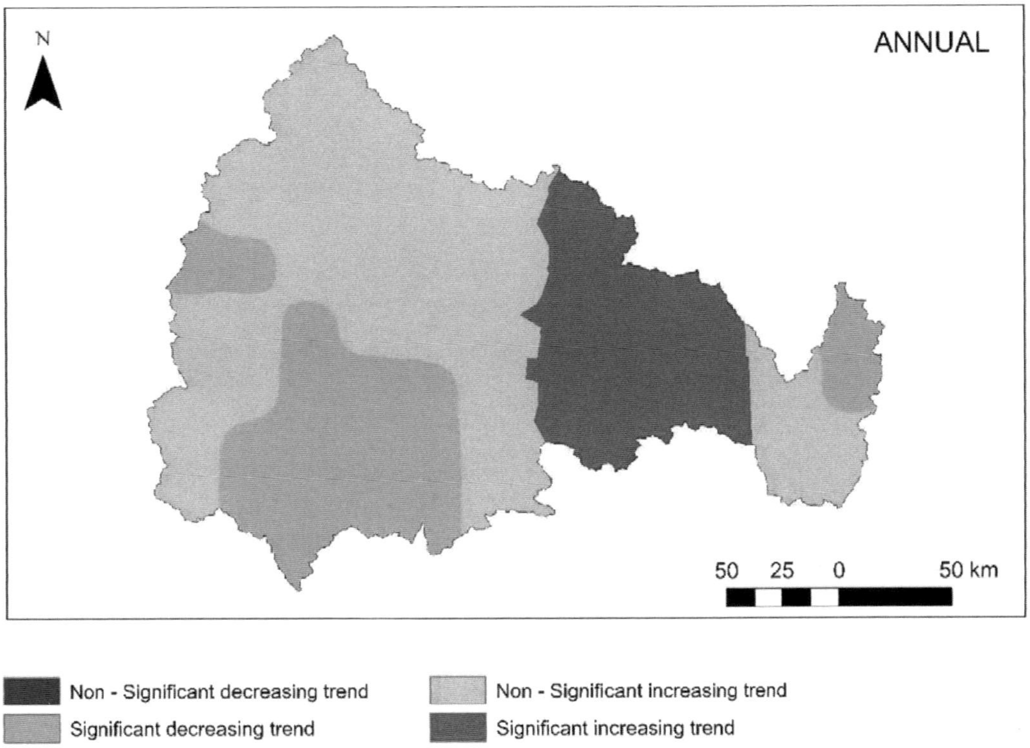

Figure 7. Annual trend of precipitation

In light of the yearly trend analysis, two sites out of a total of fifty-six have been chosen to represent the greatest and least Z values. The highest Z value, 2.386, was discovered close to Gengaraguda (lon = 82° 15', lat = 19° 15'), and the lowest Z value, -2.234, was discovered close to Gangaloor (lon = 81°, lat = 18° 45'), which were clearly indicated in figure 8 and figure 9. The plot of the time series for two stations can be seen in figure x, which is displayed below. These plots include elements such as the trend line equation and the R square value.

Murali (2022) [23] provided data consistent with an increasing trend in the amount of precipitation across Telangana. Rangarajan (2022) [31] demonstrated an increase in the amount of precipitation brought by the northeast monsoon across Tamil Nadu. The Sabarmati River basin has likewise demonstrated a tendency toward an increase in rainfall, as illustrated by Payal (2022) [32]. According to the findings of this study, the months of March, April, and September all indicate a tendency toward significantly increased rainfall across the region. The most significantly affected seasons by the large rise in the trend of precipitation are summer and the monsoon. Over the course of a year, it is effectively going to rise across the Indravathi subbasin.

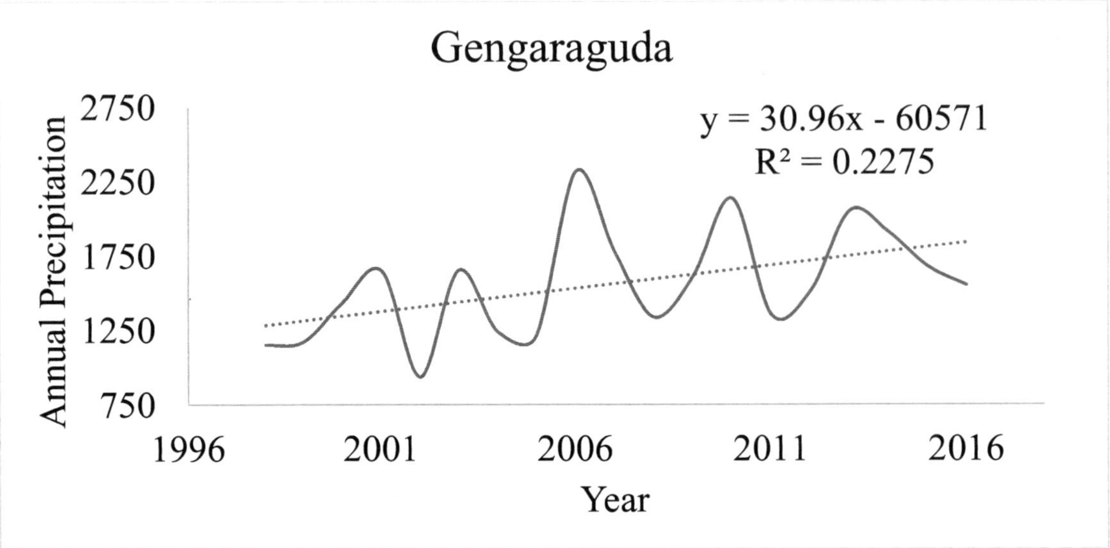

Figure 8. Annual trend of precipitation at Gengaraguda (Increasing)

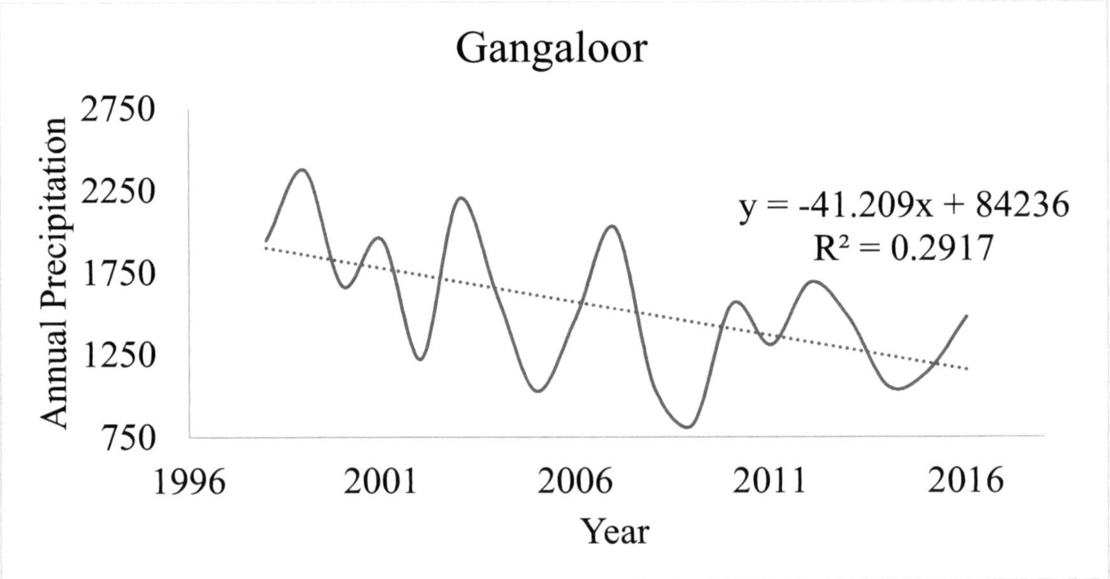

Figure 9. Annual trend of precipitation at Gangaloor (Decreasing)

4. Conclusions

This research examined the spatial variation of monthly, seasonal, and yearly rainfall in the Indravathi subbasin in India using traditional non-parametric approaches, such as the Mann-Kendall method and Sen's slope method, in order to identify long-term patterns (1998–2016). Upon examining the monthly rainfall pattern, September was determined to have the most Spatio-temporal growing trend in rainfall, whereas February had the highest Spatio-temporal falling tendency in rainfall. Nearly 80% of the basin's summer precipitation trend increased considerably, according to a study of seasonal rainfall data. The yearly precipitation in the Central eastern portions of the basin was found to be on the rise, according to statistics. It may be advantageous to properly identify rainfall trends throughout the region to raise awareness among various sectors (agricultural and industry), which are directly tied to the nation's economic growth, and to limit the dangers that these sectors face. With the aid of this study, policymakers will be able to construct a solid foundation for overall development.

Acknowledgements
The authors wish to acknowledge the Indian Meteorological Department, which has kindly provided the data, and the National Institute of Technology Tiruchirappalli for their assistance and support so this research can be completed properly. We also thank the anonymous reviewer that gave helpful comments and suggestions

References

[1] Costa A C and Soares A 2009 *Int. J. Climatol.* **29** 1956–75
[2] Santos C A G and de Morais B S 2013 *Hydrol. Sci. J.* **58** 789–96
[3] Gocic M and Trajkovic S 2013 *Glob. Planet. Change* **100** 172–82
[4] da Silva R M, Santos C A G, Moreira M, Corte-Real J, Silva V C L and Medeiros I C 2015 *Nat. Hazards* **77** 1205–21
[5] Pathak A A and Dodamani B M 2020 Model. *Earth Syst. Environ.* **6** 1357–72
[6] Allan R P, Hawkins E, Bellouin N and Collins B 2021 *IPCC, 2021: summary for Policymakers*
[7] Mirza M M Q 2003 *Clim. Policy* **3** 233–48
[8] Kundzewicz Z W, Ulbrich U, brücher T, Graczyk D, Krüger A, Leckebusch G C, Menzel L, Pińskwar I, Radziejewski M and Szwed M 2005 *Nat. Hazards* **36** 165–89
[9] Dashtpagerdi M M, Kousari M R, Vagharfard H, Ghonchepour D, Hosseini M E and Ahani H 2015 *Environ. Earth Sci.* **73** 1231–44
[10] Kundu S, Khare D, Mondal A and Mishra P K 2015 *Environ. Earth Sci.* **73** 8197–216
[11] Parry M L, Canziani O, Palutikof J, der Linden P and Hanson C 2007 *Climate change 2007-impacts, adaptation and vulnerability: Working group II contribution to the fourth assessment report of the IPCC vol 4* (Cambridge University Press)
[12] Hennessy K J and Suppiah R 1998 *Int. J. Clim.* **10** 1141–64
[13] Nicholls N and Lavery B 1992 *Int. J. Climatol.* **12** 153–63
[14] Hu Z-Z, Yang S and Wu R 2003 *J. Geophys. Res. Atmos.* **108**
[15] Zhai P and Pan X 2003 *Geophys. Res. Lett.* **30**
[16] Peterson B J, Holmes R M, McClelland J W, Vörösmarty C J, Lammers R B, Shiklomanov A I, Shiklomanov I A and Rahmstorf S 2002 *Science* **298** 2171–3
[17] Farooqi A B, Khan A H and Mir H 2005 *J. Meteorol.* **2**
[18] Buffoni L, Maugeri M and Nanni T 1999 *Theor. Appl. Climatol.* **63** 33–40
[19] da Silva V de P R 2004 *J. Arid Environ.* **58** 575–96
[20] Swain S, Sahoo S, Taloor A K, Mishra S K and Pandey A 2022 *Groundw. Sustain. Dev.* **18** 100783
[21] Bora S L, Bhuyan K, Hazarika P J, Gogoi J and Goswami K 2022 *Curr. Sci.* **122** 801
[22] Bharath A L and Venkatesh B 2022 *Indian J. Ecol.* **49** 462–9
[23] Murali A, Rangarajan S, Thattai D, Kamaraj M, Varadharajan D, Bhutia S Y and Islam M N 2022 *India II: Climate Change Impacts, Mitigation and Adaptation in Developing Countries* (Cham: Springer International Publishing) pp 351–82
[24] Sharma A, Sharma D and Panda S K 2022 *J. Hydrol.* **605** 127314
[25] Al-Abadi A M 2018 *Arab. J. Geosci.* **11** 1–19
[26] Dahe P D and Deshmukh B B 2018 Estimation of annual runoff in indravati sub basin of godavari river using statistical approach

[27] Schulzweida U, Kronblueh L and Budich R G 2019 cdo: climate data operators
[28] Mann H B 1945 *Econometrica* **13** 245–59
[29] Sen P K 1968 *J. Am. Stat. Assoc.* **63** 1379–89
[30] Hirsch R M, Slack J R and Smith R A 1982 *Water Resour. Res.* **18** 107–21
[31] Rangarajan S, Karthik Raja R, Murali A, Thattai D, Kamaraj M and Islam M N 2022 *India II: Climate Change Impacts, Mitigation and Adaptation in Developing Countries* (Cham: Springer International Publishing) pp 407–39
[32] Makhasana P and Joshi G S 2022 *Phys. Chem. Earth,* Parts A/B/C **126** 103146
[33] Reddy N M and Saravanan S 2022 *Int. J. Environ. Sci. Technol.* 1-26

IOP Conf. Series: Earth and Environmental Science 1173 (2023) 012037 doi:10.1088/1755-1315/1173/1/012037

Activating an Adequate Management & Effective Emergency Unit for The Water Supply in The Armed Conflicts Zones

T M Alnajar[1], A Fitri[2], W Almadhoun[3], M S N Hadie[4], F C Ros[5], A E Kabeel[6], and S Z H Shah[7]

[1]Faculty of Engineering, Islamic University of Gaza, P.O. Box 108 Gaza Strip
[2]Faculty of Engineering and Computer Science, Universitas Teknokrat Indonesia, Indonesia
[3]Dean of The Faculty of Engineering, Gaza University, Gaza
[4]Civil Engineering Department, Universitas Andalas, Indonesia
[5]Malaysia-Japan International Institute of Technology (MJIIT), Universiti Teknologi Malaysia, 54100 UTM Kuala Lumpur, Malaysia
[6]Faculty of Engineering, Tanta university, Egypt
[7]GAZIOSMANPAŞA University, Tokat, Turkey.

E-mail: tamernon7@gmail.com

Abstract. This study examined Gaza Strip water management after Israeli raids. This research is directly relevant to the Gaza Strip's water emergency plan. Taking immediate action helps prepare for shocks before, during, or after an emergency. This project aims to establish an updated emergency strategy for water sector emergencies. Considering Israel's Gaza raid This study is important because it addresses a vital issue that saves lives and sustains a vital industry (the water sector). In addition to other factors that support the management process, the study offers a basic method for controlling water supplies in an emergency (the Israeli aggression on the Gaza Strip). This solution is titled (decentralized supplies). This report updates a plan for managing Gaza's water supply during Israeli invasion. The report will enable decision-makers direct international organizations to improve and develop Gaza's water condition using scientific approach. The findings will benefit water experts and emergency managers.

1. Introduction

One of the greatest instances of a recent conflict zone is the Gaza Strip, where water service providers have developed on their own without the guidance of a strategy or master plan. Direct rainwater that seeps into the sands replenishes the aquifer. The overall water budget of the aquifer is also increasingly influenced by return flows from leaky supply networks, agricultural irrigation flows, and home or industrial sewage. The quantity and quality of its groundwater are significantly impacted by lateral subsurface groundwater inflows from nearby aquifers [1].

One of the most significant aspects of the environment that is regularly researched and examined as part of evaluating the state of the environment is the water sector [2,3,4]. The water problem, including the quantity and quality of the water accessible, as well as the methods for developing water work in the sector, are just a few of the significant issues that this title is related to within the context of the current study [5,6,7]. In addition to the most significant relevant institutions, civil organizations and bodies play an essential role in safeguarding water sources, enhancing their quality, and enhancing its output.

No matter how many families live there, a lot of residential structures just have one meter and one connection. There is about one meter for every 17–20 people in Gaza. Additionally, despite extensive network coverage, service delivery is inconsistent and there are significant differences in per-person supply between localities. In Gaza, just 30% of homes have access to 24-hour network water service. A little more than 15 percent of Gaza's residents have access to piped water for fewer than 10 days a month [8].

The water and sewage infrastructure in the Gaza Strip is being targeted by Israel's military operations, which result in the destruction of desalination plants for drinking water, public water supply networks, pumping and sewage treatment stations, as well as sewage and rainwater drainage networks. The effects of the invasions on Gaza extended to the water infrastructure, the most crucial of which are surface water and groundwater. The main electrical transformers that contain dangerous oils that could spill into surface or underground water and cause an environmental and health catastrophe were targeted. Additionally, the aquifer, where significant amounts of wastewater and sewage flow into the groundwater, is significantly impacted by direct targeting of sewage lines and extensions in the main and secondary streets [9]. As a result, there is an unequal distribution of water throughout much of the Strip. The owners of homes close to municipal wells or homes on the ground level receive municipal water sooner and in greater quantities than people living in remote places or on upper floors, particularly in tall buildings and towers. The projected 16 hours per day of electricity outages have a severe impact on water production and distribution, as well as restricting inhabitants' access to water. Furthermore, in some parts of the Gaza Strip, water is only provided to residents once or twice each week [10]. Additionally, the Gaza Strip has been deprived of numerous important strategic initiatives in a variety of environmental sectors due to Israel's siege. Along with restricting access to numerous pieces of machinery, tools, and apparatus necessary to improve the Strip's environmental performance [11].

The widespread violent bombing and targeting of electricity columns and lines has caused power outages, which has made all the environmental information pertaining to water more complex (resources and supplies). The shutdown of water resources, supplies, and desalination facilities, along with other significant environmental problems that will only get worse as time goes on, will have an adverse effect on people's health, the environment, as well as their economic, social, and even security situations. All prior development and enhancement plans will be scrapped or delayed as a result of the ongoing attacks, and their place will be taken by plans for relief, repair, and reconstruction. Considering the population is growing quickly and unabatedly, this indicates a significant fall and slowdown in development [11].

The research will reveal very useful guidelines for decision makers to guide international organizations to conduct better interventions to improve and develop the water situation according to appropriate scientific methodologies. In addition, the research will be of value to researchers in water topics as well as those working in the field of emergency.

2. Methodology and data collection

By developing the existing emergency plans, this study aims to find a clear approach for optimal management of water supplies in light of the Israeli aggression on the Gaza Strip. Which can move towards better management in the event of dividing the Gaza Strip into zones that are managed separately, so that the network does not stop working or the water floods in the event that the main distribution network is bombed, as well as to ensure the minimum water interruption in the event that one of the areas is bombed.

Also, this process which is called (Decentralization), greatly enhances the concept of (community participation), so that the community is one of the main elements in its management in cooperation with the concerned authorities.

The decentralization process combines with the elements of existing plans at several important points but in a somewhat different order, which will be clarified later. It is also characterized as being applied without the need for major changes on the ground.

3. Analyses and discussion

Emergency planning is the methodical process of preparing for potential occurrences, such as large crises and disasters. The plan is typically a shared document that outlines the roles and responsibilities that each participant and stakeholder will play in the multi-agency disaster response. It is a guide for organizing events, so it must take management requirements into account. In other words, it is the framework for emergency response and should outline the principles of action, collaboration, command, and communication during a civil exigency like a tragedy or significant event.

Emergency planning is the methodical process of preparing for potential occurrences, such as large crises and disasters. The plan is typically a shared document that outlines the roles and responsibilities that each participant and stakeholder will play in the multi-agency disaster response. It is a guide for organizing events, so it must take management requirements into account. In other words, it is the framework for emergency response and should outline the principles of action, collaboration, command, and communication during a civil exigency like a tragedy or significant event. The predictability of hazards and the amount of time, if any, that is required to make preparations varies greatly. However, most emergency plans include warning and related responses as two essential components.

3.1 Emergency plan frameworks

The fundamental goal of the emergency plan is to be able to maintain the bare minimum of services needed by the populace to carry out infrastructure restoration during a conflict. A movement coordination request will be made to the military forces to ensure the access to the ground for technical teams so that service providers may maintain the water and wastewater systems in working order (truce on the intervention area, avoid targeting the employees). It's also crucial to keep in mind that this approach focuses mostly on interventions for which movement coordination is necessary [12].

During war times, service providers are operating the network on a daily basis in non-military active areas and military active areas without going through this process. It is proposed that:

- In non-military active areas, the service provider acts as usual to ensure that basic services are provided to the population.
- In military active areas, a list of pre-ready coordination shall be prepared for the routine operation.
- In military active areas, coordination shall be requested to armed forces as described below to intervene in case of damages to a wash infrastructure.

3.2 Emergency procedures

It is necessary to conduct extensive national public awareness campaigns about the precautions that citizens must take in emergency situations to provide water at a minimum, without causing a state of social panic, as the concerned authorities are under a constant state of pressure and depletion that is difficult to live with in some circumstances, such as times of armed conflict. The following actions and precautions should be done in case of an emergency (water sector):

- Activating the system of communication through the various communication channels available between the various operational areas of water sector in the municipalities, the Coastal Municipalities Water Utility, the concerned parties and their working teams, and finding the appropriate tools at various levels.

- Securing the minimum and acceptable number of spare parts necessary for the maintenance and operation of water lines, wells, pumping stations and various facilities. This includes spare parts for networks and electromechanical spare parts.
- Providing all water wells and water pumping stations with a chlorine solution for water sterilization, in order to prevent pollution, which is sufficient for a period of no less than a month, and to keep chlorine tanks in a state of fullness in various areas. As well as distributing protective clothing to various areas in the Gaza Strip to be used by workers in the water sector.
- Set a station for storing chemicals needed for water wells and desalination plants.
- Urgent and comprehensive follow-up of all water well facilities, pumping stations, desalination stations, and pumping stations through contracts for maintenance services for pumps, generators and electrical systems, which are managed and implemented by engineering teams, which helps to create a state of continuous operational and functional readiness around the clock, at a rate exceeding 95%.
- Securing the readiness of water tankers, to transport water to areas where water systems may be damaged.

3.3 Proper Management

Regarding the direct Israeli attacks on infrastructure mentioned at the beginning of this study (especially during the most recent armed conflict in 2021), the aggression on infrastructure directly resulted in the destruction of a significant portion of the public water supply network in addition to desalination plants, pumping stations, wells, and water tanks as well as the primary and subsidiary water distribution network.

Operation and upkeep of several water, wastewater, and storm water facilities make up the RWU's primary activity. These facilities can be found all over the Gaza Strip. In most facilities, the daily operation of the wash facilities requires the physical presence of the operators in addition to the swift movement of the repair teams. It is suggested that regional offices (RO), which are geographically dispersed throughout the Gaza Strip, handle the daily operation and maintenance in order to ensure effective management, efficiency, and retaining the organizational structure's decentralization philosophy. The cost-effectiveness, which will be evident in reducing the number of high-level management staff, as well as the geographic dispersion and integration between the WASH facilities and networks, will decide the number of these RO.

4. Decentralization approach

Decentralized supply systems offer the possibility to provide water where centralized supply systems are not feasible due to abnormal conditions. Small-scale systems (SSS) which is one of decentralizing supply systems types, is the most typical application of small-scale systems is for community water supply. However, they have also had important applications in emergency water supply. Because emergency situations necessitate provision of clean water with limited time and resources, SSS are one of the best options to quickly provide safe water for community and camp water supply.

Piratla and Goverdhanam [13] reported that according to the comparative reliability analysis, the decentralized scenario's reliability outperformed the centralized-only scenario by an absolute margin of 5.83%. The ability of the system to meet demands under various potential hydraulic and mechanical uncertainties is referred to as reliability. For instance, in a centralized scenario, 100% reliability refers to a system that can function fully even when demand increases (up to 25%) and every pipeline in the network fails. These results clearly show that, in terms of the system's ability to meet needs under various uncertainty, decentralized supply complements conventional water supply.

The study seeks to determine an appropriate strategy for managing water supply in light of emergencies and disasters, as well as with reference to the regional water utility plan that is currently under development, through this method (decentralized water supply), in accordance with the study's objectives. In order to make the current public network the one that is primarily relied upon and with a separate management system in cases of emergency and aggression, taking into account the anticipated situation of regional water utility and its management process, the study involves a number of steps centered around the concept of decentralized water supply and its management process.

The process of decentralizing water supply covers several concepts, for example, through which it is possible to work on integrating the community within the water resources management plans and the process of implementing emergency plans, in addition to enhancing community participation in the management of water supplies during the emergency phase. It also contributes to the development of the administrative process of the water sector in the Gaza Strip from several aspects (economic, technical, environmental and operational), which makes it a strong preference for the concerned authorities to adopt and work on developing it to be the most reliable tool.

In order to clarify the steps that the study concluded within the emergency plans to manage the water sector in light of the armed conflict on the Gaza Strip under the title Decentralization Approach, a case study will be worked on, which was selected within the Jabalia Municipality authority, in the northern Gaza Strip.

4.1 Decentralization (case study)

The study proposes a solution that can be applied to be a basic system for managing water supplies in the event of an emergency, which is represented in a basic title, which is (decentralized supplies) in addition to other elements that support the management process, those elements of which part that agrees with the existing plans, This confirms the results of the study tools used (questionnaires), as it turns out that there are effective emergency plans with moderate rates that can rise to a higher and more effective level through working on the development of water supply management. The study will examine the area of influence of the Jabalia municipality in the northern Gaza Strip as a case study, so that the policy of decentralization supplies and the development of an updated emergency plan that can be extend and applied effectively during the emergency phase for whole Gaza Strip. Jabalia city is one of the most important cities in the Gaza Strip figure 1. It is a 1.5 km away from the city center. The area of Jabalia is about 18 km^2.

Figure 1. Jabalia Municipality Regions

Jabalia municipality services cover 25 official registered wells, benefiting 275,050 people living within the municipality's influence. Work will be done on dividing and developing the decentralized water supply management process based on a number of these wells so that the rest of the wells work as standby wells within the emergency plan.

Within the strategy of working with the emergency plan by adopting a schema of decentralized supply, water is distributed at the minimum limit in emergency and war situations, which is 15 l/c/d [14], which we often do not reach except in cases of complete interruption of water supply, and for which many interventions are prepared to compensate for any loss or interruption according to the supporting steps that are consistent with many points in the existing emergency plans, as follows:

- Securing the minimum and acceptable number of spare parts necessary for the maintenance and operation of water lines, wells, pumping stations and various facilities. This includes spare parts for networks and electromechanical spare parts.
- Securing the readiness of number of water tankers, with a capacity of 5-10 cubic meters each, to transport water to areas where water systems may be damaged.
- Providing the necessary fuel to operate the electricity generators in the various water facilities.
- Supplying electricity to water utilities by means of standby diesel generators.
- Distributing water through fixed/mobile water tanks, containers, and faucets, and this is supported by the provision of water bottles and bags.
- The residents are responsible for reaching the water distribution points and for bringing containers of the appropriate size for the family's daily allowance, through prior public awareness.
- Preparing the locations of the distribution stations as follows:
  - In each targeted area, water distribution stations are set up - next to a public facility, for example: a water well, a main square, a clinic, a school or other.
  - In the event of an emergency that necessitates distributing water by distribution stations, the stations are allocated to the affected areas.
  - The empty containers are distributed at the stations designated for this in the neighborhoods, and they are filled by tanks, the size of 5 cups, that move alternately.
- Water is distributed to all residents by means of mobile or fixed tanks, etc., water gallons, water bags, soft drinks, etc.
- Prepare to supply water according to the amount of 4 liters per person per day.
- Residents are responsible for accessing water distribution points and for bringing containers of adequate size for the family's daily allowance.

The following table 1 and figure 2 showing the division and distribution of zones according to the decentralized water supply system, which includes 11 zones depending on several determinants which are: the number and location of wells, the beneficiary population, the pumping rate for each well, the number of pumping hours, achieving the minimum amount of water needed in case of emergency and war, In addition to the main determinant, which is to identify the main resource and the supporting resources in the event that it stops working as a result of targeting or destroying the network. The study proposes a number of sequential steps to achieve optimal management of water supplies in light of the armed conflict on the Gaza Strip, which come in the following order: decentralized water supplies, supporting lines, water tanks, water distribution tankers. which are clarified during this part of the study.

Table 1. Decentralized Water Supply Zones.

No. of D.P	**Name of Well**		**Beneficiary population**	**Pumping Rate**	**Operating hours/day**	**L/C/d**
D.P.1	Tika Tank		30000	700	4	93
	Faisal Frigah Al-Marwah	Support Supplier				
D.P.2	Faisal Frigah		31500	150	4	19
	Al-Shifa Al-Alaween Al-Marwah	Support Supplier				
D.P.3	Al-Nazlah (New)		13300	100	4	30
	Abu Sharekh (West) Abu Sharekh (East)	Support Supplier				
D.P.4	Al-Shifa Al-Alaween		16000	230	4	57
D.P.5	Abu Sharekh (Wast) Abu Sharekh (East)		22350	250	4	44
	Al-Nazlah Al-Marwah Al-Reayah	Support Supplier				
D.P.6	Hiraa' (New)		17500	140	4	32
	Al-Zarqa	Support Supplier				
D.P.7	Abu-Hasirah Al-Bahtimi Tal Al-Zaatar		19400	120	4	24
D.P.8	Al-Moaskar		55000	380	4	27
	UNRWA Wells Yafa Arijoni	Support Supplier				
D.P.9	Al-Omari		22000	110	4	20
D.P.10	Al-Hissi		33000	150	4	18
	Al-Zain Aslia	Support Supplier				
D.P.11	Al-Fateh Al-Bosna		20000	200	4	40
	Al-Hissi	Support Supplier				

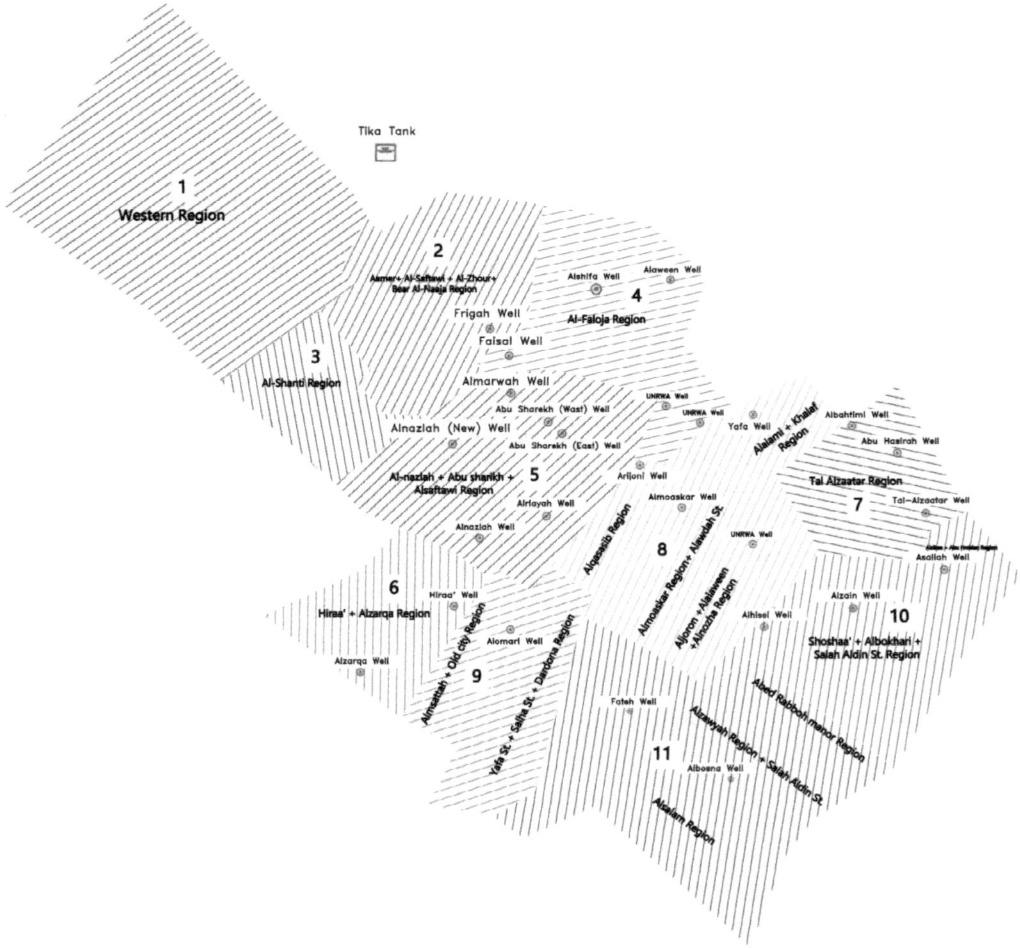

Figure 2. Decentralized Water Supply Zones.

4.2 Critical zones

After dividing the zones according to the decentralized water supply system, it was observed that there are 4 zones classified as critical zones, where there are no alternatives or sources to support them in case the network or wells are bombed and destroyed, and some of these zones also suffer from difficult access during the emergency period (The aggression on Gaza) such as the eastern regions in the event that there was a ground invasion from the eastern side, or the western regions in the event of an invasion from the western or northern side, so consideration was given to strengthening these areas within the decentralized supply policy as follows:

4.2.1 Zone 1 (western region). The western region considered as one of the critical zones, as this area supplied through the Tika reservoir alone without any alternative sources. In the normal case, if any failure occurs in the network or the pump used, the water supply will stop completely in that area, which gets worse in the event of an emergency and aggression, in the event that the network, the pump, or the tank is bombed or destroyed, an entire area with a population of about 30,000 people will not have access to water, which may cause a mass displacement of the population with the aim of reaching areas where they can obtain water. In addition to these reasons, the western region may be a place for the displaced from the eastern regions, as a result of the land invasion in the event that it occurs, so it is necessary to provide adequate quantities of water within the framework of emergency plans. According to the foregoing, the study proposes to supply the western region with a transmission line from Frigah, Faisal and Al-Marwah wells to be used as a support line in the event of a failure or targeting of the main line (Tika reservoir, Fig.3). In addition, linking the wells and alternative line through the Supervisory control and data acquisition (SCADA) system to ensure remote monitoring and controlling, which contributes to improving the emergency period management process.

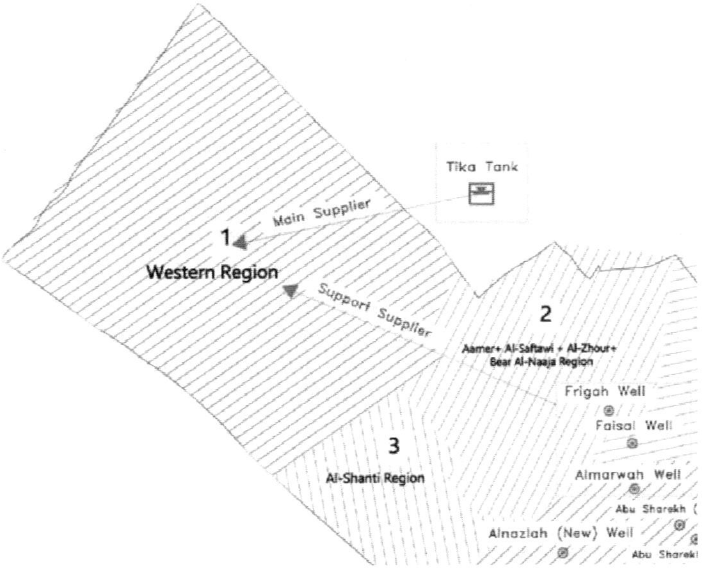

Figure 3. Zone 1 (Western Region)

4.2.2 Zone 3 (al-shanti region). Al-Shanti zone also considered as one of the critical zones, as this area is supplied through the new Al-Nazla well alone without the presence of any alternative sources. In the normal case, if any failure occurs in the network or the pump used, the water supply will stop completely in the area, which is worsening in the State of emergency and aggression. If the network, pump, or tank is bombed and destroyed, an entire area with a population of about 13,300 people will not have access to water, which may cause a mass displacement of residents with the aim of reaching areas where they can obtain water. In addition to these reasons, Al-Shanti area may be a place for displaced people from the eastern, western and northern regions, as a result of the Israeli invasion in the event that it occurs, so it is necessary to provide adequate quantities of water within the framework of emergency plans. According to the above, the study proposes to supply the Shanti zone with a transmission line from the eastern and western Abu Sharekh wells to be used as a support line in the event of a defect or targeting of the main line (the new Nazla well) as shown in figure 4.

Figure 4. Zone 3 (Al-Shanti Region)

4.2.3 Zone 7. With regard to the Tal Al-Zaatar zone, this area is considered to have a large population of about 20,000 people and may double, especially in emergency cases due to the large displacement cases from the border areas, so the study suggests the presence of a support line within the decentralized water supply so that the area is provided with a transmission line that connects it to the Al-Moaskar well as shown in figure 5.

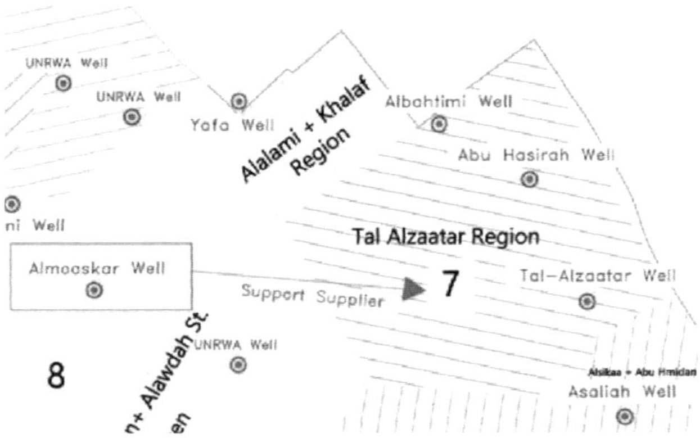

Figure 5. Zone 7 (Tal Alzaatar Region)

4.2.4 Zone 10+11. The eastern areas of Jabalia city suffer from several problems that emerge clearly during the aggression against the Gaza Strip, the most important of which is the inaccessibility of these areas as they are dangerous and militarily active areas, in addition to the presence of a large number of factories and farms that require a continuous and sufficient water supply. Zone 11 depends on Al-Fateh and Bosna wells, So the study suggests supplying Zone 11 with a transmission line from the Al-Hissi well as part of emergency plans, as well as providing Zone 10, specifically Salah El-Din Street, with a direct transmission line to ensure that water reaches the eastern border areas, as shown in figure 6. In addition, linking the wells and alternative line through the Supervisory control and data acquisition (SCADA) system to ensure remote monitoring and controlling, which contributes to improving the emergency period management process.

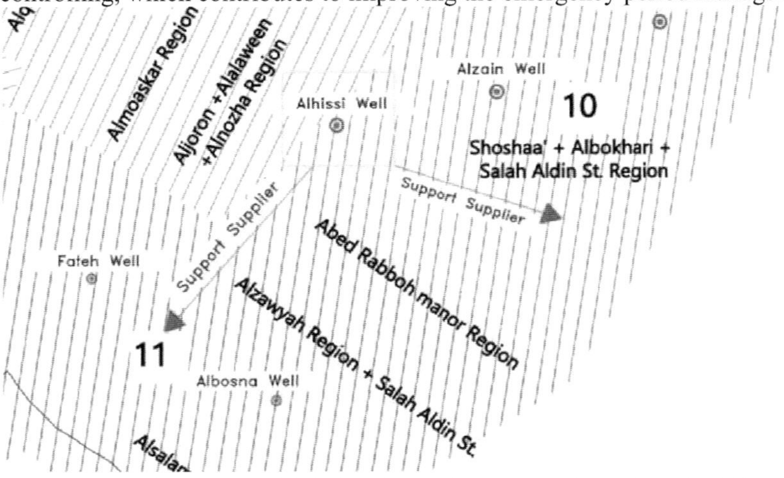

Figure 6. Zone 10+11 (Abed Rabboh manor and Salah Aldin St.)

4.3 Emergency stores

Within the preparedness stage, which is one of the most important elements and components of emergency plans, the study suggests establishing two emergency storages to be in easily accessible areas (the Alriayah and Amer region) as shown in the attached drawing see figure 7.

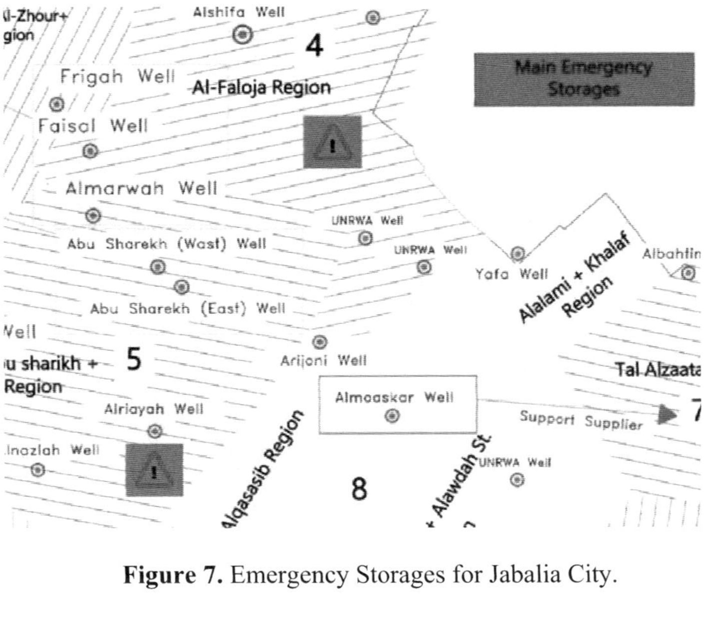

Figure 7. Emergency Storages for Jabalia City.

4.4 Water tankers distribution

Within the state of readiness tools, the drinking water distribution companies operating within the influence of Jabalia Municipality, which can be contracted directly during the period of aggression (emergency) or in advance to provide citizens with water in the targeted areas in the event of the destruction of the main lines and alternative lines and the lack of availability stations or water tanks in that area. Which is distributed as follows: Hanif Company (Western Region), Nile Company (Amer area), Al Falah Company (Abu Iskandar area), Abdel-Salam Yassin Company (Al Zain area - east of Jabalia) and Al Kahlout Company (Tal Al-Zaatar area) as shown in figure 8.

Figure 8. Water Tankers Distribution.

5. Conclusion

Based on results above, this study can conclude that:

1. The decentralization process combines with the elements of existing plans at several important points but in a different organize. It is also characterized as being applied without the need for major changes

on the ground, which will be clarified in two points: Regional Water Utility plan (RWU), and that this process depends primarily on effective management rather than a change in the network and its components.
2. Decentralized supply systems offer the possibility to provide water where centralized supply systems are not feasible due to abnormal conditions.
3. The study entails a number of steps centered around the concept of decentralized water supply and its management process, so that the current public network is the one that is basically relied upon and with a separate management system in cases of emergency and aggression, taking into account the expected situation of regional water utility and its management process.
4. The process of decentralizing water supply covers several concepts, for example, through which it is possible to work on integrating the community within the water resources management plans and the process of implementing emergency plans, in addition to enhancing community participation in the management of water supplies during the emergency phase. It also contributes to the development of the administrative process of the water sector from several aspects (economic, technical, environmental and operational), which makes it a strong preference for the concerned authorities to adopt and work on developing it to be the most reliable tool.
5. The study proposes a number of sequential steps to achieve optimal management of water supplies in light of the aggression on the Gaza Strip, which come in the following order: decentralized water supplies, supporting lines, water tanks, water distribution tankers. which are clarified during this part of the study.

References

[1] Messerschmid C. 2011 Water in Gaza: problems and prospects *Ibrahim Abu-Lughod Institute of International Studies (IALIIS)* Final Report
[2] Fitri A, Yao L, Pratiwi D, Phelia A, Dewantoro F, Safitri D, and Maulud KNA 2021 Effectiveness of a groundsill structure in reducing scouring problem at Cimadur River, Banten Province *In IOP Conference Series: Earth and Environmental Science* **880** 012026
[3] Pratiwi D, Fitri A, Phelia A, Adma NAA & Kastamto 2021 Analysis of urban flood using synthetic unit hydrograph (SUH) and flood mitigation strategies along way Halim River: a case study on Seroja street, Tanjung Senang District *In E3S Web of Conferences* **331** 07015
[4] Fitri A, Hadie MSN, Agustina A, Pratiwi D, Susarman, Pramita G & Salah HA 2021 Analyses of flood peak discharge in Cimadur river basin, Banten Province, Indonesia *In E3S Web of Conferences* **331** 08006
[5] Fitri A, Maulud KNA, Pratiwi D, Phelia P, Rossi F, and Zuhairi NZ 2020 Trend Of Water Quality Status In Kelantan River Downstream, Peninsular Malaysia *Jurnal Rekayasa Sipil (JRS-Unand)*, **16** 178-184
[6] Maulud KNA, Fitri A, Mohtar WHMW, Jaafar WSWM, Zuhairi NZ and Kamarudin MKA 2021 A study of spatial and water quality index during dry and rainy seasons at Kelantan River Basin, Peninsular Malaysia *Arabian Journal of Geosciences* **14** 1-19
[7] Fitri A, Maulud KNA, Rossi F, Dewantoro F, Harsanto P and Zuhairi NZ 2020 Spatial and Temporal Distribution of Dissolved Oxygen and Suspended Sediment in Kelantan River Basin *In 4th International Conference on Sustainable Innovation 2020–Technology, Engineering and Agriculture*, Atlantis Press: 51-54
[8] Group WB 2018 *Securing Water for Development in West Bank and Gaza: Sector Note* World Bank
[9] NIED 2021 *Environmental Impact Study of Gaza War 2021 "Baseline Study"* Final Report, Gaza: pp. 50
[10] PWA. 2017 *The water situation in Gaza* Final Report

[11] Madhoun WA and Rashid M, 2014 An overview on the mutual relationship between forests and air pollution–gain and loss *In National Conference Royal Belum (NCRB)* Gerik Perak, Malaysia
[12] CMWU, PWA and MOL Government 2015 *Water & Wastewater Sector Contingency Plan for Conflict Times (Gaza Strip)* In 3 (ed), CMWU Web Site
[13] Piratla KR and Goverdhanam S 2015 Decentralized Water Systems for Sustainable and Reliable Supply *Procedia Engineering* **118** 720-726
[14] Balfour N, Obando J and Gohil D 2020 Dimensions of water insecurity in pastoralist households in Kenya *Waterlines*, **39** 24-43

IOP Conf. Series: Earth and Environmental Science 1173 (2023) 012038 doi:10.1088/1755-1315/1173/1/012038

Analysis of extreme wind wave based on weibull and fisher tippett I (gumbel) distribution

D Pratiwi[1], R Pratama[1], R A M P[1], A Fitri [1] and F Lestari[1]
[1]Department of Civil Engineering, Faculty of Engineering and Computer Science, Universitas Teknokrat Indonesia, Bandar Lampung, Lampung 35132, Indonesia

E-mail: dian.pratiwi@teknokrat.ac.id

Abstract. Extreme waves are an important factor in coastal structures both offshore structures and onshore structure design. However, due to the difficulty of measuring ocean waves, many areas still do not have measured wave height data. Data processing is required to transform the measured wind data into wind–wave value in order to solve this issue. This study aims to analyze extreme waves based on wind data from BMKG Panjang and BMKG Radin Inten II stations. For analysis of determination of extreme waves can be done by using Fisher Tippet Type I (Gumbel) and Weibull methods. The result show that Weibull method gives the higher value than Gumbel method, with the maximum percentage of 43.04% at the 2-year return period.

1. Introduction

Extreme wave values are required when designing offshore and coastal structures, such as jetty, breakwaters, sea wall, and other onshore or offshore constructions [1,2]. Measured wave data are required to calculate the value of the extreme wave. The measured wave data is the main variable in determining the extreme wave value. The lack of offshore wave measurement data, limits stakeholders from conducting extreme wave analysis for the construction of coastal infrastructure. In order to design extreme waves in various countries, numerical analysis of the wind speed conversion data into waves may be a solution [3]. Fisher Tippett Type I (Gumbel) and Weibull methods are frequently used to determine extreme waves [4]. Both of these methods used to transform a representative wave from a recording wave into an extreme wave with a specific return period [5].

In addition to measured wave data, wind speed data also includes determinants of surface wave height (swell) [6]. Wind data measurements are usually carried out on land, while in wave forecasting, the required wind data is wind data obtained from recording winds at sea level. Generally, wind data measurements were carried out on land. While the required wind data for wave forecasting comes from the measurement of winds at sea level. Therefore, it is necessary to transform wind data on the land that is nearest to the study area in wave forecasting. Finally, using one of the necessary factors, fetch effective, the wind data is transformed into generated wave data [7].

In the Lampung region, Lampung Bay is one of the key areas for economic growth, tourism, and maritime transportation. A study of extreme waves in this region is needed to predict the sustainability of this region as a center for economic growth, tourism, and maritime transportation. In this study, wind data obtained from two stations, that is close to the study area, BMKG Panjang and far from study area, the BMKG Radin Inten II, is used to analyze extreme waves in the area. This is carried out in order to compare extreme waves based on wind data from stations far from study area, BMKG Radin Inten II and close to study area Panjang.

IOP Conf. Series: Earth and Environmental Science 1173 (2023) 012038 doi:10.1088/1755-1315/1173/1/012038

2. Reseach method

This research is located in the Lampung Bay area, precisely in the Panjang Harbour area figure 1. Daily wind data from 2011 to 2022 were gathered from BMKG Radin Inten II and BMKG Panjang. The analysis used in this study is the analysis of wind data into wave data for 12 years of recording, analysis of representative waves and determination of extreme waves using Gumbel and Weibull methods. The results of extreme waves based on wind data from BMKG Panjang and BMKG Radin Inten II will be compared and analyzed for suitability.

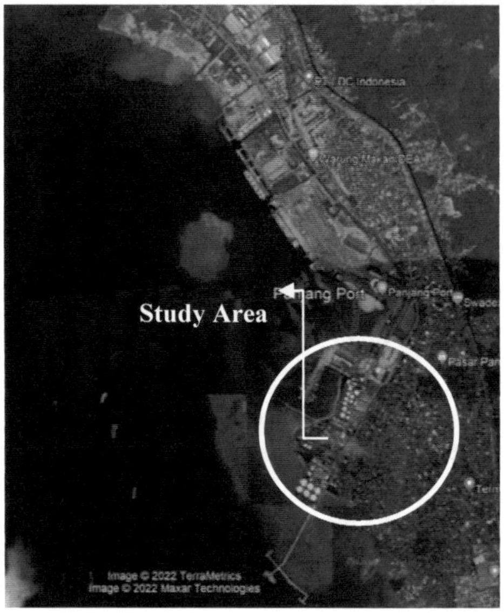

Figure 1. Study area.

2.1. Wind-wave analysis using wind data

2.1.1. Wind stress factor analysis. The wind data that has been calculated up to this point is sea level wind speed data. Next, wave forecasting is carried out based on the results of the conversion. The formulation used to determine the conversion as follows [8]:

$$U = 2.16U_s^{7/9} \quad (1)$$

The wind data that has been calculated up to this point is sea level wind speed data. Where U_s is the wind speed measured by the ship and U is the corrected wind speed. Since the required wind data is at sea level and the measured wind data is wind data on land, the measured wind data is transformed into sea surface wind data. The transformation is carried out by the relationship between wind speed at sea and on land figure 2 which is given by the relationship $R_L=U_W/U_L$. The wind stress factor (U_A) is determined after calculating the value of U using the following formula [8]:

$$U_A = 0.71U^{1.23} \quad (2)$$

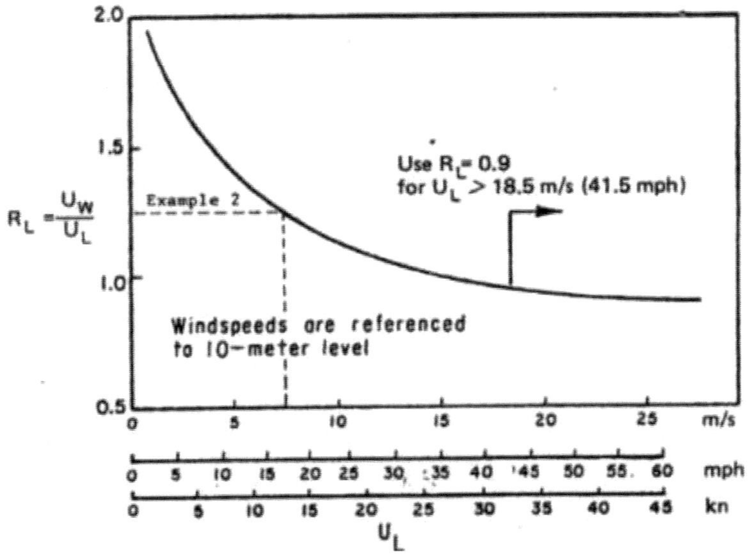

Figure 2. The relationship R_L=U_W/U_L[8].

2.1.2. Fetch. Fetch describes two dimensionless equations that define the characteristic development of wave height and wave period. The determination of the fetch value in this study is given as follows [9,10]:

$$F_{eff} = \frac{\sum X_i cos\alpha}{\sum cos\alpha} \quad (3)$$

The average of effective fetch defines as F_{eff}. X_i is the length of the fetch segment measured from the wave observation point to the end of the fetch. A is deviation on both sides of the wind direction, using increments of 5° to 45° angles on both sides of the wind direction. In this study the fetch used can be illustrated as follows:

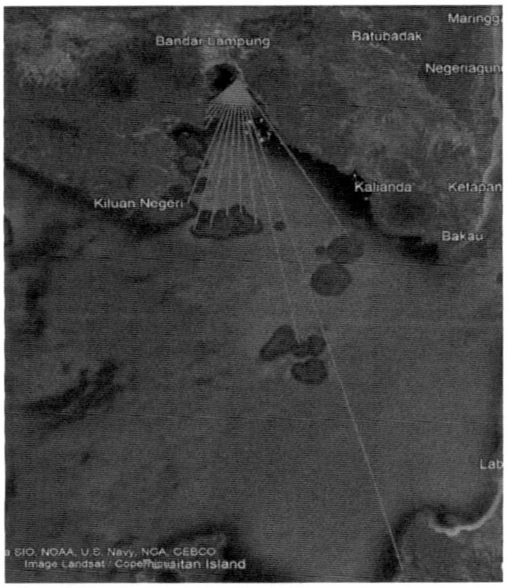

Figure 3. Effective fetch determination in Lampung Bay.

The effective fetch values in the Lampung Bay area, especially at Panjang Harbor are shown in the following table:

Table 1. Effective fetch at Panjang Harbor.

α (°)	Cos α	Xi (km)	Xi cos α
45	0.707	1.444	1.021
40	0.766	1.671	1.280
35	0.819	1.783	1.461
30	0.866	46.268	40.069
25	0.906	7.437	6.740
20	0.940	8.834	8.301
15	0.966	128.961	124.567
10	0.985	68.028	66.995
5	0.996	36.787	36.647
0	1.000	33.894	33.894
5	0.996	34.601	34.469
10	0.985	37.650	37.078
15	0.966	39.164	37.830
20	0.940	33.581	31.556
25	0.906	9.352	8.476
30	0.866	14.274	12.362
35	0.819	12.413	10.168
40	0.766	10.605	8.124
45	0.707	9.319	6.590
Σ	**16.902**	**536.066**	**507.626**

Therefore, the effective fetch value in Lampung Bay, specifically in the area of Panjang Harbor is 30.033 km.

2.1.3. The determination of wind-wave value

The following formula presents the findings of the calculation of waves in the deep sea based on SPM 1984 [8].

$$H_{m_0} = 5.112 \times 10^{-4} U_A F^{1/2} \tag{4}$$

$$T_m = 6.238 \times 10^{-2} (U_A F)^{1/3} \tag{5}$$

Where H_{m0} is the Wind–Wave, T_m is the period of the wave. Due to the influence of the wind, wind–waves will be acquired in this analysis. Furthermore, an analysis of the maximum wind–waves per year is carried out in a 12-year range.

2.2. The analysis of representative wind–waves

At this stage, statistical analysis is carried out regarding the distribution of the maximum calculated wave using the chi square method. The statistical distribution used are Chi Square. The formula for the chi-square is given by [11,12]:

$$X^2 = \sum_{i=1}^{k} \frac{(EF-OF)^2}{EF} \tag{6}$$

Where the EF is the expected value and OF observed value. The conditions that must be achieved are $X^2<X_{cr}^2$ in order for the chosen frequency distribution to be considered acceptable. By calculating the significant level with degrees of freedom, the X_{cr}^2 value can be determined.

2.3. Determination of extreme waves using fisher tippet type I (gumbel) and weibull methods
Determination of extreme waves is carried out by two methods, Gumbel and Weibull method. Where these two distributions are given as follows[8]:

Gumbel distribution

$$P(H_s \leq \widehat{H_s}) = e^{-e^{-\left(\frac{\widehat{H}_s-B}{A}\right)}} \tag{7}$$

Weibull distribution

$$P(H_s \leq \widehat{H}_s) = 1 - e^{-e^{-\left(\frac{\widehat{H}_s-B}{A}\right)^K}} \tag{8}$$

Where $P(H_s \leq \widehat{H}_s)$ is the probability that show $\widehat{H}_s$ not exceeded. H is the representative wave height. $\widehat{H}$ is the wave height with certain value. A is the scale parameter. B is the location parameter. K is the shape parameter. The following formula is used to obtain the values associated with the shape parameter[13]:

$$k = \left(\frac{0{,}9874}{\frac{\sigma}{v}}\right)^{1{,}0983} \tag{9}$$

With σ is deviation standard, υ is the wind speed. In this study, H_s (H representative) is the parameter that used in wave distribution, therefore υ value replace with H_s. furthermore, the probability assigned to each wave height is determined as follows [8]:

Weibull method

$$P(H_s \leq H_{sm}) = 1 - \frac{m-0{,}2-\frac{0{,}27}{\sqrt{K}}}{N_T+0{,}2+\frac{0{,}23}{\sqrt{K}}} \tag{10}$$

Gumbel method

$$P(H_s \leq H_{sm}) = 1 - \frac{m-0{,}44}{N_T+0{,}12} \tag{11}$$

With $P(H_s \leq H_{sm})$ is the probability of representative wind wave height to m that is not exceeded. H_{sm} is the wind wave is the wave in m sequeance. N_T is the number of wave occurrences during wave recording (may be greater than representative wave). Then, find the value of ym given by the following formula [8]:

Weibull method

$$y_m = [-\ln\{-\ln F(H_s \leq H_{sm})\}]^{\frac{1}{K}} \tag{12}$$

Gumbel method

$$y_m = -\ln\{-\ln F(H_s \leq H_{sm})\} \tag{13}$$

With F is the probability of wave height. Further, In order to determine the wave count, linear regression analysis is used. The analysis given in the following [8]:

$$H_m = \widehat{A}y_m + \widehat{B} \tag{14}$$

$\widehat{A}$ and $\widehat{B}$ are the estimated scale parameters from linear regression analysis. Next define significant waves for multiple return periods. With the formula presented as follows[8]:

$$H_{sr} = \widehat{A}y_r + \widehat{B} \tag{15}$$

While, y_r being expressed in the following formula[8]:
Gumbel method

$$y_r = -\ln\left(-\ln\left(1 - \frac{1}{LT_r}\right)\right) \tag{16}$$

Weibull method

$$y_r = (\ln(LT_r))^{\frac{1}{\kappa}} \tag{17}$$

With H_{nr} is the significant height with the return period T_r. T_r is the return period in year. K is data length in year and L is the average number of events per year with the value L=N/K. Confidence interval calculations are necessary for the analysis of extreme waves. The confidence interval in this study is provided by [8]:

$$\sigma_{nr} = \frac{1}{\sqrt{N}}[1 + \alpha(y_r - c + \varepsilon \ln v)^2]^{1/2} \tag{18}$$

$$\alpha = \alpha_1 e^{\alpha_2 N^{-1,3} + \kappa\sqrt{-\ln v}} \tag{19}$$

$$\sigma_r = \sigma_{nr}\sigma H_s \tag{20}$$

With σ_{nr} is normalized deviation standard of significant wave height with return period T_r. N is amount of significant wave height data. α_1, α_2, e, ε, κ value is empiric coefficient. σ_r is the error standard of significant wave height with return period T_r and σH_s deviation standard of significant wave height data. Then, wave height will be compared with data from BMKG Panjang and BMKG Radin Inten II.

3. Result and discussion

In the study there are 3 main discussions, including the results of wind data analysis into waves, representative wave testing and determination of extreme waves in Lampung Bay using the Gumbel and Weibull methods.

The results of the analysis of wind data into calculated waves with data sourced from BMKG Panjang and BMKG Radin Inten II are presented in figure 4 and figure 5 below:

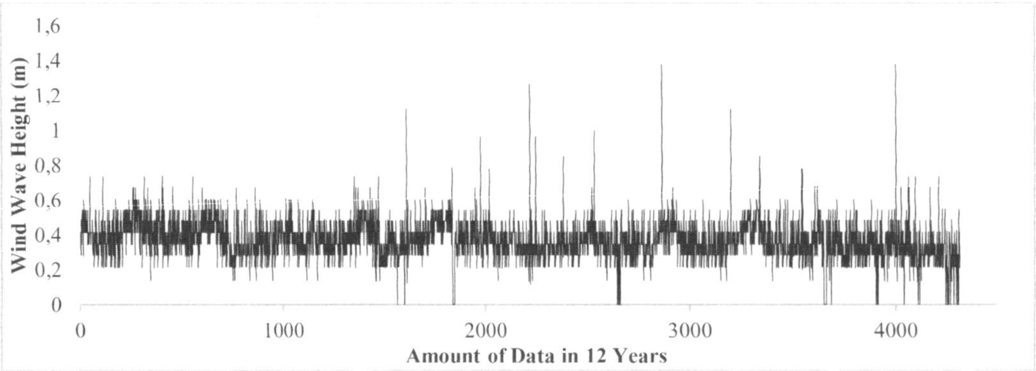

Figure 4. The relationship between wave height and amount data for 12 years with wind data source from BMKG Radin Inten II.

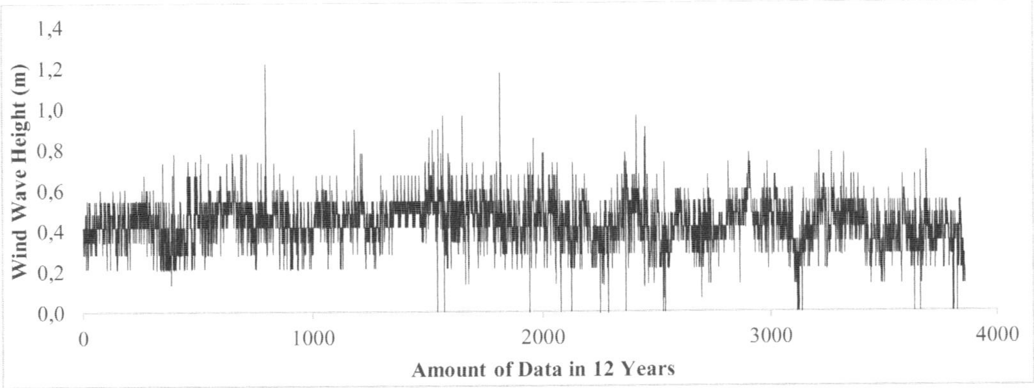

Figure 5. The Relationship between Wave Height and Amount data for 12 years with Wind data source from BMKG Panjang.

A maximum wave analysis is then conducted using wind-wave data from 2011 to 2022 based on the results of the wave analysis in figure 4 and figure 5. The maximum wind–wave results or representative waves are listed in the following table:

Table 2. Representative wind –waves based on wind data at BMKG Radin Inten II and BMKG Panjang stations.

Period	Representative Wind Wave/H_{m0} (m)	
	BMKG Panjang	BMKG Radin Inten II
2011	0.732	0.732
2012	0.777	0.732
2013	1.216	0.670
2014	0.893	0.732
2015	0.962	1.117
2016	1.168	1.262
2017	0.962	0.998
2018	0.893	1.379
2019	0.777	1.117
2020	0.777	0.777
2021	0.777	1.379
2022	0.777	0.732

The maximum representative waves determined from the recording data of BMKG Radin Inten II and BMKG Panjang are 1.379 m and 1.216 m, respectively, based on the analysis of wave data from 2011 to 2022. While the average representative waves are 0.969 m and 0.892 m respectively. The outcomes of sample waves are then used as the basis for estimating extreme waves. Previous, the distribution of the wave data distribution must be validated. The test used is the Chi-Square method. With the results of testing with wind data sources from BMKG Radin Inten II and BMKG Panjang and giving results 0.5 and $X^2 < X_{cr}{}^2$. This shows that the distribution of the data used meets the requirements and can be used in determining extreme waves with a certain return period.

The Gumbel and Weibull Methods were used to determine the maximum wave height in Lampung Bay. The results of the extreme waves are presented in figure 6 and figure 7 respectively.

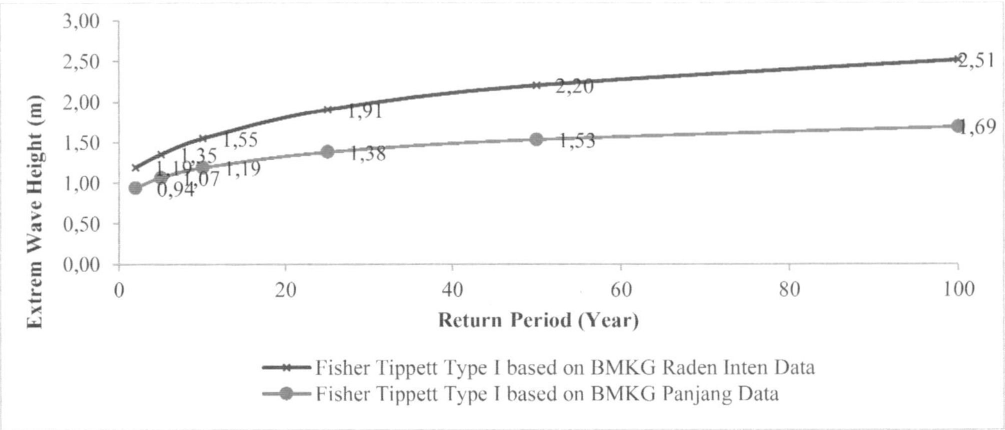

Figure 6. Comparison between extreme wave heights and certain return period with gumbel's method.

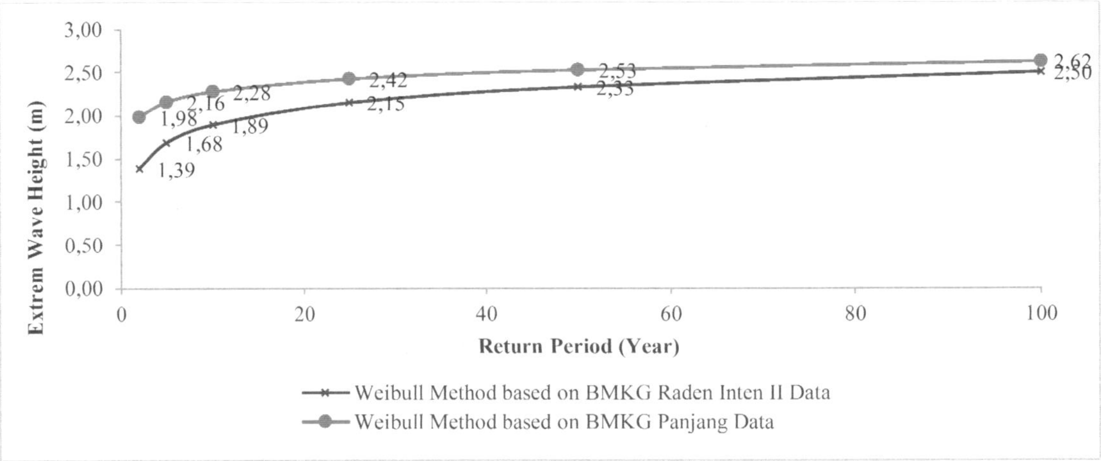

Figure 7. Comparison between extreme wave heights and certain return period with weibull's method.

The results indicate that the extreme wave value of the Weibull method, which uses data from BMKG Radin Inten II and BMKG Panjang, provides a higher value than the Gumbel method. The Weibull method, which used recording data from the BMKG Panjang, produced the highest value. The results obtained 1.98 m, 2.16 m, 2.28 m, 2.42 m, 2.53 m, and 2.62 m for the return periods of 2 years, 5 years, 10 years, 25 years, 50 years, and 100 years, respectively. The Gumbel method's results revealed a higher value at the BMKG Radin Inten II station than at the BMKG Panjang station. This is due to differences in wind data, where at BMKG Panjang station there are some wind data that are not recorded.

The percentage of extreme waves using the Weibull method based on wind data from the BMKG Panjang and BMKG Radin Inten II stations is shown in figure 8.

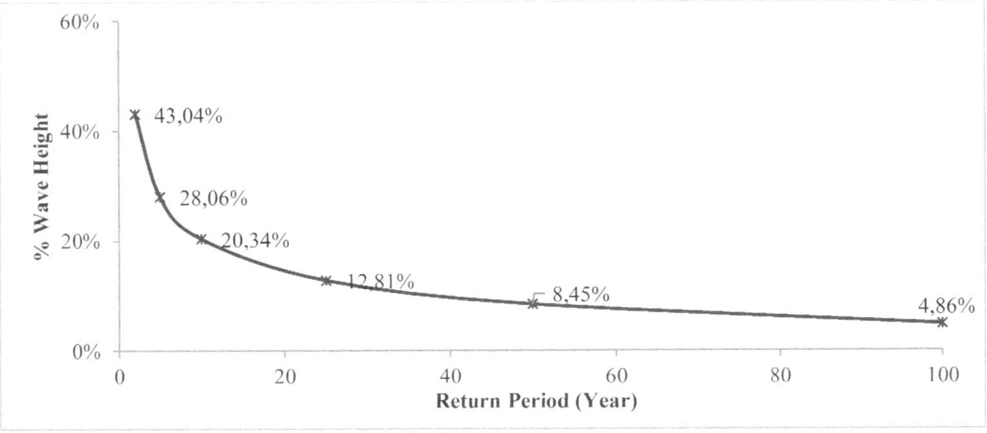

Figure 8. Percentage of difference in extreme waves based on wind data at BMKG Panjang against BMKG Radin Inten II.

Figure 8 shows the percentage of extreme waves based on wind data from BMKG Panjang against BMKG Radin Inten II. Extreme waves based on wind data at the BMKG Panjang were 43.04%, 28.06%, 20.34%, 12.81%, 8.45% and 4.86% with a return period of 2, 5, 10, 25, 50 and 100 years respectively to the BMKG Radin Inten II station.

Basically, the value of extreme waves with a certain return period is needed in determining the elevation planning of onshore and offshore infrastructures, both of coastal protection and coastal infrastructure [14]. The value used in the elevation of an infrastructure is the maximum value obtained from the wave transformation method into extreme waves. Based on the analysis's results, Weibull method gives higher value than Gumbel method. While the wind data which is transformed into extreme waves using Weibull method, BMKG Panjang data gives higher results than the data sourced from BMKG Radin Inten II. This is due to the reason that BMKG Radin Inten II station is farther away from the research location than BMKG Panjang station.

4. Conclusion

According to the study's findings, it is feasible to draw the following conclusions:

1. Several analyses are required to determine the calculated severe waves, including the examination of wind data, the selection of annual representative waves, and tests of the suitability and data distribution. Based on the results of the wave transformation analysis, the highest maximum values are at an altitude of 1.379 m and 1.216 m, for the average representative wave height of 0.969 m and 0.892 m based on wind recording data sourced from BMKG Radin Inten II and BMKG Panjang, respectively.
2. The results of wave analysis using the Weibull and Gumbel methods, show that the highest extreme wave values are found in the Weibull method. This value can be used as a reference in determining the elevation of coastal buildings and coastal protection buildings.
3. Extreme waves based on wind data sourced from BMKG Panjang give higher results than data sourced from BMKG Radin Inten II at 2, 5, 10, 25, 50 and 100 year return periods. This happens because the distance between BMKG Panjang and the research location is closer than the distance from BMKG Radin Inten II station to the research location.
4. The percentage ratio of extreme waves based on wind data sourced from BMKG Panjang gives higher results than BMKG Radin Inten II, with a maximum percentage of 43.04% at the 2-year return period.

Acknowledgment
Thank you for Universitas Teknokrat Indonesia for contributing to this research process for completion of this paper.

References

[1] Imani H, Abbaspour M, Kameranzad B, and Amirinia G 2014 Assessment of Extreme Wave Height in Chabahar Region, North of Gulf of Oman in *11th Int. Conf. Coasts, Ports Mar. Struct. (ICOPMAS 2014)* (Tehran, Iran) pp.24–26
[2] Niroomandi A, Ma G, Lou X Y S, and Xue P 2018 *E*xtreme Value Analysis of Wave Climate in Chesapeake Bay *Ocean Eng.* **159** 22–36
[3] Moeini M H, Etemad-Shahidi A, and Chegini V 2010 Wave Modeling and Extreme Value Analysis Off The Northern Coast of The Persian Gulf Appl *Ocean Res.* **32** 209–218
[4] Park S B *et al* 2020 Extreme Value Analysis of Metocean Data for Barents Sea *J. Ocean Eng. Technol* **34** 26–36
[5] Muzathik A M, Nik W B W, Samo K B, and Ibrahim M Z 2011 Ocean Wave Measurement and Wave Climate Prediction of Peninsular Malaysia *J. Phys. Sci* **22** 77–92
[6] Dobrynin M, Murawski J, Baehr J, and Ilyina T 2015 Detection and Attribution of Climate Change Signal in Ocean Wind Waves *J. Clim* **28** 1578–1591
[7] Karimpour A, Chen Q, and Twilley R R 2017 Wind Wave Behavior in Fetch and Depth Limited Estuaries *Sci. Rep* **7** 1–8
[8] CERC 1984 *S protection manual: Volume I and II* (Washington: US Army Corps of Engineering)
[9] Hwang P A 2016 Fetch- and Duration-Limited Nature of Surface Wave Growth Inside Tropical Cyclones: With Applications to Air-Sea Exchange and Remote Sensing *J. Phys. Oceanogr* **46** 41–56
[10] Hwang P A and Fan Y 2017 Effective Fetch and Duration of Tropical Cyclone Wind Fields Estimated From Simultaneous Wind and Wave Measurements: Surface Wave and Air–Sea Exchange Computation *J. Phys. Oceanogr.* **47** 447–470
[11] Farooq M, Shafique M, and Khattak M S 2018 Flood Frequency Analysis of River Swat Using Log Pearson type 3, Generalized Extreme Value, Normal, and Gumbel Max Distribution Methods *Arab J. Geosci.* **11** No. 9*)*
[12] Pratiwi D, Fitri A, Phelia A, Adma N A, and Kastamto 2021 Analysis of Urban Flood Using Synthetic Unit Hydrogaph (SUH) and Flood Mitigation Strategies Along Way Halim River a case study on Seroja street, Tanjung Senang District in *Proc. ICDMM 2021* (France: E3S Web of Conferences) **331** 07015
[13] Mahmood F H, Resen A K, and Khamees A B 2020 *Wind characteristic analysis based on Weibull distribution of Al-Salman site, Iraq Energy Reports* (Amsterdam: Elsevier) Vol. 6 Iss. 3 pp.79-87
[14] Cousineau J and Murphy E 2022 Numerical Investigation of Climate Change Effects on Storm Surges and Extreme Waves on Canada's Pacific Coast *Atmosphere* **13** 311

IOP Conf. Series: Earth and Environmental Science 1173 (2023) 012039 doi:10.1088/1755-1315/1173/1/012039

Analysis of drainage network capacity in Rawa Jaya Area, Ilir Timur I, Palembang

S Nisumanti[1], R Baniva[1] and Afrizal[1]

[1]Universitas Indo Global Mandiri, Jalan Jendral Sudirman No. 629. Palembang. South Sumatera. Indonesia

E-mail: sartika.nisumanti@uigm.ac.id

Abstract. Palembang is a lowland which is dominated by swamp areas.. The problem of flooding in this area stems from a drainage network system that is almost cut off, causing high rainfall and water not flowing properly into the river. This study aims to analyze the drainage network system's performance based on the channel's capacity to the channel's condition in analyzing the hydrology using the probability distribution method. The frequency suitability test was carried out using Smirnov-Kolmogorov and Chi-Square. The results showed that the rainfall was 116.74 mm/hour and the design flood discharge (Qr) at the ten-year return period was 3.684 m3/second. The carrying capacity of the discharge channel (Qs) is smaller than the value of Qr. Indicates that the existing drainage channels cannot accommodate the flood discharge, resulting in flooding. The channel at point 4 leads to the discharge channel; the capacity value of Qs is 4.0296 m3/second, while Qr is 3.684 m3/second. The calculation results can be estimated that the channel discharge capacity is smaller than the planned flood discharge, so the drainage channel capacity cannot accommodate flood flows.

1. Introduction

Palembang is a lowland area with many swamps and tributaries. It always causes rain in several areas to be inundated [1]. Rapid population growth in Palembang City will increase the need for space and land for burning, as well as supporting facilities and infrastructure. It turns open land or swamps into settlements, resulting in the loss of the function of swamp land as water catchment areas [2]. Increasing flood points will pose a threat to infrastructure [3]. The impact of changes in the function of swamp land is that the soil will be saturated so that the falling rainwater cannot be absorbed into the soil, which causes rainwater to flow and can cause flooding. Several areas in Palembang City experience flooding almost every year during the rainy season [4]. One of the areas experiencing these problems is Kampung Pahlawan, problems such as the state of the drainage network in the Rawa Jaya area, almost all of the channels are not well connected. If it rains, flooding often occurs in several areas behind UIN Raden Fatah. It is certainly very troubling community activities.

This research was conducted to determine the volume of drainage discharge [5,6] and the planned flood discharge of the drainage channel at the Rawa Jaya location.. This research is expected to assist the authorities in dealing with flood inundation in this area.

The drainage system serves to reduce excess water from the area so that the channel can be used properly. The drainage system must be maintained to prevent flooding. Drainage is an effort to control groundwater quality involving channels and drainage not only for surface water, but also for groundwater [8].

Rain features include:

a. Rain duration is obtained from the results of data collection on the rain gauge automatically.

b. Rain intensity is the amount of rain in a high unit of time [9]. Rain volume intensity is based on the duration and frequency of rain. At the same time, the time of rain is related to the time of concentration. Intensity is used to analyze the hydrology of the drainage system.

c. The rain curve is a graph of the relationship between the intensity and duration of rain.

d. The concentration time is used to determine the time of water flow and the farthest point from the flow area to the control point.

Rainfall analysis is 10-year data; frequency analysis is calculated using the equation in table 1

Table.1 Frequency analysis parameter equation [10,11].

Parameter	Formula
Average value	$\overline{X} = \frac{\sum_{i}^{n} -_1 x_i}{n}$
Standard deviation	$Sd = \sqrt{\frac{\sum_{i}^{n} (X_i - \overline{X})^2}{n-1}}$
Slope coefficient	$Cs = \frac{n\sum_{i}^{n} -_1 \{(X_{i)} - X\}^3}{(n-1)(n-2)Sd^3}$
Coefficient of kurtosis	$Ck = \frac{\frac{1}{n}\sum_{i}^{n} {}_{-1}\{(X_i) - \overline{X}\}^4}{Sd^4}$
Coefficient of variation	$Cv = \frac{Sd^4}{\overline{X}}$

While the frequency distribution equation used for hydrological analysis is shown in Table 2.

Table.2 Frequency Distribution Equation [12, 13, 14]

Distribution	Formula
Normal distribution	$Xt = \overline{X} + K_T S$
Normal Log distribution	$X_T = 10\overline{\log x} + K_T.S$
Log-Pearson distribution III	$X_T = 10\overline{\log x} + K.S$
Gumbel Distribution	$X_T = \overline{X} + \frac{(Y_T - Y_n)}{S_n} S$

Determine the distribution of statistical parameters from the previous event data, such as Cs, Cv, and Ck. The analysis uses rain data from the rain gauge post. The conditions are the selective distribution of Gumbel Type I -11.1396 < CS < 1.1396 and CK = 5.4002, Distribution of Log Person Type III CS $\neq 0$

The frequency suitability test is used to find the truth between the observation results and the distribution model that is expected or obtained theoretically and to find out the truth of the hypothesis (accepted/rejected) [15].

a. Smirnov-Kolmogorov test [15]

The Smirnov – Kolmogorov compatibility test is a non-parametric test that does not use a special distribution function for use in the study area.. This test determines the maximum vertical distance between observations and theory based on the sample distribution. The maximum distance equation is:

$$Dn = \max \mathrm{I}P(x) - (Po(x)\mathrm{I} \tag{1}$$

b. Chi-Square Test (SNI 2415:2016)

The chi-square test determines whether the selected distribution can represent the statistical distribution of the analyzed sample with the equation.

$$\frac{k}{X(i=1)} = \frac{(oi - Ei)^2}{Ei} \tag{2}$$

2. Materials and methodology

2.1. Study area

This research was conducted in a drainage channel in the Rawa Jaya area, Pahlawan Village, Ilir Timur I District, Palembang City. The location is shown in figure 1.

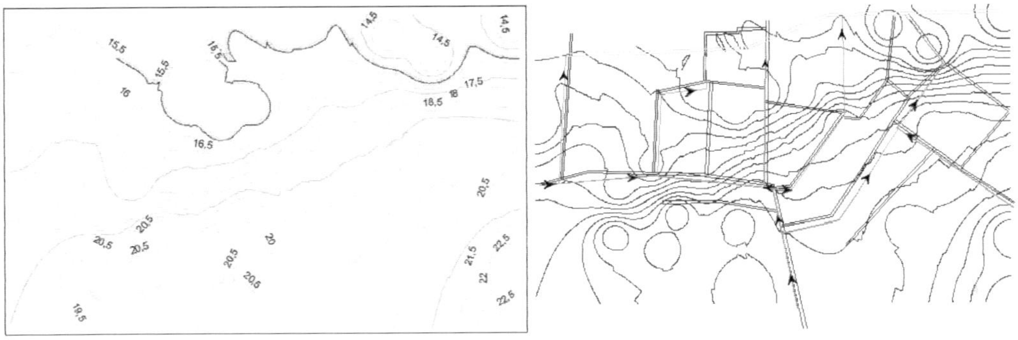

Figure 1. Research location in the Rawa Jaya Area.

2.2. Flow diagram

The flow diagram of this research can be seen in figure 2.

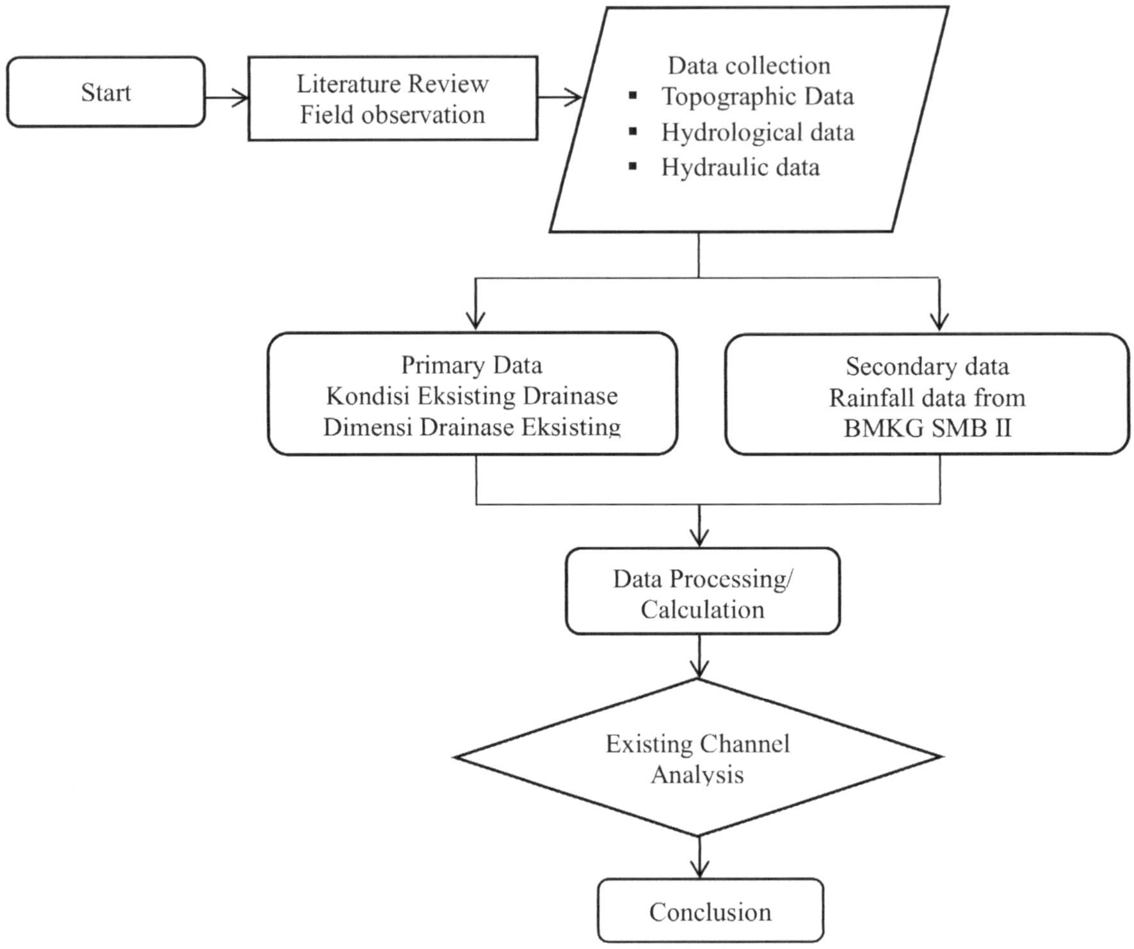

Figure 2. Flow diagram.

2.3. Method of collecting

The stages of this research go through several stages, namely:

a. Data Collection

Primary data is taken directly from the field, namely drainage measurements such as cross-section, longitudinal section, elevation, and data collection on drainage conditions. At the same time, secondary data in the form of literature and rainfall data were obtained from BMKG SMB II Palembang.

b. Hydrological Analysis [10]

Hydrological analysis is used to obtain the planned rainfall intensity, the initial stage begins with calculating the annual and monthly average rainfall [16], analyzing the frequency of rainfall with probability distribution methods, namely Normal, Log-Normal, Gumbel, and Log-Person Distributions. Distribution III, can be calculated by the Mononobe equation [17].

c. Frequency Conformity Test [11]
The frequency fit was carried out using the Smirnov-Kolmogorov test and the Chi-Square. Test.
d. The hydraulic analysis begins by calculating the flow velocity and drainage discharge capacity [18].

3. Result and discussion

3.1. Hydrology analysis

The Hydrological analysis of the data used is rainfall data for ten years obtained from the Palembang Meteorology, Climatology and Geophysics Agency Palembang rainfall station. The average rainfall data is obtained from the rain data, as shown in figure 3.

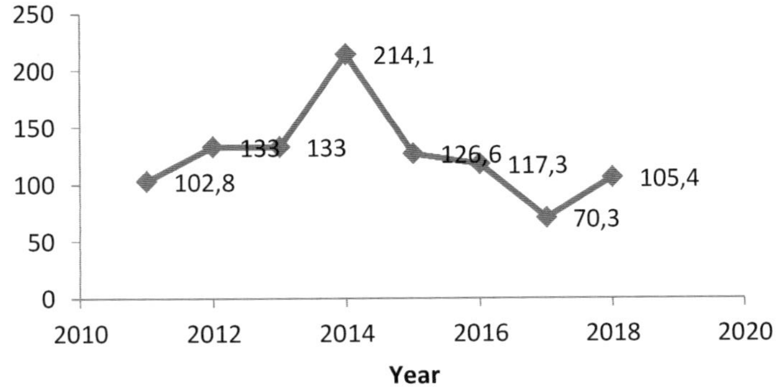

Figure 3. Daily maximum rainfall

Figure 3 shows the intensity of rainfall for ten years with the highest intensity of rain, due to the intensity of rainfall reaching the highest value of 214.1 mm and the average rainfall data obtained is 121.95 mm.

3.2. *Rain Frequency Analysis*

Rain frequency analysis was carried out to obtain the planned rainfall through the probability distribution calculation method, including the mean, standard deviation (s), coefficient of variance (Cv), Skewness coefficient (Cs), and Kurtosis coefficient (Ck) [11]. The results of data calculations as shown in table 3

Table 3. Rain frequency analysis.

	Distribution Normal	Distribution Log-Normal	Distribution Gumbel	Distribution Log Pearson III
$(\overline{X})$	121.95	2.070	121.95	2.079
S	37.326	0.122	37.326	0.122
Cv	0.306	0.059	0.306	0.059
Cs	1.676	0.486	1.676	0.486
Ck	4.615	2.679	4.615	2.679

From the results of the statistical parameter test, which is close to Log Person III, the boundary conditions for determining the distribution are shown in Table 4

Table 4. Suitable distribution type.

No	Type Distribution	Condition	The calculation results	Description
1	Normal	Cs = 0, Ck = 3	Cs=1.676 Ck = 4.615	Not
2	Log Normal	Cs = 3 Cv= 1.8; Cv = 0.6	Cs= 0.486 Cv = 0.059	Not
3	Gumbel	Cs≤1.1396; Ck = 5.4002	Cs= 1.676 Ck = 4.615	Not
4	Log Pearson III	Alternative	Cs= 0.486 Ck = 2.679	Approach

Table 4 shows that the type of distribution to determine the closest distribution is the Pearson log distribution III.

3.3. Spread match test

The results of calculating the suitability of the frequency distribution of the data sample to the probability distribution function used the Chi-Square test and the Smirnov-Kolmogorov test. Determination of the number of classes (K) obtained the calculation results 5, X = 35.93, the value of the Degree of Saturation (DK) 3, and Ei = 2. The results of the Chi-Square calculation are shown in Table 5

Table 5. Chi-square test results.

Class	Probability (%)			Oi	Ei	$\frac{(Oi-Ei)^2}{Ei}$
1	52.325	< x <	88.275	1	2	0.5
2	88.275	< x <	116.225	4	2	2
3	116.225	< x <	160.175	4	2	2
4	160.175	< x <	196.125	0	2	2
5	196.125	< x <	232.075	1	2	0.5
Amount				10	2	7

Based on the results in table 5, the Chi-Square value is 7, while the critical Chi-squared (Cr) value is 7.815, so the calculated chi-square value is smaller than the Cr value, indicating that the distribution hypothesis being tested is acceptable.

While the Smirnov-Kolmogorov compatibility test uses an average rainfall value of 121.95 mm, the distribution match test results are shown in Table 6.

Table 6. Smirnov test results – kolmogorov.

Rain (mm)	M	$P(x) = \frac{m}{n+1}$	P(x<) = 1-P (x)	D
121.1	1	0.0909	0.9091	0.0202
133	2	0.1818	0.8182	0.0404
133	3	0.1818	0.8182	0.0404
126.6	4	0.2727	0.7273	0.0606
117.3	5	0.3636	0.6364	0.0808
115.2	6	0.4545	0.5455	0.1010
105.4	7	0.5454	0.4545	0.1212
102.8	8	0.6364	0.3636	0.1414
101.8	9	0.7273	0.2727	0.1616
70.3	10	0.8182	0.1818	0.1818
		Dmaks		0.1818

Based on the compatibility test carried out with two methods, namely the Chi-Square test and Smirnov Kolmogorov, the distribution of rain data in Rawa Jaya follows the Log Person distribution III.

3.4. Rainfall plan

Rainfall data for ten years obtained from BMKG SMB II data aims to analyze the frequency by testing the distribution of the Normal, Log-Normal, Gumbel, and Log-Pearson III methods. whereas in determining the amount of rainfall the planned return period starts from 2, 5, 10, 25, and 100 years as shown in table 7.

Table 7. Return period calculation based on rain data.

Return Period (Years)	Rain characteristics (mm) according to Probability							
	Normal		Log Normal		Gumbel		Log Pearson III	
	K_T	X_T	K_T	X_T	K_T	X_T	K_T	X_T
2	0	121.95	0	117.517	-0.136	116.891	-0.082	114.863
5	0.84	153.304	0.84	148.807	1.059	161.461	0.809	147.508
10	1.28	169.727	1.28	168.393	1.849	190.965	1.322	170.412
20	1.64	183.164	1.64	186.321	2.607	219.262	1.907	200.839
50	2.05	198.468	2.05	209.074	3.588	255.892	2.306	224.670
100	2.33	208.919	2.33	226.190	4.324	283.344	2.679	249.492

From table 7, The calculation results of some of the best methods are the Pearson III Log distribution by producing rain plans of 114 mm, 147 mm, 170 mm, 200 mm, 224 mm, and 249 mm.

3.5. Hydraulics analysis

Hydraulic analysis is calculated to determine the cross-sectional capacity to water discharge which begins with calculating the V and Qs values; the results on table 8.

Table 8. Rate speed V and Q

Drainage	V	Qs
8 - 6	3.999	2.399
7 – 1	2.060	0.6798
6 until the drainage drain	3.619	1.52
5 until the drainage drain	3.638	1.528
4 until the drainage drain	5.037	4.0296
3 until the drainage drain	2.890	0.867
2 until the drainage drain	3.099	1.108
1 – 2	1.458	0.219

Table 8 shows the channel discharge capacity (Qs) on channel 4 to the drainage channel of 4.0296 m3/s which is greater than the planned flood discharge value (Qr) of 3.684 m3/s so that the existing drainage channel can accommodate the flood discharge that planned. But in other channels, the Qs value is smaller than Qr's. This shows that the existing canal cannot accommodate flood discharge, resulting in inundation or flooding. One of the problems with this drainage is that the existing canal's dimensions cannot accommodate runoff. In addition, the drainage conditions are damaged, and no maintenance is carried out; in the channel, there is a lot of garbage which causes the water flow to be blocked.

4. Conclusion

The results of research and calculations as well as analysis can be concluded that: Rainfall is quite high with rainfall intensity (i) 116.74 mm/hour and a 10-year return period Qr value of 3,684 m^3/second. The carrying capacity of the channel discharge is smaller than the value of the planned flood discharge (Qr). This indicates that the existing drainage channels cannot accommodate flood discharges, resulting in inundation or flooding. The channel at point 4 leads to the drainage channel, the discharge capacity value (Qs) is 4.0296 m^3/second, and the design flood discharge value (Qr) is 3.684 m^3/second so that the existing drainage channel can accommodate the planned flood.

Reference

[1] Lumuan M D, Sujono J and Supraba I 2019 Pola Penanganan Genangan Air Hujan Dengan Pendekatan *Jurnal Teknisia vol.* **XXII** No. 2 p 400–409
[2] Sjarief A W P and Lasminto U 2020 Analysis of water surface profile and flood discharge of Cijangkelok river *IOP Conf. Ser. Mater. Sci. Eng.* **930** 012079
[3] Faqih N and Lestarini W 2020 Redesign of road drainage system in jalan Banyumas Km 7 Selomerto, Wonosobo with normal discharge method *J. Phys. Conf. Ser.* **1511** 012063
[4] Cely-Calixto N J, Carrillo-Soto G A and Bonilla-Granados C A 2020 Optimization of a storm drainage network using the storm water management model software in different scenarios *J.*

Phys. Conf. Ser. **1708** 012030

[5] Hidayat R, Setiyadi S and Hutabarat L E 2021 Evaluation of road drainage capacity to improve optimized road performance in Kebon Pala Area East Jakarta *IOP Conf. Ser. Earth Environ. Sci.* **878** 012051

[6] Imaadiiddiin M H, Saud I, Azis S K, Wahyudi H and Adiningtyas T 2021 The Effect of Urban Drainage on the Determination the Time of Storage in the City Development Area *IOP Conf. Ser. Earth Environ. Sci.* **930** 012043

[7] Yuswo M, Wahyudi S I, Soedarsono, Boogard F C and Boer E 2022 Drainage System of Tegalsari Polder for Handling Flood and Tide in Tegal City Indonesia *IOP Conf. Ser. Earth Environ. Sci.* **955** 012008

[8] Suripin 2004 *Sistem Drainase Perkotaan yang Berkelanjutan* (Yogyakarta: Pustaka Andi Offset)

[9] Harahap, R. K.Jeumpa, dan E.M. Silitonga (2020). *Drainase Permukiman:Prinsip Dasar dan Aplikasinya* (Medan: Yayasan Kita Menulis)

[10] Bambang T 2014 *Hidrologi Terapan* (Yogyakarta: Pustaka Beta Offset)

[11] Hadisusanto N 2011 *Aplikasi Hidrologi* (Yogyakarta: Yogya Media Utama)

[12] Puspita N 2017 Analisa Kebutuhan Kolam Retensi Bandara Atung Bungsu Kota Pagar Alam *Jurnal Teknik vol* **2** *No. 2 p 116-129*

[13] Hasmar H H A 2011 Aplikasi Hidrologi (Yogyakarta: Yogya Media Utama)

[14] Badan Standar Nasional 2016 *Tata Cara Perhitungan Debit Banjir Rencana, SNI 2415:2016*

[15] Nisumanti S, Aurdin Y, Bendung D A S and Buah D A S 2019 Analisis Multikriteria Untuk Memetakan Daerah Rentan Banjir pada Sub Das Buah, Palembang *Jurnal Tekno* vol **2** No. 1 p 64-71

[16] Mulyono 2010 *Penataan Drainase Perkotaan* (Semarang: Pustaka Graha Ilmu)

[17] Haris V T, Putri L D and Soehardi F 2020 Analysis of Urban Whale Drainage Capability Pekanbaru to the Maximum Intensity of Rainfall *IOP Conf. Ser. Earth Environ. Sci.* **469** 012011

[18] Sarminingsih A, Hadiwidodo M, Rezagama A, Sausan K S and Nurullah 2021 Hydraulic analysis of Semarang River in supporting the drainage channel as water tourism *IOP Conf. Ser. Earth Environ. Sci.* **896** 012010

IOP Conf. Series: Earth and Environmental Science 1173 (2023) 012040 doi:10.1088/1755-1315/1173/1/012040

Adsorption using fly ash from stam power plants for acid mine drainage treatment

S Siswati[1], H Pratama[2] and M Giatman[2]

[1]Faculty of Public Health, Andalas University, Padang, Indonesia
[2]Master Program of Mining Engineering Pembangunan Nasional Veteran University Yogyakarta, Indonesia

E-mail: siswati@ph.unand.ac.id

Abstract. Acid mine drainage has a negative impact on the environment. Adsorption is an effective and efficient method for acid mine drainage treatment, besides that adsorption does not have a negative impact on the environment. This study aims to analyze the ability of fly ash from a steam power plant as an adsorbent in acid mine drainage treatment. Using experimental methods and batch adsorption mechanisms, with variations of adsorption parameters carried out including adsorbent dose, stirring time, and stirring speed. The test results show that fly ash is very effective in acid mine drainage treatment. Where there is an increase in pH and a decrease in the concentration of Mn and Fe metals in acid mine drainage to match the quality standard. The increase in pH was in the range of 6.3 to 7.4 with the highest effectiveness of 54.05%. In the reduction of Mnmetalthe highest effectiveness was 78.14%, while in Fe metal the highest effectiveness was 85.83%. Fly ash can be used and is very useful in reducing the impact on public health.

1. Introduction

Acid mine drainage is one of the wastes produced by the mining industry. Acid mine drainage is formed from a leachate, seepage, or drainage that is affected by the natural oxidation of sulfide minerals in rock, where the rock is exposed during the mining process[1].The formation of acid mine drainage is characterized by a low pH (1.5 – 4), high concentrations of dissolved metals (iron, aluminum, cadmium, copper, lead, zinc, arsenic and mercury), high acidity values (50 – 1500 mg/l CaCO3), high sulfate value (500 – 10,000 mg/l), salinity value (1 – 20 mS/cm) and low dissolved oxygen concentration[2].

In general, the impact of mining on the environment is a decrease in land productivity, increased soil density, erosion and sedimentation, the occurrence of soil movements or landslides, disturbance of flora and fauna, disruption of public health and the impact on microclimate change. Impacts and hazards that threaten health are still being felt in places that were once mined, because people can be exposed to mining waste and chemicals that are still attached to the ground and if not managed properly. Acid mine drainage has a negative impact on public health and the environment. Acid mine drainage can reduce water quality, both surface water and ground water. If it is channelled into the river, it will have an impact on the people living along the river and will disrupt the health of the people living along the river and will disturb living things on land and in waters. Mining threatens health in various ways including dust, chemical spills, toxic fumes, heavy metals and radiation can poison miners and cause health problems throughout their lives.

Surface and groundwater at mining sites can be polluted for years to come. As water is used up, land can dry out and become unusable for farming or grazing livestock. The long-term damage from contaminated water will last much longer than the short-term economic benefits derived from mining activities. Acidic water is mined so that it is not dangerous requires management, because it is necessary to do good management of mining water by treating it.

If not managed properly, acid mine drainage has a negative impact on the environment. Acid mine drainage treatment can be done with active treatment and passive treatment, active treatment is done by adding chemicals continuously, while passive treatment is the most effective and efficient because it takes place naturally and continuously. Adsorption is the most commonly used method because it has a simpler and more economical concept [3]. The adsorption method has several advantages, including the relatively simple process, relatively high effectiveness and efficiency, and no side effects in the form of toxic substances [4], besides that adsorption does not have a negative impact on the environment.

In general, adsorption is divided into physical adsorption and chemical adsorption. Physical adsorption is reversible, because the adsorbed molecules are not strongly attached to the surface so that the bound ion atoms can be released again. Chemical adsorption begins with physical adsorption where the particles adhere to the surface of the adsorbent and form chemical bonds. The chemical bond that occurs is very strong binding gas or liquid molecules to the solid surface so it is very difficult to be released again (irreversible), this adsorption involves a force that is much greater than physical adsorption. Based on physical adsorption, the batch system is one of the adsorption mechanisms by immersing the adsorbent in the solution to be adsorbed by weight and observing changes in quality at certain intervals.

Steam power plants produce residues, one of which is fly ash, where the use of fly ash that has not been maximized has a negative impact on the community around the location of the steam power plant. Fly ash is a promising alternative as conventional adsorbent [5], because there is a large amount of unburned carbon and has a good adsorption capacity [6]. This study aims to analyze the effectiveness of fly ash adsorption from a steam power plant as an adsorbent in acid mine drainage treatment. Characterization of fly ash by comparing the diffraction pattern before and after adsorption using x-ray diffraction (XRD). It is hoped that the results of the adsorption effectiveness test will be very useful in overcoming the impact on public health.

2. Methods

This study uses fly ash as an adsorbent for acid mine drainage treatment, fly ash used is a residue from a steam power plant located in West Sumatra. The acid mine drainage used comes from one of the mining industries, the collection location is in the sump area and the water intake applies the wastewater sampling method [7]. Sampling using purposive sampling method, namely the technique of determining the sample with certain considerations [8]. This research was conducted on a laboratory scale with experimental methods, the tools used were a hot plate magnetic stirrer and a stirring bar as a stirring medium in the adsorption process. Referring to the quality standards of mining activity waste [9], the maximum limit for Fe metal is 7 mg/l, Mn is 4 mg/l, while the pH is in the range of 6-9. Water quality testing is carried out using the atomic absorption spectrophotometry (AAS) method of testing for iron (Fe) [10], the method of testing for manganese (Mn) using the AAS method [11] and the method of testing the degree of acidity using a pH meter [12].

2.1. Characterization of fly ash

X-ray diffraction (XRD) is a non-destructive rapid analytical technique used for phase identification of crystalline materials. The data generated by XRD is in the form of x-ray diffraction intensity which is diffracted at 2θ angles. Each pattern that appears on XRD represents a crystal plane that has a certain orientation [13]. The characterization of fly ash was carried out by XRD testing before adsorption, after the adsorption process then XRD testing was carried out again, then the comparison of fly ash diffraction patterns before and after the adsorption process was obtained.

2.2 Adsorption

Adsorption is done by mixing fly ash as an adsorbent with acid mine drainage. Before being used in the adsorption process, the adsorbent was prepared with the aim of removing the moisture content and opening the pores on the adsorbent, the preparation was carried out in a dry oven at a temperature of 110oC for 17 hours [14]. After that, the adsorbent is measured on a watch glass and weighed according to a predetermined dose, then mixed with 250 ml of acid mine drainage in a beaker glass. Stirring was carried out at a speed of 150 rpm, then stirred with variations in stirring time according to the adsorption design. The following is the adsorption design in table 1.

Table 1. Adsorption design.

Adsorbent Dose	A1 (10 grams), A2 (14 grams), A3 (17 grams)	
Stirring Speed	T1 (150 rpm)	
Stirring Time	S1 (30 min), S2 (60 min), S3 (90 min)	
A1T1S1	A2T1S1	A3T1S1
A1T1S2	A2T1S2	A3T1S2
A1T1S3	A2T1S3	A3T1S3

After the adsorption process, it was filtered using whatman paper. The sample was then subjected to AAS test to determine the concentration of the remaining heavy metals and the increase in pH that occurred with a pH meter. Analysis of the effectiveness of heavy metal absorption and increase in pH was carried out using the following equations .

$$1 - \left(\frac{Ct}{Co}\right) x\ 100\% \qquad (1)$$

$$1 - \left(\frac{Co}{Ct}\right) x\ 100\% \qquad (2)$$

Co is the concentration before (mg/l) and Ct is the concentration after (mg/l), the results obtained are the percentage of the effectiveness of heavy metal absorption and the effectiveness of increasing pH.

3. Results and discussion

3.1. The test results of x-ray diffraction

The results of X-ray diffraction (XRD) of fly ash before adsorption, the characteristics of fly ash minerals are crystals. The dominant minerals are quartz (SiO_2) and mullite ($Al_6Si_2O_{13}$), before adsorption, quartz appears at an angle of 2θ = 20.82; 26.6; 32.29; 42.53; 49.95; 64.52; 68.05; and 75.05. Mullite appears at an angle 2θ = 16.42; 26.6; 30.95; 33.2; 35.22; 39.29; 40.85; 42.53; 48.12; 53.95; 57.56; 60.65; 64.52; 70.63; and 75.05. The diffraction pattern of fly ash before and after adsorption can be seen in figure 1 and figure 2.

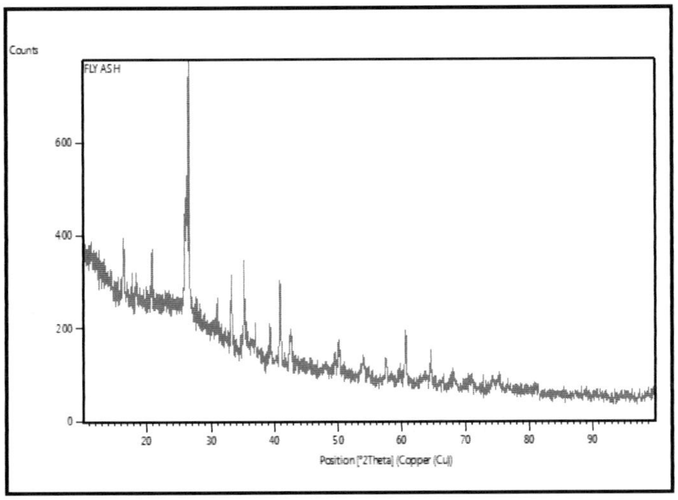

Figure 1. Fly ash diffraction pattern before adsorption.

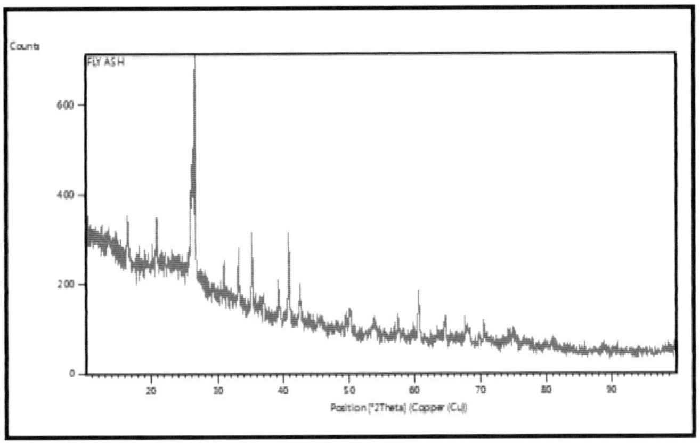

Figure 2. Fly ash diffraction pattern after adsorption.

After the experiment, fly ash showed a new peak, namely hematite (Fe_2O_3) at an angle of 2θ = 33.17; 39.28; 40.83; 53.92; 57.53; 81.07; and 88.76.Grossularite ($Ca_3Al_2Si_3O_{12}$) at angle 2θ = 35.22; 36.98; 53.92; 57.53; 74.95; and 88.76.Augite ($Ca(Mg,Fe)Si_2O_6$) at angle 2θ = 31.01; 35.22; 39.28; 40.83; 42.52; 50.07; 60.60; 70.59; and 74.95.Halloysite ($Al_2Si_2O_5(OH)_4.2H_2O$) at angle 2θ = 31.01; 35.22; 39.28; 40.83; 42.52; 50.07; 53.92; 57.53; 64.55; 67.95; 70.59; 74.95; 81.07; and 88.76.Manganite ($MnO(OH)$) at 2θ = 26.58; 39.28; 40.83; 53.92; 60.60; 67.95; 74.95; and 81.07.

Based on the results of the XRD fly ash analysis, it was seen that there were new peaks after the adsorption experiment with acid mine drainage was carried out. The results of this study are in line with[15]and[16]where the appearance of these new peaks indicates that there has been a transfer of mineral content from acid mine drainage to the adsorbent, it also indicates the occurrence of an adsorption process between silica and alumina minerals with heavy metals. Silica and alumina make fly ash an aluminisilicate mineral which has a polar side with an active hydroxyl group (-OH) when in aqueous solution.

The hydroxyl group is a component that plays an important role in the adsorption process between heavy metals and fly ash. Metal ions that are positively charged cause metal ions to be absorbed on the polar side of the material, so that when the active hydroxyl group is reacted with acid mine water, the -OH group will automatically escape from its structure and directly bind to heavy metals.

3.2. Adsorption using fly ash

The initial test results showed that the concentration of Fe and Mn in acid mine drainage was below the quality standard, namely Fe 8.86 mg/l and Mn 7.03 mg/l. Adsorption was carried out by stirring using a hot plate magnetic stirrer, then after adsorption a test was carried out to see changes in the concentration of Fe and Mn metals in acid mine drainage. The results of the adsorption of heavy metal concentration of Fe can be seen in figure 3.

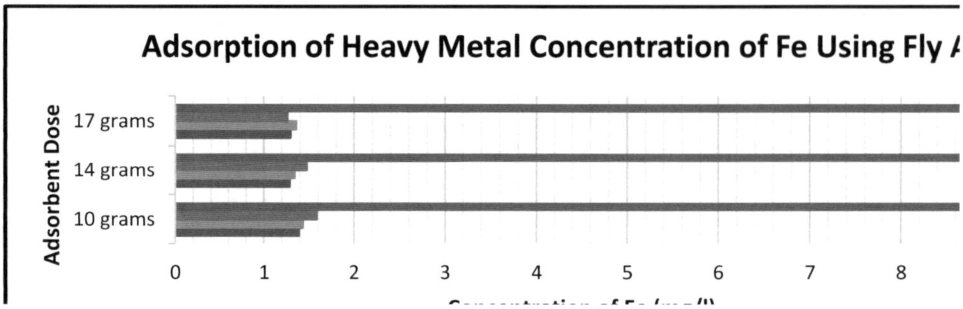

Figure 3. Adsorption of heavy metal concentration of Fe using fly ash.

Based on figure 3 above, the absorption of heavy metal Fe from 8.86 mg/l decreased to 1.26 mg/l. At a time of 30 minutes of stirring with 10 grams of fly ash the dose decreased to 1.58 mg/l, then with 14 grams the dose decreased to 1.47 mg/l and 17 grams of dose the highest decrease was 1.26 mg/l. Stirring with a time of 60 minutes and giving a dose of fly ash 10 grams decreased to 1.43 mg/l, given a dose of 14 grams the concentration decreased to 1.34 mg/l, then increased to 1.36 mg/l at a dose of 17 grams.90 minutes of stirring time with 10 grams of fly ash dose decreased to 1.39 mg/l, then decreased by 1.29 mg/l with the use of a dose of 14 grams, and increased to 1.3 mg/l at a mixing dose of 17 grams. For the results of the adsorption of heavy metal concentrations of Mn can be seen in figure 4.

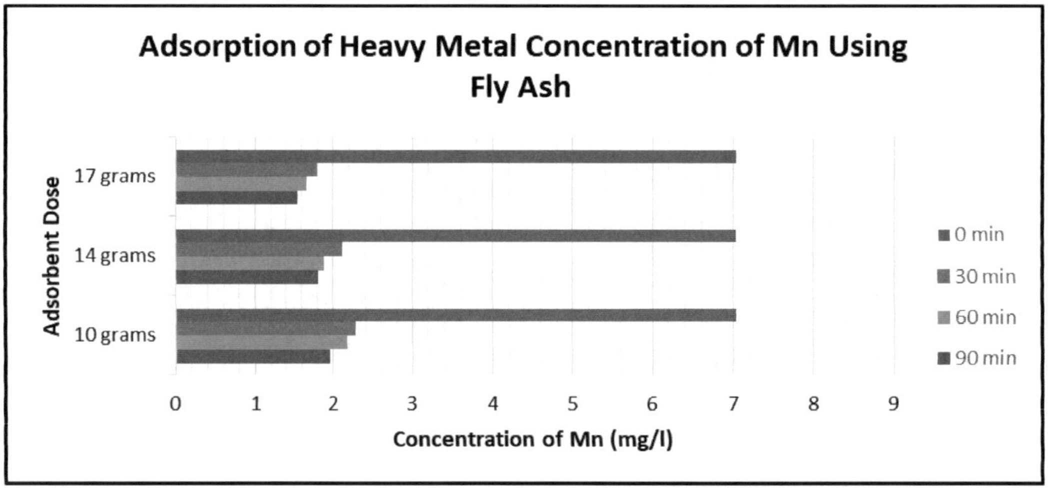

Figure 4. Adsorption of heavy metal concentration of Mn using fly ash.

Figure 4 shows the absorption of heavy metal Mn with an initial concentration of 7.04 mg/l. At the time of stirring for 30 minutes and the dose of fly ash 10 grams decreased to 2.27 mg/l, decreased to 2.11 mg/l at a dose of 14 grams, and again decreased at a dose of 17 grams to reach 1.79 mg/l.60 minutes of stirring time and 10 grams of fly ash dose decreased to 2.18 mg/l, and 2.11 mg/l at a dose of 14 grams, then at 17 grams the dose decreased to 1.66 mg/l. Stirring 90 minutes with 10 grams of fly ash dose decreased to 1.95 mg/l, when the use of 14 grams the dose decreased to 1.8 mg/l, then 17 grams of dose reached the highest decrease to 1.54 mg/l. The results of adsorption on increasing pH can be seen in figure 5.

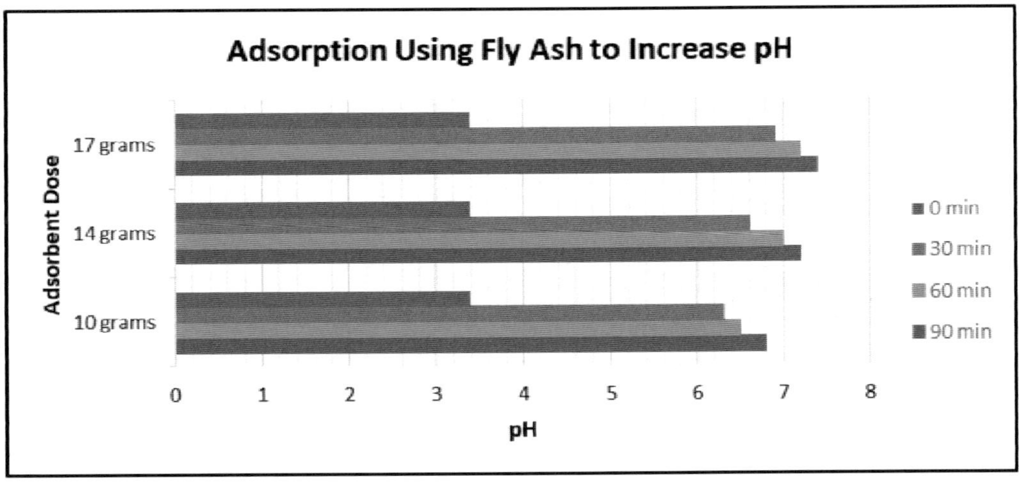

Figure 5. Adsorption using fly ash to increase pH.

Figure 5 above shows an increase in pH before and after the adsorption process.The increase in pH was proportional to the stirring time and the addition of the fly ash dose, namely the initial pH of 3.4 increased with a range of 7.4.At a stirring time of 30 minutes with the use of a dose of 10 gram fly ash, the pH increased to 6.3, then increased to 6.6 with a 14 gram dose, then 17 grams the dose reached 6.9.Stirring 60 minutes the increase occurred sequentially, each 10 gram dose of fly ash increased to 6.5, 14 gram dose increased to 7, then 17 gram dose increased to 7.2.90 minutes of stirring and 10 grams of fly ash dose increased to 6.8, then with a dose of 14 grams increased to 7.2 and the highest increase occurred at a dose of 17 grams to 7.4.

3.3. Adsorption effectiveness

Based on the results of adsorption using fly ash from a steam power plant, the effectiveness of fly ash's ability to absorb heavy metals Fe and Mn, as well as increasing pH in acid mine drainage was calculated.The effectiveness of absorption using fly ash can be seen in table 2.

Table 2. Effectiveness of adsorption using fly ash.

Dose	Fe			Mn			pH		
	30 min	60 min	90 min	30 min	60 min	90 min	30 min	60 min	90 min
10 grams	82.15%	83.86%	84.28%	67.68%	68.98%	72.28%	46.03%	47.69%	50%
14 grams	83.38%	84.91%	85.48%	70.02%	73.24%	74.48%	48.48%	51.43%	52.78%
17 grams	85.83%	84.62%	85.3%	74.63%	76.47%	78.14%	50.72%	52.78%	54.05%

Table 2 above describes the effectiveness of heavy metal absorption and pH increase using fly ash.The lowest absorption effectiveness of Fe metal was 10 grams with 30 minutes of stirring time, while the highest absorption was with 17 grams of stirring and 30 minutes of stirring time reaching 85.83%. For Mn metal the lowest effectiveness was 10 grams of dose and 30 minutes of stirring time, for the highest effectiveness on Mn metal was 17 grams with 90 minutes of stirring time reaching 78.14%. The effectiveness of increasing the pH ranged from 46.03% to 54.05%, the lowest pH increase was 30 minutes of stirring time and 10 grams of dose, while the highest was 90 minutes of stirring time and 17 grams of dose.The average effectiveness of absorption using fly ash can be seen in figure 6.

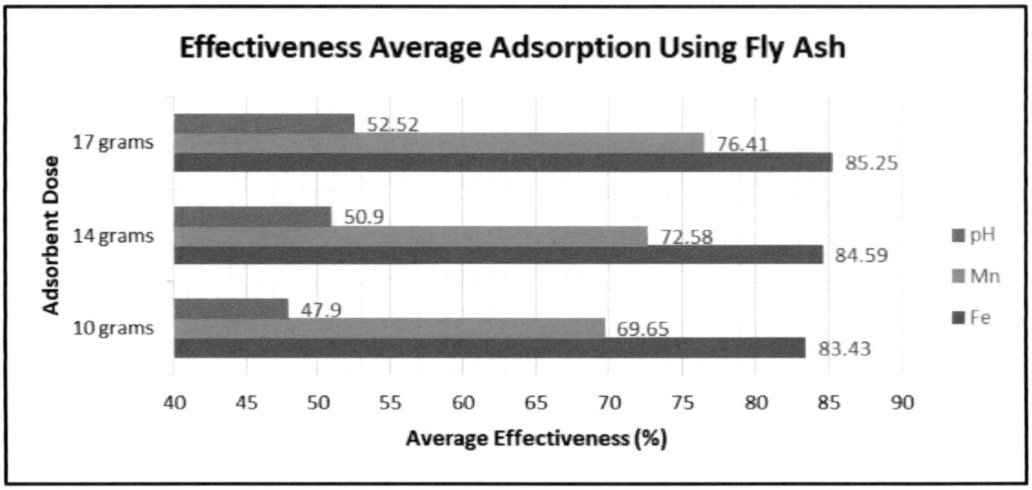

Figure 6. Effectiveness average adsorption using fly ash.

Based on figure 6 above, it shows the average effectiveness of the absorption of heavy metals Fe and Mn, as well as increasing the pH using fly ash. At a dose of 10 grams, the average effectiveness of increasing pH was 47.9%, while the average absorption effectiveness of heavy metal Mn was 69.65% and heavy metal Fe reached 83.43%. Giving a dose of 14 grams, the average effectiveness of increasing pH is 50.9%, for the average effectiveness of absorption of heavy metals Mn and Fe are 72.58% and 84.59%. The use of 17-gram dose of fly ash, the highest average effectiveness of increasing pH reached 52.52%, the average effectiveness of absorption of heavy metals Fe and Mn was also the highest, namely 76.41% and 85.25%.

The average absorption of heavy metal Fe is better than heavy metal Mn, this is because the electronegativity of Fe is greater than Mn. The absorption of Mn metal will be better at high pH and slow down if the pH is less than 8[17].The adsorption of heavy metals Fe and Mn runs faster if the dose is greater, the pH is more alkaline due to the presence of H^{+}ions on the surface of the adsorbent which causes heavy metal ions to compete with each other[18]. Water conditions in mining areas greatly affect public health, and the results of this study can be used to address the impact of labor on public health [19].

4. Conclusion and recommendation

Fly ash from PLTU is very effective as an adsorbent for acid mine drainage treatment. The results of the XRD test showed that there were new peaks in the fly ash before and after adsorption, due to the transfer of minerals from the acid mine drainage to the adsorbent, this indicated the success of the adsorption process. The best effectiveness of fly ash is at a dose of 17 grams, where the average effectiveness of increasing pH reaches 52.52%, the average effectiveness of absorption of heavy metal Mn reaches 76.41%, and Fe metal has the highest average effectiveness reaches 85.25%. The results of this study will be very useful for protecting the health of the people who work and live in the Mine area.

Reference

[1] Linggasari S, Siri H T and Salsabilita N 2018 Soda ash calculation to neutralize acid mine water pthere is tin ore mining in the nibung area of PT Kobatin, Bangka Belitung province *J.*

STTNAS **2018** 167–174
[2] Oktafia N, Madiutomo N and Solihin 2016 Pattern of Spread of Potential Acid Mine Water Actuality in Coal Mines in the Loajanan Samarinda Block, East Kalimantan *Pros. Tek. Pertamb. Univ. Islam Bandung* **2** 31–40
[3] Tangio J S 2013 Adsorption of lead metal (Pb) using water hyacinth biomass (Eichhornia crassipes) *J. Entropi* **8** 500–506
[4] Safrianti I, Nelly W and Zaharah T A 2012 Adsorption of lead (ii) by cellulose of nitric acid-activated rice straw waste: the influence of ph and contact time *J. of equatorial chem.* **1** 1–7
[5] Kulkarni S J, Dhokpande S R and Kaware J P 2013 Studies on flyash as an adsorbent for removal of various pollutants from wastewater *Int. J. Eng. Res. Technol.* **2** 1190–1195
[6] Yao Z T 2015 A comprehensive review on the applications of coal fly ash *Earth-Science Rev.* **141** 105–121
[7] National Standardization Agency 2008 *Metoda Pengambilan Contoh Air Limbah SNI 6989.59:2008* (Jakarta: BSN)
[8] Sugiyono D 2013 *Quantitative, Qualitative, and Action Research Methods* (Bandung: Alfabeta)
[9] Menteri Negara Lingkungan Hidup 2003 *Keputusan Menteri Negara Lingkungan Hidup Nomor 113 Tahun 2003 tentang Baku Mutu Air Limbah Bagi Usaha dan atau Kegiatan Pertambangan Batubara* (Jakarta: MENLH)
[10] National Standardization Agency 2009 *SNI 6989.4:2009 How to Test Iron (Fe) By Atomic Absorption Spectrophotometry (SSA)* (Jakarta: BSN)
[11] National Standardization Agency 2009 *SNI 6989.5:2009 How to Test Manganese (Mn) By Atomic Absorption Spectrophotometry (SSA)* (Jakarta: BSN)
[12] National Standardization Agency 2019 *SNI 6989.11:2019 How to Test The Degree of Acidity (pH) Using a pH Meter* (Jakarta: BSN)
[13] Widyawati N 2012 *Analisis Pengaruh Heating Rate terhadap Tingkat Kristal dan Ukuran Butir Lapisan Tipis Bzt yang Ditumbuhkan dengan Metode Sol Gel* (Surakarta: Universitas Sebelas Maret Surakarta)
[14] Putri R D, Taufiq I and Norokhim 2019 Radionuclide analysis on Fly Ash and Bottom Ash of the Teluk Sirih Power Plant using a gamma spectrometer *J. Physics Unand* **8** 387–393
[15] Gobel A P, Nursanto E and Ratminah W. D 2018 Effectiveness of Coal Fly Ash Utilization as Adsorbent in Neutralizing Acid Mine Water in Banko Mining Settling Pond PT. Bukit Asam (Persero), Tbk J. *Miner. Energi dan Lingkung.* **2** 1–11
[16] Siregar I T N 2019 *Analisis Netralisasi Air Asam Tambang Menggunakan CaO dan Fly Ash Pada PT. Alamjaya Bara Pratama Kabupaten Kutai Kartanegara, Provinsi Kalimantan Timur* (Yogyakarta: Universitas Pembangunan Nasional "Veteran" Yogyakarta)
[17] Stumm W and Morgan J J 2012 *Aquatic chemistry: chemical equilibria and rates in natural waters, 126 ed.* (New York: John Wiley & Sons Inc.)
[18] Paradise M 2021 *Adsorpsi Logam Berat Besi (Fe) dan Mangan (Mn) Menggunakan Komposit Claystone, Zeolit, dan Arang Aktif Tempurung Kelapa pada Sampel Air Asam Tambang Batubara* (Yogyakarta: Universitas Pembangunan Nasional "Veteran" Yogyakarta)
[19] https://hesperian.org/wp-content/uploads/pdf/id_cgeh_2010/id_cgeh_2010_21.pdf, pertambangan dan kesehatan, acsess 27 Agust 2022

IOP Conf. Series: Earth and Environmental Science 1173 (2023) 012041 doi:10.1088/1755-1315/1173/1/012041

Performance Index of Anai I Irrigation-Area, Padang Pariaman

Junaidi[1], Radhiatul Ulfa Jf[1], and Mas Mera[1]
[1] Civil Engineering Departement, Universitas of Andalas, Padang

Email: mas_mera@eng.unand.ac.id

Abstract. The irrigation area of Anai River I which has an area of 6,764 hectares is one of the main irrigation-area in West Sumatra Province. The objective of the present study is to determine the performance index of the main irrigation network (main building, primary canal and secondary canal) of the Batang Anai I Irrigation-Area based on the Minister of Public Works and Public Housing Regulation No. 12/PRT/M/2015, and to determine the level of service of it based on the MASSCOTE method. The method based on Permen PU No.12/PRT/M/2015 examines six aspects. They are physical infrastructure, planting productivity, supporting facilities, personnel organization, documentation and farmer association (WUA). The MASSCOTE method is based on the Rapid Appracial Procedure (RAP) which has four indicators in its assessment, namely; indicators of service, farmer association, human resources and of channel-operation modernization. Based on the Minister Regulation, the Batang Anai I is in the moderate category (76.85%) in the physical performance. Meanwhile, based on the MASSCOTE method, it is in the good category (3.22 in 4-scale) in the level of service. Based on both those categories, the management of the irrigation area only needs the operation and maintenance action.

1. Introduction

Batang Anai Irrigation Area is one of the mainstay irrigation areas in West Sumatra Province. This irrigation area is one of the largest central technical irrigation areas owned by the Province of West Sumatra with a service area of 13,604 hectares. This irrigation area is divided into 2, namely Batang Anai I with a service area of 6,764 hectares and Batang Anai II with a service area of 6,840 hectares. These two irrigation areas play an important role in increasing agricultural yields, achieving the target of national food needs, and improving the economy of farmers in Padang Pariaman Regency. This study aims to determine the performance index of the Batang Anai I Irrigation Area in Padang Pariaman Regency on the Main Irrigation Network (main building, primary canal and secondary canal) based on the Minister of PUPR Regulation No. 12/PRT/M/2015, and the level of service based on the MASSCOTE method.

IOP Conf. Series: Earth and Environmental Science 1173 (2023) 012041 doi:10.1088/1755-1315/1173/1/012041

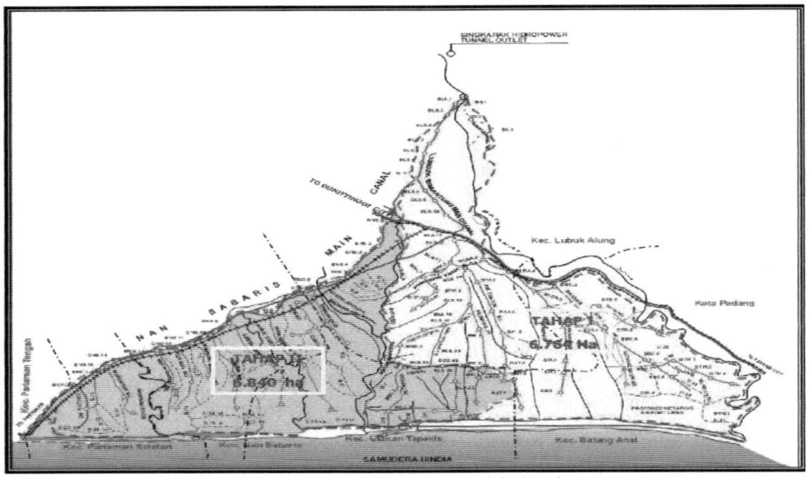

Figure 1. Layout of Batang Anai irrigation area.

2. Methodology

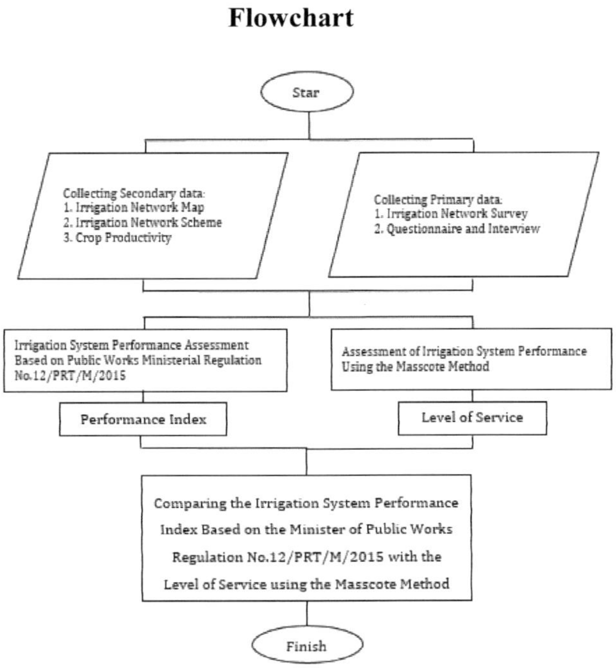

Figure 2. Flowchart

2.1 Based on the Minister of Public Works Regulation No.12 PRT/M/2015 method

Based on the Minister of Public Works Regulation No.12/PTR/M/2015 concerning Guidelines for Operation and Maintenance of Irrigation Networks, it explains that in the evaluation of irrigation network performance there are 6 indicators as aspects of irrigation area performance assessment that are monitored and reviewed, namely; Aspects of Infrastructure Conditions, Planting Productivity, Aspects of Operation and Maintenance Supporting Facilities, Aspects of the Organization of Personnel Implementing Maintenance Operations, Documentation Aspects, Aspects of Institutional Conditions of Water User Farmers' Associations (WUA). The irrigation performance review is carried out every year. The assessment is the sum of the values of all assessment parameters with numbers 0 to 100, with the following performance categories; very good with a score of IN> 90%, good with a value of 90>IN>80%, moderate with a value of 80%>IN>60%, less (Bad) = IN< 60%. For the assessment of each aspect can be seen in Table 1.

Table 1. Rating quality of Minister of Public Works Regulation No.12/PTR/M/2015 Method

ASPECT	Maximum value
1. Aspects of Infrastructure Conditions	45
2. Planting Productivity	15
3. Aspects of Operation and Maintenance Supporting Facilities	10
4. Aspects of the Organization of Personnel Implementing Maintenance Operations	15
5. Documentation Aspects	5
6. Aspects of Institutional Conditions of Water User Farmers' Associations	10
TOTAL	**100**

2.2 Based on MASSCOTE Method

The MASSCOTE (Mapping System and Service for Cannal Operation Technique) method is a method that has the aim of obtaining service and operating solutions for irrigation water users who participate in deciding water management policies to achieve agreed service levels or that need to be improved (Winardi, 2019). The first step of the MASSCOTE approach is to carry out a diagnosis that has a role in performance bottlenecks in an irrigation system, namely by evaluating the RAP (Rappid Appraial Procedure) which will provide an overview of the condition of the related irrigation system from the results of mutual assessment of the four main indicators, namely service indicators, WUA Farmers indicators, indicators of operator human resources, and indicators of modernization of channel operations (Burt, 2001). Based on this method, the performance assessment of Irrigation Areas is divided into 4 categories of Level of Service, namely as follows; very good with a score of 4, good with a score of 3-3.9, less good with a score of 2-2.9, and bad category with a value of 1-1.9. For the assessment of each indicator can be seen in Table 2.

Table 2. Rating quality of MASSCOTE Method

ASPECT	Maximum value
1. Service Orders for Irrigation Water Users	28

2. Secondary Channel	20
3. Irrigation Management HR and WUA/GWUA	20
4. Irrigation Channel Operation Modernization Indicator	8
TOTAL	76

3. Result And Discussions

3.1 Performance Index Based on the Ministerial Regulation No.12 PRT/M/2015 Method

From the results of the direct assessment in the field and the distribution of questionnaires, the results of the performance index assessment are as shown in Table 3.

Table 3. Performance index based on Ministerial Regulation No.12 PRT/M/2015 Method

		DESCRIPTION	VALUE OF PART MAX	VALUE OF CONDITIO N QUALITY	CONDITION INDEX AVAILABLE %	CONDITION INDEX MAX %
		1	3	4	5	6
I.		PHYSICAL INFRASTRUCTURE		SUM	34,98	45,00
1.		Main Building		Sub total	11,46	13,00
	1.1.	Weir	100,00	3,36	84,06	4,00
	a.	Mercu	20,00	0,68	85,00	
	b.	Wing	15,00	0,51	85,00	
	c.	Weir Floor	20,00	0,64	80,50	
	d.	Closing Embankment	20,00	0,68	85,00	
	e.	Bridge	5,00	0,17	85,00	
	f.	Operatin Board	10,00	0,34	85,00	
	g.	Ruler	5,00	0,16	82,00	
	h.	Safety rail	5,00	0,17	85,00	
	1.2.	Weir doors and gears can be operated	100,00	6,30	90,00	7,00
	a.	Intake door	50,00	3,15	90,00	
	b.	Drain door	50,00	3,15	90,00	

DESCRIPTION			VALUE OF PART MAX	VALUE OF CONDITION QUALITY	CONDITION INDEX	
					AVAILABLE %	MAX %
	1.3.	Mud Bags & Drain Door	100,00	1,80	90,00	2,00
	a.	Good Mud Bag Building	35,00	0,63	90,00	
	b.	Mud bag has been cleaned	30,00	0,54	90,00	
	c.	The Mud Bag Drain & Gears are operable	35,00	0,63	90,00	
2.	Carrier Channal			Sub Total	7,49	10,00
					83,3	
	2.1.	The capacity of each channel is sufficient to carry the maximum required/planned discharge	100,00	4,16	83,28	5,00
	2.2.	The embankment height is sufficient to avoid overflow at any time during operation	100,00	1,64	82,17	2,00
	2.3.	All channel repairs have been completed	100,00	1,69	84,30	3,00
3.	Building on the carrier channel			Sub total	7,43	9,00
					82,14	
	3.1.	Regulatory Building (Share / Share Tap / Tap) is complete and functioning	50,00	1,53	76,29	2,00
	a.	Every time and every regulatory building needs Main and Secondary Saluran	50,00	0,76	76,33	1,00
	b.	In each tertiary tapping	50,00	0,76	76,25	1,00
	3.2.	Discharge measurement can be carried out with the operation plan of the Irrigation Area	33,33	2,21	88,33	2,50
	a.	In the Intake Building (Weir / intake)	35,00	0,83	95,00	
	b.	At each control building (share/share tapping/tapping)	30,00	0,64	85,00	
	c.	In each tertiary tapping	35,00	0,74	85,00	
	3.3.	Complementary buildings are functional and complete	50,00	1,56	78,00	2,00
	a.	On the main and secondary channels	50,00	0,80	79,75	
	b.	There are no blockages in siphon buildings, culverts, bridges, gutters, cross-drains	50,00	0,76	76,25	

		DESCRIPTION	VALUE OF PART MAX	VALUE OF CONDITION QUALITY	CONDITION INDEX AVAILABLE %	MAX %
	3.4.	All repairs have been completed	100,00	2,13	85,94	2,50
	a.	Repair of regulatory (share/share tapping/tapping)	60,00	1,26	83,75	1,50
	b.	Ruler, liter scale and water level mark	10,00	0,21	85,00	0,25
	c.	Operation board	20,00	0,45	90,00	0,50
	d.	Complementary buildings	10,00	0,21	85,00	0,25
4.		Sewer and its Building		Sub total	2,55 85,00	4,00
	4.1.	All sewers and structures have been constructed and listed on the maintenance list and have been repaired and are functioning	75,00	2,55	85,00	3,00
	4.2.	There is no problem of flooding inundated	25,00	0,85	85,00	1,00
5.		Entrance / Inspection		Sub total	3,40 85,00	4,00
	5.1.	The entrance to the main building is in good condition	50,00	1,70	85,00	2,00
	5.2.	The Inspection Path and walkway along the channel have been improved	25,00	0,85	85,00	1,00
	5.3.	Every building and maintained canal can be reached easily	25,00	0,85	85,00	1,00
6.		Office, Residential and Warehouse		Sub total	2,65 57,50	5,00
	6.1.	The office is adequate for:	50,00	1,85	92,50	2,00
	a.	Ranting/Pengamat/UPTD (Setingkat Satker Balai PSDA / UPT/Cab PU Kab/Kota).	50,00	0,95	95,00	1,00
	b.	Mantri/Juru (Setingkat Korlap Balai PSDA/ Mantri Pengairan)	50,00	0,90	90,00	1,00
	6.2.	Adequate housing for:	50,00	0,80	80,00	1,00
	a.	Ranting/Pengamat/UPTD (Setingkat Satker Balai PSDA / UPT/Cab PU Kab/Kota).	50,00	0,40	80,00	0,50

		DESCRIPTION			VALUE OF PART MAX	VALUE OF CONDITION QUALITY	CONDITION INDEX AVAILABLE %	MAX %
	b.	Mantri/Juru (Setingkat Korlap Balai PSDA/ Mantri Pengairan)			50,00	0,40	80,00	0,50
6.3.		Adequate warehouse for:			33,33	-	-	2,00
	a.	Kantor Ranting/Pengamat/UPTD			50,00	-	-	1,00
	b.	Bangunan utama (BD)			25,00	-	0,00	0,50
	c.	Skot Balok dan perlengkapan dibangunan lain			25,00	-	0,00	0,50
II.		PLANTING PRODUCTIVITY (previous year)				SUM	14,16	15,00
1.		Fulfillment of water needs (Factor K)			100,00	9,00	100,00	9,00
2.		Realization of planting area		(e	100,00	3,29	82,34	4,00
		Standard Area (Ha)	3413	(a				
		Planting Season	Planting Realization (Ha)					
	-	MT. I	3413					
	-	MT. II	3413					
	-	MT. III	1605					
		Jumlah I,II,III	8.431,00	(b				
		IP Maks (%)	300,00	(c				
		Cultivation Index (IP) available = (b)/(a)x100%	247,03	(d				
		Presentase Raelisasi LuasPercentage of Realized Planted Area = (d)/(c)x100%	82,34	(e				
3.		Rice Productivity		(c	100,00	1,86	93,08	2,00
		Average rice productivity (ton / ha)	4,77	(a				

	DESCRIPTION		VALUE OF PART MAX	VALUE OF CONDITION QUALITY	CONDITION INDEX AVAILABLE %	CONDITION INDEX MAX %
	Existing rice productivity (ton / ha)	(b 4,44				
	Productivity Percentage Rice = (b)/(a)x100% If the existing rice production > average production, the percentage of rice productivity (c) is written as 100%	(c 93,08				
III.	SUPPORTING FACILITIES			SUM	5,55	10,00
1.	Equipment Operation & Maintenance			Sub total	3,00 66,67	4,00
	1.1.	Basic tools for routine maintenance	50,00	2,00	100,00	2,00
	1.2.	Personnel equipment for operations	25,00	1,00	100,00	0,50
	1.3.	Heavy equipment for mud cleaning and embankment maintenance	25,00	-	-	1,50
2.	Transportation			Sub total	1,00 33,33	2,00
	2.1.	Ranting/Pengamat/UPTD (Motorcycle)	50,00	1,00	100,00	1,00
	2.2.	Mantri/Juru (Motorcycle)	25,00	-	-	0,50
	2.3.	DPA (Motorcycle)	25,00	-	-	0,50
3.	Branch/Observer/UPTD office equipment			Sub total	1,55 77,50	2,00
	3.1.	Basic furniture for office	*50,00*	0,75	75,00	*1,00*
	3.2.	Office tools	*50,00*	0,80	80,00	*1,00*
4.	**Communication tool**			Sub total	- -	***2,00***
	4.1.	Adequate communication network for Branch/ Observer/UPTD - Balai PSDA - Bag. mop. Activity	*100,00*		-	*2,00*
IV.	**PERSONALITY ORGANIZATION**		**SUM**		**12,03**	***15,00***

No.	DESCRIPTION	VALUE OF PART MAX	VALUE OF CONDITION QUALITY	CONDITION INDEX AVAILABLE %	CONDITION INDEX MAX %
1.	**Operation & Maintenance Organization has been structured with clear boundaries, responsibilities and duties.**		Sub total	**5,00** **100,00**	*5,00*
	1.1. Ranting/Pengamat/UPTD	*40,00*	2,00	100,00	*2,00*
	1.2. Mantri/Juru	*40,00*	2,00	100,00	*2,00*
	1.3. PPA	*20,00*	1,00	100,00	*1,00*
2.	**Personnel**		Sub total	**7,03** **57,06**	***10,00***
	2.1. Quantity/Amount according to requirement	*50,00*	3,13	74,50	*4,00*
	a. Mantri/Juru	*25,00*	0,67	67,00	*1,00*
	b. PPA	*75,00*	2,46	82,00	*3,00*
	2.2. > 70% PPA for Civil Servants (if => 70 % value of part 100 %)	*100,00*	-	-	*2,00*
	2.3. Everyone understands Operation and Maintenance	*33,33*	3,90	*96,68*	*4,00*
	a. Ranting/Pengamat/UPTD	*25,00*	1,00	100,00	*1,00*
	b. Mantri/Juru	*50,00*	2,00	100,00	*2,00*
	c. PPA	*25,00*	0,90	90,05	*1,00*
V.	**DOCUMENTATION**	**JUMLAH**		**5,00**	***5,00***
1.	**Irrigation Area Databook**	***100,00***	2,00	2,00 **100,00**	***2,00***
2.	**Maps and pictures**		Sub total	**3,00** **100,00**	***3,00***
	2.1. Wall data in the office	*40,00*	1,20	100,00	*1,00*
	2.2. Executor Image	*30,00*	0,90	100,00	*1,00*
	2.3. Network Schematic (implementing & building)	*30,00*	0,90	100,00	*1,00*
VI.	**WATER USING FARMERS ASSOCIATION (WUA)**	**SUM**		**5,13**	***10,00***
A.	Sum WUA = 75 unit			70,25	

	DESCRIPTION			VALUE OF PART MAX	VALUE OF CONDITION QUALITY	CONDITION INDEX AVAILABLE %	MAX %
B.	Sum GWUA	= 7	unit				
C.	Sum IWUA	= 1	unit				
1.	**GWUA / IWUA is already a legal entity**			***100,00***	1,50	**100,00**	***1,50***
2.	**Institutional Condition GWUA / IWUA**			***100,00***	0,36	**72,50**	***0,50***
	- Develop	*(100 %)*					
	- In Progress	*(60 %)*					
	- Undeveloped	*(30 %)*					
3.	**Ulu-ulu/WUA/GWUA Meeting with Branch/Observer/UPTD**			***100,00***	0,80	**40,00**	***2,00***
	- 1/2 month	*(100 %)*					
	- once a month	*(60 %)*					
	There is irregular	*(40 %)*					
	- There isn't any yet	*(0 %)*					
4.	**WUA actively participates in surveys/network searches**			***100,00***	1,00	**100,00**	***1,00***
5.	**WUA participation in network repair and disaster management**			***100,00***	0,76	**38,00**	***2,00***
6.	**WUA dues are used for network repair**			***100,00***	-	**0,00**	***2,00***
7.	**WUA Participation in Planting Planning and Water Allocation**			***100,00***	0,71	**71,00**	***1,00***

From the calculation results obtained a recapitulation of the assessment as in Table 4.

TABLE 4. Recapitulation of Performance Index Assessment Based on the Minister of Public Works Regulation

ASPECT	Maximum value	Existing Value %
1. Aspects of Infrastructure Conditions	45	34,98

2. Planting Productivity	15	14,16
3. Aspects of Operation and Maintenance Supporting Facilities	10	5,55
4. Aspects of the Organization of Personnel Implementing Maintenance Operations	15	12,03
5. Documentation Aspects	5	4,50
6. Aspects of Institutional Conditions of Water User Farmers' Associations	10	5,13
TOTAL	**100**	**76,35**

From the table above, it can be seen that the total Irrigation System Performance Index in Batang Anai I is 76.85% which is included in the Medium Performance category.

3.2 Service Level of Irrigation Areas Based on the Masscote Method

Data analysis based on the MASSCOTE approach with Evaluation with the Rapid Appraisal Procedure (RAP) provides a good indication of the constraints and problem areas in an irrigation system. Indicators of the assessment carried out are related to operational, management and institutional procedures, hardware of water delivery systems and services. The assessment was carried out based on field observations of the respondents of irrigation Operations and Maintenance Officers as operators and farmer respondents as beneficiaries of irrigation water services which were presented in a questionnaire to conduct cross-assessment/assess each other between respondents in irrigation management in Batang Anai I Region. contained in the main indicator obtained from the sub-indicator which is the median value of the response of Operations and Maintenance Officers and Farmers where in this case Operations and Maintenance Officers and Farmers conduct a Cross Assessment of each question in the assessment Sub-Indicator. Table 5.

Table 5. Assessment of the Service Level of Irrigated Areas based on the Masscote Method

NO.	Main Indicator Name	Value (1-4)	Max. Value.	Survey Value x Quality	Total Quality Value	Data Information Source
1	Service Based on Orders for Irrigation Water Users					
	Realization of Water Delivery Service for Individual Ownership Units (eg for rice fields)	3,25	4,00	35,75	11,00	Last Delivery
	Water Delivery Service Plan for Individual Ownership Units (eg for rice fields)	3,11	4,00	34,25	11,00	Office Questions
	Realization of water delivery services at the most downstream point in the system operated by the operator	3,18	4,00	54,00	17,00	Last Delivery

NO.	Main Indicator Name	Value (1-4)	Max. Value.	Survey Value x Quality	Total Quality Value	Data Information Source
	Service Plan Water delivery at the most downstream point in the system operated by the Operator	3,01	4,00	51,25	17,00	Office Questions
	realization of Water Service Delivery from Main Channel (Primary) to Level II Channel (secondary)	3,22	4,00	14,50	4,50	Main Channel (Primary)
	Water Service Delivery Plan by Main Channel (Primary) to Level II Channel (secondary)	3,22	4,00	14,50	4,50	Office Questions
	User Orders for Irrigation Water from Operator-operated (secondary) Channel Systems	3,50	4,00	14,00	4,00	Last Delivery
Total Value I		22,50	28,00			
2	Secondary Channel					
	Gauge building hardware/hardware (on Secondary Channel)	3,25	4,00	6,50	2,00	Level II Channel
	Divider of Secondary Channel	3,33	4,00	10,00	3,00	Level II Channel
	Irrigation Management Operator Communication on for Secondary Channel	2,91	4,00	32,00	11,00	Level II Channel
	General Condition of Secondary Channel	2,80	4,00	14,00	5,00	Level II Channel
	Operation on Secondary Channel	2,83	4,00	14,15	5,00	Level II Channel
Total Value II		15,12	20,00			
3	Irrigation Management HR and WUA/GWUA					
	Employees/Operators	3,11	4,00	29,50	9,50	Employee
	WUA	3,54	4,00	23,00	6,50	WUA
	Mobility and Size of Operations Staff	4,00	4,00	4,00	1,00	Office Questions
	Computer for management records	3,00	4,00	3,00	1,00	Office Questions
	Computer for channel control	3,00	4,00	3,00	1,00	Office Questions
Total Value III		16,64	20,00			
4	Irrigation Channel Operation Modernization Indicator					
	(existing) water delivery service capabilities, to	3,50	4,00	10,50	3,00	n/a

NO.	Main Indicator Name	Value (1-4)	Max. Value.	Survey Value x Quality	Total Quality Value	Data Information Source
	support irrigation modernization					
	Changes in Channel Operations needed to support irrigation modernization	3,33	4,00	10,00	3,00	n/a
Total Value IV		6,83	8,00			
TOTAL Value (I+II+II+IV)		61,10	80,39			
AVERAGE PERCENTAGE (%)INTERNAL INDICATORS		=	(61,10/76)X100		=	80,39
ABSOLUTE VALUE		=	(80,39*4)/100		=	3,22

From the recapitulation table of the performance assessment of irrigation areas in D.I Batang Anai I with RAP evaluation shows a performance index of absolute value (Level of Service) is 3.22, then the performance index in the existing condition is in the **Good classification.**

4. Conclusion

The performance of the Irrigation Areas in the Batang Anai Irrigation Area based on the Ministerial Regulation No. 12/PRT/M/2015 is in the **medium** category (80%>IN>60%) in actual terms and is in the **good** category (3-3.9) in the service is based on the Masscote method so that for its management only **operation and maintenance** need to be carried out.

References

[1] Fachrie, S. M., Samsuar, S. and Achmad, M. (2019) "*Performance Assessment of the Main Irrigation System for the Bantimurung Irrigation Area, Maros Regency*", Agritechno Journal, 12(1), pp. 66–77. doi: 10.20956/at.v12i1.187.

[2] KPUPR (2017) "*Performance module of the irrigation system training for the operation and maintenance of irrigation at the 2017 interpreter level*".

[3] Nurrochmad, F. (2016) "*Performance Analysis of Irrigation Networks*", Agritech Journal, 27(04). doi: 10.22146/agritech.9860.

[4] Government of the Republic of Indonesia, (2015),"*Regulation of the Minister of Public Works and Public Housing No.12/PRT/M/2015 concerning Exploitation and Maintenance*". Government of the Republic of Indonesia. Jakarta.

[5] Syarifudin (2010) "*Performance Analysis of Irrigation System Case Studies in the Upper Cisangu Irrigation Area in Lebak Regency, Banten Province*", Master's Thesis in Water Resources Management, Bandung Institute of Technology.

[6] Mulyadi, Soekarno, I. and Natasaputra, S. (2014) "Irrigation Performance Assessment by Permen PU no.32/2007 and Masscote Methode Approach *with Rapid Appraisal Procedure (RAP) Evaluation at Barugbug Irrigation Area – West Java*", Jurnal Irigasi, Vol 9, No.2, Oktober 2014.

[7] Mubarok, C., Wahyudi, S. I. and Asfari, G. D. (2017) "*Irrigation Performance Assessment Based on the Pupr Ministerial Regulation No.12/Prt/M/2015 Approach and the Masscote*

Method with Rapid Appraisal Procedure (Rap) Evaluation in Irrigation Areas", Proceedings National Innovation in Smart City Development, 1(1), pp. 230–236.

[8] Winardi, W., Azmeri, A. and Masimin, M. (2020) "*Study of Irrigation System Performance in the Pandrah Irrigation Area of Bireuen Regency*", Journal of Civil Engineering and Planning Archives, 3(2), pp. 158–165. doi: 10.24815/jarsp.v3i2.16566.

[9] Permatasari, T. I., Sayekti, R. W. and Ismoyo, M. J. (2021) "*Study of Determining the Level of Irrigation Service Using the MASSCOTE Method Based on the Rapid Appraisal Procedure (RAP) as a Support for Irrigation Performance Decisions in the Makam Sukowon Irrigation Area*", Journal of Water Resources Technology and Engineering, Department Irrigation Engineering, Faculty of Engineering, University of Brawijaya, 1(2), pp. 622–634.

Detection of flood inundation in serayu watershed using landsat-8 and sentinel-2 satellite imagery

Suroso[1], A Zikri[1], P B Santoso[1] and Ardiansyah[2]

[1]Civil Engineering Department, Faculty of Engineering, Jenderal Soedirman University
[2]Agricultural Engineering Department, Faculty of Agriculture, Jenderal Soedirman University

E-mail: azani.zikri@mhs.unsoed.ac.id

Abstract. Floods are natural disasters that occur routinely almost every year in various regions in Indonesia, especially on the island of Java. According to the National Disaster Management Authority (BNPB), in 2001-2021 there were 5542 floods in Java. Determining the area of flood inundation is very important to know the potential and risk of the impact of flooding in an area. The purpose of this study is to develop a method of detecting flood inundation areas based on Landsat-8 and Sentinel-2 satellite imagery data using the Normalized Difference Water Index (NDWI) parameter. The Maximum Likelihood Classification (MLC) method was chosen to classify inundated areas and non-flooded areas. The Landsat-8 and Sentinel-2 satellite images were selected during the wet months that usually rain, January, February, March, November and December in 2020. The results show that the Landsat-8 satellite can detect water inundation better than the Sentinel-2 satellite. The results are influenced by the difference in recording time between the two satellites, the presence of clouds that cover the research location, and Landsat-8 imagery data that does not cover the entire location area.

1. Introduction

Indonesia is known as an archipelagic country with a tropical climate. Tropical climate is a climate in an area that has an average temperature per day that is higher than other climates. Areas with a tropical climate like Indonesia only have two seasons, the wet season which usually occurs in November-April, the transitional season in May-July, and the dry season which usually occurs in August-October [1-4]. This season changes due to the apparent motion of the sun and the movement of the monsoon wind. However, at this time, the change of seasons, especially in Indonesia, often does not match the time due to the effects of global warming [5].

In the rainy season, on average, Indonesia gets high and even very high rainfall accumulation. For example, according to Indonesian Agency for Meteorology, Climatology, and Geophysics (BMKG), in January 2021, most regions in Indonesia will experience an accumulation of rainfall exceeding 300 mm per month. This high rainfall can be at risk of natural disasters, one of which is a flood disaster.

Flood is a natural disaster that occurs because water overflows onto land that is usually dry. This overflow of water to the mainland can be triggered by various factors, such as high rainfall intensity, reduced catchment areas due to land use changes, the shape of the watershed, and factors caused by human behavior.

Flood is one of the most frequent natural disasters in Indonesia. According to BNPB data, the number of flood events from 2001-2021 reached 5542 events. The frequency of flood events based on Provinces in Java Island, Central Java Province is at the top with 1776 events followed by East Java, West Java, Banten, Jakarta and Yogyakarta as shown in figure 1. Based on the frequency of annual flood events, the number of flood events on the island of Java changes from year to year with the highest incidence in 2020 with 749 events which can be seen in figure 2.

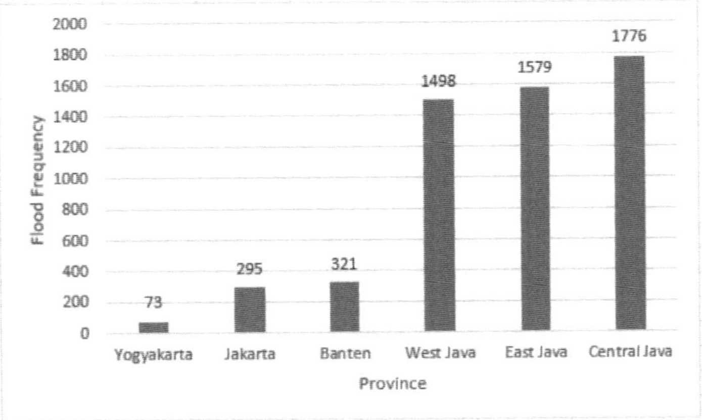

Figure 1. Flood Frequency by Province

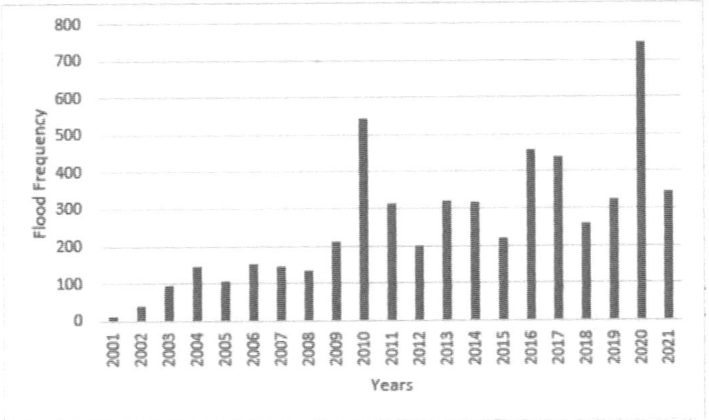

Figure 2. Annual Flood Frequency

Serayu watershed is a watershed that located in Central Java. The districts covered by the Serayu watershed are Cilacap, Banyumas, Purbalingga, Banjarnegara, and Wonosobo. According to BNPB data, flood events in five districts of the Serayu watershed coverage in 2001-2021 were 343 incidents with 60 of them occurring in 2020.

Under these conditions, the flood inundation area needs to be mapped. with the aim that flooding in the future can be minimized or prevented in certain ways. However, mapping flood inundation areas is not an easy thing if it is done directly in the field. Mapping directly in the field will require a lot of money, time, and resources so that the data obtained is accurate and confirmed. With the development of technology, one of the innovations in mapping is using remote sensing.

One of the systems in remote sensing is the satellite system. Based on its character, the data generated from the recording of the satellite system has three characters, spatial character, spectral character, and temporal character. Spatial character is remote sensing data with the smallest area that can be recorded by the sensor. Spectral characteristics are remote sensing data recorded at certain wavelengths. Temporal characteristics of a satellite imagery are the properties of satellite imagery that can record an area repeatedly within a certain time[6-8].

2. Methods

2.1. Study area

This research was conducted in the Serayu watershed which is located in five districts in Central Java, Cilacap, Banyumas, Purbalingga, Banjarnegara, and Wonosobo Regencies as shown in figure 3. The Serayu watershed has an area of 3738 km2 with downstream in the south towards the Indian Ocean. The western Serayu watershed is bordered by Cilacap, the northern part is bordered by Pemalang, Pekalongan, and Batang. Meanwhile, the eastern part of the Serayu watershed is bordered by Temanggung and the southern part of the Serayu watershed is bordered by Kebumen.

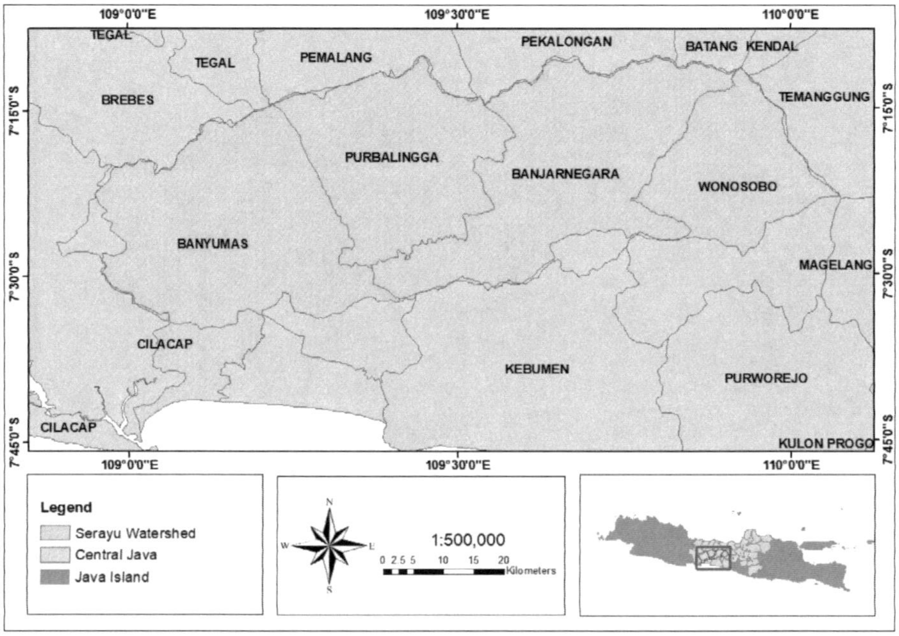

Figure 3. Study Area Map

2.2. Data set

The satellite image data that used in this study are image data from the Landsat-8 and Sentinel-2 satellites, which were downloaded from the EOS (Earth Observing System). Landsat-8 has Onboard Operational Land Imager (OLI) and Thermal Infrared Sensor (TIRS) sensors. Landsat-8 provides spatial resolution data of 30 meters (visible, NIR, SWIR), 100 meters (thermal), and 15 meters (panchromatic). The area of 1 Landsat-8 data scene is 185 km x 185 km every 16 days of re-recording. The Landsat-8 image data used is on path 120 row 065 on January 3, February 4, March 7, November 18, December 20, and June 27, 2020. Meanwhile, Sentinel-2 has 13 image data bands of which 4 bands have a resolution of 10 m, 6 bands of 20 m resolution, and 3 bands of 60 m resolution. The area per scene of

Sentinel-2 data is 290 km x 290 km with 10 days of re-recording. Sentinel-2 image data used is image data on tiles 49MBN, 49MBM, 49MCN, and 49MCM on January 2, February 1, March 17, November 7, December 27, and August 9, 2020.

Land Use and Land Cover (LULC) data is needed for training samples of satellite image data in this study The LULC is data from MODIS (Moderate Resolution Imaging Spectroradiometer) in 2020. This LULC MODIS is a product of the Terra and Aqua satellites which have a pixel size of 500 meters, geographical dimensions of 1200 km x 1200 km and have 13 layers. The layer that will be used in this research is layer type 1, namely the annual Geosphere-Biosphere program classification which has 17 classes [9-14].

2.3. Image data processing

NDWI is created based on satellite image data. Furthermore, the classification of water inundation and not water inundation is carried out using the MLC method on land cover data. Points (grids) with a distance of 1km x 1km are created in the study area for model training as shown in figure 4. The estimation of the flood inundation area is done by comparing the classification model of inundation during the flood with the classification model of inundation before the flood.

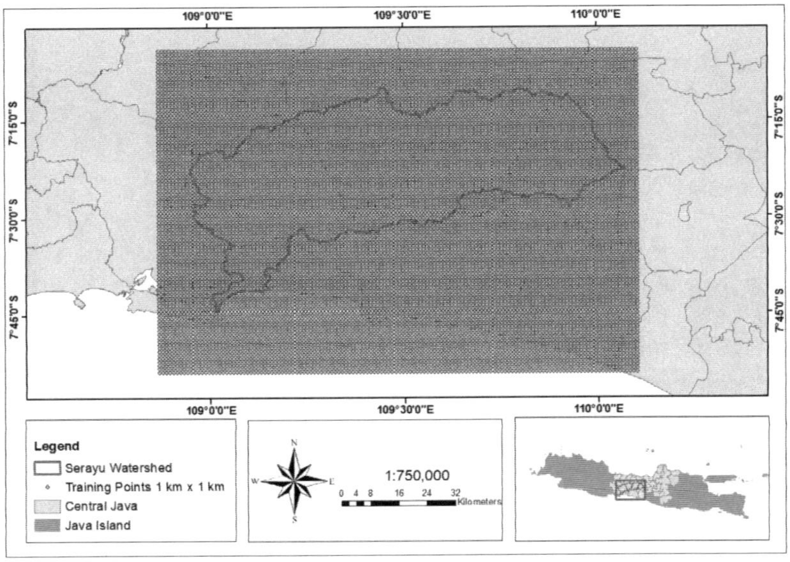

Figure 4. Sample training point.

2.4. Normalized difference water index (ndwi)

Normalized Difference Water Index (NDWI) is an index used to compare the level of wetness in satellite imagery. NDWI uses two bands in satellite imagery, the visible band which is used to assess water bodies and the Near Infrared (NIR) band which is used to emphasize the biomass content. The NDWI index value ranges from -1 to 1, which means the higher the NDWI value, the higher the wetness. The NDWI equation for Landsat-8 and Sentinel-2 can be seen in equation (1) and (2) [8][1].

$$NDWI_{Landsat-8} = \frac{Band\ 3\ (green) - Band\ 5\ (NIR)}{Band\ 3\ (green) + Band\ 5\ (NIR)} \quad (1)$$

$$NDWI_{Sentinel-2} = \frac{Band\ 3\ (green) - Band\ 8\ (NIR)}{Band\ 3\ (green) + Band\ 8\ (NIR)} \quad (2)$$

2.5. Maximum likelihood classification (mlc)

The Maximum Likelihood Classification (MLC) method is one of the supervised classification methods which is a step for transforming multispectral satellite images into special element classes[4]. The MLC method can compare and calculate the average value of the diversity between classes and existing bands. The level of accuracy of the MLC method has a fairly high value. The MLC method is the most widely used method for classification in remote sensing. In the MLC method, the training area must be used to be able to see the statistical characteristics of each category that want to classify [3]. The calculation of the MLC method can be seen in the equation (3).

$$P\ (i\mid x) = \frac{P\ (x\mid i)\ .\ P(i)}{P\ (x)} \quad (3)$$

Where:

P (i|x) = Conditional probability of a class i, which is calculated with the stipulation that vector x
without conditions.

P (x|i) = The conditional probability of the vector x, which is computed with an unconditional class.

P (i) = Probability of a class i that emerges from an image.

P (x) = The probability of the vector x.

2.6. Image data evaluation

The criteria used to evaluate the classification model for inundation and non-water areas are Critical Success Index (CSI), Probability of Detection (POD), and False Alarm Ratio (FAR). In evaluating the classification model, a sample point will be selected from the results of the classification model and will be compared with the actual condition (LULC). There are four types of combinations that will be used:

- HIT (H), where the results of the classification model and the actual condition data are inundation.
- FALSE (F), where the results of the classification model show inundation, but the actual condition is that there are no inundation.
- MISSED (M), where the results of the classification model there is no inundation, but the actual condition shows there is a inundation.
- NULL, where the results of the classification model and the actual conditions do not contain water inundation.

After knowing the combination of sample points, then it will be known how good the classification model is against the actual conditions using POD, CSI, and FAR criteria [2]. the classification model will be better if the POD and CSI values are getting closer to the value 1 and the FAR value is getting closer to the value 0. The calculation of POD, CSI, and FAR can be seen in the equation (4), (5), and (6).

$$POD = \frac{H}{(H + M)} \quad (4)$$

$$CSI = \frac{H}{(H + M + F)} \quad (5)$$

$$FAR = \frac{F}{(H + F)} \quad (6)$$

3. Result

3.1. Flood Inundation Area Detection Using Landsat-8

The water inundation area classification model is carried out using a training point sample of 1 km x 1 km on the NDWI map. The NDWI Landsat-8 map can be viewed in figure 5. The model was made using the maximum likelihood classification method on land cover maps (LULC). The results of modeling water inundation clasification can be seen in figure 6. The water inundation area classification model needs to be evaluated to find out how much confidence the water inundation area classification model has to its original condition. This evaluation was carried out on the classification model of the water inundation area at the study site. The table of evaluation results is presented in table 1. Flood inundation is detected by comparing the water inundation classification model with the water inundation classification model on 27 June 2020 which will get the area of the flood inundation from the difference between both. The resultant area of the flood inundation area is presented in table 2.

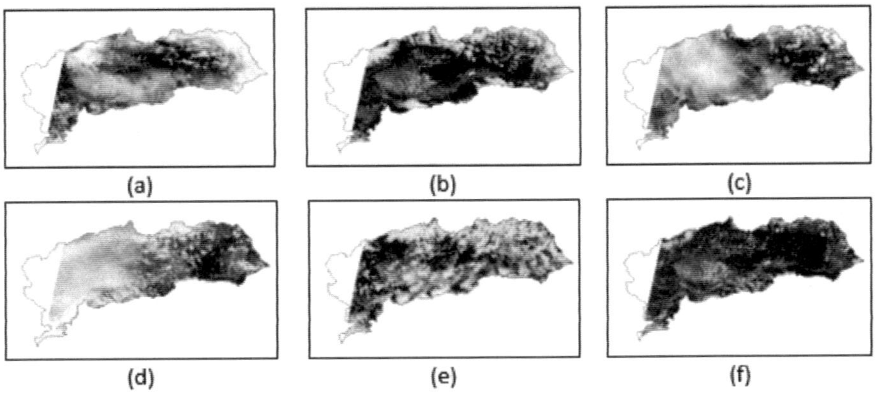

Figure 5. NDWI Landsat-8 map (a) 03/01/20 (b) 04/02/20 (c) 07/03/20 (d) 18/11/20 (e) 20/12/20 (f) 27/06/20

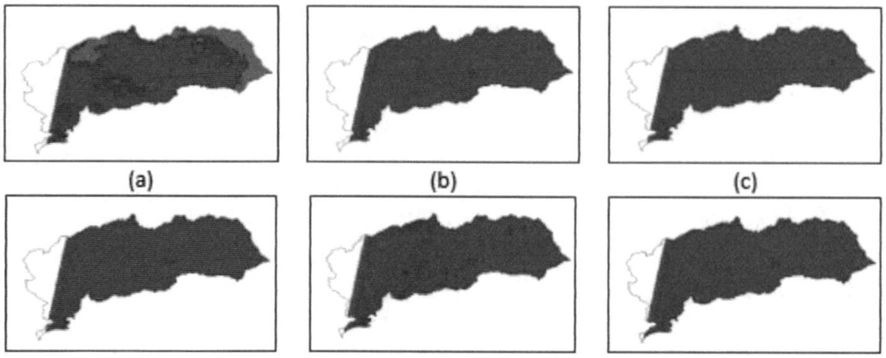

Figure 6. Landsat-8 water inundation classification model (a) 03/01/20 (b) 04/02/20 (c) 07/03/20 (d) 18/11/20 (e) 20/12 /20 (f) 27/06/20

Table 1. Evaluation of the water inundation classification model using Landsat-8 image

Date	H	F	M	NULL	POD	FAR	CSI
03/01/2020	5	696	12	2534	0.294	0.993	0.007
04/02/2020	4	8	13	3207	0.235	0.667	0.160
07/03/2020	1	10	16	3281	0.059	0.909	0.037
18/11/2020	3	1	14	3272	0.176	0.250	0.167
20/12/2020	6	72	12	3279	0.333	0.923	0.067
27/06/2020	3	5	15	3361	0.167	0.625	0.130
Average					0.211	0.728	0.095

Table 2. Area of flood inundation using Landsat-8 image

Date	Number of Flood Grids	Flood Inundation Area (m2)	Flood Inundation Area (Ha)
03/01/2020	771165	694048500	69404.85
04/02/2020	11177	10059300	1005.93
07/03/2020	8456	7610400	761.04
18/11/2020	929	836100	83.61
20/12/2020	79987	71988300	7198.83

The evaluation of the inundation model of Landsat-8 image data shows an average critical success index of 0.095, this means that the success of the Landsat-8 image data model in detecting inundation areas is 9.5%. The area of the flood inundation was found to be the widest area of 69404,85 Ha on January 3, 2020.

3.2. Flood inundation area detection using sentinel-2

In detecting flood inundation areas using Sentinel-2 imagery, the 49MBN, 49MBM, 49MCN, and 49MCM tiles are combined into one unit. Sentinel images that have been merged will be modeled using a 1 km x 1 km trainning point on the NDWI map of land cover (LULC)

using the MLC method. The NDWI Sentinel-2 map can be viewed at figure 7 and modeling of water inundation classification is presented in figure 8. The results of the water inndation classification model then evaluated to determine the level of confidence. Evaluation of the classification model of the water inundation using the Sentinel-2 image can be seen in table 3.

Sentinel-2 imagery on August 9, 2020 was chosen as comparison data to analyze flood inundation. Flood inundation detection is carried out using the grid value of the water inundation classification model will be subtracted from the grid value of the water inundation classification model on August 9, 2020 as comparison data. The flood inundation area on Sentinel-2 image can be seen in table 4.

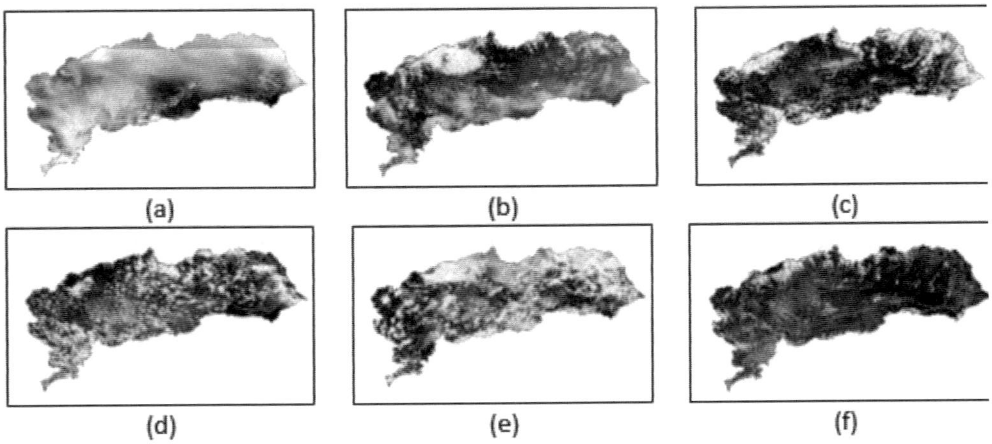

Figure 7. NDWI Sentinel-2 map (a) 02/01/20 (b) 01/02/20 (c) 17/03/20 (d) 07/11/20 (e) 27/12/20 (f) 09/08/20

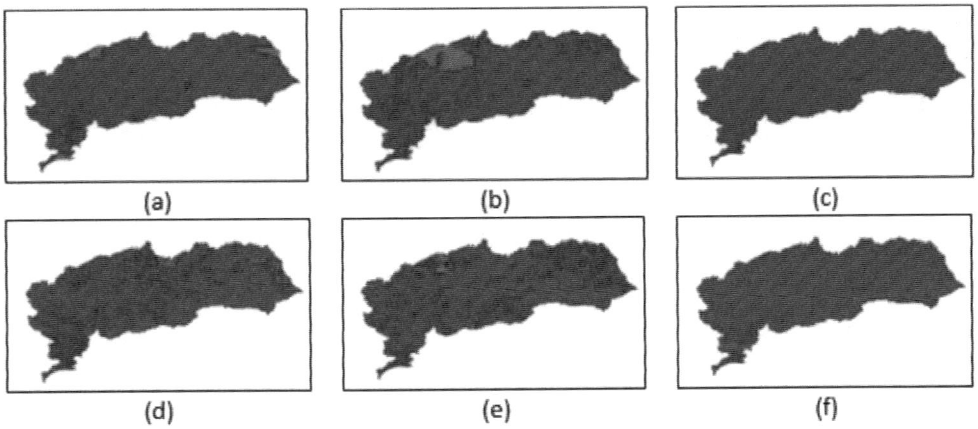

Figure 8. Sentinel-2 water inundation classification model (a) 02/01/20 (b) 01/02/20 (c) 17/03/20 (d) 07/11/20 (e) 27/12 /20 (f) 09/08/20

Table 3. Evaluation of the water inundation classification model using Sentinel-2 image

Date	H	F	M	NULL	POD	FAR	CSI
02/01/2020	5	124	13	3592	0.278	0.961	0.035

01/02/2020	6	380	12	3336	0.333	0.984	0.015
17/03/2020	3	5	15	3711	0.167	0.625	0.130
07/11/2020	5	314	13	3402	0.278	0.984	0.015
27/12/2020	5	367	13	3349	0.278	0.987	0.013
09/08/2020	5	10	13	3706	0.278	0.667	0.179
Average					0.269	0.868	0.065

Table 4. Area of flood inundation using Sentinel-2 image

Date	Number of Flood Grids	Flood Inundation Area (m2)	Flood Inundation Area (Ha)
02/01/2020	1381892	138189200	13818.92
01/02/2020	3805521	380552100	38055.21
17/03/2020	25195	2519500	251.95
07/11/2020	3280137	328013700	32801.37
27/12/2020	3644503	364450300	36445.03

The evaluation of the inundation model of Landsat-8 image data shows an average critical success index of 0.065, this means that the success of the Landsat-8 image data model in detecting inundation areas is 6.5%. The area of the flood inundation was found to be the widest area of 38055.21 Ha on January 2, 2020.

4. Conclusions

From the results of this study it can be concluded:

1. Satellite image data can be used to detect flood inundation areas. The problem with Landsat-8 and Sentinel-2 image data in detecting flood inundation areas is the presence of clouds that cause images of the earth's surface not to be recorded clearly. In addition, Landsat-8 image data does not completely cover the study area which will affect the inundation area detected during the classification process.
2. The results of the evaluation of Landsat-8 data showed that the average probability of the NDWI classification model being able to detect inundation areas was 9.5% with an average error of 72.8%. Meanwhile, for Sentinel-2 image data, the average probability of the NDWI classification model being able to detect inundation areas is 6.5% with an error rate of 86.8%.

Acknowledgments

Financial support for completing this research is graciously provided by the Universitas Jenderal Soedirman.

References

[1] McFeeters, Stuart K 1996 The use of the Normalized Difference Water Index (NDWI) in the delineation of open water features *Int. J. of remote sensing* **17.7** 1425-1432

[2] Brasil Neto, Reginaldo Moura, et al 2020 Evaluation of the TRMM product for monitoring drought over Paraíba State, northeastern Brazil: a statistical analysis *Remote Sensing* **12.14** 2184

[3] Abdulrazzak, Mohamed, et al 2019 Flash flood risk assessment in urban arid environment: case study of Taibah and Islamic universities' campuses, Medina, Kingdom of Saudi Arabia *Geomatics, Natural Hazards and Risk* **10.1** 780-796.

[4] Kseňak, Ľubomír, et al 2022 Assessment of the Usability of SAR and Optical Satellite Data for Monitoring Spatio-Temporal Changes in Surface Water: Bodrog River Case Study *Water* **14.3** 299
[5] Setiyono H, Bambang A N B, Helmi M & Yusuf M 2022 Effect rainfall season on coastal flood in Semarang City, Central Java, Indonesia *International Journal of Health Sciences* **6** 7584–7595
[6] Cavallo, Carmela, et al 2021 Continuous monitoring of the flooding dynamics in the Albufera Wetland (Spain) by Landsat-8 and Sentinel-2 datasets *Remote Sensing* **13.17** 3525
[7] Uddin, Kabir, Mir A Matin, and Franz J Meyer 2019 Operational flood mapping using multi-temporal Sentinel-1 SAR images: A case study from Bangladesh *Remote Sensing* **11.13** 1581
[8] Kaplan, Gordana and U Avdan 2017 Object-based water body extraction model using Sentinel-2 satellite imagery *European Journal of Remote Sensing* **50** 137-143.
[9] Suroso, Putudewi A 2021 Impact of land use changes on the water availability in Ciwulan watershed, West Java. In IOP Conference Series: Earth and Environmental Science (Vol. 653, No. 1, p. 012031). IOP Publishing.
[10] Suroso, Wahyuni A D & Santoso P B 2021 Prediction of Streamflow at the Pemali catchment area using Shetran model *IOP Conf. Ser.: Earth Env. Sci.* **698** 012009
[11] Suroso, Wahyuni R P 2021 Estimating of soil moisture using shetran model at Cisanggarung catchment area *IOP Conf. Ser.: Earth Env. Sci.* **653** 012028
[12] Rahmat F P 2021 Groundwater depth prediction using Shetran model in Citarum River basin. *IOP Conf. Ser.: Earth Env. Sci.* **653** 012029
[13] Suroso, Krishna K P & Santoso P B 2020 Impact of Land Use Land Cover Changes on River Discharge at Brantas Catchment Area using SHETRAN Model *IOP Conf. Ser.: Mat. Sci. and Eng.* **982** 012037
[14] Suroso, Nadawaty R & Santoso P B 2020 Detection of Flood Inundation in the Progo Watershed using the SHETRAN Model *IOP Conf. Ser.: Mat. Sci. and Eng* **982** 012038

IOP Conf. Series: Earth and Environmental Science 1173 (2023) 012043 doi:10.1088/1755-1315/1173/1/012043

Effectiveness analysis of East Sunter polder system for flood control management using HEC-HMS & HEC-RAS software

A Novrima[1] and E Anggraheni[1]

[1]Civil and Environmental Engineering Department, Faculty of Engineering, University of Indonesia, Depok 16424, Indonesia

E-mail: adesha.novrima@gmail.com

Abstract. DKI Jakarta is the capital city of Indonesia well known as a city with the most rapid development and growth in the country. This indicates that with the increase in the population of Jakarta, the need for a considerable change of land use cannot be denied. Environmental problems will get worse over time and have resulted in flooding in Jakarta which has been the concern of the local government every year. A polder system is stated to be effective in resolving flood problems especially in Jakarta that cannot drain water by gravity. The goal is to give out recommendation that is based on the existing inundation map of East *Sunter II Petukangan* polder system using HEC-RAS-Ras Mapper by adding pump system and storage area. The result will indicate the effectivity of the East *Sunter II Petukangan* polder system, as well as to give out recommendation effectiveness to the flood control management using HEC-RAS.

1. Introduction

In every rainy season, the downstream area of Jakarta becomes one of the flood-prone areas in the region. Most of the flood-prone area in Jakarta is a natural floodplain and those areas are in the flat area or in the coastal area of Jakarta [1]. Moreover, the population of DKI Jakarta by September 2020 was 10.56 million people. Compared to the previous census result, the population of DKI Jakarta continues to increase. Within the next 10 years since 2010, the population of Jakarta has increased by around 954 thousand people, or average of 88 thousand people each year [2]. This indicate that with the increase in the population of Jakarta, the need for a considerable change of land use cannot be denied. This will affect the hydrological response of an area associated with rapid urbanization which have greatly altered a large proportion of the city, and one of the crucial effects of urbanization is the intensification of basin response and increase of peak flows enhancing the flood risk during wet season [3]. Another driving factor is the rate of land subsidence. Land subsidence in DKI Jakarta is influenced by several main factors, including the number of buildings that increase the load on the soil surface, there are still many people who use groundwater as the main source of clean water, and geological structures that are dominated by alluvial [4]. Consequently, with the condition of flooding in the northern area of Jakarta which is due to the flat contours and the coastal area of Jakarta that is located below the sea level, discharge of water in Jakarta cannot be drained by gravity. Therefore, to overcome the above-stated problem in the area, a drainage system was created—a polder system [5, 6]. For the polder system to fulfil its purpose effectively, it is necessary to have a proper analysis related to the polder system to provide flood protection in DKI Jakarta, in which the polder system chosen to be analyzed is the East *Sunter II Petukangan* polder system, where the area consists of several industrial factories and area of

Sedayu City shop houses, as well residential clusters of *Sedayu,* which is one of a vital spot in the city. Most industrial area where the center of economic activities takes place is in North and East Jakarta reaching up to 810 industries [7].

Flooding can be defined as the inundation of an area due to overflowing water that exceeds the water disposal capacity in an area that causes physical, social, and economic losses [8]. Indonesia is one of the countries whose territory is prone to natural disasters, especially hydro-meteorological hazards which are caused by climate and weather [9]. This disaster damage infrastructure and significantly disrupt the economic stability of the community [8, 9]. With the risk of the damage stated, regional development planning still highlights only the rapid development of industry, commercial areas, and elite settlements. There is no directive towards the interest of environmental conservation just yet [10]. It is necessary to build and develop a unified polder system so that the water elevation, discharge, and the volume of water can be controlled. By 2020, the government of DKI Jakarta has built and developed 36 polder system and 11 polders is on the status of still under development [6].

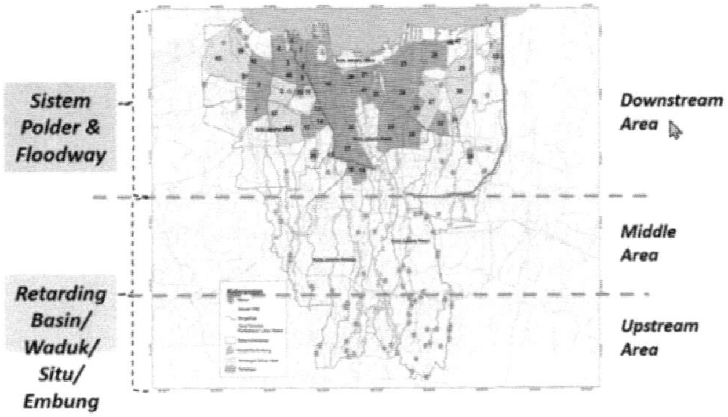

Figure 1. Jakarta's area is divided into 3 main areas. Polder systems are applied in the downstream.

2. Research objective

The main objective of the research is to obtain analysis of the effectiveness of East S*unter II Petukangan* polder system.

Figure 2. The location of this research focus on East Sunter II Petukangan (no. 4) with research area of 256.23 ha.

The first phase of the research is to analyze the existing condition output of East *Sunter II Petukangan* polder system using HEC-HMS and HEC-RAS, then to give out recommendation output that is based on the existing inundation map of the system in HEC-RAS-Ras Mapper by adding pump system and storage area. By comparing both results outputs, the research can proceed to analyze the East *Sunter II Petukangan* polder system's recommendation effectiveness to the flood control management.

3. Methodology

Two software were used to assist the simulation: hydrological analysis using HEC-HMS and hydraulic analysis using HEC-RAS (RAS Mapper). The two-software issue different outputs which will later be combined for further flood mapping with existing conditions output and recommendation output. In the hydrological modelling, the data that will be collected are the rainfall distribution map, design rainfall, land coverage (CN value), topography, and the watershed schematic. Hydraulic modeling uses software assistance namely the Hydrologic Engineering Center- Hydrologic Modeling System (HEC-HMS). In which, the output of the hydrological simulation is the unit hydrograph. The hydraulic modeling uses the Hydrologic Engineering Centre-River Analysis System (HEC-RAS). The input includes the hydrograph (Peak flow), pump design of the polder system, Digital Elevated Model (DEM) for the terrain, and drainage system characteristics in the area.

The flow used in the simulation is a steady flow condition. In the Steady Flow analysis, the flow does not change with time and only includes the maximum discharge for each existing cross section. This is only done because the research only analyzes the maximum flooding conditions that occur in the Petukangan system at a specified time. East *Sunter II Petukangan* polder system's existing inundation mapping can then be obtained to further find its recommendation. The goal is to obtain a reduction in inundation depth and area.

4. Data and result

4.1. Delineation and sub-catchment boundary

Based on the results of urban delineation, the boundary digitization was carried out using ArcGIS software to obtain the boundary shown in figure 3. The boundaries made are based on the default polder boundary but with additional adjustments that adjust the road boundaries in the area. According to the geometric calculation, the area of the Petukangan Boundary is 256.23 Ha.

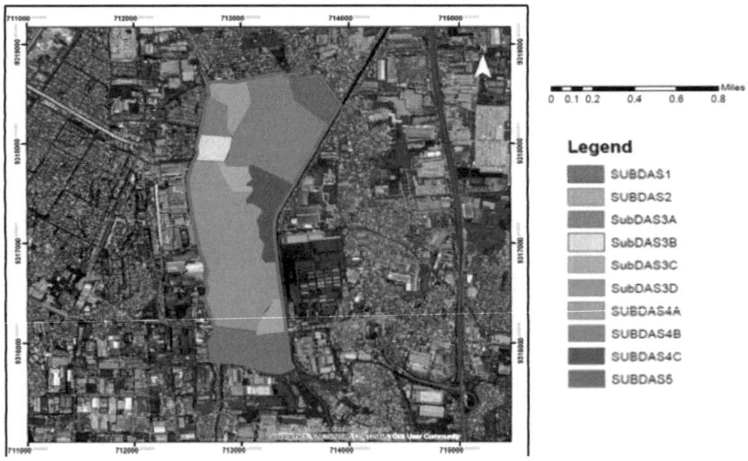

Figure 3. Petukangan boundary and 10 sub-catchment boundaries obtained from urban delineation.

The sub catchments are divided depending on the river flow and road boundary. Initially, 5 Sub-catchments are created accordingly. However, due to the conditioning of the artificial ponds which will be simulated into storage area, sub-catchment 3 is divided into 4 sections, and sub-catchment 4 is divided into 3 sections to adjust the schematic of the river system inside the *Petukangan* polder boundary. Figure 3 displays the total of 10 sub-catchments that are created.

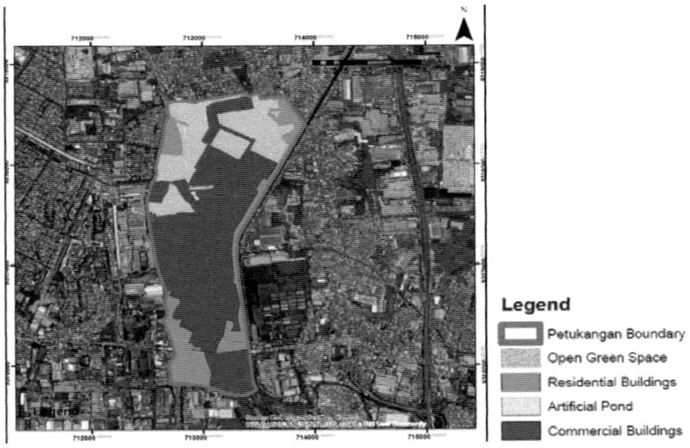

Figure 4. Petukangan System landuse data as the basis to obtain CN value for each sub-catchment.

Figure 4 illustrates the land coverage of the system. Each land use has its own CN value, and each sub-catchment has its own weighted CN value according to the percentage of each land use. This CN value will be used as the input in the HEC-HMS simulation run. The *Petukangan* boundary, according to the surveyed and satellite data, consist of 4 main land use: ponds, residential area, buildings (factories and commercial area), and open space. The most dominant of the land use is the commercial buildings and infrastructure, followed by open green spaces, residential area, and then the ponds. The impervious cover takes up 76,98% of the catchment area.

4.2. Rainfall & distribution
In calculating the design rainfall for this area, rain stations located around DKI Jakarta are plotted based on their coordinates into the Arc-GIS Base map. The nearest Station is the *Pulogadung* Station, and no Thiessen polygon affect the boundary. Hence, only *Pulogadung* Station will be included in the Rainfall design calculation shown in table 1. In calculating the rainfall design, the maximum annual data recorded by the rain station is used.

Table 1. Pulogadung maximum rainfall data station 2011-2020.

Year	Max (mm)
2011	105
2012	95
2013	144
2014	177
2015	176
2016	150
2017	102
2018	135
2019	121
2020	165

After obtaining the maximum overall data each year, it is necessary to measure the dispersion of the rainfall data. The value will be compared with the requirements needed to determine the type of distribution that will be used in calculating the rainfall design. according to the calculation, the appropriate type of distribution is the Gumbel distribution. Rainfall of 25 years will be calculated with the Gumbel Method. Since only 1 Station affects the Petukangan Boundary, a reduction factor is needed. Hence, the Rainfall Design of 25 years after multiplied by the reduction factor is 206.22 mm. In addition, smirnov-kolmogorov and chi-squared consistency distribution test is also conducted, and error tolerance of the rainfall data is accepted. Table 2 concludes the Gumbel calculation and its reduction factor.

Table 2. Gumbel distribution calculation and reduction factor.

25-Year	
n	10
Average	137
Stdev	30.54
Ytr (25)	3.20
Yn	0.53
Sn	1.09
Ktr	2.44
Xtr (Initial)	211.65 mm
Reduc. Factor	0.97
Xtr (Final)	206.22 mm

4.3. Existing inundation simulation

Figure 5 illustrates the sub-catchment division, and the river reach in the Petukangan system, with the *Cakung* drain as the main reach. The conditions are simplified to produce a schematic flowing from the upstream of the Cakung drain, to the outlet of the Petukangan system, in which this schematic will be a reference for the HEC-HMS software to process the rainfall design. After entering all the inputs in the HEC-HMS application, a discharge simulation can be performed, where each of hydrologic element's peak discharge are obtained. The peak discharge for the downstream (outlet) occurred a value of 54,9 m^3/s.

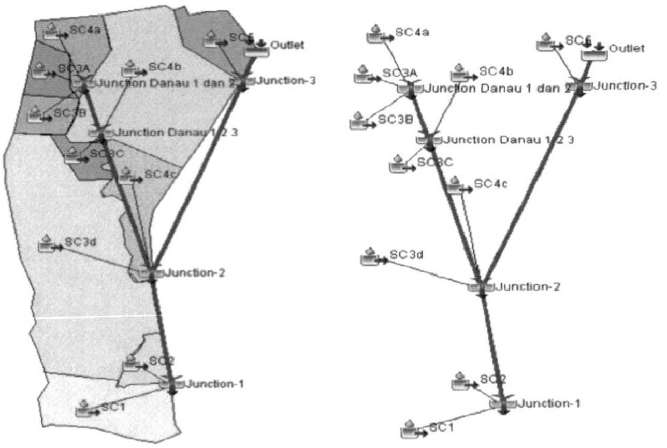

Figure 5. Existing condition of Petukangan system schematic.

In performing inundation simulations for this research, the RAS Mapper application is used to simulate the existing conditions of the Petukangan system. The terrain used is a DEM Map obtained from Lidar and Topographic analysis. An accurate topography model and the creation of geometric data elements that represent how water will move through the river system are the basis of a solid river hydraulics model. River Geometry is a depiction of the Petukangan system which is included in the RAS-Mapper application.

Figure 6. HEC-RAS simulation output of existing inundation in Petukangan system.

Figure 6 illustrates the condition of inundation due to rain with a 25-year return period that occurs in the Petukangan system. The area that is most affected by the inundation is the commercial buildings or factories located on the upper side of Petukangan system (located in subcatchment 4). The area consist of several industrial factories and area of Sedayu City shop houses, as well as the Springwood Cluster of Sedayu.

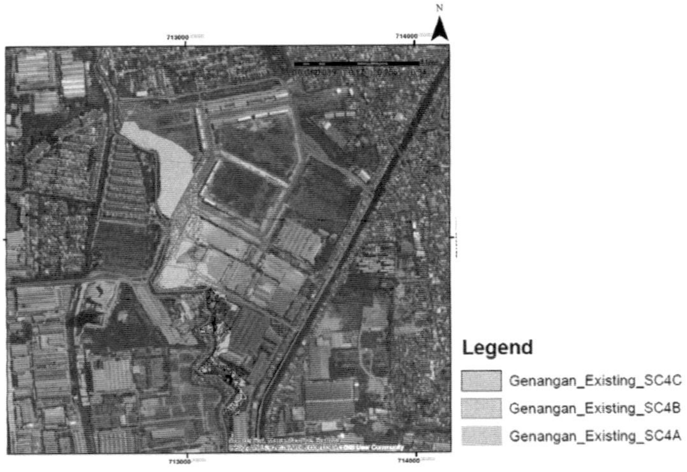

Figure 7. Exported inundation result; showing that the inundated area is located in subcatchment 4.

According to the field geometric calculation that has been exported to Arc GIS shown in figure 7, the area of the water surface of the Petukangan system's inundation is 15,14 Ha or inundated 6% of the total Petukangan area. The inundation result are briefed into table 3.

Table 3. Area of inundation, area affected, max. depth, and average depth of affected area.

SC	Area of Inundation	Sub-Catchment Area Affected	Max. Depth	Average Depth
	(Ha)	(%)	(m)	(m)
SC4A	4.97	5.28%	2.18	1.24
SC4B	6.6	7.01%	2.29	1.07
SC4C	3.57	3.79%	2.44	1.04

Overall, the average depth of the inundation is 1.11 meters, and maximum depth reached up to 2.3 meters. There needs to be a recommendation to decrease the inundated area and the depth of the inundation by modelling the storage area and addition of pumping system along the channel to reduce water that has accumulated in the inundated drainage network.

4.4. Recommendation simulation

According to the Western Java Environmental Management Project (WJEMP), pumping operation as a recommendation can be given out, where the pumping station has been designed to pump water out of the overflow drain. To give out recommendation, another HEC-HMS and HEC-RAS simulation needs to be conducted in the addition of pumping station and the geometric modelling of the artificial pond or water storage in the Petukangan System. The addition of reservoir or storage area in the system will create a schematic, as illustrated in figure 8.

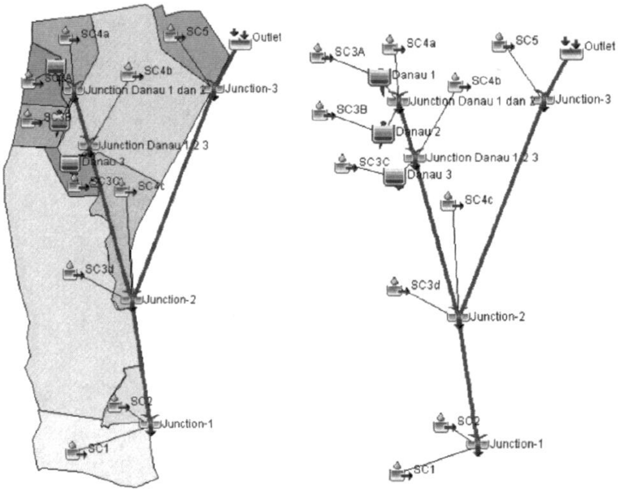

Figure 8. Recommendation schematic by adding storage area into the schematic.

Using the same parameters on each component, the author run another HEC-HMS and HEC- RAS simulation. in this case, the simulation will have an addition of a storage area and pumping station on the inundated area. the table below is the initial and recommended dimension of the 3 artificial ponds to be input into the HEC-HMS simulation.

Table 4. Existing and recommendation of pond dimension and parameter.

Pond No.	Existing		Recommendation	
	Area (Ha)	Elev. (m)	Area (Ha)	Elev. (m)
Pond 1	0.26	1.63	0.5	1.3
Pond 2	0.33	1.31	0.34	1.31
Pond 3	0.34	1.32	0.34	1.32

The maximum discharge downstream for the recommended Petukangan system is 53.9 m^3/s. In the HEC-RAS modelling, storage area and the pump will be added as the additional geometric features in the geometric data shown in figure 9. Table 4 shows the modeling recommendation input for the storage area.

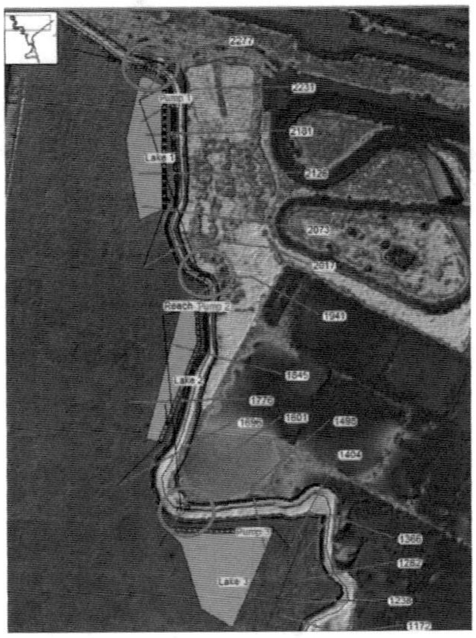

Figure 9. Modelling of 3 pumping station (circled red) and 3 storage area (blue area) in the HEC-RAS geometry.

The pump will help the discharge or water to flow faster. During conducting iterations and trial & errors, the best possible outcome is taken. Table 5 is the outcome which have a better result after several iterations. Each pump has the capacity and on/off elevation.

Table 5. Pump capacity and elevation

Pump	Elev. On	Elev. Off	Capacity
Pump 1	2.5 m	0 m	3.5 m^3/s
Pump 2	2.5 m	0.4 m	3.5 m^3/s
Pump 3	1.5 m	0 m	2.5 m^3/s

Figure 10 is the inundation result after modelling the pumping station and the storage area.

Figure 10. Recommended inundation results after given storage area and pump station.

Although the inundation is not completely gone, there is a significant reduction. In terms of the remaining inundation can use a solution by providing embankments so that water does not come out of the river when it is in an overflow condition. Moreover, the model used is steady flow, where the inundation will be permanent hence it will be difficult to remove all the inundation thoroughly. Figure 11 shows the inundated area and depth after given pump and storage area.

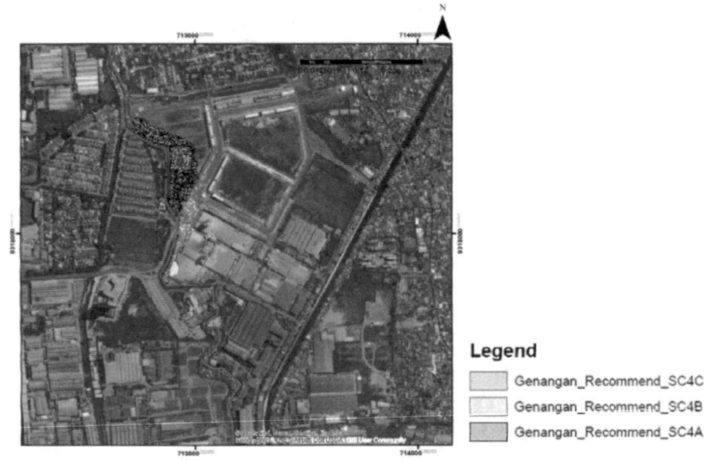

Figure 11. Exported inundation result after given storage area and pump station.

According to the field geometry computation above, the result shows that each Sub-catchment that area affected by inundation has a decrease in its area and depth, where the inundated area now only covers 7.53 Ha of the Petukangan Area. In terms of its maximum depth, it decreased to 1.4 meters and its average depth decreased to 0.54 meters.

Table 6. Recommended result of inundation area, maximum depth, and average depth of affected area.

SC	Inundation Area	Sub-Catchment Area Affected	Max Depth	Average Depth
	(Ha)	(%)	(m)	(m)
SC4A	4.12	4.38%	1.1	0.46
SC4B	2.53	2.69%	1.4	0.59
SC4C	0.88	0.93%	1.74	0.59

5. Analysis and discussion

According to the field geometry computation above, the result shows that each Sub-catchment that area affected by inundation has a decrease in its area and depth. According to the inundated area obtained from HEC-RAS, figure 12(a) shows the inundation of the existing condition, and figure 12(b) shows the recommendation scenario. Side by side, the difference in the contour is visible.

(a) (b)

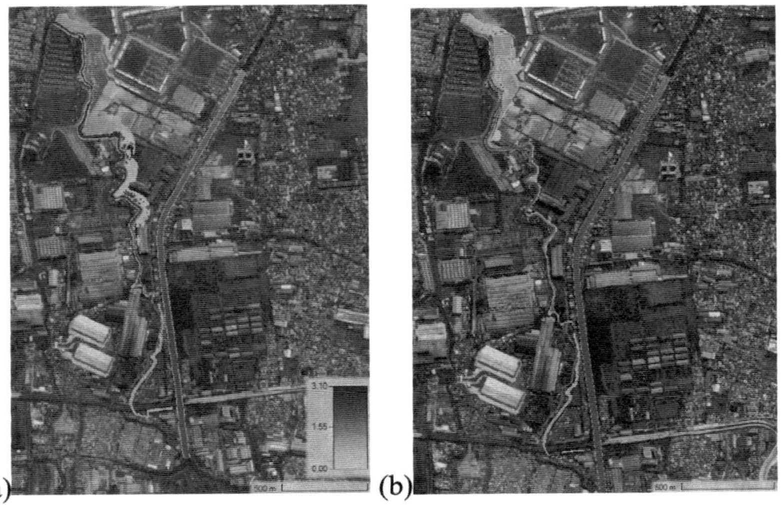

Figure 12 (a) Existing Inundation Condition (b) Recommended Inundation Condition.

A decrease in inundation area and depth is visible due to the installation of the three pump units and storage areas in the *Petukangan* Region. The percentage change of area reduction can be seen in table 7, where overall, the percentage of area reduction is 50.26%. There is also a decrease in inundation depth, in which both its maximum and average depth has a positive trend in reducing the inundation. Overall, the maximum depth reduction reached up to 38.64%, as shown in table 8, while the average depth reduction of 50.90% is as shown in table 9.

Table 7. Existing comparison of inundation area between existing and recommendation scenario.

SC Name	Area of Inundation (Ha)		Percentage of area reduction (%)
	Existing	Recommendation	
SC4A	4.97	4.12	17.10%
SC4B	6.6	2.53	61.67%
SC4C	3.57	0.88	75.35%
Overall	**15.14**	**7.53**	**50.26%**

Table 8. Existing comparison of maximum depth between existing and recommendation scenario.

SC Name	Maximum Depth (m)		Percentage of depth reduction (%)
	Existing	Recommendation	
SC4A	2.18	1.1	49.54%
SC4B	2.29	1.4	38.86%
SC4C	2.44	1.74	28.69%
Average	**2.30**	**1.41**	**38.64%**

Table 9. Existing comparison of average depth between existing and recommendation scenario.

SC Name	Average Depth (m)		Percentage of depth reduction (%)
	Existing	Recommendation	
SC4A	1.24	0.46	62.90%
SC4B	1.07	0.59	44.86%
SC4C	1.03	0.59	42.72%
Average	**0.97**	**0.69**	**50.9%**

6. Conclusion

According to the inundation mapping of 25-year return period, inundation occurred on 15.14 Ha of the whole *Petukangan* (256.23 Ha), which takes up 6% of the *Petukangan* system. the average depth of the inundation is found to be 0.97 m. Floods inundated the area of *Sedayu* City shophouses and the Springwood Cluster. Additionally, flooding also inundated several factory points operating in the area.

The modeling of the storage area system for artificial ponds and three pumping stations in areas inundated by floods can reduce inundation by 50.26%, where the inundation area is reduced to 7.53 Ha, and the average depth is reduced by 30 cm (from 0.97 m to 0.69 m).

Reducing the inundation area and depth is considered effective because overall, the inundation aspect is greatly reduced in terms of its area, the average depth aspect (reduced by 50.90%), and the maximum depth (reduced by 38.64%).

7. References

[1] Yastrib M, Tajri R and Rifdah S 2021 *IOP Conf. Ser.: Earth Environ. Sci.* **737** 012026

[2] Badan Pusat Statistik (BPS) *Data Statistik Provinsi DKI Jakarta* Retrieved from: https://jakarta.bps.go.id

[3] Federica R, Burlando P and Vollmer D 2016 Exploring the hydrological impact of increasing urbanisation on a tropical river catchment of the metropolitan Jakarta, Indonesia. *Sustainable*

Cities and Society: Elsevier **20** 210-221

[4] Cyntia, and Putu I 2018 Analisis penurunan muka tanah DKI Jakarta dengan metode differential interferometry synthetic aperture radar (DINSAR) *J. Ilmu dan Inovasi Fisika* **2** 88-89

[5] Ignatius S, Soeryanto H, Anggraheni E and Sutjiningsih D 2019 *IOP Conf. Ser.: Mater. Sci. Eng.* **669** 012039

[6] Noviadriana D et al 2020 *IOP Conf. Ser.: Earth Environ. Sci.* **437** 012028

[7] Badan Pusat Statistik Provinsi DKI Jakarta *Jakarta Dalam Angka.* Retrieved from: https://jakarta.bps.go.id/publication/2020/04/27/20f5a58abcb80a0ad2a88725/provinsi-dki-jakarta-dalam-angka-2020.html

[8] Rahayu et al 2009 *Banjir dan Upaya Penanggulangannya* (Bandung: Pusat Mitigasi Bencana (PMB-ITB))

[9] Badan Nasional Penanggulangan Bencana (BNPB) *Potensi Ancaman Bencana* Retrieved from: https://www.bnpb.go.id/potensi-ancaman-bencana

[10] Raharjdo P N 2014 7 Penyebab banjir di wilayah perkotaan yang padat penduduknya *J. Air Indonesia* **7** 205-213

Enhanced digital water level measurement for irrigation channel

J A Loekito[1], R Y Tallar[1], O Pattipawaej[1], M C Sutandi[1], T O Sihombing[1], P P Yapinus[1], J Chandra[1], M Tanubrata[1], H Lesmana[1], L Karsten[1], Tonny[1], Efferiki[1], P Madvi H A S[1] and F P Purba[1]

[1] Department of Engineering, Maranatha Christian University, Jl. Prof. Drg. Surya Sumantri, M.P.H. No. 65A, Bandung, 40164, Jawa Barat, Indonesia.

Email: jimmy.loekito@gmail.com

Abstract. Indonesia needs to improve its irrigation system by implementing respectable technology. Recently, Indonesian irrigation systems have intended to switch from analog to digital technology. However, it appeared that Indonesia was having some issues applying digital technology in many areas. For instance, delivering water from water sources to irrigation subversives requires a computer infrastructure. A manual water level monitoring tool has been used in irrigation canals to gauge the water level that flows through them. The major goal of this project is to design a digital measuring device for water level and water temperature in the irrigation channel. The appropriate information must be entered into the irrigation channels. An experimental investigation has been conducted into the accuracy of digital instrument design for detecting water level and water temperature in irrigation channels. The creation of the digital water level measurement design will be tested against manual measurement to determine the percentage difference. The percentage difference between the traditional water level measurement and the digital water level measurement is the anticipated outcome. It is possible to infer that it is crucial to use digital tools, especially when determining the water level in irrigation channels.

1. Introduction

One of the crucial tasks in civil engineering projects is building irrigation systems. The estimate of water availability and the construction of storage or reservoirs suited to the water balance should have allowed for the identification of the requirement for water for agriculture. Demands for irrigation have increased over time as a result of the growth and development of modern agriculture. Indonesia needs to improve its irrigation system by implementing respectable technology. Recently, Indonesian irrigation systems have intended to switch from analog to digital technology. However, it appeared that Indonesia was having some issues applying digital technology in many areas. For instance, delivering water from water sources to irrigation subversives requires a computer infrastructure. A manual water level monitoring tool has been used in irrigation canals to gauge the water level that flows through them. The major goal of this project is to design a digital measuring device for water level and water temperature in the irrigation channel. The appropriate information must be entered into the irrigation channels [1].

Through the provision of guaranteed irrigation water along with improved agricultural supporting services, diversification, soil conservation, drainage facilities, roads, and other facilities in the irrigation project area, the irrigation project seeks to increase agricultural production, particularly rice, to create employment opportunities, and to improve the living conditions of farmers [2].

The scope of the design work for the irrigation project will be limited to the precise design of each separate major irrigation system, including weirs, main and minor canals, and associated structures in the project area, respectively. A weir is a type of structure that is placed across a stream to direct water flow to irrigation areas by increasing the water level to maintain the necessary water height. The reservoir's purpose is to hold water, but the weir's role is to enable uninterrupted flow through the weir.

Basic studies should be carried out in order to plan and construct the weir and irrigation system in general. The primary information gathered is a topographic map, a stream profile, and a cross-section. The state of the water level can also be observed using stream flow data from gauging stations. The water level that flows through an irrigation channel was measured at this stage using a manual water level monitoring instrument. Water-saving agriculture refers to a farming practice that is able to take full advantage of the natural rainfall and irrigation facilities [3].

2. Literature review

Several parts and sensors were employed in the development of this water level monitoring instrument, among others, and will be discussed in this area of the literature review [4]. These elements and sensors play a significant role in the creation of tools. The outcomes of this study will be determined by the output findings generated by the sensors, processed by the microcontroller, and shown on a 2x16 LCD.

2.1. Microcontroller

In order to take input signals, process them, and then produce output signals in accordance with the program written into the microcontroller, it is a chip known as an IC (Integrated Circuit). This microcontroller receives its input signal via sensors that may collect data from its immediate surroundings. The actuator receives the output signal, which might have an impact on the surrounding area. In light of this, it can be said that a microcontroller may be compared to the brain of a machine or product that can communicate with its surroundings.

Because a microcontroller has a microprocessor, memory, input/output (I/O) lines, and other complementing devices on a single chip, it is also sometimes referred to as a computer. The microcontroller processes data at a significantly slower rate than a personal computer. While the working speed of the microcontroller typically runs from 1-16 MHz, the speed of the microprocessor on the commonly used PC has reached the order of GHz. Similar to a microcontroller, a PC's RAM and ROM have a capacity that may reach the order of Gbytes as compared to the order of bytes or Kbytes for a microcontroller. Despite having a much slower data processing rate and less memory than a personal computer, a microcontroller nevertheless has the potential to be used in a wide range of applications, especially considering its small size.

2.2. Ultrasonic sensor

A sensor with the ability to transform electrical quantities into sound and physical or acoustic qualities into electrical quantities is known as an ultrasonic sensor. Reflected sound waves serve as the foundation for how this sensor operates. This ultrasonic sensor may be used to determine whether an object is there or how far away it is by using the sound wave's reflection off of a certain field. It is possible to determine and estimate the object's distance using a certain frequency of reflection. Because this sensor employs ultrasonic waves or ultrasonic sound as feedback, it is known as an ultrasonic sensor.

Ultrasonic waves are sound waves with a frequency of 20,000 Hz, which is quite high. Certain ultrasonic noises are audible to certain animals, including dogs, cats, bats, and dolphins, but they are inaudible to the human ear. It is possible for ultrasonic sound to travel through solids, liquids, and gases. On the surface of a solid, ultrasonic sound reflects with a reflectivity that is virtually identical to that of ultrasonic sound on the surface of a liquid. Therefore, in this investigation, the height of the water below the water's surface will be measured using ultrasonic sensors. However, materials like textiles and foams are easily permeable to ultrasonic sound waves, which leads to less than optimal measurement results.

IOP Conf. Series: Earth and Environmental Science 1173 (2023) 012044 doi:10.1088/1755-1315/1173/1/012044

The temperature of the surrounding area must also be considered when using this sensor to determine the depth of the water. Even a little temperature variation may result in inaccurate measurements since the speed of soundwaves is not constant.

2.3. LCD (liquid crystal display)

The LCD (Liquid Crystal Display) has the ability to display numbers, characters, and symbols more clearly. Low current consumption is a benefit of LCD panels. The dot matrix LCD (Liquid Crystal Display) is an LCD module utilized in this study to display the values obtained from sensing the water level. The character viewer (LCD) part of the dot matrix M1632 LCD (Liquid Cristal Show) module is used to display characters. This LCD also has an integrated microprocessor in the form of a module, which is located at the LCD's back. This microcontroller controls communication between the LCD and the microcontroller used with the LCD module within the LCD as well as the display on the LCD screen. The LCD M1632 is an LCD module with a 2x16 display layer, or 2 rows × 16 columns.

2.4. Air temperature sensor (DHT11)

The measurement instrument used for this study must carefully take input ambient temperature into consideration because temperature depends on location. In this investigation, the temperature was measured via a DHT11 sensor. This DHT11 sensor comprises two sensors: a temperature sensor and a humidity sensor, both of which produce digital data as their output. The information from this sensor's output will be utilized as the microcontroller's input value. The temperature value will then be shown on the LCD layer using this output after further processing. Temperature and humidity may both be measured simultaneously by a sensor called the Sensos DHT11. This DHT11 sensor offers a very high level of stability and a calibration feature that is quite precise. A thermistor of the NTC (Negative Temperature Coefficient) type is included in this sensor and will be used to monitor temperature. The DHT11 also features an 8-bit microprocessor that processes the two sensors and a resistive type humidity sensor. The outcomes will be transmitted by this microcontroller in a single-wire bi-directional fashion to the output pin. A calibration function with a respectable degree of accuracy is present on each DHT11 sensor. The inbuilt sensor will detect the signal while the calibration coefficient is being put into the OTP program memory.

2.5. Power supply

An electrical device known as a power supply serves the purpose of supplying voltage and electric current to other components that need those two things in order to operate. Basically, a power supply needs an external power source, which will be transformed into the precise power source required by different electronic devices. In this investigation, a DC power supply was employed to provide the microcontroller unit with the necessary 5 volts of power. This power supply unit's ability to produce at least 1 amp for the current rating is adequate for the measuring instrument to function correctly.

3. Methods and materials

An experimental investigation has been conducted into the accuracy of digital instrument design for detecting water level and water temperature in irrigation channels. The creation of the digital water level measurement design will be tested against manual measurement to determine the percentage difference. This study used V-notch discharge measurement for the discharge measurement (figure 1). Figure 2 depicts the manual water level measurement. Figure 3 shows the manual method for visually measuring the water level. technology for measuring digital water levels.

Figure 1. V-notch discharge measurement.

Figure 2. Water level measurement.

Figure 3. Manual water level measurement.

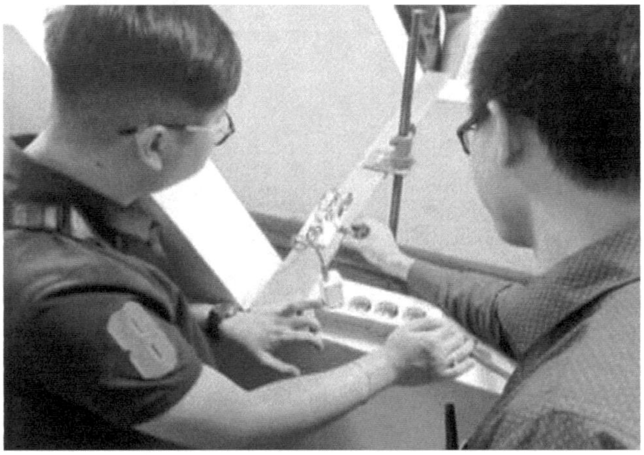

Figure 4. Digital water level measurement.

3.1. Measurement device block diagram

This diagram shows how input data from sensors is obtained, processed by the microcontroller, and measured by using a straightforward equation. After completing the equation's calculation, the microcontroller unit will display the solution on an LCD screen.

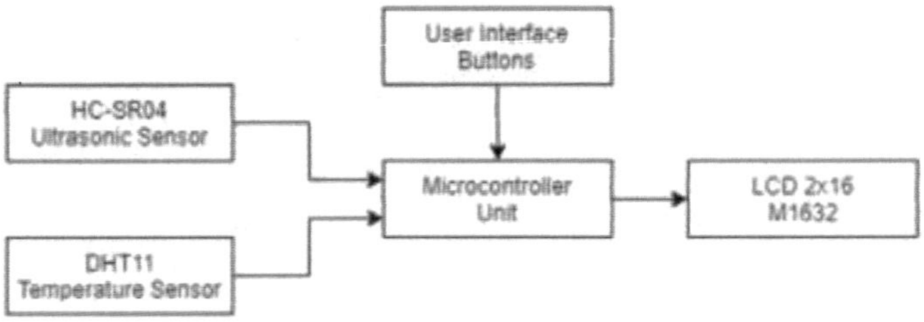

Figure 5. Measurement device block diagram.

$$d = \frac{t * 0.348}{2} \quad (1)$$

The HC-SR04 sensor measures the ultrasonic soundwave duration in microseconds and measures depth (d) in centimeters. Since an ultrasonic sound wave travels back and forth, the equation (1) must be divided by two to accurately represent depth. Please take note that the speed of a soundwave at ambient temperature is 348.369 m/s, hence this equation only applies at that temperature. The d=(t*0.0348)/2 temperature sensing unit is very important for this obvious reason, because the constant value stated in this equation is approximately 0.0348 and must be accounted for temperature variations.

3.2. Rating curve measurement

A rating curve in an irrigation system is a graph showing discharge in relation to stage at a specific point on the irrigation system, often at gauging stations where the discharge is determined using a flow measurement across the irrigation channel. Several discharge measurements are taken over a variety of irrigation channel stages or water levels. The rating curve is a crucial tool in irrigation channels because a good stage-discharge relationship at the gauging station is crucial to the accuracy of discharge data values. In this study, the accuracy of the digital measuring device for water level and water temperature in the irrigation channel was evaluated using the percentage difference. If the calculated discharge is greater than the given discharge, reduce the gate opening, otherwise increase it, and repeat the process until the difference between the calculated and the given discharge is small [5]. Based on that method as well, the results can be seen on table 1, table 2, table 3 and table 4.

Table 1. The percentage difference for Q50%.

Experiment	%Difference
1	0.1
2	0.1
3	0.1
4	0.2
5	0.2

Table 2. Percentage difference for Q10%.

Experiment	%Difference
1	0.3
2	0.3
3	0.2
4	0.3
5	0.3

Table 3. Percentage difference for Q10%.

Experiment	%Difference
1	0.3
2	0.3
3	0.2
4	0.3
5	0.3

Table 4. Percentage difference for temperature.

Experiment	%Difference
1	0
2	0.01
3	0
4	0
5	0.01

4. Result and discussion

The significant development offers utility decision makers concrete learning to speed their adoption of digital solutions and address pressing water concerns by examining how digitalization is affecting the water industry, particularly in irrigation projects. The result displays the percentage difference between analog and digital methods for measuring water levels. It is possible to infer that it is crucial to use digital tools, especially when determining the water level in irrigation channels. Figure 6 shows a rating curve.

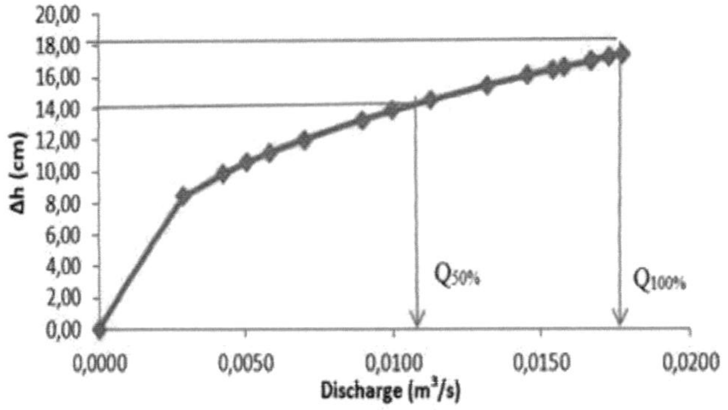

Figure 6. Rating curve.

Depending on the size and form of the stream, discharge rating curves can be simple or complicated. Stage and discharge are frequently related through empirical research using repeated observations. To establish the rating curve for a certain stream, these data are plotted against the contemporaneous stage. Furthermore, to construct the stage discharge connection throughout the complete range of stream flow data, several discharge measurements are required.

5. Conclusions

In the decision-making process, the sensitivity analysis was the final stage where the input data were slightly modified in order to observe the impact on the results [6]. If the ranking did not change, the results were said to be robust. In short, we needed to see whether the small variations in the weights would change the decision. If not, we would be reassured that our choice was reasonable. Precision and performance are the primary operating criteria for irrigation gates and other metrics. Due to the continuously changing water level as well as outside disturbances, the water level data monitoring based on a typical measurement frequently produces incorrect findings. Additionally, in some circumstances, it's necessary to measure the discharge flow in addition to the water level in order to manage water excess or shortage. This work suggests a precise non-contact water measuring system based on a microcontroller, an ultrasonic sensor (HC-SR04), and an air temperature sensor (DHT11) to detect the water level in a channel without any touch in order to address these problems. The speed of ultrasonic emission while measuring distance is impacted by the variation in air temperature. At a gauging station, the channel conditions downstream from the gauge determine the stage-discharge relation for open-channel flow. Therefore, a fundamental understanding of the channel characteristics that govern the stage discharge relation is essential for creating rating curves.

Acknowledgements
The authors are grateful to Maranatha Christian University and some students who provided assistance with this project.

References

[1] Grag S K 1998 *Irrigation Engineering and Hydraulics Structures 2nd edition* (London: The publishing company)
[2] Polak P, Nanes B and Adhikari D 1997 A low-cost drip irrigation system for small farmers in developing countries *J. of the American Water Resources Association* **33**(1) 119–124
[3] Deng X-P, Shan L, Zhang H and Turner N C 2006 Improving agricultural water use efficiency in arid and semiarid areas of China *Agricultural Water Management* **80**(1–3) 23–40
[4] International Commission on Irrigation and Drainage 2015 *Agricultural Water Management for Sustainable Development* Retrieved from http://www.icid.org/ar_2015.pdf
[5] Swamee P K 1992 Sluice-gate discharge equations *J. of Irrig. and Drain. Eng.* 118(1) 56–60
[6] Tallar R Y and Suen J P 2016 Aquaculture water quality index: A low-cost index to accelerate aquaculture development in Indonesia *Aquac. Int.* **24** 295–312

IOP Conf. Series: Earth and Environmental Science 1173 (2023) 012045 doi:10.1088/1755-1315/1173/1/012045

Backwater flow analysis due to the construction of the check dam series on Aie Dingin River using HEC-RAS

D Daoed[1], Sunaryo[1] and F Arsyad[2]

[1]Civil Engineering Department, Engineering Faculty, Andalas University, Padang, West Sumatra
[2]Undergraduate Student, Civil Engineering Department, Andalas University
Author Emails

E-mail: darwizaldaoed@gmail.com

Abstract: River profiles with steep slopes and loose-grained soil types will easily be carried away by river currents. One way to reduce the sediment rate is by constructing a check dam or a series of check dams. However, the effect of a series of check dams can influence each other, due to backwater flow. On the one hand, it is advantageous because the flow velocity is reduced, but it will raise the water level downstream of the dam and increase the height of the embankments and ponds. Based on the foregoing, an analysis of the backwater flow was carried out using HEC-RAS. The results obtained are that water backflow occurs in two water structures, namely the Koto Tuo and Koto Pulai check dams. The backwater flow is quite long, almost reaching the previous check dam. Where the length of the backwater flow is only up to the final threshold and it is advisable to observe local sedimentation behavior due to backwater flow.

1. Introduction

The river which is located in an oval-shaped watershed has many tributaries in the upstream area as well as steep and undulating topography, which has the potential to produce high water discharge and flow velocity [1]. This condition will have a bad impact on the river because there will be scouring or erosion. Also the collapse of the river bank and changes in the river channel [2,3,4]. This phenomenon is experienced by several watersheds (DAS) along the Bukit Barisan on the island of Sumatra, especially the watershed in Padang City. Besides that, the influence of climate with heavy rainfall from June to December every year. This is a wet month with precipitation (R>150 mm. Floods cause several areas along the coast to experience inundation, particularly when it rains in the afternoon to evening. Where the effect of sea level rise or tide occurs twice a day. One of the events of high-intensity rain on March 22, 2016 (R=286 mm) caused flooding and the irrigation dam collapsed and was damaged in Batang Aie Dingin or Aie Dingin river. Also flooding on 18 August 2021 and 29 September 2021 drowned the Aie Pacah, Maransi, and around Banda Luruih areas. Although the latter is concentrated rain in the upstream areas which are still in the form of rice fields and protected forests [5].

Observing the behavior and profile of Aie Dingin river which is located in the Aie Dingin watershed in Padang City, where the river has a base slope that varies from very steep, steep to gentle and relatively straight. The combination of the river profile and high rainfall intensity makes the river vulnerable to scouring and sedimentation. Based on this, the government has built multi-functional water structures, both as energy absorbers, preventing sediment transport along rivers, and at the same time for irrigation and tourism. This building is also known as a weir or check dam. There are two check dams built on this river, namely in Koto Pulai and Koto Tuo which are located sequentially or in series. Also between these dams is a highway bridge that has pillars and abutments. These two buildings have different functions but need to have strength against scour. Therefore, there may be backwater flow caused by river damming, so it is necessary to study how far the influence of water level rise due to weirs and its effect on the height and dimensions of river embankments is carried out.

Backwater is the rate of flow being restrained due to obstructions, causing the water level upstream of the building to rise to a certain distance. Barriers can be, weirs, dams, bridges, or the confluence of two or more rivers and at the confluence of rivers and shores. The effect of building barriers on the river in addition to going into the river upstream is also useful for holding sediment. To predict the length of the water level rise, the HEC-RAS software and direct step method can be used [6,7,8,9].

HEC-RAS is an application that focuses on the hydraulic analysis of a river, in contrast to HEC-HMS which focuses on hydrological analysis. The analysis carried out by HEC-RAS is an analysis of steady flow, unsteady flow, and also sediment transport. Steady flow is a flow in which the fluid flowing in any part is constant. Even this flow is divided into 2 namely Uniform flow and Non-Uniform Flow. Meanwhile, unsteady flow is a flow in which the fluid flowing in any part is not constant. HEC-RAS requires some data to run the analysis including geometry data, river data, rain data, and so on [10].

2. Study location and methodology

2.1. Study location

Figure 1. Check the dam located in the Aie Dingin river.

The research was conducted on the Aie Dingin river (Batang Aie Dingin) at latitude 0°50'48.20"S and longitude 100°21'26.60"E for the Koto Pulai check dam and latitude 0°50'26.25"S and longitude 100°22'03.14"E for the Koto Tuo check dam. Where this water structure is separated by the Padang By-Pass (Highway) bridge which is approximately 0.51 km from the Koto Pulai weir. Practically, this weir is also used in addition to controlling sediment and also protecting bridge pillars from scouring. The following Figure ***1***. Check the dam located in the Aie Dingin river shows the location of the building.

2.2. Methodology

The research method can be used as a theoretical approach through the direct step method and the standard step method. Both are standard methods of predicting gradual changes in flow. The planned flood discharge is estimated from the prevailing rainfall in the study area and determines the catchment area of rainfall with the help of Arc GIS. There are three stations for recording rainfall in the watershed, namely Batu Busuak, Koto Tuo, and Gunung Nago. Furthermore, the planned flood discharge is estimated for several return periods. While the channel cross-section is obtained by BWSS-V in the form of secondary data. After the design flood discharge and river geometric data are obtained and inputted into the HEC RAS software by determining the length of the study area first. Besides that, the calculation of the length of the water level rise is also carried out using direct-step methods. This is done so that it can be done as a comparison to the output of the software and theory. Finally, a discussion of the two outputs was carried out, and conclusions.

3. Result and discussion

3.1. Hydrology analysis

In the study area, only three representative and complete rainfall observation stations were obtained, namely Koto Tuo Station, Batu Busuak, and Gunung Nago. Data collected from 2011-2020. Although there are other stations, they are located on plains and far from the watershed. The results of the analysis show that the average maximum daily rainfall for 10 years of observation is 156 mm, the skewness coefficient Cs =1.03, variation coefficient Cv = 0.15 kurtosis coefficient Ck = 5.17 (leptokurtic). So that the CH distribution of maximum daily rainfall during 10-year observations from three observation stations satisfies the Pearson III log equation. Then the rainfall intensity with several return periods is obtained from the estimated maximum daily rainfall plan using the Log Pearson III equation and the hourly rainfall intensity is approximated by the Mononobe equation. The results of the calculation of the area of the watershed are 122.5 km^2, the slope average is 78.5%, the river length is 20.5 km and the upstream and downstream elevations are 1440 m and 12 m from the mean sea level.

Furthermore, to obtain the planned discharge, it is carried out using the rational method with the following Table 1. Rainfall intensity and planned flood discharge:

Table 1. Rainfall intensity and planned flood discharge.

Return Period (T) (year)	R (mm)	I (mm/hour)	C	A (km^2)	Q (m^3/s)
	175.5	41.3	0.3	122.5	422.4
10	190.8	44.9	0.3	122.5	459.3
25	194.9	45.9	0.3	122.5	469.2
50	223.4	52.6	0.3	122.5	537.7
100	237.3	55.8	0.3	122.5	571.1

It can be seen in the table-1 that the planned flood discharge for the 10, 25, 50, and 100-year periods increased from 422.4 m^3/s to 571.1 m^3/s. This increase is quite large, namely, around 150 m^3/s. Therefore, for the next simulation, a 100-year discharge is used.

3.2. Result of simulation

The backwater flow simulation was carried out using HEC-RAS software under conditions without and with check dams. Check dam locations are at station 9 and station 37. Meanwhile, the planned flood discharge is used for a return period of 100 years. In Figure *2.* Longitudinal profile of flow without a check dam, the following shows the results of the modeling with HEC-RAS 4.1.0 without check dams.

The image below Figure 2. Longitudinal profile of flow without a check dam and Figure 3. 3D water level profile without check dam shows the water level along the entered cross-section data and the difference in water level in each cross-section can be seen. Also showed the cross section at station 9 and station 37 where the check dam is located and the state of the flow. Next, the results after the check dam were built at stations 9 and 37 as Figure 4. Longitudinal profile of flow with check dam and Figure 5. 3D water level profile with check dam. In the figure, it is clear that the water level is rising upstream of each check dam. Especially at a distance of more than 100 m from the weir. It also shows changes in flow level along the study area.

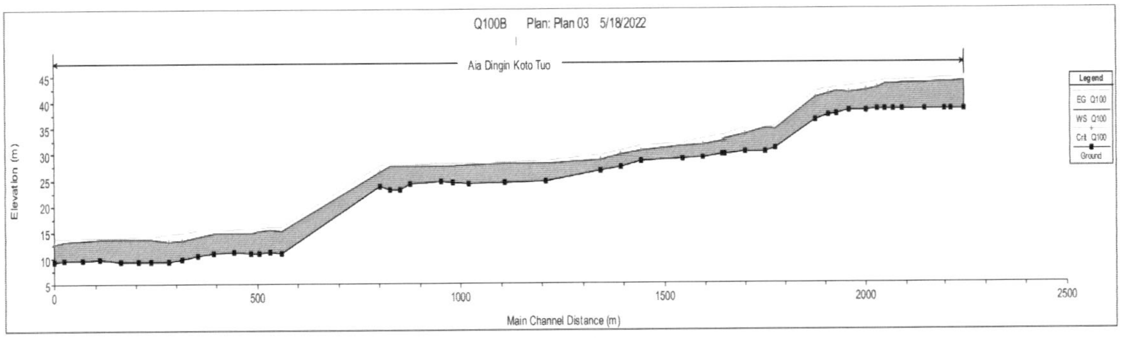

Figure 2. Longitudinal profile of flow without a check dam.

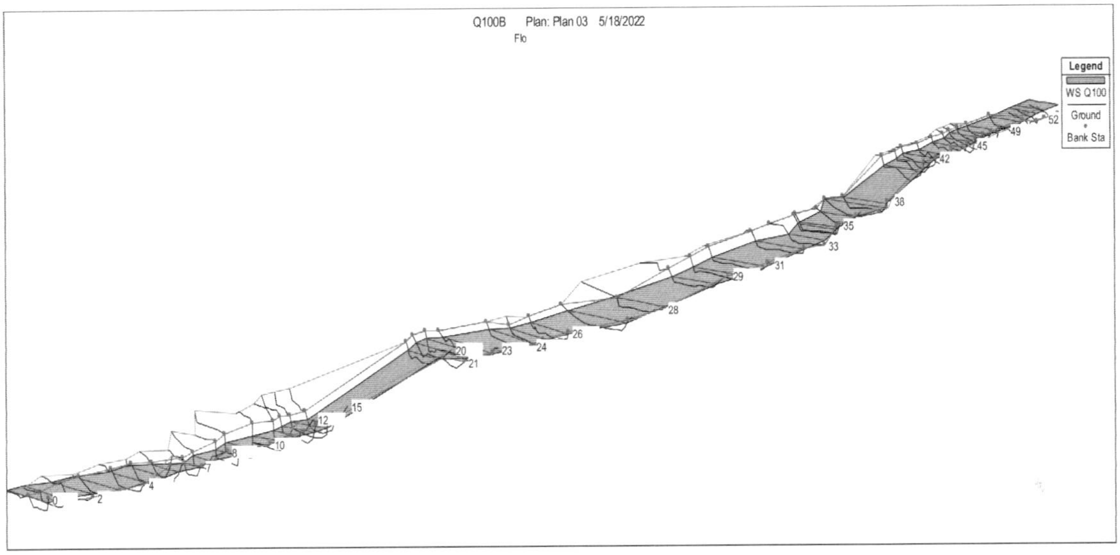

Figure 3. 3D water level profile without check dam.

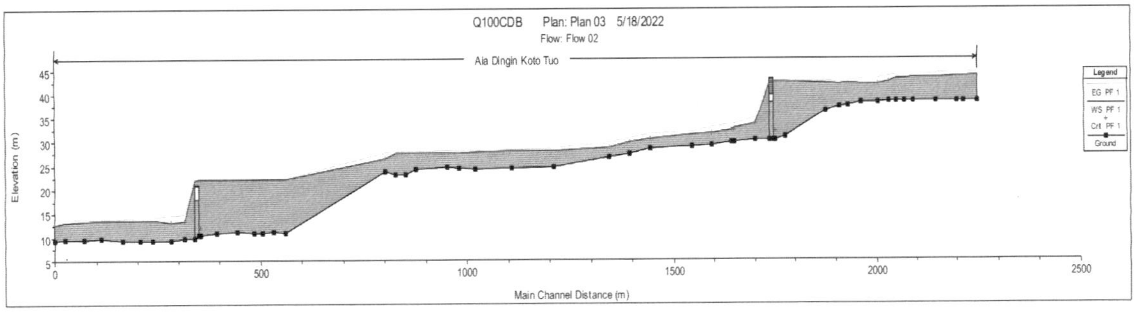

Figure 4. Longitudinal profile of flow with check dam.

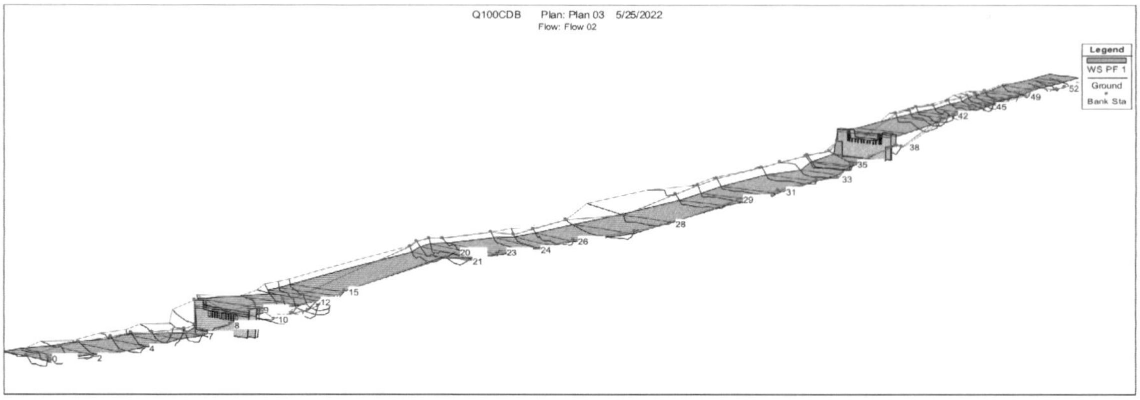

Figure 5. 3D water level profile with check dam.

3.3. Discussion

From the results of running HEC-RAS, it can be described the water level along the study area, where there are several stations with a distance between stations of 100 m. Graphically, there is a difference in water level between the theoretical analysis with the direct step method and the output of HEC-RAS. This can be seen from the two scenarios, namely without a check dam or with a check dam. This difference is very real from the two segments of the check dam locations that were built, namely station 9 and station 37. The following is an illustration of the water level rise that occurred in fFigure 6. The water level in the area at the Koto Pulai check dam and Figure 7. The water level in the area at the Koto Tuo check dam, following.

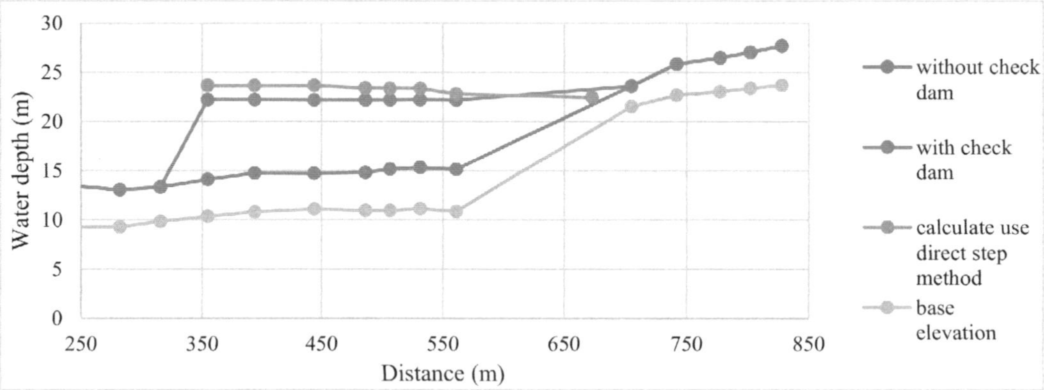

Figure 6. The water level in the area at the Koto Pulai check dam.

Figure 6 shows the water level without a check dam is a blue line, while the red line is a graph with a check dam. Where the water depth near the bulkhead was initially around 15 m to 22 m (equal to the water outlet height in the weir). The length of the effect of rising water levels upstream increases. Initially, the profile length was about 150 m to 447.7 m from the check dam. When viewed at the normal depth, the flow tends to increase to the same as the normal depth without check dams. Although the length of the water rise did not reach the Koto Tuo check dam, this was due to the steep slope of the river and the sudden change in slope at station 26. The shape of the flow profile in the longitudinal direction in the case of obstacles with check dams on a steep slope is relatively not too long, but on the contrary, the length of the water level rise will be long in the case of a rather mild and mild slope. In the case of the Koto Pulai check dam, the slope is quite sharp and the effect is only up to the By-Pass bridge pillar.

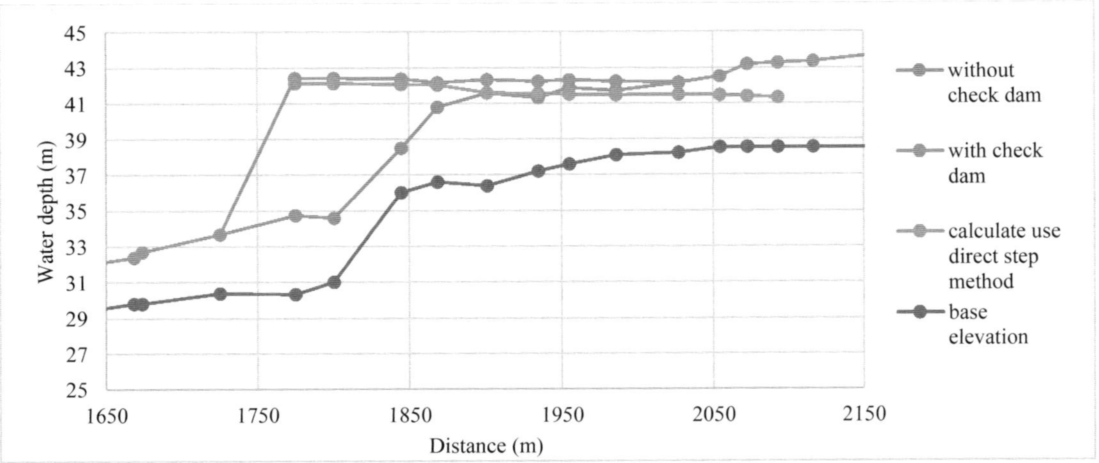

Figure 7. The water level in the area at the Koto Tuo check dam.

Meanwhile, the water level upstream from the Koto Tuo check dam is slightly shorter than that from the Koto Pulai check dam, which is 252 m. However, the flow depth near the check dam is deeper. Where without a check dam is about 35 m and becomes 42.5 m after being dammed.

In general, it is seen that the reduced length of the rise effect is the result of the slope of the riverbed [11,12,13]. Where a large slope will affect the flow velocity. The steeper will give speed and may be closer to the critical speed. So the length of the effect of rising water levels is getting shorter. Another impact will be to increase the dimensions of the embankment on the left and right of the river [14,15,16,17].

To confirm the results of the HEC-RAs output, analysis, was also carried out using the direct step method. The results of the estimation of the length of the rise from the two check dams. Where it can be seen that the depth and length of the rise in water level or backflow are close to the value of HEC-RAS. Although the length of the increase is slightly different. When compared to the results of the analysis with software and calculations based on direct and standard stages, the length of the rise in the upstream area is slightly different, from 447 m to 351 m at the Koto Tuo weir and 252 m to 342 m. Where this is most likely due to rounding and different approaches. There is also a steep slope at the location of 550-800 m with a slope of 7.5%. Likewise, upstream of the Koto Tuo check dam, there is a steep slope of 3.6%. Indirectly it is a natural function as a barrier or groundsill building. This is in line with the results of studies from [6,7,8,16,18,19,20] there is a decrease in energy due to waterfall buildings. This needs to be studied further if the length of the profile extends to this area.

Theoretically, the existence of a check dam is very beneficial in controlling the sediment rate and the position of the bridge pier which is close to the Koto Pulai check dam can avoid local scouring. On the other hand, it will accelerate the accumulation of discharge material. In addition, it is necessary to anticipate the planned flood discharge that is more than the estimated 100-year discharge, because climate change with high intensity in the watershed can occur.

4. Conclusion

The length of the water level rise due to the construction of the Koto Pulai check dam did not reach the Koto Tuo check dam location. This means that the distance between check dams is quite safe against rising water levels. The existence of the Koto Pulai check dam can raise the water level to the location of the bridge pillars, thereby reducing the impact of local erosion.

Validation of the match between the HEC-RAS outputs using the direct method to the length of the longitudinal profile looks almost the same. The estimated differences are due to rounding, Manning roughness, and the assumed cross-sectional shape. The side effect of rising water levels upstream of the weir resulted in higher dimensions of the river embankment. The effect of a large base slope at each upstream of the check dam increases annual sedimentation and accelerates maintenance and dredging work.

Acknowledgments
This publication was funded by Andalas University through a publication grant from the Faculty of Engineering, LPPM - Andalas University, and I would like to thank the Head of the Civil Engineering Department, Vice, Dean-1, Dean, and the Rector for their help.

References

[1] Daoed D, et al 2018 Evaluation of drought vulnerability on watersheds in West Sumatera Province by using Cropwat-8 and GIS *Int. J. Adv. Sci. Eng. Inf. Technol* **8**(6) 2443-2449
[2] Daoed D, et al 2021 *Study of the effects of check dam construction on the Limau Manih river using GIS in E3S Web of Conferences* (EDP Sciences)
[3] Daoed D, et al 2016 Predictions of Vulnerability Flood and Flood Prone Areas in Watershed West Sumatra Province using Arc-GIS and Category Value. *International Journal of Earth Sciences and Engineering* **09** SPL No 03 in 3rd International Conference on Earth Sciences and Engineering (ICEE 2016) p.274-279
[4] Roni Y 2022 *The Effect of Land Use Changes on Flow Discharge and Drainage Capacity of Padang City, (Indonesia language: Pengaruh Perubahan Tataguna Lahan Terhadap Debit Aliran dan Kapasitas Drainase Kota Padang*) (Padang: Andalas University)
[5] Kodoatie Robert J S 2002 *Flood Causes and Control Methods From an Environmental Perspective (Indonesia language: Banjir Beberapa Penyebab dan Metoda Pengendaliannya Dalam Perspektif Lingkungan)* (Yogyakarta: Pustaka Pelajar) p.349
[6] Subramanya K 2009 *Flow in Open Channels 3 Edition ed. 2009* (Singapore: MC Graw Hill International Edition) p.548
[7] Andrew C, Morfett J, 1986 *Hydraulics in Civil* Engineering (London, UK: Allen & Unwin (Publisher) Ltd)
[8] Darwizal D 2010 *Hydraulics and Applied for Open Channels,(Indonesia language: Hidrolika dan Terapan untuk Saluran Terbuka)* (Padang: CV Ferila) p.113
[9] Hidayat B, Daoed D and Aprilindo A 2017 Analysis of flood inundation mapping and community knowledge in flood disaster mitigation in Maransi housing, Koto Tangah sub-district, Padang city,(Indonesia language: Analisis pemetaan genangan banjir dan pengetahuan masyarakat dalam mitigasi bencana banjir di perumahan maransi kecamatan Koto Tangah Kota Padang) in *Proc. 4th ACE* (Padang:Andalas University)
[10] Utami T, Purwadi O T, and Susilo G E 2016 *W*ay Besai River Cross Section Design Through Increasing River Capacity Using HEC-RAS, (Indonesia language: Desain Penampang Sungai Way Besai Melalui Peningkatan Kapasitas Sungai Menggunakan Softwere HEC-RAS) *Jurnal Rekayasa Sipil dan Desain* **4**(2) 185-196
[11] Daoed D, et al. 2015 *S*tudy of the Batang Bayang watershed and additional flow from Batang Sikabau to meet the water needs of the Batang Bayang irrigation area, (Indonesia language: Kajian Das Batang Bayang dan tambahan debit aliran dari Batang Sikabau untuk memenuhi kebutuhan air daerah irigasi Batang Bayang) *Jurnal Rekayasa Sipil (JRS-Unand)* **11**(2) 71-80
[12] Tsai C W 2005 Flood routing in mild-sloped rivers—wave characteristics and downstream backwater effect *Journal of Hydrology* **308**(1-4) 151-167
[13] CR A and Thatikonda S 2020 Study on backwater effect due to Polavaram Dam Project under different return periods *Water* **12**(2) 576
[14] Azinfar H, 2010 *Flow resistance and associated backwater effect due to spur dikes in open channels* (Saskatoon: University of Saskatchewan)
[15] Azinfar H and Kells J A 2009 Flow resistance due to a single spur dike in an open channel *Journal of Hydraulic research* **47**(6) 755-763

[16] Liro M 2019 Dam reservoir backwater as a field-scale laboratory of human-induced changes in river biogeomorphology: A review focused on gravel-bed rivers *Science of the Total Environment* **651** 2899-2912
[17] Hidayat H, et al. 2011 Discharge estimation in a backwater affected the meandering river *Hydrology and earth system sciences* **15**(8) 2717-2728
[18] Darwizal D, Istijono B, and Hakam A2014 Integration of Small Watersheds to Meet Water Needs Using Supplementary Channels, (Indonesia language: Integrasi Daerah Aliran Sungai Kecil untuk Memenuhi Kebutuhan Air Menggunakan Saluran Suplesi) in *Pertemuan Ilmiah Tahuanan HATHI 31* (Padang: HATHI Pusat)
[19] Muhammad K A 2019 *Identification of damage to the old building due to the March 22, 2016 flood and stability test on the redesign of the Koto Tuo waterfall, (Indonesia language: Identifikasi kerusakan bangunan lama akibat banjir 22 Maret 2016 dan uji stabilitas pada desain ulang bangunan terjun Koto Tuo)* (Padang: Universitas Andalas)
[20] Siallagan B J N 2018 *Analysis of the Effect of Back Water on the Lepan River Flood, Langkat Regency, (Indonesia language: Analisa Pengaruh Back Water Terhadap Banjir Sungai Lepan Kabupaten Langkat)* (Medan: Universitas Sumatera Utara)

IOP Conf. Series: Earth and Environmental Science 1173 (2023) 012046 doi:10.1088/1755-1315/1173/1/012046

Effects of COVID-19 on transportation services and travel behaviour in Malaysia

L Gungat[1], D R Deh[1] and E E Putri[2]

[1]Civil Engineering Programme, Faculty of Engineering, Universiti Malaysia Sabah, Sabah 88400, Malaysia.
[2]Civil Engineering Department, University of Andalas, Padang Indonesia

E-mail: lillian@ums.edu.my

Abstract. Movement restrictions implementation due the COVID-19 pandemic affected people's mobility and and transportation sectors globally. Tourism, retail, transportation, and hospitality also suffered significant consequences that indirectly impacted a country's economic stability. Hence, this paper presents the Malaysian experience with the effects of the pandemic on transportation services and travel behaviour. Online sources, secondary data and a questionnaire survey were employed to collect the relevant data. These data were further analysed and presented in this paper. In general, transportation sectors involving air, rail and public bus showed decreased demand and services. A severe drop in services was observed in air transportation which is more than 97%. Most people prefer not to use public transportation to avoid the risk of COVID-19 infection. Noticeable changes in travel behaviour and mode preference during and pre-pandemic are observed, particularly on trips related to work/school. Key pandemic factors concerning transportation were fear of infection and social distance. Hence, most people prefer to use a private vehicle. The changes in transportation service and travel behaviour result in severe losses in the transportation sector in Malaysia.

1. Introduction

Coronavirus disease or COVID-19 outbreak on December 2019 at Wuhan, China was a shocking news to the whole world [1]. In June 2020, COVID-19 destroyed more than 450,000 lives and the estimated number cases had crossed more than 8,400,000 individuals. Imported cases of COVID-19 appeared in other Asian countries such as Thailand, Japan, the Republic of Korea, and Singapore around mid-January 2020. Millions of people have been quarantined to avoid the virus spread. At the same time, many countries have decided to enact specific mandatory guide for example implementation of social distancing and lockdown to hinder futher spreading of the highly transmissible COVID-19 virus. It became a mandatory norm in most countries to avoid physical contact and minimise interaction between individuals (i.e., physical distancing). This mandatory control procedure affected the whole transportation industry.

The lockdown policy implementation to decrease the possible infection of COVID-19 worldwide has impacted the mobility of people. A study by [2] reported that the mobility of air transport in the European region decreased by more than 89%. However, the pandemic had less effect on cargo traffic due to the need for medical supplies in the fight again illness [2]. Regarding transport services it was reported that the ridership in some major cities in China, USA, Iran and UK dropped by 80% [3]. The policies on the social distancing set by the government to minimise COVID-19 transmission had changes the transport services. Hence, in order to synchronise these changes, public transport operators globally have opted to reduce their service duration, cancel specific services and shut certain stations. Based on report significant increase in freight transport during the lockdown due to panic buying and started to stock up on foods and necessities at home in China [4]. Nevertheless, people choose to shop online to avoid COVID-19 infection. As a result, this rises the freight transport demand [4]. The trend of the pandemic impact on freight transport differs in some places. For example, in Colombia, the freight transport trips were reduced by 38%, possibly due to the closed down of non-essential factories during the mandatory quarantine phase [5]. The new norm has caused the decline in urban travel globally, and all public transportation modes have experienced severe losses in terms of cost. As a result, all these reductions affected economy globally. The United Nations Conference on Trade and Development (UNCTAD) reported that the global economy could lose more than 4 USD trillion in the tourism sector, which is directly related to the transport system [6].

In Malaysia, many sectors have been impacted by the extended movement restriction orders, which have been in effect since March 2020, and as expected, these measures have not been economically beneficial. Tourism, retail, transportation, and hospitality sectors suffered significant consequences. Work from home and online teaching implementation during pandemic had changed individual travel behaviour. Hence, a study on the effects of the COVID-19 pandemic on transportation services and travel behaviour based on Malaysian experiences is necessary in order to understand its effect.

2. Methodology

This study uses online sources, secondary data resources and a questionnaire survey to understand the pandemic's impact on transportation services and individual mobility. The data collection using online sources and secondary data was conducted by skimming through the related websites, reviewing and analysing the previous study's findings. An survey was carried out randomly via online to study the individual travel behaviour in Kota Kinabalu, Sabah. Whereas for other places in Malaysia, secondary data from previous research were analysed. Structured interviews with the public transport operators at Kota Kinabalu, Sabah, were conducted to identify the challenges encountered by the operators in adapting to a new norm of public transport service during the pandemic.

3. Effects on transportation services in Malaysia

Movement restrictions during the pandemic caused traffic and mobility changes, especially in major towns with high population density. The impact could be seen in land and air transportation services, whereby a dramatic drop in the quantity of public transportation users across Malaysia.

3.1. Rail transportation

Rail transportation is the main transportation in metropolitan such as Kuala Lumpur. During the pandemic, rail transport services were also affected. For example, rail transportation services in the Klang Valley area in 2020 have reduced drastically with an average reduction of 48% [7]. Figure 1 shows the ridership comparison for various rail transport services in the Klang Valley area in year 2019 and 2020. It can be observed that the Kuala Lumpur International Airport (KLIA) Express and KLIA Transit service show the highest reduction, which is 67%. These lines provide a direct transport connection between KLIA airport and the city centre, which was hit mainly by the pandemic. The reduction in ridership is followed by Kelana Jaya LRT (60%), Intercity train (55%) and KTM Commuter train (53%). The reduction in train ridership had affected the operation's sustainability and annual revenue. These happened due to the movement restriction during the first lockdown in March 2020. Besides that, another reason could be due to the fact that users were scared of virus infection when using public transport.

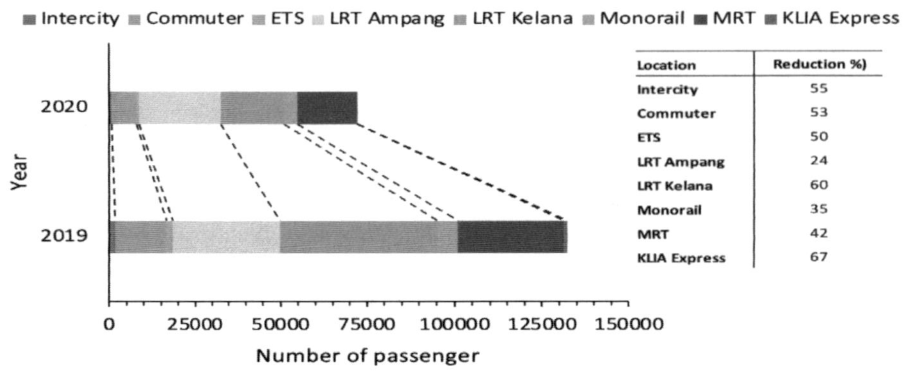

Figure 1. Ridership of train services in Klang Valley area in 2019-2020. Source: [7]

3.2. Air transportation

Air transportation and tourism industries are extremely affected when the COVID-19 cases suddenly increase in Malaysia. As stated in the annual report by MAHB (2020) shown in table 1, about 98% reduction in passengers for all international flights in April 2019 and 2020 for the main KLIA. This is even worse for the KLIA 2, which was 99.7% in April 2019 and 2020. In addition, many international and domestic flights were canceled due to the movement control policy, reducing the number of flight operations. At the same time, the airline industry must strictly follow the standard operating procedure (SOP) in handling passengers. Hence, reducing flight services indirectly increases travel costs as the airline industry needs to sustain its operation.

Table 1. Passenger traffic at KLIA('000) as of April 2020.

Airport	2020	2019	YOY(%)	2020	2019	YOY(%)
	April	April		Year to date		
KLIA Main total	49	2440	-98.0	5171	9412	-45.1
International	37	1954	-98.1	4064	7607	-46.6
ASEAN	10	753	-98.6	1472	2896	-49.2
Non-ASEAN	27	1201	-97.8	2592	4711	-45.0
Domestic	12	486	-97.5	1107	1806	-38.7
KLIA2 total	8	2821	-99.7	5592	10993	-49.1
International	3	1836	-99.7	3597	7186	-49.9
ASEAN	1	1069	-99.9	2149	4159	-48.3
Non-ASEAN	2	767	-99.7	1448	3028	-52.2
Domestic	5	985	-99.5	1995	3806	-47.6

Source: [8]

3.3. Public bus transportation

The execution of Movement Control Order (MCO) causes travel changes in daily activities such as working in the office, attending classes, shopping and etc. This also caused a significant reduction in public transport users. Based on a report, the overall ridership of Rapid Bus in KL dropped by 66% [9]. Furthermore, a significant reduction in ridership has recorded especially in big cities such as Kuala Lumpur, Penang and Kuantan. The study report shows the highest ridership reduction for Sunway BRT by 79.5%, followed by MRT feeder buses which is 75.7%. Moreover, ridership for public buses which are Rapid KL, Rapid Penang and Rapid Kuantan's also reported a decline at 65.1%, 50.8%, and 62%, respectively particularly during MCO [9].

This situation may negatively impact the public transport operators if not tackled reasonably. Malaysian attitude on public transport service during the COVID-19 pandemic and public transport usage is illustrated in figures 2 and 3, as reported by [10] in their study. Both figures are much related to the safety perception of virus infection while traveling by public transport. About 90.6% respondents are worried to take public transport during the pandemic. However, 32.7% may take the public transport if necessary. Sudden decline of public transport usage can be seen in March and April 2020 whereby most public transportation were shut down during MCO.

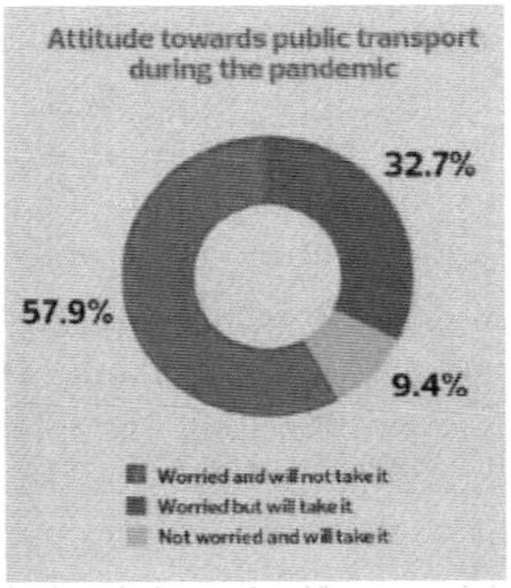

Figure 2. Malaysia's attitude towards public transport during the pandemic. Source: [10].

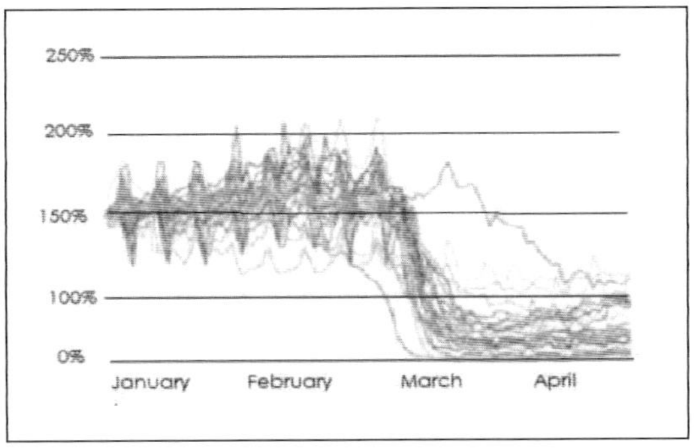

Figure 3. Effect of COVID-19 pandemic on public transit usage. Source: [10].

A survey finding regarding the people's perception of public transportation usage during the pandemic at Kota Kinabalu, Sabah (figure 4) shows that most people agreed that the probability of being infected by COVID-19 is higher (71%) when using public transportation. Based on the survey, only 5% disagreed that there is a high chance of being infected using public transport. According to [11], an enclosed environment and crowded area create favourable conditions for spreading epidemic infections, where the COVID-19 virus can spread over a large area through droplet and air circulation. Hence, there will be a high risk of infecting healthy person in indoor, inadequately ventilated areas and crowded. Previous research reported that the COVID-19 virus droplets is feasible to travel at 10 m/s speed with a concentration about 108 copies/mL. Hence, there is a high risk of an asymptomatic patient spreading the virus if they cough while on bus without wearing any protection such as mask during peak hours.

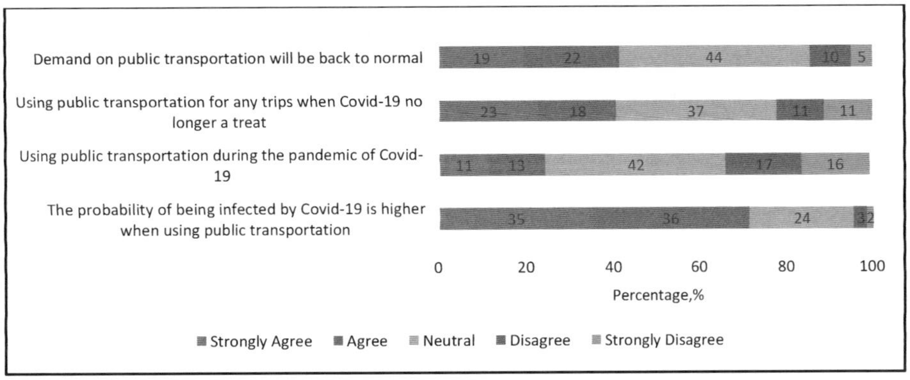

Figure 4. Respondent perception of COVID-19 pandemic effects on public transportation usage.

The findings on personal opinion on using public transportation during a pandemic indicate that only 24% of the respondents strongly agreed and agreed with the statement. In contrast, about 42% voted for neutral, and the outstanding respondents (33%) strongly disagreed and disagreed. The finding indicates that public transport user perceived commuting via public transport is risky during the pandemic, thus result in lesser ridership in comparison with the before COVID-19 . In contrast, 41% of the respondents strongly agreed to use public transportation for any trips when COVID-19 is no longer a threat. Public transport users, particularly the frequent traveler, find it safer with strict SOP. Respondent's perspective regarding the demand for public transportation, whether it will be back to normal or otherwise, indicates that 41% of the respondents are strongly agreed and agreed with the statement. The effect of the decrease in transportation related business is expected to recover and return to normal business when vaccination reaches the level of herd immunity.

Besides that, public transport operators are suffering a sudden decrease in terms of demand and fare revenues due to the infectivity risk within enclosed environment of public transport. Regulations to limit or suspend public transport service during pandemic also affected the public transport operators. Under common operating conditions of public transportation, passenger density and frequency of service are the economic indicators that drive its efficiency gains compared to other modes of transport. Nevertheless, passenger density during the pandemic decreased due to the implementation of social distancing in public transportation. Besides a drop in commuters, the higher cost of regularly cleaning and disinfecting public transportation and facilities could put an additional financial burden on public transport operators. The challenges in public transport operation are caused by low demand on public transport and distruptions in the economic stability for public transpotion daily operations. Based on the interview with the public transport operators in Kota Kinabalu Sabah, MCO implementation and restrictions on various daily activities cause substantial dropped on demand for public transport services. Apart from that, afraid of COVID-19 virus infection could be another reason why people chose not to use public transport and tend to use e-hailing services like Grab, which is more private and safer even though the price is more pricey compared with public transport. Therefore, the decrement of demand for public transport has caused severe consequences to the financial instability encountered by public transport operators. The operating costs of public transport needs a long term solution as it is not depend on demand only. An immediate adjustment of supply and production costs is not a possible solution in response to the transportation bussiness shockwave. Therefore, bus operator needs recovery time for capital cost of infrastructure, and vehicle maintenace.

4. Effects on individual travel behaviour

Based on previous research, human-mobility decrement showed substantial impact on lessening COVID-19 infection and death [12-13]. Therefore, if a person changes their travel related behaviour helpfully when consider a high risk of COVID-19 infection in daily commute, the probablity of spreading the virus can be decreased. Understanding on how COVID-19 pandemic affected the travel behaviour based on Malaysian experience is crucial to identify some key-pandemic factors. Travel is part of daily life, and each person has different reasons for travelling. Some people travel for work, school or college, recreation, shopping and others.

4.1 Primary purpose of travelling

The purpose of travel refers to the main motive to perform outdoor trips daily due to a specific requirement. Based on a survey study conducted by [14] to 435 respondents on individual travel behaviour in the COVID-19 pandemic phase, most respondents stated that their primary purpose of travelling was to work. Nevertheless, changes in primary reasons to travel are noticeable during a pandemic. Figure 5 compares the primary reason for travel before and during the pandemic. Travelling to work and school are the two primary reasons that reduced during the pandemic from 57% to 49% and 22% to 12 %, respectively. This is due to the fact that the public and private sectors opted for work from home policies implementation during the movement restriction. According to the survey done by Jobstreet's Malaysia, about 67% of Malaysian companies support the limited movement policy and commanded staff to work from home during pandemic [15]. However, the work implementation of work from home policies is depends on the nature of bussiness.

Similarly, decreased travel to school could be explained by the fact that during MCO, higher learning institutions adopted asynchronous and synchronous e-learning whereby all are conducted online. Travel for other activities also declined because gathering activities were not allowed during the lockdown. Interestingly, in the COVID-19 pandemic phase, the primary outdoor trip for shopping was raised to 29% compared to 12% before the pandemic. The increase in shopping trips is likely due to the change of workplace to online for most companies. Restriction for dining at restaurants during the MCO also contributed to an increase in shopping trips. [16] reported a similar finding related to the change in primary trip. Since dining outside was not allowed, shopping for necessities was more frequent. This explains the increase in shopping activities during COVID-19.

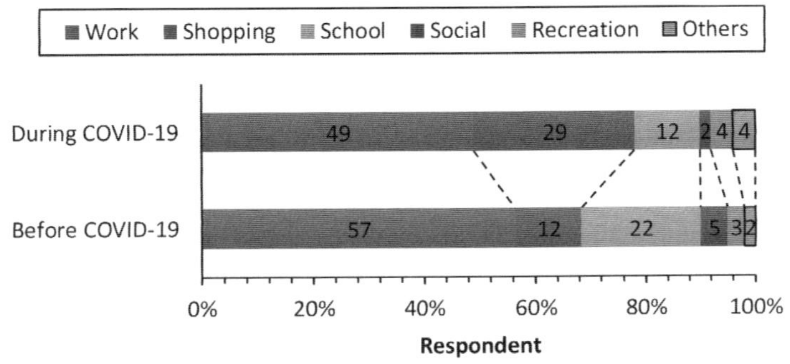

Figure 5. Comparison of the primary purpose of travelling before and during COVID-19. Source: [14]

4.2 Travel mode preference for outdoor trips

Each person has a different purpose and mode preference for travelling in their routine life. Some may prefer to use a private car, public transport, taxi/grab, motorcycle or walking. During the pandemic, a study by [14] shows the Malaysian travel mode preference for different activities (figure 6). Personal cars are the dominant preference with an average of 85% for all activities as they are more comfortable and reduce the risk of infection. However, for recreational/sports activities, about 17% prefer to use bicycles and walk.

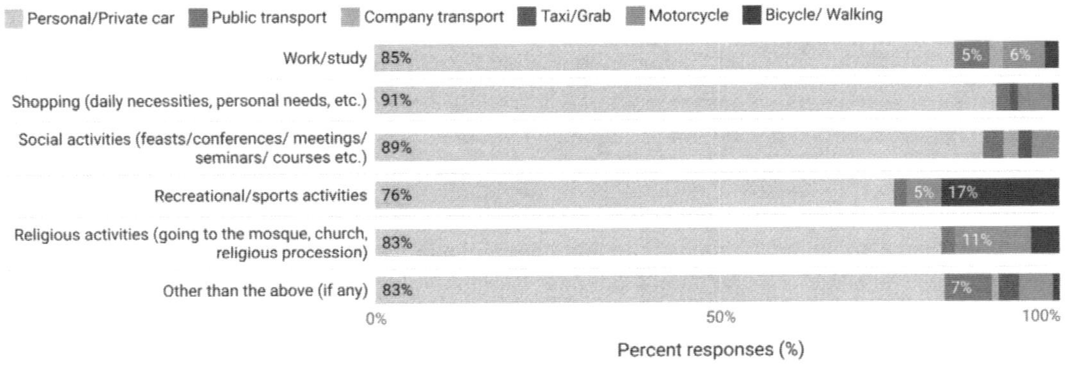

Figure 6. Travel mode of choice for various activities during pandemic. Source: [18]

A survey was conducted at Kota Kinabalu Sabah, on the travel mode preference by using public transport before and during COVID-19 for various trip purposes which are work/class, grocery shopping and social/recreational/sport as shown in figure 7. This data illustrates the public user perception of public transport usage during pandemic and pre-pandemic, whereby more people prefer to use private vehicles.

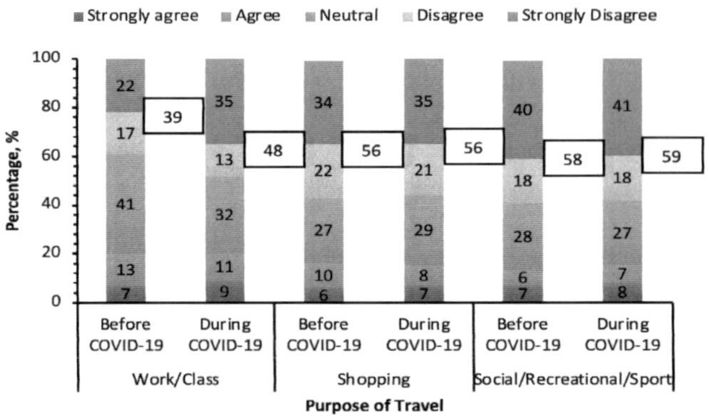

Figure 7. Travel mode preference by public transport for various trips during COVID-19 at Sabah, Malaysia.

As presented in figure 7, it can be deduced that more than 50% of the respondents strongly disagreed and disagreed with using public transportation for any of their essential trips during the pandemic. About 39% does not agree to use public transport to work/school before the pandemic, and this value increases to 48% during the pandemic. In terms of using public transport for shopping and social/recreation/sport, there were no noticeable changes between the pre and during COVID-19 pandemic. According to [17], the increasing number of respondents that decided to stop using public transportation immediately are due to public transport users do not feel safe and hence, choose private cars. In addition, an online report on March 2020, stated that simulation study on journey limitation measures in the city of Bogotá, Colombia shown reduction of user about 87 % [18]. The simulation was conducted for a bus rapid system named Trans Milenio. During pandemic, the users' mode of choice are influenced by pandemic-related factors, which were infection concerns, wearing face masks, social distance and cleanliness. Research findings by [14] on the influence of pandemic-related factors on Malaysian respondents indicated that concern for infection and wearing face masks are the highest scored (83% and 81%). Most respondents also prioritise social distancing and cleanliness when selecting the travel mode preference. Similar findings were reported by other research findings outside Malaysia, whereby car owners preferred to use private cars for commuting from home to work during the pandemic, indicating that people prioritised their health and comfort [19].

5. Discussion and conclusions

In the history of medicine, the emergence of the COVID-19 virus has challenged medical experts and even world leaders to deal with an outbreak that can spread quickly through air and contact. This incident has never been expected in Malaysia and worldwide. The virus's existence and spreading mechanism require a community to reduce contact and social distancing. Based on the Malaysian experience, the type of transportation that was severely affected as air transportation. International flights and local flights were cancelled. As of April 2020, more than a 90% reduction for all flights at KLIA. The second worst hit transportation was rail transportation. Rail transportation ridership reduced tremendously due to the less movement and fear of contact, as well as the implementation of strict SOP on the maximum capacity. Reduction of the passenger was also observed in public transportation such as busses. Most passengers are worried about using public transport during pandemic due to the high infection risk.

Undeniably, the COVID-19 pandemic has caused an adverse impact not only on public health but also on other sectors such as transportation, business, tourism and education. As a result, it affects people's travel behaviour. The likelihood of respondents travelling for work/class, shopping, recreation, sport and religious purposes during a pandemic shows reduced travel to work/class. However, travelling for shopping had increased. During pandemic, people prioritised pandemic-related factors when selecting the travel mode preference and purpose of travelling. Some key pandemic-related factors concerning transportation were fear of infection and social distance. Hence, most people prefer to use private cars. A previous study in Malaysia stated that car owners significantly prioritised pandemic-related factors when deciding their travel mode from commuting from home to work compared to non-car owners [14]. Hence, it can be concluded that Malaysia experienced a tremendous reduction in demand for transportation services during the COVID-19 pandemic and affected travel behaviour due to strict movement policy and social distancing by the government.

References

[1] Cheng S, Chang Y, Fan Chiang Y, Chien Y, Cheng M and Yang C et al 2020 First case of coronavirus disease 2019 (COVID-19) pneumonia in Taiwan *J. of The Formosan Medical Assoc.* **119 (3)** 747-751

[2] Nižetić S 2020 Impact of coronavirus (COVID-19) pandemic on air transport mobility, energy , and environment: A case study *Int. J. of Energy Res.* **44 (13)** 10953–61

[3] Gkiotsalitis K and Cats O 2021 Public transport planning adaption under the COVID-19 pandemic crisis: a literature review of research needs and directions *Transport Reviews* **41 (3)** 374-392

[4] Ho S J, Xing W, Wu W and Lee C C 2021 The Impact of COVID-19 on freight transport: Evidence from China *Method sX* ***8*** 101200

[5] Arellana J, Márquez L and Cantillo V 2020 COVID-19 outbreak in Colombia: An analysis of its impacts on transport systems *J. of Adv. Transp.* 1-16

[6] United Nations Conference on Trade and Development 2021 Global economy could lose over $4 trillion due to COVID-19 Impact on tourism https://unctad.org/news/global-economy-could-lose-over-4-trillion-due-covid-19-impact-tourism

[7] Ministry of Transport 2020 Malaysian Transportation Statistics 2020 Putrajaya: Ministry of Transport Malaysia

[8] MAHB Annual Report 2020 URL: https://mahb.listedcompany.com/ar.httml

[9] Karim K March 2020 Rapid bus passengers drop by 66 percent due to MCO New Straits Times Retrieved January 18, 2022, from https://www.nst.com.my/

[10] Sham R, Xuen C K, Ting L Y and Hye M N 2022 Public transport, tidership and safe travelling environment during COVID-19 Pandemic *Environment-Behaviour Proceedings Journal* **7 (19)** 339-346

[11] Chen L, Ban G, Long E, Kalonji G, Cheng Z, Zhang L and Guo S 2021 Estimation of the SARS-CoV-2 transmission probability in confined traffic space and evaluation of the mitigation strategies *Environmental Science and Pollution Research* **28 (31)** 42204-16

[12] Hadjidemetriou G M, Sasidharan M, Kouyialis G and Parlikad A K 2020 The Impact of government measures and human mobility trend on COVID-19-related deaths in the UK *Transportation research interdisciplinary perspectives* **6** 100167.

[13] Wei W E , Li Z , Chiew C J, Yong S E, Toh M P and Lee V J, 2020 Presymptomatic transmission of SARS-CoV-2—Singapore January 23–March 16, 2020 *Morbidity and Mortality Weekly Report* **69 (14)** 411

[14] Dias C, Abd Rahman N, Abdullah M and Sukor N S A 2021 Influence of COVID-19 Mobility-Restricting Policies on Individual Travel Behavior in Malaysia *Sustainability* **13 (24)** 13960

[15] Jobstreet 2022 67% of Malaysian Companies Required Staff to Work from Home. Available online: https://www.jobstreet.com.my/en/cms/employer/laws-of-attraction/inspirations/67-of-malaysian-companies-required-staff-to-work-from-home/ (accessed on August 16 2022).

[16] Abdullah M, Dias C, Muley D and Shahin M 2020 Exploring the impacts of COVID-19 on travel behavior and mode preferences *Transp. Res. Interdiscip. Perspect.* **8** 100255

[17] Przybylowski A, Stelmak S and Suchanek M 2021 Mobility behaviour in view of the Impact of the COVID-19 Pandemic—Public Transport Users in Gdansk Case Study *Sustainability* **13 (1)** 364

[18] Arredondo P A, Sánchez P K, Redacción A D V P, Guaithero P N S, Guaithero P N S, Martínez P S, Martínez P S & Martínez P S 2020 Bogotá, desolada en el primer día de simulacro de aislamiento por el COVID-19. Voice of America. https://www.vozdeamerica.com/coronavirus/bogota-desolada-primer-dia-simulacro-de-aislamiento-covid-19.

[19] Shakibaei S, De Jong G C, Alpkökin P and Rashidi 2021 The Impact of the COVID-19 pandemic on travel behavior in Istanbul: A panel data analysis *Sustain. Cities Soc.* **65** 102619

Developing pedestrian evacuation path parameters based on the requirements of Indonesia National Agency of Disaster Management and the Indicators of the Global Walkability

F Rifwan[1, 2], Yosritzal[1], Purnawan[1] and Yossyafra[1]

[1.] Civil Engineering, Universitas Andalas
[2.] Civil Engineering, Universitas Negeri Padang

E-mail: fitrarifwan5163@ft.unp.ac.id

Abstract. The tsunami incident in 2004, in Aceh, was killed around 230,000 people and destroyed more than 1,000 buildings. Many accidents also occurred related to the earthquakes and tsunamis in Indonesia. For example, when the big earthquakes happened in Padang City, people evacuated themselves by using cars or motorbikes (motorized movement). It created a problem; and became worse when the movement was not effective. All research related to evacuation movement suggested that walking is better than moving using vehicles. Thus, people who will evacuate by walking should be supported by good and sufficient infrastructure, such as the path. Up to recent time, there has been no research conducted in relation to finding the solution this problem. Therefore, this article reports on the research that focused on the parameters of the path design. The validity and reliability of the test, were measured by experts. The research respondents filled the form and their responses were taken as the data analyzed to get the parameters, indicators, and levels based on the adjustment of BNPB (National Agency of Disaster Management) and GWI (Global Walkability Indicators) guidelines. In conclusion, the four parameters: walking path mode conflict, availability of walking path, amenities, and obstruction were found.

1. Introduction

Based on the evacuation management, a tsunami requires massive handling. Treatment can be done before, during, and after the event. The management focuses on increasing capacity in order to reduce the combination of hazards, exposures, and vulnerabilities that may increase risks cause by tsunami. Designing facilities is a form of management for helping the evacuation process. The design is based on a thought-out planning.

Planning of the facilities should be effective in order to achieve the goal, namely: to be able to reduce the risk of a tsunami. Effectiveness requires coordination from related parties involving many agencies, from the central government to its lower levels. Planning also requires analyses, such as identification of affected risk tsunami-prone areas, the capacity of evacuation routes provided, and forms of traffic behavior. It is a form that is influenced strongly by land use conditions and the amount of transportation activity on the land [1]. In this section, the foundations are transportation support (mode support), traffic management, and the availability of shelters [2].

The foundations described above are forms that are closely related to the smoothness of evacuation which explains the traffic during the process. The expected traffic in this process is the movement from the dangerous zone to the safe zone before the tsunami comes. Aspects which related to it are the time and location of safety and adequate capacity for shelters. Currently, the problem is that the facilities that support the evacuation are still not sufficient in some areas, such as Padang, one of the big cities in Indonesia. The city is dense with people living and working at around 5 km from its coastline making the city in the tsunami red zone. This means that the danger potential is high [3].

Padang city is geographically located on the coast and is one of the tsunami-prone areas in Indonesia. Since 2004, the threat of a tsunami has always been a hazard that cannot be avoided due to the geographical conditions described previously. This geographical condition is the worst scenario in research related to the tsunami simulation in Padang City which explains that the time to reach land is 37 minutes and the effective evacuation time for the community is only 17 minutes [4].

When tsunami comes, the people (Padang Citizens) always get panic, and almost all rush to evacuate to a safe place. This process they go through with various forms of movement, such as: motorized and un-motorized. However, the process is not effective because no one can get to a safe place on time [5]. This is due to the increasing volume of vehicles that exceed the road capacity. In this process, vehicle users (motorized) also use sidewalks that should be reserved for pedestrians instead. This causes more problem as the pedestrians (un-motorized) are not able to move either. Generally, walking should be an effective mode during the evacuation because it does not conflict with the other modes; or there should be separate facilities for other modes [5].

Evacuation events in the last three earthquake occurrences (2005, 2007, and 2009) in Padang City show that people rely more on vehicles than on foot as a mode of evacuation because the nearest evacuation site does not yet exist and people wish to seek refuge at the safe zone in Bypass as soon as possible [6]. On the other hand, motorized vehicles (motorcycles and cars) are inefficient modes considering the unpredictable traffic conditions during the evacuation [7].

Essentially, the city's problems related to tsunami consists of four parameters, namely hazard, vulnerability, infrastructure, and agriculture (low capacity). These parameters will be taken as the basis for determining the priority of handling and management in the form of the Risk Prioritization Index (RPI) [8]. The top priorities for handling evacuations, specifically for tsunamis, are routes and shelters (safe temporary shelters) [2]. Evacuation infrastructure is critical because if the infrastructures have issues, it can lead to internal problems in the evacuation process [9].

Currently, the number and capacity of the evacuation routes in Padang City is not able to accommodate the large influx of movement during an actual evacuation. Furthermore, since 2013, the number of shelters in the city of Padang, has not been sufficient to cover the red zone area, so additional shelters are needed [10]. Research conducted in 2014 found that the combination of the number of buildings planned to be used as shelters is not sufficient and is only able to accommodate 75% of the population who settle or live in tsunami-prone areas in Padang City [11].

Therefore, previous evacuation processes and results of previous research show that having sufficient evacuation infrastructures is essential in solving this issue. Pathways-related research can be the starting point to the solution. The path in question is an evacuation route for pedestrians because walking is an effective mode of evacuation [5].

Research on pedestrian evacuation route is still macro, which means that there is no detailed design of evacuation routes that can be used efficiently. The research still discusses a lot about the selection of effective routes and evacuation management policies. In addition, the National Agency of Disaster Management (BNPB), which a party related to disaster management affairs in Indonesia, is also still issuing general standard evacuation routes as shown in table 1 [12].

Table 1. BNPB parameters for evacuation routes.

Parameters
1. Paths should be leading to temporary evacuation shelter (TES) or final evacuation shelter (TEA). It is accessible from all road access. It should not be narrow.
2. There should be no obstacles such as parking areas and street vendors along the path.
3. The path surface should be paved, and flat. It is not perforated, and not slippery
4. Lighting should be provided on the path
5. The path is equipped with signs on the strategic placement.

The parameters for evacuation route displayed in table 1. are still not focused on pedestrians. The pedestrian parameters are instead listed in the Global Walkability Index (GWI) Indicators, as shown in table 2 below. However, these parameters are the requirements for designing paths to be used for to normal conditions or daily routines, not for emergency situations, such as an evacuation process [13].

Table 2. Global walkability indicators (GWI).

Indicators
1. Walking Path Mode Conflict
2. Availability of Walking Path
3. Availability and Grade Crossing Safety
4. Amenities
5. Obstructions
6. Motorist Behavior
7. Disability Infrastructure
8. Security from Crime

Tables 1 and 2 are the fundamental references in designing the method and data analysis for this research

2. Research methodology

2.1. Data collected

The data were collected using a form that consists of columns. The columns comprise the parameters in table 1 that has been matched with indicators in table 2. The form was filled by four experts in transportation, infrastructure, and disaster and by linguists using a range of scores of 1-5. The experts filled out different forms, one part is related to the content and the other part related to the language. The aspects in the language forms consist of 7 parts [14].

2.2. Analysis

The form filled by the experts went through a validity testing using the Aiken formula as follows[15], [16].

$$V = \frac{\Sigma s}{[n(c-1)]} \tag{1}$$

S = r – lo if the highest rating scale is at 5
r = number given by expert (score range 1-5)
lo = the lowest validation score, in this analysis = 1
n = number of raters, in this analysis = 3 linguist and 3 content experts
c = the highest validation score, in this analysis = 5

In equation (1), The V value was matched by adjusting the Table Values of Right Tile Probabilities (P) for Selected Values of The Validity Coefficient (V) or the validation coefficient [15].

The reliability test was done using Interclass Coefficient Correlation (ICC). ICC analysis was done using Software Statistical Package for the Social Sciences (SPSS) and the interpretation table of the guideline value table [17,18]. The equation used is as follows:

$$ICC = \frac{MS_R + MS_E}{MS_R + (k-1)\, MS_E} \tag{2}$$

ICC = The reliability value is matched with table 7.

MSR = Mean Square Indicator value in the ANOVA test results table.
MSE = Residual Mean Square Value in the ANOVA test result table.
k = Number of Experts who filled out the assessments in tables 5, 6, and 7.

The research method conceived in achieving the objectives can be seen in the flowchart of figure 1 below.

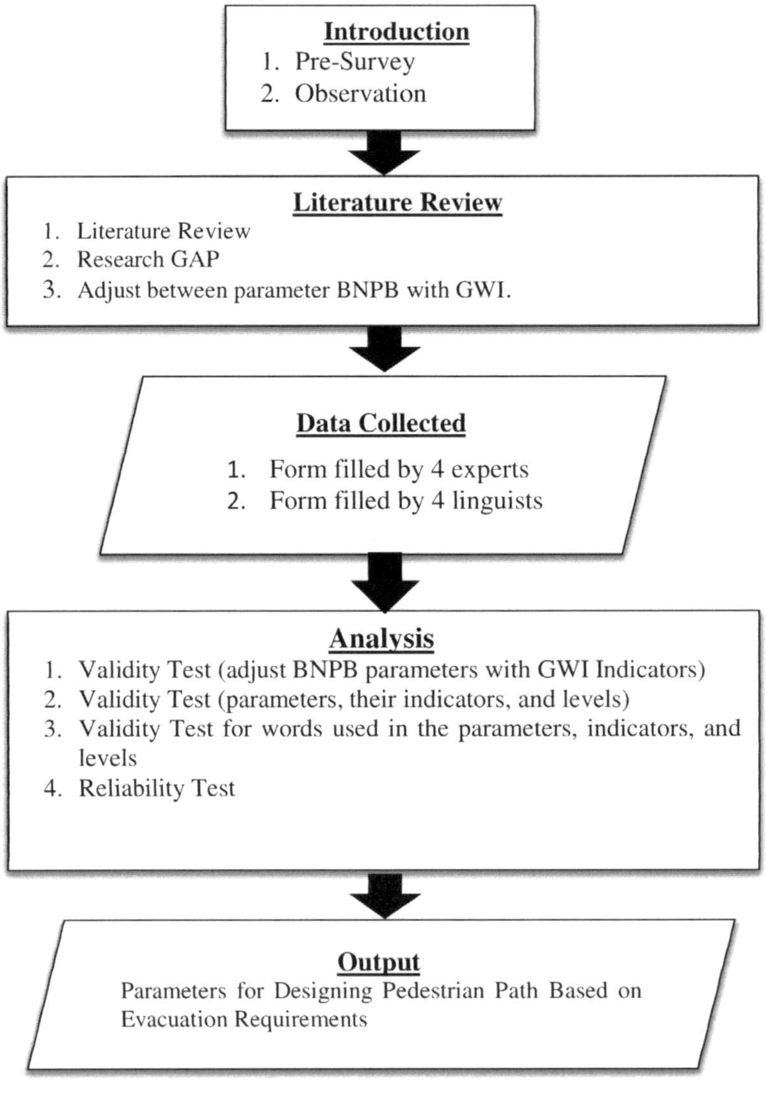

Figure 1. Research flow-chart.

3. Data and analysis

3.1. Results of validity test for adjustment of GWI indicators with BNPB parameter, firstly

Table 3 shows the results of the validity test related to the adjustment of the GWI indicators with the BNPB parameters. It produced 1 invalid parameter.

Table 3. Results of the first validity test for adjustment of GWI indicators with BNPB parameter.

Parameters	Experts Score								∑s	n(c-1)	V	Valid/Invalid [a]
	1	s	2	s	3	s	4	s				
A[b]	3	2	4	3	5	4	5	4	13	16	0.81	Not Valid
B[c]	4	3	4	3	5	4	5	4	14	16	0.88	Valid
C[d]	5	4	5	4	5	4	4	3	15	16	0.94	Valid
D[e]	5	4	5	4	5	4	4	3	15	16	0.94	Valid

[a] If V > 0.88, the parameter is valid
[b] Walking Path Mode Conflict is adjusted with Paths, meaning that it should lead to temporary evacuation shelter (TES) or final evacuation shelter (TEA). It should be accessible from all roads, and it should not be narrow.
[c] Availability of Walking Path is adjusted with Paths, meaning that it should lead to temporary evacuation shelter (TES) or final evacuation shelter (TEA). It should accessible from all roads, it should not be narrow, the path surface should be paved and flat, it should also not be perforated and slippery.
[d] Obstruction is adjusted with Paths, meaning that it should lead to temporary evacuation shelter (TES) or final evacuation shelter (TEA). It should be accessible from all roads, it s should not be narrow, and there should not be obstacles such as parking areas and street vendors along the path.
[e] Amenities is adjusted with Paths, meaning that it should lead to temporary evacuation shelter (TES) or final evacuation shelter (TEA). It should accessible from all roads, it should not be narrow, lighting should be provided on the path, and the path should be equipped with signs on the strategic placement.

3.2. Results of validity test for adjustment of GWI indicators with BNPB parameter, secondly
Table 4 shows the second validity results. All adjustments are valid.

Table 4. Results of the second validity test for adjustment of GWI indicators with BNPB parameter.

Parameters	Experts Score								∑s	n(c-1)	V	Valid/Invalid [a]
	1	s	2	s	3	s	4	s				
A	5	4	5	4	5	4	5	4	16	16	1.00	Valid
B	4	3	5	4	5	4	4	3	14	16	0.88	Valid
C	5	4	5	4	5	4	5	4	16	16	1.00	Valid
D	5	4	5	4	5	4	5	4	16	16	1.00	Valid

[a] When V > 0.88, the parameter is valid.

3.3. Results of validity test for adjustment of the parameters with its indicators and levels
Tables 5, 6, 7 and 8 show the results of adjusted parameters with the indicators for each level. The Adjustment of all parameters (Walking Path Mode Conflict, Availability of Walking Path, Obstruction, and Amenities) with each indicator and level is valid.

Table 5. Validity test results for walking path conflict mode with indicator of each level.

Parameter A[a]	Experts Score								∑s	n(c-1)	V	Valid/Invalid [b]
	1	s	2	s	3	s	4	s				
Level 1[c]	5	4	5	4	5	4	5	4	16	16	1.00	Valid
Level 2[d]	5	4	5	4	5	4	5	4	16	16	1.00	Valid
Level 3[e]	4	3	5	4	5	4	4	3	14	16	0.88	Valid
Level 4[f]	5	4	5	4	5	4	5	4	16	16	1.00	Valid
Level 5[g]	4	3	5	4	5	4	5	4	15	16	0.94	Valid

[a] Parameter Walking Path Mode Conflict. Indicator= Connectivity of each path segment to TES.
[b] When V > 0.88, the parameter is valid.
[c] Not connected to TES.
[d] Indirectly connected to TES, separated by a building entrance.
[e] Indirectly connected to TES, not separated by a building entrance.
[f] Directly connected to TES, separated by a building entrance.
[g] Directly connected to TES, not separated by a building entrance.

Table 6. Validity test results for the availability of walking path with indicators and each level.

Parameter B[a]	Experts Score								∑s	n(c-1)	V	Valid/Invalid [b]
	1	s	2	s	3	s	4	s				
Level 1[c]	4	3	5	4	4	3	5	4	14	16	0.88	Valid
Level 2[d]	5	4	5	4	4	3	4	3	14	16	0.88	Valid
Level 3[e]	5	4	5	4	5	4	5	4	16	16	1.00	Valid
Level 4[f]	5	4	5	4	5	4	5	4	16	16	1.00	Valid
Level 5[g]	5	4	5	4	5	4	5	4	16	16	1.00	Valid

[a] Parameter Availability of Walking Path. Indicators = Surface, Width, and Height of Path .
[b] When V > 0.88, the parameter is valid.
[c] Hole on the surface, width (<150 cm), height > 15 cm.
[d] Hole on the surface, width (±150 cm), height 10-15 cm.
[f] No hole on the surface, width (250 cm), height 10 -15 cm.
[g] No hole on the surface, width> 250 cm, height 10 -15 cm.

Table 7. Validity test results for obstruction with indicators and each level.

Parameter C[a]	Experts Score								∑s	n(c-1)	V	Valid/Invalid [b]
	1	s	2	s	3	s	4	s				
Level 1[c]	5	4	5	4	5	4	5	4	16	16	1.00	Valid
Level 2[d]	4	3	5	4	4	3	5	4	14	16	0.88	Valid
Level 3[e]	5	4	5	4	5	4	5	4	16	16	1.00	Valid
Level 4[f]	5	4	5	4	5	4	5	4	16	16	1.00	Valid
Level 5[g]	5	4	5	4	5	4	5	4	16	16	1.00	Valid

[a] When V> 0.88, the parameter is valid.
[b] Parameter Amenities. Indicators = evacuation signs, lighting, and disability infrastructures.
[c] No signs, lighting, and disability infrastructures.
[d] No signs and ramp for wheelchair, but lighting ready.
[e] No ramp for wheelchair and lighting, but installed signs ready.
[f] No lighting, but ramp for wheelchair and sign installed.
[g] Lighting, ramp for wheelchair and signs ready.

Table 8. Validity test results for amenities with indicators and each level.

Parameter D[a]	Experts Score								∑s	n(c-1)	V	Valid/Invalid [b]
	1	s	2	s	3	s	4	s				
Level 1[c]	5	4	5	4	5	4	5	4	16	16	1.00	Valid
Level 2[d]	5	4	5	4	5	4	5	4	16	16	1.00	Valid
Level 3[e]	5	4	5	4	5	4	5	4	16	16	1.00	Valid
Level 4[f]	5	4	5	4	4	3	4	3	14	16	0.88	Valid
Level 5[g]	5	4	4	3	5	4	4	3	14	16	0.88	Valid

[a] When V> 0.88, the parameter is valid
[b] Parameter Obstruction. Indicator = Width of Obstruction
[c] Obstruction covers all path.
[d] Obstruction covers path, accessible space < 1 m.
[e] Obstruction covers path, accessible space = 1 m.
[f] Obstruction covers path, accessible space > 1 m.
[g] No obstruction.

3.4. Results of validity test for adjustment of the languages on the parameters, indicators, and levels, firstly.

The result of the validity test is displayed in table 9. In the table, there are two assessment aspects which not valid by filled linguists [14].

Table 9. Results of first validity test for adjustment of the languages on the parameters, indicators, and levels.

Aspects	Experts Score						∑s	n(c-1)	V	Valid/Invalid [a]
	1	s	2	s	3	s				
1[b]	4	3	5	4	5	4	11	12	0.92	Valid
2[c]	5	4	4	3	4	3	10	12	0.83	No Valid
3[d]	4	3	5	4	5	4	11	12	0.92	Valid
4[e]	5	4	5	4	5	4	12	12	1.00	Valid
5[f]	4	3	5	4	4	3	10	12	0.83	No Valid
6[g]	5	4	5	4	5	4	12	12	1.00	Valid

[a] When V > 0.92, the aspect is valid
[b] Interpretation of the language, commonly used.
[c] The directness of the language used for measuring instruments (it does not have a double meaning).
[d] Clarity of language (designed based on the elements, such as the subject, predicate, and description).
[e] The objectivity of the language (the language can be understood individually and does not require the opinion of others).
[f] Word order in the sentence (the position of the subject, predicate, and adverb in the sentence).
[g] Concentration on understanding the language (the longer you concentrate on understanding the language, the less valid it is).

Due to the two invalid parameters as shown in table 9, a second validity test is required. The results are presented in table 10 below.

3.5. Results of validity test for adjustment of the languages on the parameters, indicators, and levels, secondly

Table 10 displayed the second validity test. All aspects are valid. All parameters, indicators, and levels are valid and they can be used commonly.

Table 10. Results of the second validity test for adjustment of the languages on the parameters, indicators, and levels.

Aspects	Experts Score						∑s	n(c-1)	V	Valid/Invalid [a]
	1	s	2	s	3	s				
1	4	3	5	4	5	4	11	12	0.92	Valid
2	5	4	5	4	4	3	11	12	0.92	Valid
3	4	3	5	4	5	4	11	12	0.92	Valid
4	5	4	5	4	5	4	12	12	1.00	Valid
5	5	4	5	4	5	4	12	12	1.00	Valid
6	5	4	5	4	5	4	12	12	1.00	Valid

[a]When V > 0.92, the aspect is valid

3.6. Result of reliability test for adjustment of BNPB parameters with GWI indicators

The reliability test was done to complete the validity results. Table 11 shows the reliability results along with the explanation.

Table 11. Results of reliability test for adjustment of BNPB parameters with GWI indicators.

	Value	Determination
Cronbach Alpha	0.667	*acceptable* or *sufficient*, it means that the reliability of the measuring instrument of the adjustment can be accepted. It matches the research objective.[a]
Anova	0.436	There is an agreement among the experts in assessing the parameter adjustment because the number is 5% (the significance probability).
ICC	0.667	The adjustment is reliable. It means that the parameters can be used repeatedly because it falls under the ICC category score: fair.[b]

[a]Value of Cronbach Alpha[19].
[b]ICC Table[18].

3.7. Result of reliability test for adjustment of parameters with indicators, and levels
The Adjustment of Parameters with indicators and levels was tested for the reliability. Table 12 shows the test results.

Table 12. Results of reliability test for adjustment of parameters with indicators, and levels.

	Test	Value	Interpretation
Parameter A with the indicators, and levels.	Cronbach's Alpha	0.5	Its category is acceptable or sufficient. It means that the assessment of all experts on each indicator and level for parameter A can be trusted [19].
	Anova	0.195	The agreement of all experts in assessing each indicator and level is acceptable for parameter A. It is because the value of significance is greater than 5%.
	ICC	0.5	All experts are in an agreement. Their assessments were tested in the fair category on the ICC Table[18].
Parameter B with the indicators, and levels.	Cronbach's Alpha	0.556	Same explanation with the interpretation above.
	Anova	0.426	
	ICC	0.556	
Parameter C with the indicators, and levels.	Cronbach's Alpha	0.667	Same explanation with the interpretation above.
	(Sig.) Anova	0.426	
	ICC	0.667	
Parameter D with the indicators, and levels.	Cronbach's Alpha	0.556	Same explanation with the interpretation above/
	Anova	0.426	
	ICC	0.556	

3.8. Result of reliability test for using language in the parameters, indicators and levels
The next reliability test was analyzed for language in the parameters, indicators, and levels. Table 13 shows the results.

Table 13. Results of reliability test for using language in the parameters, indicators, and levels.

	Value	Interpretation
Cronbach Alpha	0.732	All linguists have an agreement on the assessment of language aspects. The value category is relatively high.[a]
Anova	0.647	There is no difference in the significant score of each expert. The probability value is higher than 5%.
ICC	0.750	The linguists agree with the given score and they reached an agreement to classify the parameters in the good category.[b]

[a]Value of Cronbach Alpha[19].
[b]ICC Table[18].

4. Discussion

All analyses results for the parameters, indicators, and levels for pedestrian path based on evacuation needs show that they can be commonly used and displayed in table 14.

Table 14. Parameters, indicators, and levels for pedestrian path based on evacuation needs.

Parameters	Indicators	Score
Walking Path Mode Conflict	not connected to TES	1
	indirectly connected to TES, separated by a building entrance	2
	indirectly connected to TES, not separated by a building entrance	3
	directly connected to TES, separated by a building entrance	4
	directly connected to TES, not separated by a building entrance	5
Availability of Walking Path	hole on the surface, width (<150 cm), height > 15 cm	1
	hole on the surface, width (±150 cm), height 10-15 cm	2
	no hole on the surface, width (±200 cm), height 10 -15 cm	3
	no hole on the surface, width (250 cm), height 10 -15 cm	4
	no hole on the surface, width > 250 cm, height 10 -15 cm	5
Amenities	no signs, lighting, and disability infrastructures	1
	no signs and ramp for wheelchair, but lighting ready	2
	no ramp for wheelchair and lighting, but installed signs ready	3
	no lighting, but ramp for wheelchair and sign installed	4
	lighting, ramp for wheelchair and signs ready	5
Obstruction	obstruction covers all path	1
	obstruction covers path, accessible space < 1 m	2
	obstruction covers path, accessible space = 1 m	3
	obstruction covers path, accessible space > 1 m	4
	no obstruction	5

5. Conclusion

The development of parameters outlined by BNPB requirements and GWI indicators reveals that designing pedestrian paths must be guided by parameters. Based on the evacuation process, the design should include these four parameters: Walking Path Mode Conflict, Availability of Walking Path, Amenities, and Obstruction.

References

[1] Yossyafra Y, Fitri N, Sidhi R P, Yosritzal Y and Mazni D I 2020 Assessment of road traffic performance of the Tsunami evacuation road in Padang Municipality area based on the traffic volume simulation approach *E3S Web of Conferences* **156**

[2] Lindell M K 2013 *Evacuation planning, analysis, and management (*United States: Texas A&M University Hazard Reduction & Recovery Center) p. 121–150

[3] BNPB 2014 *Indeks Risiko Bencana*, Cetakan 1. Citeureup-Sentu: Direktorat Pengurangan Risiko Bencana Deputi Bidang Pencegahan dan Kesiapsiagaan,. [Online]. Available: http://inarisk.bnpb.go.id/irbi

[4] Yosritzal, Kemal B M and Siddik F 2016 Estimation of coverage area of tsunami shelters in Padang *Proceeding of National Conference of Applied Sciences, Engineering, Business and Information Technology (ASCNITech)* (Padang: Politeknik Negeri Padang) pp. 1–8.

[5] Rifwan F 2012 *Studi evaluasi efektifitas penggunaan jalur evakuasi pada zona berpotensial terkena bencana tsunami di Kota Padang* (Padang: Universitas Andalas)

[6] Alhadi Z 2014 Kesiapan Jalur da Lokasi Evakuai Publik menghadapi Risiko Bencana Gempa dan Tsunami di Kota Padang *Humanus* **13** 35–44

[7] Septima R, Munzir S and Salmawaty S 2017 Modeling and simulation of traffic flows on inclined road during evacuation process of the volcano disaster with finite difference method *Jurnal Natural* **17** 8

[8] Purnawan P, Bachtiart V S and Kurniati T 2018 Development risk analysis method for tsunami disaster *MATEC Web of Conferences* (Bandung: EDP Sciences) **197** 10005

[9] Adhi I, Hutama W and Nakamura H 2022 Disaster Evacuation Route Choices for Informal Settlements: A Conceptual Framework Accessed: Sep. 29, 2022. [Online]. Available: https://doi.org/10.34436/sjse.3.1_16

[10] Rahmadhani N, Suprayogi A and Sabri L 2013 Analisis Aksesibilitas Shelter Evakuasi Tsunami Di Kota Padang Berbasis Sistem Informasi Geografis *Jurnal Geodesi Undip* **2** 1–11

[11] Kurniawan A A, Sutikno S and Rinaldi 2014 Evaluasi Kapasitas Shelter Evakuasi Untuk Bencana Tsunami Di Kota Padang Berbasis Sistem Imformasi Geografis (SIG) *Jurnal Online Mahasiswa (JOM) Bidang Teknik dan* **1**

[12] BNPB 2014 *Pedoman Perencanaan Jalur dan Rambu Evakuasi Tsunami*, vol. 53, no. 9.

[13] Leather J, Fabian H, Gota S and Mejia A 2011 Walkability and pedestrian facilities in Asian cities state and issues Mandaluyong, Number 17, 2011.

[14] Kabir S M S 2016 Preparing Questionnaire in *Basic Guidelines for Research* pp. 181–201 [Online] Available: https://www.researchgate.net/publication/325846992_PREPARING_QUESTIONNAIRE

[15] Aiken L R 1980 Content validity and reliability of single items or questionnaires *Educ. Psychol. Meas* **40** 955–959

[16] Aiken L R 1985 Three coefficients for analyzing the reliability and validity of ratings *Educ. Psychol. Meas.* **45** 131–141

[17] Widhiarso W 2005 *Mengestimasi reliabilitas* (Yogyakarta: Fakultas Psikologi UGM) pp. 6–17

[18] Koo T K and Li M Y 2016 A Guideline of selecting and reporting intraclass correlation coefficients for reliability research *J. Chiropr. Med.* **15** 155–163

[19] Taber K S 2018 The use of cronbach's alpha when developing and reporting research instruments in science education *Res. Sci. Educ.* **48** 1273–1296

Public Preferences of Using Public Transport During the *Covid-19* Pandemic and the Implementation of Level 3 Community Activity Restrictions Policy in Padang City

Yosritzal[1*], Titi Kurniati[1], Jihan Meiriza Putri[1]

Department of Civil Engineering, Faculty of Engineering, Andalas University, Padang, West Sumatra 25163, Indonesia

*yosritzal@eng.unand.ac.id

Abstract. The Covid-19 pandemic has had a significant influence on the usage of public transport, in this case, a decrease in the use of public transport services. The decreased use of public transport was because the government created a new social system called social distance, followed by the Enforcement of Restrictions on Community Activities (PPKM). This study aims to perform utility and probability modelling and determine the most significant factor in influencing the preferences of the community in using public transport during the Covid-19 pandemic and the implementation of the PPKM level 3 policy in Padang City. Primary data was collected using online and offline questionnaires. The questionnaire used the expressed preference technique on eight alternatives with three attributes and two levels. The attributes include the difference in travel time, the cost of travel, and health protocols. The best model was $Y_{(AUK-AUO)} = 0.656 + (-0.038) X_1 + (-0.056) X_2 + 0.644 X_3$ with R-Square = 0.174 referring to all characteristics as a whole influence 17.4% of Y. According to the sensitivity test results, the travel cost difference is the most influencing variable in people's decisions to use traditional public transportation during the covid-19 pandemic and the implementation of the PPKM level 3 policy.

1. Introduction

Padang city has a relatively high population, with a population aged over 20 years, as many as 616,748 in 2020 [1]. Demand for cheap and accessible public transportation services is also high not only for those who have private vehicles *(Choice Group)* and for people who have no choice except public transportation *(Captive Group)* [2]. However, the Covid-19 pandemic has had a very influential impact on the use of public transportation, one of which is a decrease in the use of public transportation services. The decreasing use of public transportation during the pandemic was because the government declared a new social system, namely social distancing, and followed up with the Enforcement of Restrictions on Community Activities (PPKM) policy [3]. The PPKM is a policy issued by the Indonesian government in 2021, which aims to minimize the spread of the COVID-19 virus and restore the economic sector in society [4]. The PPKM policy results in limited access to public transportation services. This study investigates the public perception of the use of public transportation services during the COVID-19 pandemic after implementing the Level 3 Community Activity Restriction (PPKM) policy in Padang City.

2. Research Method

2.1. *Data Collection*

Primary data were collected using online questionnaires distributed through the researcher's social media and offline questionnaires to obtain a sample of public transport users so that the samples collected were diverse. Secondary data in this study is a study of literature from journals and supporting books related to this research. The number of samples from this study was determined using the Slovin formula based on the total population of Padang City residents aged over 20 years, namely 616,748 people in 2020 [1], so the number of samples from this study is 100 respondents.

The sampling method used in this study is *Non-Probability Sampling*, a sampling that is not known which individual from the population will be selected as a sample. The offline sampling technique in this study is the *Accidental Sampling* technique. *Accidental Sampling* is a sampling procedure that selects samples from people who are the easiest to meet or coincidentally meet with the researcher and are believed to represent the study [6]. At the same time, the online sampling technique in this study is *Simple Random Sampling,* where *Simple Random Sampling* is a method used to randomly select samples from the population so that each member of the population has an equal chance of being taken as a sample [5].

2.2. *Questionnaire Design*

This questionnaire contains questions related to the socio-economic characteristics of the respondents, including gender, age, monthly income before and during the COVID-19 pandemic, and the number of private vehicle owners. The next questions are related to the characteristics of the respondent's travel before the covid-19 pandemic (before March 18, 2020) and after the PPKM level 3 policy (October 18, 2021), including reasons for using transportation modes, frequency of use of transportation modes, costs incurred for using modes of transportation, waiting time of public transportation modes, as well as the origin and destination of the trip using this mode of transportation. The last questions are *stated preferences* questions. The attributes used in this study include travel time attributes, travel fare attributes, and health protocol attributes. The attributes offered have two categories for each, which can be seen in Table 1.

Table 1 Attribute offered

Attribute	Description
Travel time difference	10 minutes 20 minutes
Travel fare difference	Rp. 7.500, - Rp. 10.000, -
Health protocol	Implementing health protocols Not implementing health protocols

Table 2 Alternative research questionnaire

	Travelling Time Difference	Travel Fares Difference	Safety Against Health Protocol
Alternative 1	10 minutes	Rp. 7.500, -	Implementing health protocols
Alternative 2	10 minutes	Rp. 7.500, -	Not implementing health protocols
Alternative 3	10 minutes	Rp. 10.000, -	Implementing health protocols
Alternative 4	10 minutes	Rp. 10.000, -	Not implementing health protocols
Alternative 5	20 minutes	Rp. 7.500, -	Implementing health protocols
Alternative 6	20 minutes	Rp. 7.500, -	Not implementing health protocols
Alternative 7	20 minutes	Rp. 10.000, -	Implementing health protocols
Alternative 8	20 minutes	Rp. 10.000, -	Not implementing health protocols

Table 1 shows an independent variable (X) consisting of travel time, the difference in travel fares, and health protocols. The alternatives offered consist of 8 kinds of alternatives. These alternatives are calculated based on the number of attributes in the stated preference technique. If attribute (a) is stratified into the level (n), then a combination of choices (n^a) is needed [7]. The combination of choices in this study can be seen in Table 2.

Then, respondents will choose the combination offered using a point rating technique with a four-point semantic scale, namely:
(1) Don't want to use conventional public transportation
(2) Do not want to use conventional public transportation
(3) Want to use conventional public transportation
(4) Want to use conventional public transportation

2.3. *Processing and Data Analysis*
In this study, the primary data obtained from questionnaires that respondents have filled out is analyzed using the binomial logit model to transform the semantic scale into a numeric scale. After data transformation, data analysis was performed using multiple linear regression analysis. The steps in conducting data analysis are as follows:

A. Categorize the dependent variable with a predetermined rating point
B. Forming several models of multiple linear regression analysis *(multiple linear regression analysis)*. With multiple linear regression equation, namely [8]:

$$Y = a + b_1X_1 + b_2X_2 + \ldots + b_nX_n \quad (1)$$

Where :
Y = Dependent variable (predicted value)
$X_1, \ldots, X_n$ = Independent variable
a = Constant
b = Regression coefficient

C. Statistical tests were conducted to check the significance of the model, such as the coefficient of determination test (R-Square), partial significance test (T-test), and simultaneous significance test (F-test).
D. The test result was used to select a model that fits the predetermined criteria and can be used for further testing.
E. Then a sensitivity test on each attribute of the survey was conducted.

3. Result and Discussion

3.1. *Characteristics of Respondents*
Characteristics of respondents are various characteristics of respondents to research subjects, such as gender, age, main occupation, income before and during the covid-19 pandemic, as well as vehicles owned by families, which can be seen in Table 3 and Table 4.

Table 3 shows that the majority of respondents in this study were female, aged 20-29 years old, and a student. Table 4 shows that the Covid-19 pandemic resulted in a fairly high percentage increase in average income/pocket money < Rp. 1.000.000, - / month by 10.83%. Moreover, it can also be seen that most respondents own a motorcycle as much as 1 unit, and the majority do not own a car.

Table 3 Characteristics of respondents

Characteristics	Category	Frequency (%)
Gender	Male	35,38
	Female	64,62
Age	20 – 29 years old	81,59
	30 – 39 years old	6,50
	40 – 49 years old	5,78
	50 – 59 years old	5,42
	> 60 years old	0,72
Main job	Private (Entrepreneur)	5,05
	Housewife	2,89
	Employee (PNS, TNI/POLRI)	15,16
	College student	67,87
	Etc	9,03

Table 4 Characteristics of respondents. (Continuation of table 3)

Characteristics	Category	Frequency (%)	
		Before	During
Respondent's income/pocket money in the last year before and during covid-19 every month	< Rp. 1.000.000, -	38,27	49,10
	Rp. 1.000.000, - to Rp. 3.000.000, -	43,32	32,85
	Rp. 3.000.001, - to Rp. 5.000.000, -	8,30	8,66
	Rp. 5.000.001, - to Rp. 7.000.000, -	6,86	6,86
	Rp. 7.000.001, - to Rp. 9.000.000, -	2,89	1,44
	> Rp. 9.000.000, -	0,36	1,08
Characteristics	**Category**	**Frequency (%)**	
		Motorcycle	**Car**
Family-owned private vehicle	0	5,42	44,77
	1	38,63	39,35
	2	31,05	10,47
	3	20,94	3,97
	4	2,89	0,72
	5	0,36	0,36
	> 5	0,72	0,36

3.2. Respondent's Travel Characteristics

3.2.1. Private Transportation

Table 5 shows that the reason for using private transportation was the most chosen by respondents before the covid-19 pandemic, which was more cost-effective. Meanwhile, during the covid-19 pandemic, after the PPKM level 3 policy, avoiding direct contact with other people. The frequency of using private transportation during the pandemic for the frequency of 1-3 times/week experienced the highest increase of 10,16%; for the frequency of 4-6 times/week, there was an increase of 1,90%. The frequency of use 7-9 times/week decreased by 2,74%, and for private transportation more than nine times/week, the highest decrease was 9,31%. Then there was a significant decrease in the percentage of respondents with an average travel cost of Rp. 100.000 - Rp. 200.000 during the covid-19 pandemic. Furthermore, it can be seen that before and during the covid-19 pandemic, most respondents used motorbikes for campus purposes and used cars for shopping centers.

Table 5 Travel characteristics of respondents using private transportation

Characteristics	Category	Frequency (%)			
		Before	**During**		
Reason for use	More cost-effective	14,44	11,44		
	Reduce congestion	4,77	4,40		
	Short travel time	11,72	10,23		
	More practical	12,26	11,33		
	More comfortable	14,31	11,00		
	No waiting times	12,13	8,69		
	Save energy	3,68	3,19		
	Safer	7,22	-		
	Reduce air pollution	0,54	1,21		
	Personal wish	9,54	8,25		
	More flexible	9,40	8,80		
	Avoid direct contact with other people	-	11,55		
	Avoid jostling with other people	-	9,90		
Frequency of use	1 – 3 times/week	20,83	30,99		
	4 – 6 times/week	25,00	26,90		
	7 – 9 times/week	23,21	20,47		
	> 9 times/week	30,95	21,64		
Average monthly expenses	< Rp. 100.000, -	22,62	22,22		
	Rp. 100.000, - to Rp. 200.000, -	33,33	30,99		
	Rp. 200.001, - to Rp. 300.000, -	14,29	15,79		
	Rp. 300.001, - to Rp. 400.000, -	9,52	11,11		
	Rp. 400.001, - to Rp. 500.000, -	6,55	6,43		
	Rp. 500.001, - to Rp. 600.000, -	4,17	4,09		
	> Rp. 600.000, -	9,52	9,36		
		Frequency (%)			
		Motorcycle		**Car**	
		Before	**During**	**Before**	**During**
Travel destination	Shopping center	21,86	21,29	22,34	23,03
	Worship place	15,81	16,27	10,99	10,79
	Campus	23,95	23,68	15,02	15,47
	Cafe/Restaurant	14,19	14,35	12,82	13,31
	Recreation	8,84	8,61	16,48	15,11
	Office	6,28	6,22	10,26	11,15
	Others	9,07	9,57	12,09	11,15

3.2.2. Public Transportation

Table 6 shows that the reason for using public transportation was the most chosen by respondents before the covid-19 pandemic, which was more practical. Meanwhile, during the Covid-19 pandemic, the PPKM level 3 policy is more cost-effective. The frequency of using public transportation during the pandemic for the frequency of 1-3 times/week experienced the highest increase of 12,43%; for the frequency of 4-6 times/week, there was an increase of 3,19%. The frequency of use 7-9 times/week, the highest, decreased by 9,28%, and for the use of public transportation more than nine times/week, there was a decrease of 6.34%. Then there was an increase in the percentage of respondents with an average travel cost of <Rp. 50.000/month and Rp. 50.000 - Rp. 100.000/month during the Covid-19 pandemic.

Table 6 Travel characteristics of respondents using public transportation

Characteristics	Category	Frequency (%)	
		Before	During
Reason for use	No private vehicle is available	9,64	11,35
	More cost-effective	18,47	22,16
	Reduce congestion	3,61	-
	Short travel time	1,20	5,41
	More practical	19,28	14,05
	Save energy	17,67	16,22
	Personal wish	9,24	9,19
	It is safer because it applies health protocols	-	2,16
	Do not have a driving license yet	5,62	6,49
	Reduce air pollution	-	2,70
	Safer	2,81	-
	More flexible	5,62	3,24
	Do not have a private vehicle	4,42	2,16
	There is a policy of limiting the number of passengers	-	2,70
	Can socialize with other people	0,80	-
	Cannot drive a motorized vehicle	1,61	2,16
Frequency of use	1 – 3 times/week	47,00	59,43
	4 – 6 times/week	27,00	30,19
	7 – 9 times/week	14,00	4,72
	> 9 times/week	12,00	5,66
Average monthly expenses	< Rp. 50.000, -	26,00	34,91
	Rp. 50.000, - to Rp. 100.000, -	29,00	31,13
	Rp. 100.001, - to Rp. 150.000, -	22,00	12,26
	Rp. 150.001, - to Rp. 200.000, -	10,00	5,66
	Rp. 200.001, - to Rp. 300.000, -	5,00	6,60
	> Rp. 300.000, -	8,00	9,43

Table 7 Public transportation waiting time

	Waiting time							
	Frequency (%)							
	< 5 minutes		5 – 10 minutes		11 – 15 minutes		> 15 minutes	
	Before	During	Before	During	Before	During	Before	During
City transport	46,88	31,43	32,81	44,29	12,50	18,57	7,81	5,71
Trans Padang bus	21,05	21,43	36,84	42,86	28,95	30,95	13,16	4,76
Conventional taxi bike	48,39	41,94	32,26	41,94	3,23	6,45	16,13	9,68
Online taxi bike	28,81	26,15	59,32	66,15	3,39	1,54	8,47	6,15
Conventional taxi	15,79	18,18	31,58	36,36	36,84	36,36	15,79	9,09
Online taxi	25,00	24,14	58,33	62,07	4,17	10,34	12,50	3,45

Table 7 shows that the waiting time for conventional public transportation and motorcycle taxis during the covid-19 pandemic is more extended than before the covid-19 epidemic. Meanwhile, the waiting time for Trans Padang Bus, Online Ojek, Online Taxi and Conventional Taxi during the COVID-19 pandemic is the same as before the Covid-19 pandemic.

Table 8 Origin of public transportation trip

	Origin of travel													
	Frequency (%)													
	Shopping center		Worship place		Campus		Cafe/Restaurant		Recreation		Office		Etc	
	B	D	B	D	B	D	B	D	B	D	B	D	B	D
City transport	19,30	20,00	6,14	3,18	38,60	40,00	14,91	15,24	4,39	3,81	4,39	2,86	12,28	14,29
Trans Padang bus	24,49	19,05	6,12	7,14	26,53	30,95	12.24	7,14	6,12	4,76	2,04	0,00	22,45	30,95
Conventional taxi bike	10,53	5,26	5,26	10,53	47.37	47.37	0,00	5,26	0,00	5,26	0,00	0,00	36,84	26,32
Online taxi bike	19,13	15,63	3,48	5,21	32,17	39,58	20,87	19,79	6,96	5,21	5,22	4,17	12,17	10,42
Conventional taxi	13,64	5,88	4,55	11,76	27,27	47,06	9,09	5,88	9,09	11,76	4,55	0,00	31,82	17,65
Online taxi	20,45	20,59	4,55	5,88	29,55	32,35	18,18	20,59	9,09	5,88	6,82	0,00	11,36	14,71

Table 8 shows that the campus is the origin of most respondents' trips using public transportation before and during the covid-19 pandemic.

Table 9 Public transportation travel destination

	Travel Destination													
	Frequency (%)													
	Shopping center		Worship place		Campus		Cafe/Restaurant		Recreation		Office		Etc	
	B	D	B	D	B	D	B	D	B	D	B	D	B	D
City transport	21,05	17,95	6,14	3,42	39,47	38,46	14,04	16,24	3,51	4,27	4,39	7,69	11,40	11,97
Trans Padang bus	25,64	16,33	7,69	8,16	28,21	30,61	7,69	8,16	2,56	2,04	10,26	10,20	17,95	24,49
Conventional taxi bike	8,33	6,45	16,67	6,45	37,50	45,16	4,17	6,45	0,00	3,23	8,33	12,90	25,00	19,35
Online taxi bike	17,43	14,15	2,75	3,77	33,94	41,51	20,18	18,87	7,34	3,77	9,17	8,49	9,17	9,43
Conventional taxi	28,57	7,69	7,14	11,54	7,14	26,92	21,43	11,54	7,14	15,38	0,00	3,85	28,57	23,08
Online taxi	20,00	19,05	6,67	4,76	22,22	30,95	17,78	19,05	13,33	4,76	11,11	4,76	8,89	16,67

Table 9 shows that most respondents used public transportation, Trans Padang buses, conventional motorcycle taxis, online motorcycle taxis and online taxis to campus before and during the covid-19 pandemic while using conventional taxis to shopping centres before the covid-19 pandemic and heading to campus during the covid-19 pandemic.

3.3. *Community Probability to Use Public Transportation Post PPKM Level 3 Policy*

3.3.1. *Data Compilation*

The data compilation process was carried out using the multiple linear regression method. The scale value was assigned as the variable used, and the difference in travel time, the difference in travel fares, and health protocols were placed as independent variables. The general equation for the utility function of the linear model for all research samples is the number of 277 respondents who get the regression model as follows: $Y_{(AUK-AUO)} = -0{,}033 + (-0{,}025)X_1 + (-0{,}020)X_2 + 0{,}600X_3$, from the regression results above, we get the model equation with an R-Square value of 0.138. The results of the p-value test can be seen that partially the difference in travel time (X_1) and health protocol (X_3) has a significant effect on the dependent variable (H1 is accepted). In contrast, the travel fare difference variable (X_2) has no significant effect on the variable. Dependent (H_1 is rejected). The significance value of F obtained from the multilinear regression day is $3{,}683 \times 10^{-71} < 0{,}05$ (Sig. α), where H_1 is accepted simultaneously with the variables of the difference in travel time (X_1), the difference in travel fares (X_2) and health protocols (X_3). Those variables significantly influence people's probability of using public transportation after the PPKM level 3 policy.

Respondents using public transportation before the covid-19 pandemic and private transportation after the PPKM level 3 policy have the regression model as follows: $Y_{(AUK-AUO)} = -1{,}142 + (-0{,}011)\ X_1 + 0{,}072\ X_2 + 0{,}914\ X_3$. The regression model has an R-Square value of 0,348. The p-value test results show that the health protocol variable (X_3) significantly affects the dependent variable (H_1 is accepted). In contrast, the travel time difference variable (X_1) and the travel fare difference (X_2) have no significant effect on the dependent variable (H_1 is rejected). The significance value of F obtained from the multilinear regression day is $1{,}949 \times 10^{-06} < 0{,}05$ (Sig.α), so the H_1 is accepted. It means that simultaneously the variables for the difference in travel time (X_1), the difference in travel fares (X_2) and health protocols (X_3) has a significant influence on people's probability to use public transportation after the PPKM level 3 policy.

The number of respondents using public transportation before the covid-19 pandemic and after the PPKM level 3 policy, which was 100 respondents, obtained the following regression model: Y(AUK-AUO) $= 0{,}656 + (-0{,}038)\ X_1 + (-0{,}056)\ X_2 + 0{,}644\ X_3$, from the regression results above, the model equation with an R-Square value of 0.174 is obtained. The p-value test shows that all independent variables partially influence the dependent variable (H_1 is accepted). The significance value of F is $7{,}267 \times 10^{-33} < 0{,}05$ (Sig.α), so H_1 is accepted. It means that simultaneously, the variable travel time difference (X1), travel fare difference (X_2), and health protocols (X_3) have a significant effect on people's probability to use public transportation after the PPKM level 3 policy.

The equation of the utility function of a linear model for private transport users before the covid-19 pandemic and public transport after the PPKM level 3 policy is as follows: $Y_{(AUK-AUO)} = -0{,}652 + (-0{,}018)\ X_1 + (-0{,}016)\ X_2 + 1{,}182\ X_3$, with an R-Square value of 0,501. The p-value test shows that partially the health protocol variable (X_3) has a significant effect on the dependent variable (H_1 is accepted). In contrast, the travel time difference (X_1) and the travel fare difference (X_2) have no significant effect on the dependent variable (H1 is rejected). The significance value of F is $9{,}008 \times 10^{-07} < 0{,}05$ (Sig.α), so H1 is accepted. It means that simultaneously the variable travel time difference (X_1), travel fare difference (X_2), and health protocols (X_3) have a significant influence on people's probability to use public transportation after the implementation of PPKM level 3 policy.

The utility function for private transport users before the covid-19 pandemic and after the PPKM level 3 policy is as follows: $Y_{(AUK-AUO)} = -0{,}389 + (-0{,}018)\ X_1 + (-0{,}001)\ X_2 + 0{,}539\ X_3$, with an R-Square value of 0,118. The results of the p-value test show that the difference in travel time (X_1) and health protocols (X_3) has a significant effect on the dependent variable (H_1 is accepted), while the travel fare (X_2) has no significant effect on the dependent variable (H_1 is rejected). The significance value of F is $5{,}570 \times 10^{-35} < 0{,}05$ (Sig.), so H_1 is accepted. It means that simultaneously the variable difference in travel time (X_1), the difference in travel fares (X_2) and health protocols (X_3) has a significant influence on people's probability of using public transportation after the PPKM level 3 policy.

3.4. Sensitivity Analysis

The sensitivity analysis of the travel time shows that if the difference in travel time is zero, then public transportation users will prefer to use conventional public transportation (AUK) compared to online public transportation (AUO) with a probability value of 0.656 or 65.6%. The more significant the difference in travel time, the smaller the probability for people to use conventional public transportation.

The sensitivity analysis of the travel fares shows that if the difference in travel fares is zero, then public transportation users will prefer to use conventional public transportation (AUK) compared to online public transportation with a probability value of 0.656 or 65.6%. The more significant the difference in travel fares, the fewer chances of people using conventional public transportation.

The sensitivity analysis of health protocols shows that if conventional public transportation implements health protocols, the probability value of people using conventional public transportation after the PPKM level 3 policy will be greater, which is 63,8%. So, people prefer to use conventional public transportation (AUK) after the PPKM level 3 policy compared to online public transportation (AUO) if conventional public transportation implements health protocols.

4. Conclusion

The study developed a utility model of public transportation in Padang City after the implementation of the PPKM level 3 policy. The chosen model from several simulated models is $Y_{(AUK-AUO)} = 0,656 - 0,038\ X_1 - 0,056\ X_2 + 0,644\ X_3$ with X_1 being the difference in travel time, X_2 is the difference in travel fares, and X_3 is a health protocol.

The sensitivity test on all research attributes shows that the probability of using conventional public transportation was smaller when travel time and the travel fares of conventional public transportation were greater than online public transportation. As for the attributes of the health protocol, it is generic, which means that the probability for people to use conventional public transportation is more significant if conventional public transportation applies the health protocol.

Based on this research, the attribute that has the most significant influence on people's probability of using conventional public transportation is the attribute of the difference in travel fares.

5. References

[1] Statistik, B. P., 2021. *Penduduk Menurut Kelompok Umur, 2010-2020.* [Online] Available at: https://padangkota.bps.go.id/indicator/12/664/1/penduduk-menurut-kelompok-umur.html
[2] Warpani, S. P., 2002. *Pengelolaan Lalu Lintas dan Angkutan Jalan.* Bandung, Indonesia: Penerbit ITB.
[3] Adda, J., 2016. Economic Activity and the Spread of Viral Diseases: Evidence from High-Frequency Data. *The Quarterly Journal of Economics, 131(2),* pp. 891-941.
[4] Muchtar, A., 2021. Kilas Balik 1 Tahun Bersama Pandemi Covid-19. Bandung, Jawa Barat: CV. Media Sains Indonesia.
[5] Sugiyono, 2012. *Metode Penelitian Kombinasi (Mixed Methods).* Bandung: CV. Alfabeta.
[6] Sugiyono, 2013. *Metode Penelitian Kuantitatif dan R&D.* Bandung: CV. Alfabeta.
[7] Yosritzal, 2006. Review Pendekatan Stated Preference Dalam Beberapa Penelitian Transportasi Di Kota Padang. *Simposium IX FSTPT,* pp. 126-135.
[8] Uyanto, S. S., 2009. *Pedoman Analisis Data Dengan SPSS Edisi 3.* Yogyakarta: Graha Ilmu.

Accessibility Towards West Sumatra Tourism Area

Jihan Melasari[2], Purnawan[1], Yosritzal[1] and Yossyafra[1]
[1] Andalas University, Padang, Indonesia
[2] Putra Indonesia University, Padang, Indonesia

Email: jihan_melasari@upiyptk.ac.id

Abstract. Accessibility was identified as the main factor driving tourist travel. The transportation sector is one of the strategic elements in tourism. Transportation in tourism is a means to achieve tourist destinations and also a means of movement in tourist destinations. Transportation is one of the important indicators in the development of tourist destinations and the selection of destinations by tourists. West Sumatra Province is one of the provinces in Indonesia which is rich in tourist attractions and is also one of the tourist destinations that is quite taken into account not only at the national level but also at the international level. West Sumatra has been designated as one of the main national tourist destinations. This research identifies accessibility to existing tourist areas in West Sumatra and analyzes areas that have the highest and lowest accessibility and explains the existing condition of the availability of public transportation to provide services in terms of providing good accessibility to the tourist zone. This research uses a quantitative approach method to present numerical data through the normal distribution method to make each accessibility parameter into three classifications namely low, medium, and high, and also uses a qualitative approach that aims to investigate, describe and explain the existing conditions of transportation facilities to tourist zone which is the object of research using primary data. Through a qualitative approach, a detailed description of the condition of the availability of routes and public transportation will be produced which will be analyzed based on the results of mapping in the study area.

1. Introduction

Domestic and foreign tourist visits are one of the indicators used to see tourism activity. Based on the summary data attached to the West Sumatra tourism development master plan for 2014-2025, the number of tourist visits continues to increase. Based on data from the Central Statistics Agency (BPS) of West Sumatra Province in 2018, foreign tourist arrivals were 83,133, and in 2019 became 95,187 with an increase of 14.5%, domestic tourist visits in 2018 were 8,073,070 and in 2019 became 8,366,550 with an increase in the number of visitors 3.5%. To continue to increase the interest and number of tourist visits, both foreign and domestic, it is necessary to consider aspects of the carrying capacity of transportation to tourist areas and one of the important things to note is the level of accessibility in the tourist zone. The transportation sector plays a very important role in tourism development and can significantly increase tourist visits to countries, therefore transportation is very important in the tourism policy agenda [1].

Therefore, to support tourism, it is necessary to integrate the public transportation system to create smooth accessibility to the tourist zone as stated in the Strategic Plan of the Ministry of Transportation for 2020-2024. Integrated mobility aims to increase multimodal integration to make public transportation an attractive alternative to private transportation [2]. The main tourist destinations in West

Sumatra, in general, can be reached via land transportation, but public transportation to the tourist zone has not been an option used or become part of the service of the tourist zone at certain times during peak tourist visits on holidays there are still frequent traffic jams to the tourist zone. tourist zone [3]. So it is necessary to plan an integrated transportation system that is expected to improve public transportation services for tourists as a solution to provide services in terms of providing good accessibility to tourist zones.

In Alfonzo's theory, it is stated that five hierarchies influence decision-making to travel which can also be taken into consideration in planning the provision of transportation facilities and infrastructure consisting of feasibility, accessibility, security, comfort, and pleasure [4]. The main components in tourism integration consist of proximity to tourist attraction objects (ODTW), accessibility in the form of road network availability and availability of transportation modes, supporting facilities in the form of hotels; restaurants; and support networks, as well as good institutions from the government; manager; investors; and local community involvement [5]. The availability of paths and modes of connection is a supporting aspect of geographical proximity aspect.

It is also mentioned that in addition to availability, affordability of connecting transportation modes is also an aspect that must be considered in tourism integration [6,7]. In this case performs component specifications, stating that the aspect that needs to be considered in the availability of public transportation modes is the similarity of the path so that accessibility between ODTW will be easier [8]. However, transportation problems also need to be taken into account in terms of accessibility, because transportation problems that occur will automatically slow down the travel time for achieving ODTW.

2. Methodology

This research uses a quantitative approach method to present numerical data through the normal distribution method as a tool to make each accessibility parameter into three classifications, such low, medium, and high [9]. Accessibility parameters are made using data obtained from the survey, the data is sorted from the lowest data to the highest data. Then the mean and standard deviation of each of these data will be obtained which will be used to make a classification of accessibility. Classification of data using the normal distribution method can be seen in table 1 as follows:

Table 1. Normal distribution method

Accessibility	Parameter	Accessibility value
Low	$x < m+1sd$	1
Medium	$m - 1sd \leq x \leq m+1sd$	2
High	$x > m-1sd$	3

where:

- x is the value of the accessibility parameter
- m indicates mean
- sd indicatcs standard deviation

In this research, data collection was carried out using the observation method, which could be done using primary and secondary survey techniques. This stage is carried out through a field survey and network integration is done by using the Google Maps application to find out the distance with the shortest route to the airport. This paper intends to interpret the level of accessibility to top tourist areas in West Sumatra and the availability of public transportation to the area because one of the main components in the integration of tourism is the availability of connecting roads accompanied by modes of transportation [10].

3. Result and discussion

The literature review results that components that must be contained in a tourist area include accessibility which is how easy it is to achieve ODTW by tourists from their place of origin with benchmarks covering the condition of land, sea, and air transportation infrastructure; number and type of transportation facilities to ODTW; and the frequency of transportation to ODTW [11]. The results of Yang's research stated that most domestic and foreign tourists will use public transportation during their tours so the provision of public transportation facilities and infrastructure in tourist areas will affect the development of tourism [12]. Part of the transportation analysis for tourist areas includes adjusting the type of mode with the characteristics of the tourist area according to visitor demand. One of the important characteristics is the availability of multimodal which can access many activities of tourists such as to the beach, lodging, and others that can be accessed by taxi, motor vehicle, bus, bicycle, walking, using boat or boat. Given this diversity, the transportation system must encourage intermodal connectivity [13].

In general, many studies on transportation planning for tourism have not focused on the general integration system towards the tourist zone [12-18]. Therefore, support and the need for integrating the public transportation system to provide accessibility to the tourist zone. So the researchers first reviewed the level of accessibility and availability of public transportation modes in the tourist zone of West Sumatra. The factors that influence the choice of public transportation modes by tourists are accessibility, availability, affordability, comfort, and safety. Tourists need transportation that can provide quality services with an emphasis on accessibility-based transportation [19].
Factors that are also related to the selection of long-distance travel routes are direct directions, safety, congestion levels, and distance [20]. Figure 1 explains the distance from Minang Kabau International Airport to leading tourist destinations in West Sumatra, with 29 destinations spread across 19 districts and cities.

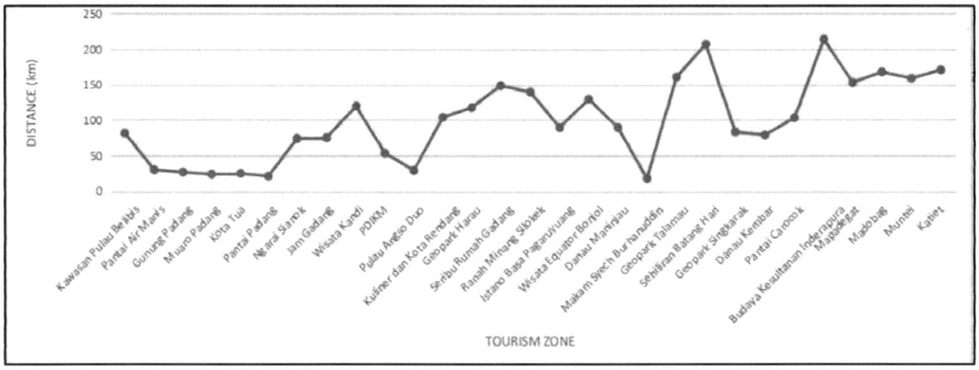

Figure 1. Distance to tourist zone (entry gates: airports).

Distance data (km) were analyzed using the normal distribution method to classify accessibility from airports to tourist areas, the mean value was 101.8 and the standard deviation was 59.5. Then the classification was obtained as shown in table 2.

Table 2. Classification of accessibility value

Accessibility	Parameter (km)	Accessibility value
High	< 40.3	3
Medium	40.3 – 160.3	2
Low	x > 160.3	1

The calculation in table 1 results in 6 areas with high accessibility levels (21%), 18 areas included in the medium accessibility levels (62%), and 5 areas with low accessibility levels (17%). This high accessibility is found in the capital area of the Province of West Sumatra which is close to the airport. But more than half of the area is at an intermediate level of accessibility.

Table 3. Availability of public transportation

	DESTINATION TYPE	CITY/ DISTRICTS	TOURISM ZONE	TRANSPORT AVAILABILITY					
				AKDP (Mini Bus)	DAMRI KSPN	TAXI	RENT CAR	TRAIN	SHIP
1.	*(Marine and Cultural Tourism)*	Padang and surrounding	Pantai Air Manis	x		√	√ (D)	√ (D)	x
			Gunung Padang	x		√	√ (D)	√ (D)	x
			Muaro Padang	x		√	√ (D)	x	x
			Kota Tua	x		√	√ (D)	x	x
			Pantai Padang	x	√ (D)	√	√ (D)	x	x
		Kabupaten Pesisir Selatan dan sekitarnya	Pantai Carocok	√ (I)	√ (D)	x	√ (D)	x	x
			Wisata Budaya kesultanan Inderapura	√ (I)	X	x	√ (D)	x	x
D		Kabupaten Padang Pariaman dan sekitarnya	Pulau Angso Duo	x		x	√ (D)	√ (I)	x
2.	*(Geopark)*	Kota Bukittinggi dan sekitarnya	Ngarai Sianok	√ (I)		x	√ (D)	x	x
			Jam Gadang	√ (I)	√ (L)	x	√ (D)	x	x
		Kabupaten Agam dan sekitarnya	Danau maninjau	√ (I)	X	x	√ (D)	x	x
		Kabupaten 50 kota dan sekitarnya	*Geopark* Harau	√ (I)	X	x	√ (D)	x	x
		Kota Payakumbuh	Wisata Kuliner dan Kota Rendang	√ (I)	X	x	√ (D)	x	x
		Kabupaten Pasaman Timur	Wisata *Equator* Bonjol	√ (I)	X	x	√ (D)	x	x

		Kabupaten Pasaman Barat	*Geopark* Talamau	√ (I)	X	x	√ (D)	x	x
3	(*Ancient Minangkabau Tourism)*	Kota Padang Panjang dan sekitarnya	Pusat Dokumentasi dan Informasi Kebudayaan Minangkabau	√ (I)	X	x	√ (D)	x	x
		Kabupaten Tanah Datar dan sekitarnya	Istano Basa Istano Basa Pagaruyuang	x	√ (D)	x	√ (D)	x	x
		Kabupaten Solok	*Geopark* Singkarak	√ (I)	√ (D)	x	√ (D)	x	x
			Danau Kembar	√ (I)	X	x	√ (D)	x	x
		Kota Solok	Pulau Belibis	√ (I)	X	x	√ (D)	x	x
		Kabupaten Solok Selatan	Saribu Rumah Gadang	-	X	x	√ (D)	x	x
4	(*Coal Mining Heritage Tourism)*	Kota Sawahlunto	Wisata Kandi	√ (I)	X	x	√ (D)	x	x
		Kabupaten Sijunjung	Ranah Minang Silokek	√ (I)	X	x	√ (D)	x	x
		Kabupaten Dharmasraya	Wisata Sehiliran Batang Hari	x	X	x	√ (D)	x	x
5	(*Whitesand Tourism)*	Kabupaten Kep. Mentawai	Mapadegat	x	X	x	x	x	√ (D)
			Madobag	x	X	x	x	x	√ (D)
			Muntai	x		x	x	x	√ (D)
			Katiet	x	√ (D)	x	x	x	√ (D)

Where :

- D indicates direct, transportation takes tourists directly to their destination
- I indicate indirectly, The available transportation cannot take tourists directly to tourist destinations and requires connecting transportation

4. Conclusion

This research is part of the introduction of further research that will be carried out in tourism development that is supported by an integrated tourist transportation system, it must begin with an understanding of the existing conditions of accessibility and availability of public transportation to tourist areas. Of the 29 top tourist areas in west Sumatera, there are 17 with high accessibility, 62 % have medium accessibility, and 21% have low accessibility. Based on the existing condition matrix of the availability of public transportation, it is known that there are still tourist areas that have not been served by public transportation and require connecting transportation.

References

[1] Seetanah B Khadaroo J 2009 An analysis of the relationship between transport capital and tourism development in a dynamic framework. Tour. Econ. 15 (4).785e802.

[2] Merkert R, Bushell J and Beck M J 2019 Collaboration as a service (CaaS) to fully integrate public transportation – Lessons from long-distance travel to reimagine mobility as a service. Transportation Research Part A, (XXXX), 1–16. https://doi.org/10.1016/j.tra.2019.09.025.

[3] Hesna Y, Suraji A, Istijono B, Hidayat B and Ophyandri T 2016 Kajian Kapasitas Infrastruktur : Suatu upaya peningkatan pariwisata Sumatera Barat. Simposium I jaringan perguruan tinggi untuk pembangunan infrastuktur Indonesia.

[4] Alfonzo M A 2005 Environment and behavior https://doi.org/10.1177/0013916504274016

[5] Utama I Gusti Bagus Rai 2015 Pengembangan Wisata Kota sebagai Masa Depan Indonesia.

[6] Postma Albert 2002 *An approach for integrated development of quality tourism* Dublin: Dublin institute of technology.

[7] Ferreira Joao and Cristina Estevao 2009 Regional Competitiveness of Tourism Cluster: A Conceptual Model Proposal. Portugal: University of Beira Interior.

[8] O A Ajala 2013 Tourism and Integrated Development: A Geographic Perspective. Malaysia Journal of Society and Space Volume 09 Issue 1.

[9] Sulistiyono 2012 Kategorisasi berdasarkan interval nilai dengan metoda distribusi normal. Jakarta.

[10] Utama I Gusti Bagus Rai 2015 Pengembangan Wisata Kota sebagai Masa Depan Indonesia.

[11] Musenaf 1996 Manajemen Usaha Pariwisata Indonesia. Jakarta: PT Toko Gunung Agung.

[12] Yang Y 2010 Analysis Of Public Transport For Urban Tourism In China Master Of Arts In Transport Policy And Planning Yao R Kwesi E and Adjei C 2018 Public transport mode preferences of international tourists in Ghana : Implications for transport planning. Travel Behaviour and Society 11 June 2017 1–8. https://doi.org/10.1016/j.tbs.2017.11.002

[13] Janic M Reggiani A 2001 Integrated transport systems in European Union: An overview of some recent developments. Transport Reviews 21 (4), 469-497.

[14] Ibrahim M 2003 Improvements and integration of a public transport system: the case of Singapore. Cities 20 (3), 205-216.doi:10.1016/S0264-2751(03)00014-3.

[15] Solecka K and Jacek Ż 2014 Integration of the urban public transportation system with the application of traffic simulation, 3(July), 259–268. https://doi.org/10.1016/j.trpro.2014.10.005

[16] Cheng Y H and Chen S Y 2015 Perceived Accessibility, Mobility, and Connectivity of Public Transportation Systems. Transportation Research Part A: Policy and Practice, 77, 386-403. doi:10.1016/j.tra.2015.05.003

[17] Andriani I 2018 Integrasi transportasi dalam mendukung kawasan destinasi wisata Tanjung Kelayang Kabupaten Belitung, 16(5), 27–42.

[18] Merkert R, Bushell J, and Beck M J 2019 Collaboration as a service (CaaS) to fully integrate public transportation – Lessons from long-distance travel to reimagine mobility as a service. Transportation Research Part A, (XXXX), 1–16. https://doi.org/10.1016/j.tra.2019.09.025

[19] Reilly J Williams P and Haider W 2010 Research in Transportation Economics Moving towards more eco-efficient tourist transportation to a resort destination: The case of Whistler, British

Columbia. Research in Transportation Economics, 26(1), 66–73. https://doi.org/10.1016/j.retrec.2009.10.009
[20] Dallen J 2007 The challenges of diverse visitor perceptions: Rail policy and sustainable transport at the resort destination. Journal of Transport Geography

Development of Traffic Delay and Queue Metamodel on Landslide Zone Based on Simulation Using VISSIM

Purnawan[1,a)] **and Ridhola Alfathira**[2]

[1,2]*Department of Civil Engineering, Faculty of Engineering, Andalas University, Padang, West Sumatra 25163, Indonesia*

[a)] Corresponding author: purnawan@eng.unand.ac.id

Abstract. Landslides occur due to instability in the soil triggered by vibrations, high rainfall and topographical conditions of an area. Landslide is a natural disaster that often occurs in West Sumatra. This landslide certainly has an impact on traffic around the area. Congestion is the impact caused by landslides due to falling chunks of material onto the road. Lack of traffic control during landslides makes travel ineffective and inefficient. Therefore, the traffic control and its impact on possible landslides were tested using a VISSIM microscopic simulation software by setting the queue and delay parameters as the outputs. Based on this output of simulations, the metamodel was developed using regression analysis. This model could be used to predict the queue and delay of traffic because of landslide at road network. Two traffic control conditions that were simulated, open and close lane is used if the width of the landslides covering the road is more than one lane. Two-way traffic control in landslide zone is used when the width of the landslides covering the road is less than one lane.

1. Introduction

Geographically, Indonesia is located between two continents and two oceans that have a tropical climate, making Indonesia prone to natural disasters. One of the natural disasters that often occur in Indonesia is landslides. Landslides occur due to instability in the soil triggered by vibrations, high rainfall and topographical conditions of an area. According to Christady, landslides show the movement of soil and rock down a slope due to the direct influence of Earth's gravity. Landslide is also defined as the movement of slope-forming material caused by shear failure along one or more landslide areas [1]. The slope of the slope and the length of the slope are important factors that affect the occurrence of landslides. West Sumatra Province is traversed by the Bukit Barisan plateau, which makes West Sumatra Province vulnerable to landslides. Several areas in West Sumatra Province are prone to landslides, the landslide-prone roads are spread on national roads, provincial roads and district or city roads in areas with hilly topography.

According to Khairah, there are 6 forms of landslides based on the volume of soil covering the road that can be seen in Figure 1 [2]:

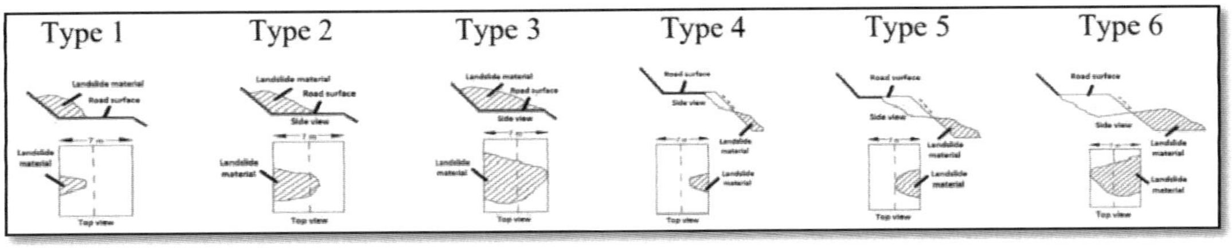

Figure 1. Landslides based on the volume of soil covering the road

In forms 1 and 4, landslide material can be seen only cover less than or equal 1/2 of the road. So, the remaining roads can still be used by a two way road in the landslide zone with limited speed. In forms 2 and 5, landslide materials have started to cover 1/2 the road or more. However, vehicles can still pass by alternately or by opening and closing the road system with low speed. In forms 3 and 6, the landslide material has completely covered the road. The landslide had caused the road body to collapse and break, so that the road at that time could no longer be passed.

The average cause of landslides is the high intensity of rainfall causing the slope to collapse onto the road. As a result, traffic in the area is congested. Therefore, traffic control and transportation modeling are needed and important to anticipate the effects of landslides on traffic.

Predicting traffic flow by making a transportation model can support an effective traffic management. Modeling has an important role in transportation planning. Transportation modeling is an element in planning as a requirement in an effective planning system where the results of the transportation modeling can be used to solve transportation problems [3]. VISSIM is a software that can be used for microscopic simulations. VISSIM is used to analyze road and traffic networks. Various effectiveness calculations can be entered in this VISSIM, such as speed, delay, queue, travel time and stop [4].

According to J. Sprinkle, meta models are a powerful tool to represent structure, behavior and other conditions by utilizing patterns from running results, which can later be visualized through mathematical equations [5]. Meta models are also used to minimize time in the simulation process, by performing a predictive analysis of a model. Based on the benefit of metamodel, this simulation developed queue and delay metamodel.

2. Reseacrh Method

2.1. Simulation Scenario Models

Scenarios are determined based on landslide events that often and have occurred in West Sumatra with minimum and maximum landslide size conditions. The length of the road used is 1000 meters, with road width of 8 meters. For each scenario, 5 simulations were performed by changing the traffic volume from 400 vehicles/hour to 2000 vehicles/hour. The following is a simulation scenario contained in Table 1 :

Table 1. Simulation Scenario Models

Traffic Control	Landslide Zone	Landslide Dimension		Vehicle Composition (%)				
		Length (m)	Width (m)	Motorcycle	Car	Pick Up	Bus	Truck
Open and Close Lane	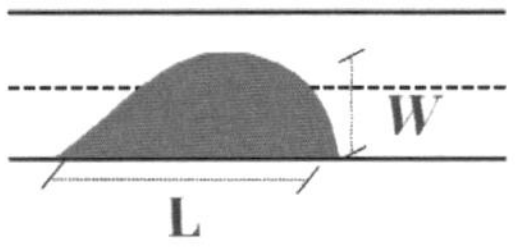	7	5					
		12	6					
				38	46	4	7	5
Two Way in Landslide Zone	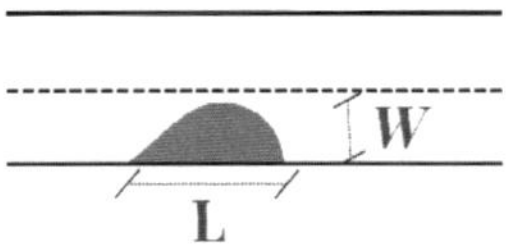	6	2					
		9	4					

2.2. Vissim Simulation

In the process of using VISSIM to perform traffic simulations, input data such as traffic volume and vehicle speed are required which will be used and processed into a simulation model and will be analyzed through this program. the following are the stages of the VISSIM simulation:

- Input background and creating a road network (links)
- Make connectors between roads (connectors)
- Input vehicle speed
- Input vehicle composition
- Determine the route of travel
- Entering the composition of the travel route
- Enter the number of vehicles
- Perform calibration and validation
- Running simulation

2.3. Driving Behavior

The driving behaviour data that used for simulation is shown in Table 2. This data was developed based on the condition of traffic at landslide zone.

Table 2. Driving Behavior

Parameters	Value
Desired position at free flow	any
Overtake on same lane	on
Distance standing	0.2 m
Distance driving	0.6 m
Average standstill distance	0.6 m
Additive part of safety distance	1.00
Multiplicative part of safety distance	1.00

2.4. Meta Models

Linear Regression is one of the most widely used meta modeling methods. Linear regression is used to see the effect between variables which will later be displayed in a mathematical modeling equation. Regression analysis is measuring the relationship between two or more variables which is expressed in the form of a relationship or function. A strict separation is needed between the independent variable and the dependent variable, usually symbolized by x and y. in regression there must be a specified variable and a determining variable in other words, there is a dependency between one variable and another variable. The two variables in the regression are usually causal or cause and effect that is mutually influential. Thus, regression is a form of a certain function between the dependent variable (y) and the independent variable (x) or it can be stated that the regression is a function y=f(x) [7]. Here is the theoretical model of Multiple Linear Regression.

$$Y_{ei} = a + b_{1i}x_{1i} + b_{2i}x_{2i} + \ldots + b_{ni}x_{ni} \quad (1)$$

Where:
Yei : number of observed variables
Xni : the observed nth independent variable
a : observed constant
b : coefficient which states effect

3. Result and Discussion

3.1. Vissim Simulated Result

a) Open and Close Traffic Control

From simulation using VISSIM, there are the queue and delay results from traffic control open and close lane scenario, the value of queue and delay is as shown in Figure 2 and Figure 3.

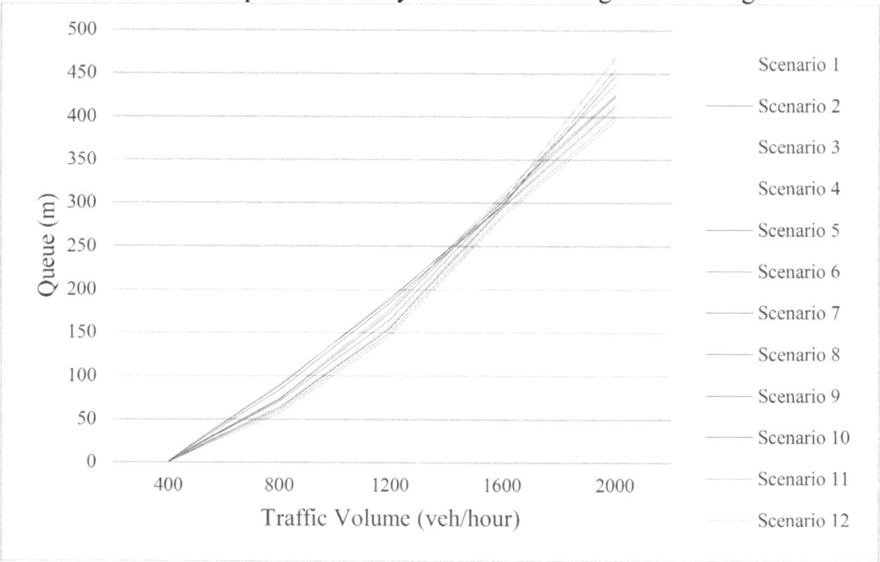

Figure 2. Queue of Traffic Control Open and Close Lane

In Figure 2 shows a comparison of the queue length of each scenario in the traffic control open and close lane. It can be seen that the graphs have a comparable relationship. Where, the higher the traffic volume, the longer the queue that will be generated. In addition, the dimensions of the avalanche also affect the queue length in the area. The larger the dimensions of the avalanche, the fewer roads that can be passed, so the queue will be longer.

From the graph, scenario 1 with a landslide dimension of 7 x 5 m has a remaining road of 3 meters for vehicles to pass. Scenario 1 has the lowest queue length, which is 0,3 meters at a volume of 400 vehicles/hour and 396 meters at a volume of 2000 vehicles/hour. While the highest queue length is in scenario 12 with a landslide dimension of 12 x 6 m, and the rest of the road is 2 meters for vehicles to pass. The queue length is 1 meter at a volume of 400 vehicles/hour and 468 meters at a volume of 2000 vehicles/hour.

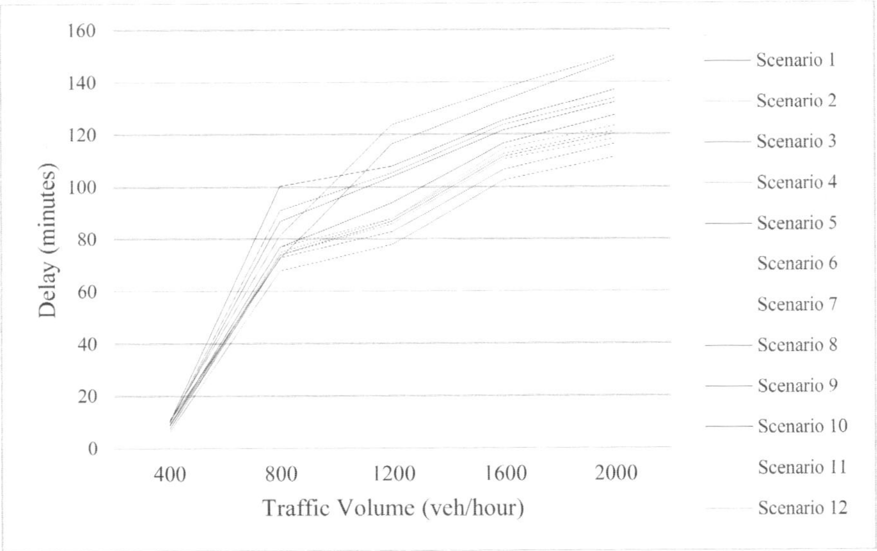

Figure 3. Delay of Traffic Control Open and Close Lane

Figure 3 shows a comparison of the delays for each scenario in in the traffic control open and close lane.

It can be seen that the graphs have a comparable relationship. Where, the higher the traffic volume, the longer the delay will be. In addition, the dimensions of the avalanche also affect the queue length in the area. The larger the dimensions of the avalanche, the fewer roads that can be passed, so the delay will be longer.

From the graph, scenario 1 with a landslide dimension of 7 x 5 m has a remaining road of 3 meters for vehicles to pass. Scenario 1 has the lowest delay, which is 6,32 minutes at a volume of 400 vehicles/hour and 111,26 minutes at a volume of 2000 vehicles/hour. While the longest delay is in scenario 12 with a landslide dimension of 12 x 6 m, and the rest of the road is 2 meters for vehicles to pass. The delay is 10,74 minutes at a volume of 400 vehicles/hour and 149,96 minutes at a volume of 2000 vehicles/hour.

b) Two Way on Landslide Zone Traffic Control

After simulated with VISSIM, there are queue and delay results from traffic control two way in landslide zone in Figure 4 and Figure 5:

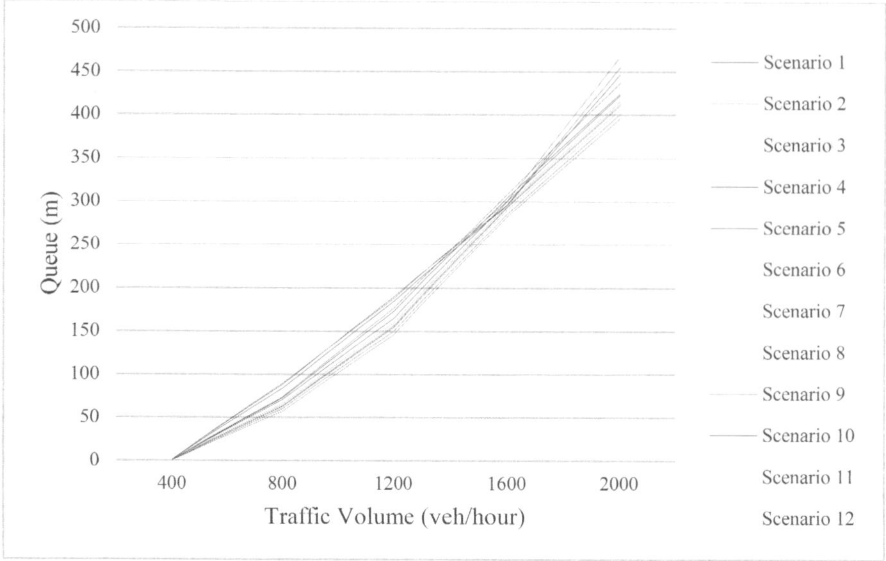

Figure 4. Queue of Traffic Control Two Way In Landslide Zone

Figure 4 shows a comparison of the queue length of each scenario in the traffic control two way in landslide zone. It can be seen that the graphs have a comparable relationship. Where, the higher the traffic volume, the longer the queue that will be generated. In addition, the dimensions of the avalanche also affect the queue length in the area. The larger the dimensions of the avalanche, the fewer roads that can be passed, so the queue will be longer.

From the graph, scenario 1 with a landslide dimension of 6 x 2 m has a remaining road of 6 meters for vehicles to pass. Scenario 1 has the lowest queue length, which is 0 meters at a volume of 400 vehicles/hour and 74 meters at a volume of 2000 vehicles/hour. While the highest queue length is in scenario 12 with a landslide dimension of 9 x 4 m, and the rest of the road is 4 meters for vehicles to pass. The queue length is 0 meters at a volume of 400 vehicles/hour and 93 meters at a volume of 2000 vehicles/hour.

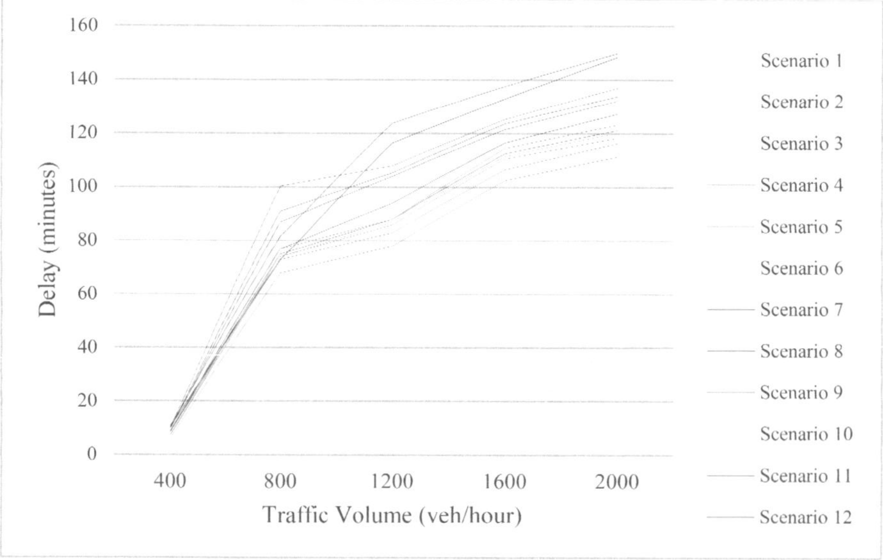

Figure 5. Delay of Traffic Control Two Way In Landslide Zone

Figure 5 shows a comparison of the delays for each scenario in the traffic control two way in landslide zone.

It can be seen that the graphs have a comparable relationship. Where, the higher the traffic volume, the longer the delay will be. In addition, the dimensions of the avalanche also affect the queue length in the area. The larger the dimensions of the avalanche, the fewer roads that can be passed, so the delay will be longer. From the graph, scenario 1 with a landslide dimension of 6 x 2 m has a remaining road of 6 meters for vehicles to pass. Scenario 1 has the lowest delay, which is 0,03 minutes at a volume of 400 vehicles/hour and 3,75 minutes at a volume of 2000 vehicles/hour. While the longest delay is in scenario 12 with a landslide dimension of 9 x 4 m, and the rest of the road is 4 meters for vehicles to pass. The delay is 0,09 minutes at a volume of 400 vehicles/hour and 4,76 minutes at a volume of 2000 vehicles/hour.

3.2. Meta Models

a) Open and Close Lane Traffic Control

Based on the simulation, metamodel for analysis of queue and delay has been developed. The alternative models are shown in Table 3 and Tabel 4.

Table 3. Queue Meta Model of Open and Close Lane Traffic Control at Landslide Zone

No	Equation	R Square	F count	F table	T count		T Table
					X1	X2	
1	Y = -228.534 + 0.458 X1	0.801	233.224	4.01	15.272	-	2.00172
2	Y = 143.893 + 0.916 X2	0.004	0.231	4.01	-	0.481	2.00172
3	Y = -282.691 + 0.458 X1 + 1.031 X2	0.806	118.304	3.16	15.344	1.214	2.39357

Note : Y = queue (m), X1 = Traffic volume (veh/hr), X2 = size of landslide area (m^2)

Based on Table 3 above, it can be seen that those result meet the requirements for F count > F table and T count > T table are equation number 1. So, the best equation is obtained for the length of the queue for Traffic Control Open and Close Lane is Y = -228.534 + 0.458 X1.

Table 4. Delay Meta Model of Open and Close Lane Traffic Control at Landslide Zone

No	Equation	R Square	F count	F table	T count		T Table
					X1	X2	
1	Y = -39.061 + 0.137 X1	0.890	471.474	4.01	21.713	-	2.00172
2	Y = 46.480 + 0.760 X2	0.034	2.054	4.01	-	1.433	2.00172
3	Y = -80.799 + 0.137 X1 + 0.794 X2	0.928	366.260	3.16	26.562	5.430	2.39357

Note : Y = delay (sec, X1 = Traffic volume (veh/hr), X2 = size of landslide area (m^2)

Based on Table 4 above, it can be seen that those that meet the requirements for F count > F table and T count > T table are equations number 1 and 3. Then between the two equations the largest R square value is taken, which is equation number 3. So, the best equation is obtained for the adjustment delay Traffic Control Open and Close Lane is Y = -80.799 + 0.137 X1 + 0.794 X2.

b) Two Way on Landslide Zone Traffic Control

Queue and delay meta model of traffic control open and close lane can be seen in the Table 5 and Table 6 :

Table 5. Queue Meta Model of Two Way Traffic Control at Landslide Zone

No	Equation	R Square	F count	F table	T count		T Table
					X1	X2	
1	Y = -37.138 + 0.049 X1	0.561	73.995	4.01	8.602	-	2.00172
2	Y = 15.272 + 0.165 X2	0.001	0.072	4.01	-	0.267	2.00172
3	Y = -41.055 + 0.049 X1 + 0.174 X2	0.562	36.562	3.16	8.542	0.422	2.39357

Note : Y = queue (m), X1 = Traffic volume (veh/hr), X2 = size of landslide area (m^2)

Based on Table 51 above, it can be seen that result that meet the requirements for F count > F table and T count > T table are equation number 1. So, the best equation is obtained for the length of the queue for Traffic Control Two Way in Landslide Zone is Y = -37.138 + 0.049 X1.

Table 6. Delay Meta Model of Two Way Traffic Control at Landslide Zone

No	Equation	R Square	F count	F table	T count		T Table
					X1	X2	
1	Y = -1.764 + 0.003 X1	0.792	233.224	4.01	14.865	-	2.00172
2	Y = 1.088 + 0.015 X2	0.005	0.231	4.01	-	0.515	2.00172
3	Y = -2.118 + 0.003 X1 + 0.016 X2	0.797	118.304	3.16	14.915	1.170	2.39357

Note : Y = delay (sec, X1 = Traffic volume (veh/hr), X2 = size of landslide area (m^2)

Based on Table 6 above, it can be seen that those that meet the requirements for F count > F table and T count > T table are equation number 1. So, the best equation is obtained for the adjustment delay for Traffic Control Two Way in Landslide Zone is Y = -1,764 + 0,003 X1.

4. Conclusion

Based on the result of metamodel development, the metamodel of queue and delay has been developed. Two independent variable that used during development, traffic volume and landslide size. Best metamodel was selected from two scenarios: (a) Open and close lane traffic control at landslide zone, (b) Two-way traffic control at landslide zone. It was concluded that traffic volume is always used as independent variable in all metamodel, landslide size is only included in delay metamodel at open and close traffic control. All meta model has negative constant, because the simulation input for traffic volume and landslide size is beginning at relatively high values.

References

[1] Christady Hary Hardiyatmo 2012. *Landslide soil and erosion.* (Indonesia: Gajah Mada University Press. Yogyakarta) (*In Bahasa*)
[2] Khairah M 2021. *Traffic management at landslide disaster in West Sumatera and its impact to passenger and drivers*. Vol. 708, No. 1 (IOP Conference Series: Earth and Environmental Science) p 012047
[3] Tamin Ofyar Z 2000. *Transportation Planning and Modelling pemodelan transportasi*. (Indonesia: Institut Teknologi Bandung) *(In Bahasa)*
[4] VISSIM 5.30-05 User Manual
[5] J. Sprinkle, B. Rumpe, H. Vangheluwe, G. Karsai 2014. *"Metamodelling", in Model-Based Engineering of Embedded Real-Time Systems* p. 57-76
[6] Gustavsson F.N 2007. *New Transportation Research Progress*. (New York: Nova Science Publishers, Inc)
[7] Kurniawan R 2016. Regression Analysis. Prenada Media. *(In Bahasa)*

Land use and land cover change analysis of flood prone area using remote sensing data and machine learning in Malang Raya, East Java, Indonesia

M I Hariyono[1], D Ramdani[1], F E S Silalahi[1], A A Kurniawan[3], N Indriasari[2], and M Buswari[4]

[1] Research Organization for Earth Sciences and Maritime - BRIN, Indonesia 16911
[2] Research Organization for Aeronautics and Space - BRIN, Indonesia 13710
[3] Department of Public Works Water Resources of Malang Regency, Indonesia 65163
[4] Brawijaya University, Faculty of Economics and Business, Indonesia 65145

E-mail: moch053@brin.go.id

Abstract. A natural disaster is one type of disaster that requires holistic handling. One of the natural disasters that often occurs is flooding. A flood is overflowing a large amount of water beyond its normal confines. Floods often cause an increased vulnerability of fatalities, so disaster mitigation efforts are needed. To reduce flood risk, this study uses remote sensing data to analyze changes in land use and land cover. The research location is Malang Raya, one of the locations prone to flooding in East Java. The method analyzes the results of processing satellite image data with the machine learning tools available at Google Earth Engine (GEE). This method provides convenience in accelerating the object classification in satellite image data. The study results indicate that using satellite image data analysis and machine learning processes is beneficial in accelerating object classification and approaching the actual conditions, making it easier to make a policy in handling disaster risk reduction.

1. Introduction

Flood disaster is one of the most common natural disasters. Floods occur when water overflows and inundates the land [1]. Flood conditions in the downstream area can be caused by the presence of fairly high-intensity of rain. Significant damage to nature can be caused by catastrophic floods [2]. Flood prevention and control measures need to be taken to reduce the effects of flood disasters, such as damage to natural resources, infrastructure, agriculture, and others [3]. Under these conditions, flood susceptibility analysis is needed for future flood prevention and mitigation strategies [4]. One of the analyzes can be done by observing changes in land use and land cover (LULC). The change in LULC can have an impact on environmental conditions in a disaster-prone location or something that can make the location more vulnerable to flooding.

Changes in LULC are one of the issues that are widely discussed in many studies. Information related to LULC is needed to implement sustainable development, ecosystem conservation and environmental management [5]. Multi-temporal data is needed to facilitate the analysis of a study area that is easily obtained from remote sensing data. We can use satellite image data from remote sensing technology to analyze LULC changes. Previous studies have stated that changes in LULC can affect flood frequency [6], [7], flood volume [8], and flood propagation [9].

It is a challenge in planning and managing flood disasters because it requires information and understanding related to LULC information in flood-prone areas. The development of remote sensing technology and Geographic Information System (GIS) application nowadays can assist in flood analysis of multi-year LULC changes, especially for hydro-logical studies [10], [11], modelling and predicting floods [12], [13], as well as assisting in analysing all relevant information to obtain the flood area hazard zone [14] – [16].

This study aims to analyze LULC changes using remote sensing data to reduce the risk of flooding. For the LULC classification process, machine learning algorithms accelerate the classification of remote sensing data. The Google Earth Engine (GEE) platform uses the LULC class classification process from satellite image data. GEE is a free cloud-based platform for geospatial analysis with supercomputer capabilities for data processing and complex calculations [17], [18]. The results of the classification are used to measure and map temporal changes in land use/cover in the study area from 2016 to 2021.

2. Material and methods

2.1 Study area

The research location is the Malang Raya area, also known as the Malang Metropolitan Area (Figure 1). This area is located in East Java, Indonesia, which includes three administrative areas: Malang City, Batu City, and Malang Regency. The choice of this location is the object of research because Malang Raya is unique, it is a location where there are many tourist attractions with a topography dominated by highlands, but floods often occur, especially in downstream areas.

This location is prone to geological and hydro-meteorological disasters. Geographically, Malang Raya is located between 112° 06' East Longitude to 112° 57' East Longitude and between 112° 06' South Latitude and 8° 26'South Latitude. Topography Malang Raya comprises a plateau area surrounded by several mountains and lowland or valley regions from 250 to 526 meters above sea level. This plateau is categorized as limestone hills (Kendeng Mountains) in the south from 0 to 650 meters above sea level. The Tengger-Semeru area is in the east part, stretch from north to south from 500 to 3,600 meters above sea level. Kawi – Arjuno is in the west part from 500 to 3,300 meters above sea level. Another, the mountain Buring, is located east of Malang Regency. Mountainous topography with hilly slope conditions makes the Malang area a great area in great demand for residences and resting places. At the same time, the rivers that flow in the Malang Raya area are the Brantas, Amprong, and Bango [19], [20].

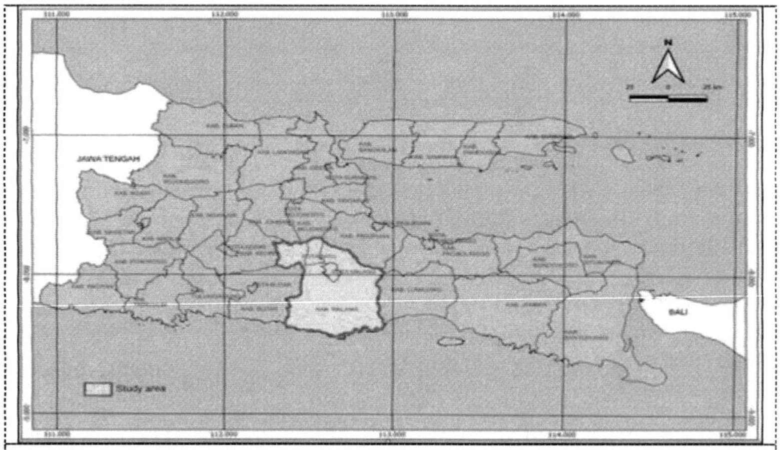

Figure 1. The Malang Raya area

The area of Malang Raya is approximately 3,800 km^2 (table 1), located in the mid-south region of East Java Province. Malang Raya is one of the areas with the most extreme topography in Indonesia, especially in the eastern part. There are many mountains in the research location, while the lowland areas are mainly located on the south side in the form of the coast of the South Java Sea.

Table 1. Malang Raya Administrative Region (Source: Statistics of Malang and Malang Regency).

No	Administrative Region	Area in km^2
1	Malang City	145.28
2	Batu City	202.3
3	Malang Regency	3,534.86
	Malang Raya	3,882.44

2.2 Data preparation.
The data used in this study are Sentinel-1, Sentinel-2 and NOAA VIRRS data in 2016 and 2021 and SRTM (table 2). A period of 5 years is sufficient to represent the analysis of changes in LULC in the research area. Sentinel-1 and Sentinel-2 data have an accuracy of 10m, enough to be used as LULC change analysis data. The data already exists on the Google Earth Engine (GEE), so that we can carry out the classification process with the machine learning algorithm in GEE. We use the platform of GEE as a tool for the remote sensing data classification process by applying the machine learning algorithms.

This study uses the LULC classes: buildings, roads, waters, bare land, and high and low vegetation. The data used for the analysis process is Sentinel imagery, obtained free of charge and available from the GEE platform. Sentinel imagery covers the research area, Malang, Raya East Java, about 3,882.44 km^2. The choice of using this data is because the data provided by GEE has been corrected for atmospheric effects so that it can improve the detection of changes in LULC surface objects.

Table 2. Data series of respective years.

No	Data	Acquisition Year	Band	Resolution (m)
1	Sentinel-1	2016	VV, VH	10
2	Sentinel-1	2021	VV, VH	10
3	Sentinel-2	2016	Multispectral	10
4	Sentinel-2	2021	Multispectral	10
5	NOAA VIIRS	2016	Average DNB radiance values	500
6	NOAA VIIRS	2021	Average DNB radiance values	500
7	SRTM	2000	slope	30 m

Many methods have been developed for analyzing land use/cover changes, one of which is machine learning algorithms. In this study, the LULC class classification process uses machine learning technology from the GEE platform. In the analysis, we use remote sensing technology and GIS to analyze changes that occur temporally to changes in the land surface. The analysis is projected in a map view of the classification results and a table. The remote sensing data for 2016 and 2021 are processed separately for LULC class classification: buildings, roads, waters, bare land, high vegetation, and low vegetation. These results analyzed the changes that occurred over five years. The data processing method shows below (figure 2).

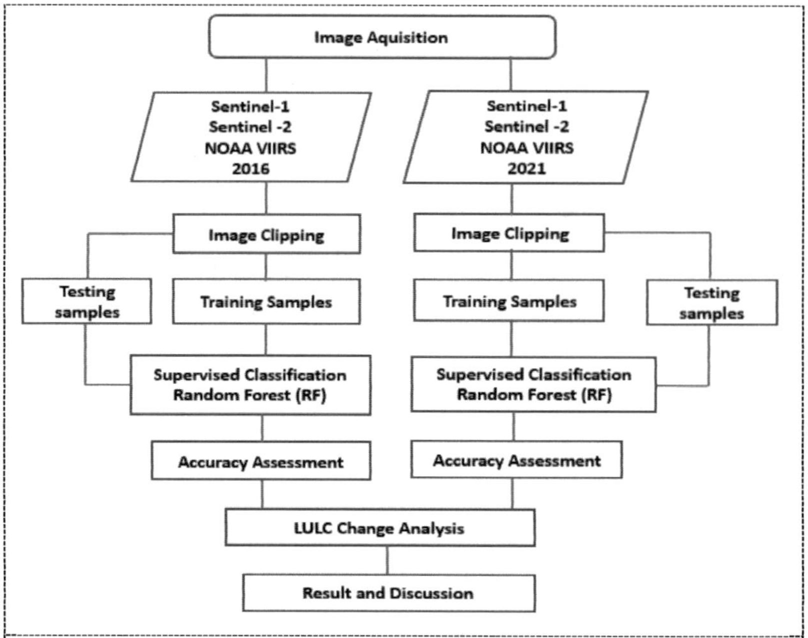

Figure 2. GEE platform data processing

In addition to using Sentinel 2 Bands: B1, B2, B3, B4, B8A, B11 and B12, they also use a combination of these bands, namely the New Built-up Index (NBI), the Build-up Or Bareness Index (BOBI), Modified Normalized Difference Water Index (MNDWI), the Normalized Difference Built-up Index (NDBI), the Normalized Difference Vegetation Index (NDVI), the Ratio Residence-area Index (RRI), and Soil salinity Index (SI), NOAA VIIRS: Night-Time Light (NTL), and Vegetation Adjusted NTL Urban Index (VANUI) (the combination of NT and NDVI), Sentinel 1 GRD: VH, VV, Variance of VV and Variance of VH (texture metrics from the Gray Level Co-occurrence Matrix polarization VV and VH), and SRTM using slope data.

Sentinel 1, Sentinel 2 and NOAA VIIRS data are retrieved over five years in 2016 and 2021 with cloud coverage removed. Subsequently, the image was cropped according to the needs of the research location with the vector data format.

This study uses a pixel-based supervised classification method for the LULC classification process. A sampling of the area is done manually based on the visual appearance of the surface in the available image data. The number of samples for each class and the selection of sampling locations significantly affect the classification results. The sample for each class is then used 80% as the training area and 20% for the test. This study uses Random Forest (RF) classification method with the number of trees parameter 500, variables Per Split 6 and the other default prices in GEE. RF performs better for detecting built-up land and other classes in the land area of 81 km^2 compared to Minimum Distance Accuracy or Support Vector Machine [18]. Accuracy is obtained using the confusion matrix, including the overall accuracy and Kappa. The RF variables' importance is calculated after the training process to determine the contribution from predictors during training.

3. Result and discussion

The LULC classification using Random Forest from this confusion matrix produces an overall accuracy of 1 and Kappa 1 for 2016 and 2021. As for the results of Random Forest Variable Importance, it shows in figure 3.

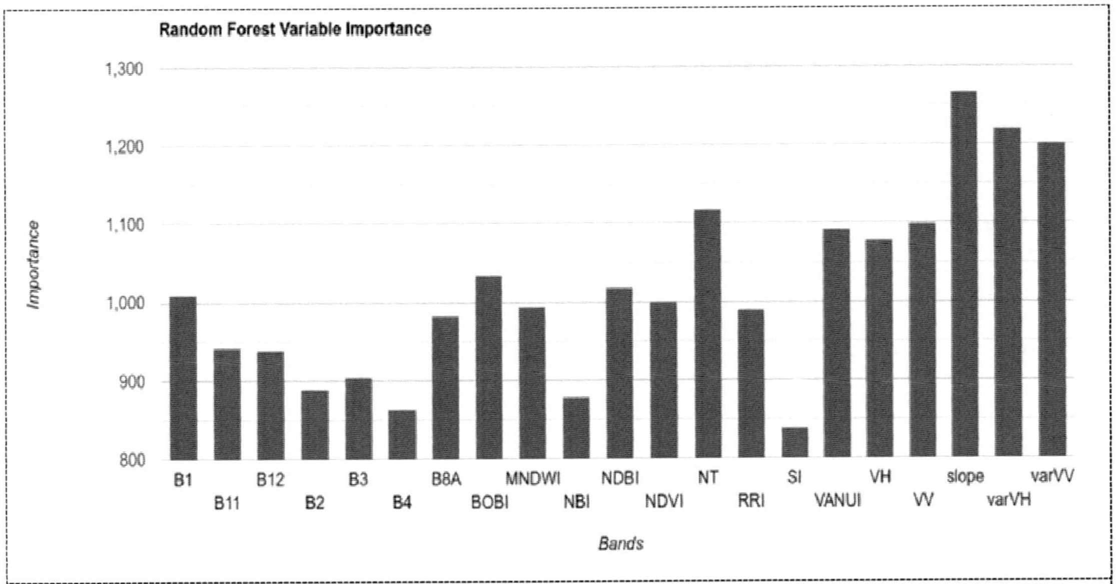

Figure 3. Random forest variable importance

The results of the LULC classification shows below (figure 4). The LULC in 2016 (figure 4a) show that most of the Malang Raya area is dominated by high and low vegetation, especially in the Malang City area, which is dominated mainly by buildings. The LULC in 2021 (figure 4b) shows a change from 2016, namely an increase in the class of settlements, especially in the Malang Regency area. These changes resulted in a reduction in high and low vegetation objects.

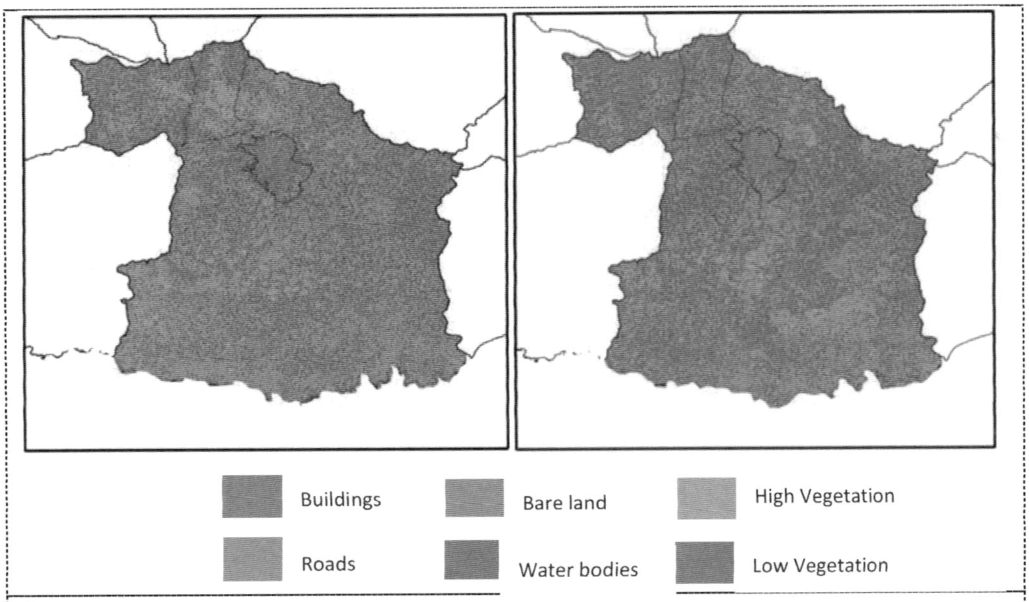

Figure 4. LULC classification results in 2016 (a) and 2021 (b) for the Malang Raya area

The image data used has a pixel size of 10 m. This affects the results of land cover that has a size of less than 10 m. The road class at the research location is not quite visible from the classification results

IOP Conf. Series: Earth and Environmental Science 1173 (2023) 012051 doi:10.1088/1755-1315/1173/1/012051

because it has a road width that is mostly less than 10m. For changes in the area of each class, we can see in table 3. Flood events often occur in Malang City by looking at the land use changes in table 1. In figure 1, the orange color appears, which indicates land use changes in settlements in Malang City.

Table 3. LULC classes 2016 and 2021

Land Classes	2016 (km^2)	2021 (km^2)
Buildings	1.159,597	1.348,604
Roads	493,126	499,597
Bare Land	419,732	436,390
Water Bodies	401,025	431,093
High Vegetations	585,209	483,679
Low Vegetations	748,404	614,728

The analyzes and information in the field show that most of the disaster-prone locations in Batu City are in the Bumi Aji District (figure 5). This area has the highest elevation compared to other sub-districts in Batu City. The hilly contours of the land caused the flood disaster at this location. The plantation area had not applied the terrace cultivation method (*terasering*), and there were no trees to limit soil erosion by reducing the water flow energy, so it was also prone to landslides.

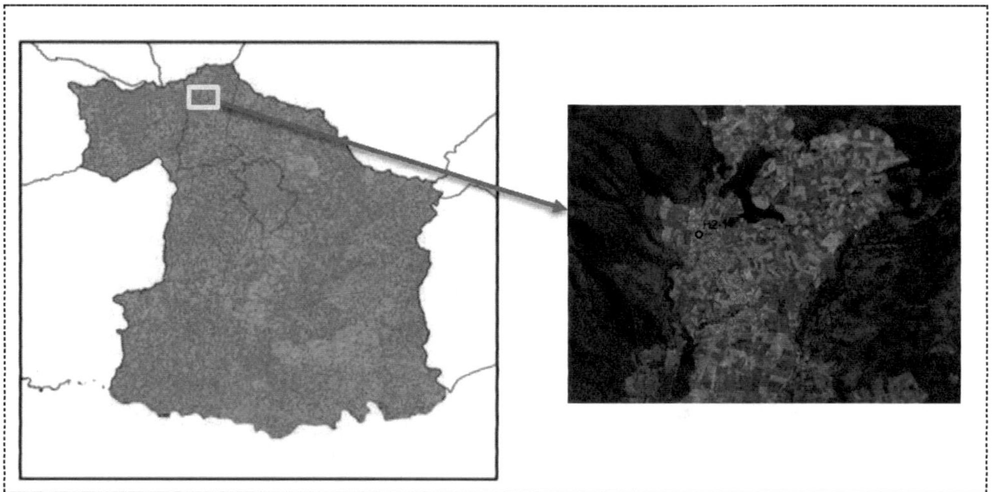

Figure 5. The Flood-prone Area in Batu City

Flood-prone locations in Malang City are primarily located in the city center (figure 6). Inundation of water often occurs in these locations. Overflowing waterways usually cause flood disasters because they cannot accommodate large volumes of water due to high rainfall intensity. Malang Regency is the largest compared to Malang city and Batu city in the coverage area of the Greater Malang area.

Figure 6. The Flood-prone Area in Malang City

Flood-prone locations in Malang Regency occur in certain villages, mainly in the southern part of Malang. This area has a lower contour or land elevation than the northern part of Malang (figure 7).

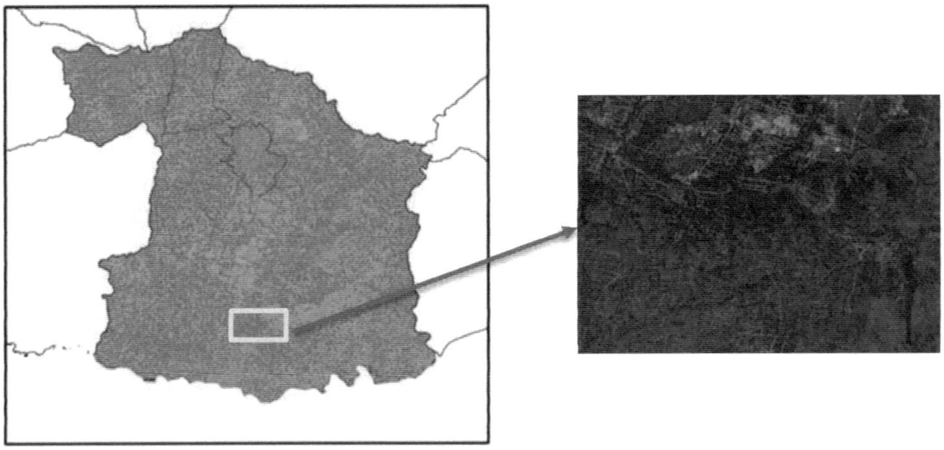

Figure 7. The Flood-prone Area in Malang Regency

4. Conclusions

Based on the analysis of LULC classification, data processing shows a change in land use in Malang Raya from 2016 to 2021, mainly for settlements. Many of land covers with vegetation classes have changed functions into residential areas, and this is one of the triggers for a flood disaster where the water catchment area is reduced. Regional governance is needed so the rapid building development does not impact the risk of natural disasters, especially floods.

5. Acknowledgements

We would like to thank the Center for Disaster Research, Research Organization for Earth Sciences and Maritime - National Research and Innovation Agency (BRIN) for providing this research opportunity and budgetary support. From *Daftar Isian Pelaksanaan Anggaran* (DIPA) Research Organization for Earth Sciences and Maritime - BRIN No. SP DIPA-124.01.1.690501 year 2022, on 17 November 2021.

6. References

[1] Apel H, Aronica G T, Kreibich H, and Thieken A H 2009 Flood risk analyses - How detailed do we need to be? Natural Hazards vol. **49**, no. 1, pp. 79–98
[2] Modarres R, Sarhadi A, and Burn D H 2016 Changes of extreme drought and flood events in Iran *Global and Planetary Change* vol. **144**, pp. 67–81
[3] Huang X et al. 2008 Flood hazard in Hunan province of China: An economic loss analysis *Natural Hazards* vol. **47**, no.1, pp. 65–73
[4] Yu J, Qin X, and Larsen O 2013 Joint Monte Carlo and possibilistic simulation for flood damage assessment *Stochastic Environmental Research Risk Assessment* vol. **27**, pp. 725–735
[5] Saleem A, Corner R and Awange J 2018 On the possibility of using CORONA and Landsat data for evaluating and mapping long-term LULC: Case study of Iraqi Kurdistan *Applied Geography,* vol. **90**, pp. 145–154
[6] Montanari A and Brath A 2004 A stochastic approach for assessing the uncertainty of rainfall-runoff simulations *Water Resources Research*, vol. **40**, pp. 1–11
[7] Teixeira Z, Teixeira H, and Marques J C 2014 Systematic processes of land use / land cover change to identify relevant driving forces : Implications on water quality *Science of the Total Environment*, vol. **470–471**, pp. 1320–1335
[8] Miller J D, Kim H, Kjeldsen T R, Packman J, Grebby S and Dearden R 2014 Assessing the impact of urbanization on storm runoff in a peri-urban catchment using historical change in impervious cover *Journal of Hydrology,* vol. **515**, pp. 59–70
[9] Di Baldassarre G, Castellarin A and Brath A 2009 Analyse des effets de la surélévation des levées sur la propagation des crues: Exemple du Fleuve Pô, Italie *Hydrological Science Journal,* vol. **54**, pp. 1007–1017
[10] Broxton P D, Zeng X, Sulla-Menase D, and Troch P A 2014 A Global Land Cover Climatology Using MODIS Data *Journal of Applied Meteorology and Climatology*, vol. **53**, pp. 1593–1605
[11] Wanders N, Bierkens M F, de Jong S, de Roo A and Karssenberg D 2014 The benefits of using remotely sensed soil moisture in parameter identification of large-scale hydrological models *Water Resources Research,* vol. **50**, pp. 6874–6891
[12] Haq M, Akhtar M, Muhammad S, Paras S and Rahmatullah J 2012 Techniques of Remote Sensing and GIS for flood monitoring and damage assessment: A case study of Sindh province, Pakistan *The Egyptian Journal of Remote Sensing and Space Science,* vol. **15**, pp. 135–141
[13] Elbialy S, Mahmoud A, Pradhan B, and Buchroithner M 2014 Application of spaceborne synthetic aperture radar data for extraction of soil moisture and its use in hydrological modelling at Gottleuba Catchment , Saxony, Germany *Journal of Flood Risk Management*, vol. **7**, pp. 159–175
[14] Sun X *et al.* 2012 Sensitivity of digital soil maps based on FCM to the fuzzy exponent and the number of clusters *Geoderma*, vol. **171–172**, pp. 24–34
[15] Akgun A 2012 A comparison of landslide susceptibility maps produced by logistic regression , multi-criteria decision , and likelihood ratio methods : a case study at İ zmir , Turkey *Landslides*, vol. **9**, pp. 93–106
[16] Hariyono M I, Kurniawan A A, Rokhmatuloh and Tambunan M P 2022 The use of airborne Lidar data to analyze flood disaster area: Case study of Sekarbela Subdistrict, Mataram *IOP Conference Series : Earth and Environmental Science* **950** 012086. doi:10.1088/1755-1315/950/1/012086
[17] Patel N N *et al. 2015* Multitemporal settlement and population mapping from Landsat using Google Earth Engine *International Journal Applied Earth Observation and Geoinformation*, vol. **35**, pp. 199–208, 2015.

[18] Rudiastuti A W, Lumban-Gaol Y, Silalahi F E S, Prihanto Y and Pranowo W S 2022 Implementing Random Forest Algorithm in GEE: Separation and Transferability on Built-Up Area in Central Java, Indonesia *International Journal on Informatics Visual,* vol. **6**, pp. 74–82
[19] Regency M and Figures I N 2022 *Kabupaten Malang dalam angka.*
[20] Statistik B P and Malang K 2022 *Catalog : 1102001.3573. Kota Malang dalam angka*

Population distribution analysis in Banda Aceh City for tsunami disaster risk reduction

I Amri[1], R R Alami[2] and A Serlia[3]
[1] The Graduate School of Universitas Gadjah Mada, Bulaksumur, Yogyakarta 55281, Indonesia
[2] Faculty of Social & Political Science, Universitas Gadjah Mada, Bulaksumur, Yogyakarta 55281, Indonesia
[3] Regional Office of the National Land Agency of Aceh Province, Lamgugob, Banda Aceh 23115, Indonesia

E-mail: ikhwan.amri@mail.ugm.ac.id

Abstract. Areas affected by the 2004 Indian Ocean Tsunami have recovered following a long-term reconstruction process, including Banda Aceh City, Indonesia. Using data from the Statistics Indonesia, this study aims to analyze population dynamics in Banda Aceh City, starting from 2004 (pre-tsunami), 2005 (shortly after the tsunami), as well as 2010 and 2020 (post-reconstruction phase). Population dynamics are calculated at the sub-district (*kecamatan*) level by observing trends and changes in size and rank based on population density and population growth rate values. Some of the tsunami-affected sub-districts have experienced population recovery to pre-disaster levels, while population growth in safe areas is relatively high in line with development directions to the east and south of the city. However, disaster risk reduction tasks will be even more challenging as the population in coastal areas is growing at an alarming rate. Therefore, the government needs to design a comprehensive policy to minimize the population's exposure to the tsunami hazard.

1. Introduction

As an area located on the Ring of Fire, Indonesia has a relatively high potential for a tsunami disaster, whereas the source mechanism comes from earthquakes, volcanic eruptions, or underwater landslides. Historical records since 416 AD proved that tsunamis repeatedly occur around subduction zones, and many of them destroy human civilization significantly [1]. In the last two decades, several tsunami disasters in Indonesia caused devastating impacts, including two recent tsunamis triggered by non-seismic factors [2,3]. This fact highlights the importance of seeking more serious disaster mitigation actions to minimize the tsunami risk.

Population is often considered one of the important elements in various risk models [4]. Theoretically, the level of risk will be higher along with the density of the population. Many works of literature also pay attention to the high rate of urbanization in disaster-prone areas, including coastal areas [5,6]. In Indonesia itself, the population concentration is predominantly found in the lowlands facing the sea. Spatial planning that ignores local hazards can create excessive damage when disasters occur.

The 2004 Indian Ocean Tsunami in Aceh Province has sufficiently shown that the combination of high hazard and vulnerability in an area could highly trigger a devastating impact, as the event itself was classified as one of the deadliest disasters of all century. The worst intensity of damage was found in Banda Aceh, the center of government and economy in Aceh Province, where half of the total built-up land area was damaged. About a third of the population became victims of the ferocity of the tsunami waves, while more than 65,000 people were displaced to safer areas. Major damages were reported to infrastructures and public services, thus having implications for the emergence of long-term crises in people's lives [7].

Post-disaster reconstruction and rehabilitation play a crucial role in restoring city functions and socio-demographic conditions. Based on the development of the theory of random growth-locational fundamentals, there are two possible scenarios in the context of changes in population during the recovery process [8]. First, a disaster-affected area will experience permanent impacts or even continuous depopulation due to the out-migration process, both influenced by individual risk perceptions and regulatory instruments that regulate the prohibition of living in-situ. Second, a population explosion can occur following the rebuilding of settlements and infrastructure, as well as a long-term economic investment. The latter scenario assuredly recreates challenges in disaster risk reduction efforts.

By taking a case study in Banda Aceh City, this article aims to analyze population dynamics in each sub-district (*kecamatan*) unit. The periods selected include 2004 (before the tsunami), 2005 (shortly after the tsunami), 2010 and 2020 (post-tsunami). The data in 2004 became the basis for determining whether the condition of the population in the affected area had recovered as it was before the tsunami.

2. Method

2.1. Study area

Banda Aceh is the capital city of Aceh Province, Indonesia. This area is surrounded by Aceh Besar Regency and is directly adjacent to the Malacca Strait and the Indian Ocean. It consists of nine sub-districts (*kecamatan*): (1) Meuraxa, (2) Kuta Raja, (3) Jaya Baru, (4) Syiah Kuala, (5) Kuta Alam, (6) Baiturrahman, (7) Banda Raya, (8) Lueng Bata, and (9) Ulee Kareng. The area of the city is about 59 km^2, but the scope of the spatial planning area is about 61 km^2.

Banda Aceh is characterized by a low-lying flat topography associated with fluvio-marine depositional environments. This area is situated as a discharge area of the Krueng Aceh Basin, surrounded by two strike-slip fault segments that are part of the Great Sumatra fault. Due to intensive urban development in the coastal plains of Banda Aceh, coastal hazards, including the tsunami generated by submarine earthquakes from the west coast of Sumatra Island, potentially threaten the urban population and disrupt the existence of major socio-economic activities.

The 2004 Indian Ocean Tsunami caused fatal damage in large parts of Banda Aceh City. The height of the tsunami waves was estimated at 10 meters, causing complete destruction in an area about 2-3 km wide along the Banda Aceh coast (Figure 1). The west coast was the one that suffered the most losses. Meuraxa Sub-district and Kuta Raja Sub-district can be said to be completely destroyed at that time [9]. Significant losses were also reported in other sub-districts, namely Jaya Baru, Kuta Alam, Syiah Kuala, and Baiturrahman. Meanwhile, three sub-districts that are farther from the shoreline were not significantly affected, such as Banda Raya, Lueng Bata, and Ulee Kareng. The regions, including the surrounding suburbs, are designed to be the new development centers as they are relatively safe from tsunami hazard.

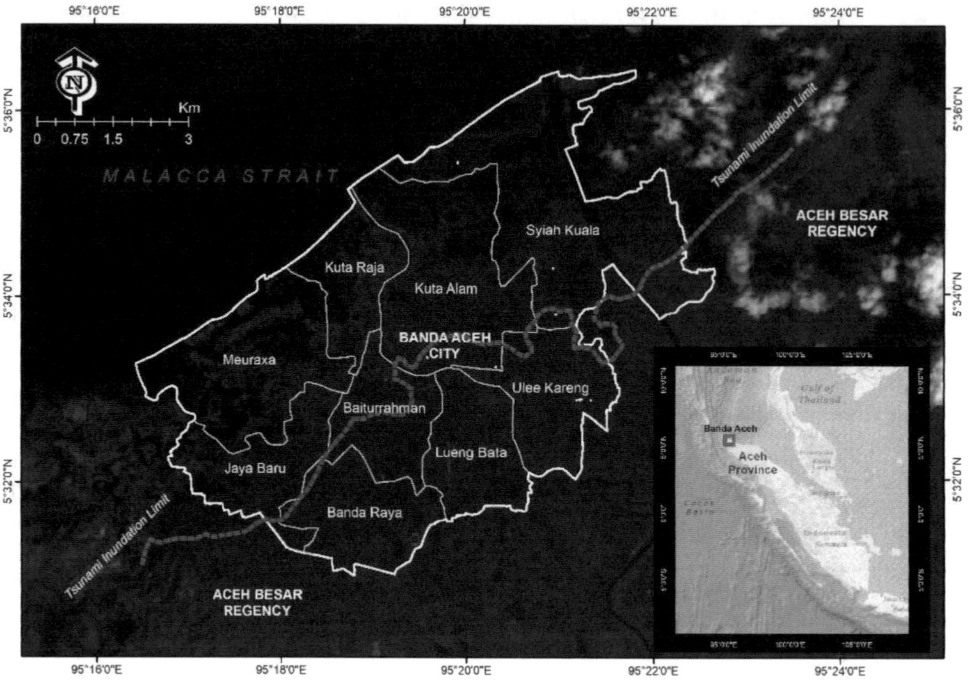

Figure 1. Damage to the 2004 tsunami in Banda Aceh and its surroundings.

2.2. Data sources

This study uses population data per sub-district in 2004, 2005, 2010, and 2020 collected from the annual publication of the Statistics Indonesia (Table 1). A person is defined to be part of the population if living in an area for at least six months or less than six months but intends to stay. The Population Census is the main source of population data in Indonesia. It takes place every 10 years and is usually held in years ending in 0 (zero). The Statistics Indonesia also coordinates the Intercensal Population Survey every year ending in 5 (five). Specifically for the Aceh and Nias regions, a Population Census was carried out in 2005 to support the provision of basic data for post-tsunami reconstruction and rehabilitation planning. Other sources of population data are projections calculated annually based on Population Census and Intercensal Population Surveys. The data for 2004 acts as a comparison to the post-disaster population condition.

Table 1. Population size data in Banda Aceh City.

Sub-District	Area (km^2)	Population Size			
		2004	2005	2010	2020
Meuraxa	7.26	34,592	2,221	16,484	26,861
Kuta Raja	5.21	21,632	2,978	10,433	15,291
Jaya Baru	3.78	21,305	12,340	22,031	25,939
Syiah Kuala	14.24	32,590	25,418	34,850	32,969
Kuta Alam	10.05	54,718	35,033	42,217	42,505
Baiturrahman	4.54	37,715	33,582	30,377	32,513
Banda Raya	4.79	23,995	24,257	20,891	25,228
Lueng Bata	5.34	19,232	19,284	23,592	24,336
Ulee Kareng	6.15	19,319	22,768	22,571	27,257
Total	61.36	265,098	177,881	223,446	252,899

2.3. *Estimation of population dynamics*

The main population metrics calculated in this study are population density (*PD*) and population growth rate (*PGR*). The *PD* formula refers to the crude density, as shown in equation (1). It is obtained by dividing the total population (*P*) by the total land area in square kilometers (*A*). Meanwhile, *PGR* is calculated to analyze changes in population between two time periods (*t*) using the geometric method, as shown in equation (2). If the *PGR* is positive, there is an increase in population in the final year (P_t) compared to the initial year (P_0). Conversely, if the *PGR* is negative, there is a reduction in population in the final year (P_t) compared to the initial year (P_0).

$$PD = \frac{P}{A} \tag{1}$$

$$PGR = \left(\frac{P_t}{P_0}\right)^{\frac{1}{t}} - 1 \tag{2}$$

Population dynamics are calculated at the sub-district level by observing trends and changes in size and rank based on *PD* and *PGR* values. This analysis can be useful for estimating whether an area affected by a disaster can rebound after going through a recovery period. While in safer areas, there is also the possibility of significant population changes as people begin to become more aware of natural hazards.

3. Results

3.1. *Population density*

According to the obtained results of population density (*PD*) in Banda Aceh which is divided into several main periods from 2004 to 2020, most of the sub-districts in the area had been through dynamic changes throughout the year. The major changes were directly caused by the tsunami disaster that occurred on 26 December 2004. A population density map of Banda Aceh from 2004 to 2020 in Figure 2 represents the dynamic changes of *PD* in each sub-district. The map uses the *PD* data of Banda Aceh provided by the local authorities showing the inclines and the declines of *PD* in the study area. Most of the sub-districts experienced a temporal decline in *PD* from 2004 to 2005 due to the tsunami disaster that generally destroyed the western coastal region of Banda Aceh. Some sub-districts that experienced a significant decline in *PD* value from 2004 to 2005 included Meuraxa with the *PD* value of 4,764.74 people/km^2 declining to 305.92 people/km^2, Jaya Baru with the *PD* value 5,636.24 people/km^2 declining to 3,264.55 people/km^2, Kutaraja with the *PD* value of 4,152.02 people/km^2 declining to 571.59 people/km^2, and Kuta Alam with the *PD* value of 5,444.58 people/km^2 declining to 3,485.87 people/km^2.

As the reconstruction programs were conducted immediately after the disaster, during the second period from 2005 to 2010, some sub-districts with the most significant declines were rapidly progressing and gaining higher values of *PD*. For example, the *PD* value in Jaya Baru, Meuraxa, Kuta Raja, and Kuta Alam reached 5,828.31 people/km^2, 2,270.52 people/km^2, 2,002.50 people/km^2, and 4,200.70 people/km^2, respectively. Furthermore, during the third period from 2010 to 2020, most sub-districts in Banda Aceh City were mostly experiencing the inclines of population density indicating that the population growth of this city was also growing rapidly. Some of the obtained *PD* values in the sub-districts were even higher than their previous values in 2004 during the pre-tsunami period. For example, the highly populated Jaya Baru which experienced a major decline in *PD* value after the tsunami disaster in 2020 had a *PD* value of 6,862.17 people/km^2. This value was higher than in 2004 when it was only 5,636.24 people/km^2. The only sub-district that experienced a decline in *PD* value during this most recent period was Syiah Kuala, which previously had a *PD* value of 2,447.33 people/km^2 and slightly declined to 2,315.24 people/km^2 in 2020. However, despite the decline in the area, the *PD* value of Syiah Kuala in 2020 was still higher than the one in 2004, which was 2,288.62 people/km^2.

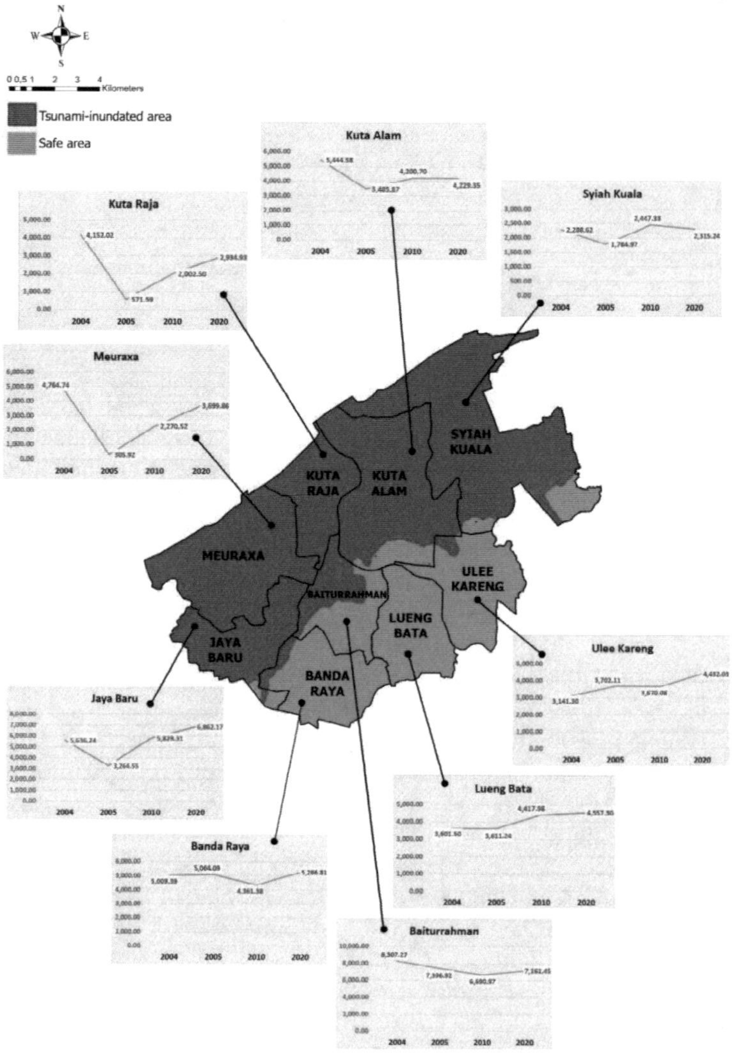

Figure 2. Population density map of Banda Aceh.

Figure 3 shows another visualization of the dynamic changes occurring from 2004 to 2020. From 2004 to 2005, the graphic shows some significant declines in *PD* in several sub-districts, specifically Jaya Baru, Kuta Alam, Meuraxa, and Kutaraja. During this period, Jaya Baru that used to be the second most populated area in Banda Aceh dropped to the sixth rank in 2005. It also happened to Meuraxa which dropped its rank from fifth to ninth in 2005. Most of these areas are located in the western coastal region of Banda Aceh that was heavily devastated by the tsunami disaster. The disaster significantly reduced the numbers of people living in these areas, whether due to the excess death toll as the direct disaster effect or due to the higher number of the surviving residents that decided to move to safer areas. Some sub-districts in Banda Aceh during the first period (2004–2005) were also going through an incline in terms of their *PD*, specifically in Lueng Bata, Ulee Kareng, Banda Raya, and Syiah Kuala. Three of the total four sub-districts with inclined *PD* are located outside the tsunami inundation zone. Meanwhile, the *PD*'s rank of Baiturrahman tended to be more stable during the first period compared to the other eight sub-districts in Banda Aceh.

During the second period (2005–2010), several sub-districts with previous declines experienced progressive changes, including Jaya Baru and Meuraxa. Jaya Baru gained back the second rank as the most populated sub-district despite being in the sixth rank in 2005. Meuraxa also experienced an incline from the ninth rank in 2005 to the eighth rank in 2010. In contrast, Banda Raya and Ulee Kareng showed declines in the *PD* rank. However, several sub-districts remained in the same rank during this period, such as Syiah Kuala in the seventh rank, Kuta Alam in the fifth rank, and Baiturrahman in the first rank.

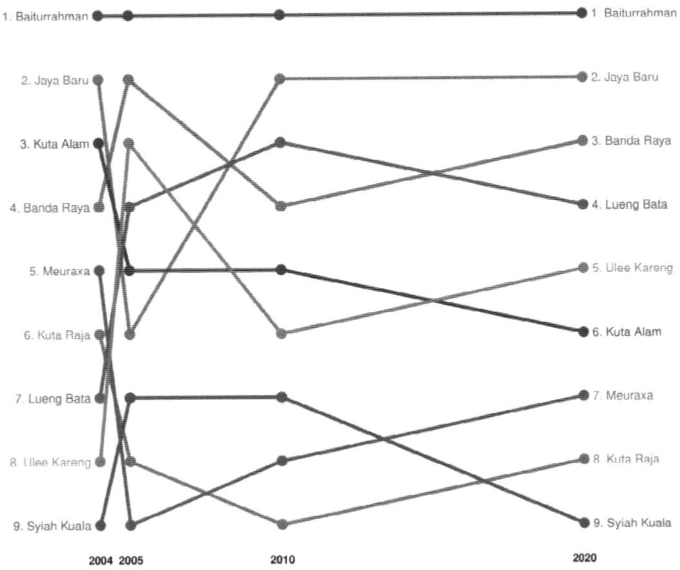

Figure 3. Rank in population density.

In the third period of 2010–2020, the sub-district of Baiturrahman still ranked as the most populated area in Banda Aceh City. It indicated that the tsunami disaster that happened in 2004 did not significantly change the *PD*'s rank in the sub-district. The remaining *PD* value was still higher than the other sub-districts in Banda Aceh, even though the *PD* might slightly drop in each period. It is in line with the location of Baiturrahman in the center area of the city that provides many fundamental activities for the residents. Jaya Baru was the second area with the highest *PD* even though it significantly dropped in 2005 due to the tsunami disaster in 2004. Jaya Baru's recent rank was then sequentially followed by Banda Raya, Lueng Bata, and Ulee Kareng that are considerably safer due to their more distant location to the tsunami inundation zone of the city. In line with this result, these areas are potential for further improvement in the future urban development.

3.2. *Population growth rate*

Population growth rate (*PGR*) expresses the increase or decrease in population size over a certain period in percent. This calculation complements the previous results, representing how rapidly population changes are closely related to demographic processes. Figure 4 shows the variation in *PGR* values in each sub-district. In general, the *PGR* curve in all tsunami-affected areas has a similar appearance, except for Baiturrahman. The tsunami-affected areas experienced negative population growth from 2004 to 2005. The heaviest losses were found on the west coast of Banda Aceh, particularly Meuraxa (-93.58%) and Kuta Raja (-86.23%). In contrast, the *PGR* value in areas not affected by the tsunami was positive, with the highest being in Ulee Kareng (17.85%).

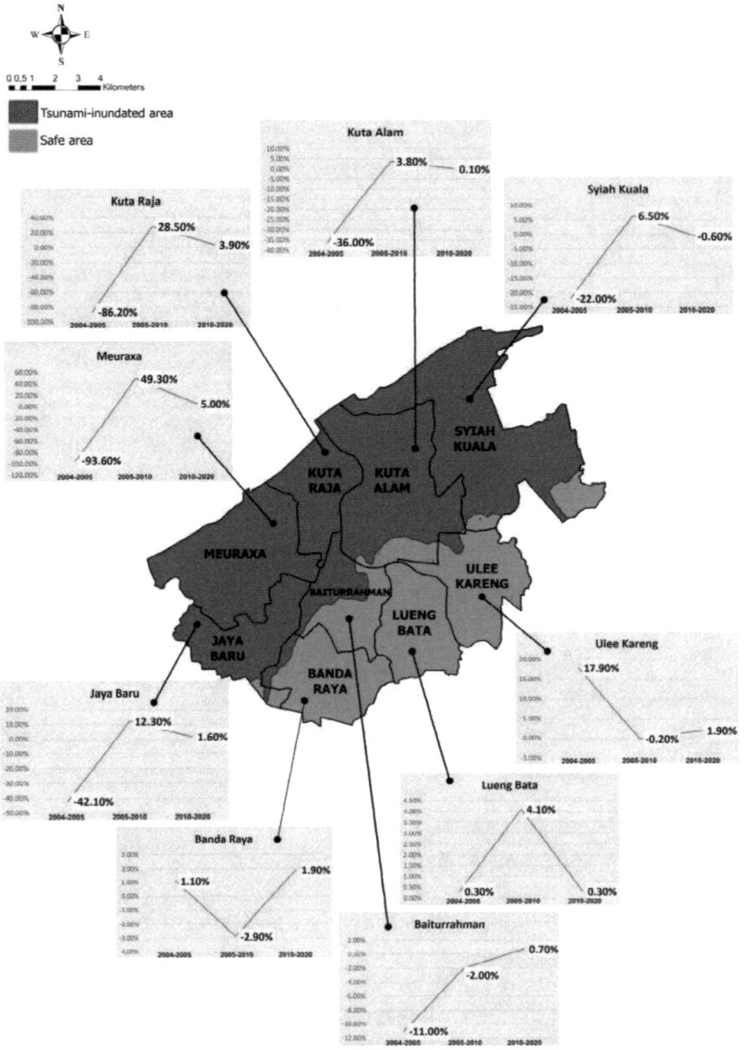

Figure 4. Population growth rate map of Banda Aceh.

The rapid increase in *PGR* values in tsunami-affected areas started from 2005 to 2010 is in line with the reconstruction and rehabilitation activities. There is a tendency that *PGR* was worth more in areas that got the most damage. One example is the *PGR* value in Meuraxa which reached 49.31%. Meanwhile, negative population growth was found in Baiturrahman (-1.99%), Banda Raya (-2.94%), and Ulee Kareng (-0.17%). In the 2010-2020 period, all sub-districts had positive *PGR* values, except in Syiah Kuala. However, growth rates in most sub-districts tended to be lower than in the previous period. Banda Raya and Ulee Kareng, two areas that in the previous period had negative *PGR* values, experienced a rather significant growth increase, where the percentage growth of both became 1.90%.

Figure 5 shows the dynamics of the *PGR* in Banda Aceh based on the growth rate rank. Some significant changes that happened from the first period (2004–2005) to the second period (2005–2010) were caused by the tsunami disaster. Specifically, areas in the first to fourth ranks dropped into the last ranks regarding their population growth, including Ulee Kareng, Banda Raya, Lueng Bata, and Baiturrahman. On the other hand, the rest of the areas in the last ranks became in the top six of the *PGR*. Jaya Baru, Kuta Raja, and Meuraxa were in the three last ranks of the population growth in the first period. The three areas became in the top three ranks in the second period. Kuta Alam stayed in the sixth rank as the first period.

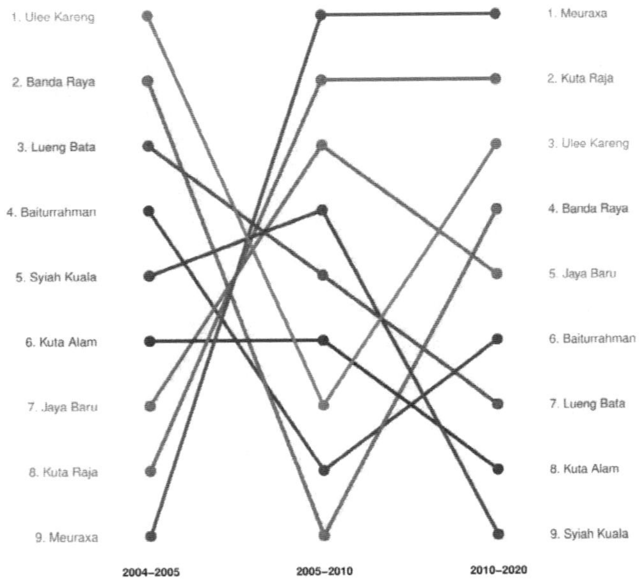

Figure 5. Rank in population growth rate.

In the third period (2010–2020), some areas still maintained the same rank as before, such as Meuraxa and Kuta Raja in the first and second ranks. Besides, Ulee Kareng and Baiturrahman, and Banda Raya went to higher ranks than before. The areas were in the three last ranks for the second period but became the third, the fourth, and the sixth ranks in the last period, respectively. The rest of them fell into lower ranks in the third period.

The most interesting part of the data is the newest result of the population growth rate ranks in the third period compared to the first period. After being hit by the tsunami, some affected areas significantly tended to experience rapid population growth during the recovery phase. For instance, Meuraxa was in the very last rank in the first period as it was the most affected area by the tsunami, but then it had the highest population growth rate in the second and last period. Kuta Raja was ranked eighth out of nine in the first period. It then became in the second rank for the next period.

4. Discussion

According to the obtained result of the population density (*PD*) and population growth rate (*PGR*) of Banda Aceh from 2004 to 2020, the city had experienced several major declines and inclines in population growth as the impact of the tsunami that occurred on 26 December 2004. The obtained *PD* values in 2004 and 2005 show that most sub-districts in Banda Aceh experienced a major decline in *PD* as the devastating effect of the tsunami disaster, especially in the western coastal area of the city.

There are at least four sub-districts with the most major *PD* decline, including Jaya Baru, Meuraxa, Kuta Raja, and Kuta Alam. This condition is estimated to be happening due to the excess death toll caused by the tsunami disaster and also as the indirect effect of the residents deciding to move to areas with safer shelters during the early post-disaster stage.

However, from 2005 to 2010, the devastated sub-districts were rapidly recovering and experiencing significant inclines of population growth. These areas were gaining back the residents' interest to reside despite their locations in the tsunami inundation region. The increases in *PD* in most sub-districts were still ongoing in 2010 and 2020, although the *PGR* appears to be decreasing slightly. However, in 2020, some sub-districts had already reached higher *PD* values compared to 2004, and some others had somewhat different values of *PD*. This trend indicates that Banda Aceh is still growing and getting more populated each year. It is in line with the recent *PGR* of Banda Aceh that was predominantly positive. The uncontrolled population growth is potentially affecting many fundamental aspects of the city. Therefore, the current demographic conditions of Banda Aceh have to be taken into account seriously by the local authorities.

Various studies reveal the factors of population growth in the former tsunami inundation zone of Banda Aceh. Syamsidik et al. found that tsunami hazard is no longer a major consideration in the options of the place of residence [9]. The motivation of residents to live in the hazard zone is more influenced by the price and rent of land, distance from the workplace, and social relations factors. Research by Siregar et al. focused on tsunami survivors, where people who own land in coastal areas have a greater tendency to resettle in the area than those who do not own land [10]. In addition, the hazard zone is also in demand by migrants who cannot afford to live in urban inland due to increasingly expensive property prices [11].

Demographic changes have a persistent relationship with the dynamics of urbanization. After the 2004 tsunami, the new spatial plan initially designed a strict ban on development in coastal areas heavily affected by the disaster. However, the policy had drawn controversy in the community. As a result, the construction of settlements and urban facilities was allowed to return there with certain restrictions. Urban expansion took place rapidly during the reconstruction period, in line with the population explosion [12,13]. Figure 6 shows the development of human civilization in the form of urban land use in high-risk environments. The existence of activity centers induces urban (re)development in the surrounding area (neighborhood effect). As a result of the long-term recovery, the built-up land distribution in 2020 is increasingly similar to before the disaster.

Figure 6. Changes in Banda Aceh's urban physical condition from Google Earth imagery.

Based on the existing urban expansion trend, the coastal area of Banda Aceh is predicted to grow in the next few years [14]. Without proper policy intervention, this scenario will also be followed by uncontrolled population growth. The consequence of continued population growth in the hazard zone is an increase in social vulnerability. This situation is contradictory to one of the objectives of the "build back better" strategy, namely to prevent future vulnerabilities from the same hazard [15].

One of the risk reduction measures is implemented through urbanization control. To date, Banda Aceh's spatial planning has accommodated various aspects of tsunami mitigation, such as spatial assessment of disaster susceptibility, construction of evacuation routes and shelters, and strict development provisions in tsunami-prone areas. Unfortunately, the implementation of regional regulations related to disaster management has not run optimally [13,16]. Coastal wetlands or green open spaces, as tsunami natural barriers, must be protected from all anthropogenic activities that can degrade them. Active participation of local communities and strong coordination among stakeholders determine the success of disaster risk reduction actions.

Community capacity assessment is no less crucial according to the risk mitigation framework. Oktari et al. evaluated the level of preparedness for coastal hazards in Banda Aceh [16]. Ironically, they found that the community's preparedness index value was low, especially on the early warning and resource mobilization parameters. Another major concern is the existence of migrants who lack local hazard knowledge or disaster literacy. The community's ability to deal with the tsunami hazard should be improved in several ways, for example, by promoting disaster socialization with formal or informal methods and conducting regular tsunami drills.

Referring to changes in population size in Banda Raya, Lueng Bata, and Ulee Kareng, population growth in the safe zone is considered to be increasing rapidly. This condition is one of the effects of development priorities following the establishment of new activity centers in urban planning, especially the new main center in Lueng Bata. There are far more trade, office, and housing activities than during the pre-tsunami period. Some areas in Aceh Besar adjacent to Banda Aceh have even transformed into an urban landscape. The urban growth process in Banda Aceh horizontally has reached a saturation level so that suburban areas are predicted to experience rapid regional transformation [17]. Due to the threat of urban sprawl in floodplain landforms and around inland active faults, it is also necessary to develop appropriate planning and strategies to minimize the consequences of multi-hazards.

5. Conclusion

This study discusses population dynamics in Banda Aceh related to past disasters using sub-district level data in four non-consecutive years (2004, 2005, 2010, and 2020). Of the six sub-districts affected by the tsunami, two sub-districts have experienced population recovery regarding the value of *PD*, namely Jaya Baru and Syiah Kuala. The population rank in the safer sub-district moved up significantly in line with development priorities to the east and south of the city. Meanwhile, the population in Meuraxa and Kuta Raja, the two areas most affected by the tsunami, has not fully recovered. Interestingly, the *PGR* values in these two sub-districts have been in the highest ranks in recent periods.

High-risk areas are feared to be hotspots for population growth and urbanization. The positive benefits of long-term reconstruction have overshadowed the shock events of the past. This research calls for serious attention from policy-makers to avoid the re-emergence of vulnerability to similar hazard. Further disaster risk reduction strategies need to be promoted and consolidated in development planning to achieve sustainability.

Understanding the linkage between population dynamics and disaster risk profiles contributes prominently to resilience in urban environments. A more comprehensive analysis should be carried out to overcome the limitations of this study. Demographic components, such as fertility, mortality, and net migration, need to be considered in analyzing population changes, as well as being the basis for predicting future social vulnerability. In addition, the inclusion of population dynamics in risk reduction strategies will be more useful if it considers the population compositions or particular characteristics that are more sensitive in capturing marginal groups. This result can be complemented with analyzes of population changes in other regions to draw valuable lessons from the past disaster.

References

[1] BMKG 2019 *Katalog Tsunami Indonesia Tahun 416-2018* (Jakarta: BMKG)
[2] Mutaqin B W, Amri I, and Aditya B 2020 Pola kejadian tsunami dan perkembangan manajemen bencana di Indonesia setelah Tsunami Samudra Hindia Tahun 2004: Sebuah tinjauan *Jurnal Lingkungan dan Bencana Geologi* **11** 73–86
[3] Titov V V 2021 Hard lessons of the 2018 Indonesian tsunamis *Pure Appl. Geophys.* **178** 1121–1133
[4] Tenerelli P, Gallego J F, and Erhlich D 2015 Population density modelling in support of disaster risk assessment *Int. J. Disaster Risk Reduct.* **13** 334–341
[5] Neumann B, Vafeidis A T, Zimmermann J, and Nicholls R J 2015 Future coastal population growth and exposure to sea-level rise and coastal flooding – A global assessment *PLoS ONE* **10** e0118571
[6] Rudiarto I, Handayani W, and Setyono J S 2018 A regional perspective on urbanization and climate-related disasters in the northern coastal region of Central Java, Indonesia *Land* **7** 34
[7] JICA 2005 *The Study on the Urgent Rehabilitation and Reconstruction Support Program for Aceh Province and Affected Areas in North Sumatra (Urgent Rehabilitation and Reconstruction Plan for Banda Aceh City)* (Banda Aceh and Tokyo: JICA)
[8] Brata A G, de Groot H L F, and Rietveld P 2014 The impact of the Indian Ocean Tsunami and the Nias Earthquake on the spatial distribution of population in Northern Sumatra *Bull. Indones. Econ. Stud.* **50** 101–121
[9] Syamsidik, Oktari R S, Munadi K, Arief S, and Fajri I Z 2017 Changes in coastal land use and the reasons for selecting places to live in Banda Aceh 10 years after the 2004 Indian Ocean Tsunami *Nat. Hazards* **88** 1503–1521
[10] Siregar M I, Dawood T C, Fakhruddin, Rizki C Z, Fitriyani, and Reza M 2019 Why tsunami

survivors remain living in the same location? Evidences from Lampulo Village, Banda Aceh *IOP Conf. Ser. Earth Environ. Sci.* **273** 012060

[11] McCaughey J W, Daly P, Mundir I, Mahdi S, and Patt A 2018 Socio-economic consequences of post-disaster reconstruction in hazard-exposed areas *Nat. Sustain.* **1** 38–43

[12] Amri I, Prabaswara I W, and Giyarsih S R 2020 Spatio-temporal analysis of post-disaster built-up expansion in Banda Aceh City, Indonesia *2020 6th International Conference on Science and Technology (ICST)* 01–06

[13] Amri I and Giyarsih S R 2022 Monitoring urban physical growth in tsunami-affected areas: a case study of Banda Aceh City, Indonesia *GeoJournal* **87** 1929–1944

[14] Burhan I M, Achmad A, Rizkiya P, and Zainuddin 2020 Forecasting the land-use change of urban coastal area in Banda Aceh and its impact on urban sustainability using Landusesim Cellular Automata simulation model *Aceh International Journal of Science and Technology* **9** 120–131

[15] Dube E 2019 The build-back-better concept as a disaster risk reduction strategy for positive reconstruction and sustainable development in Zimbabwe: A literature study *Int. J. Disaster Risk Reduct.* **43** 101401

[16] Oktari R S, Syamsidik, Idroes R, Sofyan H, and Munadi K 2020 City resilience towards coastal hazards: An integrated bottom-up and top-down assessment *Water* **12** 2823

[17] Amri I and Giyarsih S R 2020 Quantifying urban physical growth types in Banda Aceh City after the 2004 Indian Ocean Tsunami *E3S Web Conf.* **200** 07003

Acknowledgments

The authors would like to thank Badan Pusat Statistik Kota Banda Aceh for their support in providing population data. This research was funded by the Master Program in Disaster Management, Graduate School of Universitas Gadjah Mada.

IOP Conf. Series: Earth and Environmental Science 1173 (2023) 012053 doi:10.1088/1755-1315/1173/1/012053

Upstream Brantas Watershed management strategies for flood mitigation (a review: Batu District)

D A Handayani[1], A Kurniadi[1] and F Bahar[1]

[1] Disaster Management, Universitas Pertahanan Republik Indonesia

E-mail: diah.ayuretnani@gmail.com

Abstract. The floods that occurred in Batu in November 2021 were the result of improper management of the upstream area of the Brantas river. This paper reviewing the parties involved in s watershed management strategies with floods. We start by reviewing the watershed management by the stakeholder in upperment areas Brantas. Then we discuss the mitigation strategies' role in invoicing the negative impact of floods. This research uses qualitative descriptive methods through secondary data collection and literature studies on previous research on management strategies for the upstream area of the Brantas River. This result of the analysis showed that the watershed manages by several leading sectors. The good carrying capacity of the upstream area of the Brantas river is a collaboration of management by the stakeholders. That ability to generate the desired output of basic resources to reduce the risk of several flooding disasters, while maintaining the desired environmental quality. Efforts to develop the regional management can be pursued through community empowerment activities which in practice are carried out through various forms of cooperation/partnerships including community groups. That ability to generate the desired output of basic resources to reduce disaster risk of more severe flooding disasters, while maintaining the environmental quality.

1. Introduction

Flash flood forcer three sub-district and nine villages in Batu City on Thursday, November 4th 2021, at 14.00 WIB. Most of the affected areas are settlements on the border of the Brantas River [1]. Floods and landslides often forced Batu City are indicated as a result of the conversion of protected forest land on the slopes of the upper reaches of the Brantas River into annuals / swallowing fields on the slopes of Mount Arjuno and Mount Anjasmoro. A forest fire in 2019 occurred on Mount Arjuno, resulting in trees in a forest area of 300 ha drying and easily falling and drifting in the dry season that they accumulate in the river valley.

Floods, erosion, landslides in the rainy season and prolonged droughts in the dry season, are closely related to the mishandling of watershed land management. The upper reaches of the Brantas River have weathering soil properties in the form of volcanic deposits on the cliffs or left and right of the river are easily eroded if exposed to sufficiently fast water currents. Public awareness is still low about controlling forest and land management in a sustainable manner, and forest and critical land restoration programs have not been implemented optimally. floods caused by natural factors and human activities are related to the utilization of natural resources, which leads to the degradation and destruction of watersheds. The intensity of rainfall that is very high, accompanied by declining watershed conditions, changes in land use by humans, poor drainage, and government social institutions and communities that are not yet large are the main factors in flooding [2].

Critical and very critical land in Batu Area, the upstream Brantas watershed shows a very severe figure of 51.76 percent. In other words, more than half of Brantas upstream have been damaged. The leading in cause of disasters is environmental damage, especially in the upper reaches of the river area as a water catchment area [3]. The implementation of disaster management in situations where there is a potential for disaster due to land destruction and loss of infiltration a done through mitigation efforts. Mitigation is a series of effort measures to reduce disaster risk, both physical development and awareness and improvement of disaster-facing capabilities [4]. Flood mitigation activities can be carried out through the management of the upstream area of the watershed by paying attention to the implementation of spatial planning in the development and implementation of piercing to increase community capacity.

Water resource management is carried out in a comprehensive, integrated, and environmentally sound manner is a complex problem and involves all parties, users, and managers, so it is inevitable that joint efforts are needed to start using the one river basin, one plan and one integrated management [5]. From the explanation above, this paper discusses the strategy of watershed management areas using Batu flood study. We start by reviewing the watershed management by the stakeholder and the community in upperment areas Brant. Then we discuss mitigation strategies' role in invoicing the negative impact of floods in Batu

2. Research method

This research uses qualitative descriptive methods through secondary data collection, documents, interview and literature studies on previous research on management strategies for the upstream area of the Brantas River in Batu District. The subject of this study is the Community and Stakeholders related to the management of the upstream region of the Brantas River. Data analysis is carried out by systematically searching and compiling the data obtained organizing data into categories, describing them into units, synthesizing, collecting into patterns, and choosing which ones are important. It will be studied and make conclusions so that you and others can understand them [6]. The data is organized and describes watershed management strategies and threats in the upstream area to improve flood disaster mitigation efforts. The literature study illustrates how the management of the upstream region of the Brantas watershed can be an effort to mitigate flood disasters in Batu District.

3. Result and discussion

3.1. *Upstream Brantas Watershed management strategy*

The Brantas River area is the second largest WS in Java Island, located in East Java Province at 110°30' E to 112°55' E and 7°01' E to 8°15' E. Brantas River km2 has a length of ± 320 km. It has a river area of ± 14,103 which covers ± 25% of the area of East Java Province or ± 9% of the area of Java Island. Upstream watershed management aims to provide policy direction related to water use, natural resource use, land use and spatial planning policies [3]. Watershed ecosystems are usually divided into upstream, middle, and downstream areas. Biogeophysically, the upstream area is a conservation area, has a higher drainage density, with a slope greater than 15%, not a flooded area, the regulation of water use is determined by drainage patterns, and the type of vegetation is generally forest stands. The upstream watershed ecosystem is an important part because it has a function of protecting all aspects of the watershed, namely in terms of throleon of water management, and local groups of residents [7]. River basin management refers to operational policies as an essential reference in carrying out natural resource management activity. Activeional policy directions in water resource management for WS Brantas include the fields of 1) Conservation, 2) Utilization of water resources, 3) Control of water damage, 4) Water resource information systems, and 5) Empowerment and community participation. The upstream management of the Brantas watershed is the responsibility of all elements of society. Collaboration on the management of upstream Brantas is carried out in partnership with the Regional Government of Batu City and related institutions, non-governmental organizations [3]. The organization the Brantas Project is the large level of sedimen significantion in reservoirs due to erosion that upstream erosion Brantas Source Watershed, which is much larger than expected. At that time there were still no perceived problems associated with fluctuations in water discharge. Erosion prevention efforts have been implemented by several government institutions, including initiatives from the Brantas Project which were then handed over to PJT 1 as the manager. Reforestation efforts, assistance to the community (farmers) to create conservation buildings (terraces, drop structures, gully plugs, dividing channels, terrace reinforcement etc.), and encourage farmers to apply soil and water conservation efforts in their agricultural practices [5].

Stakeholders at the East Java Provincial level related to the hydrological management of the Brantas Source Watershed are the Regional Environmental Impact Control Agency (Bapedalda), the Forestry Service, the Irrigation Public Works Office, the Health Office, the Energy and Mineral Resources Service. Meanwhile, several other agencies and agencies that are also indirectly related to watershed issues are the Agriculture Office, the Livestock Service and the Community Empowerment Agency (Bapemas). The next group is central agencies or institutions operating in areas such as the Brantas River Basin Center (BBWS), Perum Jasa Tirta I (PJT), Perum Perhutani KPH Malang, and the Brantas Watershed Management Center (BP DAS). Another party that plays a role in this field in Batu City is the Environmental Services Program (ESP)-USAID East Java, an international NGO that has activities at the national level [3, 5].

Table 1. Sources of Natural Resources Information Systems at WS Brantas

Sector	Watershed Management	Flood Control	Water Quantity Management and Utilization	Water Quality Management	River Environmental Management	others
members	BBWS Brantas	BBWS Brantas	BBWS Brantas	BBWS Brantas	BBWS Brantas	Department of Industry and Commerce
	BPDAS-HL	Dept. of PSDA	Dept. of PSDA	DLH Prov. East Java	Dept. of PSDA	BPS
	Perhutani	PJT-I	PDAM	PJT-I		Ministryof ESDM
		Satkorlak Prov.	PLN	Ministry of Health		Ministry of PUPR
		Bappeda Prov./Kab	Industry			Ministry of LHK
		BNPB BPBD	PJT-I			Bappeda Prov. East Java
						BMKG and others

Source: Kep.Men. PU 2378/KPTS/M/2020 Strategy of Watershed Managemen Natural Source Brantas River, 2020

IOP Conf. Series: Earth and Environmental Science 1173 (2023) 012053 doi:10.1088/1755-1315/1173/1/012053

The management of the water resource information system at WS Brantas is organized by the Government, Regional Governments and River Basin Managers who are authorized, to provide information on these water resources and are obliged to maintain the accuracy, correctness, and timeliness of data and information [8]. Stakeholder non-govemmental among other Kelompok Tani Hutan, Lembaga Masyarakat Desa Hutan, IPPHTI, Pro-Fauna, Persatuan Pengussaha Hotel dan Restauran, and many others have main role and function in coordination and restoration in uppstream Brantas River areas. The Brantas river basin has great potential for the development of leading sectors, especially in the fields of industry, trade in services, tourism, agriculture and plantations. This potential, if not managed properly, will cause environmental problems towards the use of space in the future, due to the disruption of protected areas and nature conservation areas due to the development of residents and their activities. Area management is one of the efforts to prevent greater losses in the event of a disaster. This condition is aggravated by the condition of the community in managing agricultural land on the slopes of the valley path which is not balanced with soil buffer plants or stand crops [3]. Sumber Brantas Watershed is one out of five sub catchments of the Upper Brantas River, situated in Batu District, East Java, Indonesia and covering an area about 174 km2. Recently, the hydrology of the watershed was considered to be degraded in terms of the quality and quantity of water yield. Fast floods were more often during the rainy seasons, while droughts were more severe during the dry seasons [5].

3.2. Strategy of management critical land in upper river

The Brantas Source Watershed is one of the most critical of the approximately 29 watersheds in East Java. Almost half of this watershed area is included in the category of critical land [8, 9]. The most prominent environmental issues in this region are (a) land use from forests to vegetable crops, (b) decreased water quantity and quality, and (c) land degradation [5]. That way it is necessary to have an area management strategy that is carried out comprehensively by all parties with environmental insight and sustainability. In the upstream area of the river, there has now been a change in the function of the area. Along Tulungrejo to Brantas Source, Bumiaji. The conversion of forest areas causes a decrease in the work of water catchment areas. Many of these areas are agricultural land, one of which is also driven by the conversion of productive land for tourism and other uses such as hotels and housing. As a result, some people began to open up forest areas for agriculture. This shows that there are factors of economic inequality, especially access and management of land, thereby encouraging the expansion of new agricultural spaces [10].

Sumber Brantas Village is a village in upper Batu, their community-based collaboration has been formed in environmental management with Tahura R. Soerjo and Perum Jasa Tirta I. This collaboration program aims to increase public awareness to be involved in preserving the Brantas watershed and the forest that surrounds the Sumber Brantas village, the collaboration and their role activity shown in table 2 Several factors that support the implementation of the collaboration program are: 1) awareness of Brantas' strategic value; 2) an understanding of roles and responsibilities; 3) cultural aspects [6-13]. The strategy carried out based on spatial planning is the existence of the Brantas Source Arboretum, R. Soerjo Forest Park and the Perum Perhutani Protected Forest Area of KPH Malang. The arboretum is a garden of collection of biological resources for science, but also has the functions of regulating water systems, controlling erosion, forming a microclimate, and tourist attractions. The purpose of developing the Brantas Source Arboreum (ASB) determined by Perum Jasa Tirta includes (a) for the preservation of the Brantas River spring (water catchment area), (b) collection of types of highland plants, (c) for research and education facilities, and (d) for recreational facilities containing elements of environmental education [14].

Table 2. The role and function of stakeholders involved in restoring land functions upstream of the Brantas watershed in Sumberbrantas Village

No	Stakeholder	Role	Function
1	Brantas Watershed Management Center (BPDAS-HL)	Facilitator, planning upstream management activities of the Brantas watershed.	implementing RHL and Soil and Water conservation
2	Perum Jasa Tirta (PJT) 1	Development of upstream watershed management and the sustainability of the implementation of water resources management.	implementing RHL and Soil and Water conservation
3	Taman Hutan Raya R. Soerjo	Maintaining water catchment areas for river continuity	Protection of buffer systems and maintaining biodiversity
4	The Community	Mediator between agencies and the community	Support the progrqm of stakeholder work, maintain the forest area around the village, and maintain the Brantas river area flowing through the village

Source: Cahyani, 2019

The Protected Forest Area (KHL) managed by Perum Perhutani Unit II East Java covers 11 (eleven KPH). The area of HL area in the KPH Malang area is 69,372 ha or 59% of the total forest area in KPH Malang. Administratively, the area of protected forest areas in Malang Regency is 61,343.90 ha and 8,028.1 ha in Batu City. Protected forests in the Malang KPH area which is the headwaters of the Brantas watershed as a protection area as well as a source of water and a source of livelihood for the surrounding community. Protected forests in the Malang KPH area which are the headwaters of the Brantas watershed and as a source of water need to be preserved so as not to pollute the water surface of the Brantas River which is a good source of raw water for the community containing elements of environmental education [13].

The Tahura Raden Soerjo area has many springs. Some of the springs found in the tahura area: the Brantas river spring located on Mount Anjosmoro in the village of Sumber Brantas which belongs to the Brantas watershed area. The forest area of Tahura Raden Soerjo covers the area of the East. This forest area combines a protected forest area of 22,908.3 ha and an Arjuno Lalijiwo nature reserve covering an area of 4,960 ha, with an area located in the stone city covering an area of 5,342.50 ha. The development and management of forests in the Tahura R. Soerjo area must include 3 (three) aspects, namely: 1) aspects of community empowerment, 2) aspects of sustainability of the function and benefits of forest resources 3) aspects of management professionalism [13]. Optimization of natural resources in various activities or other aspects is indispensable for increasing the resilience of the country. The use of an environmentally friendly economy that can be managed properly by the community is one of the steps to optimize Indonesia's natural resources. In general consesus on scientific literature that forest ecosystems play a significant role in efforts to prevent or at least reduce the risk of natural disasters. Biodiversity including plants and wildlife as well as being protected is a state wealth that must be maintained and preserved. The eruption of Mount Merapi still continues to occur periodically, so in its management requires a strategy to adjust to the conditions and the level of damage [14].

3.3. Flood disaster risk reduction through the management of the upper Brantas River area

Flooding is a natural phenomenon that is an accumulation of several factors including river conditions, upstream conditions, conditions of cultivated areas and sea tides. The potential for the threat of flood disasters is likely due to the state of damaged river bodies, damage to catchment areas, violations of regional spatial planning, violations of laws increased, development planning is less integrated, and community discipline is relatively low [15]. Land management in the upstream area of the river is the main factor causing flooding. The upstream area has a function as a buffer area and water catchment. The increasing case of forest land use to agricultural land is the cause of damage to land function and ecological function such as the loss of catchment areas and the increase in sedimentation in riverbeds [16, 17]. In general, the area of Batu City that is hit by flash floods is flat with undulating hills. The flattening area is an inundation ar,ea while the sloping to slightly steep hills are the flow track of flash flood flows. The source of the avalanche and the logs coming from the upper reaches is an area with a morphology of steep to very steep hills. Avalanches occur on slopes with a slope of more than 60° [1]. BPBD Batu and BPPTKG claim that the main cause of flash floods in Batu City in 2021 is the result of the remaining forest fires on Mount Arjuno in 2019 which also impacted protected forest areas and conservation forests in the upper reaches of the Brantas river. As an effort to restore the area, greening and reforestation activities have been carried out around springs and upstream of the watershed and in forest areas. This activity was realized because of good coordination between stakeholders and the community [6].

In table 1 we know that in flood control by BBWS Brantas, Department of PSDA, Company of Jasa Tirta I, BPBD, Satkorlak Province. Flood risk management strategy through construction activities, monitoring and maintenance of embankments, monitoring and controlling sediments and maintaining watersheds. The government implemented the strategy in collaboration with local communities. The community has the same opportunity to play a role in the process of planning, implementing and supervising water resource management. The role of the community in can be realized through a community-based Early warning system to reduce risks due to flooding [18,19]. Floods that hit the Batu City area were controlled by water overflows due to increased water volume during high rainfall and the presence of deposition upstream of material in the form of wood, boulder rocks and avalanche material. Overflow occurs due to the swater supply the waterways and rapids in the high upper reaches and flows and empties into the valleys. In addition, several points of soil movement were also found upstream which can be the cause of water overflow [1]. Watershed Brantas has experienced regional criticism since 1970, and has not been resolved until now. Almost half of the Brantas river watershed is included in critical land [16]. The most prominent environmental issues in this region are a) land use change from forests to vegetable crops, b) decreased water quantity and quality, and c) land degradation [5]. The Brantas River Basin has a relatively high level of erosion and sedimentation hazards, especially in the Central Watershed, Ringin Bandulan, Kondang Merak and Brantas watersheds in the upper and middle parts, caused by many slopes collapses, especially in areas with steep slopes [3]. The protected forest area under the management of Tahura Raden Soerjo and Perum Perhutani and the production forest under Perum Perhutani which is located in the Brantas Source Watershed is an infiltration area of 111 main springs in the Batu City area. Deforestation that has occurred in protected forest areas and in prodion forest areas has resulted in changes in forest cover, which has an impact on the malfunctioning of catchment [5]. The above conditions foster awareness from all parties to carry out rehabilitation of damaged forests and lands (RHL) in order to repair and restore the function and productivity of these natural resources. The activity is directed as a planned and integrated national-scale movement, involving various related parties, both governments, private and the wider community.

There is a critical land of 38.78% and a very critical 6.25% of the total area of Batu City. This shows that land use is unsustainable and there is still a lack of efforts to rehabilitate forests and critical lands in the area [9]. Poorly managed forests can pose hazards and various risks to forest-dependent life. Some natural disasters that can arise due to disturbed forest ecosystems or damaged is the risk of cliff material collapse, floods and landslides, and debris flows [14]. The main cause of disasters is environmental damage, especially in the upper reaches of the river basin (WS) as a catchment area. The above conditions foster awareness from all parties to carry out rehabilitation of damaged forests and lands (RHL) in order to repair and restore the function and productivity of these natural resources. Perum Perhutani KPH Malang with LMDH (Lembaga Masyarakat Desa Hutan), and also Tahura

Raden Soerjo with MPA (Masyarakat Peduli Api) actively socializing about sustainable forest cultivation and management to reduce the risk of floods and landslides.

The activity is directed as a planned and integrated national-scale movement, involving various related parties, both governments, private and the wider community. Mitigation is one of the efforts to reduce disaster risk, which is carried out by making buildings/structuring human resource capabilities that aim to avoid and or reduce the risk of disasters [18]. Disaster risk reduction is a cost-effective investment in reducing the impact and damage that may be caused in the future. The management of Brantas River water resources information resources is organized by the Government, Local Governments and River Basin Managers in accordance with their authority. The management of information sources is useful in increasing the role of stakeholders, so as to achieve comprehensive flood control. The government is also trying to overcome the flood problem as a whole through structural and non-structural efforts. Structural efforts were carried out, among others, through river normalization, canal flooding and interconnections to lower water level elevations. The construction of reservoirs and flood retention reservoirs is carried out to minimize flood discharge. In an effort to prevent the overflow of floods at a certain height, it is carried out by building and maintaining embankments [19]. The operation of reservoirs upstream and the regulation of water levels in dams and weir along the Brantas River by the Public Service Company (PJT I) through the operation of reservoirs, monitoring of dam water gates and sluices, exploring riverbed degradation and repairing embankments [3, 19].

With the collaboration between the government, NGO, Volunteer and the community have a purpose to restore the function of uppstream area of Brantas watershed back to normal. The activity carried out such as technical and vegetative or structural and non-strucural strategies. Non-structural strategies pursued by the government in handling floods as a whole are: 1) improvement of early warning systems; 2) floodplain management through greening, spatial planning, reforestation and erosion control; 3) Implementation of education through public information, counseling, law enforcement and determination of environmental policies. The development of an early warning system is one of the efforts to reduce disaster risk in a non-structural manner centered on communities living in flood-prone areas from downstream to upstream. Structural and nonstructural strategies are carried out to reduce the amount of damage caused by floods. The disaster risk reduction strategy by paying attention to the sustainable environment that has been implemented in the upstream of the rocky area is the management of forest areas that function as catchment areas and prevent erosion. Forests also have an important role to play in mitigation efforts to reduce disaster risk. The destruction of forests can increase people's vulnerability to hydrometrological disasters, so that they can carry out disaster management efforts by reducing risks [20]. In fact, naturally an ecosystem including forest ecosystem has its resilience or the ability to recover (resilience) itself after experiencing disturbances. Ecosystems need human assistance or intervention in order to preserve [21]. The community can also build cooperation with business institutions, non-governmental organizations, or other institutions outside the community to synergize with each other in disaster risk reduction [22]. The existence of synergy between the community, the Government and related stakeholders/institutions can improve the management activities of upstream watersheds.

References

[1] Geological Agency *flash flood inspection report in Batu City, East Java Province center for volcanology and geological disaster mitigation, ministry of energy and mineral resources 2020* Retrieved from: https://vsi.esdm.go.id/index.php/gerakan-tanah/kejadian-gerakan-tanah/3840-laporan-pemeriksaan-banjir-bandag-di-kota-batu-provinsi-jawa-timur

[2] Wiwin B, Gravitiani E dan Mujiyo 2017 Flood mitigation efforts in the samin sub-watershed through the development of disaster resilient communities *J. Teknologi Lingkungan* **18** 241-250

[3] Minister of Public Works of the Republic of Indonesia 2010 *Decree of the minister of public works number 268 of 2010 concerning the pattern of water resources management of the Brantas River area* (Jakarta: PUPR)
[4] President of The Republic Of Indonesia 2007 *Undang-Undang Nomor 24 Tahun 2007 of Penanggulangan Bencana* (Jakarta: Government of Republic Indonesia)
[5] Widianto D S, Sudarto, and Lestariningsih 2010 *Implementation of rapid hydrological assessment (RHA)* (Bogor: World Agroforestry)
[6] Cahyani I G 2019 *Collaboration of Stakeholders on Returning the Function of Brantas Watershed in Effort to Reduce the Risk of Natural Disaster (Study in Sumberbrantas Village, Bumiaji District, Batu City)* (Malang: Brawijaya University)
[7] C. Asdak *2006* The carrying capacity of water resources as a consideration for spatial planning *J. Tek. Ling P3TL-BPPT* **7** 16-25
[8] Minister of Public Works of the Republic of Indonesia 2020 *Decree of the Minister of Public Works Number 237 of 2020 concerning the Pattern of Water Resources Management of the Brantas River Area* (Jakarta: PUPR)
[9] Hirijanto, Suprapto B and Hargono E 2014 Critical Land Analysis for Water Resources Conservation in Batu Malang City Area *Jurnal Rekayasa Sipil* **2** 68-81
[10] Anonymous 2021 *Floods in Batu City, Climate Disasters and Spatial Planning.* Retrieved from: https://www.walhi.or.id/banjir-di-kota-batu-petaka-bencana-iklim-dan-tata-ruang
[11] Kurniawan P V, Sudjoko A and Antoni 2018 Collaboration to care for Brantas (an environmental communication study in Sumber Brantas Village, Batu city, East Java) *CHANNEL* **6** 69-82
[12] Baskara M, Mundandar A and Samingan T 1998 *Landscape planning for Sumber Brantas arboretum as outdoor recreation area* (West Jakarta: Buletin Taman dan Lanskap Indonesia)
[13] Sulistyaningsih T S and Sunarto 2013 *Model Collaboration Partnership for Watershed Green Open Space Management for National Strategic River Sustainability (study in the Brantas watershed of East Java* Retrieved from: https://pemerintahan.umm.ac.id/files/file/tri%20sulis%20STRANAS.pdf
[14] Moos C, Bebi P, Schawarz M, Stoffel M, Sudmeier-Rieux K and Dorren L 2018 Ecosystem-based disaster risk reduction in mountains *Earth-Science Reviews* **177** 497–513
[15] Rohmi F 2022 *Psychological Preparedness of Communities in Flood-Prone Areas of Sitiarjo Sumbermanjing Wetan Village* Retrieved from: https://ejournal.umm.ac.id/index.php/keperawatan/article/view/3932/4409 **7** 88-93
[16] Waskito, N T 2013 Organizational Culture Model in Brantas Watershed Management *Humanity* **8** 75-82
[17] Utomo P A 2009 *Estimation of Distribution of Flash Flood Prone Areas Sub-Watersheds Of Brantas Source Watersheds Of Batu City Application: Genriver Model & Geographic Information System* Retrieved from: http://repository.ub.ac.id/id/eprint/128112/
[18] Yuliani F and Aulia R 2018 Peatland Restoration Method in The Context of Peatland Fire Disaster Mitigation and Community Empowerment *Sosio Informa.* **4** 448-460
[19] Anonymous 2006 *LPM Presentation* (Yogyakarta: Perhutani BPKH XI Jawa Madura).
[20] Khairullah S I and Eldina F 2016 *Community Perception of the Function of Mangrove Forests in Disaster Risk Reduction Efforts* (*JIKA*) p 110-119
[21] Haryadi, Sunarto and Sugiyarto 2020 *Strategi Pengelolaan Kawasan Hutan Taman Nasional Gunung Merapi (TNGM) Provinsi DIY* (Surakarta: National Seminar on Biology and Science Education) p 343-348
[22] Habibullah 2013 Community-Based Disaster Management Policy: Disaster Preparedness Villages and Disaster Resilient Villages *Informasi* **18** 133-150

IOP Conf. Series: Earth and Environmental Science 1173 (2023) 012054 doi:10.1088/1755-1315/1173/1/012054

Disaster management and its importance: a review paper of Dakshina Kannada and Kodagu districts, Karnataka, India

M C Jayaprakash[1],D Venkat Reddy[2] and B Istijono[3]

[1]Department of Civil Engineering, Mangalore Institute of Technology and Engineering, Moodabidri, Affiliated to VTU& Approved by AICTE, India.
[2]Retd -HOD of Civil Eng.-National Institute of Technology -Karnataka -NITK, Adjunct Professor, Dr. Babasaheb Ambedkar Technological University, LONERE-402103.
[3]Professor of Water Resources Engineering and Management, University of Andalas, Padang, Indonesia.

E-mail: jayaprakash@mite.ac.in

Abstract. An extreme occurrence within the earth's system that causes human fatalities, injuries, and property destruction is referred to as a "disaster" from a social perspective. Disaster management is one of the most critical topics around the world. Many countries are getting affected because of its high magnitude impact on economy and loss of life. The Karnataka government must always be on guard and ready to meet problems due to the frequent natural disasters like landslides, floods, droughts, and human-induced dangers. In order to lessen the effect of disasters on the community, the Government of Karnataka has built enhanced institutional structures for disaster risk reduction. The main aim of the review paper of this, is to create awareness about the situations occurring due to any form of disasters and to develop preparedness in younger generation to face disaster situations.

1. Introduction

According to science, at least one or more geological phenomena or earth system processes are actively, continuously, and cyclically involved in every natural disaster that has ever occurred on the earth surface of the world. These natural occurrences or processes are referred to as "Geohazards" because they pose a risk to people, other living things, their property, and the environment (Rashmi Jain et al, 2019). Even though disaster is a common term yet people are not aware how to be prepared and safe. On the other hand, even though the NGOs and government are providing facilities. As a citizen of India, it is our duty to take initiative for helping our society (Raj Nirmal et al, 2020).

The ideas of Disaster Risk Management (DRM) must be incorporated into every sector of growth. This would guarantee that all construction projects - roads, airports, canals, hospitals, schools, bridges, and railways—are constructed in accordance with the necessary standards and enhance the resilience of the communities they aspire to serve. This underlines the requirement to ensure that all infrastructure development complies with the highest requirements for disaster safety. Such a tactic is wise and will be profitable in the long run. It is essential that disaster risk factors be taken into account in all public investments. Such prospects are represented in India by the "housing for everyone" programme and "smart cities" efforts. The Karnataka State Disaster Management Plan (KSDMP) has been created and updated annually with the assistance of the State Executive

Committee in order to streamline and direct the state's disaster management actions (SEC). The KSDMP has been supplying pertinent information to the parties involved in disaster response (KSDMP 2020-21). Along with KSDMP, each district's DDMP preparation is a "dynamic document" that it will continually be revised in accordance with district best practices and knowledge base (DDMP 2022-23).

1.1 Paradigm shift

A paradigm change in disaster management in India has been institutionalised through the DM Act of 2005 and the NPDM of 2009. This represents a transition from a focus on emergency reaction to one on proactive prevention, mitigation, and preparedness. While natural disasters cannot be prevented, the NPDM emphasises that they may be averted from becoming catastrophes with the right mitigation and risk reduction methods. Hazards combine with social, physical, economic, and environmental vulnerabilities to create disaster risk. The NPDM recommends a multifaceted strategy including the following for disaster risk reduction and mitigation:

- Implementing risk mitigation strategies across all development initiatives.
- Initiating mitigation initiatives via cooperative efforts of the Central & State Governments in designated high precedence regions.
- Promoting and helping mitigation efforts at the State level.
- Examining natives' understanding of disasters and coping techniques.
- Giving heritage structure protection the weight, it deserves.

1.2 Natural Hazards

The Disaster Information Management Methodology of DesInventar uses a "universally recognised categorization system that divides catastrophes occurring from natural hazards into five primary groups and is utilised internationally for the Sendai targets monitoring"

1. **Geophysical:** Geological processes or phenomena that may result in property destruction, loss of livelihoods or other health effects, disturbance of social or economic order, or environmental harm. Some of these processes have significant contributions from hydro-meteorological aspects.
2. **Hydrological:** Events brought on by variations in the regular water cycle and/or overflowing of water bodies as a result of favorable wind conditions.
3. **Meteorological:** Short-lived/small to meso-scale atmospheric events that occur (in the spectrum from minutes to days).
4. **Climatological:** long-lived meso- to macro-scale processes that result in certain events (in the spectrum from intra-seasonal to multi-decadal climatic variability).
5. **Biological:** Exposure to pathogenic microorganisms, bioactive, and toxic substances that may result in loss of life, illness, injury, or other health effects, property damage, loss of livelihoods and services, social and economic disruption, or environmental damage is an example of an organic process or phenomenon that is spread by biological vectors.

2. Kodagu at a glance

One among the smallest district in Karnataka is Kodagu which is also known as Coorg. It is the most beautiful hill station in south-west Karnataka region. This place is also called the coffee cup of India. The main source of income for this place is agriculture and tourism. The town capital of Kodagu district is Madikeri with an estimated population of 30000.

Torrential rains and floods accompanied by landslides in Kodagu district have brought agriculture to a standstill. The district's economy is based on agriculture, which produces rice, pepper, cardamom, and coffee. In 2017–18, 45,000 growers produced 1,16 lakh tonnes of Arabica and Robusta coffee. Kodagu contributes significantly to the production of fragrant soft drinks and provides 40% of the country's coffee. Farmers in Kodagu anticipated a big crop in 2018–19, but excessive rain and flooding damaged over 60% of the district's berries, and topsoil washed away in the majority of places.

and no longer be able to discern the farmed terrain. More than 150 landslide-prone locations have been identified by the Geological Survey of India in Kodagu alone.

Due to its proximity to the shore, Kodagu Prefecture has a typical tropical climate with mild to moderate humidity. High humidity, frequent downpours, and a chilly summer are some of its characteristics, making it recognised to be fairly pleasant and healthful. A monsoon season that lasts from early June to late September occupies a large portion of the year. Even "certain portions of the district receive a large quantity of rain" in October and November after the monsoon. According to an analysis of data from the previous ten years, the district's westernmost and most heavily wooded Madikeri taluk has the most rainfall (an average of 3302.46 mm), and its easternmost and driest portions (average 2105.22 mm). The taluk in Somvarpet is the region of the district that is farther north and has less forest cover; it is also close to the Plains.

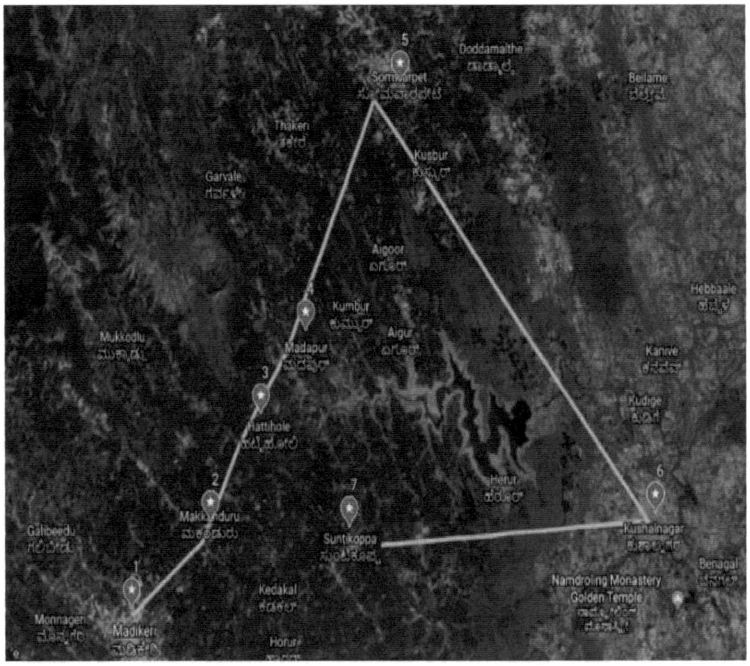

Figure 1. Disaster prone locations in Kodagu District

3. Dakshina Kannada at a glance

The headquarters of Dakshina Kannada are located at Mangaluru. Mangaluru Taluk is traversed by the Netravati River and the Gurupura River, the two principal rivers in the region. The district and Mangaluru in particular are well connected by road, rail, sea, and air. The Konkan Railway and National Highway 66 (Kanyakumari to Mumbai) both travels north to south along the district's coastline. From west to east, NH 75 traverses the district as it travels from Mangaluru to Bangalore.

Landslide occurrence is a frequent and recurring phenomenon in Western Ghats especially during monsoon. A number of landslides occurred in the isolated mounts and hill ranges of Western Ghats during 2018 and 2019 even at present, resulting in loss of lives and extensive damage to roads, dwelling units and other properties even death, uprooting of lot of trees and disrupting traffic movements. The disastrous event created panic amongst the general public and generated serious concern among the administrators, scientists and engineers. District administration carried out rapid assessment of the landslide event as well as the previous events along the three important road corridors of Dakshina Kannada and the adjoining districts, Karnataka, from during January 2019. The various studies taken were to find out type, causative factors, magnitude of the landslides for arriving control and corrective measures and also to identify critical landslides and slopes that warrants detail study and workout long term to control or preventive measures.

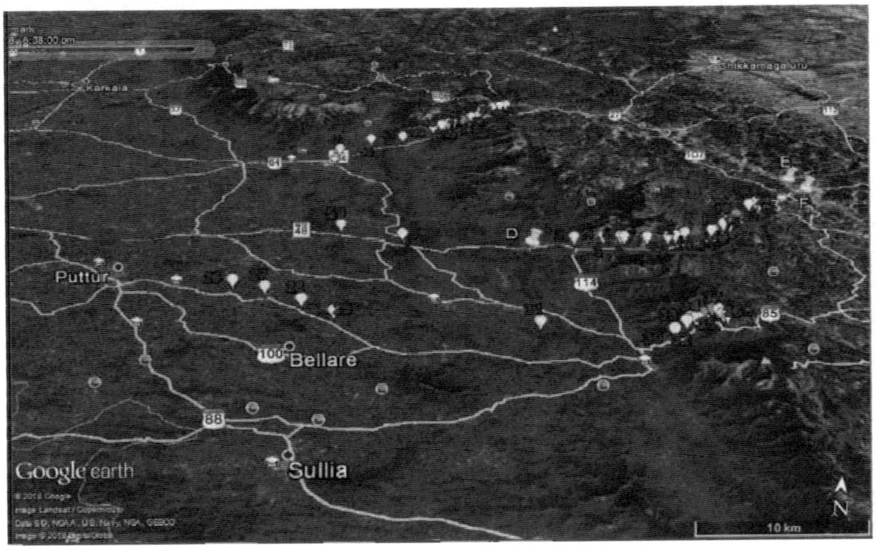

Figure 2. Map of slide/slide prone locations along NH 48 and NH 234 and SH 85

4. Priorities; Targets

India is a member to the Sendai Framework, a voluntary, non-binding 15-year agreement that acknowledges the state's important role in disaster risk reduction but also that, the duty must be shared with other stakeholders, such as local Government, the corporate sectors, and others.

The following four areas of action are prioritised under the Sendai framework:

1. Being aware of catastrophe risk.
2. Increasing the effectiveness of catastrophe risk management.
3. Making investments to reduce catastrophe risk to increase resilience.
4. Improving catastrophe readiness to provide an efficient response and "Build Back Better" in terms of recovery, rehabilitation, and reconstruction.

Figure 3. Sendai Framework for Disaster Risk Reduction (DRR) - 7 Global Targets

5. Conclusion

- Public have lack of knowledge about how to deal with a disaster.
- Children should be taught about coping with a disaster during the primary education itself.
- Disaster preparedness should be added to the curriculum.
- Disaster prone areas need special drainage systems.
- Disaster preparedness should be done as an activity in rural areas.

Despite being a cross-cutting characteristic, the accountability framework now includes a specific thematic action area for social inclusion, along with a list of related thematic sub-areas of action.

The catastrophe serves as a serious wake-up call for the local populace and the government to protect the ecology and stop the overuse of natural resources. The whole Western Ghats have been divided into three separate categories of ecologically sensitive zones, each of which has varied degrees of restrictions on development, according to the 'Gadgill' commission report. By protecting natural resources from overexploitation, these proposals might help the Western Ghats' ecological deterioration. The Western Ghats ecology continued to deteriorate as a result of the government's "negligence to issue" these regulations, which made it easier for similar calamities to happen.

References

[1] Alam E. 2016 Earthquake and tsunami knowledge, risk perception and preparedness in the South-East Bangladesh *Journal of Geography & Natural Disasters* **6(1)** 1-7

[2] Balachandar. S., Reddy Vivek. B., Shankar Neeraj., and Flood 2017 Disaster Management in India–A Case Study on Chennai Floods *Journal of Civil Engineering and Environmental Technology* **4(2)** 166-170

[3] Chaudhuri Sabuj Kumar and Chakraborty Sarthak. 2018 University Libraries and Community Preparedness for Natural Disasters: A Participatory Based Action Plan, Conf. paper

[4] D.N Vinutha, K Shridhara Raje and M. R Janardhana 2014 Landform studies and geomorphological mapping of a part of Coorg district, Karnataka state *International journal of Geology, earth and Environmental science* ISSN: 2277-2081.

[5] DDMP2022-23. District Disaster Management Plan for Dakshin Kannada District. [ONLINE] Available at: https://drive.google.com/file/d/1y06CvjGjp5fj4rU3lVksmJ4X2UNgg_QN/view. (2022-23)

[6] KSDMP2020-21. Karnataka State Disaster Management Plan. [ONLINE] Available at: https://ksdma.karnataka.gov.in/storage/pdf-files/Plans/Volume-I_30-08-2020_compressed.pdf. (2020-21)

[7] Megokedono Vakha, M Nagaraju, K.G Asha Manjari 2016 Land use land cover mapping using Geo-informatics of Kodagu District, Karnataka *International Journal of Geomatics and Geosciences* **7** 1

[8] NDMA 2019 National Disaster Management Authority of India. [ONLINE] Available at: https://ndma.gov.in/en/maharashtra.html. (Accessed 27 November 2019).

[9] Raj Nirmal, Lokande Sheetal and Jain Palak, A Study on Awareness and Challenges in Natural Disaster Preparedness in Karnataka *International Journal of Advances in Engineering and Management (IJAEM)* **2**, Issue 1, pp: 1216-1224 (2020)

[10] Rashma Jain, Sreelakshmy Madhusoodhanan, Arun S Nair, Mohd. Azmal and Niriksha Rani 2019 Disaster Analysis of Kodagu District using Geomatics Technology *International Journal of Engineering Research & Technology (IJERT)* **8** Issue 06

[11] Sakthivel R, M. Janaki and Manikandan N 2021 GIS Based Flood Susceptibility Mapping Using Frequency Ratio Model in Dakshina Kannada, Karnataka *International Journal of Remote Sensing & Geoscience (IJRSG)* **10** Issue 6

Development of post-disaster rehabilitation and reconstruction information system

T A Sundara[1,2], E Usada[3], H Saputra[1], L Arief[4] and M Hamidi[4]

[1]Metamedia University, Padang, Indonesia
[2]Universiti Kebangsaan Malaysia, Bangi, Malaysia
[3]Telkom Institute of Technology, Purwokerto, Indonesia
[4]Andalas University, Padang, Indonesia

Email: tri.sundara@stmikindonesia.ac.id

Abstract. Information is a crucial aspect in disaster management. A reliable information systems play an important role in every stages of disaster activities: pre-disaster, disaster response, and post-disaster. This study aims to develop a post-disaster information system related to rehabilitation and reconstruction activities. Development of the system adopt System Development Life Cycle (SDLC) methodology: requirement, analysis, design, and implementation. This rehabilitation and reconstruction information system was developed for a web platform using various web technologies, such as Apache web service, MySQL database, Yii framework, JQuery, OpenMaps, among others. The information system has been successfully developed according to user requirements. It has been tested and implemented in post-disaster rehabilitation and reconstruction activities in Ministry of General Working and Public Housing.

1. Introduction

Information is a crucial aspect in disaster management. Natural disasters, such as earthquakes, tsunamis, floods, and others will cause serious damage and affect the people affected by the disaster. Therefore, reliable information related to disaster locations, residents affected by disasters, and others is an important aspect in disaster management activities.

Disaster activities consist of:

- Pre-disaster / mitigation
- Disaster response
- Post-disaster/rehabilitation and reconstruction

Each stage of the disaster activity requires different pieces of information in order to be carried out properly. Information systems can be used for disaster management at these stages.

In pre-disaster resilience, information related to the early warning system and disaster mitigation is needed. At the disaster response stage, information on disaster maps, disaster logistics, and reliable and up-to-date disaster human resources is more important information. Meanwhile, in the post-disaster stage, information regarding rehabilitation and reconstruction activities is more valuable information.

This study aims to develop a post-disaster information system related to rehabilitation and reconstruction activities. The system was developed according to the analysis of user needs and implemented in rehabilitation and reconstruction activities by the Ministry of General Working and Public Housing.

2. Disaster response information system

Information systems play an important role in exchanging data related to the management of disaster activities. Sakurai has mapped out the role of information technology for disaster management according to the stage of the disaster.

In the pre-disaster stage, information systems can be used to reduce risks and make preparations for disasters. Risk reduction can be done by real-time risk monitoring using information technology. The disaster phase can also be empowered through information technology, such as simulations using Virtual Reality (VR) or building a knowledge repository about disasters.

In the event of a disaster, response and recovery activities can be empowered through the information system. In the response stage, the information system can be used to provide situation information. Meanwhile, in the recovery phase, information systems can be empowered to manage recovery activities, such as rehabilitation and reconstruction.

Rehabilitation is the improvement and restoration of all aspects of public or community services to an adequate level in post-disaster areas with the main objective of normalizing or running normally all aspects of government and community life in post-disaster areas.

3. Methodology

This paper will discuss the development of information systems for post-disaster rehabilitation and reconstruction. The development of the information system uses the System Development Life Cycle methodology (SDLC). In the methodology, development is divided into stages:

1. Requirements analysis
2. Software design such as architectural design
3. Software development
4. Testing and Deployment

3.1. Requirements analysis

Prior to system development, analysis was carried out to capture user requirements. Users in this system include the Ministry of PUPR, local governments, residents affected by disasters, in addition to other users who need information related to disaster response activities. Based on user needs, essential use cases have been extracted as shown in table 1.

Table 1. Essential functionalities obtained through requirements analysis.

Component	Functionality	Goal
Back-end services	Authorized Mechanism	To enhance the reliability of obtained data
	Disaster-affected Residents Information	To provide data on disaster-affected residents and mapping them into rehabilitation and reconstruction projects
	Rehabilitation and Reconstruction Information	To provide data on post-disaster rehabilitation and reconstruction project activities
Front-end services	Geographic Information System	To display the base map and overlaid disaster-prone zone, risks, and/or facilities

3.2. Architectural design

The core information system of the disaster response information system is built based on an integrated disaster database. These databases include demographic databases, facilities and infrastructure databases, disaster HR databases, disaster logistics databases, disaster location databases, in addition to other databases required or provided by other entities and communicated through the Application Programming Interface (API). Figure 1 shows the architecture of the information system.

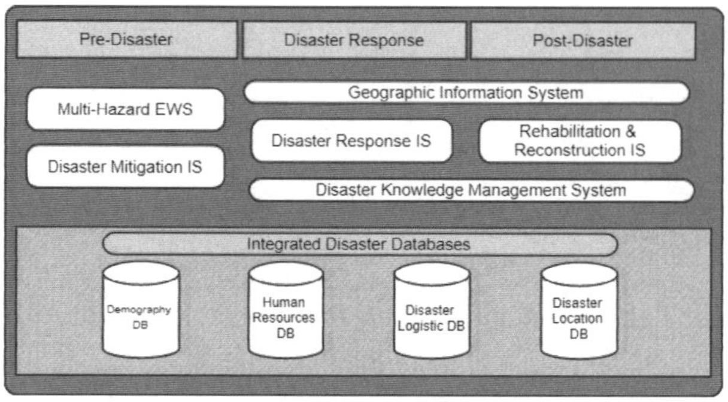

Figure 1. Core information system architecture.

The information system is developed according to the needs of each stage of disaster management. At the mitigation and preparedness stage, a multi-hazard Early Warning System and a disaster response information system are needed. At the emergency response stage, an emergency response information system and a disaster information map are needed. Meanwhile, at the stage of rehabilitation and reconstruction, an information system on rehabilitation and reconstruction is required. The information system for emergency response and rehabilitation and reconstruction will establish a Disaster Knowledge Management System.

3.3. System development

This rehabilitation and reconstruction information system was developed for a web platform. As a web service, Apache is used while MySQL is used as database. For back-end development, the Yii framework is used, while JQuery is mostly used for front-end development. OpenMaps is mainly used to build GIS functionalities. Various other web technologies have also been utilized in the system development environment.

4. Result and discussion

The system is implemented in the development environment according to user requirement. A web-based information system impemented and deployed for Ministry of General Work and Public Housing.

4.1. Implementation

Various features according to user requirements are applied to the information system, including: disaster-affected residents, rehabilitation and reconstruction projects, GIS, and others.

4.1.1. Disaster-affected residents. The essential information in the disaster response information system is about the people affected by the disaster. Disaster-affected residents are people who are victims of earthquake, tsunami, or liquefaction disasters who previously lived in the red zone area as referred to in the applicable zoning regulations. Many of the affected residents lost their homes due to the disaster. In such conditions, the handling can be rehabilitation of residential residents (in situ) or relocation (ex situ). For this reason, detailed information about the resident is needed, including personal data, family, occupation, previous residence, and relocation options. Figure 2 shows an excerpts of disaster-affected residents data in the information system.

Figure 2. Disaster-affecter residents information.

4.1.2. Rehabilitation and reconstruction projects. Post-disaster rehabilitation and reconstruction activities aim to restore people's lives after a disaster event. These activities include the rehabilitation or construction of damaged or destroyed buildings due to disasters. Development activities include: permanent residentis, satellite permanent residences, independent permanent residences, residential infrastructure (water, sanitation, road, drainage), public facilities, and community-based regional infrastructure development. All rehabilitation and reconstruction activities can be monitored through the dashboard as shown in the figure 3.

Figure 3. Dashboard for monitoring rehabilitation and reconstruction projects.

The detail of each projects also present in the information system. Figure 4 shows the details of a permanent residence project. Master plan / site plan and residential development progress can be seen in the system.

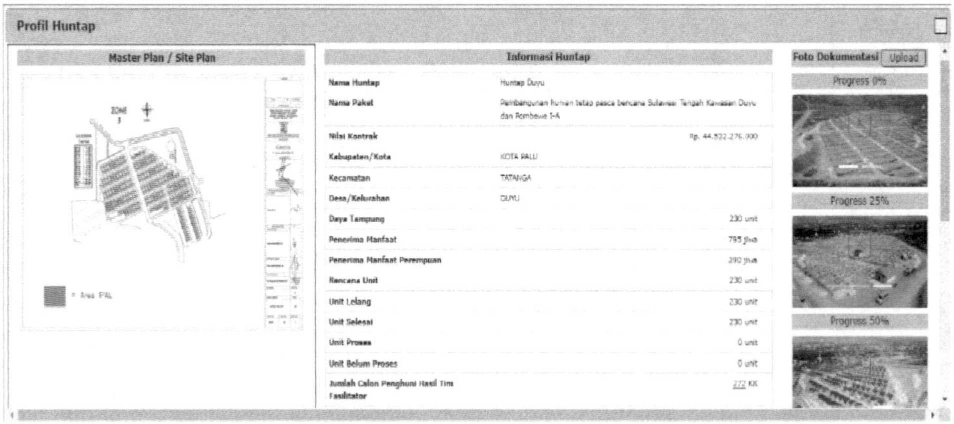

Figure 4. Details of a development of housing unit project.

4.1.3. Geographic information system (GIS). GIS is an important element in the post-disaster rehabilitation and reconstruction information system. It can provide spatial data related to ongoing rehabilitation and reconstruction activities. Various essential information can be presented on it to give an overview of the activity.

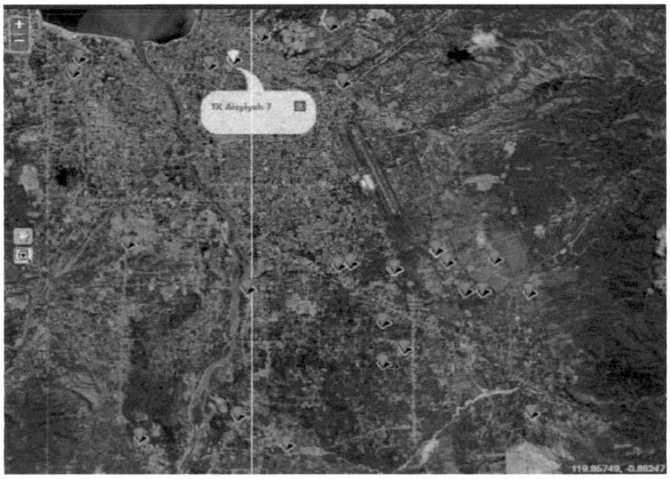

Figure 5. Geographic information system.

Figure 5 provides an example of the location of the construction of public facilities in post-disaster rehabilitation and reconstruction activities. GIS provides an intuitive experience of the distribution of activities.

4.1.4. Disaster prone zone. A disaster prone zone also implemented on the geographic information system. It becomes the basis for the preparation of the master plan for the recovery and redevelopment of the post-disaster area and the preparation of the spatial revision, especially in the relocation area by taking into account the zoning classification and post-disaster spatial directions.

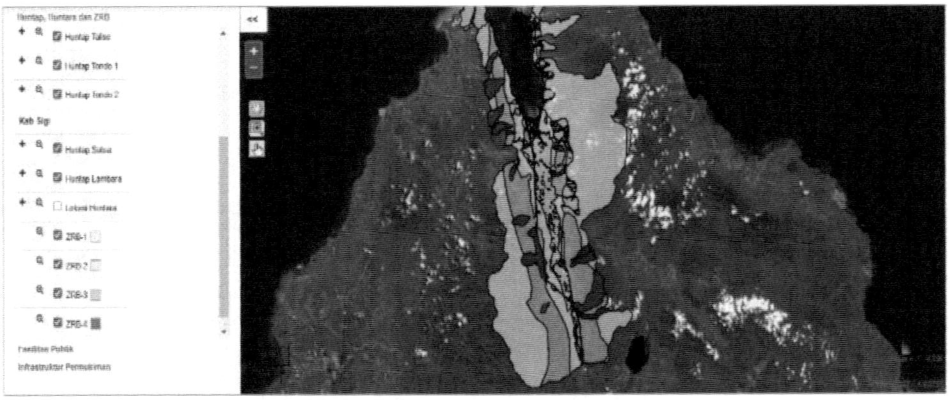

Figure 6. Disaster prone zone.

Disaster Prone Zones (*Zona Rawan Bencana, ZRB*) are divided into 4 categories, namely ZRB-1, ZRB-2, ZRB-3, and ZRB-4. ZRB 4's post-disaster spatial directive is no redevelopment and new construction allowed. The use of space in ZRB 4 is prioritized for the function of protected areas, green open spaces, and monuments. Therefore, residential units in this zone are recommended to be relocated immediately. Spatial direction after the ZRB 3 disaster is not allowed for the construction of new residential functions as well as important and high-risk facilities.

4.1.5. Housing units. Shelters are still being built to move disaster-affected residents from disaster-prone zones to new areas that are safe from disasters. Some residents choose to move to permanent residential areas, while others choose to move to independent permanent residences or satellite permanent residences.

The relocation of independent shelters is relocation for affected residents previous disaster (earthquake, tsunami or liquefaction) living in a disaster-prone red zone construction of temporary housing on prepared land/land with proof of legal ownership.

The relocation of the satellite huntap is relocation for affected residents previous disaster (earthquake, tsunami or liquefaction) living in a disaster-prone red zone construction of temporary housing on land/land that has been determined by the local government..

Residential areas are still built as a residential complex for residents. The rehabilitation and reconstruction information system provides a GIS for the shelter as shown in figure 7.

Figure 7. Map of housing units.

4.1.6. Relocation. Relocation is an effort to move part or all of an activity the following facilities and infrastructure to support these activities from one place to another in order to enhance the safety factor, feasibility, legality of use with due regard the relationship between the displaced and the natural environment and construction at the destination. The rehabilitation and reconstruction information system will record the origin and destination of the residents so as shown in figure 8.

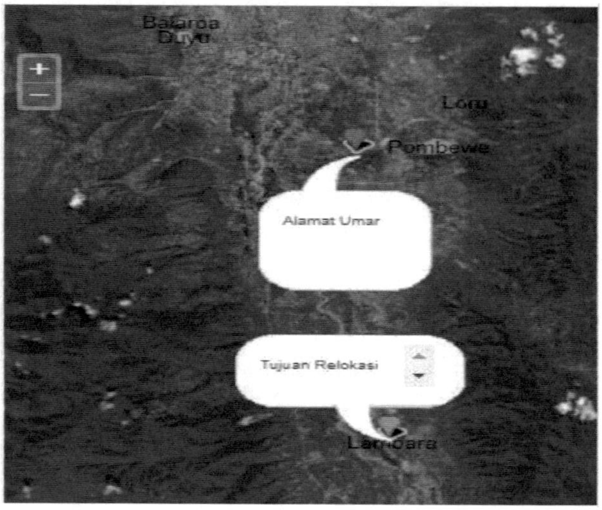

Figure 8. Relocation of residents.

The progress of relocation and occupancy by residents is then mapped on the permanent housing being built. Figure 9 shows details of the progress of occupancy by residents. The red box indicates that the housing unit has been occupied. Thus, the progress of development and relocation activities can be monitored and evaluated.

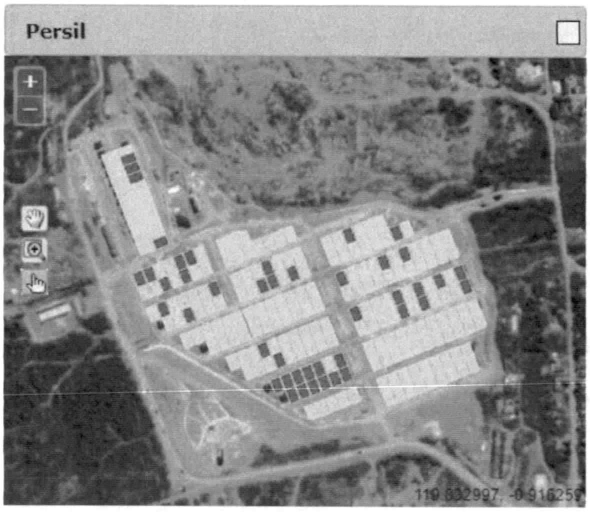

Figure 9. House occupancy.

4.2. Evaluation

The rehabilitation and reconstruction information systems has been implemented in Ministry of General Working and Public Housing. The information obtained from and through the system, such as disaster-affected residents, ongoing rehabilitation and reconstruction projects, permanent housing, infrastructure, and public facilities, could be used to evaluate the achievements of the rehabilition and reconstruction projects.

This achievement could be measured through key performance indicators (KPI), such as: percentage of housing unit built, percentage of infrastructures and public facilities built according to resilience standards. This could be measured with the following equation:

$$KPI = \frac{\sum Realization\ of\ Facilites\ Built}{\sum Target\ of\ Facilities\ Built} \tag{1}$$

The information has been made available for general public. While not all features that have been developed are presented in this paper, nevertheless there are still many features that still need to be developed according to user feedbacks.

5. Conclusion and recommendation

The development of a disaster response information system using the SDLC methodology has met the defined user requirements. The information systems provide data of disaster-affected residents to rehabilitation and reconstruction activities. Many of which supported by relevant geographic information systems

The system has been implemented for Ministry of General Work and Public Housing. Based on the feedbacks and suggestions from the users, future work on the rehabilitation and reconstructions information systems should improve user experience and bringing new functionalities. For example, it could be redesigned to facilitate more user-friendly interaction. Thematic maps could be included from other relevant sources. So, all information from various sources and every stage of disaster management could be integrated with a holistic strategy.

Acknowledgments

The authors would like to thank the Ministry of General Working and Public Housing, Metamedia University, World Bank, and Asian Development Bank for their support in the research.

References

[1] Sakurai M and Murayama Y 2019 *Prog. Disaster Sci.* **2** 100012
[2] Yu Z, Liang R, Wang Y and Song X 2011 *Energy Procedia* **16** 1199
[3] Silva L D O, Bandeira R A D M and Campos V B G 2017 *Transp. Res. Procedia* **27** 1137
[4] Jeong S and Cheong T 2012 *Trop. Cyclone Res. Rev.* **1** 207
[5] Shimokawa S, Fukahori H and Gao W 2016 *Procedia - Soc. Behav. Sci.* **216** 481
[6] Stute M, Maass M, Schons T, Kaufhold M A, Reuter C and Hollick M 2020 *Int. J. Disaster Risk Reduct.* **47**
[7] Jayawardene V, Huggins T J, Prasanna R and Fakhruddin B 2021 *Prog. Disaster Sci.* 12, 100202
[8] Nespeca V, Comes T, Meesters K and Brazier F 2020 Int. J. Disaster Risk Reduct. **51**101887
[9] Pribadi K S, Abduh M, Wirahadikusumah R D, Hanifa N R, Irsyam M, Kusumaningrum P and Puri E 2021 *Int. J. Disaster Risk Reduct.* **64** 102424
[10] Khorram-Manesh A, Goniewicz K and Hertelendy A J 2021 *Am. J. Emerg. Med.* **45** 623
[11] Wang C, Qin F and D J S R 2020 *Comput. Commun.* **151** 548
[12] Shigeyasu T and Matsumoto N 2020 *Internet of Things (Netherlands)* **11** 1
[13] Paul S K, Naim J and Mou M A 2022 *Nature-Based Solut.* **2** 100024
[14] Saraswati G 2014 *Procedia - Soc. Behav. Sci.* **135** 112

[15] Uddin K andMatin M A 2021 *Prog. Disaster Sci.* **11** 100185
[16] Busayo E T, Kalumba A M, Afuye G A, Olusola A O, Ololade O O and Orimoloye I R 2022 *Environ. Sustain. Indic.* **14** 100175
[17] Chan S W, Abid S K, Sulaiman N, Nazir U and Azam K 2022 *Heliyon* **8**
[18] Vankova L, Krejza Z, Kocourkova G and Laciga J 2021 *Procedia Comput. Sci.* **196** 708
[19] Twomlow A, Grainger S, Cieslik K, Paul J D and Buytaert W 2022 *Int. J. Disaster Risk Reduct.* **77** 103067
[20] Gerster J, Boret S P, Morimoto R, Gordon A and Shibayama A 2022 *Int. J. Disaster Risk Reduct.* **77** 103085
[21] Yamashita R 2022 *SSRN Electron. J.* **78** 103150
[22] Ayenew T, Tassew S F and Workneh B S 2022 *African J. Emerg. Med.* **12** 246
[23] Talubo J P, Morse S and Saroj D 2022 *Environ. Challenges* **7** 100511
[24] Tantri F andAmir S 2022 *J. Saf. Sci. Resil.* **3** 288
[25] Haryana I K, Fikriyah V N and Yulianti N V 2013 *Procedia Environ. Sci.* **17** 434
[26] Nefros K C, Kitsara G S and Photis Y N 2018 *IFAC-PapersOnLine* **51** 360
[27] Poonam, Rana N, Champati ray P K, Bisht P, Bagri D S, Wasson R J and Sundriyal Y 2017 *Geomorphology* **284** 41
[28] Alam R R R, Adityawan M B, Farid M, Chrysanti A, Widyaningtias and Kusuma M A 2021 *IOP Conf. Ser. Earth Environ. Sci.* **708** 012003
[29] He Y, Zhang D and Fang Y 2017 *Int. J. Disaster Risk Reduct.* **25** 101
[30] Kohl P A, Orourke A P, Schmidman D L, Dopkin W A and Birnbaum M L 2005 *Prehosp. Disaster Med.* **20** 355
[31] Sanjaya J, Triatmadja R and Triatmodjo B 2018 *J. Civ. Eng. Forum* **4** 179
[32] Wargadalam R, Nakanishi H, Vidyattama Y, Black J and Suenaga Y 2021 IOP Conf. Ser. Earth Environ. Sci. **630** 012023

The pattern of material supply chain in post-disaster reconstruction: a case study in Indonesia's disaster-prone area

Putranesia[1,2], T Ophiyandri [1], F A Ismail [1] and B Hidayat [1]

[1]Dept. of Civil Engineering, Faculty of Engineering, Andalas University, Indonesia.
[2]Dept. of Quantity Surveying, Faculty of Engineering and Planning, Bung Hatta University, Indonesia

E-mail: putranesia@bunghatta.ac.id

Abstract. The government, the community, or non-governmental organisations can all participate in the reconstruction process after a disaster. The management of material procurement and logistics may entail participation from several stakeholders. It may be crucial to a project's success in reconstruction. A reliable supply chain is required to guarantee an appropriate building supply. The supply chain for building materials significantly impacts the course and results of reconstruction after a disaster. This study intends to map the post-disaster restoration process's formation of the construction supply chain pattern. In order to detect potential issues and disturbances that could contribute to the failure of the post-disaster restoration process, the supply chain pattern for building materials is being mapped. The path of the supply chain for construction materials is undoubtedly significantly impacted by post-disaster environmental circumstances. Because it can impact project prices and timelines, the choice of building material suppliers is crucial in construction supply chain management. During the reconstruction process in disaster-affected areas, this research used qualitative approaches by conducting surveys and questionnaires to gather data and facts.

1. Introduction

The impacts of damage following Indonesia's natural disasters, such as the earthquake in Yogyakarta in 2006, which devastated hundreds of thousands of homes, with an estimated 154,000 dwellings demolished and 260,000 homes significantly damaged [1], earthquakes in Padang and Mentawai in 2009 which caused damage to places where people lived with a total of 279,432 housing units [2] and then the earthquake in Donggala, Palu in 2018 with the impact of damage caused as many as 68,235 houses damaged [3], and earthquake in Lombok, Nusa Tenggara Barat that impact 33.926 house damage [4]. As a result, many people do not get proper housing and even lose their homes. This condition requires Efforts to answer the need for rehabilitation, reconstruction, and re-procurement of residential houses. These activities are known as post-disaster reconstruction process activities. The post-disaster reconstruction process involves various elements and parties working well, quickly, and accurately. Local community involvement in rehabilitating residential houses and damaged infrastructure reconstruction requires building materials and many resources. However, the available supply is relatively insufficient for the existing collection [5]. The post-disaster reconstruction process can be carried out by various parties, governments, non-government institutions, or non-governmental organisations. Several post-disaster reconstruction processes that have been carried out have implemented community-based and project-based (contractor) patterns, such as in Aceh [6], Lombok [7,8], and Palu [9].

One of the critical factors for the success of a post-disaster reconstruction can be seen in the success of the housing construction program for disaster-affected communities [8,9,10]. The supporting factor for successfully implementing the post-disaster reconstruction process is the availability of construction material supplies. Management of material procurement and logistics for the success of a construction project is an essential factor in the success of a construction project. According to Bajomo, Ogbeyemi, & Zhang [11], material procurement often faces cost overruns, overscheduling, delays in material supply, and loss of productivity.

The adequacy of the availability of construction materials cannot be separated from a well-running supply chain of construction materials. The supply chain of construction materials dramatically influences the process and outcome of post-disaster reconstruction. The construction material supply chain is a relationship that intensively involves the activities of the parties involved in the construction network, from the material's source to the construction project's completion [12]. One of the fundamental problems in the construction supply chain is the availability of construction materials, where the supply process (procurement) is very vulnerable to disruption. Under normal conditions, of course, the problem of supplying/procuring construction materials is not a significant problem. However, during the reconstruction period, where an uncertain situation arises. The environment is in crisis [13], and construction materials' supply/procurement (procurement) in the post-disaster period can become a fundamental problem. Post-disaster environmental conditions certainly significantly affect the course of the construction material supply chain. The selection of construction material sources is essential in construction supply chain management because it can affect project costs and schedules.

2. Methodology and concept

2.1. Methodology

Research methodology using The Research Union Process developed by Saunders [14]. A pragmatism paradigm research philosophy is used to map the construction material in disaster-prone areas. Then through a research approach followed by an inductive approach using case study investigations and field surveys. We conducted a comprehensive literature review and questionnaire survey as part of our qualitative research. Questionnaires were sent out in cross-sectional time to the parties involved. A literature review allowed us to compare our findings with those of other scientists. Resilience assessment can start with a hierarchical and inductive evaluation [15].

2.2. Concept

The pattern of the supply chain of construction materials is known by carrying out a mapping of the supply of materials during the post-disaster reconstruction process. The mapping of the supply chain pattern of construction materials is intended to identify problems and disturbances that may arise so that they can become one of the factors causing the failure of the post-disaster reconstruction process. Post-disaster environmental conditions certainly significantly affect the course of the construction material supply chain. The selection of construction material sources is essential in construction supply chain management because it can affect project costs and schedules. Mapping (the patterns) begins with collecting related data and seeking answers to research questions. The mapping is done by collecting data through interview methods and distributing questionnaires to parties involved (contractors, consultants, construction implementers, and suppliers/building material stores) to rehabilitate and reconstruct damaged buildings.

In the Padang and Pasaman, a supply chain survey for building materials was carried out. The city of Padang was chosen because of its 2009 significant earthquake, which significantly damaged nearby buildings at the time, particularly community residential homes[2], and because it was anticipated that the Mentawai segment movement would cause another significant earthquake in the area[16]. The

Pasaman region was also picked since it was a disaster-affected area due to the earthquake that happened at the start of 2022. The selection of the two parts represents areas that are at risk of experiencing disasters, namely the coastal zone (Padang) and the mountainous zone (Pasaman). By visiting building supply stores in the two towns and gathering information on materials that eventually became the primary materials used for the implementation of the rehabilitation and reconstruction process, the survey of the supply chain pattern of construction materials in the cities of Padang and Pasaman was carried out. Construction material data collection is based on the use of materials in the community's rehabilitation and reconstruction in the restoration, rebuilding, and procurement of residential buildings affected by the disaster. Materials that are the object of the survey are divided into two main categories: (1) natural materials and (2) industrial materials. Natural materials in question are sourced and used raw materials directly from nature without using industrial processing processes. The natural materials are sand, river stone, bricks, and gravel / crushed stone. Meanwhile, what is meant by industrial materials are materials that require industrial processing to be used as construction materials. These materials are iron, zincalume mild steel, woven wire, cement, GRC Board (Glass-Reinforced Concrete), and roof covering materials (zinc, onduline, and others).

3. A literature study and result

3.1. Literature study

Management of material procurement and logistics for the success of a construction project is an essential factor in the success of a construction project. Material procurement often faces cost overruns, overscheduling, material supply delays, and productivity loss [11]. Previous researchers developed several supply chain models for the procurement of construction materials. Le, Jarroudi, Dao, & Chaabane developed an integrated complexity model of the construction supply chain with a third-party logistics partnership pattern (Third-Party Logistics)[17]. The advantage of using a third-party logistics partnership is that it can save costs in terms of providing logistics warehouses and reducing transportation costs offered by suppliers to improve construction logistics performance. Further, Bajomo developed a dynamic construction material supply chain (CMSC) model based on a causality system in the Engineering, Procurement, and Construction (EPC) industry[11]. Imbalances related to supply/market conditions pose risks and affect the construction of supply chain management strategies [18]. A construction project's supply chain structure can be determined, based on the supply chain's main components, including delivery, method, operational elements, and functions[19]. On the other hand, Construction projects are often unique and time-limited, requiring a high degree of customisation and involving many Engineer-to-Order (ETO) components [20]. Under the same conditions, the construction supply chain (CSC) has unique characteristics, such as being a temporary organisation that requires high coordination efforts to align the processes of supply chain actors [21].

Reflecting on the rehabilitation/reconstruction carried out in Aceh and Nias, there are two methods of providing reconstruction: a contractor-based approach and a community-based approach [10]. In addition to Palu and Lombok, a community participation-based approach to restoration and reconstruction as well as resettlement (procurement of housing units) was made there [7,8]. Large-scale housing projects, in this case, can be equated with the post-disaster reconstruction and rehabilitation process. The role and ability of suppliers become an evaluation of the prevailing supply/market conditions [18]. Rehabilitation/reconstruction of community residential houses carried out by the community on a self-managed basis. This activity uses a community-based approach and can be interpreted as a deep concern for the reconstruction program [22, 6].

3.2. Result

Based on community-based settlement rehabilitation/reconstruction activities, the building materials supply chain has a particular pattern. The community-based construction material supply chain pattern has six variations: (1) Supply chain patterns for building materials for design and construction, (2) Supply-chain Patterns for Construction Materials Owner as a source of materials, (3) The supply chain pattern for the procurement of construction materials is mixed Owner and Contractor, (4) The Supply-chain Pattern for Construction Materials Owner as the principal provider, (5) The Supply-chain Pattern for Construction Materials Contractor serving as the primary provider, (6) The Supply-chain Pattern for Construction Materials a subcontractor who assists the main contractor.

Figures 1, 2, 3, 4, 5 and 6 below explains the pattern of the community-based construction material supply chain.

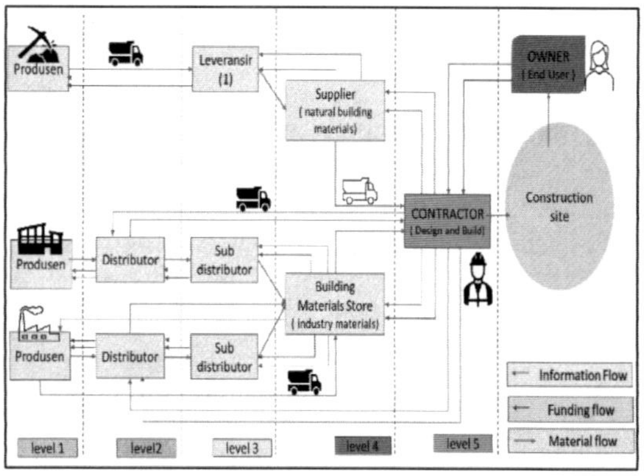

Figure 1. Supply chain patterns for building materials for design and construction

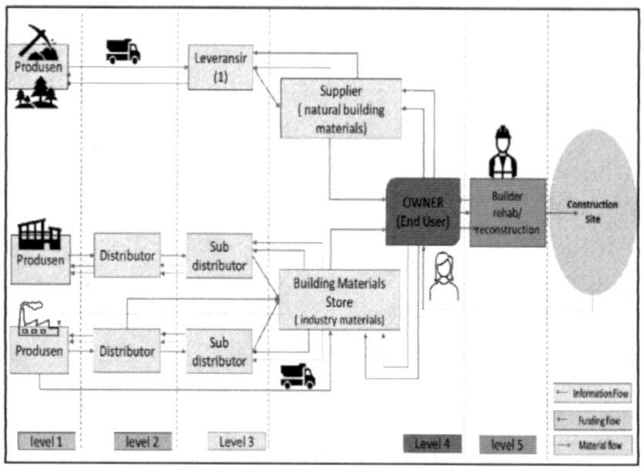

Figure 2. Supply-chain Pattern for Construction Materials Owner as a source of materials

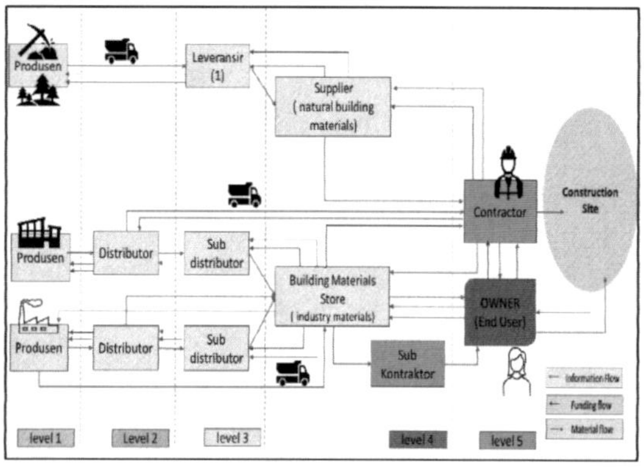

Figure 3. The supply chain pattern for the procurement of construction materials is a mixed Owner and Contractor

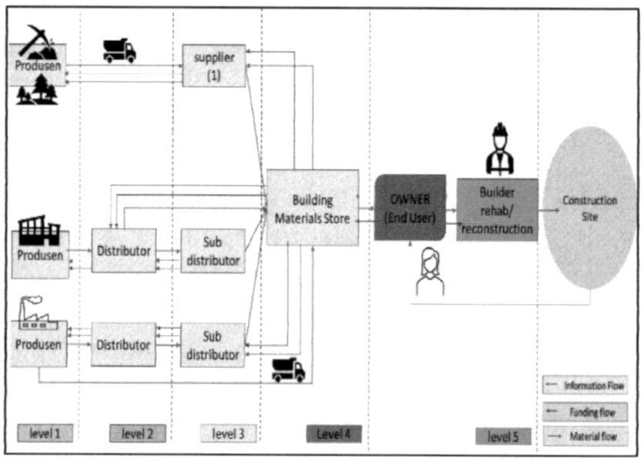

Figure 4. The Supply-chain Pattern for Construction Materials Owner as the principal provider

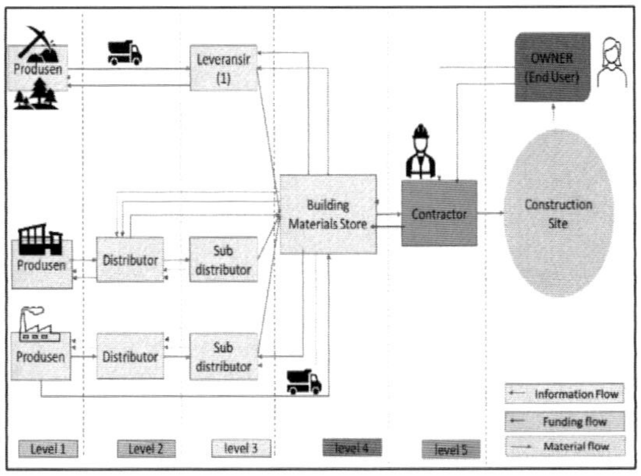

Figure 5. The Supply-chain Pattern for Construction Materials Contractors serving as the primary provider

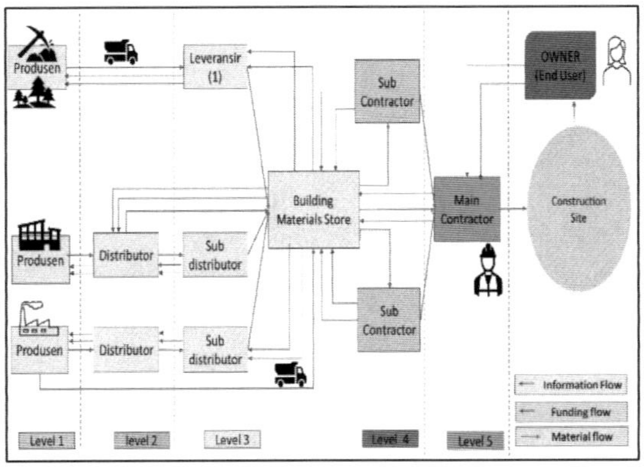

Figure 6. The Supply-chain Pattern for Construction Materials is a subcontractor who assists the main contractor

Based on the appointment of contractors by the Government (Contactor-Based) to carry out rehabilitation and reconstruction of community residential houses, two patterns of building material supply chains were found ; (1) The pattern of the supply chain for building materials based on contractors, (2) The supply chain for building materials based on contractors, with supporting subcontractors. Figures 7 and 8 below explain the contractor-based construction material supply chain pattern.

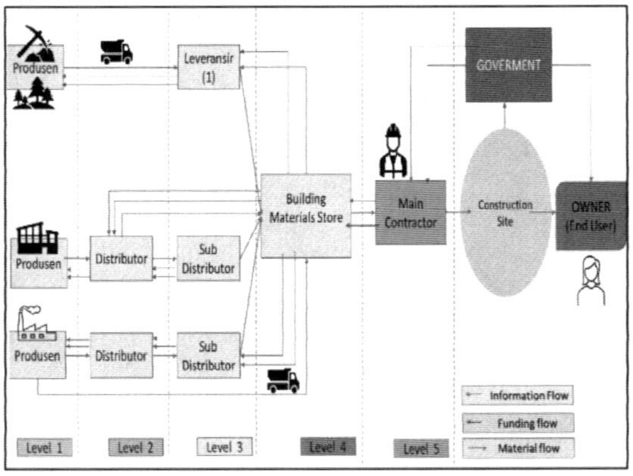

Figure 7. The Supply-chain Pattern for Construction Materials Contractors serving as the primary provider

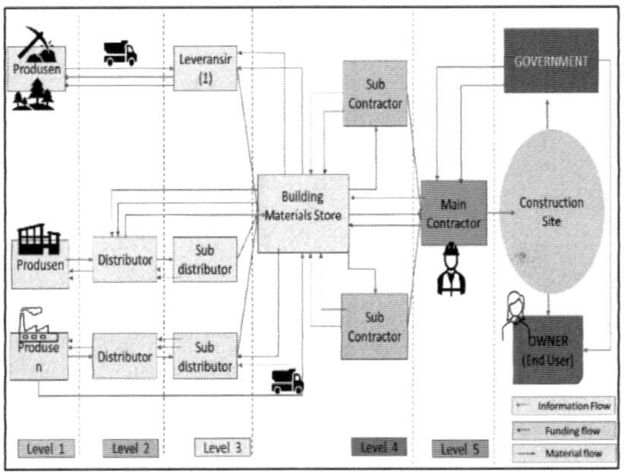

Figure 8. The Supply-chain Pattern for Construction Materials is a subcontractor who assists the main contractor

4. Conclusion

The placement of categories of materials used in the implementation of rehabilitation, reconstruction, and procurement (resettlement) as "Disaster Emergency Materials" is based on efforts to place these materials as materials needed in decision-making and policies to carry out post-disaster reconstruction, where the post-disaster situation and conditions covered as an uncertain and critical situation (emergency). According to the observed patterns, it was discovered that the supply chain pattern based on the community was more variable than the supply chain pattern based on contractors, raising concerns that the design would be subject to interference. It is possible to conduct more studies to ascertain the resilience of the developed supply chain pattern and then locate a flexible supply chain pattern for critical situations in different disaster zones. And This research is also expected to open up further research opportunities to find temporal patterns in the supply chain of disaster-resistant construction materials.

References

[1] BAPPENAS 2006 Penilaian Awal Kerusakan dan Kerugian Bencana Alam di Yogyakarta dan Jawa Tengah (Jakarta: Bappenas)

[2] L. H. P. BNPB 2009 *Hilang Meninggal Luka Berat Luka Ringan Mengungsi* (Jakarta: BNPB) 0211–9

[3] Pemerintah Provinsi Sulawesi Tengah 2018 *Laporan Finalisasi Data dan Informasi Bencana Gempa Bumi , Tsunami dan Likuifaksi di Sulawesi Tengah Per Tanggal 20 Des 2018* (Sulawesi Tengah: Pemerintah Provinsi Sulawesi Tengah)

[4] BNPB 2016 Info bencana BNPB *Info Bencana Inf. Kebencanaan Bln. Teraktual* 1–4

[5] Chang Y, Wilkinson S, Brunsdon D, Seville E and Potangaroa R 2011 An integrated approach: managing resources for post-disaster reconstruction *Disasters* **35** 739–765

[6] M. C. an A. D. Ophiyandri T 2015 Critical success factors for community-based post-disaster housing reconstruction projects in the pre-construction stage in Indonesia *Int. J. Disaster Resil. Built Environ.* **6** 102–116

[7] Paramastri M A 2021 Community Development, Local Wisdoms, and Ineffective Government Aid: The Case of Lombok Post-Disaster House Reconstruction Project *Glob. South Rev.* **2** 136

[8] Pribadi K S, Pradoto R G, Hanafi E A and Rasmawan I M A B 156 Lombok earthquake, one year later: Housing sector recovery *E3S Web Conf.* **156** 1–10

[9] Hidayah N, Satyarno I and Saputra A 2020 Housing rehabilitation and reconstruction in Central Sulawesi post-2018 earthquake *Proc. The 1st Geosciences and Environmental Sciences Symposium (ICST) 2020* (E3S Web Conf.) **200** 03004

[10] Ophiyandri T, Amaratunga D, and Pathirage C 2013 Risk management for community-based post-disaster housing reconstruction project In: International Post Graduate Research Conference *8th-10th April 2013,* (Salford, UK: IPGRC) p 1077-1087

[11] Bajomo M, Ogbeyemi A, and Zhang W 2022 A systems dynamics approach to the management of material procurement for Engineering, Procurement and Construction industry *Int. J. Prod. Econ.* **244** 108390

[12] Pryke S 2009 *Construction Supply Chain Management: Concepts and Case Studies*, This editi. John Wiley & Sons Ltd, The Atrium, Southern Gate, Chichester, West Sussex, PO19 8SQ, (United Kingdom: Blackwell Publishing Ltd)

[13] Shrivastava S P and George 1982 Disaster Containment Strategies *J. Bus. Strategy* **3** 21–33

[14] Saunders A, Philip K L, and Thornhill 2019 *Research Methods for Business Students, Eighth edi.* (New York : Pearson) Print edition typeset in 9.5/12 ITC Slimbach Std by Pearson CSC

[15] Rus K, Kilar V and Koren D 2018 Resilience assessment of complex urban systems to natural disasters: A new literature review *Int. J. Disaster Risk Reduct.* **31** 311–330

[16] Damayanti C, Yamko A K, Souisa C J, Barends W and Naroly I L P T 2020 Pemodelan Segmentasi Mentawai-Pagai: Studi Kasus Gempa Megathrust di Indonesia *J. Geosains dan Remote Sens.* **1** 105–110

[17] Le P L, Jarroudi I, Dao T M and Chaabane A 2021 Integrated construction supply chain: an optimal decision-making model with third-party logistics partnership *Constr. Manag. Econ.* **39** 133–155

[18] Seth D, Nemani V K, Pokharel S and Al Sayed A Y 2018 Impact of competitive conditions on supplier evaluation: a construction supply chain case study *Prod. Plan. Control* **29** 217–235

[19] Forghani M A, Dadkhah S and Dadras M 2022 Sustainable development based on the modeling process to identify and determine the structure of the supply chain of construction projects and in line with regional planning (Case study: Qeshm city in Iran) *Geogr. (Regional*

Planning)
[20] Braglia M, Dallasega P and Marrazzini L 2020 Overall Construction Productivity: a new lean metric to identify construction losses and analyse their causes in Engineer-to-Order construction supply chains *Prod. Plan. Control* **0** 1–18
[21] Dallasega P, Rauch E and Linder C 2018 Industry 4.0 as an enabler of proximity for construction supply chains: A systematic literature review *Comput. Ind.* **99** 205–225
[22] Ophiyandri T, Amaratunga R, and Pathirage C 2010 Community based post disaster housing reconstruction: Indonesian perspective *Int. J. Disaster Resil. Built Environ.* 1–17

Landfilling as individual adaptation for coastal flooding (*Rob*) countermeasures in North Central Java

S N Purnomo[1], W Widiyanto[1] and Suroso[1]

[1]Civil Engineering Department, Engineering Faculty, Universitas Jenderal Soedirman, Jl. Mayjen Sungkono KM 5, Purbalingga, Central Java, Indonesia, 53371

E-mail: wahyu.widiyanto@unsoed.ac.id

Abstract. The northern coastline of Central Java is vulnerable to coastal floods. In recent years, the area impacted by coastal flooding on the northern shore of Central Java has grown and encroached further onto the mainland. This precipitated the "sinking" of Central Java's northern coast. The deterioration of the north shore of Central Java has a social and economic impact on the local community. Every year floods along the northern shore of Central Java cause growing property losses, particularly for the region's residents. This forces the local community if they wish to continue living in the area, to adapt to the severe conditions in their immediate environment. Each member of the local community undertakes the adaptation process of filling the soil to raise the elevation of their home and the infrastructure surrounding it. This study demonstrates that using landfilling to elevate the base elevation of dwellings and road infrastructure is a successful adaptation technique; nevertheless, it will incur large expenses for the people, as the landfilling process would need to be repeated frequently due to sea level rise.

1. Introduction

Floods are a type of natural disaster that can result in fatalities, property damage, functional damage, and disruptions to vital infrastructure, transportation, and communication [1]. Indonesia, which consists of more than 17,000 islands, has the second-longest coastline in the world at 99,083 kilometers, making it susceptible to coastal flooding. Climate change, which influences extreme weather and sea level rise (SLR), tsunamis induced by tectonic earthquakes on the seafloor, and land subsidence caused by natural compaction, groundwater pumping, and construction loads are all factors that contribute to coastal flooding [2–5].

On May 23-25, 2022, coastal flooding, known as *rob*, affected the majority of the northern shore of Central Java. *Rob* is the Javanese (one of the local languages in Indonesia) term for the persistent and temporary flooding that occurs along the northern coast of the island [6]. On that date, the districts of Brebes, Tegal, Pemalang, Pekalongan, Batang, Kendal, Semarang, Demak, Jepara, Pati, and Rembang were badly impacted by coastal flooding. However, according to history, the flood in the area did not only happen on that date, but it happened virtually every day.

To overcome the tidal floods that occurred on the north shore of Central Java, the central and provincial governments have made numerous infrastructure-related measures, including the construction of dikes and pumping stations [7]. However, local people feel that this is insufficient to solve the problem of tidal flooding along the northern shore of Central Java. An effort to protect themselves, their homes, and the environment from the coastal flooding that happens in their surroundings, has prompted residents of various locations to build individual adaptations. Several

IOP Conf. Series: Earth and Environmental Science 1173 (2023) 012057 doi:10.1088/1755-1315/1173/1/012057

initiatives have been made to combat the phenomena of tidal flooding in various regions [8]. In Semarang, Central Java, there have been numerous analyses of the population's efforts to adapt to land subsidence and coastal flooding [6,9,10]. Migration or relocation of residents impacted by coastal flooding is one of them, as is raising the level of the floor of the house by landfilling if local residents wish to preserve their homes. As an effort to adjust the population on the north coast of Central Java, this study will explore raising the elevation of the house's floor through landfilling in the residence or infrastructure surrounding the residence.

2. Study area and method

The study area in this study was conducted along the northern coast of Central Java, as shown in Figure 1.

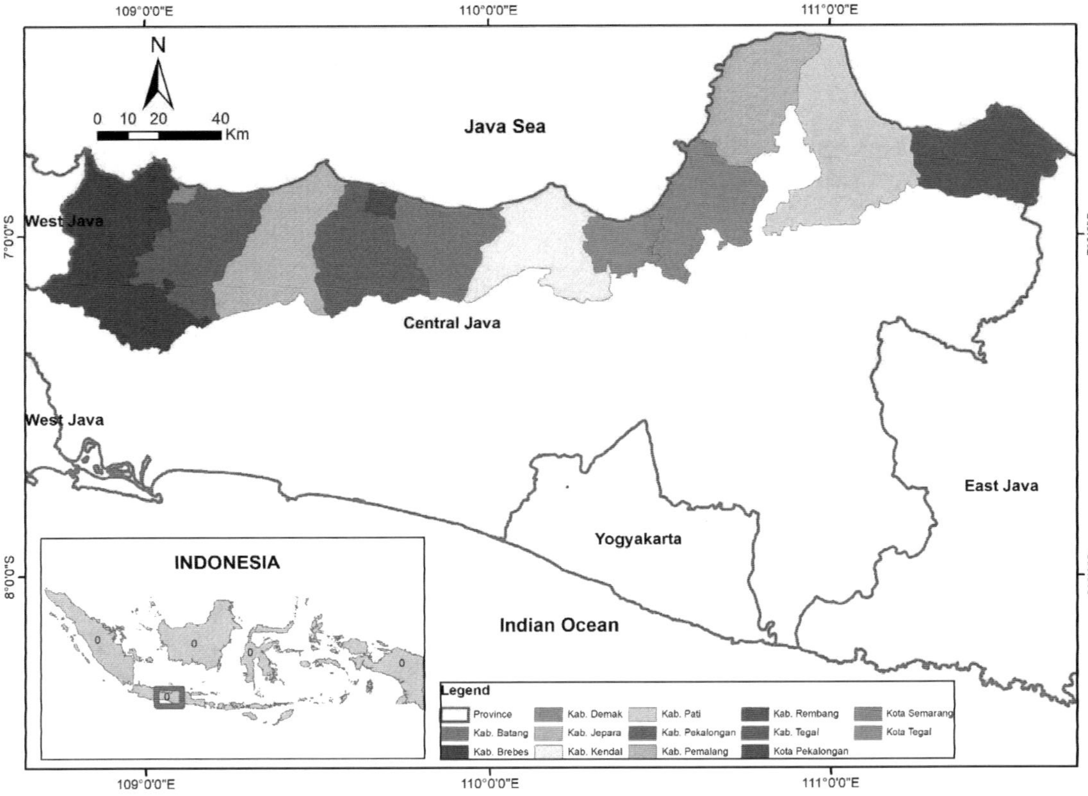

Figure 1. Areas experiencing coastal flooding along the northern coast of Central Java, which is directly adjacent to the Java Sea.

The research method used in this paper is a quantitative descriptive method so that an objective description of the condition of the area on the north coast of Java will be obtained. In order to support the research method, a field survey was conducted between 22 and 24 July 2022 along the coast of Central Java, encompassing the districts of Brebes, Tegal, Pemalang, Pekalongan, Batang, Kendal, Semarang, Demak, Jepara, Pati, and Rembang. This field survey was very helpful in determining the individual adaptation measures performed by communities on the north shore of Java, which are impacted by coastal flooding on the north coast of Central Java.

The field survey was done to collect primary data and create a comprehensive picture of the condition of flood-prone coastal areas. Observing the particulars of local conditions, measuring and estimating inundation heights, and recording the post-coastal flooding that had happened throughout the northern shore of Central Java, from the westernmost to the easternmost coast, comprised the field survey. The surveyor team also conducted interviews on the damage, inundation depth, and inundation duration that happened on the north shore of Central Java to get further information. The survey results are then examined and supplemented with social media news and data.

3. Finding and discussion

Adaptation refers to the process and the outcome, which leads to many interpretations and much debate [11]. In general, humans will attempt to adapt to harsh environmental situations [12]. Coastal flooding has impacted residents of the north shore of Central Java for decades. Many of them eventually gave up and moved to locations that were safer from coastal flooding, but many residents continued to try to survive and remain in their homes despite the high costs that would be spent as a result of the adaptation process.

The survey undertaken by the surveying team indicates that flooding and inundation along the northern coast of Java are not identical. The north coast of western Java is subjected to tidal waves every day on average, making the region more vulnerable and susceptible to more severe flooding damage than the north coast of eastern Java. The western part of the north coast of Java, where the city of Tegal is located, is practically daily flooded due to high tides. Typically, *rob* in the northern city of Tegal begins at 2:00 p.m. This is in contrast to Pati, which is situated on the eastern part of Java's northern shore. In Pati, coastal flooding only occurs under extreme conditions, as was the case from May 23 to May 25, 2022. So that Pati is one of the areas that, although being located on the north coast of Java, suffered very less damage in comparison to other coastal areas.

According to the results of interviews with local inhabitants, there are a number of sites where flooding and inundation are becoming more prevalent. In recent years, Pekalongan City has experienced an increase in the frequency and severity of tidal flooding, as floods and inundations are rapidly penetrating the mainland. An identical event occurred in Semarang. There are various plausible causes for the widespread flooding and inundation that happened on Java's northern shore, including land subsidence [4,5].

This tidal flood resulted in enormous material losses because it resulted in the inundation of houses and road infrastructure located in areas affected by the tidal flood. The owners of the properties that were flooded virtually every day eventually abandoned them, while housing prices in the region decreased since no one wanted to buy them. Figure 2 depicts a description of the homes affected by the coastal flooding.

Figure 2 demonstrates that coastal flooding has a devastating effect on life along the northern shore of Central Java. In addition to causing damage to homes of people, practically daily flooding also causes significant damage to the pavements of several road infrastructures. Local governments in a number of regions have taken measures to elevate the elevation of roads to protect them from inundation during tidal flooding. However, this did impact the homes surrounding the elevated road. Eventually, the ground and floor elevations of the houses placed along the road are lower than the road elevation. The result is that, during a tidal flood, water coming from the road will enter people's homes, but the higher road elevation from the residents' dwellings will block the water that has entered each house, so prolonging the inundation period in each home.

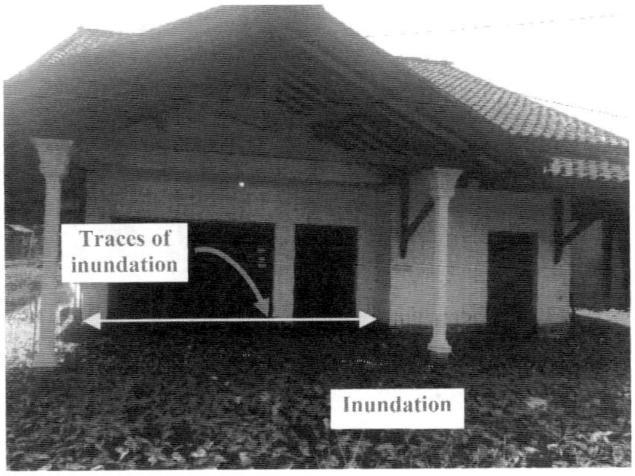

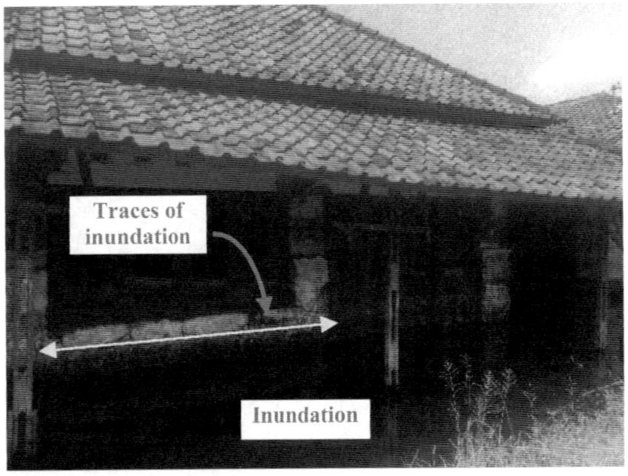

Figure 2. Residents' houses were flooded due to the impact of coastal flooding in Blancanan, Pekalongan City, and their private vehicles were damaged as a result of being submerged in floodwaters during coastal flooding.

Based on field measurements, observations, and interviews, the height of floods and inundation along the northern shore of Java range from 5 centimeters to over 100 centimeters in certain regions, particularly during the flood of May 23-25, 2022. Although practically every day is affected by flooding and inundation, many residents prefer to remain in the area for a variety of reasons. Local residents who have remained in their homes despite the tidal waves are attempting to elevate the base elevation of their dwellings by filling the soil within their homes and the infrastructure surrounding their homes. This landfilling approach is regarded as effective in Sayung, Demak, and Central Java [13] and mostly along the north coast of Central Java, as shown in Figure 3.

Figure 3. The process of landfilling as an individual adaptation effort due to coastal flooding on the north coast of Central Java.

Considering the reality of sea level rise, people who desire to preserve their homes will typically attempt to raise the elevation of their homes through frequent landfilling. Raising the elevation of a house's floor will be followed by raising the elevation of the house's walls and roof for persons with a reasonably decent economic standing. The road infrastructure near the house is also inundated and destroyed as a result of coastal flooding, thus just elevating the level of the house is evidently insufficient. Figure 4 depicts residences whose owners raised the ground floor as a kind of adaptation to the near-daily flooding and inundation on the northern coast of Java.

Figure 4. In an effort to adapt to coastal flooding on the northern shore of Central Java, homeowners are landfilling and elevating the base of their homes.

Landfilling materials include soil, sand, stone, rubble left over from the demolition of home walls or floors, mixed materials, splits, asbestos fragments, and many others. The cost of landfilling material fluctuates. Obviously, the price will increase with the quality of the requested materials. To lower the expense of landfilling, residents typically use asbestos fragments, or debris from the demolition of walls or floors of houses mixed with soil. As an illustration, the price of asbestos fragments used for backfill in Central Java Province is approximate Rp. 80,000.00 (US$ 5.17) per cubic meter, while the price of debris left over from the demolition of structures is Rp. 100,000.00 (US$ 6.46) per cubic meter, and the price of a landfill service is Rp. 20,000.00 (US$ 1.29). The finance requirements for land filling for a very simple house constructed on an area of 72 m^2 and requiring a floor elevation rise of up to 0.8 m are depicted in Table 1.

Table 1. The finance requirements illustration for landfilling for a very simple house.

No	Work Item	Volume	Price Rp/m^3	Price US$/$m^3$	Total Price (Rp/m^3)	Total Price (US$/$m^3$)
1	Volume of landfill	57.60	80,000	5.17	4,608,000	297.79
2	Labor	57.60	20,000	1.29	1,152,000	74.30
		Total			5,760,000	372.10

Meanwhile, the monthly regional minimum salary for Central Java in 2022 is IDR 1,813,011.00 (US$ 117.18). Due to the rise in sea level, landfilling is seen as highly costly if it is performed regularly.

Although it is regarded as a successful method for preventing coastal flooding, this landfilling procedure requires the attention of the local government, as landfilling sites are widely dispersed throughout the northern shore of Central Java. Due to the vast amount of land utilized for landfilling along the coast of Central Java, it is anticipated that this practice may harm the ecology. In order to prevent disruption of the current ecosystem, it is necessary to impose regulations on the quarrying of soil for landfilling.

4. Conclusions

People who prefer to preserve their properties on the north coast of Central Java must adapt to coastal flooding by utilizing landfills. This landfilling is deemed highly beneficial for the local community, but it is very expensive, as the trend of sea level rise, and raising the elevation of the road around their homes will be followed by the elevation of their home and the landfilling of the infrastructure surrounding their home. Although deemed profitable, landfilling also will have an impact on the local ecosystem due to the enormous number of people and expansive region involved. Therefore, additional research must be conducted on quarry backfill soil material.

Acknowledgments

The authors are grateful for the financial support provided by the Research and Community Service Institution (LPPM) Universitas Jenderal Soedirman in particular, the Riset Dasar UNSOED program under contract T/692/UN23.18/PT01.03/2022. In addition, the authors would like to thank the Hydraulic and Environment Laboratory of the Department of Civil Engineering at Universitas Jenderal Soedirman for providing laboratory facilities.

References

[1] Kalyanapu A J, Burian S J and McPherson T N 2009 Effect of land use-based surface roughness on hydrologic model output *J. Spat Hydrol* **9** 51–71.
[2] Erkens G, Bucx T, Dam R, DeLange G and Lambert J 2015 Sinking coastal cities *Proc. Int. Assoc. Hydrol Sci* **372** 189–98.
[3] Taherkhani M, Vitousek S, Barnard P L, Frazer N, Anderson T R and Fletcher C H 2020 Sea-level rise exponentially increases coastal flood frequency *Sci. Rep.* **10** 1–17.
[4] Lo W C, Purnomo S N, Sarah D, Aghnia S and Hardini P 2021 Groundwater modelling in Urban Development to achieve sustainability of groundwater Resources: A Case Study of Semarang City *Water.* **13** 1395.
[5] Lo W C, Purnomo S N, Dewanto B G and Sarah D 2022 Integration of numerical models and

inSAR techniques to assess land subsidence due to excessive groundwater Abstraction in the *Coastal and Lowland Regions of Semarang City Water* **14** 201.
[6] BuchoriI, Pramitasari A, Sugiri A, Maryono M, Basuki Y and Sejati A W 2018 Adaptation to coastal flooding and inundation: Mitigations and migration pattern in Semarang City, Indonesia *Ocean and Coastal Management.* **163** 445–55.
[7] Andreas H, Abidin H Z, Gumilar I, Purnama T and Yuwono B 2016 Adaptation and Mitigation of Land Subsidence in Semarang. In: International Symposium on Earth Hazard and Disaster Mitigation (ISEDM) AIP Conf Proc 1857, 060005-1-060005-7. 2017.
[8] Uy N Takeuchi Y and Shaw R 2011 Local adaptation for livelihood resilience in Albay, Philippines *Env. Hazards.* **10** 139–53.
[9] Harwitasari D and vanAst J A 2011 Climate change adaptation in practice: People's responses to tidal flooding in Semarang, Indonesia *J. Flood Risk Manag.* **4** 216–33.
[10] Suryanti E D, Marfai M A 2008 Adaptation of the coastal area of Semarang to the danger of tidal flooding (Rob). *J. Kebencanaan Indonesia* **1** 335–46.
[11] Simonet G 2010 The concept of adaptation: Interdisciplinary scope and involvement in climate change *SAPIENS Surv Perspect Integr Environ Soc.* **3** 9.
[12] Ilardo M and Nielsen R 2020 Human adaptation to extreme environmental conditions *Curr Opin Genet Dev* **53** 77–82
[13] Rudiarto I, Rengganis H, Sarasadi A and Caesar E 2020 The Effectiveness of Strategy Adaptations on Tidal Flood in the Coastal Areas of Sayung, Demak, Central Java, Indonesia *IOP Conf. Ser.: Earth Environ. Sci.* **448** 012090

A study of community slope risk and awareness in Penang, Malaysia

F Ahmad[1], S I Azmi[2], V Rajan[1], N Hashim[3], M Azmi[1] and E E Putri[4]

[1]School of Civil Engineering, Universiti Sains Malaysia, Malaysia
[2]Centre for Global Sustainability Studies, Universiti Sains Malaysia, Malaysia
[3]Faculty of Maritime Studies, Universiti Malaysia Terengganu, Malaysia
[4]Civil Engineering Department, Universitas Andalas, Padang Indonesia

E-mail: cefahmad@usm.my

Abstract. Malaysia faces a high probability of experiencing landslide disasters, particularly during rainy weather. Copious rain would result in landslide prone areas with low slope strength; thus increasing the possibility of landslide disasters occurring and threatening the safety of surrounding communities. The locations of this study are in the state of Penang that has been listed as landslide prone areas, namely Mukim 17 at Batu Ferringhi, as well as Tanjung Bungah and Tanjung Tokong, two areas which are in Mukim 18. This research aims to map community slope risk areas and to evaluate the understanding, knowledge, and awareness of the community on slope risk and landslides. There are two methods involved, namely GIS and survey. The findings show that the estimated community slope risk mapping is 24.71% and 20.62% of inhabitants living in high risks landslide zones in Mukim 17 and 18. 62 respondents participated in the survey and the result analysis shows that more than 50% have a strong fundamental knowledge of slope risks, while less than 50% are aware of possible landslide risks, and 42.6% will always obey an evacuation order. Overall findings show that most of the community does have a general awareness of slope risks and landslides.

1. Introduction

As stated by the United Nations Office for Disaster Risk Reduction (UNDRR), a disaster can be defined as "a serious disruption of the functioning of a community or society involving widespread human, material, economic or environmental losses and impacts, which exceeds the ability of the affected community or society to cope with using its own resources" [1]. Any tragic event with great loss stemming from events like earthquakes, floods, catastrophic accidents, fires, or explosions is considered a disaster [2]. A disaster is sudden, drastic, and normally occurs without any alarm or warning [3]. Most disasters have a natural origin, but some disasters are man-made as well. Compared to natural hazards resulting from natural disasters, man-made disasters are known for outcomes that arise from man-made hazards in which the threats that have human intent elements, neglect, or error that include a man-made device or system failure that could be considered crimes [4]. As defined by the International Federation of Red Cross and Red Crescent Societies (IFRC), man-made disasters are incidents resulting from environmental or technological emergencies triggered by people that occur in or near human communities. Environmental degradation, emissions, and incidents such as industrial, technical, and transport typically involving the manufacture, use, or transportation of hazardous materials may also be included as factors of man-made disasters [4].

Landslides are unpredictable types of natural disasters that can occur anywhere in the world. Slope failures occur due to weakened self-retainability of the earth which may cause landslides. Triggering events like earthquakes, heavy rain, or upsetting the inclined surface through excavation will facilitate the occurrence of landslides. Any structure that is built above the bottom or grown will result in structural failure leading to landslides because soil deformity can lead to toppling or cracking. People's failure to escape in the event of the collapse of a slope occurring in areas of human settlements can increase the rate of causalities and property damage. Failure of stability or steadiness of slope is a serious concern of geologists because of its collapsing susceptibility.

Landslides in Malaysia are reported to usually occur between October and January, a period that coincides with the most amount of rainfall [5, 6]. In recent years, Malaysia has experienced several landslide disasters resulting from heavy rainfall. Penang is one of the states in Malaysia that has been facing numerous major and medium landslide disasters. The Penang slope failures phenomenon is now at an alarming stage that requires serious consideration. Indeed landslides have become common incidences in Penang Island as featured by the media each year. Massive hill area developments are increasing the risk of slope failure in the surrounding community. Malaysia has lost millions of Ringgit in the post-recovery process through various landslide tragedies. Hence, preparedness for slope failures is important, as such incidents could be entirely preventable.

Most landslide incidents occur in populated areas and present major hazards to people and property. The community living in the risk zone should be prioritized, as there are no parties in the world that can guarantee the safety of the community in the event of a landslide. The community needs to be able to appreciate the risks, and should always be prepared, for the possible occurrence of landslides. Thus, the community's knowledge, understanding, and awareness of landslides must be evaluated. Apart from that, the community slope risk map also needs to be developed as the community should always be aware of their surroundings. The slope risk map would highlight the level of the risk zone by category. The research is located at Batu Ferringhi, Tanjung Bungah, and Tanjung Tokong. These locations have been selected due to the lack of research that has been conducted on landslide risk assessment towards the community over the past 5 (five) years. Therefore, the research aims to map the community slope risk area and evaluate the understanding, knowledge, and awareness of people towards slope risks and landslides. This paper also discusses factors triggering landslides, impacts of landslides, disaster management as well as government actions about landslide issues.

2. Literature review

2.1. Factors triggering landslide

There are 3 (three) phases of slopes, (i) stable, (ii) marginally stable, and (iii) actively unstable. The stable slope is the stability of the margin is high enough to handle all of the destabilizing forces. Meanwhile for marginally stable slope is when destabilizing forces reach a certain level of activity that will likely cause the slopes to fail at some point in the future. The actively unstable slope happens when destabilizing forces continuously or intermittently produce movements [7, 8].

These 3 (three) phases provide the context for identifying and classifying the causal factors according to their effect on preparatory and triggering causal factors [9, 7]. Causal factors of the preparatory are those that make the slope prone to motion without actively initiating it, and thus help to place the slope in a moderately stable state. The representation of a ground crack is a series of mass movements on the soil. They are usually clustered or parallel in plan and have a few centimetres in width and several meters in length, which distinguishes them from the much shorter cracks in desiccation. A popular indicator of imminent slope failure is the development of cracks and any change in their widening intensity. The presence of surface subsidence or disruption and rises are also good indicators of future collapse [9].

Then, triggering causal factors will activate movement that causes causal factors to change the slope from being marginally stable to significantly unstable [9, 7]. Several external causes of slope instability lead to a greater shear stress together with several internal causes leading to a lowering in shear resistance, starting with the general term of the safety factor (Fs) provided by the ratio of resistant stresses (τ f) to shear stresses (τ), where Fs = τ f / τ [10, 7, 11]. Nonetheless, some factors influence all terms in determining the safety factor. The effect of multiple contributory factors on the safety factor of a slope often differs with time due to a variety of reasons. Short-term changes in the safety factor are shown in various analyses that may occur due to seasonal level influences on the groundwater rates, whereas longer-term indicate weathering effects or improvements in groundwater conditions over the longer period. Slope instability also may not be necessarily due to one causal factor as highlighted [9].

Man's actions can cause a natural process to accelerate or slow down, slope failure. There are many causes made by man's actions including (i) removal of lateral support by cutting slopes for roads and other structures, quarrying, removing retaining walls, and lowering reservoirs, (ii) adding weight to slopes such as landfills, stockpiles of ore or rock, waste piles, construction of heavy buildings and other structures, and water leaking from pipelines, sewers, canals, and reservoirs, (iii) burdening the slopes by adding weight such as landfills, stockpiles of ore or rock, waste piles, construction of heavy buildings, and other structures, and water leaking from pipelines, sewers, canals, and reservoirs, (iv) vibrations from explosives, machinery, road and air traffic, (v) decreasing of the underlying aid by extracting granular and soluble content, extraction, failure of strength or degradation and freezing of the underlying substance, (vi) regional tilting which raises angles to the hill, (vii) lateral water pressure in cracks, water freeze in cracks, and hydration of minerals and mobilization of residual tension, and (viii) volcanic cycles that alter soil and rock pressures, such as magma chamber inflation or deflation, changes in lava-lake temperatures, and increasing ground tremors could create landslides [9, 7].

2.2. *Impacts of landslide*

Landslides have adversely affected both humans and wildlife. There are reports of human and animal casualties and the destruction of human property every year. This has disrupted people's normal lives. The scale of the damage caused each year by landslide hazards has grown rapidly, so solving these issues have become more costly. The research about landslides' effects on man and nature is significant, as human interference with nature has become more evident in the last decade. The following paragraphs address some of the impacts of the landslides on man and nature in general.

According to Madaan [12], landslides can lead to economic crises. Landslides also can result in property destruction. If the collapse is severe, it could destroy the city's and country's economies. The area affected usually needs to implement recovery efforts after a landslide. This rehabilitation means a massive outlay of capital. It is a well-known fact that the economic development of any area depends largely on the area's available surface communication which is transportation. Landslides can destroy surface networks in hilly areas all over the country. Each year, the extent of the damage caused by landslides is very high and solving such issues becomes more costly. Landslides can cause serious damage to property due to the forceful flow of mud, debris, and rocks. A single landslide can decimate infrastructures such as roads, railways, leisure destinations, buildings, and communication systems [12]. The roads might remain impassable for two to three days after the incidents occurred.

Besides that, landslides can also claim the lives of people and animals. Communities that live at the foot of hills and mountains have a greater risk of experiencing landslides that cause fatalities. A large collapse brings with its huge rocks, heavy rubble, and thick dirt that is capable of killing lots of people [12]. In Malaysia, there are 60 cases of landslides, and 611 lives were lost from 1900 to 2009, although these figures do not consider unrecorded events, particularly between 1900 and 1990. Records from 1989 to 2009 indicate that there were 56 landslide occurrences, 33 were injured and 541 were killed. The average figure is around 2.8 landslides per year and 1.7 landslides occurring each year with the loss of human injuries and properties [13].

Generally, people have faced threats by staying under the shadow of collapsing mountains and hazards by building houses in old and collapsed landslide ruins. Landslides have resulted in both public and government bearing direct and indirect damages. Direct loss consists of loss of human lives and injuries, disruption to forest land, loss of property, and expenditure on measures of rehabilitation apart from relief. The psychological losses experienced include depression as the pain that the affected people have to endure is unmeasurable.

Furthermore, the devastation caused by landslides leaves behind unsightly rugged environments and affected the beauty of the landscape. The downhill layer of dirt, rock, and rubble will cover the ground used by the community for agricultural or social purposes. Landslides can also affect river ecosystems as sliding downhill dirt, debris, and rock can obstruct the natural flow of the rivers. Changes in water quality might occur because of natural water flow intrusion and adversely affect river species, such as fish, shrimp, and crab. Communities that rely on river water for household and agricultural activities also would suffer if water flow is disrupted [12]. Then, the cleaning process would become difficult as uneven, steep, and hilly soil surfaces would prevent the use of machinery. In some extreme cases when cracks exist on the soil, the ground cannot even be used as pastures as it is unsafe for both animals and humans.

2.3. Disaster management
The IFRC [14] describes disaster management as the coordination and management of resources and commitments to tackle all humanitarian elements of emergencies in particular preparedness, response, and recovery to reduce the impact of disasters. Figure 1 shows the flow of disaster management by IFRC where the first phase of disaster management is to prepare and reduce the risk of disaster by responding to disasters, then the second phase covers the process of recovery and reduces risk as well. IFRC focuses on the local community as they are the first people that will respond to a disaster. IFRC assists the communities in disaster preparedness in order to reduce their vulnerability to a disaster.

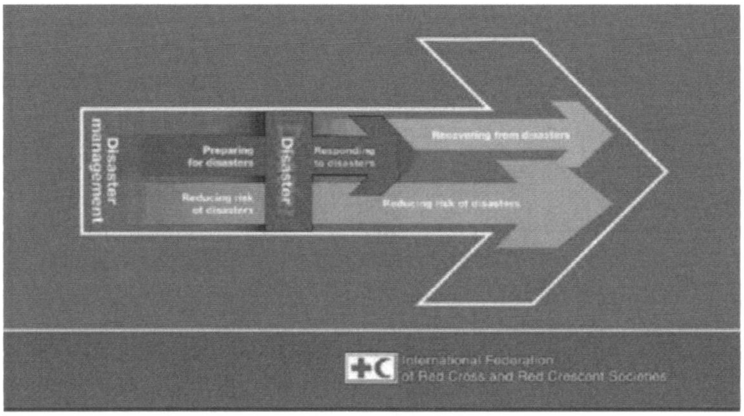

Figure 1. The flow of disaster management by The International Federation of Red Cross and Red Crescent Societies [14]

Disaster management in Malaysia is under the responsibility of the National Security Council (MKN). The responsibility of MKN in disaster management began in 1997 until 2015. MKN provides a mechanism for the recognition of the roles and responsibilities of agencies involved with disaster management. Disaster incidents that are under this directive include natural disasters, such as storms, fires, droughts, coastal erosion, landslides, or accidents triggered by storms and heavy rain [15]. The roles of MKN are now taken over by the National Disaster Management Agency (NADMA). Currently, NADMA acts as a focal point in disaster management in Malaysia together with the state and other federal government agencies [15, 16]. NADMA's roles and responsibilities include coordination of disaster risk reduction programs, public awareness plan delivery, operation of disaster relief trust funds, and deployment of the Special Malaysia Disaster Assistance and Rescue (SMART) team to carry out Search and Rescue (SAR) operations [15]. Generally, NADMA adopts the 4 (four) main strategies consisting of prevention and mitigation, preparedness, response, and recovery [15]. The mechanism cycle of Malaysia's disaster risk reduction is shown in Fig. 2 which consists of 4 (four) main stages. The following paragraph thoroughly defines the terminology for disaster risk reduction including disaster prevention, preparedness, response, and recovery.

Figure 2. Malaysia's disaster risk reduction cycle

Disaster prevention is described as the principle of collaborating in activities aimed at preventing or minimizing future adverse impacts by actions taken beforehand which are the activities designed to protect against the effects of disasters. According to the United Nations International Strategy for Disaster Reduction (UNISDR) [17], prevention is the outright shunning of a hazard's negative impact and connected disasters. World Confederation for Physical Therapy (WCPT) also stresses that while it is impossible to prevent all accidents, improvise risk reduction, evacuation plans, environmental preparation, and standards of design, will reduce the risk of loss of life and prevention of injuries to people. The Hyogo Framework was implemented in 2005 as a ten-year global plan for natural disaster risk reduction with the involvement of 168 countries [18]. The framework provides guiding principles, goals actions, and realistic means to achieve vulnerable communities' disaster resilience [19].

Then, as defined by the UNISDR [17], preparedness is the information and capacities developed by governments, skilled response and recovery organizations, communities, and people to effectively anticipate respond to, and live through, the impacts of doubtless, imminent, or current hazard events or conditions. According to the International Committee of the Red Cross (ICRC), disaster preparedness is applied to mitigate the effects of disasters, either natural or man-made disasters. It is practiced with the combination of risk reduction strategies and avoiding emergency scenarios saving people's lives and livelihoods in any disaster situation that enables the affected community to return to normal conditions within a minimal period [4].

Meanwhile, a response is the provision of emergency services and public assistance during or immediately after a disaster in order to save lives, cut back health impacts, guarantee public safety, and meet the fundamental subsistence of individuals of the folks affected [17]. The distinction between this period and the corresponding recovery stage is not readily seen, mainly on urgent and short-term needs. Current solutions such as a temporary supply of housing and water can continue well during the recovery phase. Stabilizing survivors' physical and emotional health as well as rescuing them from immediate danger are the primary goals of disaster response or relief, which go hand in hand with rescuing the deceased and restoring essential services including water and power. For this level of disaster management, multi-agency collaboration is essential to reduce the impact of a catastrophe and its long-term consequences, such as relief efforts like evacuation, rehabilitation, food and water supply, emergency health treatment, illness, and injury prevention, reconstruction of vital services such as telecommunications and transportation, and temporary shelter delivery [4].

After the occurrence of a landslide disaster, the affected communities are often still vulnerable and most of them are really in need of assistance in recovering. Disaster recovery is when the government and local communities are assisted in rebuilding and recovering from crises, as well as in preventing conflict from returning. In doing so, recovery seeks not only to catalyse long-term development activities but also to build on previous humanitarian efforts to ensure that their contributions become assets for development [19]. Thus, several disaster recovery programmes can be implemented by related agencies to assist the affected communities. Those programmes include (i) rebuilding infrastructure such as homes, schools, hospitals, and roads, (ii) health care and rehabilitation center, (iii) developing activities such as building human resources for health, and (iv) policy and practice development to avoid or mitigate similar situations in the future [9].

2.4. Government strategies and actions over landslides issues

There are lots of strategies and actions taken by the government to address landslide issues. The purpose of these strategies and actions is to lessen the number and effects of natural disasters such as landslides. Malaysia's absence of a database of slopes and landslides is one of the explanations why successful land-use preparation, management, and systematic risk assessment decision making has been challenging. One of the first steps taken, therefore, by the Slope Engineering Division is to collect data for effective slope control by cataloging slopes. A risk evaluation developed based on a ranking system known as Slope Management and Risk Tracking (SMART) has been performed. The analysis of the spatial study is broken down into two stages. Firstly, geomorphological mapping works, and secondly hazard map and paperwork development. Light Detection and Ranging (LiDAR) technology is used for the production of digital terrain models (DTM) in geomorphological mapping works. Shortly, these slopes will be provided with more data quantitative analysis, and risk assessment [20]. Using Geographic Information Systems (GIS) in landslide hazard assessment is also one of the strategies and actions taken by the government to overcome landslide issues. According to Burrough [21], GIS is a powerful set of tools for collecting, storing, retrieving at will, transforming, and displaying spatial data from the real world for a particular set of purposes. GIS is also a strong set of tools to collect, store, retrieve at will, convert, and interpret real-world spatial data for specific purposes. Many previous researchers, engage thoroughly with the benefits of using GIS to determine landslide risk as compared to conventional techniques [21, 22]. GIS's benefits in determining the possibility of landfills include a much broader set of approaches to achievable hazard assessment.

To enable the public to take precautionary measures to reduce their vulnerability, landslide alert systems are introduced to counter landslip threats. This system will assist builders, contractors, and others who are likely to suffer landslip losses. The warning also advises the relevant government agencies and disaster management and organizations to take appropriate actions, such as temporary shelter opening, search and rescue activities, school closing, and relief work. One can break the Landslide Warning (LW) model into two forms, (i) to track landslide-prone areas and (ii) to sensitive positions where significant landslides occur. Slope movement monitoring is conducted in real-time using a surface monitoring device to assess the slope surface motions [20]. Then, to overcome the rainy monsoon seasons, slope inspection for all federal roads was carried out by the Public Work Department (PWD) technical staff with the assistance of the district engineering team. Apart from that, regular inspections of hill slopes are also carried out by local authorities. These efforts are made in order to maintain and reduce the risk of landfill slides. As reported by PWD, a spending cost of nearly RM 140 million per annum is incurred for routine maintenance and minor repairs to road slopes [20]. Thus, regular inspections are needed to be consistently carried out as the maintenance and repair costs could be reduced.

Guidelines and manuals promoting routine work and processes are also good strategies to overcome a disaster. Several guidelines have been created to assist engineers in carrying out their research, especially those at PWD. Guidelines implemented by other existing organisations outlining proper procedures have been adapted with minor changes. For example, the slope maintenance guidelines for Malaysia were adapted with minor changes from the Hong Kong Guidelines for Slope Maintenance. Inspection plans for slopes like personnel and inspection frequency are provided in the guidelines. In 2009, the Town and Country Planning Department, a federal agency, drafted expansion on hilly terrain and highland guidelines to verify that man-made slopes are well developed, correctly engineered, well built, and maintained accordingly [20].

3. Methodology

This research involves 2 (two) different methods, (i) community slope risk mapping process by using IDRISI Selva version 17.0 software and (ii) community slope risk evaluation by using a survey.

3.1. Community slope risk mapping

The research was conducted in Mukim 17 (Batu Ferringhi) and Mukim 18 (Tanjung Bungah and Tanjung Tokong). The areas are located in the Penang Island Northeast District. The district covers half of the northeastern area of the island. The locations have been selected because the focus of this research is mainly on community living and utilising hillside areas in their daily routine. The mapping process of community slope risk used is IDRISI Selva version 17.0 software to produce a geographical information system (GIS)-based Landslide Susceptibility Mapping (LSM).

3.2. Community slope risk evaluation

A survey was used as the evaluation method. Questionnaires were distributed among the communities living in Mukim 17 (Batu Ferringhi) and Mukim 18 (Tanjung Bungah and Tanjung Tokong). The total number of participants for this survey is 62 respondents. The questionnaire used comprises 4 (four) sections, namely (i) Section A: Demographic profile of respondents, (ii) Section B: The state of the communities' environmental knowledge, (iii) Section C: The state of the communities' environmental review, and (iv) Section D: The state of the communities' environmental action and practice. All questions are closed-ended questions. The respondents were required to answer all questions in each section. The data obtained were analysed using Microsoft Excel software.

4. Results and discussion

This section presents the results and discussion of the findings from the analysis of community slope risk mapping and survey. The community slope risk mapping categorises areas from highly risky to lowly risky of the community living in Mukim 17 (Batu Ferringhi) and Mukim 18 (Tanjung Bungah and Tanjung Tokong). Both mappings were interpreted directly from the IDRISI Selva. The community's understanding, knowledge, and awareness of people towards slope risk and landslide were analysed according to the questionnaire's sections namely demographic profile, environmental knowledge, environmental review, and environmental action and practice of respondents. The result and discussion of the community slope risk mapping and community survey are explained in the following subsections.

4.1. Community slope risk mapping in Mukim 17 (Batu Ferringhi)

The image in Fig. 3 is the landslide risk map in Mukim 17 (Batu Ferringhi). The black line represents the boundary of Mukim 17. From the figure, it can be seen that the red region occupied the most area, followed by the purple, green, orange, and lastly yellow regions. The 5 (five) colours shown in Fig. 3 represent the level of risk of the zones, namely red for the very low-risk zone, yellow for the low-risk zone, green for the moderate-risk zone, orange for the high-risk zone, and purple for the very high-risk zone. Figure 3 indicates that the area of Mukim 17 is generally in a very low-risk zone the landslide at the same time, but there are also specific areas in Mukim 17 with very high landslide risk.

In order to increase the accuracy of the risk map, the analysis done by the software also included several parameters such as areas, population, and percentage of the population for each region. The analysis for these parameters is summarised in table 1. From the table, the largest area in Mukim 17 represented by the colour red covers approximately 34.89 km^2. While the purple region area which indicates very high risk covers about 14.01 km^2. The third largest area covered by green comprises an area which is 5.81 km^2 and is indicated as a moderate risk area. From Fig. 3, this green region can be seen in the coastal area. The next region is the area in orange indicating a high-risk zone of 1.971 km^2, while the smallest area is yellow, a low-risk zone covering about 0.02 km^2.

According to City Population [23], the population density per square kilometre for Mukim 17 is 545 persons/km^2 and the total population predicted through the risk map would be 30,902 persons. Thus, based on the risk map population, the total population affected in the very high-risk zone is 7,635 persons, which is 24.71 % of the total population. For the high-risk zone, 1,074 persons are estimated to be affected which is 3.48 % of the total population. While the total risk, which is a combination of high risk and very high risk, is 28.19 % of the population will be exposed to landslides risk.

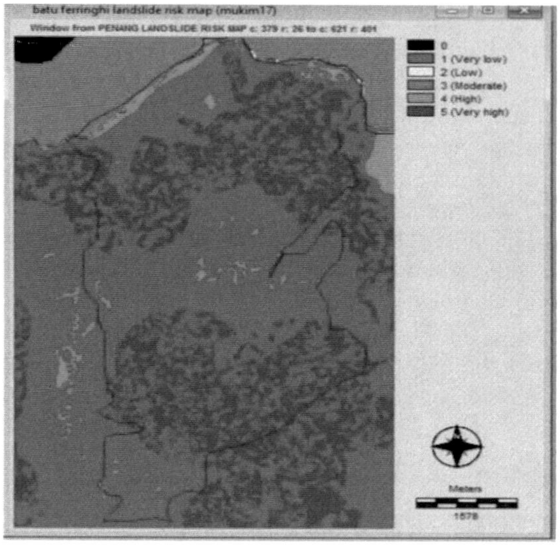

Figure 3. Batu Ferringhi landslide risk map

Table 1. Summary of area and population affected due to landslides in Mukim 17.

No.	Colour	Area (km^2)	Population (person)	Population (%)	Zone
1	Red	34.890	19,015	61.53	Very low risk
2	Yellow	0.020	11	0.04	Low risk
3	Green	5.810	3,166	10.25	Moderate risk
4	Orange	1.971	1,074	3.48	High risk
5	Purple	14.010	7,635	24.71	Very high risk

4.2. Community Slope Risk Mapping in Mukim 18 (Tanjung Bungah and Tanjung Tokong)

Figure 4 shows the landslide risk map in Mukim 18 located in the areas of Tanjung Bungah and Tanjung Tokong. Based on the figure, it can be seen that the red region occupies the most area followed by the purple region, the orange region, the green region, and finally the yellow region. As mentioned in the above section, 5 (five) colours are used to represent the level of risk of the regions, (i) very low, (ii) low, (iii) moderate, (iv) high, and (v) very high. As shown in the figure, most areas in Mukim 18 have a very low risk of landslide occurrences but at the same time, certain places do have a very high risk of experiencing landslides.

Based on analysis done by the software, the largest area which is the very low-risk zone (red) area covers approximately 21.14 km^2 of the total area in Mukim 18. This is followed by the very high-risk zone (purple) which is the second largest area and followed by the high-risk zone (orange), moderate risk zone (green), and low-risk zone (yellow) with an area of 6.10 km^2, 1.51 km^2, 0.81 km^2, and 0.023 km^2, respectively.

In 2015, the population density per square kilometre in Mukim 18 is 3,342 persons/km^2 [23]. Hence, based on the given value, the total population predicted through the risk map is 98,866 persons. Thus, the total population affected in the very high-risk region is 20,386 persons, which is 20.62 % of the total population. For the high-risk region, 5,046 persons are estimated to be affected which is 5.10 % of the total population. Hence the total risk, which is a combination of high risk and very high risk is 25.72 % of the population will be exposed to landslides risk.

The affected area and population for each region were summarised in Table 2.

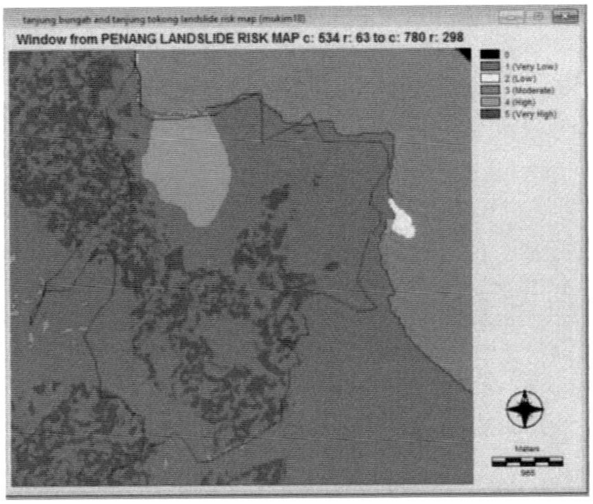

Figure 4. Tanjung Bungah and Tanjung Tokong landslide risk map

TABLE 2. Summary of area and population affected due to landslides in Mukim 18 (Tanjung Bungah and Tanjung Tokong).

No.	Colour	Area (km^2)	Population (person)	Population (%)	Zone
1	Red	21.14	70,650	71.46	Very low risk
2	Yellow	0.023	77	0.08	Low risk
3	Green	0.810	2,707	2.74	Moderate risk
4	Orange	1.510	5,046	5.10	High risk
5	Purple	6.100	20,386	20.62	Very high risk

4.3. Demographic profile of respondents

This section represents the findings of Section A of the questionnaire. Section A comprises respondents' information such as gender, citizenship, age, and occupation. From the analysis, the percentage of females and males involved in this survey is 61.3% and 38.7% respectively. Most of the respondents involved in this survey are Malaysians with a percentage of 93.5% and the balance is non-Malaysian respondents. 54.8% of respondents were aged between the ages of 25-64 years old, while the other 45.2% are between 14-24 years old. In terms of occupation, 70.9% of the respondents are students, followed by 19.4% who are private sector employees, 3.2% are government sector employees and the balance of 6.5% are unemployed or retired. Table 3 shows the distribution of the respondents according to demographic profile.

Table 3. Distribution of respondents according to demographic profile.

n = 62	Demographic Profile	Percentage (%)
Gender	Male	38.7
	Female	61.3
Citizenship	Malaysian	93.5
	Non-Malaysian	6.5
Age	15-24 years old	45.2
	25-64 years old	54.8
	65 years old and above	0
Occupation	Government Sector	3.2
	Private Sector	19.4
	Students	70.9
	Unemployed/Retired	6.5

4.4. The state of the communities' environmental knowledge

This section covers Section B of the questionnaire. Section B consists of 3 (three) questions, (i) the type of natural disaster that results in damage to the environment, property, and human lives, (ii) the potential of landslides areas, and (iii) the type of landslides. The purpose of these questions is to investigate the respondents' understanding, knowledge, and awareness regarding landslide hazards and risks.

The first question is on respondents' opinions regarding the type of natural disaster that results in damage to the environment, property, and human lives. There are about 93.6% (58) of them agreed that landslides do cause damage to the environment, property, and human lives. While the other 4.8% (3) respondents answered not sure, and 1.6% (1) respondents answered no. The graph trend for landslides is also almost the same as that of floods, where most of the respondents agreed with the statement and only 3.2% (2) of them answered no. In the case of drought and heavy rain, the respondents' agreement level decreased to 62.9% (39) respondents for both. While 19.4% (12) and 27.4% (17) of respondents answered no to drought and heavy rain. The remaining respondents who answered not sure about both drought and heavy rain are 17.7% (11) and 9.7% (6) respectively. Overall, many respondents are aware of the landslide's risk factors and hazards due to natural disasters and some of the respondents are lacking knowledge related to the risk factors or damages caused by natural disasters.

The next question revealed the respondents' opinion on the potential of landslides occurring in different types of areas. The options given for this question are flat land, gentle slope, steep slope, mountain, and sea. Most respondents (85.5% (53)) agree that steep slopes are potential landslide areas followed by mountains (80.6% (50)) of respondents. These results show that most respondents are aware and alert to the potential areas of landslides occurring. However, there are still 8.1% (5) and 11.3% (7) respondents who disagreed with this, and the rest have no idea whether both areas are potential landslide areas. While for a gentle slope, the percentage of agreeing is 48.4% (30) respondents, 24.2% (15) respondents disagreed, and about 27.4% (17) of respondents are undecided on whether a gentle slope area is a potential landslide area. The percentage of respondents who disagree that flat land and sea areas are potential landslide areas is 56.4% (35) and 53.2% (33). These results indicate the respondents know the geographical aspects of landslide areas. However, there are also a few respondents who agreed that flat land and sea areas are potential landslide areas with 30.6% (19) of the former and 24.2% (15) of the latter, respectively. While the rest are unsure and have no idea about both areas. Therefore, it could be assumed that several people in the community still do not have any knowledge or awareness regarding the risk factors and hazards of natural disasters as they could not recognise potential landslide areas.

The next question is about respondents' opinions regarding the type of landslides. The number of respondents' opinions about mud flow is about 41.9% (26) of respondents, which is the highest compared to other types of landslides. Meanwhile, around 35.3% (22) of respondents have chosen debris flow as their answer. Based on the results, many respondents do know the types of landslides, however there are also quite a several respondents who still lack knowledge about landslides.

4.5. The state of the communities' environmental review

This section presents the responses in Section C of the questionnaire. There are 4 (four) questions in Section C, (i) respondents' awareness of the risks of landslides, (ii) respondents' opinion in considering landslides as a threat, (iii) respondents' opinion on whether the speed of landslides able to save people's lives and (iv) potential risk for people living near the slope. These questions identify the level of knowledge and awareness of the risk factors and potential hazards of natural disasters such as landslides.

There are about 32.3% of the respondents responded to always being aware of the risks of landslides. Approximately 29% of the respondents responded to being aware sometimes and 24.2% of them responded to often being aware of the risks of landslides. However, almost 8.1% and 6.5% of the respondents are never and are rarely aware, of the risks of landslides. The lack of awareness of the respondents would place them in a dangerous zone that could threaten their lives. The reason people are not aware of the risks of landslides is due to the lack of knowledge and awareness regarding landslides. Overall, the results look quite dangerous because there are only 32.3% of the respondents are always aware of landslide risks, while the rest are considered lacking in awareness. To overcome this situation, people should learn and understand more about landslides risks with the help and support from the government and local authorities, as well as the surrounding communities in order to prepare themselves for handling and facing landslides, including the early detection of landslides.

The next question is about the respondents' opinions on considering landslides as a threat. The analysis revealed about 33.9% of the respondents responded that they often consider landslides as a threat and approximately 29% of them always consider landslides as a threat. While about 21% of the respondents stated that they sometimes consider landslides as a threat and another 9.7% and 6.5% of them never and rarely consider landslides as a threat. In general, people that never and rarely consider landslides as threats need to change their mindset as they will put their lives in danger. Enhancing people's knowledge and awareness are critical issues that need to be solved by the government and local authorities.

The respondents also have been asked whether the speed of landslides can save people's lives. In fact, the speed of landslides depends on several factors such as landslide material and movement. Particularly, the height and gradient of the land will determine the speed of a landslide movement. The steeper the slope, the more likely the components to gush downhill. Based on the analysis, about 54.8% of the respondents have responded that the ability or possibility to save people's lives is very less or not at all with the speed of the landslide flow. Yet, there are still 22.6% of respondents who said it is possible to save people's lives, and another 22.6% of them are not sure. Overall, the result shows that there are still people with lack knowledge regarding the speed of landslides and the ability or possibility of saving people's lives during landslides incidences.

Besides that, the respondents also have been evaluated for the potential risk of living near the slope. The findings exposed many respondents to a lack of knowledge about the risk of people living near the slope. However, most respondents strongly agreed with these two (2) potential risks namely landslides causing injury and the possibility of death by being buried alive or other reasons. In addition, the possibility of being emotionally disturbed has also been agreed to by the respondents. The least number of responses given is for strong disagreement regarding other potential risks, which are disease due to landslides and possibilities of illness due to supply shortage. Besides that, there are also a high number of neutral responses for the potential risk of being exposed to disease due to landslides and possibilities of illness due to supply shortage. Overall, the respondents show that they are quite aware of the risks and effects of landslides.

4.6. The state of the communities' environmental action and practice

This section presents the respondents' responses in Section D which consists of 3 (three) questions namely (i) the availability of landslide awareness campaign, (ii) the actions taken by local authorities to curb landslide issues, and (iii) obedience to the evacuation order related to any natural disaster occurrence. These questions identify the knowledge and awareness level among respondents towards the risk and hazards of natural disasters like landslides. Additionally, the questions also would enhance the awareness of respondents as the questions might increase their curiousity about natural disasters and landslides, especially on steps that should be taken when handling the disaster.

The analysis exposed approximately 77.4% of the respondents who say that they never, rarely, and sometimes notice or attend the landslide awareness campaign. The other 12.9% and 9.7% of them say that they often see or attend landslide awareness campaigns. The result indicates that most people are not aware of or do not care about the landslide awareness campaign. Another possible reason is that the government or local authorities only conduct a few campaigns in the community area of residence. It is dangerous, particularly for those with a lack of knowledge regarding the risk and hazard of landslides as this will make them vulnerable during a disaster.

The respondents also have been asked about the frequency of actions taken by local authorities to curb landslide issues. This question identifies the awareness of respondents regarding the actions taken by local authorities to curb landslide issues. The analysis revealed about 62.9%, believe that local authorities only take action sometimes and 14.5% of them said that local authorities always take action to curb landslide issues. The other 12.9% of the respondents said local authorities often take action to curb landslide issues, while the rest said local authorities rarely (6.5%) and never act (3.2%) to curb landslide issues. Most of the respondents' impressions of the frequency of actions taken by local authorities to curb landslide issues are "sometimes" instead of "often" or "always". This indicates that they do feel that local authorities must be more proactive in issues relating to landslides. Furthermore, the involvement of local government in landslide issues would eventually create some awareness regarding the risk and hazard as well as enhance responsiveness during the incidence of natural disasters like landslides.

In terms of communities' compliance with evacuation orders in the event of any natural disaster, 41.9% of the respondents responded to will always obey the evacuation orders given. Followed by, 22.6% and 14.5% of them who had responded to will sometimes and often obeyed the evacuation orders. The remainder which is around 11.3% and 9.7% respectively say that they never and rarely obey evacuation orders in the event of any natural disaster. Based on the result, indicates that some people in the community lack knowledge and awareness regarding the effects of natural disasters on them or their families. Therefore, in order to increase the knowledge and awareness level of society about natural disasters, people must be educated by way of organising interesting activities or programmes that could attract them to attend and participate. This might prevent them from being victims of natural disasters like landslides.

5. Conclusion

The community slope risk map for each location has identified that both Mukim 17 and Mukim 18 consist of very high-risk and high-risk zones that can affect thousands of people in the community residing there as well as those using the infrastructure in those areas as part of their daily routine. It is predicted that 24.71% out of 30,902 persons will be affected in the very high-risk zone in Mukim 17. While in Mukim 18, it is predicted that 20.62% out of 98,866 persons would be affected in the zone of very high risk. Thus, more than 50% of the population in Mukim 18 and Mukim 17 only face minimal risk from landslide occurrence. However, more precautionary and safety measures are needed to take into account those residing in very high-risk and high-risk zones. It is recommended to implement more actions and strategies on slope usage to prevent losses of human lives, economic losses due to infrastructure damages, and burdensome reconstruction costs.

The evaluation of the understanding, knowledge, and awareness of the community towards slope risk and landslide of the 62 respondents involved in this study shows that most of the communities have a general idea of landslides. In order to enhance their level of understanding, knowledge, and awareness of landslides and slope risk, local authorities should organise campaigns, programmes, or exhibitions that are related to landslides and slope risk. These events should be attended by all communities and the involvement of those residing in very high-risk and high-risk zones should be made compulsory. Local authorities should also conduct training sessions on community emergency responses in the event of a disaster. Local authorities could also organise workshops that require community leaders to attend. By implementing these programmes, the level of communities' understanding, knowledge, and awareness of landslides and slope risk would be enhanced, thus helping the government to address the situation. Therefore, sufficient understanding, knowledge, and awareness of landslides and slope risk would help in reducing the incidence of landslides and slope risk as early signs of impending landslides can be identified by the communities at risk.

Acknowledgement
The authors would like to acknowledge the Universiti Sains Malaysia for funding this research through the Research University grant scheme Project Code (8014094).

References

[1] Abdullah C H 2013 Landslide risk management in Malaysia *WIT Transactions on the Built Environment* **133** 255–265
[2] Aronoff S 1989 Geographic information systems: A management perspective *Geocarto International* **4:4**, 58
[3] Azimi M A, Zakaria S S and Majid T A 2019 Disaster risks from economic perspective: Malaysian scenario *IOP Conf. Series: Earth and Env. Sci.* **244** 012009
[4] Bamgbose O A, The Law and Practice on Disaster Issues, (Safari Books Ltd, Onireke, 2018), pp. 445.
[5] Burrough P A 1986 Principles of geographical information systems for land resources assessment. Geocarto International **1:3** 54
[6] World Bank Group 2021 *Historical monthly temperature and precipitation for Malaysia during the time period 1991-2020* World Bank Group: Climate Change Knowledge Portal. Available from https://climateknowledgeportal.worldbank.org/country/malaysia/climate-data-historical
[7] Center for Excellence in Disaster Management and Humanitarian Assistance 2019 *Malaysia Disaster Management Reference Handbook* Available from https://reliefweb.int/report/malaysia/malaysia-disaster-management-reference-handbook-june-2019
[8] Center for Excellence in Disaster Management and Humanitarian Assistance 2016 *Malaysia Disaster Management Reference Handbook* Available from https://reliefweb.int/report/malaysia/malaysia-disaster-management-reference-handbook-2016
[9] City Population 2010 *The population of districts and parishes (mukims) of Malaysia by census years.* Available from https://www.citypopulation.de/en/malaysia/mukim/admin/
[10] Crozier M J 1986 *Landslides: causes, consequences and environment.* (London: Croom Helm) pp. 252
[11] International Federation of Red Cross and Red Crescent Societies, About disaster management. Available from https://www.ifrc.org/en/what-we-do/disaster-management/about-disaster-management/ (accessed on 21 June 2021).
[12] S. Maadan, Causes, Effects and Types of Landslides, Earth Eclipse. Available from https://eartheclipse.com/natural-disaster/causes-effects-and-types-of-landslides.html (accessed on 21 June 2021).
[13] Margarint M C and Niculita M 2017 Landslide type and pattern in Moldavian Plateau, NE Romania," in *Landform dynamics and evolution in Romania*, edited by M. Radoane and A. Vespremeanu-Stroe (Springer Cham, Switzerland) pp. 271–304
[14] Nikita, Natural and Man-Made Disaster and their Impact on Environment. Biology Discussion. Available from https://www.biologydiscussion.com/essay/natural-and-man-made-disaster-and-their-impact-on-environment/7134 (accessed on 21 June 2021).
[15] J. I. Nkpadobi 2009 *Stability of slope cuts in the Dinding schist of the Ukay Perdana area, Selangor Darul Ehsan, Malaysia* (Malaysia: Master thesis, Universiti Malaya)

[16] Physiopedia Contributors, Disaster Management. Physiopedia. Available from https://www.physio-pedia.com/Disaster_Management (accessed on 21 June 2021).
[17] Popescu M E 2002 *Landslide causal factors and landslide remediatial options* 3rd International Conference on Landslides, Slope Stability and Safety of Infrastructures (CI-Premier PTE LTD, Singapor) pp. 61–81
[18] Terzaghi K, Peck R B and Mesri G 1996 *Soil Mechanics in Engineering Practice third edition* (New York: John Wiley and Sons Inc.) pp. 534
[19] The Star 2018 Malaysia among countries especially prone to landslides. Available from https://www.thestar.com.my/news/nation/2018/12/04/msia-ranks-highly-for-landslides-country-experienced-185-occurrences-annually-in-past-10-years
[20] United Nations International Strategy for Disaster Reduction 2009 *UNISDR Terminology on Disaster Risk Reduction* Available from https://www.undrr.org/publication/2009-unisdr-terminology-disaster-risk-reduction
[21] United Nations Office for Outer Space Affairs, Risks and Disasters. Available from https://www.un-spider.org/risks-and-disasters (accessed on 21 June 2021).
[22] World Confederation for Physical Therapy 2016 *WCPT Report: The role of physical therapists in disaster management* Retrieved from https://world.physio/sites/default/files/2020-06/Disaster-Management-Report-201603.pdf
[23] Zimmermann M and Keiler M I 2015 International frameworks for disaster risk reduction: Useful guidance for sustainable mountain development? *Mountain Research and Development* **35(2)** 195–202

The Importance of Village Resilience Assessment in Village Based Programs to Support Regional Resilience

Virana Fatwa Nurmala[1], Berton Suar Pelita Panjaitan[2], Adi Subiyanto[3,]

[1]Defense University of the Republic of Indonesia, [2]National Disaster Management Agency, [3]Defense University of the Republic of Indonesia

Email :viranafatwa@gmail.com,

Abstract. Village Resilience Assessment is a process of assessing resilience that is carried out by the community independently to determine the level of resilience. This assessment framework adopts village and urban village disaster resilience standards in SNI 8357:2017 and adapts the DROP (Disaster Resilience of Place) model. The methodology in this study uses a qualitative method with a descriptive analysis approach. Assessment of village disaster resilience produces several outputs, such as Village Resilience Index Values, Village Resilience Levels, and recommendations for action to increase village capacity. The Village Resilience Index value consists of three levels, namely Tangguh Utama <83.33, Tangguh Madya 58.3 – 83.33, and Tangguh Pratama <58.3. The results show that the Village resilience assessment is important because the Village Resilience Assessment is a tool to measure the level of village resilience in a simple way so that it will make it easier for village communities to carry out action plans to reduce vulnerability and increase capacity. disaster resilience village which aims to map the conditions and initial description of the level of risk, level of vulnerability and level of village capacity in dealing with disaster threats.
Keywords: Disaster Resilient Village, Assessment of village resilienc, Collaboration

1. Introduction

The territory of the Unitary State of the Republic of Indonesia has geological, geographical, hydrological, demographic and sociological conditions that make it vulnerable to disasters, both natural and social disasters. Disaster events, both extensive and intensive, have resulted in suffering, an increase in the number of persons with disabilities and loss of life, loss and damage to assets of individuals, private and state.

The focus of disaster management in sustainable development is disaster risk management which aims to reduce, avoid and transfer the impact of hazards through activities and measures for prevention, disaster mitigation and preparedness. Community-based disaster risk management that places the community as an important actor is the main strategy of disaster risk management initiatives to build disaster resilience. This is through strengthening the smallest government units that can directly target the community, namely in the village area. Based on data from the Directorate General of Population and Civil Registration (Dukcapil) of the Ministry of Home Affairs, in Indonesia there are 83,381 villages spread over 34 provinces throughout Indonesia and 53. 000 of them are included in disaster-prone villages. This means that more than 51 million families in Indonesia live in disaster-prone areas.

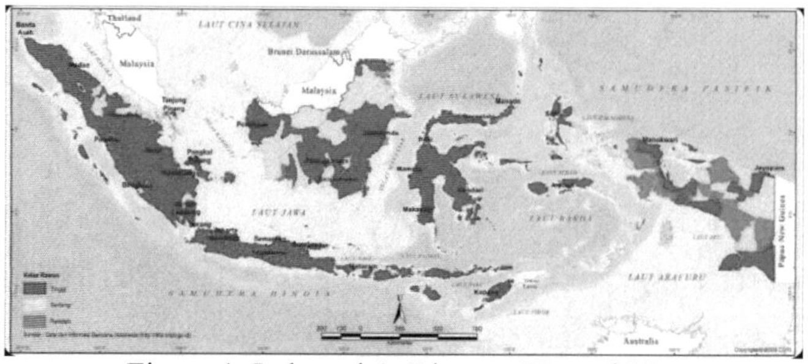

Figure 1. Indonesia's Disaster Hazard Map

Development of a community-based disaster management strategy in order to strengthen the community at the village level, namely the Disaster Resilient Village Program (destana) which was initiated in 2012 by the Directorate of Preparedness, Deputy for Prevention of the National Disaster Management Agency. The program with the title Disaster Resilient Village is a community-based risk management program with the hope that the community will not only become the object of the process but become the subject and can be actively involved in reviewing, analyzing, handling, monitoring and evaluating disaster risk reduction efforts in their area by maximizing disaster risk reduction. existing local resources.

The disaster resilient village program developed by the National Disaster Management Agency includes several activities, one of which is the Village Resilience Assessment. This village resilience assessment activity is one of the main activities in the village disaster resilience program. However, to what extent is the assessment of village resilience so important, and how is the process of assessing village resilience to support the creation of a disaster-resilient village.

2. Methodology

This study uses a qualitative method with a descriptive analysis approach. Qualitative research is used to explore and deepen a socio-environmental phenomenon that cannot be quantified. This study uses data sourced from primary data (interviews) and secondary data (documents). The subject under study is the Village Resilience Assessment in the disaster resilient village program of the National Disaster Management Agency. The informants selected in this study used a purposive sampling technique, namely the Director of Preparedness of the National Disaster Management Agency, the Head of the Sub-directorate for Strengthening Community Resilience of BNPB, the National Destana Facilitator Team and the Staff of the BNPB Preparedness Directorate. The interview technique was carried out freely and was designed with the concept of discussion or Focus Group Discussion (FGD) with resource persons as informants. Data triangulation technique is used to validate data by checking data from various sources. The author also conducts participatory observation, namely collecting data through direct observation, feeling and being in observation activities and the life of the object of observation.

3. Research Results and Discussion

3.1 Disaster Resilient Village

Disaster is not only a humanitarian issue but also an investment in future development. Therefore, disaster management efforts are the business of all parties, namely the government, non-governmental organizations, the business world, and the active participation of the community as mandated in Law Number 24 of 2007 concerning Disaster Management.

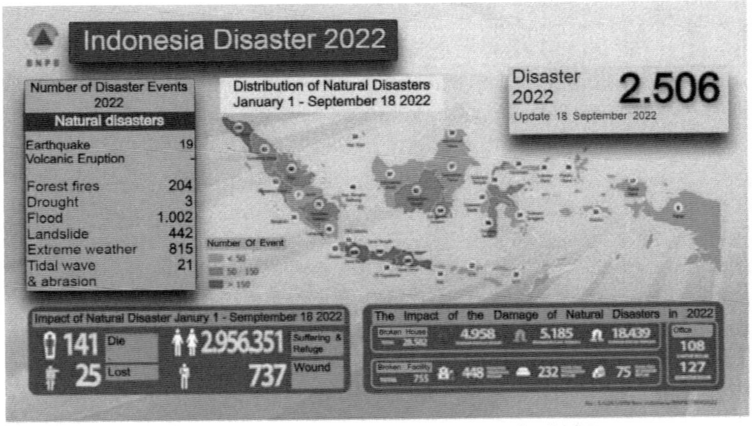

Figure 2. Indonesia Disaster Data in 2022

A disaster-resilient village is a village that has the independent ability to adapt and deal with disaster threats, as well as recover quickly from the adverse effects of disasters, if they are hit by a disaster. Disaster-resilient villages have the ability to recognize threats in their area and are able to organize community resources to reduce vulnerability and at the same time increase capacity to reduce disaster risk. The ability to recognize threats to reduce vulnerability and increase capacity is initiated by using a tool called village resilience assessment. This village resilience assessment is a tool used in one of the series of activities to form a disaster-resilient village.

Efforts to deal with the community are carried out by the village independently or by other parties to ensure its sustainability. Without government intervention, each village has had its own resilience, which we know as local wisdom. To suspend the village can not be done within a year. It takes a process to build sustainable resilience. Making disaster awareness culture into development values in the village.

The scope of activities in the disaster resilient village program includes:

1. Organizing the socialization of village activities and the introduction of the basics of disaster management to the village government, forums/institutions that already exist at the village level.
2. The Village Resilience Assessment is the initial activity in the Destana activity which aims to map the conditions and initial description of the risk level, vulnerability level and village capacity level in dealing with the tsunami threat as a baseline for the level of village disaster resilience. The implementation identification process is carried out with community participation by involving stakeholders at the village level and taking into account the representation of vulnerable groups and gender. Assessment of village resilience is carried out at the initial and final stages of village implementation to measure the impact of the resilience efforts that have been implemented.
3. Establishment of a Disaster Risk Reduction Forum (FPRB). Facilitate the establishment of a Disaster Risk Reduction Forum in the target Village. The forum must be institutionalized in a sustainable manner at the village level and have clear coordination with the local government, BPBD in handling emergency disaster response. This forum includes elements of the village-

level pentahelix. It is hoped that in the future, this forum will become part of the sustainability of the Destana program and coordinate with the Regency/City Government (BPBD). The legalization of this forum is determined through a Village Head Decree (SK).

4. Formation of Village Disaster Management Volunteers Village level Disaster Management Volunteers are selected from volunteers who are already in the village with regional, age, gender, and vulnerable groups represented through a Village Head Decree. The process of forming volunteers is carried out through a series of meetings with the active involvement of the community.
5. Preparation of community-based risk maps. Preparation of a participatory village risk map with the active involvement of the community. Communities are empowered to be able to recognize potential disaster threats, locations of gathering points, locations of strategic buildings and temporary evacuation sites in villages. Understanding the things above will help community resilience. The preparation of the risk map is carried out in a participatory manner and is accompanied by a Village Facilitator. The resulting village risk map is printed and posted at the village hall.
6. Preparation of community action plans for disaster risk reduction (RAK PRB). The process of preparing the Community Action Plan for Disaster Risk Reduction (RAK-PRB) must be carried out in a participatory manner involving elements of the village-level pentahelix based on the disaster risk map that has been prepared. This document contains a plan for priority village activities in implementing disaster risk reduction efforts within a certain period of time. The community action plan takes into account the geographical conditions and the availability of infrastructure in the village area. In order to ensure the sustainability of the community action plan for disaster risk reduction, it must be integrated into the village medium-term development plan (RPJMDes).
7. Formation of disaster-resilient families (Katana) in the village. Facilitate the selection and preparation of Disaster Resilient Families in the Village with criteria for selecting resilient families including knowledge of disaster risk, knowledge of disaster emergencies, knowledge of access to formal disaster information, knowledge of locations and evacuation routes. The selected disaster-resilient families will be provided with a family preparedness kit (e.g., a disaster preparedness bag or a first aid kit).
8. Preparation of standard operating procedures for community-based early warning. The preparation of standard operational warning procedures aims to provide guidance to the public in responding to disaster early warning information. The preparation of standard operating procedures is carried out based on local wisdom by taking into account the structure and culture and habits of the local community. This activity will also identify early warning mechanisms/tools/activities at the community level based on local wisdom and existing assets in the community.
9. Preparation of village-level Independent Evacuation Plans and evacuation trials/simulations. The process of preparing the evacuation plan document is carried out through the active role of the community. This activity includes the introduction and mapping of evacuation signs, evacuation route signs and village level temporary gathering places. In addition, every family/individual is empowered to be able to understand the threat conditions of the surrounding environment and to identify ways to save themselves in the event of a disaster. This activity includes blocking and installing evacuation route maps in crowded places. The pilot of the Independent Evacuation Plan was carried out after the Village Self-Evacuation Plan was drafted, in which the evacuation route was determined, the location points for the evacuation route signposts, location of gathering/evacuation points and standard operating procedures or evacuation implementation guidelines with their organizational structure. The preparation process involves the community as well as vulnerable groups in the community including families who are considered the most vulnerable (disabled, elderly, children under five and women) to find out when the evacuation point is reached from the location of the trial

participant's house, and the completeness and testability of the independent evacuation plan that is prepared.

10. Structural/non-structural mitigation. One of the efforts to reduce the risk and impact of disasters is through mitigation efforts, both structural and non-structural mitigation. Structural mitigation forms can be in the form of artificial or natural construction that can reduce or minimize the impact of disasters. Non-structural mitigation is all efforts/actions that are not physically related to reduce disaster risks and impacts, can be in the form of regulations, policies, socialization, government and community capacity building, training and disaster mitigation simulations. The proposed structural mitigation needs refer to the disaster risk reduction community action plan document that has been prepared. The proposed structural mitigation needs already contain detailed specifications. Structural mitigation forms can be in the form of vegetation mitigation,
11. The process of implementing structural mitigation activities is carried out through the active participation of the community where the community will be facilitated to identify artificial or natural forms of structural mitigation that can be utilized for tsunami disaster mitigation. Village communities will be facilitated and accompanied by facilitators from the planning, implementation to maintenance stages in order to maintain the sustainability of the structural mitigation that is built. The entire series of activities carried out at the village level is carried out through the method of discussion group forums with community deliberations / village deliberations with a pattern of community assistance by the Village Facilitator.

3.2 Village Resilience Assessment Process and Method

In the regional disaster resilience assessment, there are 5 components, 32 indicators and 128 questions. The questions asked started at the pre-disaster, emergency response, and post-disaster stages. In the First component, namely Basic Services, there are fourteen indicators assessed, namely Indicator 1). Implementation of Education Services, Indicator 2). Health Service Delivery, Indicator 3). Transportation Facilities and Infrastructure, Indicator 4). Provision of Clean Water, Indicator 5). Food & Nutrition Programs and Activities, Indicator 6). Empowerment of Micro, Small and Medium Enterprises and the Economy through Village-owned Enterprises, Indicator 7). Communication/Information Facilities and Infrastructure, Indicator 8). Administration and Population Services, Indicator 9). Environmental Protection and Management Regulations/Activities, Indicator 10). Religious Regulations and Activities, Indicator 11). Cultural Regulations and Activities, Indicator 12). Security Protection Regulations and Activities, Indicator 13). Regulations and Activities for Women's Empowerment and Child Protection, Indicator 14). Special Services (Disabled/Disabled). In this first component, it refers to Permendagri Number 101 of 2018 concerning Minimum Service Technical Standards for District/City Regional Disaster Affairs. The aim is that this Village Disaster Resilience Assessment can be used in all Ministries/Agencies that have similar village-based Programs. So that in the future the assessment of village resilience can be integrated with each other with other ministry/institutional programs. Regulations and Activities for Women's Empowerment and Child Protection, Indicator 14). Special Services (Disabled/Disabled). In this first component, it refers to Permendagri Number 101 of 2018 concerning Minimum Service Technical Standards for District/City Regional Disaster Affairs. The aim is that this Village Disaster Resilience Assessment can be used in all Ministries/Agencies that have similar village-based Programs. So that in the future the assessment of village resilience can be integrated with each other with other ministry/institutional programs. Regulations and Activities for Women's Empowerment and Child Protection, Indicator 14). Special Services (Disabled/Disabled). In this first component, it refers to Permendagri Number 101 of 2018 concerning Minimum Service Technical Standards for District/City Regional Disaster Affairs. The aim is that this Village Disaster Resilience Assessment can be used in all Ministries/Agencies that have similar village-based Programs. So that in the future the assessment of village resilience can be integrated with each other with other ministry/institutional programs. The aim is that this Village Disaster Resilience Assessment can be used in all Ministries/Agencies that have similar village-based

Programs. So that in the future the assessment of village resilience can be integrated with each other with other ministry/institutional programs. The aim is that this Village Disaster Resilience Assessment can be used in all Ministries/Agencies that have similar village-based Programs. So that in the future the assessment of village resilience can be integrated with each other with other ministry/institutional programs.

The second component, namely Regulations and Policies for Disaster Management, contains five indicators, namely Indicator 15). Disaster Risk Assessment, Indicator 16). Disaster Management Plan, Indicator 17). Disaster Management Training, Indicator 18). Disaster Management Institutions (Disaster Management Groups/Organizations), and Indicator 19). Assistance/Cooperation in the Implementation of Disaster Management Activities. Furthermore, the third component, namely Prevention and Mitigation which includes 3 Indicators, namely Indicator 20). Disaster Prevention, Indicator 21). Disaster Mitigation, and Indicator 22). Dissemination of Disaster Knowledge. Next is the fourth component, namely Emergency Preparedness which includes six indicators, namely Indicator 23). Hazard Warning, Indicator 24). Dissemination of Hazard Warning, Indicator 25). Evacuation Plan, Indicator 26). Evacuation Path, Indicator 27). Refugee Places, Indicator 28). Preparedness Test/Simulation. And finally, the fifth component is Recovery Preparedness which includes four indicators, namely Indicator 29). Independent Post-Disaster Assessment Mechanism, Indicator 30). Independent Early Recovery Mechanism, Indicator 31). Transparent and Accountable Self-Aid Management Mechanism, Indicator 32). Rebuilding Better

In assessing disaster resilience, villages have principles and methods. namely Self-assessment by the village based on evidence, Involving elements of the village community, Focus Group Discussion to agree on the assessment, can be assisted by the Facilitator. In addition, the participants involved in the Focus Group Discussion to verify and agree on the results of the assessment ideally are the Village Head, Village Government Apparatus, Village government division who know and understand the village administration system, BPD Representatives (Village Representative Body), village leaders, Representatives of disaster management groups/organizations in the village, representatives of social organizations, women, youth and professions (farmer groups, fishermen), representatives of people with disabilities/disabled people, representatives of people living in high-risk areas,

In carrying out the village resilience assessment process, it is necessary to be accompanied by a facilitator who has previously attended the facilitator training so that they understand the substance of the assessment, understand the evidence and sources of verification evidence, and understand the assessment procedure. The role of the facilitator here is to help participants understand the questions so that participants can provide answers according to the actual conditions of the village, ensure the availability of supporting evidence data/documents, and ensure the relevance of supporting evidence data/documents.

Implementation of the village resilience assessment has several stages that must be passed, namely:

- The first stage is preparation by coordinating with the village government with the aim of agreeing on an assessment of village resilience, identifying participants and supporting data, and determining the time and place for the village resilience assessment.
- The second stage is the collection of supporting data by ensuring that the data needed for the assessment of village resilience is available
- The third stage is the implementation of the village resilience assessment. In carrying out the village resilience assessment to ensure that participants understand the substance of the village resilience assessment, agree on the results, and ensure the availability of supporting data.
- The fourth stage is the submission of the results of the village resilience assessment by confirming and validating supporting data evidence, compiling reports, and submitting results and recommendations.

Furthermore, verification evidence according to the 5 components and 32 indicators in the village resilience assessment is described in the table below:

Table 1. Component 1: Basic Service

Indicator	Proof of Verification
1. Provision of Educational Services	Verification evidence in the form of photos of formal and non-formal education facilities and administration
2. Implementation of Health Services	Verification evidence in the form of photos of health facilities and service delivery
3. Transportation Facilities and Infrastructure	Verification proof in the form of photos of transportation facilities/infrastructure
4. Provision of Clean Water	Verification evidence in the form of photos of clean water supply facilities/infrastructure
5. Food & Nutrition Programs and Activities	Verification evidence in the form of photos of food and nutrition program activities
6. Empowering MSMEs and the Economy through BUMDES	Verification evidence in the form of photographic documents and facilities and infrastructure for micro, small and cooperative businesses for economic development
7. Communication/Information Facilities and Infrastructure	Evidence of verification in the form of information/communication service procedure documents (in MS Word/Excel/PDF format)
8. Administration and Population Services	Evidence of verification in the form of administrative and population service procedure documents (in MS Word/Excel/PDF format) and photos of activities
9. Environmental Protection and Management Regulations/Activities	environmental protection and management regulatory documents (in MS Word/Excel/PDF format) and photos of activities
10. Religious Regulations and Activities	religious regulation documents (in MS Word/Excel/PDF format) and photos of activities
11. Regulations and Cultural Activities	cultural regulation documents (in MS Word/Excel/PDF format) and photos of activities
12. Security Protection Regulations and Activities	security protection regulatory documents (in MS Word/Excel/PDF format) and photos of activities
13. Regulations and Activities for Women's Empowerment and Child Protection	documents on women's empowerment and child protection (in MS Word/Excel/PDF format) and photos of activities
14. Special Services (Disabled)	special service regulation documents for the disabled (in MS Word/Excel/PDF format) and photos of activities

Table 2. Component 2: Disaster Management Regulations and Policies

Indicator	Proof of Verification
15. Disaster Risk Assessment	disaster risk assessment document (in MS

	Word/Excel/PDF format)
16. Disaster Management Plan	Disaster Management Plan document (in MS Word/Excel/PDF format)
17. Disaster Management Training	photos or documents of disaster management training reports (in MS Word/Excel/PDF format)
18. PB Institutions (PB Group/Organization)	photos or documents of disaster management organization/group structure (in MS Word/Excel/PDF format)
19. Assistance/Cooperation in the Implementation of Disaster Management Activities	photos or reports of cooperation reports on disaster management activities with external parties (in MS Word/Excel/PDF format)

Table 3. Component 3: Prevention and Mitigation

Indicator	Verification Proof
20. Disaster Prevention	Activity reports and activity photos
21. Disaster Mitigation	Reports and photos of activities
22. Dissemination of Disaster Knowledge	Reports and photos of activities

Table 4. Component 4: Emergency Preparedness

Indicator	Proof of Verification
23. Danger Warning	photo or document of receipt of disaster warning
24. Dissemination of Danger Warnings	dissemination of flood hazard warnings, can be in the form of; (1) SOP document for disseminating disaster warnings (if any), or (2) photos of tools used for disseminating disaster warnings to the public
25. Evacuation Plan	photo of evacuation map or disaster evacuation plan document
26. Evacuation Path	photo of disaster evacuation routes and signs
27. Refugee Places (more than 1 type of disaster)	photos of disaster shelters
28. Preparedness Test/Simulation (more than 1 type of disaster)	photos of the implementation of disaster preparedness/simulation exercises

Table 5. Component 5: Recovery Preparedness

Indicator	Proof of Verification
29. Independent Post-Disaster Assessment Mechanism	Guidance Documents/mechanisms, SOPs/mechanisms for communities/villages/kels in conducting post-disaster assessments

30. Self-Recovery Mechanism	Guidelines, SOPs/mechanisms. Community/village/ reports in carrying out recovery
31. Transparent and Accountable Aid Management Mechanism	Guidelines, SOPs/mechanisms. Report on community/village/ activities in aid management
32. Rebuilding Better	Guidelines, SOPs/mechanisms. Reports on community/village/ activities in rebuilding for the better

Below is an illustration of the data input and data analysis process in assessing village/output resilience

Answers, explanations, and verification documents are inputted and written on an Excel file-based analysis tool (Village Resilience Level Calculating Software). And then it will automatically look at the calculated index value and village resilience level. The village disaster resilience assessment then produces several outputs, such as Village Resilience Index Values, Village Resilience Levels, and recommendations for action to increase village capacity. First, the Major Disaster Resilient Village (score > 83.33). This level is the highest level that can be achieved by a village that participates in this program. This level is characterized by:

- The existence of a disaster risk reduction policy that has been legalized in the form of a Village Regulation or legal instrument at the level;
- There is a disaster management planning document that has been integrated into the RPJMDes and detailed into the RKPDes;
- The existence of a disaster risk reduction forum consisting of community representatives, including women's groups and vulnerable groups, and village government representatives, which function actively;
- The existence of a village disaster management volunteer team that is routinely actively involved in capacity building, knowledge and disaster education activities for its members and the community in general;
- There are systematic efforts to conduct risk assessment, risk management and vulnerability reduction, including alternative productive economic activities to reduce vulnerability;
- There are systematic efforts to increase disaster preparedness and response capacity.

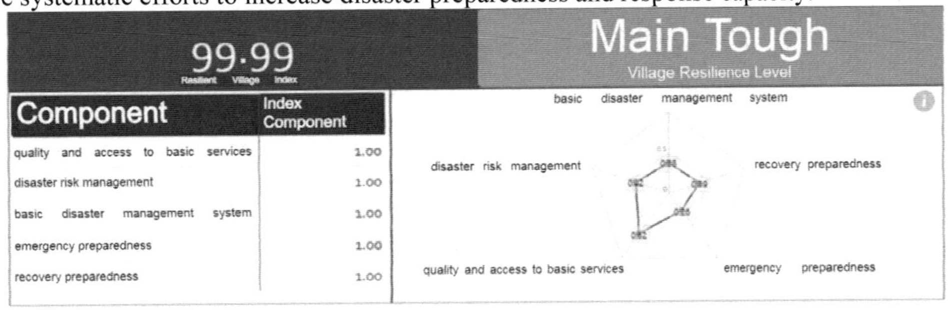

Figure 3: an example of the index value and the level of village disaster resilience in the village Cikawung ading, Cipatujah District, Tasikmalaya Regency

Second, Madya Disaster Resilient Village (score 58.33 – 83.33). This level is an intermediate level which is characterized by:

- There is a disaster risk reduction policy that is being developed at the village or sub-district level;

- The existence of disaster management planning documents that have been compiled but have not been integrated into village planning instruments;
- The existence of a disaster risk reduction forum consisting of representatives from the community, including women's groups and vulnerable groups, but not yet fully functional and active;
- The existence of a village disaster management volunteer team that is involved in capacity building, knowledge and disaster education activities for its members and the community in general, but is not routine and not very active;
- There are efforts to conduct risk assessment, risk management and vulnerability reduction, including alternative productive economic activities to reduce vulnerability, but these have not been tested;
- There are efforts to increase disaster preparedness and response capacity that have not been tested and systematic

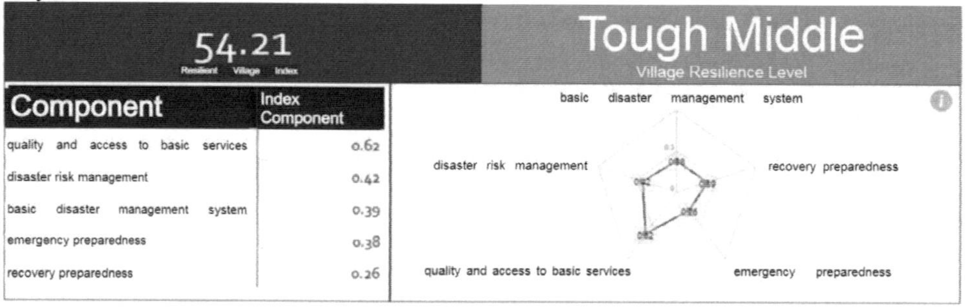

Figure 4: Examples of index values and village disaster resilience levels in Gadingsari Village, Sanden District, Bantul Regency

Third, Primary Disaster Resilient Village (score < 58.33). This level is the initial level which is characterized by:

- There are initial efforts to develop disaster risk reduction policies at the village or sub-district level;
- There are initial efforts to prepare disaster management planning documents;
- There are initial efforts to establish a disaster risk reduction forum consisting of representatives from the community;
- There are initial efforts to form a village disaster management volunteer team;
- There are initial efforts to conduct risk assessment, risk management and vulnerability reduction;
- There are initial efforts to increase disaster preparedness and response capacity;

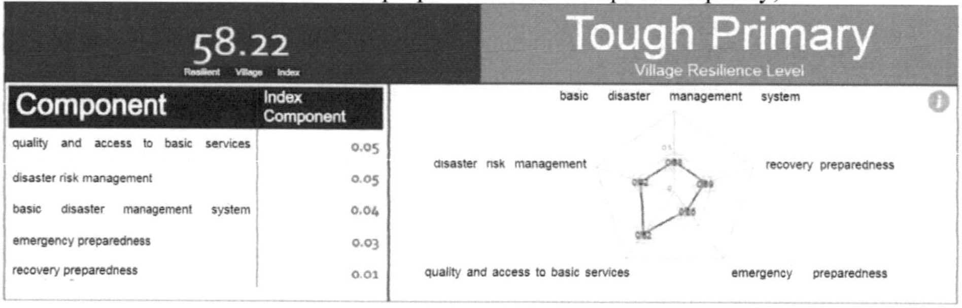

Figure 5 example of the index value and the level of village disaster resilience in Pulosarok Village, Singkil District, Aceh Singkil Regency

From the results of the index value and the level of village disaster resilience, it will produce recommendations for an action framework to increase disaster resilience. Recommendations and action plans may vary for each village, this is because the needs and shortcomings of each village to become a resilient village are different.

3.3 Assessment of Village Resilience is Important in Village-Based Programs Ministries/Institutions

In the village there is an assessment of village resilience which is a tool to measure the level of village resilience in a simple way. Apart from being a simple tool for measuring the level of resilience, these results are also used as a basis or reference in developing disaster-resilient villages. The results of this assessment present aspects that are still lacking and must be improved, so that resilient village developers can direct their efforts in a more focused and integrated manner. Currently, the village resilience assessment is a resilience assessment tool that already has Indonesian national standards (SNI). By indirectly assessing village resilience institutionally, it can be recommended for village-based development programs. With the support of Minimum Service Standards which can be used as a basis for this minimal service to be carried out for risk reduction. In addition, the Village Resilience Assessment is an initial activity in Destana activities which aims to map conditions and an initial picture of the level of risk, vulnerability level and village capacity level. in dealing with the threat of a tsunami as a baseline for the level of village disaster resilience. The implementation identification process is carried out with community participation by involving stakeholders at the village level and taking into account the representation of vulnerable groups and gender. Assessment of village resilience is carried out at the initial and final stages of village implementation to measure the impact of the resilience efforts that have been implemented.

Village resilience assessment is a concept that offers a more systematic and cross-sectoral approach as an approach to disaster risk reduction, climate change adaptation, and the humanitarian sector. This assessment framework adopts village and urban village disaster resilience standards in SNI 8357:2017 and adapts the DROP (Disaster Resilience of Place) model. This choice is based on expert recommendations. Although DROP was designed to be implemented in the United States, the commonalities of the nature of disaster resilience make it applicable to other country contexts. DROP has also been adopted and adapted in several other countries, such as in Australia, Norway, Thailand and Zimbabwe. The model assumes that natural systems interact with social systems and then create vulnerability and at the same time disaster resilience that is inherent (inherent) in the community. This inherent vulnerability is a value of the ability of a population to prepare to respond and recover from disasters, while the value of resilience is a community characteristic that may strengthen or weaken its capacity to reduce hazards, prepare for, deal with, and recover from the impact of disasters, these vulnerabilities and resilience are assumed to already exist. before the disaster [15]. The DROP model has a comprehensive framework for assessing a community's overall disaster resilience and can pinpoint specific components that can increase or decrease disaster resilience. Furthermore,

Assessment of rural disaster resilience has played an important role in increasing efforts to improve disaster resilience in line with the high number of catastrophic scale disasters (causing large amounts of life and property) throughout the world in the last two decades. Village Resilience is a process of assessing resilience that is carried out by the community independently to determine the level of resilience (SNI 8357:2017).

The objectives of the village disaster resilience assessment are as a means of education and awareness of disaster risk to the community and all parties, to provide a base line for disaster resilience in the smallest government units of the Republic of Indonesia, and to support the implementation of disaster management in the regions.

The benefits of assessing village disaster resilience are as follows:

1. For Communities in Disaster-Prone Areas
    a. Villagers understand their level of disaster resilience
    b. Availability of data to make efforts to increase village resilience

2. For Village Government
    a. The village government understands the efforts that must be made to increase resilience
    b. The Village Government has basic data for the development and improvement of basic services
    c. The Village Government has basic data in the preparation of development plans
3. For Regency/City Government
    a. Availability of evaluative data for regional development programs
    b. Availability of basic data for the preparation of regional development programs
4. For Disaster Management Organizations/Institutions
    a. Availability of basic data for community resilience program development
    b. Availability of basic data for conducting research
5. For business institutions to carry out PB
    a. Availability of data to develop CSR/Donation programs to increase village resilience
    b. Business institutions in the village understand the disaster risk that threatens the sustainability of their business.
6. For Educational and Research Institutions
    a. Availability of data for research composing materials
    b. Availability of basic data to develop community service programs

In accordance with Indonesia's 2020-2024 mission is to make Indonesia resilient in disasters. Disaster management is very multi-disciplinary and multi-disciplinary, so collaboration from all sectors is needed, especially within Ministries/Agencies. Capacity building and integration at the community level are the keys to the resilience of Indonesian society, where the village is the pivot in implementation. Various village-based capacity building programs under several ministries/agencies need collaboration with each other so as to facilitate and not burden the village in implementing these programs. Building a village-based program requires determination to build a community that is resilient in accordance with Indonesia's goal of becoming disaster-resilient in 2045.

So far, in the context of reducing disaster risk, Ministries/Agencies have various village-based programs, such as:

1. Disaster Resilient Village (Destana) – BNPB
2. Disaster Preparedness Village (KSB): Ministry of Social Affairs
3. Covid/Disaster Resilient Villages – Kemendes PDTT
4. Communities Care for Fire and Climate Village Program (Proklim) – Ministry of Environment and Forestry
5. Active Alert Village – Ministry of Health
6. Energy Independent Village – Ministry of Energy and Mineral Resources
7. Tourism Village – Kemenparekraf
8. Tangguh Coastal Village – KKP
9. Food Independent Village – Ministry of Agriculture
10. Smart Village Disaster Safe Education Unit (SPAB) – Kemdikbud
11. Child Care Village – Kemen PPA

Of all village-based programs in various ministries, institutions have the main goal of creating community resilience. With this common goal, the assessment of village resilience can be used as a unifying tool and a collaboration tool for ministries/agencies in Indonesia. So that a resilient community in all sectors can be realized. To realize this big goal, it takes the cooperation of all parties and reduces the ego-sectoral that often appears in the world of bureaucracy. This collaboration will also make it easier for the community to implement it.

Community complaints so far when they find several village-based programs from different ministries/institutions are the difficulties and confusion of the community in implementing them. This village resilience assessment will make it easier for village communities to take action to reduce vulnerability and increase capacity. With the output of the village resilience assessment in the form of recommendations, it will be easier for the community to understand and follow up and advocate for the village government to jointly implement it. Another thing is that the collaboration of ministries/agencies also aims to complement each other between village-based programs, as well as the creation of "One Data For All" as budget effectiveness and efficiency. Assessment of village resilience is also very important because it is an early data collection. A program will be successful if it departs from real data in accordance with existing conditions and according to the community itself. Meanwhile, in its implementation, it is also necessary to adapt to local wisdom in the region.

Basically the village-based program is a synergy program between actors. Therefore, there is a need for coordination between Ministries/Institutions that have village-based programs to determine the baseline that will be used as a key indicator so that it can be accommodated in the PKD indicator based on the duties and functions of each ministry. We need an umbrella that supports each other to achieve shared needs between Ministries/Institutions. What is done by the Ministry/Agency program in the village is recognized and gets a lot of contributions. Village-based programs are not just village administration programs, but we build cross-sectoral approaches. A comprehensive assessment of village resilience contains various intersecting indicators and is related to indicators related to other ministry/institutional village programs. However, there are still shortcomings, so it is necessary to establish key indicators so that the assessment that already has Indonesian standards can be used and applied in various sectors.

Meanwhile, the output of the village resilience assessment is recommendations and actions to increase disaster resilience. Examples of recommendations and actions to increase disaster resilience that have the resilience level of the Tangguh Pratama village, and the recommendations for the framework of action that need to be carried out by the village are as follows:

1. Evaluating, planning, and implementing improvements to the implementation of education services
2. Proposing the implementation of health services through the Village Musrenbang mechanism to the Regency/Municipal Musrenbang according to the minimum health service standards
3. Propose budget allocations through the village musrenbang mechanism
4. Propose cooperation with agencies/institutions.organizations for identification of needs, planning, and implementation of clean water supply programs
5. Propose food and nutrition programs and activities through the village musrenbang mechanism
6. Propose programs for empowering micro, small and medium enterprises and/or economic development through village-owned enterprises through village musrenbang
7. Propose the provision of communication and information facilities and infrastructure in the village musrenbang in accordance with the minimum service standards for communication/information facilities and infrastructure
8. Propose the administration of administrative and population services in village development planning deliberations
9. Etc

How community resilience is the basic pillar of building community welfare, the assessment of village resilience is a baseline to measure village resilience which later the village can contribute by assessing the level of independence of the village itself. After that, the toughness to be built can refer to those that have been made in synergy and harmony. Through the assessment of village resilience, all elements can collaborate and synergize to achieve a resilient Indonesia. Recently, ministries/agencies have realized the importance of assessing village resilience for all village-based programs. Therefore, ministries/agencies coordinate and agree to jointly use village resilience assessments in their programs. It's just that before that happens, the need for improvement and additional key indicators that are adjusted to the program of the respective ministries/agencies. Therefore, it is necessary to make a Policy Brief to be able to see that each program/program becomes a common framework or a common framework with the ultimate goal of 3 issues, namely:

- Disaster Risk Reduction and Climate Change Adaptation
- Economic resilience
- Adaptive and inclusive social protection

4. Conclusions and recommendations

In the regional disaster resilience assessment, there are 5 components, 32 indicators and 128 questions. The questions asked started at the pre-disaster, emergency response, and post-disaster stages. The assessment of village disaster resilience produces several outputs, such as the Value of the Village Resilience Index, the Village Resilience Level, and recommendations for action to increase village capacity. The Village Resilience Index value consists of three levels, namely Tangguh Utama <83.33, Tangguh Madya 58.3 – 83.33, and Tangguh Pratama <58.3.

Village toughness assessment is important because

- Village Resilience Assessment (PKD) is a tool to measure the level of village resilience in a simple way.
- Village Resilience Assessment is the initial activity in disaster resilient village activities which aims to map conditions and initial descriptions of the level of risk, level of vulnerability and level of village capacity in facing threats as a baseline for the level of village disaster resilience. The implementation identification process is carried out with community participation by involving stakeholders at the village level and taking into account the representation of vulnerable groups and gender. Assessment of village resilience is carried out at the initial and final stages of village implementation to measure the impact of the resilience efforts that have been implemented.
- Assessment of village resilience can be used as a unifying tool and a collaboration tool for ministries/agencies in Indonesia.
- The collaboration of ministries/agencies also aims to complement each other between village-based programs, as well as the creation of "One Data For All" as budget effectiveness and efficiency.
- This village resilience assessment will make it easier for village communities to take action to reduce vulnerability and increase capacity
- With the output of the village resilience assessment in the form of recommendations, it will be easier for the community to understand and follow up and advocate for the village government to jointly implement it.

In addition, the process and method of assessing village resilience, which is currently comprehensive, still needs to be refined again to accommodate various indicators attached to the duties and functions of other ministries/agencies.

References

[1] Law Number 24 of 2007 concerning Disaster Management
[2] Regulation of the Head of the Disaster Management Agency Number 1 of 2012 concerning General Guidelines for Disaster-Resilient Villages
[3] Regulation of the Minister of Home Affairs Number 101 of 2018 concerning Technical Standards for Basic Services on Minimum Standards for Regional Disaster Sub-District City Districts
[4] National Agency for Disaster Management. 2021. *Village Resilience Assessment Guide*
[5] Indonesian National Standard 8357:2017. Disaster-Resilient Villages and Villages
[6] Annisa, Aulia dan Dewi Liesnoor Setyowati. (2019). *"Kapasitas Masyarakat dalam Upaya Pengurangan Risiko Bencana Tanah Longsor di Desa Tempur Kecamatan Keling Kabupaten Jepara Tahun 2018."* Jurnal Edu Geography, Vol. 7 No. 1.
[7] Effendi Tajjudin Noer. 2013. *The Culture of Community Mutual Cooperation in Today's Social Changes. Journal of Sociological Thought* Vol. 2, No. 1, May 2013

[8] Kayadoe, Fretha Julian et al. (2016). "Kajian Ketangguhan Masyarakat dalam Menghadapi Bencana Banjir Bandang Way Ela di Desa Negeri Lima Kabupaten Maluku Tengah." Jurnal Dialog Penanggulangan Bencana BNPB, Vol. 7 No. 1.

[9] Rini, Ratih Setyo et al. (2021). *"Pemantauan dan Evaluasi Kegiatan Pemberdayaan dan Pengembangan Ketangguhan Masyarakat Tangguh Bencana di Provinsi Riau."* Scientific Journal of Reflection, Vol. 4 No. 3.

[10] Shaw, R.; Team, I. *Climate disaster resilience: Focus on coastal urban cities in Asia*. Asian J. environment. Disaster Management. 2009, 1, 101–116.

[11] Suryana, Ma'rif Nanang dan Sriyono. (2021). *"Evaluasi Pelaksanaan Program Desa Tangguh Bencana Desa Sambungrejo Kecamatan Grabag Kabupaten Magelang Tahun 2019."* Jurnal Edu Geography, Vol. 9 No. 1.

[12] Sanit, Maria Serlince. (2018). *"Strategi Mitigasi Bencana Banjir Bandang Berdasarkan Nilai Ketangguhan Di Kabupaten Malaka, Nusa Tenggara Timur."* Institut Teknologi Nasional Malang.

IOP Conf. Series: Earth and Environmental Science 1173 (2023) 012060 doi:10.1088/1755-1315/1173/1/012060

Numerical Modelling of Constructed Wetland using Software HYDRUS 2D

M. Adek Rizaldi[1]*, Lily Montarcih Limantara[2]*, Ery Suhartanto[2], Donny Harisuseno[2] and Nadia Karima Izzaty[3]

[1] Department of Civil Engineering, Faculty of Engineering, University of Brawijaya, Malang, Indonesia
[2] Department of Water Resources, Faculty of Engineering, University of Brawijaya, Malang, Indonesia
[3] Department of Civil Engineering, Faculty of Engineering, Universitas Indonesia, Depok, Indonesia
* adek.rizaldi@yahoo.com

Abstract. One of the problems involving water bodies is the pollution of water sources, which later becomes the parent of various social, economic, and health problems. One of these conditions can be overcome by utilizing phytoremediation through constructed wetlands. HYDRUS is a software that offers easy modelling of pollutant decay mechanisms in water bodies and constructed wetlands using finite element method. This study aims to show and analyse how HYDRUS software can model the mechanism in constructed wetland by using finite element method. The parameters observed are the time for the pollutant, namely ammonia, nitrate, and inorganic phosphorous, to reach the outlet, and the response curve of the pollutant loading on the model. From the simulation, it can be inferred that the maximum velocity of the water going through the constructed wetland is 8.11 m/day, or 811 cm/day. The dominating velocity in the wetland is around 160 cm/day, or 1.6 m/day. The response curve of the pollutant transport is also in accordance with theoretical response for impulse loading. The result yields the effectivity of simulated constructed wetland, which are 90.33% for ammonia and nitrate, and 90.26% for inorganic phosphorous. The result yields quite optimistic effectivity of the constructed wetland, which may be caused by the assumptions made in the model, which goes through simplification method. It can also be caused by the differences in plants assumption. In the HYDRUS simulation, the plant used is grass, which does not specify what kind of grass. Meanwhile, the physical simulation uses water bamboo.

1. Introduction

The increase in human activity associated with the nearby water body is one of the effects of population growth and development in a region. The amount of waste produced by human activity in the form of residential waste as well as other types of waste like industrial waste and agricultural waste is also rising naturally. The groundwater can potentially be affected by this occurrence, in addition to the water surface.

In terms of reducing pollutant in water body, one of the offered solutions is constructed wetland. It has been mentioned in many papers and researches that this solution is quite cheap, and easy to be maintained as constructed wetlands can use local plants. In addition, the pollutant substance can also be re-processed as the economic valuable biomass [1].

Water bamboo, or *Equisetum hyemale*, is one of the plants that can be utilized for constructed wetlands. According to earlier studies, water bamboo (*Equisetum hyemale*) may retain BOD_5 content up to a level of 90.34%, whereas COD content can be removed with an efficiency of 89.67% [2]. According to another study, BOD, COD, TSS, and ammonia all had the highest percentages of elimination in the artificial wetland created by *E. hyemale* without aeration: 72.58%, 82%, 78.77%, and 75.31% [2].
In order to know the effectivity of a plant in constructed wetlands, a sum amount of money and time might be needed in order to build the reactor from scratch and make physical simulations for it. The alternative would be using modeling software, one of which is provided by HYDRUS. HYDRUS is a water quality modelling software with the method of finite element. This software can model how a certain pollutant transports in water body in 3D and constructed wetlands in 2D. This study aims to show and analyze how HYDRUS can model pollutant removal in constructed wetlands.

2. Materials and Method

2.1. Literature Study

One of the effects of human activity is pollution. The quantity of waste created by human activity will rise in tandem with the amount of human activity in a given location. The sorts of pollutants that are created depend on the activity in the area around them, such as waste produced as a consequence of residential activity differing from waste produced as a result of industrial activity, agricultural waste, and other types of activity.

It is stated that raw water, which is primarily used as a source of drinking water, can disturb the nervous system, blood circulation system, and in severe cases, type of pollutant as heavy metal can cause the development of tumor and cancer cell [3]. These pollutants include acrylamide, chlorine and its derivative, nitrogen and its derivative, phosphor and its derivative, and other substance. In addition to the negative effects on health, additional studies have demonstrated that when a river is moderately polluted, economic growth in the downstream region declines by 1.4%, and when the river is heavily polluted, economic growth declines by 2% [4].

Constructed wetland is a bioremediation technology which mimics the mechanism of natural wetlands. In other words, the pollutant decay mechanism in the constructed wetland mimics the natural mechanism that occurs in the too. Essentially, the constructed wetland is situated between the sea and the land. The concept is especially important in areas where the wetland can only be wet for a short period of time per year, such as the tropics, which are distinguished by the rainy and dry seasons [2].

According to a study conducted at Xing-Qing Lake in Xi'an, China, the constructed wetland can reduce pollutant concentrations by 84.2% for COD, 53.8% for NH3-N, 47.9% for total nitrogen, 73.3% for total phosphor, and 86.6% for SS [5]. Another study in KKRBCW (Kaoping River Rail Bridge Constructed Wetland), Taiwan, found that wetland reduces pollution concentrations by 97% for TC, 55% for BOD, and more than 30% for nutrient. Taiwan has built a wetland system in a large open space with a surface area of 120 ha and a capacity of 17,000 to 19,000 m3 influent per day [6].

The vegetation used in this study is water bamboo, also known as Equisetum hyemale. The constructed wetland with peat soil media and water bamboo vegetation can process waste water in the commercial area well, reducing the COD value from 97.7 mg/l to 43 mg/l or 55.98% and the TSS value from 447.90 mg/l to 66.00 mg/l or 85.26% [7]. According to another study, the effectiveness of this vegetation in removing pollutants from hospital liquid waste was 72.58% for BOD, 82% for COD, 78.77% for TSS, and 75.31% for ammonia [8]. Then, from industrial and household waste, it was discovered that bamboo vegetation can absorb waste in amount of 89.99% [9].

An alternative to conducting experiments as mentioned above is to perform numerical modeling. There are many mathematical models that can be used for numerical calculations, and one of them is HYDRUS. Based on research that has been conducted at the University of Lisbon, Portugal, the accuracy of the HYDRUS program has been obtained which gives fairly good results of pollutant decay from numerical models and laboratory-scale physical models, namely 68% and 67%, respectively [10].

HYDRUS is the software that is being used in this research. It is used to simulate the transport of water, heat, and solutes in porous media in two- or three-dimensional way. This software numerically solves the Richards equation for saturated-unsaturated water flow and the convection dispersion equation for heat and solute transfer. For the calculation method, HYDRUS uses the finite element method [11].

$$\frac{\delta\theta}{\delta t} = \frac{\delta}{\delta x_i}\left[K\left(K_{ij}^A\frac{\delta h}{\delta x_j} + K_{iz}^A\right)\right] - S \quad (1)$$

$$K(h,x,y,z) = K_s(x,y,z)\,K_r(h,x,y,z) \quad (2)$$

θ	:	Volumetric water content	(L^3/L^3)
h	:	Pressure head	(L)
S	:	Sink parameter	(/T)
$x_i\ (i = 1,2,3)$	:	Spatial coordinates	(L)
K_{ij}^A	:	Anisotropic tensor	
K	:	Function of unsaturated hydraulic conductivity	
K_r	:	Relative hydraulic conductivity	(L/T)
K_s	:	Saturated hydraulic conductivity	(L/T)

For partial differential equations governing the transport of pollutants in a first-order sequential decay in the liquid, solid, or gas phases, it is governed by the following equation [11].

$$\frac{\partial\theta c_1}{\partial t} + \frac{\partial\rho s_1}{\partial t} + \frac{\partial a_v g_1}{\partial t} = \frac{\partial}{\partial x}\left(\theta D_{ij,1}^{w}\frac{\partial c_1}{\partial x_j}\right) + \frac{\partial}{\partial x}\left(a_v D_{ij,1}^{g}\frac{\partial g_1}{\partial x_j}\right) - \frac{\partial q_i c_1}{\partial x_i} - Sc_{r,1}$$
$$-(\mu_{w,1} + \mu'_{w,1})\theta c_1 - (\mu_{s,1} + \mu'_{s,1})\rho s_1 - \left(\mu_{g,1} + \mu'_{g,1}\right)a_v g_1 + \gamma_{w,1}\theta + \gamma_{s,1}\rho + \gamma_{g,1}a_v \quad (3)$$

c	:	Liquid phase index	(M/L^3)
s	:	Solid phase index	(M/M)
g	:	Gaseous phase index	(M/L^3)
q_i	:	The ith component of the volumetric flux density	(L/T)
μ	:	First-order decay rates for solutes in the liquid, solid, and gas phases	(/T)
μ'	:	First-order decay rates providing relationships between individual chain species	(/T)
γ	:	0th-order decay rates for solutes in the liquid, solid, and gaseous phases	(M/L^3T)
ρ	:	Soil density	(M/L^3)
a_v	:	Soil air content	(L^3/L^3)
S	:	Sink	(/T)
c_r	:	Sink concentration	(M/L^3)
D_{ij}^w	:	Dispersion coefficient for liquid phase	(L^2/T)
D_{ij}^g	:	Dispersion coefficient for gaseous phase	(L^2/T)

2.2. Research Method

Figure 1 illustrates the research design step that will be carried out in order to achieve the research goal.

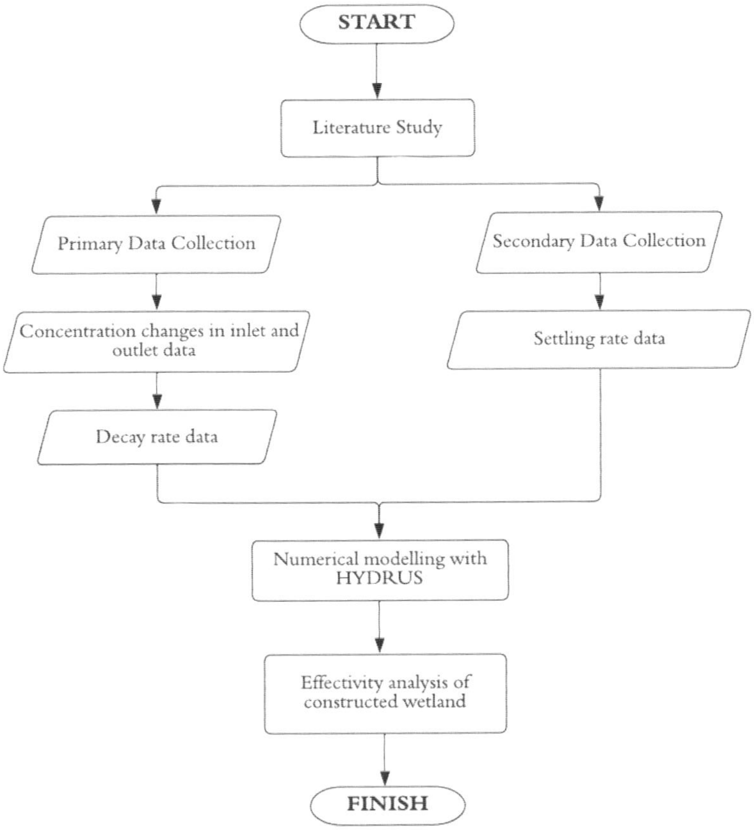

Figure 1. Flow chart of research

The first step would be literature study, which has been explained before in previous section. The next step is data gathering, which consists of two steps, primary and secondary. Primary data collected are concentration changes in inlet and outlet and also decay rate data. Secondary data collected is settling rate data. Both are then combined in order to run simulations with HYDRUS software.

3. Results and Discussion

The physical model that was previously created is used in the first step of this research. This wetland's mini reactor is made up of four reactor series units, each of which has a half-cylinder shape with a length of 1 m and a diameter of 50 cm. Every day, a sample is taken by analyzing the residence time. There are four samples of each constructed wetland reactor and one sample of the storage tank's inlet. The samples are collected over the course of 9 days, and the measurements are performed with a spectrophotometer.

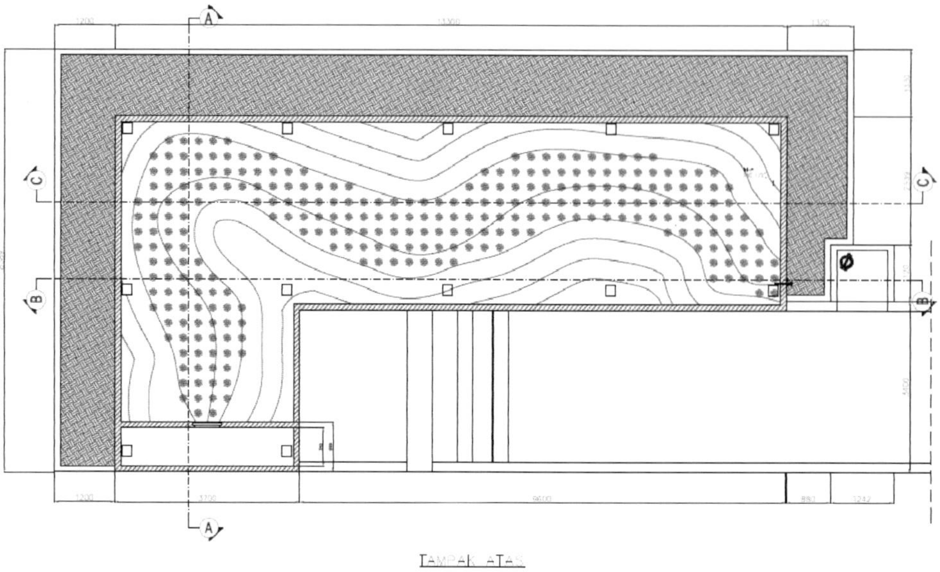

Figure 2. Constructed wetland site map

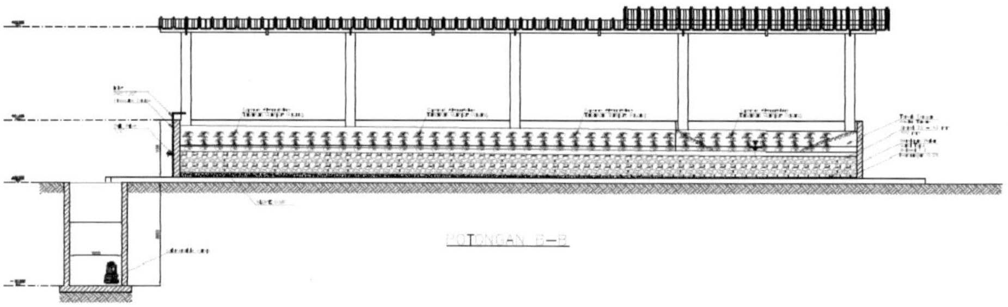

Figure 3. Constructed wetland side view

After obtaining the concentration data, the data is transformed into a curve and an exponential formulation. This step is performed to determine the first order decay from each parameter. The decay speed of each parameter is calculated based on the results of data collection and concentration measurement. According to the regression analysis results, the decay speed for each parameter is 0.026/day for DO, 0.024/day for ammonia, 0.006/day for phosphate, and 0.1214/day for nitrate.

Table 1. Recapitulation of input parameter for HYDRUS

Component	Value	Source
Inlet velocity	0.2 m/hour	Field data
Rate of phosphate decay	0.006 /day	Field data
Rate of ammonia decay	0.024 /day	Field data
Rate of nitrate decay	0.1214 /day	Field data
Rate of phosphor deposition	0.000245434 m/hour	(Gomez-Gallegos et al., 2021)
Rate of ammonia deposition	0.135 m/hour	Assumption: the same as nitrate
Rate of nitrogen deposition	0.135 m/hour	(Gomez-Gallegos et al., 2021)

The initial conditions of all modeled pollutant concentrations are set. In this case, because the actual initial conditions in the soil are not known before the contaminants flow into it, it is assumed that the initial conditions of the reactor are 0 g/m^3, which means there is no pollutant initially in the wetland.

In terms of boundary conditions, boundary conditions are set for two parameters, namely water flow and solute transport. In the water flow parameter, the boundary condition defined is the direction of the water flow from the reactor inlet to its outlet. Meanwhile, for pollutant transport, the boundary conditions are set in the form of first-type for reactor inlet and third-type for reactor outlet.

The HYDRUS model was simulated for 50 days, in order to determine the time needed for the pollutant to reach the outlet. From the simulation, it can be inferred that the maximum velocity of the water going through the constructed wetland is 8.11 m/day, or 811 cm/day. The dominating velocity in the wetland is around 160 cm/day, or 1.6 m/day.

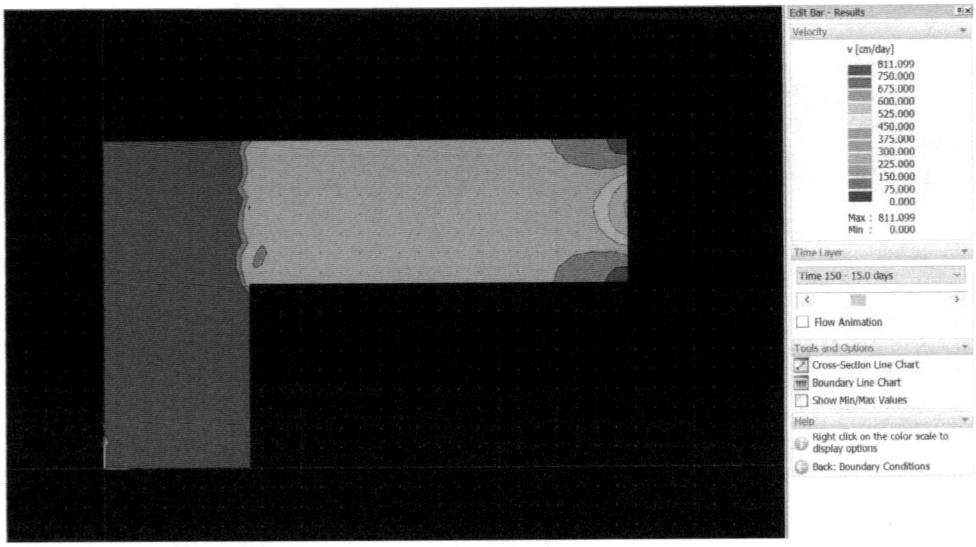

Figure 4. Velocity distribution in the wetland

Meanwhile, the direction of the velocity vector is also in accordance with the specified boundary conditions, namely from the right-hand side of the reactor all the way to the outlet, which is located as specified in the following picture.

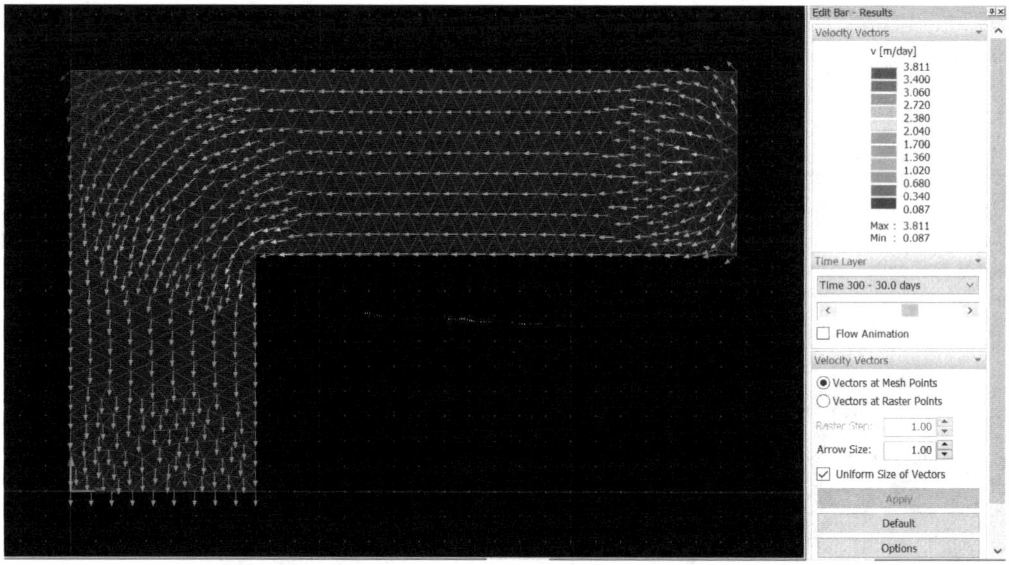

Figure 5. Velocity vector distribution in the wetland

In the constructed wetland, the amount of time required for the pollutant to reach the outlet is around 50 days, which is the same as the simulated time. Since the loading is assumed to be in the form of a pulse of 0.1 days (2.4 hours), the result graph should look like a pulse too. The following figure illustrates the three observation points distributed in the wetland.

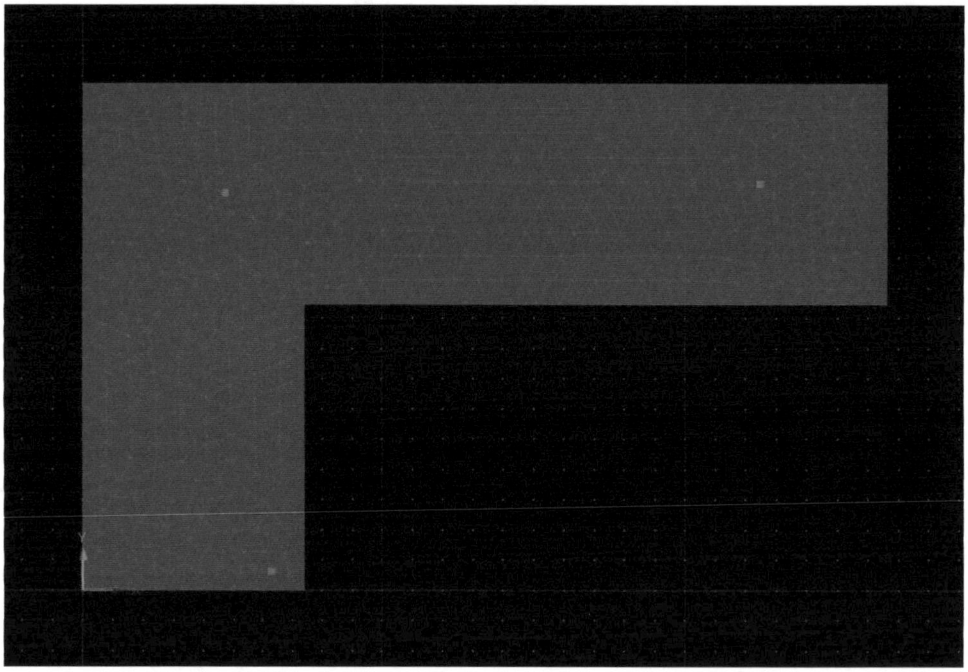

Figure 6. Observation points in the constructed wetland model

In those three randomly placed observation points, graphs of response of the constructed wetland when pollutants are being poured can be displayed. The results are as follows for nitrate, ammonia, and inorganic phosporous.

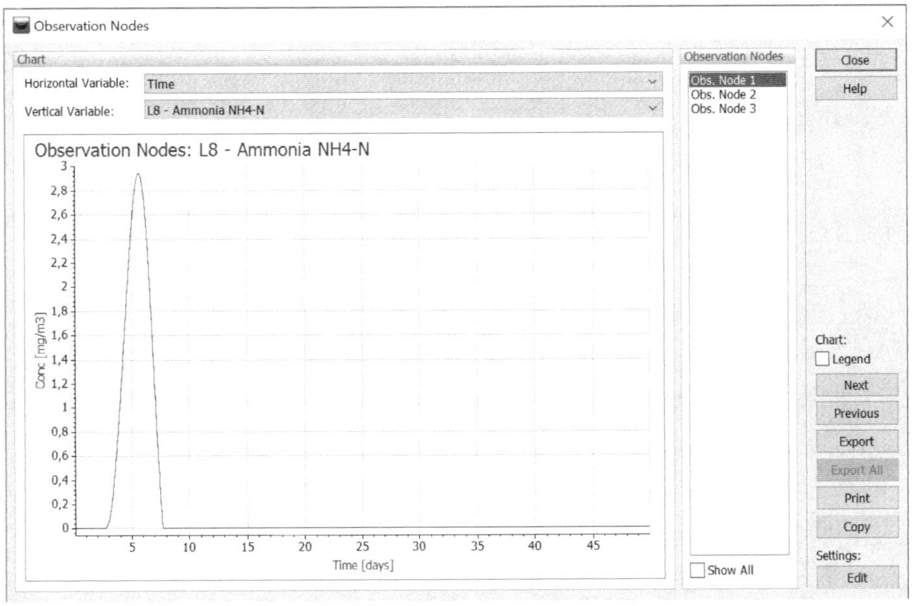

Figure 7. Ammonia response graphic at observation point 1 (near inlet)

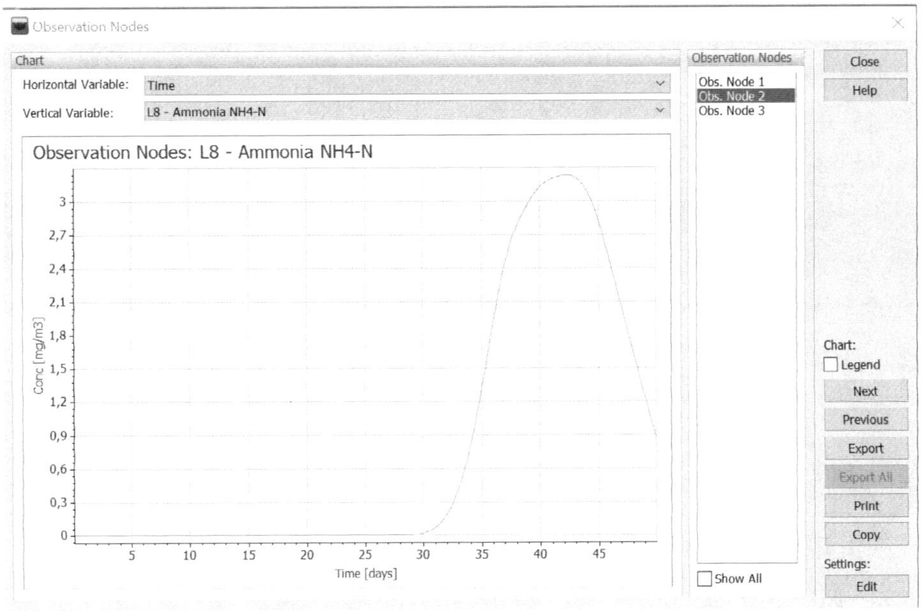

Figure 8. Ammonia response graphic at observation point 2 (near outlet)

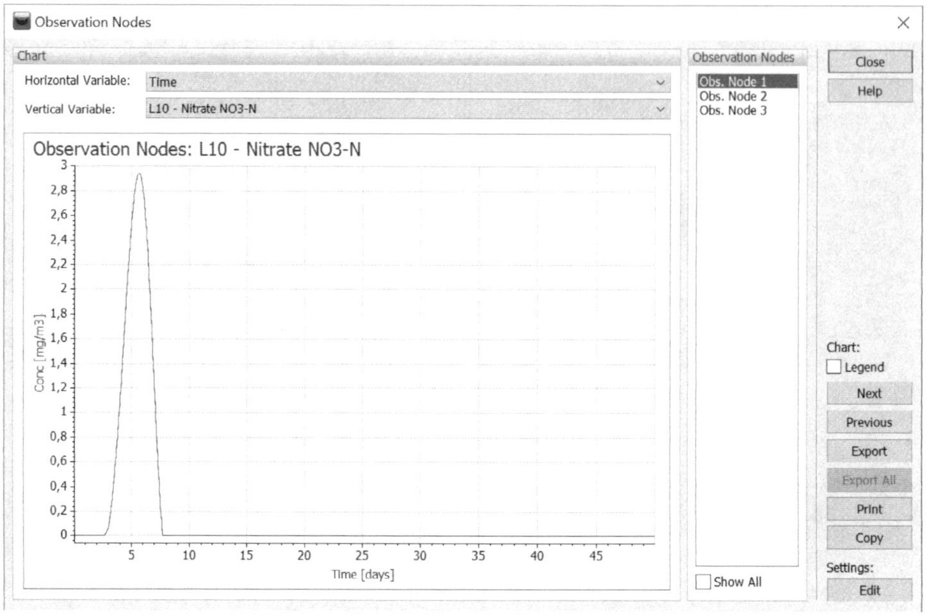

Figure 9. Nitrate response graphic at observation point 1 (near inlet)

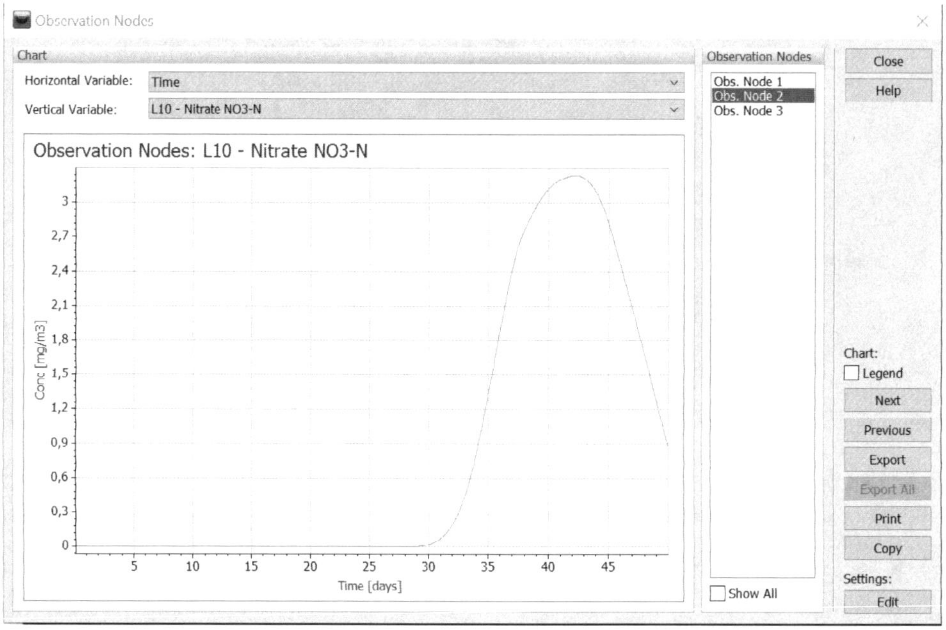

Figure 10. Nitrate response graphic at observation point 2 (near outlet)

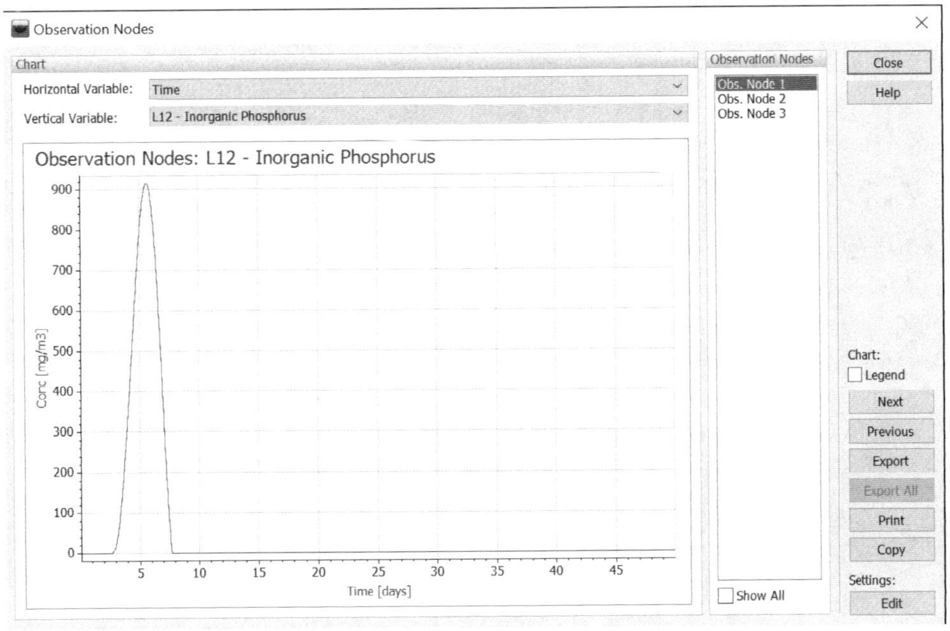

Figure 11. Phosphorus response graphic at observation point 1 (near inlet)

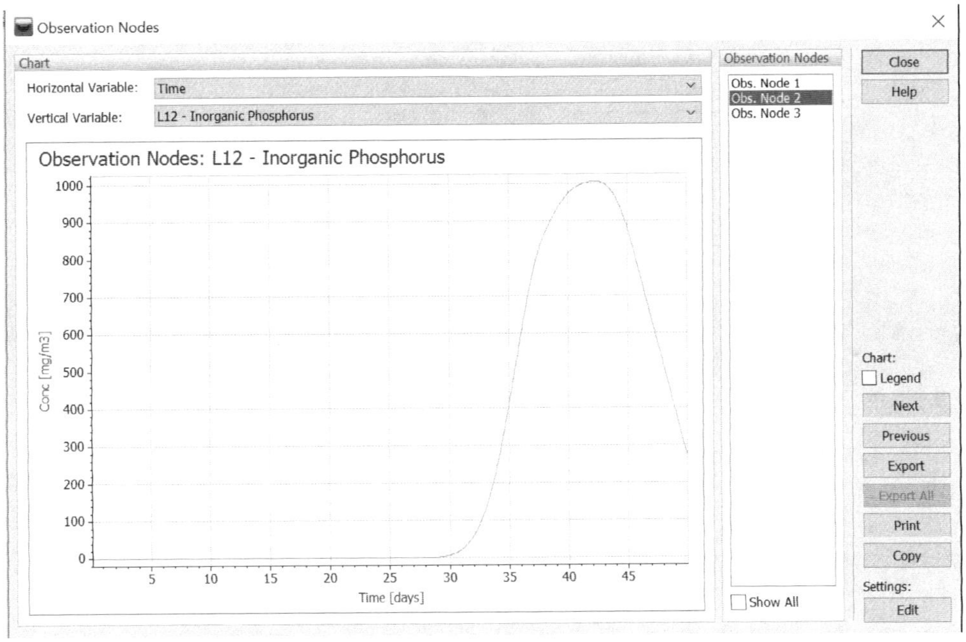

Figure 12. Phosphorus response graphic at observation point 2 (near outlet)

As it can be inferred from graphics above, the behavior of pollutants is almost the same. At observation point near the inlet, they reach the peak in around 5 hours, with the concentration of ammonia and nitrate of 2.9 mg/m^3 and inorganic phosphorous of about 910 mg/m^3. The behavior demonstrated by all pollutants corresponds with the theoretical graphs, in which the pulse loading will be followed by exponentially declining concentration [12].

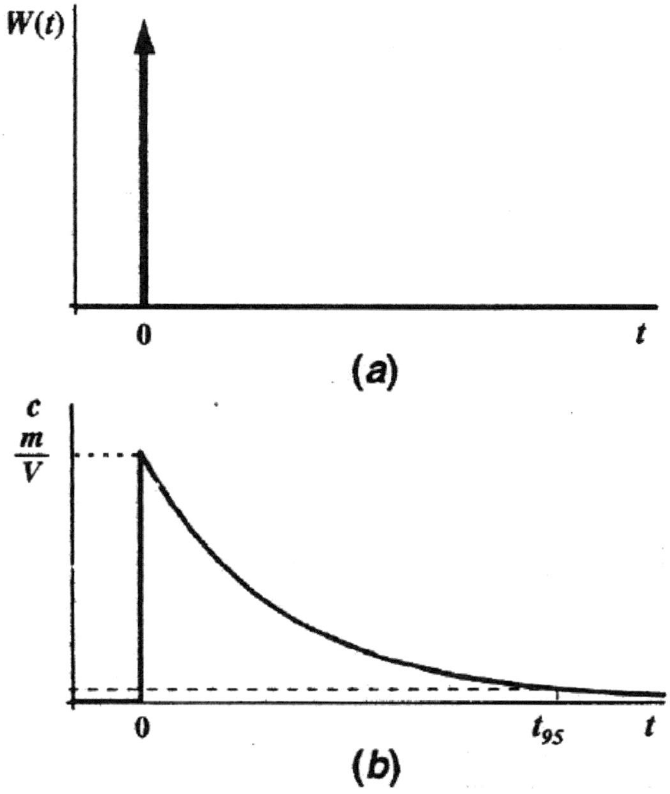

Figure 13. Input and output graphics of pulse loading

Based on the simulated constructed wetland, the effectivity of simulated wetland can be calculated. As for ammonia and nitrate, the result yields the same result of effectivity, which is 90.33%. For inorganic phosphorous, the result yields the effectivity of 90.26% removal. This is quite an optimistic number. In physical model around the world, it is that the effectivity of water bamboo is 75.31% for ammonia [4], and for phosphate is varied from 30-91% [13]. The differences in this number may be caused by the assumptions made in the model, which goes through simplified method. It can also be caused by the differences in plants assumption. In the HYDRUS simulation, the plant used is grass, which does not specify what kind of grass. Meanwhile, the physical simulation uses water bamboo.

4. Conclusions

The increase in human activity associated with the nearby water body is one of the effects of population growth and development in a region. In terms of reducing pollutant in water body, one of the offered solutions is constructed wetland. In order to know the effectivity of a plant in constructed wetlands, a sum amount of money and time might be needed in order to build the reactor from scratch and make physical simulations for it. The alternative would be using modeling software, one of which is provided by HYDRUS. This study aims to show and analyze how HYDRUS can model pollutant removal in constructed wetlands.

The result shows that the peak velocity in the wetland is around 8 m/day, with the dominating value of 1.6 m/day. The response curve for ammonia, nitrate, and inorganic phosphorous also corresponds with the theoretical response curve for impulse loading. The peak concentration for ammonia and nitrate in the outlet is 2.9 mg/m^3 and for inorganic phosphorous is around 910 mg/m^3. The result yields the effectivity of simulated constructed wetland, which are 90.33% for ammonia and nitrate, and 90.26% for inorganic phosphorous.

References

[1] NUGRAHA A. S. and WARDONO H. R. I. Efisiensi bambu air (*Equisetum hyemale*) sebagai fitoremediator kadar biological oxygen demand pada limbah cair industritahu di desa Prembun kecamatan Tambak kabupaten Banyumas. *Buletin Keslingmas*, 34(3), 2016, p. 189–194. https://doi.org/10.31983/keslingmas.v34i3.3071

[2] HASAN A. and KADARUSMAN H. Pengaruh aerasi terhadap pengolahan limbah cair rumah sakit dengan metode Constructed Wetland. *Ruwa Jurai: Jurnal Kesehatan Lingkungan*, 16(1), 41, 2022, https://doi.org/10.26630/rj.v16i1.3157

[3] DESBUREAUX S., DAMANIA R., RODELLA A.S., RUSS J., and ZAVERI E. *The Impact of Water Quality on GDP Growth Evidence from around the World*, 2019, www.worldbank.org/gwsp

[4] Suoth, A. E., & Nazir, E. (2016). Karakteristik Air Limbah Rumah Tangga (Grey Water) Pada Salah Satu Perumahan Menengah Keatas Yang Berada Di Tangerang Selatan. Ecolab., 10(2), 47–102.

[5] Cui, F., Zhou, Q., Wang, Y., & Zhao, Y. Q. (2011). Application of constructed wetland for urban lake water purification: Trial of Xing-qing Lake in Xi'an city, China. In *Journal of Environmental Science and Health - Part A Toxic/Hazardous Substances and Environmental Engineering* (Vol. 46, Issue 7, pp. 795–799). https://doi.org/10.1080/10934529.2011.571994

[6] Wu, C. Y., Liu, J. K., Cheng, S. H., Surampalli, D. E., Chen, C. W., & Kao, C. M. (2010). Constructed wetland for water quality improvement: A case study from Taiwan. *Water Science and Technology*, *62*(10), 2408–2418. https://doi.org/10.2166/wst.2010.492

[7] Siswanto, L., Darmayanti, Y., Lilis, H., & Ridwan, M. (2014). Pengolahan Air Limbah Hotel Dengan Metode Free Surface Constructed Wetland Menggunakan Tumbuhan Equisetum hymale. In *Jurnal Teknobiologi* (Issue 1).

[8] Hasan, A., & Kadarusman, H. (2022). PENGARUH AERASI TERHADAP PENGOLAHAN LIMBAH CAIR RUMAH SAKIT DENGAN METODE CONSTRUCTED WETLAND. *Ruwa Jurai: Jurnal Kesehatan Lingkungan*, *16*(1), 41. https://doi.org/10.26630/rj.v16i1.3157

[9] Nugraha, A. S., & Wardono, H. R. I. (2015). EFISIENSI BAMBU AIR (Equisetum hyemale) SEBAGAI FITOREMEDIATOR KADAR BIOLOGICAL OXYGEN DEMAND PADA LIMBAH CAIR INDUSTRI TAHU DI DESA PREMBUN KECAMATAN TAMBAK KABUPATEN BANYUMAS TAHUN 2015. *Buletin Keslingmas*, *34*(3), 189–194. https://doi.org/10.31983/keslingmas.v34i3.3071

[10] Rizzo, A., Langergraber, G., Galvão, A., Boano, F., Revelli, R., & Ridolfi, L. (2014). Modelling the response of laboratory horizontal flow constructed wetlands to unsteady organic loads with HYDRUS-CWM1. *Ecological Engineering*, *68*, 209–213. https://doi.org/10.1016/j.ecoleng.2014.03.073

[11] Šejna, M., Šimůnek, J., & Th van Genuchten, M. (2020a). The HYDRUS Software Package for Simulating the Two-and Three-Dimensional Movement of Water, Heat, and Multiple Solutes in Variably-Saturated Porous Media User Manual.

[12] Chapra, S. C. (1997). *Surface Water Quality Modeling*. The Mcgraw-Hill Companies, Inc.

[13] AL KHOLIF M., HIDAYAT S., and SUTRISNO J. Pengaruh tanaman bintang air (Cyperus Papyrus) dan bambu air (Equisetum Hyemale) dalam mengolah limbah domestik. *Serambi Engineering*, 2020.

Strategy for improving Green Open Space (RTH) based on community participation in reduce the risk of flood disaster in Jakarta

P Salsabila[1], S Maarif[1] and D A P Sari[1]

[1]Disaster Management Study Program, National Security Faculty, Republic of Indonesia Defense University. IPSC Sentul Area, Bogor, West Java, Indonesia

E-mail: puspa.salsabila@gmail.com

Abstract. Natural Floods are the main form of environmental disaster worldwide in terms of frequency, and scale is associated with a tenfold increase in costs over the last half-century. Generally, chronic flood disasters occur in several cities in Southeast Asia, including Jakarta. Misuse of spatial planning is said to be the main factor causing flooding. A community-based approach in urban planning with Green Open Space (RTH) can be a solution to reducing the risk of flood disasters. Through giving directions to the community to comply with regulations in building settlements, administering Building Permits, and consulting first with the PTSP of the nearest sub-district or sub-district before buying land, it is something that the community can do so that the percentage of land use mismatches and land use conversion does not increase.

1. Introduction

Nowadays, floods is the leader of global environmental disaster in terms of frequency, scale, and followed by the increment of cost over the last half century. Generally, high severity of flood disasters occur in several cities in Southeast Asia, including Jakarta [1]. Jakarta is located on a delta plain, where 13 rivers flow into the Jakarta bay. Therefore, naturally the location of Jakarta is vulnerable to flooding as experienced in 1996, 2002 and 2007 [2]. Although flooding in Jakarta is not a new problem, the scale of flood impacts has increased rapidly in recent decades [3]. Based on DKI Jakarta Provincial Government, from a total of 267 urban villages as many as 167 were flooded in February 2020 or approximately 60% of urban villages in Jakarta were affected by floods [4].

Physical and socio-economic factors are the two key factors that can increase the risk of flooding in Jakarta. Physical factors include land subsidence, low drainage or storage capacity in Jakarta's rivers and canals due to clogged waste, sediment eroded from upstream, and climate change. Furthermore, socio-economic factors include rapid growing of population, and changes in land use that lead to the growth of economic assets located in areas that are potentially prone to flooding [3].

As a megapolitan city, the development of various sectors, such as industry, trade, transportation, housing, and so on, in Jakarta is massive. However, this condition caused several environmental problems, such as the large number of agricultural areas being converted into residential and industrial areas, significant disruptions to the ecological and hydrological functions of the highlands and river catchments, as well an increasing rate of groundwater extraction due to the high population growth and high level of industrial and agricultural activities [5].

In addition, the flood problem in Jakarta could become more severe in the future due to climate change. Based on the study by Setianingsi4h, *et al.* shows the average sea level rise in the Java Sea from 2006-2016 of ± 6.80 mm/year. Moreover, the prediction of an increase in the severity and frequency of extreme rainfall in the 21st century can contribute to an increase in extreme river discharge [3].

On one hand, some people claim that floods in Indonesia occur because of the misuse of spatial planning. One of them is Jan Sopaheluwakan, a senior researcher from the LIPI Geotechnology Research Center, stated that Jakarta's natural conditions require Jakarta to have two areas, namely a green space for water infiltration in the south area and a blue space for storing water in the northern area of Jakarta. However, this condition has now been damaged due to a large number of buildings in almost all areas of the capital [6]. On the other hand, the use of space in Indonesia must be based on the Regional Spatial Planning (RTRW) which in its preparation refers to Law Number 26 of 2007 concerning Spatial Planning (Spatial Planning Law). If it refers to the Spatial Planning Law, then the flood disaster should not have occurred [7].

Based on the explanation above, the focus of attention in this article is the strategy of increasing green open space (RTH) based on community participation in reducing the risk of flooding in Jakarta.

2. Theoritical basis

2.1 Floods

Flooding is defined as a mass of water produced from runoff from the groundwate is relatively high and cannot be stored which naturally overflows and causes inundation. A flood becomes a disaster when the flood has caused property loss, environmental damage, and threatens as well as disrupts people's livelihoods. Flood in Indonesia have led to flooding as a disaster because it has caused loss of property even lives [7].

2.2 Spatial planning

Spatial planning is everything that is based on spatial considerations and rules. Spatial planning is the arrangement of everything related to human life in space as a place for the organization of life and livelihood physically and functionally [8]. The main considerations in physical planning are normative considerations, functional considerations and physical considerations. These three considerations are manifested in the form of a form that is not spatial or a-spatial and which is spatial or spatial. Diagrammatically the relationship between consideration and physical form can be seen in Table 1 [7].

Table 1. Consideration aspects of physical planning

Aspects of Consideration	Forms of A-Spatial	Forms of Spatial
Normative	1. Values 2. Socio-cultural patterns 3. Traditions 4. Religion and belief 5. Legislation 6. Relationships between groups or individuals 7. Tradition	1. Geographical requirements 2. Requirements for the location of the building and the environment 3. Ethnic grouping 4. Space requirements related to values and culture 5. The nature of the space of an activity
Functional	1. Scope and nature of functional activities 2. The essence of functional activities 3. Functional relationships between human groups 4. Basic needs for a particular function	1. Spatial relations between external functions 2. Spatial relationships between internal functions 3. Grouping activity functions according to their nature 4. Characteristics of the placement of functional activities 5. Space functions
Physical	1. Ability or quality of soil and space based on their basic physical properties 2. The relationship between human life and the basic physical properties of the environment 3. Basic characteristics of nature (soil, climate) 4. Basic needs for the biological physical environment	1. Macro layout 2. Circulation network 3. Space capital 4. Environmental management 5. Layout landscape layout

3. Research method

This research used qualitative analysis methods which is the literature review. The literature review is a method used to collect data or sources related to the topic raised in a writing. The literature review can be obtained from various sources, such as journals, books, documentation, the internet, and libraries [9]. The data collection process is carried out by filtering e-journals that describe flood, Green Open Space (RTH), based on the criteria determined by the author. The criteria include the year of the literature sources taken from 2012 until 2022, matching of the writing keywords, and the relevance of the results of writing and discussion. Strategies in collecting various literature journals using https://scholar.google.co.id/. In every journal that has been selected based on the criteria, a conclusion is drawn that describes the related explanation.

4. Result and Discussion

The mismatch of land use in 2019 with the planned spatial pattern in the RTRW is in the Protected Area, Green Open Space (RTH) for Cultivation, and Non-Green Open Space, with a total area of 5,098.37 Ha, where the biggest discrepancy is in the use of the RTH cultivation space pattern. for settlements, which is equal to 3,659.3 Ha. Land use mismatch with the regional spatial plan in Jakarta in 2019 can be seen in Table 2 [10].

Table 2. Land use distribution in Jakarta

Space Pattern	Land Use	Land area (Ha)
Protected area	Pond (*empang*)	3,14
	Building	17,45
	Garden/Plantation	79,04
	Residential	848,19
	Ricefield	113,15
	Pond (*tambak*)	103,5
	Mixed Plants	11,86
	Moor	94,8
	Settlement	8,38
Cultivation of Green Open Space	Building	121,56
	Residential	3.659,3
Cultivation of Non- Green Open Space	Building	1,76
	Residential	36,23

Based on these data, it can be concluded that most of the land is converted into residential or settlements. Actually, every area or zone of designation has building regulations that must be obeyed, such as the Basic Building Coefficient (KDB). KDB requires that the ground floor area of the building does not exceed the percentage set by the Regional Government. One example of a housing zone that is designated as a residence and is divided into sub-zones of very small houses, small houses, medium houses, large houses, and flat houses with KDB above 30% (thirty percent). For example, the KDB of house A is 70% meaning that if the land area is $100m^2$, then the maximum area of the ground floor is $70m^2$ and $30m^2$ is used for green open space (RTH) [11].

However, in fact most of residentials and settlements do not comply with these regulations, especially in densely populated areas where the distance between houses is too close and there is no green open space. Whereas green open space has many benefits, namely (i) direct benefits (in the sense of being fast and tangible), namely creating beauty and comfort (shady, fresh, cool) and getting materials to sell (wood, leaves, flowers, fruit); (ii) Indirect (long-term and intangible) benefits, namely highly effective air purifier, maintenance of the continuity of groundwater supplies, preservation of environmental functions of all existing flora and fauna contents (biological conservation or biodiversity). Green open space if built on a wide scale and across various land functions, can contribute to preserving natural processes and their functions, as well as supporting the sustainable growth of the city [12].

Spatial planning that needed to be encouraging is a spatial plan that is in accordance with the principles of spatial planning based on community participation. Based on community participation, it is hoped that regional and environmental development will be carried out in an integrated manner [8]. In addition, with the consideration that the community is a local expert about their own environment, so they know more about their environmental conditions than the government. This is important because the detection of signs of environmental damage that has the potential to cause flooding needs to be known by the public in advance [13].

Based on this explanation, adopting a community-based approach in urban planning with Green Open Space (RTH) community is a local expert about their own environment, so they know more about their environmental conditions than the government. community is a local expert about their own environment, so they know more about their environmental conditions than the government., which then develops into a sustainable city planning concept to be more specific into ecological city planning. In addition, it can reduce the risk of flooding in Jakarta.

The importance of the government's role in fostering public awareness about environmental care. Awareness of preserving the green environment will definitely be better if every citizen knows the function of green open space for the urban environment is to improve the quality of life and environment in the city with the aim of reducing the flood disaster in Jakarta which is getting worse every year. Directions to the community to continue to comply with regulations in building settlements, apply for Building Permits, and first consult with the PTSP of the nearest sub-district or sub-district before buying land are things that can be done by the community so that the percentage of land use mismatches and land use conversion does not increase.

5. Conclusion

Land use in DKI Jakarta shows the wider the built-up area and the narrower the green open space (RTH). This phenomenon can increase the frequency of flood events in Jakarta. To overcome this, there is a need for calls and invitations to the community to continue to comply with regulations in building settlements, administering Building Permits, and consulting first with the PTSP of the nearest sub-district or sub-district before buying land is something that can be done by the community so that the percentage of land use does not match. and land use conversion does not increase spatial planning in DKI Jakarta to consider green open space (RTH) in residence. The strategy in promoting green space regulation needs to involve all relevant parties so that appropriate spatial planning can be realized. Control over development activities also needs to be improved.

6. References

[1] Padawangi R and Douglass M 2015 Water, water everywhere: toward participatory solutions to chronic urban flooding in Jakarta *Pacific Affairs* **88** 517-550

[2] Texier P 2008 Floods in Jakarta: when the extreme reveals daily structural constraints and mismanagement *Disaster Prevention and Management: An Int. J*

[3] Budiyono Y, Aerts, J C, Tollenaar D and Ward P J 2016 River flood risk in Jakarta under scenarios of future change *Natural hazards and Earth System Sci* **16** 757-774

[4] *Cross-Year Flood Data* Retrieved From: https://pantaubanjir.jakarta.go.id/data-banjir-lintas-tahun

[5] Abidin H Z, Andreas H, Gumilar I and Wibowo I R R 2015 On correlation between urban development, land subsidence and flooding phenomena in Jakarta *Proceedings of the Int. Association of Hydrological Sci.* **370** 15-20

[6] Wiwoho L H 2014 *Banjir Jakarta Akibat Kesalahan Tata Ruang Bertahun-tahun* Retrieved from: https://megapolitan.kompas.com/read/2014/01/22/2156212/

[7] Qodriyatun S N 2020 Bencana banjir: pengawasan dan pengendalian pemanfaatan ruang berdasarkan UU penataan ruang dan RUU Cipta Kerja *J. Masalah-masalah Sosial* **11** 29-42

[8] Rozi A 2008 Perencanaan tata ruang berbasis partisipasi masyarakat dalam mitigasi bencana alam di Kelurahan Laksana Kota Banda Aceh *J. Ilmu Administrasi: Media Pengembangan Ilmu dan Praktek Administrasi* **5** 5

[9] Munandar A and Wardaningsi S 2018 Kesiapsiagaan perawat dalam penatalaksanaan aspek psikologis akibat bencana alam: a literature review *E-Journal UMM* **9** 72-81

[10] Septriana F E, Alnavis N B, Gustia R, Wirawan R R., Putri N P, Hasibuan H S and Tambunan R P 2020 Dampak perubahan penutupan lahan terhadap sistem hidrologi di Jakarta *Majalah Ilmiah Globe* **22** 51-58

[11] Karuniastuti N 2015 Bangunan ramah lingkungan *Swara Patra* **5** Retrieved from: http://ejurnal.ppsdmmigas.esdm.go.id/sp/index.php/swarapatra/article/view/110

[12] Sidauruk T 2019 Kebutuhan ruang terbuka hijau di perkotaan *J. Geografi* **4** 79-94
[13] Subrata O 2012 Pengelolaan banjir berbasis masyarakat studi kasus: Kabupaten Bojonegoro *J. Sumber Daya Air* **8** 125-140
[14] Setianingsih W, Sasmito B and Bashit N 2018) Analisis sea level rise di Laut Utara Jawa terhadap perubahan garis pantai wilayah Demak pada tahun 2006-2016 *J. Geodesi Undip* **7** 53-64

IOP Conf. Series: Earth and Environmental Science 1173 (2023) 012062 doi:10.1088/1755-1315/1173/1/012062

Estimating the Cost of Strauss Pile Retaining Wall Works Using the Cost Significant Method

Y Setiawan[1], Yelvi[1], A B Broto[1], M M Rafli[1], I Supriyadi[1] and Salimah A[1]

[1]Civil Engineering Department, Politenik Negeri Jakarta, Jl. Prof. DR. G.A. Siwabessy, Depok City, Jawa Barat. 16425

E-mail: yelvi@sipil.pnj.ac.id

Abstract. Cost estimation is an important element in a project. In the early stages of a project, information for estimates tends to be less accurate. Therefore, a cost estimation model is needed that can explain most of the projects based on the lack of information. The estimation method used is the Cost Significant Model. The aim is to determine the accuracy of the Strauss Mast Retaining Wall Work cost estimation model against the actual project cost. Data was collected using the sampling method, and 5 (five) similar project packages were obtained from 2019-2021. Produce 2 (two) Significant Cost Items that have cumulative costs (80%), namely Strauss Pile Work (X3) of 32% and Pile Work (X4) of 48%. Then it is analyzed to produce a regression equation formula using the SPSS application, and the equation is Y = 5696.59+5.270 X3-1,389X4 with a Cost Model Factor of 1.11. Estimated accuracy ranges from -7.3% to +8.4%. With an average of +0%. The accuracy rate is in class 1 of the International AACE classification, which has a lower limit of -3% to -10% and an upper limit of +3% to 15%. The estimation model is feasible to use to check the bid estimate.

1. Introduction

A retaining wall is a building that functions to stabilize the soil in certain soil conditions, especially for natural slopes and artificial slopes, as well as slopes due to soil filling [1]. Strauss foundation is a category of bore pile foundation that has a shallow depth. The use of the Strauss pile foundation is due to the fact that the project location is a densely populated environment and is considered the most suitable for the level of soil instability in the location, and the difficulty of accessing heavy equipment to enter the site is also one of the considerations for choosing this method [2]. The Cost Significant Model (CMF) is one of the models for estimating the total cost of construction. This estimate is based on past bidding data which relies more on the price that is most significant in influencing the total project cost as the basis for estimating using the program, namely Finding Multiple Linear Equations [3]. The resulting estimation model is expected to provide initial cost considerations and information for the Strauss Pile Retaining Wall Construction Project on the Ciliwung Lama River – Inspection Road and Embankment (RSCM).

Khamistan (2018) in Cost Estimation Analysis Using the Cost Significant Model Method as the basis for calculating Reinforced Concrete Bridge Construction in Aceh Tamiang district, the equation for the estimation model is Y = -19.832.201.41 + 1.948 X1 + 0.889 X2+ 4.274 X3 with a CMF of 0 .9844, the level of accuracy of the estimation results is in class 1 of the International AACE classification with a value between -3.37% to +1.69% [4]. Annisatul et al. (2021), in the Cost Estimation Model with a Cost Significant Model on the construction of houses in Tegal Regency, obtained the equation of the estimation model Y = -503,865,643 + 8,289 X4 with an accuracy rate of -5.1355% to 4.26% in the International AACE classification located in class 2 [5]. The results of this study indicate that the use of the CSM method as a cost estimation analysis method is considered appropriate and can produce appropriate results. Therefore, the Cost Significant Model method was determined to fulfill the purpose of this study, namely to estimate the initial cost of the Strauss Pile Retaining Wall Work Project.

2. Method

The research location is on the Ciliwung Lama River – Inspection Road and Embankment (RSCM). Precisely on Inspection Street, RT.07 RW.05 Kenari Village, Senen District, Central Jakarta Administrative City. This study uses a sampling technique from secondary data in the form of data from previous projects with the same type of project. Data retrieval is carried out with the following references: Collecting historical data in the form of Bill of Quantities (Quantity and Price List) sheet pile work packages for Strauss pile reinforcement, price components of work costs, and work implementation costs that are collected without Value Added Tax (VAT). Collected data are five (5) work packages whose budget is from 2019 to 2021.

Some work locations used as secondary data include :

Table 1. Job Location

Number	Job Name	Location
1	Sunter Lake Revitalization Works	Jakarta
2	Repair of DPT Simpruk PHB Permata	Jakarta
3	Cileungsi-Cibeet DPT planning project	Bogor
4	PHB Ciragil Retaining Wall Works	Jakarta
5	Kemang Express Retaining Wall	Bekasi

Source: Department of Highways and Water Resources and Document Study

This study uses cost. The data specifications used are in the form of variable identification, namely the cost of the work component as the independent variable (X) and the total cost as the dependent variable (Y).

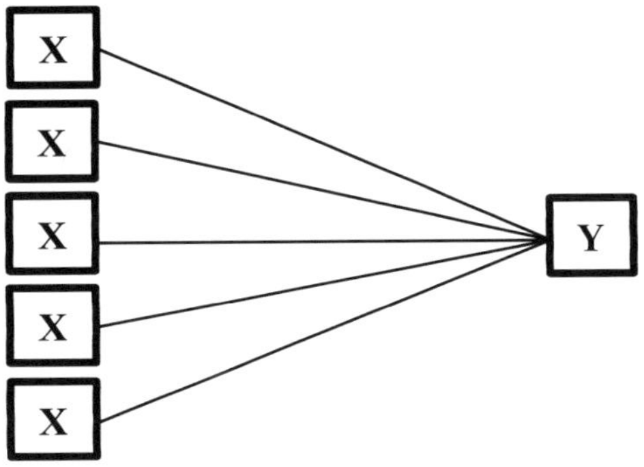

Figure 1. The Relationship between the independent variable and the dependent variable

Information :
X1 = Preparatory work
X2 = Earthwork
X3 = Strauss Pile work
X4 = Sheet Pile work
X5 = Other work

Y = Total project cost

Stages of data processing using the Cost Significant Model Method. Cost modeling formulation using multiple linear regression. Analysis of secondary data information was carried out before entering into the statistical program. The building area for each work package varies. Therefore the data is adjusted to the cost per m^2 so that it is uniform. Total cost (Y) and work component costs (X1 to X5) are divided by building area for each work package, so Y is the cost per m2 of building area, and X1 to X5 is the cost component price per m^2 of building area. Stages of analysis or data processing using the method Cost significant model as follows:

2.1. Identify variables
This study uses the price of the work cost component as the independent variable (X) and the total cost as the dependent variable (Y) [6].

2.2. Calculation of Time Value and Position Value
Calculating the data obtained in this study, namely data from projects from 2019 to 2021, will be converted to 2022 using the Future Value (FV) formula. Because the projects being reviewed are in different locations, the price at the site is also converted to the planned location price using the Construction Cost Index [7].

2.3. Identify cost significant items
Cost Significant Items are identified as the most extensive items whose percentage is equal to or greater than 80% of the total cost [6].

2.4. Test requirements analysis
To get a regression model that is Best Linear Unbias Estimator (BLUE), it is necessary to test the classical assumption. The classical assumption tests carried out in this study are the normality test, multicollinearity test, heteroscedasticity test, and autocorrelation test [8].

2.5. Data analysis
In this study, the data analysis used is linear regression analysis with the help of SPSS application software and several tests, including the coefficient of determination and the Anova test/F test.

2.6. Model accuracy
The cost estimation results using the Cost Significant Model are obtained by dividing the cost of the estimated model by the Cost Significant Factor (CMF). CMF is the average ratio of the estimated cost of the model to the actual cost. The estimation results of the Cost Significant Model are obtained from the calculation compared to the implementation costs (actual costs) of the project under review.

3. Result and Discussion

3.1. Identify variables
There are five independent variables (X) and one dependent variable (Y). Table 2 contains project data that includes the independent variable (X) and the dependent variable (Y) by m^2 of building area. Information on independent variables, namely for X1 is preparatory work, X2 is earthwork, X3 is Strauss pile work, X4 is sheet pile work, and X5 is other work.

Table 2. Variable Identification and Project Data by m^2 Building Area

Package	Year	Area (m2)	Dependent Variable		Independent Variable				
					X1	X2	X3	X4	X5
P1	2019	185	Y1	601995,94	56864,86	48730,81	187173,19	280687,39	28539,68
P2	2020	170	Y2	735715,91	69617,65	58372,94	229149,53	343635,67	34940,12
P3	2020	175	Y3	829164,86	56319,41	59607,00	265635,17	414175,74	33427,54
P4	2021	165	Y4	841742,79	79696,97	66333,33	262326,06	393387,64	39998,79
P5	2021	180	Y5	896142,50	60838,87	64830,00	286951,57	458955,29	36110,00

3.2. *Calculation of time value and position value*

Project implementation activities are carried out from 2019 to 2021 and will be converted to 2022. So, the price must be adjusted to the inflation prevailing in that year. Inflation data is obtained from the Central Statistics Agency (BPS). Time Value effect formula $F = P\ (1 + i\ )^n$. F = Projected price, P = pre-projected price, I = inflation factor.

Table 3. Inflation Data

Number	Year	General Inflation (%)
1	2019	3,13
2	2020	2,63
3	2021	2,58

source: Central Statistics Agency (BPS)

3.3. *Identify cost significant items*

Cost Significant Items are the largest cost components whose percentage is equal to or greater than 80% of the total cost. Table 4 contains the proportion of cost components.

Table 4. Proportion of Cost Components

Symbol	**Mean (Rp)**	**percentage**
Y	866.271,79	100
X_1	71.059,21	8
X_2	65.903,38	8
X_3	273.371,25	32
X_4	417.735,44	48
X_5	38.202,50	4

From table 4, it is found that the independent variables X3 and X4 have made up 80% of the total cost (Y). it can be concluded that the variables X3 (Strauss Pile work) and X4 (Sheet Pile work) have cost significant items.

3.4. *Test requirements analysis*

3.4.1. Normality test. The normality test was carried out with the Smirnov-Kolmogorov test. The data is considered normal if the significance level of each variable is more than 0.01. The test results can be seen in Table 5.

Table 5. Normality Test

Description	Symbol	Probability (Sig)
Total project cost	Y	0.20
Strauss Pile Work	X3	0.20
Sheet Pile Work	X2	0.20

The test results show that the significance level is greater than the level of effectiveness required. So it can be concluded that the variables are normally distributed.

3.4.2. Multicollinearity test. The multicollinearity test aims to determine whether or not there is a linear relationship between the variance inflation factor (VIF). If the VIF is less than ten or the tolerance value is more than 0.01, it can be concluded that there is no multicollinearity problem. The results of the research VIF are 3,806. This means that there is no multicollinearity in the regression model.

3.4.3. Heteroscedasticity test. The Heteroscedasticity test aims to determine whether there is an inequality of variance and residuals from one observation to another in the regression model. The basis for decision-making is if the significance is more significant than 0.05, then the regression model does not have heteroscedasticity problems. The study resulted in a value of 0.286 for X3 and 0.253 for X4. This means that there is no heteroscedasticity in the regression model. The heteroscedasticity test can be seen from the graph using the scatterplot test [9].

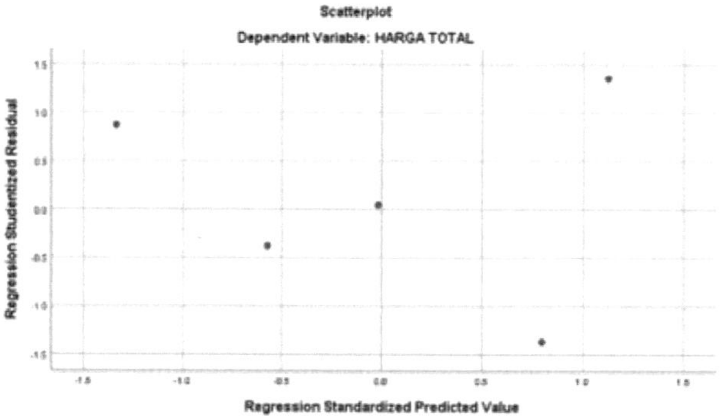

Figure 2. Scatterplot Test

The graphic image shows that the points spread above and below the number 0 on the Y axis, so there is no heteroscedasticity.

3.4.4. Autocorrelation test. The autocorrelation test aims to test for a correlation between the first period disturbance error and the previous period error in the linear regression model. If the significance level is above 0.05, it can be said that the residual is random, and there is no correlation. The autocorrelation test was carried out with the Durbin Watson test. The result of Durbin Watson is 1,325, which is more than 0.05. It can be concluded that there is no autocorrelation.

3.5. Data analysis

In this study, the data analysis used was multiple regression with the help of SPSS application software and carried out several tests, including the coefficient of determination and the Anova test/F test. From table 6, the following regression model can be obtained:

$$Y = 5696.59 + 5.270X3 - 1.389X4 \quad (1)$$

Information :
Y = Estimated Total Project Cost
X3 = Strauss Pile Work
X4 = Sheet Pile Work

Table 6. Formulation of Regression Model

Model	Unstandardized Coefficients	
	B	Std. Error
1. (constant)	5696.59	2106.572
Strauss Pile Work	5.270	.075
Sheet Pile Work	-1.389	.045

3.5.1. Coefficient of Determination. The analysis of determination aims to determine the basic percentage of variation in the dependent variable that can be explained by the independent variable. The results of the analysis are shown in Table 7.

Table 7. Coefficient of Determination

Model	R	R Square	Adjusted R square
1	1.000[a]	.9996	.9988

Based on the results of the SPSS "Model summary" output in table 7, it is known that the coefficient value or R Square is 0.9996, indicating that 99.96% of costs (Y) are influenced by factors X3 and X4. While the rest (100% - 99.96%) = 0.04% influenced by other factors.

3.5.2. ANOVA Test / F Test. Based on the results of the ANOVA test, the significance level value is 0.000 or less than 0.05, meaning that the independent variable can explain the dependent variable significantly. The independent variable can substantially explain the dependent variable.

3.6. Model accuracy

The initial stage in the regression model's accuracy is calculating the Cost Model Factor (CMF). CMF is the ratio between the estimated cost of the model and the actual cost. The estimated cost of the model is obtained by entering the cost of the Strauss pile and sheet pile work into the model obtained. The calculation results are shown in Table 8.

Table 8. Calculation of Cost Model Factor (CMF)

No	Estimated Total Project Cost /m^2 (Y) (Rp)	Actual Total Cost /m^2 (Y)(Rp)	CMF
(1)	(2)	(3)	(4)
	$Y = 5696{,}59+5{,}270\ (X_3) - 1{,}389(X_4)$		(2)/(3)
P1	656107,91	601995,94	1,09
P2	777805,21	735715,91	1,06
P3	998207,77	829164,86	1,20
P4	866941,24	841742,79	1,03
P5	1050378,59	896142,50	1,17
	Average CMF		1,11

After getting the average value of CMF, determine the estimated cost significant model. CSM estimation is obtained by dividing the model estimation cost by the average CMF. The calculation results are shown in Table 9.

Table 9. Estimated Cost Significant Model

No	Estimated Total Project Cost /m^2 (Y) (Rp)	Estimated CSM (Rp)
(1)	(2)	(3)
	Y = 5696,59+5,270 (X_3) -1,389(X_4)	(CMF=1,11) (3) = (2)/CMF
P1	656107,91	590767,30
P2	777805,21	700344,99
P3	998207,77	898798,06
P4	866941,24	780604,13
P5	1050378,59	945773,29

After getting the average CMF and calculating the CSM estimate, the next step is to calculate the level of model accuracy by estimating the CSM cost minus the actual total cost, divided by the actual total cost, then multiplied by 100%. The calculation results are shown in Table 10.

Table 10. Cost Significant Model Accuracy

Estimated CSM (Rp)	Actual total cost (Rp)	Accuracy (%)
(1)	(2)	(3)
(CMF=1,11) (3) = (2)/CMF		((1)-(2))/(2)x100%
590767,30	601995,94	-1,9%
700344,99	735715,91	-4,8%
898798,06	829164,86	8,4%
780604,13	841742,79	-7,3%
945773,29	896142,50	5,5%
	Max	8,4%
	Min	-7,3%
	Averages	0%

Based on the calculations in table 10, the model's accuracy with a significant cost ranged from -7.3% to 8.4%, with an average of 0%. Based on the International AACE classification, the accuracy of the estimated cost significant model is in class 1 and is suitable for checking bid estimates.

Table 11. Cost Estimation Classification According to AACE International

Estimation Class	End Usage (Typical purpose of estimate)	Methodology (Typical estimating method)	Expected Accuracy Range (Typical low and high range)
Class 5	Concept Screening	Equipment Factored or Parametric Models	Low -20% to -50% High +30% to +100%
Class 4	Study or Feasibility	Equipment Factored or Parametric Models	Low -15% to 30% High +20% to 50%

Estimation Class	End Usage (Typical purpose of estimate)	Methodology (Typical estimating method)	Expected Accuracy Range (Typical low and high range)
Class 3	Budget, Authorization, or Control	Semi-Detailed Unit Cost With Assemble Level Line Items	Low -10% to -20% High +10% to +30%
Class 2	Control or Bid/Tender	Detail Unit Cost with Forced Detailed Take-off	Low -5% to -15% High +5% to +20%
Class 1	Check Estimate or Bid/Tender	Detail Unit Cost With Detailed Take-Off	Low -3% to -10% High +3% to +15%

source: Christensen and Dysert (2005) [10]

4. Conclusion

The work that significantly affects the total cost of the Strauss Pile Retaining Wall Project is the Strauss Pile Work (X3) of 32% and the Sheet Pile Work (X4) of 48%. Has a percentage of 80%. The cost model for Strauss Pile Retaining Wall works using the cost significant model method is: Y = 5696.59 + 5,270X3 - 1,389X4, where Y is the cost by m2 of Strauss Pile Retaining Wall work and X is the cost of work items / significant variable by m2 work Strauss Pile Retaining Wall. The forecast model's accuracy ranges from -7.3% to +8.4%. With an average of +0%. The accuracy rate is in class 1 of the International AACE classification, which has a limit of -3% to -10% and an upper limit of +3% to 15%. The estimation model is feasible to use for checking the bid estimates.

Acknowledgments
The authors express their sincere gratitude to the Research and Community Service Unit (UP2M) of the Jakarta State Polytechnic for sponsoring this research. The authors would also like to thank all those who have supported this research.

References

[1] M. K. J. H. T. A. T. and M. 2017 Analisis Stabilitas Dinding Penahan Tanah (Sudi Kasus : Sekitar Areal PT. Trakindo, Desa Maumbi, Kabupaten Minahasa Utara)," *Jurnal Teknik Sipil ISSN: 2337:6732* **5** 285-294

[2] G. E. Dwiyanto 2020 Analisis Estimasi Biaya Dengan Metode Cost Significant Model Pada Konstruksi Jembatan Rangka Baja Di Provinsi Sulawesi Selatan 1-62

[3] C. Yuliana, A. Z. Ulimaz and R. H. Kartadipura 2020 Estimasi Biaya Dengan Menggunakan Cost Significant Model Pada Pekerjaan Pemeliharaan Jalan Di Kota Banjarbaru *ISSN : 2302-8394* **9** 24-30

[4] Khamistan September 2018 Analisis Estimasi Biaya Dengan Metode Cost Significant Model Sebagai Dasar Perhitungan Konstruksi Jembatan Beton Bertulang Di Kabupaten Aceh Tamiang *Teras Jurnal* **8** 444-454

[5] A. Muthoharoh, S. Luthfianto and I. Mirajhusnita 2021 Model Estimasi Biaya Dengan Metode Cost Significant Model Pada Pembangunan Rumah Di Kabupaten Tegal," *Rancang Bangun* **07** 9-19

[6] I. Dipohusodo, Manajemen Proyek & Konstruksi, Kanisius, 1996.

[7] R. M. Fikri and J. Sekarsari 2015 Analisis Estimasi Biaya Proyek Peningkatan Jalan Beton Di kabupaten Tangerang Dengan Metode Cost Significant Model," *Seminar Nasional Teknik Sipil ISSN : 2459 -9727* 337-348

[8] A. E. Brahmana 2009 Analisis Regresi Berganda Terhadap Faktor-faktor yang Mempengaruhi

Laju Inflasi Tahun 2006-2007," *Tugas Akhir*

[9] S. A. Muali and W. Hartono Juni 2016 Estimasi Biaya Konstruksi Bangunan Gedung Dengan Metode Cost Significant Model Studi Kasus Proyek Lanjutan Pembangunan Gedung di Universitas Sebelas Maret Surakarta," *e-Jurnal Matriks Teknik Sipil* 498-504

[10] P. Christensen and L. R. Dysert 2005 AACE International Recommended Practice," *TCM Framework : 7.3 - Cost Estimating and Budgeting* 18R-97

[11] I. N. Y. Astana September 2017 Estimasi Biaya Konstruksi Gedung dengan Cost Significant Model," *Riset Rekayasa Sipil ISSN 2579-7999* 7-15

[12] I. Nurpa'i, D. A. Susanto and N. S. Nugroho September 2020 Estimasi Biaya Menggunakan Metode Cost Significant Model Pada Pembangunan Peningkatan Jalan Studi Kasus Pembangunan Peningkatan Jalan Di Kabupaten Sukabumi," *ISSN 2715-4831*

[13] P. S. d. H. R. M. W. Poh 1995 Cost Significant Modelling - Its Potential for Use in South-east Asi," *Engineering, Constuction, and Archtictural Management.* **2** 121-139

[14] Ananda April 2022 Pondasi Strauss Pile : Pengertian, Metode, Perhitungan, dan Kisaran Harga [Online]. Available: https://www.gramedia.com/best-seller/pondasi-strauss-pile/.

IOP Conf. Series: Earth and Environmental Science 1173 (2023) 012063 doi:10.1088/1755-1315/1173/1/012063

The Role of the Government of Riau Provincial in Dealing with Forest and Land Fires

Ketsana Sengdara[1], **Achmed Sukendro**[2], **and Heridadi**[3]

[1] Student in Disaster Management Study Program, Republic of Indonesia Defense University, Indonesia
[2] Lecturer in Department of Peace and Conflict Resolution, Republic of Indonesia Defense University, Indonesia
[3] Lecturer in Department of Disaster Management, Republic of Indonesia Defense University, Indonesia

Abstract Riau is one of the provinces with a higher risk of forest and land fires in Indonesia. In this study, the researchers wanted to understand the role of the Government of Riau provincial in responding to the problems. The methodology used in this study was a descriptive qualitative research method using data collection techniques in the form of primary data from interviews and FGD as well as secondary data from previous studies and documents supporting this study. The findings of this study are the role of the Government of Riau provincial in reducing forest and land fires in Riau. This's reflected in a 90% reduction in burned area from 2016 to 2021. Not only does the Riau government work alone, but multiple parties are supporting the Riau government, namely the Indonesian Armed Forces, the Republic of Indonesia Police, the National Research and Innovation Agency, as well as other agencies that help reduce the risk of forest and land fires disasters in Riau.

1. Introduction

Riau is a province of Indonesia, it's miles are positioned in a noticeably strategic line both at gift or destiny due to the fact it's miles positioned in both the local exchange line and global exchange line through inside Association for Southeast Asian Nations (ASEAN). Riau is positioned between 01°05'00" of southern range to 02°25'00" of northern range and 100°00'00" to 105°05'00" of Japanese longitude with the subsequent border regions: North (Malaka Strait and North Sumatera Province); South (Jambi and West Sumatera Province); West (West Sumatera Province); East (Riau Island Province and Malaka Strait).

Riau Islands as Indonesia's thirty-second province in 2002 ought to be visible as a part of a broader fashion that noticed the advent of the latest subnational authorities entities seven provinces and 112 districts in total throughout the country after the end of the New Order era. Riau Province is a tropical location with most temperature via way of means of about 35,10C and minimal temperature via way of means of about 21,80° C. Riau location may be categorized as inflammable to flammable with an index of flaming index of about 0 - 330 (low - high). As a result, in each yr., there are constantly discovered numerous warm spots that motive catastrophic smog in entire regions of Riau Province. In a different case, the rainfall depth is about 1700 mm - 4000 mm/yr. Administratively, Riau Province includes 10 regencies and a pair of towns, which include 166 sub-districts and 1.846 villages. Riau's overall populace will increase to 5.877.887 human beings from 5.867.358 human beings in the year 2014. Compared to the year 2013, it will increase via way of means of 45.999 human beings (https://www.riau.go.id).

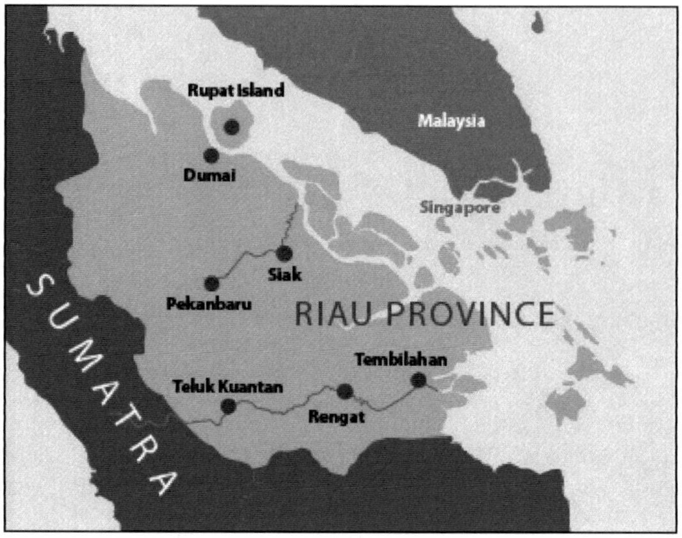

Figure 1. Map of Riau Province.
Sources: (https://www.apeopleloved.com/riau-malay-prayer-journey/)

Riau's peatlands, cover an area of approximately ±4.9 million hectares, equivalent to approximately 54.44% of Riau's land area, and consist of freshwater peat marshes and tidal flat peatlands. consists of a Swamp. Most of them are concentrated in the eastern region. This low-lying area is prone to flooding and inundation, which is common.

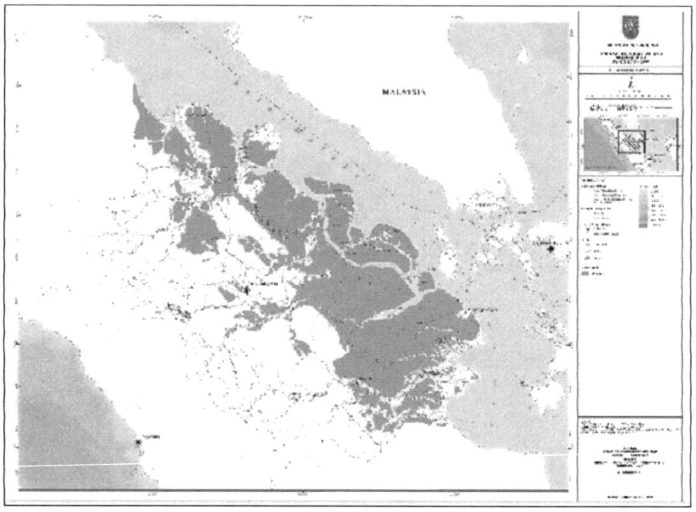

Figure 2. Peatland Map of Riau Province.
Sources: Regional Development Planning Agency, Research and Development Expose of Riau Province 2021

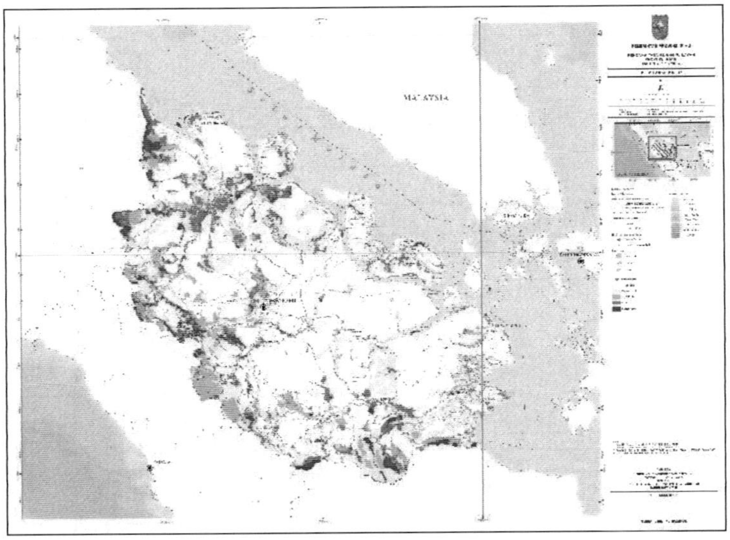

Figure 3. Critical Land map of Riau Province.
Sources: Regional Development Planning Agency, Research and Development Expose of Riau Province 2021

Last year, 2019, Riau's forest and land fire incidents captured public and international attention. The 2019 Riau wildfires weren't the worst, especially since 1997, 2004, 2014, and 2015 wildfires did more damage than his 2019 fire. All political parties agree that forest and land fires are unconstitutional in a democratic country, and the 1945 constitution of the Republic of Indonesia states that its land, water, and natural resources are under government control. For what people need for prosperity. The essence of this constitutional statement is that the state, as a state host, is committed to the best stewardship of its natural wealth, including protecting forests from forest and land fires.

Forest and land fires are also seen as a result of the government's failure to care for their citizens. Governments themselves recognize that wildfires are a multifaceted problem that requires cross-sectoral coordination to tackle. In this regard, the Government has established a coordination strategy with a bureaucratic approach through the Presidential Decree No. 16 of 2011 on Improving Forest and Land Fire Management. Forests and lands were the new fire (Budiningsih, 2017).

Forest and land fires in Riau continue every year as if the government failed to prevent forest and land fires. From another perspective, governments should be able to contain forest and land fires. Many regulations are politically crafted but very difficult to enforce. It seems that the device is not vigilant enough in these areas, despite the high number of forest fires. The new government is in a hurry to act, appearing panicked even when fires break out, but calming down when the fires are brought under control. Alert levels remain high, at least according to the hotspot index of satellite monitoring results. Communication in wildfire prevention and response is lacking due to uncoordinated human resource quality between states and regions.

No,	Districts	Hazard				
		Spacious (Ha)				Class Level
		Low	Currently	High	Total	
A	**Districts/Kebupaten**					
1	Bengkalis	57.315	171.591	616.675	845.581	High
2	Indragiri Hilir	157.928	313.147	851.696	1.322.771	High
3	Indragiri Hulu	116.878	389.537	320.388	826.803	High
4	Kampar	277.872	377.525	363.014	1.018.411	High
5	Kepulauan Meranti	42.754	120.907	197.248	360.909	High
6	Kuantan Singingi	94.124	228.313	231.903	554.340	High
7	Pelalawan	57.240	312.506	936.606	1.306.352	High
8	Rokan Hilir	24.547	229.779	677.601	931.927	High
9	Rokan Hulu	132.932	446.022	186.980	765.934	High
10	Siak	32.756	242.593	505.881	781.230	High
B	**Districts/Kota**					
11	Kota Dumai	12.345	32.386	153.259	197.990	High
12	Kota Pekanbaru	38.602	15.331	9.904	63.837	High
	Riau Provice	**1.045.293**	**2.879.637**	**5.051.155**	**8.976.085**	**High**

Figure 4. Hazard Class Level of Forest and Land Fire
Sources: Disaster Risk Assessment Document, BPBD Riau Province 2021

❖ **Literature Review**

a. *Role and Theory*

Role theory emphasizes the nature of individuals as social actors who study their behavior according to their work environment and position in society. Role theory attempts to describe the interactions between individuals within an organization, focusing on the roles they play.

Everywhere there is a role you play in life. Roles such as spouses, parents, charitable volunteers, engineers, and managers play important roles in families, communities, and work. These roles also give individuals a sense of who they are and whom they become. Within an organization, everyone ends up being an employee, subordinate, manager, department member, customer, supplier, project team member, and so on. These roles may be performed separately (e.g., as a manager in the morning, and as a subordinate in the afternoon) or concurrently (e.g., as a member of the product development team, or as a departmental representative). It will be executed. manager. Neither individuals nor organizations seem able to function without roles in which structured interdependencies organize and create a network of interlocking roles and responsibilities.

According to Soekanto (2009), the meaning of the role is a work that is carried out dynamically according to the status or position it carries. These statuses and positions correspond to social order, even in the order of action they are all adapted to different roles.

b. Theory of Mitigation

The act of moderating something or the state of moderation: The process or consequence of making something less serious, dangerous, painful, harsh, or harmful The mitigation of the suffering of punishment. It was a time when the Cold War eased, including some renegotiations of the (https://www.merriam-webster.com).

Mitigation is all the efforts and activities undertaken to reduce and minimize the impact of a disaster, including physical preparedness, vigilance, and mobilization capacity readiness and readiness

(Depdagri, 2003). Mitigation is an action taken to reduce or minimize the impact of a disaster on society (DKP, 2004). Mitigation is any effort or activity aimed at reducing the impact of a natural or man-made disaster on a nation or society (Carter, 1991).

- *Non-Structural Mitigation*

Non-structural measures tend to be more sustainable because they involve active community participation. National and regional policies should prioritize nonstructural alternatives with lower costs and fewer environmental side effects, and implement structural measures only as a last resort. (Shrestha, 2008).
Non-structural mitigation is an effort to reduce the impact of disasters through policy regulations. For example, the DM Law or the Disaster Management Law, making city spatial plans, or other activities that are useful for strengthening the capacity of citizens (https://www.gramedia.com).

- *Structural Mitigation*

The first approach to mitigating flash floods is to reduce the likelihood of landslides and debris flows and to create landslide embankments. This means taking steps to reduce the likelihood of slope failure. Measures to reduce the gradient error can be divided into three categories: Controls, Constraints, and Others.
Structural measures are physical structures to reduce or avoid the potential effects of hazards or the application of engineering practices or techniques to achieve resilience and resistance to hazards of structures or systems. Common structural measures to reduce disaster risk are dams, flood protection, breakwaters, seismic structures, and shelters.
Structural mitigation is the act of physical change or protection from disasters and hazards. An example of structural mitigation is when a family strengthens their house to make it more wind or earthquake-resistant. In addition, other examples of structural mitigation measures include creating a sandbag barrier around the house in case of potential flooding. In general, structural mitigation is the direct action that people take, build or move to better protect life and property.

❖ **Community Based Disaster Risk Reduction**

Community-Based Disaster Risk Reduction is a series of efforts to reduce disaster risk carried out in the face of disaster threats and or the application of physical and non-physical efforts carried out by community members in an actively participatory and organized manner (Direktorat Jenderal Cipta Karya, 2013). Community-Based Disaster Risk Reduction is a working perspective and methodology on disaster management that evolves and is based on communities (Indonesian Constitution No.24/ 2007) Community Based Disaster Risk Reduction is a disaster risk management process that actively involves at-risk communities to assess, analyze, and manage, monitor and evaluate disaster risk in order to reduce their vulnerabilities and increase their capacity (Haghebaert, 2012).

❖ **Disaster Risk Reduction**

According to UNISDR (2010), disaster risk reduction concepts and practices analyze the causative factors of disasters, such as reducing exposure to disasters, reducing vulnerability, prudent environmental management, and enhancing preparedness for adverse events. It is implemented through a systematic effort to address this. This is to his observations of Twig (2009), who defines disaster risk reduction as a systematic approach to identifying, assessing, and reducing disaster risk. Disaster risk reduction aims to reduce socio-economic vulnerability to disasters and to manage environmental and other hazards that cause disasters. Another objective of the disaster risk reduction efforts implemented by Widana (2019) is to reduce the damage caused by natural disaster threats such as earthquakes, floods, droughts, and cyclones through preventive measures. Disasters often follow natural disasters. The severity of a disaster depends on the extent to which the hazard impacts society and the environment. The magnitude of the impact, in turn, depends on the choices we make about our lives and our environment.

2. Research Methods

In the research entitled The Role of Government of Riau Provincial in Dealing with Forest and Land Fires, researchers used a writing method that used descriptive qualitative writing techniques. Descriptive qualitative writing techniques are writing techniques that use qualitative data presented descriptively. Qualitative data using primary and secondary data. Researchers obtained primary data through interviews with a competent caregiver for domestic work at the Indonesia Defense University and the Focus Group Discussion FGD. Meanwhile, in procuring secondary data, the researcher leveraged journals from previous studies, her research-related websites, and documents supporting that study.

This topic selection technique uses targeted selection. This means that selected resource personnel or informants meet established criteria and are directly involved in implementing disaster management. Informants involved in this study were provided by various ministries and agencies involved in implementing disaster management, including a. Director of Riau's Research and Development Planning Agency (Bappedalitbang); b. Director of Water Resources, Public Works Department, Housing and Rural Spatial Planning (PUPR-PKPP) Riau. c. Young Expert in Functional Environmental Impact Assessment, Ministry of Environment and Forestry, Riau Province. i.e. A young engineer at the National Institute for Research and Development (BRIN). e. Commander of Military Resort Command 031 Wirabima. f. Air Force Base (Lanud) Commander of TNI AU Roezmin Nurjadin Riau. G. Riau Police Chief.

3. Result And Discussion

3.1. Prevention And Mitigation

To find out how to prevent and mitigate forest and land fires (karhutla), the researcher asked several questions related to the efforts made to prevent forest and land fires in Riau province.

According to the Secretary of the Research and Development Planning Agency (Bappedalitbang) of Riau Province, disaster mitigation has been well managed. Surveillance technology is being developed in the form of real-time forest and land fire dashboards that are accessible to both the public and governments and is working with government agencies on their use of satellites. In addition, we have support from the Task Force and TNI/POLRI to help the Riau Government with forest and land fire prevention. According to Focus Group Discussion, we were joining a Domestic Work College (KKDN) between Disaster Management Master Students of the Republic of Indonesia Defense University and Riau Government provincial on 9th February 2022 there are 41 participants.

According to the Head of Water Resources, Public Works Agency for Housing Spatial Planning for Settlement and Land Areas (PUPR-PKPP) Riau, Structural mitigation to overcome forest and land fires comes in the form of peat restoration through the construction of irrigation systems (water structures) to maintain water levels. In addition, PUPR services also strive to protect peatlands by building irrigation network canals or canal gating systems.

According to the Young Expert Environmental Impact Control Functionalist, The Job Creation Act allows for the burning of firebreak construction records of up to 2 hectares. In reality, it is about the uncontrollability of existing fires, so the solution offered by the government is to provide heavy equipment support. are being implemented.

According to the Young Engineer, the National Research and Innovation Agency (BRIN), Installing sensors in hotspots can prevent forest and land fires. We also use Weather Modification Technology (TMC) to control conditions so that wetlands do not dry out even in hot weather.

According to the Commander of the Military Command Resort Command (Korem) 031 Wirabima, One of the forest and land fire prevention strategies is the village Pembina Bintara (Babinsa) program. In this program, members eat together, sleep together, and build social communication through community communication and interaction. Hope is to be independent. If someone is identified as one of those who will burn the country down, the community itself will prevent it. Korem also conducts regular simulation exercises at all times, including during the rainy season, with the aim of maintaining the operational capability of the force.

According to the TNI AU Air Base (Lanud), Commander Roesmin Nurjadin, Riau Prevention and mitigation efforts are conducted through patrol activities on hotspot information from satellites. Special fire surveillance flights, as well as commercial aircraft flights passing through Riau, can provide information on the situation on the ground due to the presence of integrated communications and coordination. In addition, the Indonesian Air Force has also set up daily Babino Pudirga to circle the air base ±5 km, visiting communities, building good relationships, and educating the public to change their behavior to be more environmentally friendly.

According to the Head of the Operational Bureau of the Riau Police, the Riau Police has developed an information system called the Nusantara Yellow Lancang Dashboard that is highly effective in combating hotspots. However, if you have limited technology, you can report hotspots or hotspots visually or directly monitor hotspots on peatlands. Then you can report to the most powerful agency in the village with access to the application. Reported information is sent to local law enforcement so they can take immediate action.

3.2. Preparedness

To find out how to prepare for forest and land fires (karhutla) disaster, the researcher asked several questions related to the efforts made to prevent forest and land fires disaster in Riau province.

According to the Secretary of the Research and Development Planning Agency (Bappedalitbang) of Riau Province, Civil protection was well managed. Programs initiated by the Ministry of Environment and Forestry, such as B. Firefighting communities and peat trading communities in villages vulnerable to forest and land fires, are designed to improve community capacity and prevent forest and land fire disasters. Born to respond quickly and cultivate the land so that it can absorb water without damage. In addition, the head of water resources at the Public Employment Agency for Urban and Rural Spatial Planning (PUPR-PKPP) Riau said knowledge of forest and land fire disaster preparedness was also carried out by the oil palm farmers department.

According to the Young Expert Environmental Impact Control Functionalist, In addition to the Fire Care Village Program, there are also CSR and Climate Village programs that are implemented in areas prone to forest and land fires (to reduce the country's CO2 emissions, this program helps prevent burns in agricultural activities try). The program is considered highly effective as it can reduce the incidence of forest and land fires by up to 90%.

3.3. Emergency Response

To find out how the efforts were made during the emergency response to forest and land fires (karhutla), the researcher asked several questions related to the efforts made to prevent forest and land fires in Riau province.

According to the Young Engineer, the National Research and Innovation Agency (BRIN) Weather Modification Technology (TMC) is not only used in the prevention and mitigation phases but also plays a role in the emergency response phase. During the emergency response phase, TMCs are used to extinguish hotspots and control the rate of fire progression to limit the spread of fires and wildfires.

According to the TNI AU Air Base (Lanud) Commander Roesmin Nurjadin Riau, He provided aircraft and/or helicopters (depending on the situation on the ground) to help conduct TMC operations as ground forces and local governments were unable to reach the area.

According to the Head of the Operational Bureau of the Riau Police, Once the yellow intimidating dashboard retrieves information about hotspots, Babinkatibnas officials share information with their closest members to confirm case reports. If Kemudida falls into the hazardous category, judging by the level of fire reported on the picket line, it will be entered on the dashboard and coordinated with the police department for firefighting efforts. If the police cannot coordinate with the police, the state will take over the firefighting work if the fire area exceeds 5 hectares.

According to the Secretary of the Research and Development Planning Agency (Bappedalitbang) of Riau Province, If the country has already burned to the ground, further investigation will be conducted to find out the cause. Additionally, we work with other sectors such as the Ministry of

Environment and Forestry (KLHK), PUPR Services, TNI/POLRI, and local governments to set up command posts in case of emergencies.

4. Conclusion

Activities related to forest and land fire prevention and control include building firebreaks and building facilities and infrastructure. The purpose of a firebreak is to divide a large fuel area into several parts/fragments so that in the event of a fire the entire fuel area or vegetation is not engulfed by the fire.

Land and forest fires are usually caused by damage to peatland hydrological/aquatic systems resulting from uncontrolled human activities such as deforestation and open burning. To prevent peat from burning, you can prevent it by building a dam around it. This dam stock is maintained near rivers and reservoirs and can pump water from rivers and other reservoirs into large peat bogs during the dry season. Activities in the fire and wildfire disaster preparedness phase are carried out through socialization and program establishment. B. Fire Protection Community.

The Governor of Riau representing the local government has set an emergency alert status. In controlling the problem, the Riau Province Task Force has carried out several blackouts, such as in Bengkalis, Siak and Rengat. In early March the Directorate of Land and Forest Fire Control will launch the Integrated Patrol for the Prevention of Forest and Land Fires and we will strengthen coordination with various parties such as MPA and other Task Force Stakeholders.

a. Socialization

The socialization of forest fire prevention is the most important initial activity in firefighting activities, and it is a work that must be done continuously. Socializing fire safety is a more economical way to reduce fire damage and casualties without the use of expensive equipment. Socialization occurs through the provision of fire information and fire prevention and control efforts. The purpose of this activity is to provide the community with fire knowledge, prevention efforts, and community preparedness education in the event of a fire.

b. Program Establishment

Community fire brigade effectiveness in fire prevention and suppression requires regular training activities to improve knowledge and skills in fire safety practices. This activity takes place at least once a year before the end of the rainy season or the beginning of the dry season.

In the event of a bushfire or swamp fire, the fire alarm center receives information from the patrol team or the community and reports to the village chief via HT radio if the fire cannot be extinguished during normal patrol operations. A community that cares about the fire. In addition, fire caretakers must mobilize all available resources to prevent the spread of fires and extinguish fires while they are still small.

REFERENCES

[1] ADRRN/UNISDR 2010 *Terminologi Pengurangan Risiko Bencana.* Indonesia: Asian Disaster Reduction and Response Network (ADRRN) dan of UNISDR Asia and the Pacific Office, Bangkok.
[2] Carter, Nick 1991 *Disaster Management: A Disaster Manager's Handbook.* ADB: Manila.
[3] Direktorat Jenderal Cipta Karya 2013 *Laporan Akuntabikitas Kinerja Instansi Pemerintah.*
[4] Haghebaert 2012 *Community Practices for Disaster Risk Reduction in Japan.* Springer: Japan.
[5] Indonesian Constitution 2007.
[6] Reed, Melissa 2015 Mitigation: *Structural and Non-structural. Public Health Preparedness.*
[7] Rizzo, J. R., House, R. J., and Lirtzman, S. I 1970 *Role conflict and ambiguity in complex organizations. Administrative Science Quarterly,* 15, 150 – 163.

IOP Conf. Series: Earth and Environmental Science 1173 (2023) 012063 doi:10.1088/1755-1315/1173/1/012063

[8] Twigg, John 2009 *Characteristics of a Disaster-resilient Community.* A guidance note for the DFID.

Internet/website:

[1] https://www.gramedia.com/literasi/mitigasi-bencana/. Accessed May 11th, 2022
[2] https://www.merriam-webster.com/dictionary/mitigation. Accessed May 12th, 2022
[3] https://www.riau.go.id/home/en/content/61/data-umum. Accessed May 12th, 2022
[4] https://www.undrr.org. Accessed May 12th, 2022
[5] UN/ISDR terminology of disaster risk reduction https://www.unisdr.org/eng/library/lib_terminology_eng%20home.htm. Accessed May 12th, 2022
[6] https://doi.org/10.24198/jwp. vli2. 11052.\ Accessed May 12th, 2022

IOP Conf. Series: Earth and Environmental Science 1173 (2023) 012064 doi:10.1088/1755-1315/1173/1/012064

Village resilience to tsunami hazard: a Desk Research on the west-coastal villages of the Regency of Serang, Province of Banten, Indonesia

D A Tiwi[1], D Ganesha[2], N A Akhirianto[3], A P Putra[1], R P Umbara[1], A F Shomim[1] and Astisiasari[1]

[1] The National Research and Innovation Agency, Geological Disaster Research Center, South Tangerang, Indonesia
[2] The National Research and Innovation Agency, Regional Research and Innovation Agency, Jakarta, Indonesia
[3] The National Research and Innovation Agency, Hydrodynamic Technology Research Center, Yogyakarta, Indonesia

E-mail: dwia001@brin.go.id

Abstract. This desk research delivers an introductory study of village resilience on the western coastline of the Regency of Serang, Province of Banten, Indonesia. It aims to preliminary characterize the level of village resilience to a tsunami hazard. This study utilizes secondary dataset, retrieved from the authorized institutions (statistical bureau and regional government) and accredited news (government bulletin and news agency). It adopts and modifies the four notable methods on resilience analysis, in order to accommodate the relevancy to study area's dataset availability. The reformulated method is developed by taking into account the variables of: vulnerability (physical, social, and economical aspects) and capacity (preparedness and mitigation measures) within a village as its unit analysis. Furthermore, SWOT analysis is conducted to present the appropriate recommendations for strengthening the resilience. The study finds that generally, tsunami resilience in study area is still considered inadequate, where among the eleven villages, seven villages set low for the Resilience Index (RI). In regards to this, SWOT analysis assembles a total of fifteen possible strategies and seventeen desk-based recommendations to promote the tsunami-resilient village.

1. Introduction

Resilience has a significant mechanism to not only reduce disaster risk but also to support sustainable development. It reflects the capability of the cross-stakeholders system in anticipating, coping, and thriving effectively in the face of disaster. Accordingly, this paper captures the prevailing level of village resilience in the coastal Districts of Anyer and Cinangka, Province of Banten. These districts support the economic development of the Regency of Serang as well as the Province of Banten through tourism and industrial sectors, though also encounter susceptibility to several geological hazards, e.g., tsunami, volcanic eruption, and submarine landslide. Both of the districts record a number of historical tsunami disasters at the cost of innumerable losses.

Based on the map of the tsunami hazard zone of Banten Province [1], the western shoreline of Serang Regency has a high susceptibility to tsunami hazard, with the potential wave height reaching >3 meters and tsunami intensity of >VII. Several tsunamis occurred on the Sunda Strait are: (1) earthquake-induced tsunamis (1722, 1852, dan 1958); (2) volcanogenic tsunamis (416, 1883, 1928, and 2018), and (3) other unknown causes (1851, 1883, 1889, and 1930) [2,3]. Figure 1 shows the impacted area of the latest Sunda Strait tsunami on December 22, 2018 [4]. Despite its hazardous circumstances, the west-coast area of Serang Regency is also directly allocated as a part of the Sunda Strait National Strategic Area Zoning Plan (RZ KSN Selat Sunda). Whereas the Sunda Strait itself potentially supports as an important gate of Indonesian Archipelago Sea Channel (ALKI), that is as an economic development corridor in Indonesia [5]. Thus, it demands critical actions pertaining to coastal resilience, in order to reduce the tsunami risk in coastal areas, as well as to strengthen the social and economic protections. Ineffectiveness in resilience contributes to the adverse effects on a disaster event. Additionally, the unknown status of resilience also aggravates the disaster management practices.

Accordingly, this desk research focuses on the eleven (11) coastal villages in the Regency of Serang, selected through a purposive sampling, with the main criteria are the west-coastal villages that record a tsunami impact, and have a tourism prospect. This paper aims to acknowledge the preliminary level of village resilience to tsunami hazard, through the currently-available secondary dataset. Moreover, by capturing the gap on its resilience, this work also aims to suggest the appropriate recommendations pertaining to resilience reinforcement, through a SWOT analysis.

This paper addresses resilience from the proxies of vulnerability & capacity, by using both of qualitative & quantitative approaches. It is conducted in two analyses by (1) mapping the resilience level with indexed-overlay analysis, and (2) mapping the tailor-made recommendations with SWOT analysis. This study adopts and modifies the four (4) notable methods on resilience analysis, in order to accommodate the relevancy to study area's dataset availability. The reformulated method is developed by taking into account the variables of vulnerability (physical, social, and economical aspects) and capacity (preparedness and mitigation measures) within each of the village as the unit analysis.

2. Methods

This desk research focuses on the village unit analysis, where eleven (11) coastal villages were selected through a purposive sampling. The criteria represent the west-coastal villages that record historical tsunami and have tourism prospects. Figure 2 shows the selected villages, that are: four (4) villages in the District of Anyar (Villages of Anyar, Tambangayam, Cikoneng, and Bandulu), and seven (7) villages in the District of Cinangka (Villages of Cinangka, Kamasan, Sindanglaya, Karangsuraga, Bulakan, Pasauran, and Umbultanjung).

2.1. Reference methods

This study utilizes four (4) notable methods that reflect the level of village resilience. And then modifies them accordingly, to the relevancy of the study area's dataset availability.

- Method-1: City Resilience Index [6]
- Method-2: Disaster Resilience Scorecard for Cities [7]
- Method-3: Resilience Index Development Towards Natural Disasters in Indonesia [8]
- Method-4: Regional Development Planning-Based on Disaster Risk Reduction [9]

Table 1 shows the modified synthesis from these four reference methods, based on the adjusted variables, parameters, and indicators, that are currently available for this desk research, as well as comprehensively subjected to the expert-judgment weighing.

2.2. Secondary Dataset

Data acquisition is retrieved from the authorized sources (statistical bureau and regional government), as well as the accredited news (government bulletin, news agency).

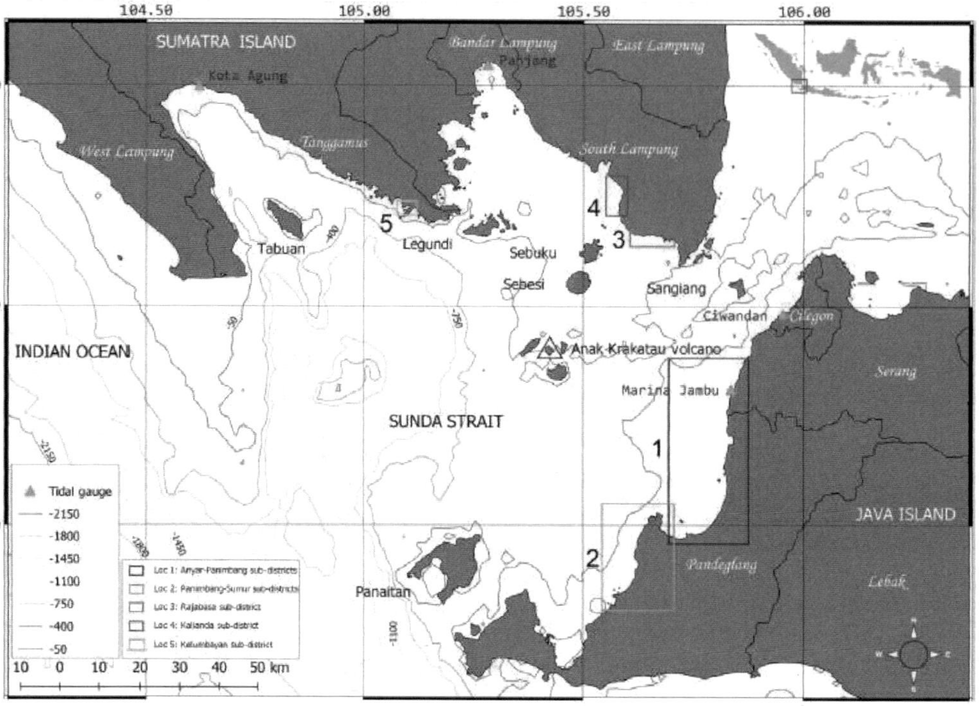

Figure 1. The impacted area of the latest Sunda Strait tsunami on December 22, 2018 [4].

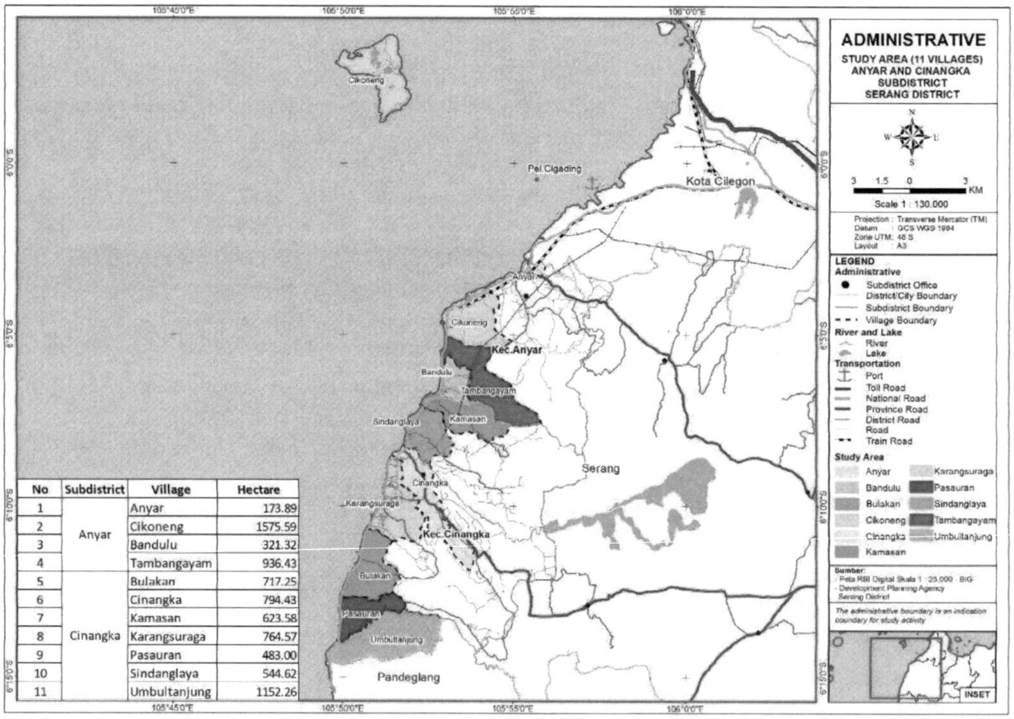

Figure 2. The Study location of eleven (11) west-coastal villages on the Regency of Serang, Province of Banten, Indonesia.

Table 1. The modified synthesis from these four reference methods, based on the adjusted variables, parameters, and indicators.

Variable	Parameter	Indicator
	Vulnerability	
Physical Vulnerability (0.4)	Built-up area (0.6)	% of built-up area (1)
	Public facility (0.4)	Number of health infrastructure (0.5)
		Number of educational infrastructure (0.25)
		Number of worship infrastructure (0.25)
Social Vulnerability (0.3)	Population density (0.6)	Level of population density (1)
	Vulnerable group (0.4)	Sex ratio (1)
Economic Vulnerability (0.3)	Productive land area (0.3)	% of rice field and plantation area (1)
	Family welfare (0.4)	% Pre-prosperity Families (1)
	Economic activities (0.3)	Number of minimarkets, shops / stalls (1)
	Capacity	
Preparedness Capacity (0.6)	Evacuation simulation drill (0.3)	Availability of evacuation guidelines / SOP (0.5)
		Implementation of evacuation simulation drill (0.5)
	Early warning (0.2)	Availability of tsunami early warning system (1)
	Disaster organization / task force (0.2)	Availability of disaster-related organization (1)
	Resource mobilization (0.1)	Availability of standby post and emergency logistics (1)
	Fund source (0.2)	Availability of village fund for disaster occurrence (1)
Mitigation Capacity (0.4)	Structural (0.6)	Availability of temporary evacuation shelter / point (0.5)
		Availability of danger, evacuation, warning, and forbiddance signs (0.25)
		Availability of breakwater structure / greenbelt (0.25)
	Non-Structural (0.4)	Availability of hazard-zone and evacuation route maps (0.3)
		Availability of tsunami disaster contingency plans / guidelines (0.4)
		Disaster-based Detailed Spatial Plan (RDTR) (0.3)

Source: Modifications from [6, 7, 8, and 9].

2.3. Analysis

Level of village resilience is analyzed through five (5) steps, i.e.: (1) Vulnerability Index (VI) analysis; (2) Capacity Index (CI) analysis; (3) Resilience Index (RI) analysis; (4) Index mapping for VI, CI, and RI; and (5) SWOT analysis for the recommendations on strengthening the resilience.

2.3.1. Vulnerability Index (VI). Vulnerability refers to a combined factors related to the physical, social, economic, and environmental aspects, that may increase the adverse impact of susceptible hazard [10]. This paper used three (3) vulnerability variables, i.e.: physical, social, and economic aspects (See Table 1 for the variables, parameters, indicators, and weighing values). Vulnerability Index (VI) is formulated as follows (Suciastuti, 2021):

$$VS_{ji} = \sum_{k=1}^{k=LV_j} x_{kji} VC_{kji} \quad (1)$$

$$VD_i = \sum_{j=1}^{j=MV_i} y_{ji} VS_{ji} \quad (2)$$

$$VI = \sum_{i=1}^{i=NV} z_i VD_i \quad (3)$$

Where:

VI = Vulnerability Index
VDi = vulnerability score on variable i
VSji = parameter score j from variable i
VCkji = indicator score k from parameter j & variable i
zi = weight of vulnerability variable i
yji = weight of parameter j in vulnerability variable i
xkji = weight of indicator k in parameter j & vulnerability variable i
NV = number of vulnerability variable

MVi = number of parameters on vulnerability variable i
LVj = number of indicators on vulnerability parameter j

2.3.2. Capacity Index (CI). Capacity refers to a collective aspect related to the available strengths and resources, within the communities, society, and organizations, in order to mitigate the risk and strengthen the resilience [10]. This paper used two capacity variables, i.e.: preparedness and mitigation aspects (See Table 1 for the variables, parameters, indicators, and their weighing values). Capacity Index (CI) is formulated as follows (Suciastuti, 2021):

$$CS_{ji} = \sum_{k=1}^{k=LC_j} u_{kji}\, CT_{kji} \tag{4}$$

$$CD_i = \sum_{j=1}^{j=MC_i} v_{ji}\, CS_{ji} \tag{5}$$

$$CI = \sum_{i=1}^{i=NC} w_i\, CD_i \tag{6}$$

Where:

CI = Capacity Index
CDi = capacity score on variable i
CSji = parameter score j from variable i
CTkji = indicator score k from parameter j & variable i
wi = weight of capacity variable i
vji = weight of parameter j in capacity variable i
ukji = weight of indicator k in parameter j & vulnerability variable i
NC = number of capacity variable
MCi = number of parameters on capacity variable i
LCj = number of indicators on capacity parameter

2.3.3. Resilience Index (RI). Resilience Index is defined as the ratio of Capacity Index to Vulnerability Index (Suciastuti, 2021):

$$RI = \frac{CI}{VI} \tag{7}$$

Where:

RI = Resilience Index
CI = Capacity Index
VI = Vulnerability Index

3. Results

Resilience Index (RI) is measured through the proxies of Vulnerability Index (VI) and Capacity Index (CI). Furthermore, SWOT analysis is conducted to extract the appropriate recommendations, pertaining to the enhancement and establishment of village resilience.

3.1. Vulnerability index (VI)

Vulnerability Index (VI) is composed of three variables of: physical, social, and economic vulnerabilities (see Figure 3a to Figure 3f). Figure 4 shows the map of Vulnerability Index and its index valuation. Here, the high VI relatively indicates a severe exposure, disruption, loss, and adversity; while the medium VI implies a notion of intermediate collapse upon a disaster event; and the low VI performs where the villages generally have the least impact upon the face of disaster.

From the Physical vulnerability, Figure 3b shows five (5) villages with the above-average score. Village of Anyar (District of Anyar) has the highest score (0.923), that is mainly supported by the highest percentage of built-up area (42% coverage). While the Village of Pasauran (District of Cinangka) has the lowest score (0.0342), that is mainly derived from its lowest percentage on built-up area (3.94% coverage).

For the social vulnerability, Figure 3d displays five (5) villages with the above-mean score. Village of Anyar (District of Anyar) has the highest score (1.000), that is mainly induced by its high density (4,603 people/km^2) and low sex-ratio (103). Meanwhile, Village of Umbultanjung (District of Cinangka) has the lowest score (0.0219), especially related to its highest sex-ratio (139).

On the economic vulnerability, Figure 3f presents five (5) villages with the above-average score. Here, Village of Sindanglaya (District of Cinangka) has the highest score (0.5531), that is promoted by both of the highest percentage of rice fields (90.5% coverage) and pre-prosperity families. Meanwhile,

Village of Cikoneng (District of Anyar) has the lowest score (0.1714), due to having the lowest percentage of productive land.

Figure 4 shows the nine (9) villages in low VI. This result implies that the west-coastal villages generally have minor exposures to a disaster impact. And thus, may possibly associated with a lesser adverse impact. The highest level of VI is on Villages of Anyar, District of Anyar (0,7790), that is acquired from the highest values on its physical- and social-vulnerabilities. This village has a high level of built-up area (42% coverage), high density (4,603 people/km^2), and low sex-ratio (103). Meanwhile Village of Umbultanjung, District of Cinangka has the lowest VI (0.1664), that is mainly related to its highest sex-ratio (139).

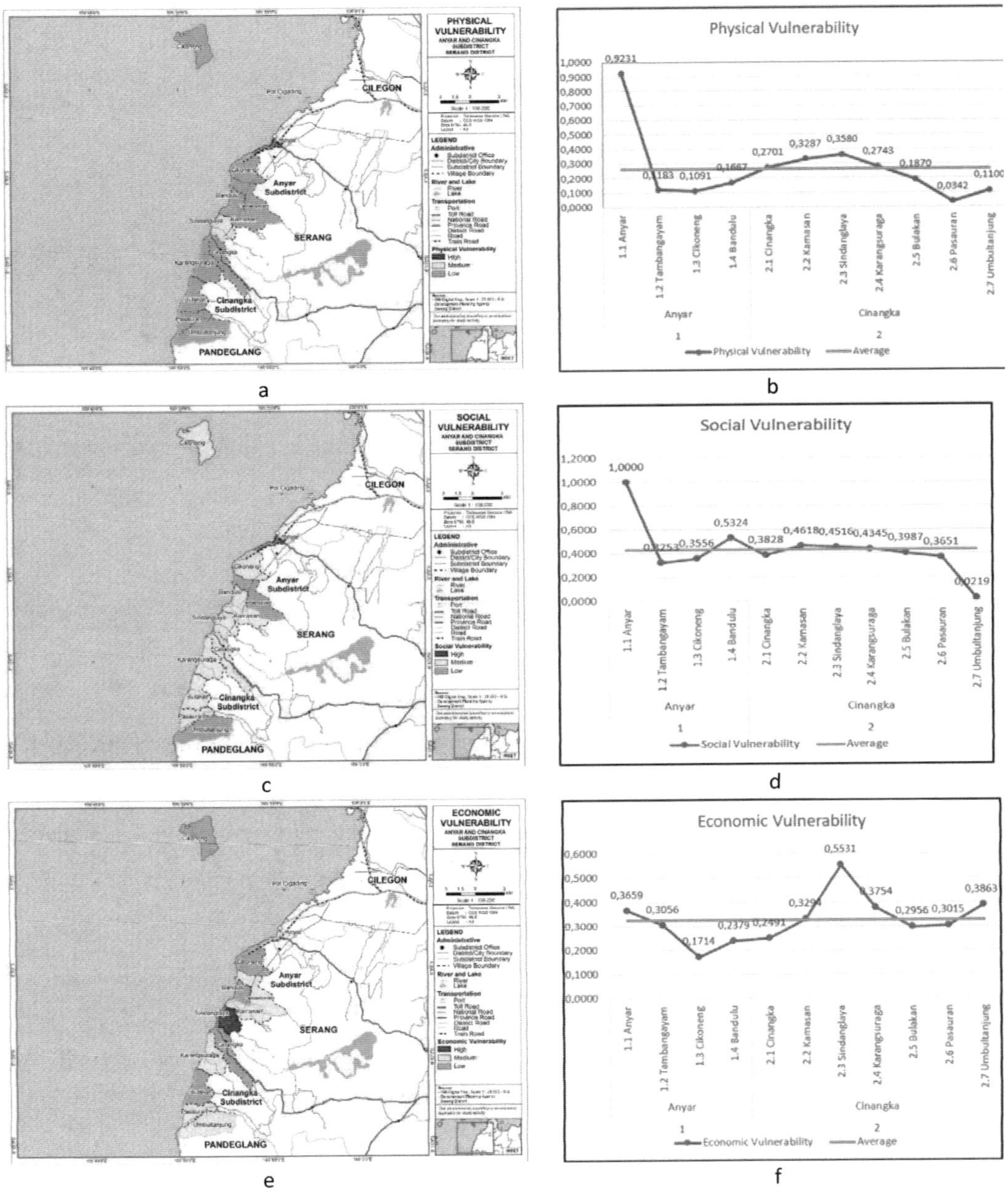

Figure 3. Variables of vulnerability in the study area, year 2020: physical vulnerability: (a) Map, (b) Statistical figure; social vulnerability: (c) Map, (d) Statistical figure; and economic vulnerability: (e) Map, (f) Statistical figure (Data Processing, 2022).

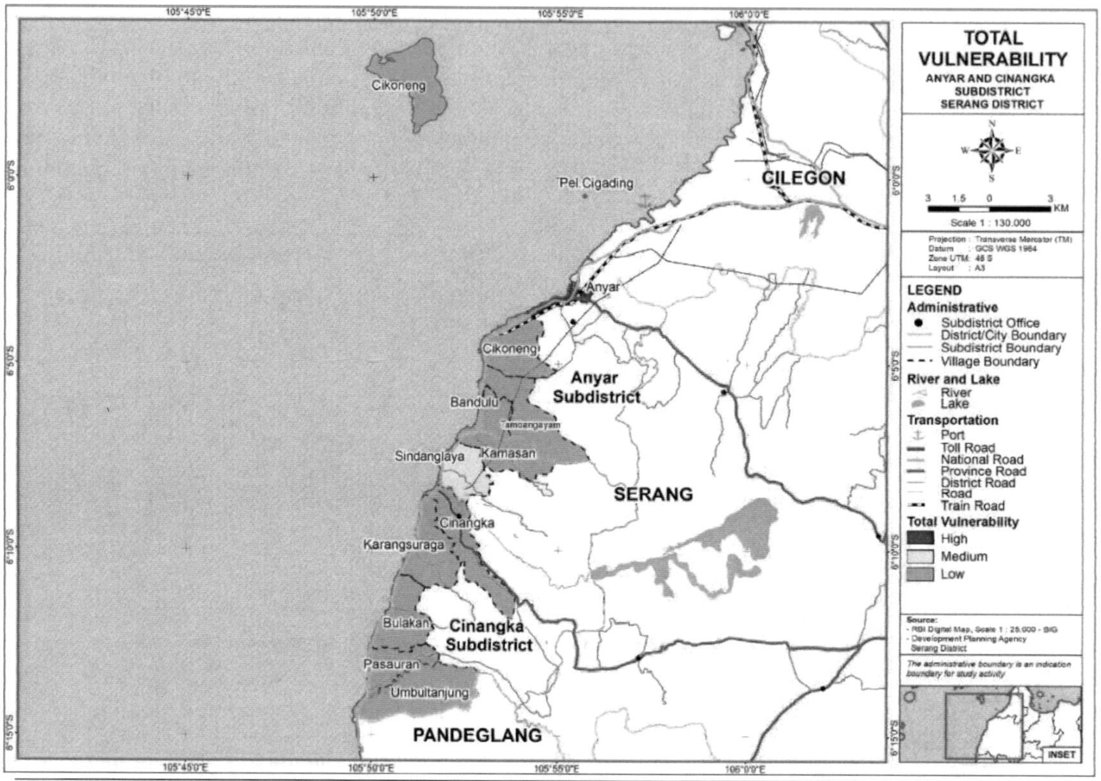

No	District	Village	Physical Vulnerability		Social Vulnerability		Economic Vulnerability		Vulnerability Index	
			Score	Class	Score	Class	Score	Class	Score	Class
1	Anyar	1.1 Anyar	0,9231	High	1,0000	High	0,3659	Medium	0,7790	High
		1.2 Tambangayam	0,1183	Low	0,3253	Low	0,3056	Medium	0,2366	Low
		1.3 Cikoneng	0,1091	Low	0,3556	Medium	0,1714	Low	0,2017	Low
		1.4 Bandulu	0,1667	Low	0,5324	Medium	0,2379	Low	0,2977	Low
2	Cinangka	2.1 Cinangka	0,2701	Low	0,3828	Medium	0,2491	Low	0,2976	Low
		2.2 Kamasan	0,3287	Low	0,4618	Medium	0,3294	Medium	0,3688	Low
		2.3 Sindanglaya	0,3580	Medium	0,4516	Medium	0,5531	High	0,4446	Medium
		2.4 Karangsuraga	0,2743	Low	0,4345	Medium	0,3754	Medium	0,3527	Low
		2.5 Bulakan	0,1870	Low	0,3987	Medium	0,2956	Low	0,2831	Low
		2.6 Pasauran	0,0342	Low	0,3651	Medium	0,3015	Medium	0,2137	Low
		2.7 Umbultanjung	0,1100	Low	0,0219	Low	0,3863	Medium	0,1664	Low

Figure 4. Vulnerability index in the districts of Anyar and Cinangka, Regency of Serang, Province of Banten, Year 2020 (Data Processing, 2022).

3.2. Capacity index (CI)

Capacity Index (CI) is composed of two variables of preparedness and mitigation (see Figure 5a and 5b). Figure 6 shows the map of Capacity Index and its index valuation. Here, the high CI relatively indicates the well-managed resources, in terms of its preparedness and mitigation measures. While medium CI implies the enabled potentials to be incorporated in strengthening resiliency. And low CI performs where villages generally have the least beneficial resources upon the face of disaster.

The study area has two (2) villages on high CI, three (3) villages with medium CI, and six (6) villages on low CI. This result indicates that generally, these west-coastal villages need to consider and encourage its capacity building to boost the resiliency. This condition requires an accelerated capacity building, especially targeted on the disaster-related programs, in order to mitigate disaster risk, enhance village resilience, and support sustainable development.

The Highest level of CI is on the Villages of Bulakan, District of Cinangka (0.760667). It is contributed by the high values on preparedness and mitigation. Meanwhile, the lowest CI (0.107333) exists on four (4) villages in the District of Cinangka, i.e.: Villages of Cinangka, Kamasan, Sindanglaya, and Umbultanjung.

Villages that have low-score in preparedness may reach medium-score on its CI, if they are able to manage the mitigation capacity, or vice-versa (e.g., on Villages of Anyar, Cikoneng and Pasauran). Here, Villages of Anyar and Cikoneng have established several mitigation measures, such as: temporary evacuation sites, evacuation routes, hazard signs and maps. However, important improvements can still be implemented on the preparedness capacity, in order to attain a high CI, such as: tsunami drill, disaster organization, and fund allocation.

Two villages with the highest CI (i.e.: Villages of Karangsuraga and Bulakan, District of Cinangka) have been categorized as Destana (Disaster-resilient Village) by the BNPB. Both of the villages are able to actively involved in maintaining their resources (e.g.: risk-informed communities and disaster management practices).

Moreover, the Village of Bulakan also has a tidal station located at the pier of Marina Jambu (06°11'21.22" S and 105°50'27.95" E). Together with the dissemination from BMKG's tsunami early warning system, this wave-height measuring tool is able to conform the upcoming wave surge or tsunami, in a more accurate manner. Moreover, both of the villages can still make an essential improvement, especially related to the non-structural mitigation capacity. In respect to this, conducting a disaster-based Detailed Spatial Plan (RDTR) is highly encouraged.

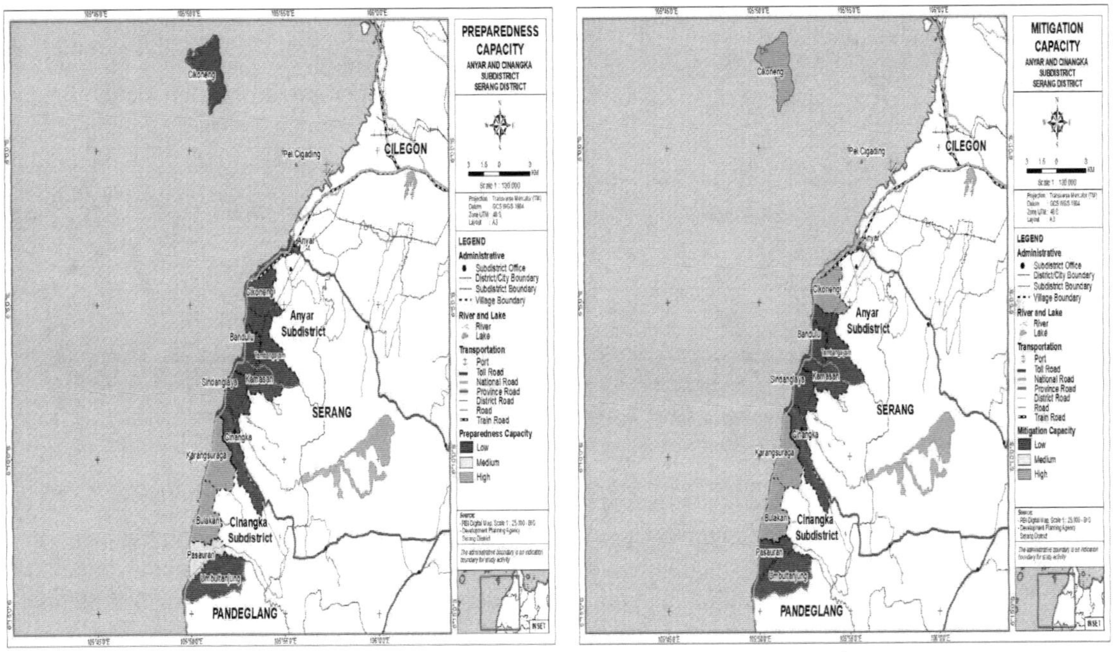

Figure 5. Variables of capacity in the districts of Anyar and Cinangka, Regency of Serang, Province of Banten, year 2020: (a) Preparedness capacity map and (b) Mitigation capacity map (Data processing, 2022).

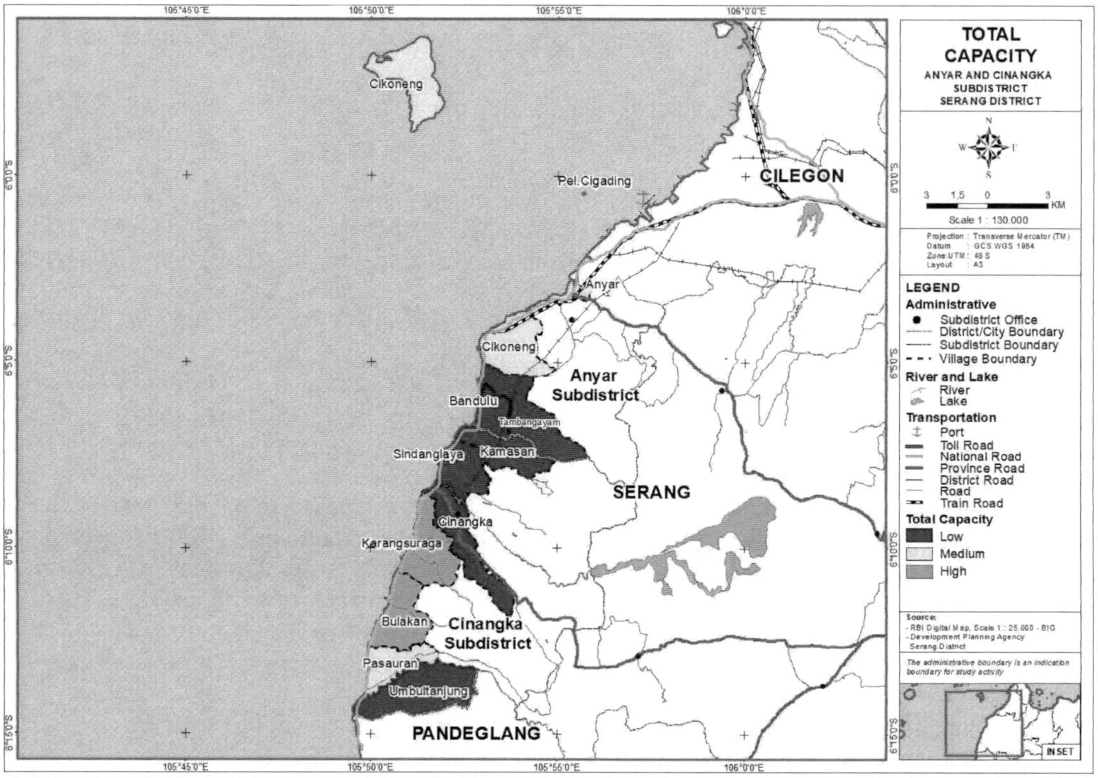

No	District	Village	Preparedness Capacity		Mitigation Capacity		Capacity Index	
			Score	Class	Score	Class	Score	Class
1	Anyar	1.1 Anyar	0.33333	Low	0.513333	High	0.405333	Medium
		1.2 Tambangayam	0.11667	Low	0.213333	Low	0.155333	Low
		1.3 Cikoneng	0.33333	Low	0.513333	High	0.405333	Medium
		1.4 Bandulu	0.11667	Low	0.213333	Low	0.155333	Low
2	Cinangka	2.1 Cinangka	0.11667	Low	0.093333	Low	0.107333	Low
		2.2 Kamasan	0.11667	Low	0.093333	Low	0.107333	Low
		2.3 Sindanglaya	0.11667	Low	0.093333	Low	0.107333	Low
		2.4 Karangsuraga	0.73333	High	0.626667	High	0.690667	High
		2.5 Bulakan	0.85000	High	0.626667	High	0.760667	High
		2.6 Pasauran	0.58333	Medium	0.093333	Low	0.387333	Medium
		2.7 Umbultanjung	0.11667	Low	0.093333	Low	0.107333	Low

Figure 6. Capacity index in the districts of Anyar and Cinangka, Regency of Serang, Province of Banten, year 2020 (Data processing, 2022).

3.3. Resilience index (RI)

Resilience Index (RI) is contributed from the two proxies of Vulnerability Index and Capacity Index. Figure 7 shows the map of Resilience Index and its valuation. Here, among eleven (11) villages, there are three (3) villages that have high scores on its Resilience Index, i.e., Villages of: Cikoneng (District of Anyar), Karangsuraga and Bulakan (District of Cinangka). This condition may generally indicate a well-managed capacity on a low-exposed vulnerability.

However, seven (7) villages still perform low on its Resilience Index. It implies that generally, the tsunami resilience in the Districts of Anyer and Cinangka is still inadequate. For example, Village of Sindanglaya, District of Cinangka has the lowest RI (0.2414). This condition entails an indication where although having a medium level of exposure for its vulnerability, the resources of its capacity is still considered insufficient. And thus, vital actions are needed for establishing and strengthening the resilience of coastal villages upon the tsunami hazard.

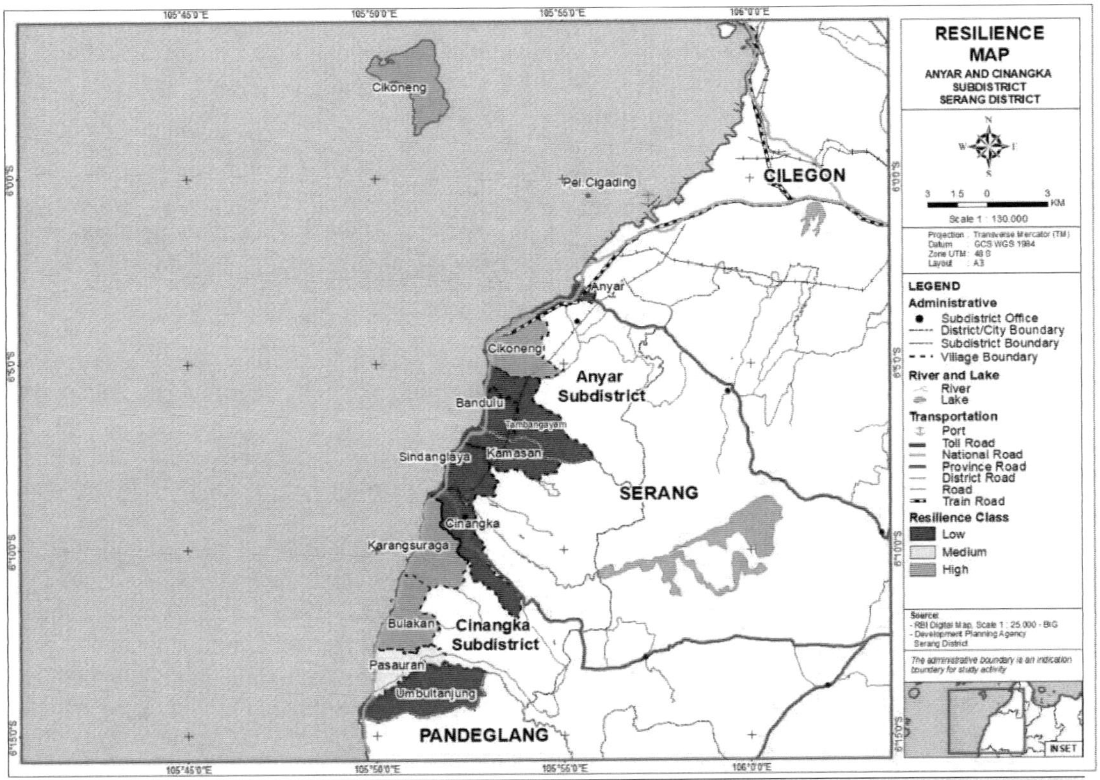

No	District	Village	Vulnerability Index		Capacity Index		Resilience Index	
			Index	Class	Index	Class	Index	Class
1	Anyar	1.1 Anyar	0,7790	High	0.4053	Medium	0.5203	Low
		1.2 Tambangayam	0,2366	Low	0.1553	Low	0.6565	Low
		1.3 Cikoneng	0,2017	Low	0.4053	Medium	2.0093	High
		1.4 Bandulu	0,2977	Low	0.1553	Low	0.5217	Low
2	Cinangka	2.1 Cinangka	0,2976	Low	0.1073	Low	0.3606	Low
		2.2 Kamasan	0,3688	Low	0.1073	Low	0.2910	Low
		2.3 Sindanglaya	0,4446	Medium	0.1073	Low	0.2414	Low
		2.4 Karangsuraga	0,3527	Low	0.6907	High	1.9584	High
		2.5 Bulakan	0,2831	Low	0.7607	High	2.6871	High
		2.6 Pasauran	0,2137	Low	0.3873	Medium	1.8129	Medium
		2.7 Umbultanjung	0,1664	Low	0.1073	Low	0.6449	Low

Figure 7. Resilience index in the districts of Anyar and Cinangka, Regency of Serang, Province of Banten, year 2020 (Data processing, 2022).

3.4. *SWOT analysis*
SWOT analysis captures four (4) aspects on the Strengths (S), Weaknesses (W), Opportunities (O), and Threats (T) form internal and external factors, pertaining to resilience support. Table 2 below displays the matrix result of SWOT analysis. The SWOT matrix assembled those four aspects to convene the appropriate strategies in promoting the tsunami-resilient village in study area.

3.4.1. *Strength (S)*
- There are 24 recorded DESTANA (disaster-resilient village, referring to the Regulation of the head of BNPB 1/2012) [11] in the Regency of Serang since 2021 [12]. Moreover, Regency of Serang held an emergency-response preparedness on the KSB (disaster-resilient village, referring to the Regulation of the Ministry of Social Affairs 128/2011 [13]) for the TAGANA (disaster-alert front) in 2020 [14, 15].
- There are about 198 personnel of TAGANA in the Regency of Serang in 2020 [16, 17], that serve emergency responses through community participation in the face of disaster [14, 15].

- Disaster-safe school (SMAB) is committed to bring awareness on the disaster risk and emergency response, as a part of the risk mitigation (Regulation of the head of BNPB 4/2012) [18]. There are at least three recorded SMAB in the Regency of Serang, i.e., Nurul Fikri Boarding School Serang, and two elementary schools in the Districts of Anyar and Cinangka [19-21].
- Tsunami drill was held in the District of Anyar on June 22, 2022. It involved hundreds of local people, TAGANA personnel, and government officers (BPBD, SAR, TNI, Police). Moreover, an earthquake-tsunami simulation was also held by the governmental authorities (BPBD Serang, TNI, and Basarnas), in the Village of Pasauran (District of Cinangka) in 2021 [22].

3.4.2. Weakness (W)
- Regency of Serang tends to have an increasing number of poverty and rate of employment. Since 2016, poverty rose for 264.47%; and in 2021, the unemployment rate reached 10.58% [23]
- Regency of Serang also has an increasing trend for population growth and density. In 2020, the population density reached 1.106 people/km^2 with the growth of 2,07/year [23]. Based on the Regulation of the head of BNPB 2/2012 [24], Regency of Serang has a medium-level vulnerability. However, from this study, Village of Anyar has a high Vulnerability Index (VI)
- Tsunami Early Warning System (TEWS) is an important aspect to the 500km-long coastline of Banten Province. Unfortunately, the only one (1) TEWS in Pasauran is already inactive due to vandalism (stolen instrument part) [25]. This may exacerbate tsunami risk on the densely populated areas, as well as on the tourism and business zones [26].
- A collaborative endeavour between government and private sector for providing an adequate and safe Tsunami Evacuation Site / Shelter (TES) is highly encouraged. At the moment, there is only one recorded shelter in the District of Labuan, Province of Pandeglang [27]. Yet, recommendation had been made by BPBD Serang Regency for the local government to improve the TES [28].

3.4.3. Opportunity (O)
- One of the priority points on the utilisation of village fund (established fund based on Regulation of the Government of Indonesia 8/2016) [29] is to support disaster risk management, based on local authority [30].
- BPBD has responsibility in a disaster risk management cycle (pre-, emergency, and post-disaster) [31]; while the social affairs service has responsibility in the disaster-related social problems and empowerment. These governmental organizations have both of the fund and program to strengthen the capacity and lessen the vulnerability
- Regency of Serang has two major industrial zones, i.e.: 4,000 Ha of West Serang Industrial Zone (on the Districts of Bojonegara, Pulo Ampel, and Kramatwatu), and 1,115 Ha of East Serang Industrial Zone (on the Districts of Cikande, Kibin, Kragilan dan Jawilan). There are a total of 145 companies in large- and medium-scales. This offers environmental collaborations on risk management and sustainable practices
- Through the Regional Regulation of the Serang Regency 8/2014 [32], the west-coastal area of the Regency of Serang was involved in the development plan of Tourism Attraction Designation (DTW). In 2021, there were 48 registered hotels (24 on the District of Cinangka, and 18 on the District of Anyar) [33].
- Universities possess an important role to not only serving academia, but also the communities. In 2021, there are at least 10 established universities in the Regency of Serang [23].
- The routine agenda between local government and community has also involved the activities on tsunami preparedness, e.g., the traditional ceremonies (e.g., Ngaruk Panjang Mulud, Ruat Laut, Bedolan Pamarayan, Sedekah Bumi, Nurun Kapal, Syahadat Bumi, Ritual Keceran, and Dalailan) [34], and also marathon race on the Indonesia's Independence Day that follows the tsunami evacuation route

3.4.4. Threat (T)

- The west coast of Serang Regency is susceptible to tsunami, that is sourced from earthquake, subduction, eruption, and landslide [35].
- The marine and coastal tourism potentials in the Regency of Serang had attracted 1.374.043 tourists in 2020, and 1.477.253 tourists in 2021 [33].
- The staged establishment for Destana and KSB is bound to the regional budget allocation. At a moment, KSB has established in the District of Anyar, Regency of Serang [36].

Table 2. Matrix of SWOT-based strategies for promoting the tsunami-resilient village in study area.

Internal Factors / External Factors	Strengths (S)	Weaknesses (W)
	1. Disaster-resilient village (Destana) & disaster-resilient village (KSB) 2. Disaster- resilient front (Tagana) 3. Disaster-safe school (SMAB) 4. Tsunami drill 5. Tsunami Early Warning System 6. Presidential Reg. 18/2020 (2020-2024 National Med-Term Development Plan) 7. Anyer Regional Planning Sector	1. Poverty and unemployment rate 2. Population density and growth rate 3. Insufficient Tsunami Early Warning System and warning sign 4. Inadequate evacuation site / shelter 5. Absentee regional contingency plan in tsunami hazard 6. Absentee coastal-area empowerment
Opportunities (O)	**S-O Strategy**	**W-O Strategy**
1. Village-fund budget 2. BPBD & Social Governance 3. Industrial- & tourism-zones 4. Universities 5. Local government-community agenda on the routine activities	1. S1 & O1. Destana replication with village-fund budget 2. S4, S6, O2–O5. Penta-helix collaboration in tsunami drill routine 3. S7, O3, O5. Disaster mitigation-based Detailed Spatial Plan (RDTR) 4. S4 & O5. tsunami drill-integrated routine agenda between local government and community	1. W4 & O3. Collaboration for improving / establishing the vertical evacuation site 2. W1 & O2. Governmental capital-assisted program, service, and fund 3. W1, O3, O4. Small enterprises accompaniment with universities 4. W5 & O2. Establishing regional contingency plan for tsunami 5. W6, O2, O5. Regional regulation for coastal-area empowerment
Threats (T)	**S-T Strategy**	**W-T Strategy**
1. Tsunami hazard 2. Tourist visits growth 3. Unestablished Destana or KSB	1. S1–S4, T1. Mapping & dissemination of Tsunami-hazard and -risk with local communities 2. S1, S2, & T3. Destana replication with local communities 3. S4, T2. Tsunami drill for/with tourist involvement	1. W3 & T1. Tsunami Early Warning System with local wisdom 2. W1 & T2. Local empowerment with handmade souvenirs 3. W2, W3, T1, T2. Memorial monument as commemoration

3.5. Desk-based recommendations

Learning from the village characteristics and conditions, exploratory recommendations can be framed and proposed, especially in terms of vulnerability reduction and capacity building. Appendix proposes desk-based recommendations to establish and enhance village resilience in the study area.

4. Conclusion

Resilience has a significant mechanism to not only reduce disaster risk but also to support sustainable development. It reflects the capability of cross-stakeholders system in anticipating, coping, and thriving effectively in the face of disaster. Resilience Index (RI) is measured through the two proxies of Vulnerability Index (VI) and Capacity Index (CI) (see Figure 7 for the map of Resilience Index)

The study area has nine (9) villages with low VI. It implies that from the level of VI, the west-coastal villages in study area generally have minor exposures to a disaster impact. And thus, may possibly associated with a lesser adverse impact. The highest level of VI is at the Villages of Anyar, District of Anyar (0,7790), that is promoted by the high level of built-up area (42% coverage), populated density (4,603 people/km^2), and low sex-ratio (103).

Moreover, the study area has six (6) villages with low CI. This condition implies that generally, the west-coastal villages preserve inefficient capacities to reach tsunami resilience. And thus, need to encourage their capacity building to boost the resilience.

The Resilience Index shows that among the eleven (11) villages, there are three (3) villages that have high scores on RI, i.e.: the Villages of Cikoneng (District of Anyar), Karangsuraga and Bulakan (District of Cinangka). It showcases an implied circumstance of a well-managed capacity on a low-exposed vulnerability. However, seven (7) villages still perform low on RI. It possibly envisages a situation that generally, tsunami resilience in the Districts of Anyer and Cinangka is still inadequate. Taking everything into consideration, prompt and vital measures are highly encouraged in order to establish and strengthen the village resilience (e.g., through vulnerability reduction and capacity building). Here, SWOT analysis assists to convene the appropriate strategies.

SWOT analysis assembled a total of fifteen (15) possible strategies, to promote the tsunami-resilient village, i.e.: four (4) S-O strategies, five (5) W-O strategies, three (3) S-T strategies, and three (3) W-T strategies. Moreover, this paper also proposes seventeen (17) desk-based recommendations, adjusted from each of the village circumstances, as catered on the Appendix. Lastly, this desk study also suggests a continued endeavor on field study, to acquire primary dataset through a Focussed Group Discussion (FGD) from the relevant stakeholders, and thus, prevails a legitimate status on village resilience to tsunami hazard.

Acknowledgements
The authors are all affiliated to The National Research and Innovation Agency (BRIN). This research was fully supported by BRIN.

Appendix

Desk-based Recommendations for Promoting the Tsunami-resilient Village in the Study Area

Recommendations \ Study Area	District of Anyar				District of Cinangka						
	Anyar	Tambang-ayam	Cikoneng	Bandulu	Cinangka	Kamasan	Sindang-laya	Karang-suraga	Bulakan	Pasauran	Umbul-tanjung
VULNERABILITY REDUCTION											
Education infrastructure & disaster knowledge dissemination	•										
Health infrastructure & emergency response	•										
Worship houses & disaster knowledge dissemination								•			•
Welfare-related program (employment, social assistance, enterprise association)		•									
Economic trade & infrastructures for emergency response	•					•					
Productive land utilization										•	
CAPACITY BUILDING											
Periodic evacuation simulation / drilling with guidelines / SOP		•		•	•	•	•		•	•	•
Community trainings & education (first aid, building capacity for women,										•	•
Tsunami early warning system	√										
Disaster-related organization, DRR forum, and volunteers	•	•		•	•	•	•				
Standby post and emergency logistics	•	•	√	•	•	•	•	√	√		
Village fund for disaster occurrence	•	•	√	•	•	•	•				
Temporary evacuation shelter / gathering point	•	•		•	•	•	•			•	•
Warning Signs (danger, evacuation route, prohibition) & Maps of hazard zone	•	•		•			•				•
Coastal defense: breakwater structure / greenbelt	•	•		•	•						•
Tsunami disaster contingency plans / guidelines											
Disaster-based Detailed Spatial Plan (RDTR)	•										

References

[1] ESDM 2009 One Map Kebencanaan (https://geoportal.esdm.go.id/kebencanaan/)

[2] BMKG 2019 Katalog Tsunami Indonesia Tahun 416–2018 per-Wilayah, (https://cdn.bmkg.go.id/Web/Katalog-Tsunami-Indonesia-416-2018-per-Wilayah.pdf)

[3] ESDM 2018 Tanggapan Kejadian Tsunami Selat Sunda, Tanggal 22 Desember 2018, (https://www.esdm.go.id/id/media-center/arsip-berita/tanggapan-kejadian-tsunami-di-selat-sunda-tanggal-22-desember-2018)

[4] Syamsidik et al 2020 The 22 December 2018 Mount Anak Krakatau volcanogenic tsunami on Sunda Strait coasts, Indonesia: tsunami and damage characteristics. *Nat. Hazards Earth Syst. Sci* **20** 549–565

[5] KKP 2020 Rencana Zonasi Kawasan Strategis Nasional Kawasan Selat Sunda Berbasis Mitigasi Bencana Webinar Pusat Riset Kelautan

[6] Arup 2015 City Resilience Index London: The Rockefeller Foundation

[7] UNDRR 2017 Disaster Resilience Scorecard for Cities – Detailed Level Assessment

[8] Kusumastuti et al 2014 Developing a Resilience Index Towards Natural Disaster in Indonesia International Journal of Disaster Risk Reduction 10 pp 327-340

[9] Muta'ali L 2014 Perencanaan Pengembangan Wilayah Berbasis Pengurangan Risiko Bencana. BPFG – UGM

[10] UNDRR 2009 UNISDR Terminology on Disaster Risk Reduction Online Glossary

[11] Regulation of the head of BNPB Number 1 2012 General Guidelines for Disaster Resilient Villages (Pedoman Umum Desa/ Kelurahan Tangguh Bencana)

[12] https://www.infobanten.id/bpbd-kabupaten-serang-kurangi-resiko-bencana/, accessed on 25 Juli 2022

[13] Regulation of the Ministry of Social Affairs Number 128 2011 Disaster Resilient Village (Kampung Siaga Bencana)

[14] https://dinsos.serangkab.go.id/baca/berita/dinas-sosial-kabupaten-serang-kuatkan-kelembagaan-karang-taruna-kecamatan-dan-desa-ajak-lebih-produktif, accessed on 25 Juli 2022

[15] https://dinsos.serangkab.go.id/baca/berita/kegiatan-pembinaan-kampung-siaga-bencana-ksb, accessed on 25 Juli 2022

[16] Social Affairs of Serang Regency 2021 Rencana Kerja Dinas Sosial Kabupaten Serang Tahun 2022.

[17] Social Affairs of Serang Regency 2021 Rencana Strategis Dinas Sosial Kabupaten Serang Tahun 2021-2026

[18] Regulation of the head of BNPB Number 4 2012 Guidelines for the Implementation of Disaster-Safe Schools/Madrasah (Pedoman Penerapan Sekolah/ Madrasah Aman dari Bencana)

[19] https://nfbs.or.id/nfbs-serang-diresmikan-sebagai-sekolah-aman-bencana-oleh-bnpb, accessed on 25 Juli 2022

[20] https://pattirobanten.or.id/membangun-ketangguhan-bencana-di-sekolah/, accessed on 25 Juli 2022

[21] https://www.atmago.com/berita-warga/antusias-warga-dalam-pertemuan-sekolah-aman-bencana-di-sdn-pasauran-1_91658238-fe7b-4b19-8ed5-825060c572d7, accessed on 25 Juli 2022

[22] https://www.bantennews.co.id/simulasi-penanggulangan-bencana-kabupaten-serang-digambarkan-diterjang-tsunami/, accessed on 26 Juli 2021

[23] BPS of Serang Regency 2021 Recency of Serang in Figures

[24] Regulation of the head of BNPB Number 2 2012 General Guidelines for Disaster Risk Assessment (Pedoman Umum Pengkajian Risiko Bencana)

[25] https://news.detik.com/berita/d-6051991/anak-krakatau-siaga-1-sirene-peringatan-tsunami-di-pesisir-serang-tak-aktif, accessed on 26 Juli 2021

[26] https://www.antaranews.com/berita/730361/banten-butuhkan-banyak-sirine-peringatan-dini-tsunami, accessed on 26 Juli 2022

[27] https://www.cnnindonesia.com/nasional/20200928132501-20-551809/alat-peringatan-dini-tsunami-banten-belum-berfungsi-3-rusak, accessed on 26 Juli 2022

[28] https://faktabanten.co.id/serang/berpotensi-tsunami-pemkab-serang-diminta-segera-bangun-gedung-shelter/, accessed on 26 Juli 2021

[29] Regulation of Government of The Republic of Indonesia Number 8 2016 Second Amendment on the State Budget-based Village Fund (Dana Desa yang Bersumber dari APBN)

[30] Regulation of the Ministry of Village, Disadvantaged Area, and Transmigration 7/2021 2021 priority use of village funds (Prioritas Penggunaan Dana Desa)

[31] Law of The Republic of Indonesia Number 24 2007 Disaster Management (Penanggulangan Bencana)

[32] Regional Regulation of the Serang Regency Number 8 2014 Masterplan of Tourism Development in Serang RegencyYear 2014-2025

[33] BPS of Serang Regency 2022 Recency of Serang in Figures

[34] http://perpustakaanbpnbjabar.kemdikbud.go.id/index.php?p=show_detail&id=3268, diakses 26 Juli 2021

[35] Geological Agency 2019 Dinamika Geologi Selat Sunda dalam Pembangunan Berkelanjutan (https://www.esdm.go.id/assets/media/content/content-dinamika-geologi-selat-sunda.pdf)

[36] https://bencana-kesehatan.net/index.php/13-berita/berita/1592-banten-bentuk-kampung-siaga-bencana, accessed on 26 Juli 2021

A conceptual framework for implementing lean construction and building information modeling (BIM) in post-disaster housing reconstruction projects

T A Syaputra[1] and A D Rarasati[1]

[1]Civil Engineering Department, Faculty of Engineering, Universitas Indonesia, Depok, 16424, Indonesia,

Email: ayomi@eng.ui.ac.id

Abstract. The occurrence of natural disasters in Indonesia every year results in massive damage to people's homes. It requires a fast and appropriate post-disaster housing reconstruction (PDHR) strategy. The building firm or contractor is among the most significant contributors; in post-disaster conditions, the contractor has to work in a project environment that is much more complicated than conventional projects, and this often results in poor contractor performance. An approach to improve contractor performance is the application of Lean Construction. In line with this linking, the function of the Building Information Modelling (BIM) tool in the application of Lean has also proven effective in conventional projects. However, studies focusing on its simultaneous application to PDHR projects are still lacking. A conceptual framework that can be applied to specific Lean and BIM implementations in PDHR projects is needed to address this issue. This paper proposes a conceptual framework by identifying and analyzing various previous studies as a proper basis. The study shows several Lean Construction and BIM factors relevant to contractor performance indicators in post-disaster housing reconstruction projects.

1. Introduction

Natural disasters in Indonesia that occur every year result in massive damage to the houses of affected communities; in 2019, there were 73,723 damaged housing units. In 2020 there were 65,743 units, and 141,795 units in 2022 [1]. It requires a fast and appropriate post-disaster housing reconstruction (PDHR) strategy. The building firm or contractor is among the most significant contributors. In post-disaster conditions, the contractor has to work in a project environment that is much more complicated than conventional projects. It often results in poor contractor performance [2]. The critical challenges that cause performance problems in recovery projects are cost overruns, usually due to delays in meeting delivery schedules and material wastage [3].

An effective method to improve contractor performance is applying the Lean Construction method. Lean is proven to be effective in supporting the improvement of contractor performance as measured against various indicators, including cost and time performance indicators [4 Lean implementation in post-disaster reconstruction aids in improving reconstruction performance, resulting in improved built environment quality and efficient workflow. Lean can also shorten the reconstruction phase by focusing on mobilization time for all stakeholders [5]. In addition, to lean, BIM is a popular choice in the construction sector for handling complex and dynamic challenges. BIM helps to improve contractor performance by managing project costs, time, and delivery, which are the most important elements in evaluating PDR projects. [6]. BIM functions such as visualization and automatic document creation can synergize with Lean principles such as reducing variability and increasing flexibility to efficiently improve construction project performance [7].

This paper begins with the background of contractor performance in PDHR, Lean Construction, and BIM, followed by the research methodology. The findings related to the factors contributing to an overview of the theory of effective lean implementation, followed by factors from the BIM function and performance. Contractors in the context of PDHR project research and report conclusions.

2. Background of Contractor Performance in Post-disaster Housing Reconstruction (PDHR) Projects, Lean Construction, and BIM

2.1. Contractor Performance in PDHR Projects,

Construction costs, time, quality, and sustainable development are all included in the definition of contractor performance as a whole, with the guiding principle that achieving one area of performance should not come at the expense of other components [27]. The success of a construction project is greatly dependent on the performance of the contractor. Improved contractor performance leads to client satisfaction and contractor reputation. A better understanding of the client's values in contract services, which are indicators of contractor performance and are grouped into Time, Cost, Quality, Procurement, Scope, Integration, Risk, Communication, and Human Resources, is necessary for the contractor to achieve the success of the reconstruction program [18]. Because the project environment is far more difficult during a disaster than it is during normal projects, poor contractor performance is frequently the result [2]. One of the critical challenges that cause performance problems in recovery projects is cost overruns, usually due to delays in meeting delivery schedules and material wastage [3].

2.2. Lean Construction

Lean construction is a method for tackling all construction jobs to reduce waste, save time, and create as many values as possible [8]. Lean Construction, in general, is the practice of using lean concepts to generate value that can satisfy the owner and improve construction efficiency [9].

2.3. Building Information Modelling (BIM)

BIM is an adjectival term used to describe tools, procedures, and technologies that are made possible by digital documenting of buildings' performance, design, construction, and operation [16]. The capacity to create representations of virtual building models using machine-readable parametric objects that display acceptable behavior for creating, analyzing and testing building designs is a feature of BIM software [7]. All project stakeholders can be involved in the facility design, building, and management process using BIM. An important strategy for enhancing the overall construction project is the implementation of BIM as a process that generates a "Building Information Model." BIM includes information that is organized, defined, and exchangeable in addition to 3D visual modeling [17].

3. Research Methodology

A literature review in this paper arrives from academic journals, conference proceedings, books, and publications from certain institutions, such as the Lean Construction Institute (LCI). Several keywords were used in the literature search, including the performance of post-disaster project contractors, application of lean-to post-disaster reconstruction projects, application of BIM to post-disaster reconstruction projects, and the relationship between the lean application and BIM. Initially, publications not containing these keywords in their abstracts were screened. Then a brief content review was conducted on the remaining papers to eliminate those that were not relevant.

The final publications selected to cover various implementations of lean and BIM approaches that affect contractor performance in post-disaster reconstruction from 1999 to 2022. This review is not restricted to a specific geographic area.

4. Results

Several indicators of lean and BIM, as well as indicators of the performance of these contractors, are revealed in these findings based on some of the literature. These indicators help to improve contractor performance in post-disaster reconstruction projects. The literature may not include all pertinent works, but the authors want to emphasize that it can demonstrate how the research trends are progressing.

4.1. Implementation of Lean Construction in PDHR

Lean implementation during post-disaster reconstruction helps to enhance reconstruction performance, resulting in higher built environment quality and efficient workflow. Lean can also expedite the reconstruction process by emphasizing stakeholder mobilization [5] and lean construction principles in conventional projects make a 6.15 % to 9.56 % reduction in total project duration [32]. By reducing the Percent Expected Time-overrun (PET) value and increasing the Percent Plan Completed (PPC) value, the majority of the project's risk factors can be minimized and mitigated, which can result in a 15.57% reduction in overall project time [33]. This study reveals how the techniques of the Lean method contribute to improving contractor performance in post-disaster housing reconstruction projects.

4.1.1. Quick Mobilization

One of the lean strategies for realizing this is to concentrate on the timing of mobilization for all stakeholders. Completing reconstruction on time is very important to reduce the impact on people affected by the disaster. The rapid mobilization that can be applied in post-disaster reconstruction, (1) Internal mobilization carried out by the Main Contractor is separate from external mobilization which refers to the location carried out by sub-contractors, meaning that all stakeholders involved in construction tasks must separate their mobilization activities from one another, (2) subcontractors are required to mobilize beforehand or concurrently with general contractors, and (3) prepared at the mitigation stage before natural disasters for initial activities such as tendering, design, and standard stages [5].

4.1.2. Pull Scheduling

Several problems, such as the supply not being delivered on time due to the difficulty accessing the road to the site, resulted in delays in construction completion time. Pull Scheduling is a Lean method of planning construction that enhances the efficiency of the reconstruction process; it can be done in several ways: (1) Create a plan to account for future requirements and shorten the time it takes to complete reconstruction work, (2) planning resource allocation and choice to clear resource pools or queues and reduce waiting times, and (3) Along with the requirements for previous activities, resource selection must also take into account the requirements for ongoing work, queues, and subsequent activities [10].

4.1.3. Just in Time (JIT)

Increasing demand for resources such as materials and labor during post-disaster reconstruction results in increased waiting times. JIT is a construction project logistics management technique to increase productivity by providing suitable materials in the correct quantity, time, and quality. Waste includes extra inventory, waiting periods, inspection delays, and time spent locating defects and is anything that does not improve the final product [11]. The implementation of JIT in reconstruction can be done by: (1) Anticipating excess material inventory at the project site that does not affect the final result, (2) Shortening the time it takes to wait, check, and find problematic materials, and (3) Controlling inventory of incoming and existing materials at the project site.

4.1.4. Prefabrication and Modularization

In PDHR projects, the use of a module system for housing components makes the reconstruction time faster. Due to the separation of resources, which is a property of modular building, it is the ideal method to offer permanent housing more quickly. Prefabrication is used in parts of a production project in an off-site plant that is moved on-site to be installed. Modularization involves dividing a space into equal parts or modules of the same size to aid in the mass prefabrication of project components [12].

4.1.5. Six Sigma (6S)
The high quality of post-disaster housing is very important, apart from anticipating future disasters, this is also a joint international consensus "Build back better". Six Sigma is a statistically based methodology for significantly reducing customer-defined defect rates to eliminate effects and errors from each process [13], thereby maintaining high quality. Six sigmas in post-disaster reconstruction can be done by maintaining an optimal resource buffer and eliminating unnecessary resources [5].

4.1.6. Last Planner System
LPS is an effort to look back at what was planned before it was carried out by the workers doing the work. The lean construction concept will replace the usual push principle (where the planned action must be carried out by field personnel) with a pull system. The principles of the IDIC are; (1) As the time to perform the work draws near, make more detailed plans, (2) Together with people who will conduct the work, make arrangements, (3) As a team, identify and eliminate impediments to planned tasks, (4) Make promises that can be made reliable, and (5) learn from damage [14]. The Last Planner System is practiced post-disaster reconstruction to communicate effectively and collaborate [28].

4.2. Implementation of BIM in PDHR
Reconstruction has been viewed as a dynamic, complicated system that poses several difficulties and issues. BIM is frequently utilized in the construction sector because it effectively addresses complex and dynamic problems. BIM supports improving contractor performance by controlling project costs, time, and delivery which are the most significant factors in evaluating post-disaster reconstruction projects [6]. The application of a BIM-based cost estimation tool can help decision-makers calculate each project's total cost-efficiently and effectively and choose the most suitable proposed option for reconstruction purposes in the shortest possible time [29]. The following are a few BIM features that help boost contractor performance in PDHR.

4.2.1. Documentation
BIM helps automatic conformity evaluation to easily compare what was done with what was planned [7]. Measure real-time productivity with the BIM interface. PDHR will improve contractor performance, especially regarding the quality of work.

4.2.2. Data Integration
In post-disaster reconstruction with complex and dynamic conditions, accurate data becomes the main requirement to control the project. Therefore, several functions of BIM can accommodate these needs; (1) BIM helps make it easier to determine the location of logistics, the layout of materials and equipment, as well as labor circulation, (2) BIM can automatically create cost estimates and schedules, allowing users to access real-time material types or specifications and project material quantities [29].

4.2.3. Visualization
In post-disaster reconstruction, visualization of information is important to make it easy for workers to understand. Some of the following BIM functions that can accommodate this are: (1) BIM makes building designs more approachable and understandable to stakeholders by providing designs with a certain level of realism, and (2) BIM visualizes the construction project progression and creation status using a building model display. Graphics convey material to workforces in the construction environment [7, 30].

4.2.4. Coordination
In complex and dynamic post-disaster reconstruction, effective and efficient coordination is an absolute must. BIM, with its automation, can accommodate these things; (1) Facilitate the coordination and forecasting needed by Manpower, Equipment, and Materials, (2) BIM facilitates the exchange of information and interoperability (between applications) between parties, and (3) BIM automates the creation of construction and construction tasks. BIM automates the creation of construction tasks and modeling the completion of previous tasks, space, information, and resources [7].

4.3. Contractors' performance indicators in PDHR projects
Contrary to usual projects, post-disaster restoration involves customers dealing with challenging and complex processes. To achieve the success of the reconstruction program, contractors need a better understanding of client values in contract services which are indicators of contractor performance [18]. Cost overruns are usually due to delays in meeting delivery schedules and material wastage [3], which is a problem that must be resolved and part of the indicators of contractor performance.

4.3.1. Time
It develops even more critically in reconstruction projects that require to be carried out in a backup because the timeliness of post-disaster reconstruction projects is essential for the involved communities, and any deviation from the programmed agenda reasons a response. Prompt after society [19].

4.3.2. Cost
Significant cost overruns can occur in post-disaster reconstruction projects. The client's forecasting policy requires the preparation of an 'impossible to exceed but not overly conservative estimate. For post-disaster reconstruction, clients value "budgetary fit" more than "lower contract prices" [20].

4.3.3. Quality
The client public has recognized the higher quality standard as the only differentiating value within the quality category that needs improvement. Although post-disaster restoration must be finished on time, quality investment is a viable method that shouldn't be sacrificed for expediency [21]. Quality control and assurance are crucial in reconstruction situations [22].

4.4. The relationship between Lean and BIM that can improve contractor performance
BIM functions such as visualization and automatic document creation can synergize with Lean principles such as reducing variability and increasing flexibility to efficiently improve construction project performance [7]. Though conceptually different from the Lean technique, which is a managerial process, and although BIM is a supporting technology, the interaction is reciprocal or two-way [31]. With this linkage, lean and BIM will improve contractor performance in PDR projects.

Figure 1 below demonstrates the conceptual foundation for applying Lean and BIM methodologies in PDHR initiatives. In particular, the relationship between the three variables involves some dependency assumptions. H1 describes the relationship between lean construction implementation affecting contractor performance, H2 describes BIM implementation affects contractor performance and H0 describes the mutually influencing relationship or synergy between lean construction and BIM affecting contractor performance in post-disaster housing projects.

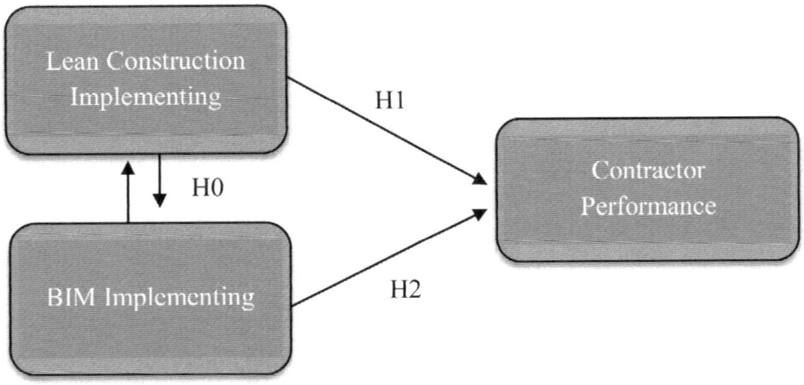

Figure 1. A framework for operationalizing the factors that contribute to implementing Lean and BIM in PDHR projects

5. Conclusion

The main objective of the current study is to propose a conceptual context for improving contractor performance through implementing lean construction and BIM. This paper states the outcomes of a literature study examining the effective implementation of lean construction and BIM in post-disaster housing reconstruction designs. Just-in-Time (JIT), Prefabrication and Modularization, Six Sigma, Last Planner System, and Visual Management. The BIM functions contributing are Documentation, Data Integration, Visualization, and Coordination. This study also determined contractor performance indicators: Time, Cost, and Quality. In this regard, the proposed conceptual framework can serve as a roadmap for construction companies to improve performance with lean construction and BIM implementation in PDHR.

References

[1] Badan Nasional Penanggulangan Bencana 2022

[2] Rotimi J O B et al 2006 The regulatory framework for effective post-disaster reconstruction in New Zealand *Int. Conf. on Post-Disaster Reconstruction Meeting Stakeholder Interests Information and Research for Reconstruction (i-REC)* pp 119 - 126

[3] Chang Y et al 2012 Resourcing for post-disaster reconstruction: a comparative study of Indonesia and China *Disaster Prevention and Management*

[4] Rosli M Tamyez P F M and Zahari A R 2020 The effects of suitability and acceptability of lean principles in waste management flow on construction project performance *Int. Journal of construction management*

[5] Mojtahedi M and Oo Bee L 2017 Possibility of Applying Lean in Post-Disaster Reconstruction an Evaluation Study *Journal of Sustainability - Special Issue on Disaster Risk Reduction and Resilient Built Environment*

[6] Baarimah A O et al 2022 A Bibliometric Analysis and Review of Building Information Modelling for Post-Disaster Reconstruction *Sustainability*

[7] Sacks R et al 2010 Requirements for building information modeling based lean production management systems for construction *Automation in Construction.*

[8] Koskela L 2000 *An exploration toward a production theory and its application to construction. Construction* Technical Research Centre of Finland

[9] Lean Construction Institute

[10] Tommelein B I D 1999 Pull-driven scheduling for pipe–spool installation *Journal of Construction Engineering and Management,* 124(4) 279-288 1999

[11] Pheng L S and Chuan C J 2001 Just-in-Time Management of precast concrete components.

Journal of Construction Engineering and Management 127(6) 494-501

[12] Hermes M 2015 Prefabrication & Modularization as a Part of Lean Construction –Status Quo in Germany *In Proc. 23rd Ann. Conf. of the Int'l Group for Lean Construction Perth Australia*

[13] Linderman K et al 2003 Six Sigma a goal-theoretic perspective *Journal of Operations Management* 21 pp 193-203

[14] Ballard G and Howell G 2010 *Lean project Building Research & Information* March 2012 pp 37-41

[15] Tezel A Koskela L Aziz Z 2017 Lean Construction in Small-Medium Sized Enterprises (SMEs) an Exploration of the Highways Supply Chain *Conf. of the Int. Group for Lean Construction (IGLC)* UK

[16] Eastman C, Sacks R M dan Paul T 2008 *BIM Handbook: A Guide to Building Information Modeling for Owners, Managers, Designers, Engineers, and Contractors*

[17] Tardif M and Smith K D 2009 *Building Information Modeling: A Strategic Implementation Guide*

[18] Aliakbarlou S Wilkinson S Costello S B and Jang H 2017 Client values within post-disaster reconstruction contracting services *Disaster Prev. Manage. Int. J* 26 3 pp 348–360

[19] Barenstein J D and Pittet D 2007 *Post-disaster Housing Reconstruction Current Trends and Sustainable Alternatives for Tsunami-Affected Communities in Coastal Tamil Nadu* (Switzerland: University of Applied Sciences of Southern Switzerland)

[20] Creedy G D Skitmore M Wong J K W 2010 Evaluation of risk factors leading to cost overrun in delivery of highway construction projects *J. Constr. Eng. Manag.* 136 pp 528–537

[21] Clinton, W J 2006 Lessons learned from tsunami recovery Key propositions for building back better *Office of the UN Secretary-General'sSpecial Envoy for Tsunami Recovery, New York*

[22] Rapp R R 2011 Disaster recovery project management Bringing order from chaos *Purdue Univ., West Lafayette, IN*

[23] Chang-Richard Y Wilkinson S Potangaroa R and Seville E 2013 Resource challenges for housing reconstruction: A longitudinal study of the Australian bushfires *Disaster Prev. Manage. Int. J.* 22 pp 172–181

[24] I A O 2015 Construction labor productivity as a correlate of project performance An empirical evidence for wall plastering activity *Civil Eng. Dimension* vol 17(1) pp 1-10

[25] Sha K and Jiang Z 2003 Improving rural laborers' status in China's construction industry *Building Research & Information* Vol. 31 No 6 pp 464-47

[26] Plebankiewicz E 2010 Construction contractor prequalification Polish clients perspective *Journal of Civil Engineering and Management* Vol 16 No 1) pp 57-64

[27] Xiao H and Proverbs D 2003 Factors influencing contractor performance: an international investigation *Journal of Engineering, Construction and Architectural Management*

[28] Demikersen S 2020 Investigating the synergy between lean construction practices and post-disaster management processes *Challenge Journal of Structural Mechanics* 6

[29] Kermanshahhachi S and Rouhanizadeh B 2018 Feasibility analysis of post-disaster reconstruction alternatives using automated BIM-based construction cost estimation tool *In Proc. of CSCE 6th International Disaster Mitigation Specialty Conference, Montreal Canadian Society of Civil Engineering* pp 13-16

[30] Gerber D et al 2010 Building Information Modeling and Lean Construction Technology Methodology and Advances From Practice *18th Annual Conf. Int. Group for Lean Construction Haifa Israel*

[31] Leite F et al 2012 BIM and Lean interactions from the BIM capability maturity model perspective A case study *Proc. 20th Annual Conf. of the Int. Group for Lean Construction*

[32] Erol H 2014 *Identifying the effects of Lean Construction principles on variability project duration* (Turkey: Middle East Technical University)

[33] Issa U 2013 Implementation of lean construction techniques for minimizing the effect of the risk on project construction time *Alexandria Eng. Journal* 697-704 p52

IOP Conf. Series: Earth and Environmental Science 1173 (2023) 012066 doi:10.1088/1755-1315/1173/1/012066

Structural mitigation of rob flood disaster through mangrove forest conservation in Indonesia coastal areas

N Adani[1], Y Subiakto[2] and S Pranoto[2]
[1]Disaster Management Study Program, National Security Faculty, Republic of Indonesia
[2]Defense University. IPSC Sentul Area, Bogor, West Java, Indonesia

E-mail: adaninadhila7@gmail.com

Abstract. Natural disasters that occur can disrupt human life and result in human casualties, environmental damage, property losses, and psychological impacts. One of the natural disasters that occur every year in Indonesia is flooding. Floods that occur due to inundation of rainwater only during the rainy season or coastal abrasion/rob. The causes of tidal flooding vary from natural conditions to human activities in coastal areas. The state of the reduced mangrove area that can potentially cause tidal flooding, the greater the impact and the more frequent the frequency. The purpose of this study is to increase awareness of the importance of structural-based mitigation in mitigating tidal flooding in coastal areas of Indonesia to the community, government and non-governmental organizations or organizations. The method in this study uses a qualitative descriptive approach using secondary data in the form of literature, books, laws, and research journals. The condition of mangrove forest areas that are increasingly critical requires synergies between ministries, namely playing a role in determining protected areas and cultivating mangrove forests, managing mangroves at the community level, especially related to forest area management and community empowerment and supported by the use of technology such as drones in massive mangrove replanting.

1. Introduction

Disaster is a disturbance to the functioning of society that has a detrimental impact on social, psychological, economic or environmental conditions and even extends to human life and usually exceeds the ability of the community to cope using their own resources [1]. As stated in Law Number 24 of 2007 concerning Disaster Management that a disaster is an event or series of events that threatens and disrupts people's lives and livelihoods caused by natural and/or non-natural factors as well as human factors, resulting in human casualties, damage environment, property loss, and psychological impact [2].

Today, various types of natural and non-natural disaster threats occur in Indonesia, when viewed from the infographic data of the National Disaster Management Agency every year, especially the number of natural disasters that occur has increased significantly [3-7]. Throughout 2021 National Disaster Management Agency recorded 3,092 events dominated by hydrometeorological disasters, namely floods with 1,298 events, followed by 804 extreme weather events, and 632 landslides [7]. Therefore, Indonesia is dubbed as a disaster laboratory where natural disasters always occur every year. When natural disasters are faced with vulnerability, it can disrupt the survival of people in all sectors who generally make a living as farmers or fishermen because of damaged or failed or failed ponds, rice fields and gardens [8].

The most frequent natural disasters that hit Indonesia every year are floods. According to Fajrianda, Bakar, Syamsidik (2017) flood disasters can be caused by several things, namely inundation of rainwater during the rainy season, tidal flooding and climate change. As a result, roads that connect between villages/kelurahan and between people's houses eventually have to be flooded, besides that in coastal areas there is land erosion that is increasingly critical [8]. With the frequency of floods that often occur in Indonesia, disaster management needs to be carried out so that these risks can be managed or even reduced. According to Worosuprojo (2012) the division of disaster management phases includes six phases, namely: prevention or mitigation phase, preparedness phase; includes early warning, emergency response, relief, rehabilitation and reconstruction. In the disaster mitigation phase, it can be divided into two approaches, namely [9]:

1) Structural Mitigation which is a physical development process by;
   a) Spatial planning with conservation of mangrove forests, coral reefs, sand dunes, and coastal forests.
   b) Infrastructure development by building earthquake-resistant houses, sea dikes, breakwater cliffs, and houses on stilts, etc.
2) Non-structural mitigation which is an activity to increase community capacity and awareness by:
   a) Training and Education,
   b) Socialization / Counseling,
   c) Field Practice/Simulation.

Structural disaster mitigation is a form of disaster management that focuses on planning and design as well as structural (physical) development as the basis for concentration in disaster management [9]. While non-structural mitigation is more focused on social problems or social impacts that will develop during and after a disaster event, the focus of attention on non-structural mitigation is looking for a regulatory system that will regulate all social aspects in a disaster-affected society [1]. Both perspectives have the same urgency with their respective portions.

With the current new paradigm, the SENDAI Framework (2015-2030), most of which is focused on the phase of disaster prevention or mitigation, which is the most significant aspect of its influence in disaster risk reduction, namely the mitigation phase, both non-structural and structural. Rif'an (2012) said that structural mitigation by spatial planning and infrastructure development in disaster-prone areas. Determination of structures and spatial plans that are suitable for areas with potential disasters can minimize casualties and environmental damage caused [1].

Indonesia, which is known as a maritime country consisting of thousands of islands, makes its territory vulnerable to tidal flooding and tsunamis in coastal areas. Currently, coastal forest areas in several areas in Indonesia, especially on the north coast of Java, Bali, East Kalimantan, Sumatra and South Sulawesi, have been degraded due to the destruction of mangrove forests and conversion of mangroves into ponds, plantations, industries, residential areas and agricultural land [10].

Mangroves can not only break waves during high tide, but can also dampen tsunamis [11]. According to Alongi (2005) the factors that determine mangroves can reduce tsunamis include: tree density, tree diameter, presence of coastal habitats (grasslands, coral reefs, hills), forest width, tree height, forest slope, soil texture, proportion of biomass , above ground level at roots, forest location, type of lowland vegetation adjacent to mangrove vegetation, tidal wave width of seawater forest, distance from tectonic events, tsunami size and velocity, and angle of incidence of the tsunami relative to the coastline [11].

But the fact is that it has reached 160-200 thousand ha/year the rate of mangrove degradation [10]. According to Muryani et al (2011), development activities are the biggest contributor to the decline of mangrove land in Indonesia by shifting the designation to ponds and agricultural areas as well as taking timber for commercial purposes [10]. According to research conducted by Murdiyarso et al., (2015) most of the deforestation of mangroves in forest areas occurs due to encroachment by communities for the development of shrimp or fish ponds [22]. Conditions of damage to mangrove forests caused by human activities in coastal areas that change mangrove forests into various other uses thus reducing the area and natural conditions such as climate change and the survival of mangroves in their habitat [12]. Meanwhile, according to Kusmana (2003) there are three main factors that cause mangrove damage, namely (a) conversion of mangrove forests that does not pay attention to environmental factors, (b) environmental pollution, and (c) excessive logging [10].

With the reduction of mangrove areas that can withstand tidal waves in coastal areas, the potential for disasters will be even greater with more frequent occurrences and a reduced ability to erode land in coastal areas, break waves, withstand strong winds so that coastal areas become submerged in tidal flooding [12]. Several tidal floods and coastal abrasion occurred in various parts of Indonesia, such as Leuge Village, Aceh [8], Tanggul Tlare Village, Jepara [10], Ketapang Raya Village, East Lombok [11], Langkat Regency, North Sumatra [12] and the City of Pariaman, West Sumatra [13].

2. Research method

The method in this study uses a qualitative descriptive approach through library research using secondary data in the form of books, laws and e-journals regarding spatial planning, especially in coastal areas by utilizing mangrove forest conservation [1, 8, 10, 11, 12, 13, 18, 21].

3. Result and discussion

In reducing the risk of tidal flooding while protecting coastal areas from salt water infiltration into the mainland, the threat of abrasion, absorbing pollutants, maintaining coastal and marine productivity, and sea breezes, structural mitigation can be done by means of structural mitigation, namely the creation of protected areas in coastal areas by developing mangrove forests and forests. beach [11]. Mangroves are a collection of plants that live in coastal areas located between the sea and land on the banks of rivers to estuaries that lead to the sea [12]. Mangrove ecosystems (mangroves) are ecosystems located in coastal areas that are affected by tides so that their roots are always flooded [10].

Mangroves have the structure and types of leaves and roots that are unique and different for each species, but in general have the ability to survive in muddy soil, making mangrove plants very suitable for coastal areas [12]. According to Sannigrahi et al (2019), in addition to having the function of local economic resilience, mangroves can also function as a barrier to sea waves in coastal areas that have the potential for coastal abrasion and tidal flooding [12].

According to Lasibani & Eni (2009) the benefits of mangrove ecosystems related to physical functions are as disaster mitigation such as coastal protection from abrasion, tidal waves (rob), can neutralize water pollution to a certain extent, hold mud and trap sediment transported by surface water currents. , dampen waves and wind storms for coastal areas, prevent seawater intrusion into the mainland, and tsunamis [10]. According to Quartel (2007) mangroves grow on a substrate that has a clay base and a density of mangrove vegetation consisting of roots, stems, and branches, for example the *Kandelia candel* mangrove species which can reduce coastal abrasion and tsunami waves [11]. Abrasion is the process of eroding land or land which mostly occurs due to the factor of sea water waves which cause sea water to easily rise to the surface and can cause a decrease in land area [10].

Structural or physical activities to mitigate tidal and tsunami natural disasters include: (1) coastal ecosystem management, (2) coastal vegetation, (3) provision of early warning systems with coastal vegetation such as mangrove ecosystems, (4) provision of health facilities and infrastructure, (5) the use of tsunami-dampening buildings, (6) the use of tsunami-friendly building constructions, (7) the provision of self-rescue facilities [14]. Adri et al (2020) research results in Japan showed that a mangrove forest with a tree density of 30 trees/100 m^2, a tree diameter of 15 centimeters and a width of 200 meters can reduce about 50% of tsunami wave energy [12].

Ecologically, it functions as protection for the terrestrial and marine ecosystems as well as habitats for various fauna, including as protection against abrasion/erosion; wave; strong winds, as a place to find food; spawn; breeding various types of fish and shrimp, controlling marine intrusion, controlling malaria, maintaining water quality (reducing water pollutant pollutants), as well as being a CO_2 absorber and producing O_2 which is relatively high compared to other forest types [20]. In addition to utilizing mangrove forest conservation, zoning is also carried out in accordance with the general guidelines for spatial planning for coastal and small islands issued by the Ministry of Marine Affairs and Fisheries which refers to the Decree of the Minister of Fisheries and Marine Affairs No: KEP.34/MEN/2002, namely determination [14]:

1) Core zone is a conservation zone in the form of protected areas and nature reserves. The function of this activity is directly related to marine or marine ecosystems and fisheries.
2) Limited utilization zone in the form of a buffer zone consisting of land and sea areas. For land areas in the form of ponds and for the sea in the form of coral reef cultivation areas, seaweed, and beach tourism activities. For areas of limited use, it is necessary to limit the types of activities that should not cause great attraction to preserve the coastal ecosystem.
3) Free zones/other zones according to their designation in the form of activities that are not directly related to the sea such as urban activities.

This zoning determination also considers several protection, accommodation and relocation activities for areas that are possible to be carried out based on the IPCC (1990), as well as implementing Government Regulation of the Republic of Indonesia No. 64 of 2010 concerning Disaster Management for Coastal Areas and Small Islands [15] and structural/physical activities to mitigate against types of sea level rise disasters [16]. In addition, structural mitigation in reducing the risk of tidal flooding in coastal areas can be done by developing a grid patterned road network, creating a green belt area, determining evacuation routes, and providing evacuation shelters [17].

Planning for coastal areas based on tidal flood and tsunami disaster mitigation in coastal areas is divided into two ways, namely structural mitigation efforts, such as coastal reclamation plans, making buffer zones with the waterfront city concept and non-structural mitigation efforts, such as making policies on natural disaster mitigation in general and tsunami disaster mitigation in particular [18]. The latest innovation is replanting trees, especially mangroves that can reduce tidal flooding and coastal abrasion, have the ability to improve air quality, store large amounts of carbon, and maintain marine life and biodiversity through the sowing of mangrove seeds which are distributed via unmanned aerial vehicles (UAV) in a fast, safe, cost-effective, and environmentally friendly way if done properly [22]. Several companies have proven the speed of unmanned aerial vehicles/drones (UAVs) in forest restoration efforts for example, Flash Forest, a Canadian-based start-up, has claimed to have the ability to plant 10,000–20,000 seed pods per day, and that their UAVs are capable of shoots 200 seed pods into the ground in 160 seconds [23-24]

However, in addition to structural mitigation, it is also necessary to carry out non-structural mitigation by providing information on tidal flood risk reduction based on mangrove forests which aims to increase public awareness and understanding of the importance of maintaining and conserving mangrove forests to minimize the impact of tidal flooding and tsunami in coastal areas and commit to planting restore mangrove areas that have begun to erode or are damaged [11]. This is evidenced by research conducted by Inghug & Sios (2019) which states that there are two reasons for the failure to protect mangroves, namely stakeholder participation and community awareness of mangrove ecosystems [19].

Structural mitigation efforts in reducing the risk of tidal flooding in coastal areas need to be supported by the implementation of policies that regulate spatial planning/zoning of disaster-safe coastal areas/land use and sanctions for violators, as well as supported by non-structural mitigation in the form of training and disaster simulations, building standardization policies. in disaster-prone areas as well as infrastructure and facilities that are resistant to tidal waves and tsunamis, development of an early warning system for rising tidal waves and tsunamis [18].

4. Conclusion

According to data, the most frequent natural disasters that hit Indonesia are floods and one of them is tidal flooding (tidal flooding) in coastal areas. The causes of this tidal flood vary from natural conditions such as climate change, the survival of mangroves in their habitats and human activities in coastal areas that turn mangrove forests into various other uses. With such conditions, coastal areas are becoming increasingly vulnerable to tidal flood disasters, so structural disaster mitigation efforts are needed to reduce the risk of tidal flooding by establishing structures, zoning, and spatial patterns in the planning area that contain activities that may and may not be carried out as well as activities which can only be done after obtaining a permit in the coastal area and also the implementation of mangrove forest conservation activities.

Mangrove management needs to involve various sectors, especially fisheries, forestry, the environment, and underdeveloped areas. Therefore it is important, collaboration and cooperation in accordance with their fields in mangrove management at the ministry level such as the Ministry of Agrarian Affairs and Spatial Planning/National Land Agency play a role in determining protected and cultivated areas, while the Ministry of Environment and Forestry and the Ministry of Maritime Affairs and Fisheries plays an important role in mangrove management at the grassroots, particularly related to forest area management and community empowerment.

And supported by the use of technology that has been innovating at this time makes it easier for us to replant mangrove land that has been depleted quickly, save time and cost with a wide range using drones. In addition, it can also be harmonized with non-structural mitigation, namely providing understanding to the surrounding community together with organizations and non-governmental organizations engaged in mangrove conservation of the importance of mangrove forests as a means of reducing tidal flood disaster risk so that it can run optimally, so as to minimize casualties as well as economic and social impacts as well as environmental damage due to tidal flooding.

References

[1] Rif'an, A A 2012 *Proceedings of the National Seminar on Remote Sensing and Geographic Information Systems*, (Surakarta: Muhammadiyah University of Surakarta) pp 577-586

[2] Law of the Republic of Indonesia Number 24 of 2007 concerning Disaster Management.

[3] National Disaster Management Agency, "Kejadian Bencana 2017" (Disaster Events in 2017), Available online from https://bnpb.go.id/infografis/kejadian-bencana-tahun-2017 [Accessed September 29, 2022]

[4] National Disaster Management Agency, "Kejadian Bencana 2018" (Disaster Events in 2018) Available online from https://www.bnpb.go.id/infografis/kejadian-bencana-tahun-2018 [Accessed September 29, 2022]

[5] National Disaster Management Agency, "Kejadian Bencana 2019" (Disaster Events in 2019), Available online from https://bnpb.go.id/infografis/kejadian-bencana-tahun-2019 [Accessed September 29, 2022]

[6] National Disaster Management Agency, "Kejadian Bencana 2020" (Disaster Events in 2020), Available online from https://bnpb.go.id/berita/bnpb-telah-selesaikan-verifikasi-data-bencana-indonesia-2020 [Accessed September 29, 2022]

[7] National Disaster Management Agency, "Catatan Refleksi Akhir Tahun Penanggulangan Bencana 2021" (Reflection Notes for the End of the Year for Disaster Management 2021), Available online from https://www.bnpb.go.id/berita/catatan-refleksi-akhir-tahun-penanggulangan-bencana-2021 [Accessed September 29, 2022]

[8] Fajrianda M, Bakar Y A, Syamsidik 2017 *Journal of Master of Disaster Science Postgraduate Syiah Kuala University*, **4**, 1, pp 1–5.

[9] Worosuprojo S 2012 *Proceedings of the National Seminar on Remote Sensing and Geographic Information Systems*, (Surakarta: Muhammadiyah University of Surakarta) pp. 3-6

[10] Utomo B, Budiastuti S, and Muryani C 2017 *Journal of Environmental Science*, **1**, 2, pp 117–123. DOI:10.14710/vol.15.2.117-123
[11] Santoro D, Yamin M, and Mahrus, M 2019 *Journal of Devotion Masters in Science Education*, **2**, 1, pp 12–16.
[12] Hamzah A H P, Anggoro S, and Puryono S 2020 *Proceedings of the 2nd National Seminar on Earth Environmental Engineering* (Yogyakarta: University of National Development "Veteran" Yogyakarta) pp 11-18
[13] Pratiwi N A, and Alhadi Z 2021 *Journal of Policy, Governance, Development and Empowerment*, **1**, 1, pp 18-29
[14] Decree of the Minister of Marine Affairs and Fisheries Number Kep. 34/Men/2002 concerning General Guidelines for Spatial Planning for Coastal and Small Islands
[15] Government Regulation No. 64 of 2010 concerning Disaster Mitigation in Coastal Areas and Small Islands
[16] Isdianto A, Citrosiswoyo W, and Sambodho K 2014 *Journal of Settlements*, **9**, 3 pp 148–157.
[17] Ahmad S N A, Mardianto D, and Hizbaron, D R 2021 *Journal of Civil Engineering and Planning*, **2**, 1, pp 1–19
[18] Purwanto, N I, Poluian R J, dan Takumansang 2017 *Spatial Journal: Urban and Regional Planning*, **4**, 3, pp 1-8
[19] Risa, N E W, Alamsyah R, Tenriwaruwaty A, Wahyuni A P, and Liswahyuni A 2021 *Tarjih: Agribusiness Development Journal*, **1**, 2, pp 32-40
[20] Marhawati M, Rahmadani A, Astry A, Gunawan G, Minarni M, Rosmaladewi R, and Khatimah U 2021 *Journal Lepa-Lepa Open*, **1**, 3, pp 386-392
[21] Salminah M, and Alviya I 2019 *Journal of Forestry Policy Analysis*, **16**, 1, pp 11-29
[22] Mohan M, Richardson G, *et al.* 2021 *Remote Sensing*, **13**, 13, pp 1-30.
[23] Kickstart, "Flash Forest: Using Drones to Plant 1 Billion Trees" Available online from https://www.kickstarter.com/projects/499734388/flash-forest-using-drones-to-plant-1-billion-trees-by- [Accessed September 29, 2022]
[24] Bullfrog Power, "Flash Forest: Revolutionizing Reforestation with Tree-Planting Drones. Bullfrog Power. 2020". Available online from https://www.bullfrogpower.com/blog-flash-forest-revolutionizing-reforestation-tree-planting-drones/ [Accessed September 29, 2022]

Utilizing information technology towards reducing forest and peatland fire risk in Riau Province, Indonesia

R J Eklou[1], I D K K Widana[1] and Suwito[2]

[1]Disaster Management Study Program, Faculty of National Security, Republic of Indonesia Defense University, Jl. Salemba Raya No. 14, Jakarta Pusat, Jakarta 10430, Indonesia
[2]Institute of Research and Community Service, Republic of Indonesia Defense University, IPSC Sentul Area, Citeureup District, Bogor Regency 16811, Indonesia

E-mail: joseph_richie@student.uns.ac.id

Abstract. Forest and peatland fires have become a seasonal challenge experienced in Riau Province. This disaster causes huge economic and environmental losses, accompanied by its alarming impact on human health. The haze as a result of land burning in this province is known to disrupt activities in neighbouring countries causing political and diplomatic tensions. To reduce the risk of forest and peatland fire, the Riau Regional Police developed the Lancang Kuning Nusantara (LKN) digital application. This study examines the general role of the application in reducing forest and peatland fire risks. A qualitative descriptive approach is adopted in analysing the primary data obtained through focus group discussion. Data from secondary sources were also obtained. Results from this study indicate that the use of the LKN digital application in Riau Province has helped immensely in the area of disaster risk communication and logistics support.

1. Introduction

Something is always burning on earth. In a global report highlighting catastrophic occurrences around the globe, the world experienced 19 disastrous wildfires in 2021 causing 128 human deaths and US$9.2 billion in economic losses. When compared to the 2001-2020 annual average data, the world experienced 11 catastrophic wildfires causing 77 human deaths and US$ 5.0 billion economic losses [1]. These figures raise a cause for alarm and attention. The UN Environment Programme (UNEP) predicts that if forest fire risks are not properly managed in the coming years, the world will steadily experience 14% increase in catastrophic wildfires by 2030, 30% increase by 2050, and 50% increase by 2100 [2]. Climate change, poor land and forest management systems are consistently linked to the driving force intensifying fire outbreaks in recent years [3,4]. Deliberate human activities mainly for economic reasons are also reported to trigger fire outbreaks [3–6].

According to the Indonesia Disaster Risk index [7], forest fires remain a major concern for the Riau Province. Peatlands constitute around 57% of the province's total land area (figure 1). This rich ecosystem could increase the risk of potential wildfires if not properly managed. The Indonesia National Disaster Management Agency (BNPB) reported that 99% of forest and peatland fires are caused by agricultural and industrial activities around these complex and vulnerable ecosystems, especially during dry seasons. BNPB claims that only 1% of these fire outbreaks are influenced by natural causes [8]. Another independent source reported that 98% of forest and peatland fires in Indonesia are caused by human-related activities [9].

In 2014, the Provincial Disaster Management Agency (BPBD) in Riau have not formulated a comprehensive strategy for managing fire risks. The regional government stated that it was unable to handle forest and peatlands fires due to the lack of heavy fire equipment and insufficient funds [10]. The approach used in managing fire risk at that time was ineffective due to a lack of hotspot monitoring system, lack of coordination and communication between stakeholders which led to an underestimation of the problem, and rare socialization activities for communities situated around forests and peatlands. In terms of scale, Riau experienced the most devastating forest and peatland fire in 2015 with an affected land area of 183,808.59 ha [11].

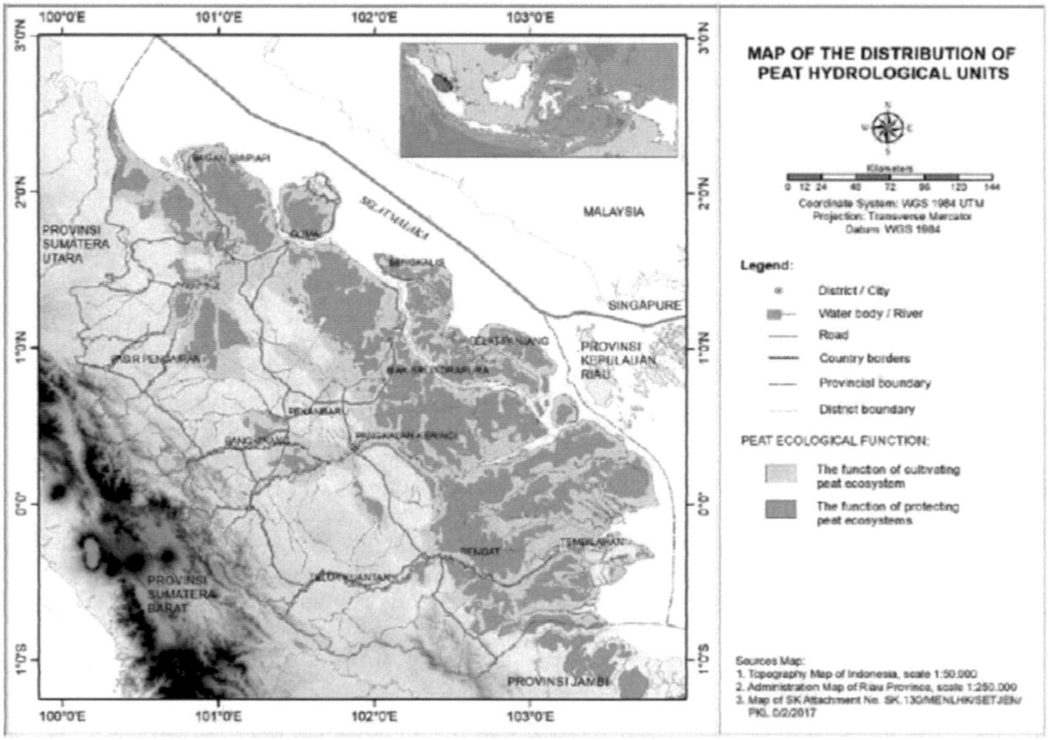

Figure 1. Riau Province Peatlands (Scale 1:50 000) [12].

In 2019, Indonesia experienced another fatal loss due to forest and land fires affecting 942,484 ha land area resulting in US$ 5.2 billion (Rp 75 trillion) economic loss [13]. The disaster affected 90,550 hectares of land area in Riau Province. To combat the seasonal occurrence of forest and peatland fires throughout Indonesia, the government passed a policy [14] towards increasing preventive measures and strengthening law enforcement.

The Riau Regional Police Force (Polda Riau) developed the Lancang Kuning Nusantara (LKN) Application in 2020. A web and android based application for managing and responding to forest and peatlands fire risk. It is intended that with the deployment of this technology, the police can detect any potential fire hotspot and take the most optimal action to prevent a fire outbreak and support suppression actions. Including Riau Province, the LKN application is utilized by the regional police force in 12 Indonesian provinces across Sumatra Island, Kalimantan Island and Bangka Belitung Islands. This article aims to analyse the general role which the application has had in Riau province towards reducing forest and peatland fire risks. Research questions to be addressed are; 1) What are the features embedded in this technology which support fire disaster risk management? 2) At what stages of disaster management are these features relevant in reducing forest and peatland fire risks? Findings from this study can provide knowledge, particularly to disaster managers in developing nations on the role of digital technologies in fighting land fires and help advance the disaster risk management field at large.

2. Literature review

2.1. *Human security and fire disaster (forest and peatland fire)*

The United Nations Development Program (UNDP) formulated a comprehensive human-centred approach which can be used to analyse threats and challenges systematically toward achieving sustainable development [15]. Disasters are one of the threats which can inflict loss and suffering on humans. The human security approach cuts across seven dimensions of security– economic, food, health, environmental, personal, community and political.

The objective of Indonesia's human security according to its national defense policies [16,17] is to create a sense of security and peace for the Indonesian people. Disasters are a threat to human security. Forest and peatland fires must be tackled with urgency to prevent colossal economic and environmental losses.

The forest and peatland fires that burnt 2,611,000 ha of land area in June-October 2015 affected 28 million people in Indonesia. This disaster caused the death of at least 19 people with nearly 500,000 people who experienced acute respiratory complications [18]. The education sector was affected because schools were forced to close. In addition, this disaster caused haze emergencies in various Indonesian provinces, including South Sumatra, Central Kalimantan, East Kalimantan, South Kalimantan, Papua, West Kalimantan, Riau, and Jambi. This situation was also referred to as a transborder disaster because the haze from Riau Province affected neighbouring countries - Malaysia, Singapore and Brunei Darussalam [19].

Based on the reviewed literature, it can be seen that forest and peatland fires not only cause economic and environmental insecurities. This disaster which is dominantly influenced by human-related activities impacts people's lives in the education, health, and transportation sectors. Forest and peatland fires also threaten the existence of flora and fauna in peat ecosystems. In regards to transborder impacts, haze as a result of land burning in Riau Province is also known to disrupt activities in neighbouring countries causing political and diplomatic tensions.

2.2. *Technology and fire disaster (forest and peatland fire)*

The Indonesia disaster management law chapter 2, article 3 places science and technology as an integral part of disaster risk management [20]. Major stakeholders in disaster management are encouraged to leverage science and technology to streamline and accelerate processes in disaster management practices. These technologies can be utilized in early warning systems, emergency response systems and recovery phases of disaster management. As a collective effort toward addressing fire disaster risks, UNESCO (United Nations Educational, Scientific and Cultural Organization) encourages non-governmental organizations (NGOs), communities and academic institutions to conduct more research on science and technology development in the area of forest and land fire mitigation. UNESCO also reported the development of certain information technology (internet, GPS, mapping) as relevant [21].

The deployment of appropriate technology is necessary and strategic in disaster mitigation efforts toward reducing the risks of forest and land fires. The outlook on disaster risk management technologies for fire and peatland fire provides five priority technologies which need to be developed to support disaster risk reduction (DRR) activities in Indonesia. These technologies focus on; zero-burning land management, fire danger rating system, water level detection, water management, and weather modification [22].

Based on the reviewed literature, it is evident that disaster-related policies in Indonesia provide an enabling system for the use and development of technologies toward achieving its DRR goals.

However, it is important to assess the effectiveness of these technologies after deployment to study if the predefined outcomes have been achieved [23].

3. Method and data

This study was conducted qualitatively using a descriptive research design. This approach was chosen because it is theoretically proven to be applicable in studies with research objectives which aim to analyse a situation, phenomenon or problem in a systematic manner [24,25]. As this article aims to analyse the general role of the LKN digital application towards reducing forest and peatland fire risks in Riau Province, this approach is useful in analysing what features are embedded in the technology, as well as when these features are relevant towards reducing forest and peatland fire risks.

The data analysed in this study were collected from primary and secondary sources. Primary data were obtained through a virtual Focus Group Discussion (FGD) with the Riau Regional Police (Polda Riau) as the main developer and operator of the LKN application. Secondary data in this study were obtained from relevant previous research, photographs, documents, statutes, and reports from the internet which are related to the addressed topic. This study uses the interactive model for data analysis [26]. This analysis technique is implemented continuously and iteratively until the obtained data is saturated. The interactive model data analysis technique consists of data collection, data condensation, data display, and drawing conclusions.

4. Results and discussion

4.1. Data flow and application system

The Lancang Kuning Nusantara (LKN) application is an innovative information technology developed to manage forest and peatland fire risks. This application system combines imagery from 4 satellites (AQUA, NOAA, TERRA, SNPP) to facilitate the real-time monitoring of hotspots, prevention of fire disasters, suppression of fire outbreaks, and efficient mobilization of resources to manage fire risks (figure 2). This application is operated by the Riau Police, however, can be accessed by the registered public which supports effective stakeholder coordination when dealing with forest and peatland fires.

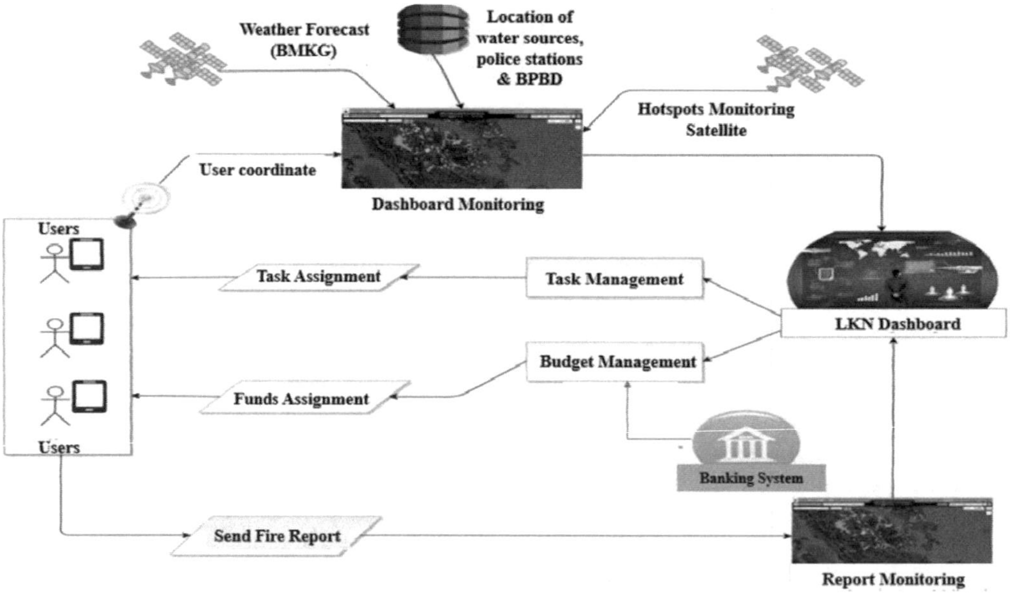

Figure 2. Data Flow Diagram.
(Source: Polda Riau, 2022)

The main features embedded in the application system are as follows;

a. Reporting feature: registered users that comprise communities, volunteers and task forces can use this feature to send direct messages regarding a spotted or potential hotspot.
b. Dashboard feature: useful in visualizing the number of reports, hotspots, available personnel around hotspots, and other useful data used in making well-informed decisions.
c. Budget management feature: the application is integrated with some bank systems which makes it useful in managing the budget allocated for responding to fire risks. This feature aids in fostering budget transparency and proper management of funds.
d. Task management feature: this feature is useful in matching affected areas with human resources for timely and effective response to any fire risk in forests and peatland areas.
e. Other features: include information on water resources, weather forecast, location of police stations and task force units, helipad locations, and other resources needed for effective and efficient management of forest and peatland fire risks.

4.2. Role of LKN application in reducing forest and peatland fire risk

In the different stages of forest and land fire risk management, the use of the LKN digital application in Riau Province has helped immensely in the area of disaster risk communication and logistics support. The communication system in this application helps the command centre in providing directions and coordination efforts to personnel and volunteers in the field. It was reported that communication is an important factor when dealing with disaster risks, one of which is through building communication networks and facilities [27]. Communication is not only needed in a disaster emergency, it is essential in every stage of disaster risk management. In pre-disaster situations, adequate information about disaster risks in an area, training and internalization of habits must also be carried out sustainably through effective disaster communication.

The LKN application also supports the effective and efficient allocation of logistics. This is made possible by providing precise and accurate information needed to execute and understand the needs of a fire extinguishing or suppression operation. Through mapping systems, the nearest resources are deployed to locations with urgent and specific needs. Logistics plays an important role in disaster risk management. It provides the link between plans and action. In actual terms, logistics provides the bridge between procurement and distribution of humanitarian aid or equipment [28].

4.2.1. Pre-disaster stage. This application practically serves as an early warning system to prevent forest and peatland fire disasters. With the satellite's continuous surveillance over the Riau Province, it can detect hotspots in real-time and directly notify the command centre to verify the data before making a decision on countermeasures. With the report feature, stakeholders including the police, village heads, and community volunteers can monitor and report forest and peatland fire incidents in their respective areas so that the wide spread of forest fires can be anticipated and handled as early as possible. The involvement of the entire community in the monitoring process confirms the concept of community-based disaster risk reduction in the Indonesian context. As these communities and villages are most likely to be affected by the devasting impacts of forest and peatland fires, it becomes reasonable that they are constantly involved in forest and peatland fire prevention measures. The LKN application is in line with the Sendai framework [29] priority of enhancing community preparedness for measured and effective response.

4.2.2. Emergency response stage. During the fire emergency response stage, the LKN application serves as an essential means of communication that facilitates the coordination of all elements required in fire suppression operations. After verification of hotspots, logistics and manpower needs can be quickly assessed for deployment through the features embedded in this application. Usually, a task force team will be dispatched to extinguish the fire in collaboration with the community and companies closest to the fire incident location. The LKN application can also be utilized in sending logistics to personnel responding to forest and peatland fires by allocating the nearest resources required in suppressing the fire. This feature aims to maximize the effectiveness and response time of the assigned task force. The application is also equipped with a weather forecast and wind direction features to help monitor field conditions during suppression operations. Moreover, prompt disbursement of operational budgets to emergency responders can be facilitated through this application system as it is connected to the banking system. The system supports budget accountability. The area of forest and peatland impacted by fire occurrences experienced a further decline from 15,442 hectares in 2020 to 8,970 hectares in 2021 (figure 3). The Riau police force reported that the LKN application had a significant contribution towards reducing forest and peatland fires since the adoption of this technology in 2020.

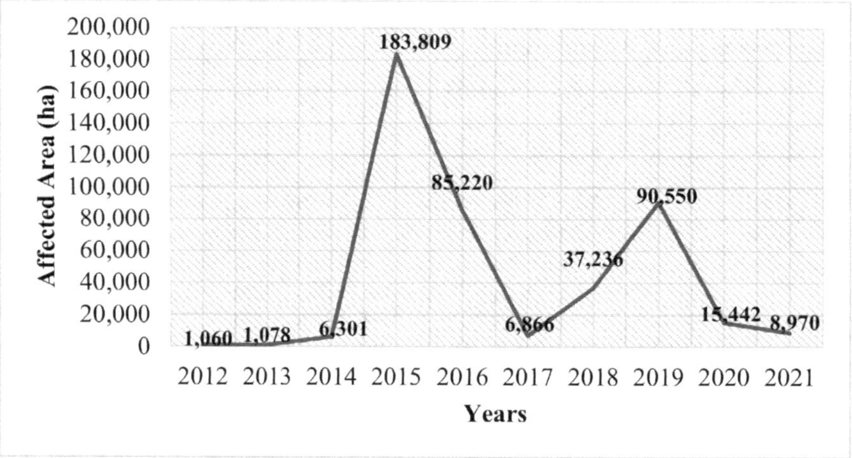

Figure 3. Ten Years Trend of Forest and Peatland Fires in Riau Province.
(Data Source: MenLHK, 2021)

4.2.3. Recovery stage. Through the LKN application, the command centre can directly monitor the success rate of fire suppression operations so that decisions can be made on whether to lift the standby alert for field personnel. The Riau police have also developed techniques in determining whether the fire was intentionally caused. The police investigate the scenes of the fire in other to analyse potential causes. An indicator of an intentional trigger of forest and peatland fire is repeated outbreaks in previously handled spots. The investigations conducted by the Riau Police aid in enforcing law and strict sanctions on perpetrators of forest and peatland fires. The LKN application has also been useful in Post-Disaster Needs Assessment (PDNA) which aims to carry out impact analysis, and estimate resources needed in the preparation of forest and peatland rehabilitation plans. The budget management and dashboard features in the application can provide some useful information needed during the PDNA. This information becomes the basis for looking at existing patterns in the field, therefore planning for a more integrated and targeted response. Results from the PDNA might recommend the construction of canals or reservoirs to help maintain soil water levels in forests and peatlands. As an incentive and community involvement strategy, the Riau Police reward individuals who have participated and provided information towards reducing forest and peatland fires. These rewards include certificates of appreciation and custom achievement pins presented by the Riau police.

5. Conclusion

The Lancang Kuning Nusantara (LKN) application is a digital innovation which is proven to have supported efforts in reducing the risks of forest and peatland fires in Riau Province. With this application, stakeholders such as the police force, local governments, village heads, business entities, communities, and volunteers can monitor and report forest and peatland fires in real-time. The application has supported efforts in managing forest and peatland fires at all stages of disaster management. In the pre-disaster stage, this technology plays an important role in mitigating fire outbreaks and monitoring forest and peatlands conditions. During the emergency response stage, this technology helps in facilitating communication to support coordination efforts needed to suppress any slight emergence of fire. In the recovery stage, it supports land rehabilitation efforts and transparent budget management. Investing in technologies that are useful in DRR activities, especially in the context of monitoring and mitigating fire disasters is a responsible action for mankind and the environment. The following recommendations are provided based on the findings from this study;

a. The local disaster management agency can support the socialization of this digital application in communities carrying out daily activities around forest and peatland ecosystems. Constant engagement with the local society can help produce local wisdom which can be integrated into the technology for optimal monitoring and management of forests and peatlands.
b. Government and academic institutions can support research which aims to understand and measure the long-term impact of forest and peatland fires on human health. Globally, research in this area is still minimal.
c. The Riau regional police can collaborate with other local and national institutions to develop additional features on the LKN application for addressing other types of hazards in the province by adopting an all-hazards approach.

Acknowledgement
The authors wish to thank the Riau Regional Police for providing insightful data and information during the focus group discussion which was useful in the writing of this paper.

References
[1] CRED 2022 *2021 Disasters in Numbers* (Brussels: CRED)
[2] UNEP 2022 *Number of wildfires to rise by 50% by 2100 and governments are not prepared, experts warn* (UN Environment Programme)
[3] Flannigan M, Cantin A S, De Groot W J, Wotton M, Newbery A and Gowman L M 2013 *Forest Ecology and Management* **294** 54
[4] Tyukavina A, Potapov P, Hansen M C, Pickens A, Stehman S V, Turubanova S, Parker D, Zalles V, Lima A, Kommareddy I, Song X P, Wang and Harris N 2022 *Frontiers in Remote Sensing* **3** 1
[5] NASA 2022 *Global fire map* (NASA Earth Observatory)
[6] UNEP 2022 *Spreading like Wildfire – The Rising Threat of Extraordinary Landscape Fires. A UNEP Rapid Response Assessment. Nairobi*
[7] IRBI 2021 *Indeks Risiko Bencana Indonesia* (Jakarta: BNPB)
[8] Nugroho S 2019 *99% Penyebab Kebakaran Hutan dan Lahan Adalah Ulah Manusia* (BNPB)
[9] Sulaeman D and Ayunda D 2020 *4 Impact of Land Preparation by Burning on the Biophysical Condition of Peatlands* (WRI Indonesia)
[10] Sutisna S, Sari D A P, and Pradana R 2018 *J. Manajemen Bencana* **4** 1
[11] MenLHK 2021 *Peta - Sipongi | Data Luas Karhutla* (SiPongi)
[12] Syahza A, Bakce D, Nasrul B, and Mustofa R 2020 *Int. J. of Sustainable Development and Planning* **15** 1119
[13] BNPB Editorial 2019 *Kerugian Kebakaran Hutan dan Lahan Sepanjang 2019 Capai Rp 75 Triliun* (BNPB)
[14] Instruksi Presiden Nomor 3 Tahun 2020 Tentang Penanggulangan Kebakaran Hutan Dan Lahan (https://peraturan.bpk.go.id/Home/Details/133045/inpres-no-3-tahun-2020, Indonesia)
[15] United Nations 2016 *Human Security Handbook: An Integrated Approach for the Realization of the SDG's*
[16] Undang-Undang Republik Indonesia Nomor 3 Tahun 2002 Tentang Pertahanan Negara (2002)
[17] Peraturan Presiden Nomor 07 Tahun 2008 Tentang Pertahanan Negara (2008)
[18] World Bank 2016 *The Cost of Fire: An Economic Analysis of Indonseia's 2015 Fire Crisis*

(Jakarta: World Bank)
[19] Yuliani F and Rahman A 2018 *Sosio Informa* **4**
[20] Undang-Undang Nomor 24 Tahun 2007 Tentang Penanggulangan Bencana (2007)
[21] Susanto D et al 2018 *Buku Panduan Mitigasi Dan Pengendalian Kebakaran Hutan Dan Lahan: Handbook for Forest and Land Fire Mitigation and Control* (Jakarta: UNESCO)
[22] Kristijono A et al 2019 *Outlook Teknologi Pengurangan Risiko Bencana Karhutla: 2019 Outlook on Disaster Risk Reduction Technologies for Fire and Land Fire* (Jakarta: BPPT)
[23] Sariani S, Yaningsih Y, and el Khairat M, JEES 2020 *J. of English Educators Society* **5** 67
[24] Kim H, Sefcik J S, and Bradway C 2017 *Research in Nursing and Health* **40** 23
[25] Kumar R 2011 *Research Methodology: A Step-by-Step Guide for Beginners, 3rd Edition* (London: SAGE Publications)
[26] Miles M B, Huberman A M, and Saldaña J 2014 *Qualitative Data Analysis A Methods Sourcebook, 3rd Edition* (California: SAGE Publications)
[27] Putri C E and Hamzah R E 2021 *J. Pustaka Komunikasi* **4** 66
[28] Yang Z, Guo L, and Yang Z 2017 *Annals of Operations Research* **283** 917
[29] United Nations 2015 *Sendai Framework for Disaster Risk Reduction 2015 - 2030*

Community-based disaster risk reduction efforts in facing the threat of abrasion disasters at Bungin Beach, Bekasi Regency during the covid-19 pandemic

S Fathoni[1], S Sutisna[1] and M Supriyatno[1]

[1]Disaster Management Study Program, Faculty of National Security, Republic of Indonesia Defense University. Salemba Raya street number 14, Central Jakarta, Province DKI Jakarta, 10430

E-mail: sfthoni93@gmail.com

Abstract. Abrasion causes coastal areas to have a very high level of dynamics and often has an impact on environmental damage, so it needs to be anticipated. In the last 2 years, the whole world has faced a crisis caused by the pandemic due to Corona Virus Disease (Covid-19). During the pandemic, the environment where coastal communities live is still threatened with abrasion. Some mangrove planting and rehabilitation activities had to be stopped to suppress the spread of the virus. The purpose of this study was to identify the role of community groups (Tourism Awareness Group) as a key factor in overcoming abrasion events during the Covid-19 pandemic at Bungin Beach. Interviews and direct observations were conducted to obtain survey data. Additional analyzes were then carried out through relevant journals and literature reviews. As a result, from 1997 to 2021 there has been abrasion on Bungin beach. The Tourism Awareness Group plays a role in post-pandemic recovery, especially in the economic sector. Through mangrove planting activities, not only to restore the economy of coastal communities during the pandemic but also to protect coastal ecosystems from the threat of abrasion on Bungin Beach.

1. Introduction

Coastal areas have very high dynamic levels. Common natural activities such as wind, waves, tides, and ocean currents strongly influence changes in coastal zones. Over time, the direction of change in Indonesia's coastal areas tends to lead to negative conditions. In terms of terminology, the dynamics that occur in coastal regions are forms of change caused by energies within the Earth (intrinsic) and energies from extraterrestrial sources (extrinsic), including those of human activity [1]. Dynamics occurring in coastal areas can lead to environmental disruption through changes in coastlines due to abrasion and sedimentation. Coastal zone dynamics can also be driven by activities that alter coastal zone function by various human activities in the region.

Bungin Beach is administratively included in the Pantai Bakti Village area in Muaragembong District, Bekasi District. The location is located in North Java, facing the Java Sea and next to Karawang Regency. The village of Pantai Bakti, based on Central Statistics Service (BPS) data for 2022, is the second largest in the Muaragembong dynasty with an area of 36.2 km^2 of which 70% is mainly ponds and ponds. rice field. Bungin Beach is one of the tourist attractions, especially for locals living in Bekasi Regency and surrounding areas. Some people have a livelihood as farmers or fishermen. This is certainly one of the factors behind the many dynamics occurring in the coastal area of the Bekasi Regency [2].

Coastal areas around the world are highly vulnerable to threats from abrasive hazards. The Bungin Beach area is not immune to a threat either. Abrasion, also called shoreline erosion, is a process of coastal erosion caused by ocean waves and is influenced by many factors that cause abrasion, in addition to abiotic factors, such as wind, tidal waves, and the type or form of coastal sediment. Abrasion can also be caused by biotic factors caused by human activities [3]. The state of Bungin Beach is a fishing settlement area and also functions as a tourist destination, so the large number of human activities taking place in the area can increase the rate of abrasion occurring in the area. area.

Based on Law of the Republic of Indonesia Number 24 of 2007 concerning Disaster Management chapter 1 paragraph (1), explains that a disaster is an event or series of events that threatens and disrupts the lives and livelihoods of people by natural or non-natural factors, including human factors themselves, resulting in the presence of victims, environmental destruction, and damage to property/objects, as well as its influence on the psychological effects of the affected community. disaster. But the law does not include abrasion as one type of natural disaster. This may be because abrasion occurs slowly (slow onset) so it is not included in the disaster classification. In other definitions and types of disaster, that abrasion is also included in the disaster classification [4]. Considering that the event of abrasion has the potential that can damage the environment and cause loss of property, the occurrence of abrasion in an area needs to be anticipated.

In the past 2 years, the world is facing a crisis caused by the existence of a pandemic due to the Corona Virus Disease 2019 (Covid-19) spread. Indonesia is also affected due to the spread of the virus to interferes with the lives and livelihoods of the people. Then Covid-19 is also designated as a type of non-natural disaster. At first, this virus was only in urban areas. However, due to the massive spread of viruses, in the end, the whole community throughout Indonesia was affected by this virus, including the coastal area of Bungin Beach. The people in the region are very vulnerable when facing the wave of pandemics that hit Indonesia in early 2020. The main problem of coastal communities due to the implementation of activity restrictions by the Government (PSBB or PPKM) in order to reduce cases of Covid-19 infection caused the breakdown of fishermen's trading chains with the market, especially those in the Indonesian small islands areas whose economy is highly dependent on the large island area/ urban area. Then the fall in the price of fish because fishermen can only sell their catch fish in a limited area. Coastal communities suffered economic losses because they did not get adequate income during the Covid-19 pandemic.

In addition, the coastal environment on Bungin Beach was still threatened by abrasion because there was still illegal logging of mangrove forests by the local community. Some mangrove planting and rehabilitation activities organized by the government had to be stopped to suppress the pace of the spread of the virus in in the coastal community's environment. If in the coastal area there is a fairly high rate of abrasion coupled with the conditions of the shelter Muaragembong coastal topography (lowland with elevation <50) is allegedly a supporting factor for the occurrence of tidal floods (rob) and seawater intrusion into community residential environments with high frequencies [5]. Of course, the coastal community will become victims. The tidal flood will inundate the houses of the community and damage the objects in their residence. Sea water chemically has a material that is corrosive/destructive which is quite high when compared to the content of fresh water.

Considering that the most affected victims of the disaster are the community itself, it is necessary to make efforts that aim to increase the awareness and capacity of the community in anticipating the possibility of disasters that occur in the area where they live. So it is expected that disaster risk can be avoided, reduced, prevented, or even eliminated. For this reason, it is necessary to implement Community Based Disaster Risk Reduction (CBDRR) which places the community as the main subject or actor in disaster management activities and disaster risk reduction. This CBDRR activity is certainly influenced by various factors that can be a supporter of the success of these activities. One factor is the socio-cultural condition in the coastal community which usually still upholds customary values or local wisdom. Disaster management efforts through the community's local wisdom approach are the easiest way to increase community capacity in dealing with disaster threats [6]. The reason is that information or material that is often used in daily life is usually easier to understand making the local knowledge of communities available as part of efforts to reduce disasters in coastal areas.

Seeing the impact caused by these threats, it is necessary to identify and analyze the danger of abrasion to reduce the impact of losses on society and the government from these threats, especially during the pandemic. The purpose of this study was to identify the role of community groups (Tourism Awareness Group) as a key factor in overcoming abrasion events during the Covid-19 pandemic at Bungin Beach. The coastal community of Pantai Bakti Village has a Tourism Awareness Community Group (Pokdarwis/TAG) Bungin Bakti Jaya which was originally formed as a driver in the management of the Tourism Area on Bungin Beach. However, researchers want to know during this pandemic the activities that have been carried out to overcome the threat of abrasion in the Bungin Beach area, Pantai Bakti Village. As well as the extent of the impact produced by community groups in preventing the threat of abrasion at Bungin Beach, Pantai Bakti Village.

2. Theoretical basis

2.1. Coastal community

Humans have the capacity to absorb things that happen around them, then followed by analyzing and interpreting them as a result of observation or experience, which can be used to predict or even as a basis for consideration in decision-making. Local knowledge of society is a concept that refers to the knowledge possessed by a group of people/society who live in certain areas for a long time. Local knowledge of society such as fishermen who live in a coastal area is usually obtained based on experience inherited from generation to generation. Local wisdom that is integrated into norms, belief systems, or culture, then expressed in long-held traditions and norms, is called local cultural wisdom [7].

Coastal communities are people who live in coastal and marine environments and engage in resource-related socioeconomic activities. Coastal communities are highly dependent on the availability and conditions of coastal and marine resources. Coastal communities are still underdeveloped and include marginalized people. Moreover, many aspects of life are unknown to outsiders, including the characteristics of coastal societies. They have different aspects of knowledge, trust, social roles, and social structures. On the other hand, even in the face of disaster threats, coastal areas do not have many options to overcome existing problems [8].

Coastal communities have characteristics that tend to be different from other communities. Judging from the anthropological perspective, this condition is based on social facts that coastal communities have unique cultural patterns due to their interaction with the environment and the resources they contain. For coastal communities, culture is a system of ideas or cognitive systems that function as guidelines for life in society, references to social behavior patterns, and a means to interpret and interpret various events that occur in their environment. This trait is very closely related to fishery activities are highly influenced by factors such as environment, season, and market, and the characteristics of coastal areas are also affected by these factors. [9]. Culture must help the ability to survive society or adjust an individual to the environment of life. Anthropological aspects that are considered important as a builder of the identity of coastal community culture, such as gender systems, patron-client relations, patterns of exploitation of fisheries resources, and social leadership are aspects that shape the socio-cultural character of coastal communities [10].

2.2. Community-based disaster risk reduction

The CBDRR activity is based on community wisdom and is one form of local intervention in leading to civil society that has a high capacity in measuring the resilience of the community towards a disaster threat. In relation to CBDRR, disaster risk reduction through awareness raising, capacity building to deal with disaster threats, and/or application of physical and non-physical efforts by community members. Various efforts are being made to reduce active, participatory, and well-organized. This activity places the community as the main subject or actor in disaster management activities and disaster risk reduction because the community is the frontline affected by the disaster [11].

In the implementation of CBDRR activities based on the principles of disaster management that exist in Law of the Republic of Indonesia Number 24 of 2007 which includes the principles of speed and accuracy, prioritization, coordination, efficiency and effectiveness, transparency and accountability, partnership, empowerment, and non-discrimination. The principle of speed and accuracy is where disaster management activities must be carried out quickly and precisely following the demands of the situation to minimize the delay in handling which will have effects on casualty or increased casualty numbers. Priority principles prioritize disaster management efforts as a priority in maintaining the safety of the human soul. The principles of coordination in disaster management are based on good coordination and mutual support among the various departments involved. The principle of effectiveness, prioritizing disaster management activities runs effectively and as efficiently as possible. The principle of transparency, where disaster management activities are carried out openly and can be accounted for by the organizers. The principle of partnership prioritizes cooperation between individuals, groups, or organizations to achieve common goals. The principle of empowerment aims to actively involve community involvement in disaster management. The principle of non-discrimination is not discriminating against treatment based on gender, ethnicity, religion, race, and any political flow when organizing disaster management.

The National Disaster Management Agency carries out the Disaster Resilient Village program which is expected to have the ability of villages to adapt and cope with various disaster threats and recover quickly from the negative effects of disasters [12]. Based on the Head of National Disaster Management Agency Regulation No. 1 of 2012 concerning Guidelines for Disaster Resilient Villages (Destana), the Destana program was developed on the basis that disaster is a common problem and therefore all aspects and elements in the community must be involved. The community is the main actor in disaster risk reduction efforts carried out in a participatory manner by mobilizing local resources. In its implementation, it is required to be based on humanity, justice, and gender equality following the principles of disaster management in Law No. 24 of 2007.

Based on the sociocultural characteristics of coastal communities tend to live in groups. One of them is the Tourism Awareness Group (Pokdarwis/TAG). TAG is one element of the stakeholder group consisting of individuals who have a strategic role in developing and managing the region's natural and cultural wealth potential. It is hoped that the area can become a tourist destination that can provide benefits to the community. TAG acts as a driver of tourism awareness and "Sapta Pesona" in the tourism area, deepens the understanding of tourism, increases the role and participation of communities in tourism development, and makes tourism development successful. Then public awareness of the environment is very necessary, especially in dealing with the dynamics that occur on the coast including the threat of abrasion [13].

2.3. Abrasion control

The status of the coastal zone, which is the transition zone between sea and land often causes coastal areas subject to pressure from a variety of phenomena originating from both land and sea. Phenomenons such as floods or erosion and human activities carried out such as the construction of settlements, clearing forests for rice fields, pond development, and so on ultimately have an impact on coastal ecosystems. The hydro-oceanographic processes from the sea can influence coastal areas such as wave thumbs, changes in current patterns, and tidal phenomena which are also possibly be strengthened by the influence of global climate change activities [14].

Natural phenomena from land that affect changes in coastline, namely land erosion and sedimentation due to tidal currents resulting in tidal floods or changes in river flows (in the estuary). Coastal erosion or abrasion is the erosion or reduction of land (beaches) due to wave activity, currents, and tides. Land compaction causes the land surface to subside and flood, altering coastlines. Coasts are said to wear away because sediment migration occurs to larger points relative to the number of sediments transported from them [15].

One form of coastal defense against the threat of abrasion naturally is with mangrove plants. Mangroves have functions, among others, as a wave and abrasion, as an intrusion damper, as a habitat for various species, and as vegetation that can be used for humans. The coastal area ecosystem becomes stable if covered by mangrove forests. Conversely, problems will arise in the coastal area if the mangrove forest environment is damaged due to human hands or by other activities that can cause damage in the form of abrasion and coastal landslides [16]. Mangrove forests function as a sediment trap to precipitate mud in tree roots so that it can be a barrier to the entry of seawater into land areas. The role of mangrove forests is to reduce coastal disasters by reducing damage caused by waves, and damage from large storms, due to tsunamis, reducing erosion and binding soil and can compensate for sea level rise [17].

2.4. Covid-19 prevention

Based on Decree HK.01.07/Menkes/413/2020 of the Minister of Health, the Republic of Indonesia about the 2019 Coronavirus Infectious Disease Prevention and Control Guidelines (Covid-19) explains that Covid-19 is one of the infectious diseases caused by severe acute respiratory syndrome coronavirus 2 (Sarscov-2). SARS-COV-2 is a new type of coronavirus that has never been identified in humans. There are at least two types of coronavirus known to cause severe illness, including Middle East Respiratory Syndrome (MERS) and Severe Acute Respiratory Syndrome (SARS). Signs and symptoms of Covid-19 infection include symptoms of acute respiratory illness such as fever, cough, and shortness of breath. The virus has an average incubation period of 5-6 days, with a maximum incubation period of 14 days. Severe cases of Covid-19 can cause dangerous illnesses such as acute respiratory syndrome, pneumonia, and even death.

Local wisdom is very important to be applied in the prevention and control of Covid-19 because the principles of local wisdom are embedded, one of which is the existence of shared values which are the main heart of a culture in society that can run well. Local wisdom can work in three ways, namely togetherness which provides the principle of "caring for the community" as an ideological basis, and spirituality which is a manifestation of the spirit to fight the threat of civilization by turning local wisdom into collective wisdom. A moral code that places culture and religion as the basis for a "code of ethics" that ensures fair, mutual cooperation, and progress-oriented life in the emergence of protocols or Standard Operating Procedures (SOPs) in society [18].

Categorically that the Covid-19 problem includes the appearance of social change in the community, decreased income, and difficulties in terms of the economy. While the government's policies in handling Covid-19 itself, one of which is through the Social Safety Net Program (JPS), as well as preventive and promotive policies directly to the entire community. Covid-19 pandemic conditions that occur today have a serious impact on the economic and social order in Indonesia. Limitations of employment due to declining business activities due to the existence of various policies, one of them being the Massive Social Restrictions Policy (PSBB) or the enactment of Limitation of Community Activities (PPKM) [19]. This disrupts various coastal community activities both in coastal economic activities including tourism activities and community activities in preventing abrasion in their area.

3. Methodology

This research takes the form of qualitative research, which is a study conducted in the situation and condition of the object experienced with the target to get an answer and also disclose various problems regarding the tasks of Tourism Awareness Groups (Pokdarwis/TAG) in preventing the threat of abrasion during the Pandemic Covid-19 period. The informant is determined using the purposive sampling technique, namely when the initial stage of the data collected comes from people who can provide information and their views on the role of tourism awareness groups in the region. The informant in this study was the management (chairman) and members of the Bungin Bakti Jaya

Tourism Awareness Group. To strengthen the results of the study, researchers also conducted additional research through journals or various literature studies related to this study to be able to enrich the results of the analysis of this study. Also carried out measurements of abrasion based on the results of research through the Landsat satellite image to get an overview of the field conditions that are happening at Bungin Beach.

4. Results

4.1. Abrasion condition at Bungin Beach

to understand the condition of the coastal land at Bungin Beach, Pantai Bakti Village in Muaragembong District, Bekasi Regency was carried out by observing using remote sensing with satellite image from Landsat 8 OLI/TIRS and Google Earth Application. Landsat 8 satellite imagery data has a spatial resolution of 30 m with a global characteristic of 12 meters of accuracy and a 90% Circular Error (CE) confidence level which is certainly closely related to geometric accuracy [20]. Satellite imagery data were taken from 1997 to 2021 to find out how much abrasion occurred within 24 years. Satellite image data processing is carried out using QGIS v3.20 software which is then obtained by the results of abrasion in 2021 as follows in table 1 and figure 1.

Table 1. Measurement of the area of abrasion and sedimentation in the Muaragembong sub-district

Years	Meter (m)	Area (km^2)		Total
	Shoreline Position (Bungin Beach)	Abrasion	Sedimentation	
1997 – 2001	-41,66	-1,89	0,81	-1,08
2001 – 2005	-8,06	-1,47	0,45	-1,02
2005 – 2009	-33,67	-2,02	0,44	-1,58
2009 – 2013	-30,06	-9,11	0,3	-8,81
2013 – 2017	105,12	-2,14	1,26	-0,88
2017 – 2021	-20,5	-1,93	1,01	-0,92
Total	-28,82	-18,52	4,25	-14,30

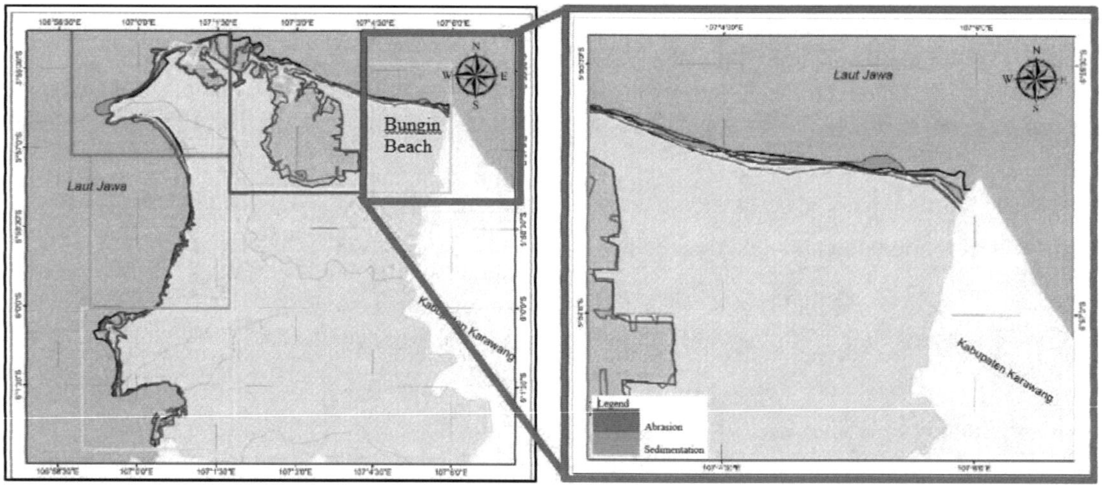

(a) (b)

Figure 1. (a) Map of abrasion and sedimentation that occurred in Muaragembong District, (b) Abrasion and sedimentation map on Bungin Beach.

The observations show that the coastal area of Muaragembong District has experienced abrasion from 1997 -2021 Muaragembong District has lost a land area of -14.30 km^2. Over the period 2017-2021, based on measurements, there was a decrease in the area of abrasion and around -1.93 $km^{2,}$ and an increase of 1.01 km^2. Planting measures implemented in 2020 failed to significantly increase the deposition area. However, the impact is quite visible, especially in the location where the planting is done because at that location the land is starting to form. In the interpretation of satellite imagery, the difference between water bodies and land can be seen from the color, such as blue for water bodies, green for areas covered by vegetation/plants, and brown for land (mud).

Abrasion that occurs is not only influenced by the dynamics of coastal areas but there is also little human intervention. In some coastal locations affected by abrasion is a pond of coastal residents who live there. But before becoming a pond, the land was previously a mangrove forest which was then converted into a pond. This change of land causes disruption of ecosystems in coastal areas. The loss of the protective balance of coastal areas causes abrasion in the region. The sea water which is now inundating the former pond area is now permanently inundating.

Sediment conditions found throughout the coastal area of Muaragembong are the result of autochthonous deposits. Residual deposits are formed as a result of the deposition process. This is caused by very intense chemical weathering, especially in tropical regions with high rainfall. Under these conditions, most rocks produce soils that lose dissolved material. Such soil morphology is called laterite soil (laterite). This type of soil is only found in and around the coastal areas of Muara Genbong, and its constituents are highly soluble in seawater, so the land is easily eroded [21].

While in several coastal locations the Muaragembong District experienced accretion or sedimentation due to its location located in the River mouth. Of course, a natural thing with a pretty good mangrove condition and also a river estuary area that carries a lot of materials and sediments carried by the river flow and then settles in the river mouth. The location of Bungin Beach is also seen experiencing abrasion so it is necessary to mitigate and prevent it so as not to exacerbate the condition of abrasion and have an impact on the survival of the people in the region. Figure 2 shows the situation at Bungin Beach.

(a) (b) (c) (d)

Figure 2. (a) River mouth at Bungin Beach (b) Mangrove condition in Bungin Beach (c) View of Java Sea from Bungin Beach (d) Condition of Bungin Beach.

Bungin Beach which is next to the Karawang Regency is separated by rivers. Although close to river mouths, it does not increase the potential for sedimentation on the coast because around the river mouth there is also no mangrove. The mouth of the river became a place for the entry of fishing vessels to be operated by the local residents. In the absence of organisms that bind sediment causing the region continues to experience abrasion. The dominant soil condition is sand so it cannot be used to plant mangroves. Efforts to overcome abrasion are carried out by making wave-retaining embankments made of sacks filled with coastal sand. Sometimes people use used tires detained with bamboo arranged along the river mouth. This is done to reduce erosion caused by the flow of river water. The absence of mangrove plants at the end of the river mouth causes the abrasion that occurs in this location to be greater than on the coast.

4.2. CB-DRR activities by Bungin Bakti Jaya tourism awareness groups

CBDRR is an approach that encourages communities to manage disaster risk at the local/lowest level. These kinds of activities require a series of efforts by interpreting the threat and risk of the disaster it faces, carrying out priority handling/reduction of disaster risk it faces, and reducing, monitoring, and evaluating its performance to reduce the disaster. But the main points of the two are the optimal implementation of the resources they have and the integral part of the daily life of the community [22].

At the community level, many community groups are tasked with disaster management efforts around their area. The purpose of using CBDRR as a disaster risk management process is to actively involve high-risk communities in researching, analyzing, managing, monitoring, and assessing disaster risk to reduce vulnerability and strengthen their capacity. A community can be a collection and order called a "Circle of Friends" (Paguyuban) with a kinship value. In other meaning, the community is also associated with social capital as a relationship between individuals, including social networks, reciprocal relationships, and mutual trust. For example, the community of one village that lives in the same geographical environment is exposed to threats and recurring disaster risks so that they have the same crisis experience. The existence of a condition of the same risk provides an opportunity for increasing a sense of the same fate.

In the scope of the Bungin Beach area, the community is in a condition where they live in an environment that faces the same threat, namely the threat of abrasion. Their region routinely experiences annual tidal floods which cause houses the closest to the sea to become submerged by tides. Conditions that occur continuously make the community must always ready to face disasters and one of the efforts is through the community or community group that had already been formed. But having many groups in a community environment tends to be less effective because it results in many overlapping interests in the community. The associations of coastal communities in the form of groups or communities in the Bungin coastal area are quite diverse. There are fishermen groups, fish farmers groups, religious groups, village groups, and so on. However, tourism awareness groups at Bungin Beach have a vital role in the community. They are the most active group in the Bungin Beach area apart from their religious group (worship and da'wah groups).

The Tourism Awareness Group (TAG) is a community group that is predicted to behave as a motivator, mover, and communicator a good way to boom the readiness and focus of the network across the vacation spot of tourist places so we can act as suitable hosts for the improvement of tourism, and feature focus of the possibilities and blessings that may be gained. Developed tourism activities to improve the economic welfare of the surrounding community [23]. The main focus of the formation of TAG is aimed at developing the economy of the community around tourist sites, another goal is to introduce, maintain and utilize the potential tourism resources that exist in each region. The word "preserving" refers to activities that aim to maintain and protect ecosystems from various threats, one of which is the threat of disasters in coastal areas, namely abrasion, and tidal flooding. Therefore, the role of the group in overcoming abrasion, apart from serving as managers of tourist areas, they are also required to maintain beach conditions so that they remain attractive to tourists.

4.3. The role of the Bungin Bakti Jaya tourism awareness group in abrasion management at Bungin Beach during the covid-19 pandemic

When the coronavirus (Covid-19) began to spread throughout the world, including in Indonesia, the Indonesian government announced the implementation of Massive Social Restrictions (PSBB) to start controlling the spread of the virus. Expanding social distancing to speed up the treatment of the coronavirus disease 2019 based on Indonesia Government Regulation No. 21 of 2020. This regulation regulates large-scale social distancing measures in response to the spread of COVID-19 by limiting activities that can lead to crowding. This allows local governments to restrict the movement of people and goods within and outside their areas as long as they have approval from the relevant ministry (Ministry of Health). The rule also states that restrictions on activities, at least during school holidays and on weekdays, include restrictions on activities in public places and/or restrictions on religious activities or facilities that may lead to crowds.

Since the emergence of regulations regarding restrictions on community activities, the Bungin Beach tourist area has been temporarily closed. This causes the disruption of several community activities on Bungin Beach which is a tourist area. People are forced to stay home and reduce contact with others to help limit the spread of coronavirus. So there is a decrease in the number of tourists who come. The Bungin Bakti Jaya Tourism Awareness Group which manages the Bungin Beach area is also affected by the regulation. Community activities that routinely monitor and plant mangroves in the context of preventing abrasion that occurs on Bungin beach are reduced. They are more focused on meeting the needs of daily life during the enforcement of Mass Social Restrictions (PSBB) or Enforcement of Restrictions on Community Activities (PPKM). Most of the people in Bungin Beach, especially those who are members of tourism awareness groups, live as fishermen and pond farmers. During this crisis condition, they are quite difficult to sell their fishery products. Due to the Covid-19 pandemic in the Bungin Beach area, the economic sector has a greater impact on the community than the health sector. However, there are still other sectors that threaten the community at Bungin Beach, namely threats to the environmental sector originating from tidal flooding and coastal abrasion.

In October 2020, the Ministry of Environment and Forestry of the Republic of Indonesia organized a program known as the Intensive Work Program (Program Padat Karya). This is based on Decree No. 353/MENLHK/SETJEN/DAS.1/8/2020 of the Minister of Environment and Forestry of the Republic of Indonesia about the Mangrove Planting Intensive Operational Plan for 2020. Economic activity in countries around the world, including Indonesia affected by the Covid-19 pandemic. The government in this regard has taken extraordinary policies and steps in the area of finance to save health and the national economy, and the National Economic Recovery Program (PEN) is one of them. The National Economic Recovery Program is a form of policy response taken by the government to maintain and prevent business activities from getting worse. In addition, the program is also expected to accelerate the national economic recovery and support state financial policies. One of the PEN programs implemented by the Ministry of Environment and Forestry (KLHK) is the Mangrove Planting Intensive activity.

Indonesia has a distribution of mangroves covering an area of 3,311,207 ha, of which 637,624 ha are in critical condition. Mangrove forests are one of the sources of livelihood for coastal communities who during this pandemic felt the most significant impact of the economic downturn. Therefore, through the Mangrove Planting Intensive Work activity, it is hoped that it can become an economic stimulus for the community around the mangrove ecosystem and at the same time accelerate the national economic recovery. This Mangrove Planting-Intensive Activity, apart from being an effort to accelerate national economic recovery, is also expected to be part of corrective measures in the 2019-2024 Working Cabinet era to make real and shape the future. This is done through efforts to control the physical impact of climate change, use sustainable economics, take sides with the community with social and community forestry as development drivers, strengthen institutional capacity at site and group levels, and safeguard ecosystems through rehabilitation/planting of mangroves.

The Bungin Bakti Jaya is the executor of the planting activities in the field. Planting implementers are community groups that carry out labor-intensive mangrove planting activities according to a simple technical design. The types of mangrove seedlings planted were ringworm (*Rhizophora mucronata*), ringworm (*Rhizophora stylosa*), Pidada (*Sonneratia caseolaris*), and api-api (*Avicennia sp.*) mangroves with a total number of 2500 seedlings to be planted. The distance between seedlings is 1x1 meter so the area used as a planting location is about 2.5 km². The location chosen is in the north of Pantai Bakti Village, Muaragembong District, Bekasi Regency, precisely in the Bungin Beach tourist area, and is about 500 m from the tourist entrance to the west. The location selection is based on the sediment condition in the form of mud and is in the tidal zone so that it is suitable for mangrove planting activities and has a high chance for the mangroves to survive. The activities of the Bungin Bakti Jaya in implementing mangrove planting are documented in Figure 3.

(a) (b) (c)

Figure 3. (a) Making planting site; (b) Planting implementation team (Bungin Bakti Jaya TAG); (c) Mangrove planting process.

Apart from the Mangrove Planting Intensive Activities organized by the Ministry of Environment and Forestry, the Bungin Bakti Jaya Group also regularly conducts independent nurseries and plantings. Bungin Beach is one of the local tourist destinations in Bekasi Regency. So the program owned by the Tourism Awareness Group aims to develop tourism in the area. According to Rahim (2012), the presence of Pokdarwis/TAG in relation to the development of tourist destinations serves as one of the driving forces to help create a favorable environment and atmosphere at the local level of the region, which together has a positive impact on the development of tourist destinations in the context of the wider region. Pokdarwis/TAG's role and contribution must be continuously supported and expanded both qualitatively and quantitatively to support the development and growth of the destination and especially to strengthen the role of municipalities in the tourism development of the region [23].

Mangrove forests in Indonesia can provide benefits to local communities as a buffer for their livelihoods [24]. Local communities harvest different types of fish from the mangrove ecosystem to provide income and food for their families. Locals still use mangrove forests for firewood, housing, furniture, and other building materials. Some fishermen use products from mangrove forests in making anchors, boat balancing joints, and coloring fishing nets. Products from these mangrove forests are harvested on a small or large scale, then contribute to local livelihoods and exports on a national scale. Visible benefits of mangrove systems include social and cultural functions. Mangrove forests support religious and spiritual values as well as aesthetic and recreational values for ecotourism activities.

Overall, tourism awareness groups do not have a direct relationship with disaster management activities. However, the Bungin Bakti Jaya Group has indirectly participated in overcoming the threat of abrasion in the Bungin Coastal area. The purpose of the group's mangrove planting is to preserve nature in the coastal area so that the natural ecosystem there is maintained. An abrasion is a form of threat that can damage the beauty and ecosystems in coastal areas. The impact will affect the economy of communities around tourist destinations if the condition of the tourist beach is damaged due to abrasion. Especially during this pandemic, the number of local tourists has decreased, so the community's economy is slightly disrupted. However, this pandemic status will not last forever. So the role of Tourism Awareness Group as the manager is to keep the beach condition in good condition so that when the pandemic begins to subside, Bungin beach does not lose its charm.

5. Conclusion

Based on the results of coastline measurements, it can be seen that during 2017-2021 there was a decrease in the land area due to abrasion, which was around -1.93 km^2 and increased by 1.01 km^2. The planting measures implemented in 2020 were not able to significantly increase the deposition area. However, the impact is quite visible, especially at the planting site because the soil is starting to form in that location. Abrasion events that occur in the Bungin Beach area can be prevented through the active role of the local community. The role of the Bungin Bakti Jaya tourism awareness group to overcome the threat of abrasion on the coast of Bungin Beach is of them through planting and monitoring mangroves. The existence of the National Economic Recovery Program (PEN) organized by the Ministry of Environment and Forestry during the pandemic period was enough to help the coastal community of Bungin Beach during the implementation of the PSBB. Mangrove planting activities carried out during the Covid-19 pandemic aim to restore the economy of the affected coastal communities. In addition, mangrove planting activities also prevent abrasion that still occurs during the pandemic. Although the role of the tourism awareness group is not directly related to disaster management activities, the planting activities carried out by these community groups to preserve the mangrove ecosystem are closely related to efforts to prevent abrasion in the coastal area of Bungin Beach. Mangrove planting activities also aim to maintain local tourism potential, so that when the pandemic ends, the economy of communities around tourist destinations can recover quickly.

Acknowledge

I would like to thank those who supported the researchers in launching this scientific article. To Andalas University and the Republic of Indonesia Defense University for the implementation of ICDM3 in 2022. To the Bungin Bakti Jaya Tourism Awareness Group for their willingness to provide the data needed for this research. To Dr. IDK Kerta Widana as Secretary of the Disaster Management Study Program who has supported this research so that it could be presented at a conference. To Professor Sobar Sutisna and Mr. Makmur Supriyatno as author supervisors. I hope this research can be a source of reference to support CBDRM in the coastal regions of Indonesia.

References

[1] Putranto A 2020 Tipologi, Dinamika, dan Potensi Bencana Alam Di Kawasan Pesisir Kabupaten Tulungagung *EKOLOGIA: Scientific Journal of Basic and Environmental Sciences* **20** 14–23

[2] Bekasi Regency Central Bureau of Statistics 2022 *Kecamatan Muara Gembong Dalam Angka 2021* (Bekasi: Bekasi Regency Central Bureau of Statistics)

[3] Wisyanto 2019 Analisis Bahaya Abrasi Di Wilayah Kabupaten Banggai Kepulauan *Jurnal ALAMI: Journal of Disaster Risk Reduction Technology* **3** 21–31

[4] Mildawati M, Aritonang R T E Y, Pradana A G B, Roswanto, Hastuti S, Suarjat J, Nastiti A D, Kridanta S J, Nurpatria I and Winarso U T 2018 *Modul DIKLAT Dasar Manajemen Bencana* (Jakarta: National Disaster Management Authority)

[5] Maulani A, Taufiq-SPJ N and Pratikto I 2021 Perubahan Lahan Mangrove di Pesisir Muara Gembong, Bekasi, Jawa Barat *J. Mar. Res.* **10** 55–63
[6] Herlina M 2019 *Kearifan Lokal Untuk Mitigasi Bencana Pada Masyarakat Rawan Bencana Gempa, Tsunami, Longsor, Banjir Di Kabupaten Pesisir Barat Provinsi Lampung* (Semarang: Universitas Negeri Semarang)
[7] Zamzami L and Hendrawan 2011 Kearifan Budaya Lokal Masyarakat Maritim Untuk Upaya Mitigasi Bencana Di Sumatera Barat *Andalas University Scientific Articles* **1** 37–48
[8] Fatmasari D 2016 Analisis Sosial Ekonomi Dan Budaya Masyarakat Pesisir Desa Waruduwur, Kecamatan Mundu, Kabupaten Cirebon *Scientific Article Syekh Nurjati State Islamic Institute of Cirebon* **1** 144–66
[9] Sidiq S S 2019 *Sosiologi Masyarakat Pesisir: Buku Ajar* (Pekanbaru: Taman Karya)
[10] Kusnadi 2010 Kebudayaan Masyarakat Nelayan *JELAJAH BUDAYA: Center for Preservation of History and Traditional Values, Ministry of Culture and Tourism of the Republic of Indonesia* **1** 1–9
[11] Ministry of Public Works of the Republic of Indonesia 2013 *Pedoman Teknis Pengurangan Risiko Bencana Berbasis Komunitas (PRB-BK)* (Jakarta: Directorate General of Human Settlements, Ministry of Public Works of the Republic of Indonesia)
[12] National Disaster Management Agency 2022 *Tentang Pengelolaan Risiko Bencana Berbasis Komunitas* (Jakarta: Directorate of Preparedness of the National Disaster Management Agency)
[13] Putrawan P E and Ardana D M J 2019 Peran Kelompok Sadar Wisata (Pokdarwis) Dalam Pengembangan Pariwisata Di Desa Munduk Kecamatan Banjar Kabupaten Buleleng *Locus Scientific Magazines, Faculty of Social and Political Sciences, Panji Sakti University* **11** 40–54
[14] Pinto Z 2015 Kajian Perilaku Masyarakat Pesisir yang Mengakibatkan Kerusakan Lingkungan (Studi Kasus di Pantai Kuwaru, Desa Poncosari, Kecamatan Srandakan, Kabupaten Bantul, Provinsi DIY) *JWL: Regional and Environmental Journal* **3** 163–74
[15] Damaywanti K 2013 Dampak Abrasi Pantai terhadap Lingkungan Sosial (Studi Kasus di Desa Bedono, Sayung Demak) *Proceedings of the 2013 National Seminar on Natural Resources and Environmental Management* **1** 363–7
[16] Pratiwi N A and Alhadi Z 2021 Efektivitas Pelestarian Mangrove Sebagai Sarana Pengurangan Resiko Bencana Di Kota Pariaman *PGDE: Journal of Policy, Governance, Development and Empowerment* **1** 18–29
[17] Soedarmo S P K 2018 *Pelestarian Hutan Mangrove dan Peran Serta Masyarakat Pesisir* (Jakarta: Ministry of Environment and Forestry of the Republic of Indonesia)
[18] Rahmawati Y, Santaufanny F F A, Hati E M, and Roziqin A 2021 Kearifan lokal dalam menghadapi COVID-19: Studi kampung tangguh di Jawa Timur *Jurnal Masyarakat Indonesia* **47** 1-12.
[19] Ikmal N M and Noor M 2021 Kebijakan Pemerintah Indonesia Dalam Penanganan Covid-19 *Journal of Research and Development of Central Java Province* **19** 155–66
[20] Dianovita and Mahendra R 2014 Kajian Ketelitian Geometri Citra Landsat 8 Level 1T *Proceedings of the 2014 National Seminar on Remote Sensing* **1** 254–60
[21] Nugraha R B A, Syaharani L, Iska R, Mulyana D, Wahyudin Y, Purbani D, Jayawiguna H, Triyono, Setiawan A and Fajar P 2019 The impact of land used changes on mangrove forest and shoreline dynamic in Muara Gembong, Bekasi, West Java *IOP Conf. Ser.: Earth Environ. Sci.* **241** 012-018
[22] Lassa J, Eko Teguh Paripurno, Jannah N M, Pujiono P, Magatani A, Pristianto H, Sudira C, and Parlan H 2014 *Community Based Disaster Risk Management (CBDRM) Guide* (Jakarta: Indonesian Disaster Management Society (MPBI))
[23] Rahim F 2012 *Buku Pedoman Kelompok Sadar Wisata* (Jakarta: Ministry of Tourism and Creative Economy of the Republic of Indonesia)

[24] Murdiyarso D, Purbopuspito J, Kauffman J B, Warren M W, Sasmito S D, Donato D C, Manuri S, Krisnawati H, Taberima S and Kurnianto S 2015 The potential of Indonesian mangrove forests for global climate change mitigation *Nature Climate Change* **5** 1089–92

IOP Conf. Series: Earth and Environmental Science 1173 (2023) 012069 doi:10.1088/1755-1315/1173/1/012069

Potential debris flow after earthquake in Mount Talamau Pasaman district and West Pasaman district

F Bari[1], B Istijono[1], R Yuhendra[2], A Hakam[3], M Noer[1] and T Ophiyandri[3]

[1] Post Graduate School, Universitas Andalas, West Sumatra, Indonesia.
[2] Water Resources Development Provincial Office of West Sumatra Province, Indonesia.
[3] Department of Civil Engineering, Universitas Andalas, West Sumatra, Indonesia.

E-mail: bistijono@eng.unand.ac.id

Abstract. The 6.1 Richter Scale earthquake on February 25, 2022 in Pasaman district and West Pasaman district caused landslides around Mount Talamau with an avalanche outcrop of 1,200 hectares, resulting in debris flow in a number of tributaries in rivers of Bt. Nango, Bt. Lampang, Bt. Timbo Abu, Bt. Tobi, Bt. Sopan, Bt. Muaro Tayo in Talamau subdistrict, West Pasaman district and the rivers of Bt. Haluan, Bt. Tampa, Bt, Malampah, Bt. Kilangan in Malampah subdistrict, Pasaman district. The potential for debris flow can still occur when the rainfall intensity exceeds 50 mm/day with a discharge of 107 m^3/sec and a debris volume of 5,000,000 m^3 in one of 10 tributaries around Mount Talamau. The estimated total volume is estimated at 24 million m^3 currently deposited in the sediment zone of 1,200 hectares of landslide outcrop area.

1. Introduction

On Friday, February 25, 2022, there was an earthquake centered in West Pasaman district. The foreshock earthquake occurred at 08:35:51 WIT with the epicenter on land at coordinates 0.14° north latitude and 99.99° east longitude, with a magnitude of 5.2 at a depth of 10 km, 18 km east of the Indian ocean. The United States Geological Survey released that the earthquake occurred at coordinates 99.984° east longitude and 0.222° north latitude, magnitude (M 5.0) at a depth of 10 km. The main earthquake occurred on Friday, February 25, 2022 08:39:29 WIT with the epicenter on land at coordinates 99.98° east longitude and 0.15° north latitude, about 17 km northeast of West Pasaman district, with a magnitude of M 6,2 at a depth of 10 km. After the main earthquake, there were several aftershocks of lower magnitude [1]. The Pasaman earthquake killed 10 people, injured 465 people and damaged 3,312 houses. In addition, this earthquake also displaced about 16,000 people [2].

The earthquake that occurred has caused landslides around Mount Talamau, Malampah subdistrict, Pasaman district and Talamau subdistrict, West Pasaman district. The tributaries around Mount Talamau are included in the Masang-Pasaman watersheds, the rivers namely: Bt. Haluan, Bt. Tampa, Bt. Malampah, Bt. Kilangan in Malampah subdistrict, Pasaman district and the rivers of Bt. Nago, Bt. Lampang, Bt. Timbo Abu, Bt. Tobi, Bt Sopan, Bt Muaro Tayo in Talamau subdistrict, West Pasaman district, shown in figure 1.

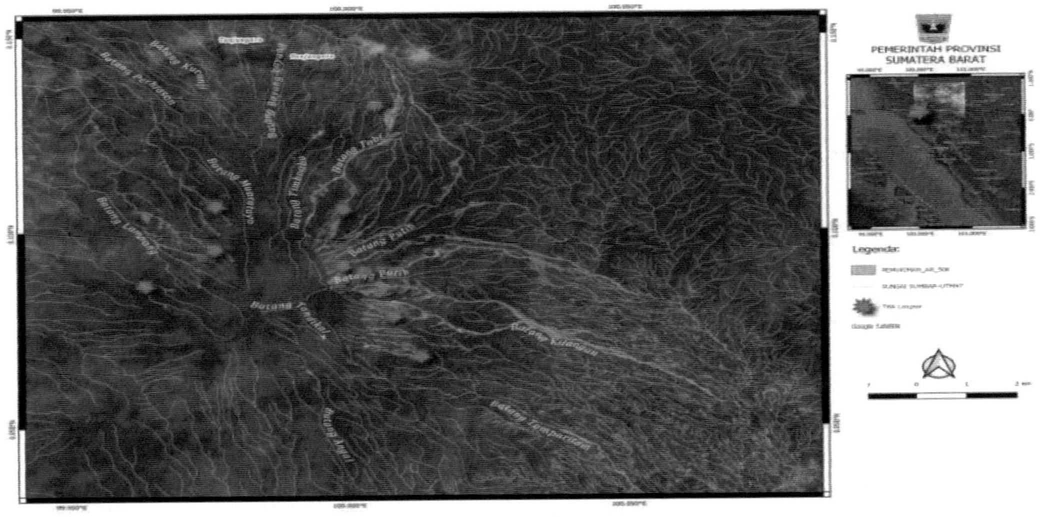

Figure 1. Study area.

2. Methodology

This study was conducted by:

a. Literature review, describing globally the theories and basics of analysis that will be used in solving existing problems, both to analyze the factors and supporting data as well as technical analysis of the causes and potential for debris flow around Mount Talamau.

b. Collecting primary data (photogrammetric field identification) and secondary data; Sentinel Image data can be downloaded at https://earthexplorer.usgs.gov/ Mount Talamau position 0.0500 north latitude - 0.1500 north latitude; 99.9500 east longitude – 100.050 east longitude as of March 4, 2022, March 24, 2022 and March 29, 2022; Data Digital Elevation Model AP_26031_FBD_F7190_RT1.dem 12m resolution (ALOS JAXA).

c. Determine the factors causing debris flow and analyze the slope instability mapping and the potential for debris flow.

d. Mapping potential slope stability areas. Modern methods of slope stability use rainfall data, geological maps, digital terrain data, and geographic information systems in the estimation and provisional prediction of the distribution of potential slope instability. Slope stability mapping method ranges from pure empirical, empirical-probability basis, and analytical, the results of which are described in the geographic information system base-map.

3. Literature review

3.1. Definition of galodo / debris flow

Galodo/debris flow is a natural phenomenon of sediment movement/transportation above the slope of the ground surface which is divided into two (2) types, namely: the influence of gravity and other mass movements which are influenced by fluid dynamics forces and their movements such as the movement of individual particles.

The movement of individual particles consists of bed load, suspended load and wash load in river discharge streams, deserts or sediments on the coast. The movement of mass sediment transport consists of: a. erosion and landslides; b. debris avalanches (cliffs); c. pyroclastic flows (volcanic eruptions/lava), and d. debris flow (*galodo*)

In this case, the term is a sedimentary disaster where all particulate matter comes from clay soil types and boulder rocks.

There are four basic geophysical mechanisms for massive sediment movement, as follows:

a. Landslide (landslips): the collapse of a relatively dry land surface mass with rock types and their combination due to the force of gravity.
b. Cliff avalanche (debris avalanches): sudden and rapid movement of mixed soil and rock on steep slopes.
c. Pyroclastic flow (pyroclastic flow): lava flow due to volcanic eruptions
d. Debris flow: flow that occurs in particles that spread in water or move together to flow in large volumes even at a slope of 30 (tends to be flat).

Takahashi (2001) classifies subaerial mass flow consisting of material grains (various dimensions of particles from sandy to mountainous blocks.) Subaerial mass sediment movement which is closely related to sediment disasters is divided into fall, flow and shear (slip).

The movement of sediment masses sometimes triggers serious natural disasters. The phenomenon of this destructive force has a large impact on human volume (magnitude) and mobility and is very important for sediment disaster mitigation.

3.2. Debris flow and its causes

Debris flow is a rapidly descending mass movement of muddy slopes and high density in alluvial fan rock types. Debris flow often occurs due to the process of mass movement of sediment from steep slopes to the river network system in each steep mountain area with the influence of the time of occurrence of rain intensity [6].

The most dangerous sediment disasters are debris flows and cliff avalanches with high flow velocity, impact force, and river channel length, which are very difficult to predict. This is the cause of the avalanche that began with the 6.1 Richter Scale earthquake on February 25, 2022 at Mount Talamau, Pasaman and West Pasaman District. So that the circumference of Mount Talamau experienced an avalanche outcrop of 1,200 ha. The impact of debris flow results in human losses, damage to housing, roads, irrigation networks, clean water, agriculture, livestock and forests.

The division of the debris flow path is divided into three zones, namely: the initiation zone; transport zone; deposition zone. In general, the debris flow area is at the beginning of the 200-450 slope zone, while the slopes > 450 have a thin layer of soil (discontinuous) that is prone to sliding. The initial sliding magnitude (volume) varies widely around 10 m^3, which will trigger a major debris flow. In addition, in the final phase, the debris flow expands into large rock sliding (debris avalanches). Avalanches debris triggers debris flow in the transport zone at the upstream of the river (stream channel). In certain cases, at the beginning of the sliding it can inhibit the flow of river discharge (natural dam / erosion dam). So that rapid erosion hits and causes the formation of debris flow waves in the upstream of the river. In the transport zone, the slope of the debris flow is around 100, which consists of non-eroded riverbeds (bedrock) or on eroded cliffs/riverbeds that flow up to the deposit zone, as shown in figures 2, 3 and 4.

Figure 2. Starting zone of debris flow Bt. Kilangan river, Malampah subdistrict, Pasaman District. Source: Water Resources Development Provincial Office of West Sumatra.

Figure 3. Transport zone of Bt. Timbo Abu River, Talamau subdistrict West Pasaman district. Source: Meteorology, Climatology, and Geophysical Agency Padang Panjang.

Figure 4. Bt. Timbo Abu River deposit zone, Talamau subdistrict, West Pasaman district. Source: Meteorology, Climatology, and Geophysical Agency Padang Panjang.

Figure 5. Slope instability after the earthquake upstream of the Bt. Timbo Abu River, Talamau subdistrict, West Pasaman district. Source: Meteorology, Climatology, and Geophysical Agency Padang Panjang.

3.3. *Slope instability*
Slope instability is usually caused by debris flow events (figure 5). In the area of unconsolidated material layers, the riverbed (bedrock) undergoes weathering of the soil layer.

Shallow landslides separate and condense into debris flows, which are thin, longitude and irregular in shape from various viewpoints that are several meters wide and long and only slightly parallel to the longitude axis. At the edges of the thicker erosion, it shows that the mass of erosion is the shape of the lens in the cross-section of the upstream and downstream. Several prominent tree roots were exposed at the bottom of the collapsed surface; this shows the roots visible from the edge of the avalanche outcrop. The erosion area layer consists of sediments of small rocks, bedrock, boulder, cobble mixed in a fine-grained sandy arrangement.

The contribution of rain of water infiltration affects the slope instability of Mount Talamau, sliding occurs around the peak and valley slopes (figure 5) with shallow sliding on steep slopes. The increasing thickness of the soil and the incorporation of groundwater often have an effect on the instability of the slopes around the valley. Although the seismic effect is not a separate factor for the occurrence of landslides, the strength of the earthquake influence can also be the cause of slope instability and have an impact on shallow erosion that drives the flow of debris or debris avalanches [7].

4. Results and Discussion

4.1. *Result*
Identification of the total erosion outcrop covering an area of 1,200 hectares along Mount Talamau with a sediment thickness of 5 – 15 m on the rivers of Bt. Haluan, Bt. Tampa, Bt. Malampah, Bt. Kilangan in Malampah subdistrict, Pasaman district and the rivers of Bt. Nango, Bt. Lampang, Bt. Timbo Abu, Bt. Tobi, Bt. Sopan, Bt. Muaro Tayo in Talamau subdistrict, West Pasaman district.

The results of the identification of field photogrammetric analysis with drones and comparisons before the occurrence of debris flow resulted in sediment thickness reaching 5-15 meters.

The area of debris flow deposits in the river of Bt. Kilangan, Malampah subdistrict, Pasaman district is about 100 hectares with an average thickness of 5 m, with an estimated debris volume of 5,000,000 m^3 of sediment, from the resulting outcrop area (debris avalanches) of about 140 hectares.

Estimated total volume of avalanche outcrop area of 1,200 hectares with a collapse thickness of 2 m is estimated at 24 million m^3 currently deposited in the sediment zone, as shown in figures 6, 7 and 8.

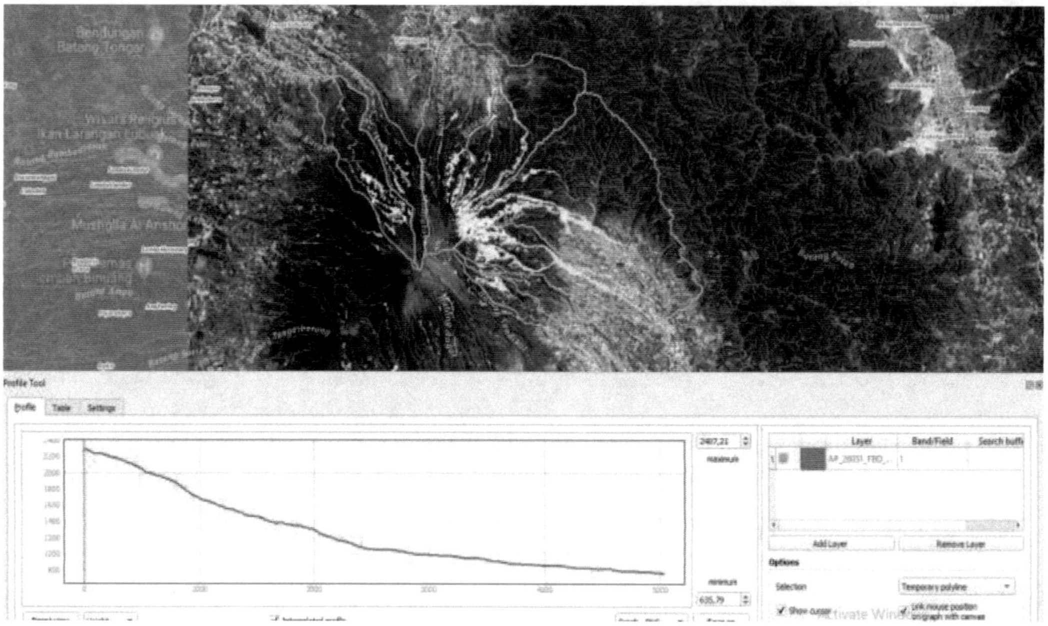

Figure 6. Slope of the Bt. Timbo Abu River, Talamau subdistrict, West Pasaman district.

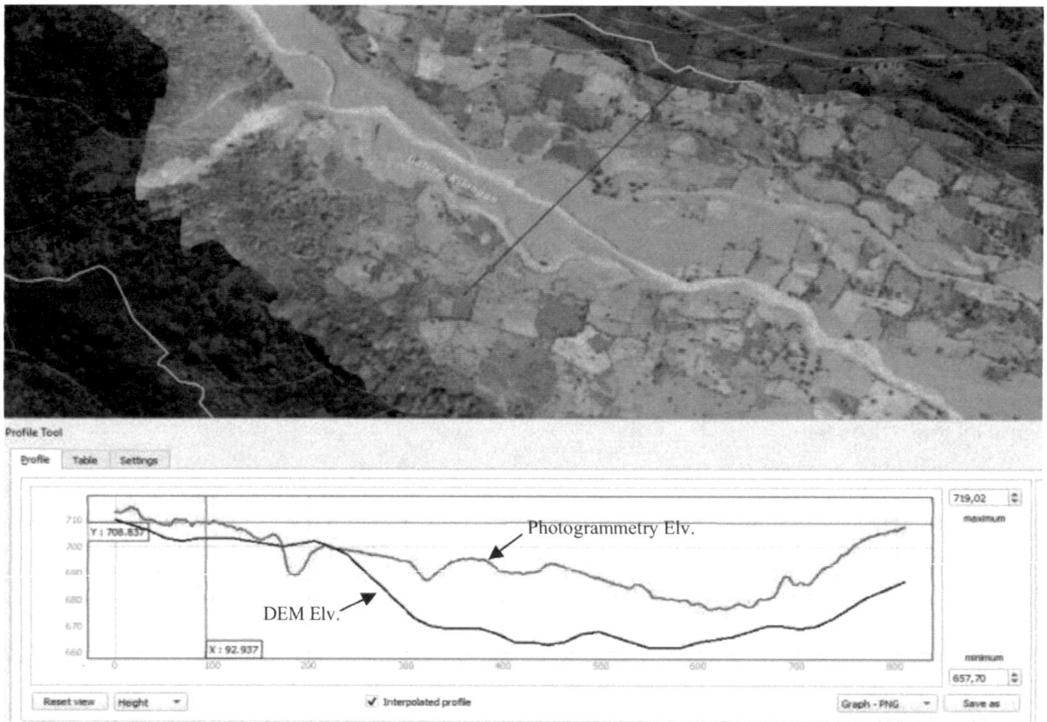

Figure 7. Comparison of the thickness of the debris flow deposits in Bt. Kilangan river, Malampah subdistrict, Pasaman district. Source: Water Resources Development Provincial Office of West Sumatra, captured March 5, 2022.

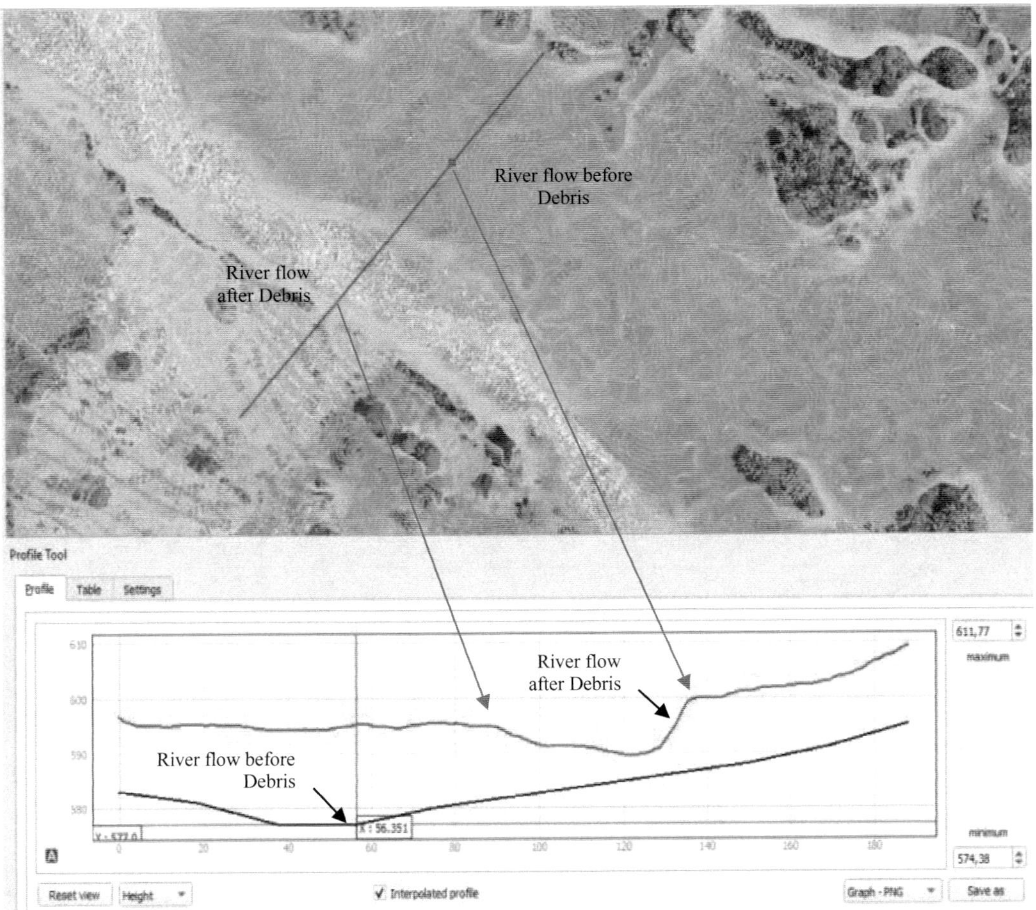

Figure 8. Contour lines per 25 cm in the Bt. Kilangan zone, Malampah subdistrict, Pasaman district, (100.03525 east longitude; 0.07498 north latitude).

Table 1. Rain intensity and river discharge Bt. Kilangan.

No	Return Period (Year)	Empirical Method	Method of Limit Hydrograph Synthesis (HSS)
		Rational	Nakayasu
1	2	68.524	25.381
2	5	85.324	31.173
3	10	95.709	34.753
4	25	107,078	38.673
5	50	117.016	42.099
6	100	125.889	45.158
7	200	134.378	48.085

Table 2. Characteristics of the Bt. Kilangan watershed.

Characteristic	Value	Unit
Water Shed (km²)	7.1	km²
Main River Length	7.5	km
Net Rain (Effective), Ro (mm)	1.0	mm
Hydrograph Parameter (α)	1.474	
Flow Coefficient	0.70	
Time Lag, tg -> L < 15	0.861	hour
tr= (0.5 sd. 1.0) tg -> 1	0.861	hour
Tp = tg+0.8*tr	1.549	hour
T 0.3= a*tg	1.269	hour
0.5.T0.3	0.634	hour
1,5.T0.3	1.903	hour
2.T0.3	2.537	hour
Tp+T0.3	2.817	hour
Tp+T0.3+1.5T0.3	4.720	hour
Qp= A*Ro/(3,6*(0.3*Tp+T0.3))	0.794	m³/s/mm

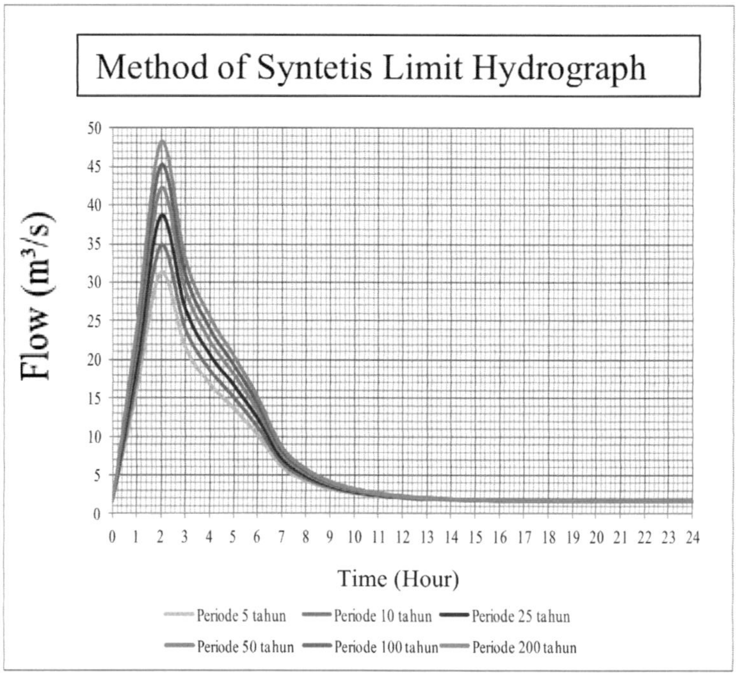

Figure 9. Synthesis unit hydrodraft of nakayasu.

Based on the above calculations (tables 1 and 2), one of them is Bt. Kilangan River with obtained results:

a. The volume of flood discharge for a period of 25 years is 107.07 m^3/s (Rational) and 38.67 m^3/s (Nakayasu), as shown in figure 9
b. The sedimentation volume of the sediment zone debris is about 5,000,000 m^3. Estimated total volume of avalanche outcrop area of 1,200 hectares with a collapse thickness of 2 m is estimated at 24 million m^3 currently deposited in the sediment zone.

4.2. Discussion

4.2.1. Causes of debris flow. Based on the sentinel image map dated March 29, 2022, there is an avalanche outcrop on Mount Talamau that is exposed to an area of 1,200 hectares. The potential threat of debris flow is still very active when the rainfall concentration reaches more than 50 mm/day at the peak of Mount Talamau. Factors that are susceptible to landslides in mountain areas are included in the high-risk category (figure 10).

The 6.1 Richter Scale earthquake on Mount Talamau, which is also a high ground movement area, is the main cause of debris flow.

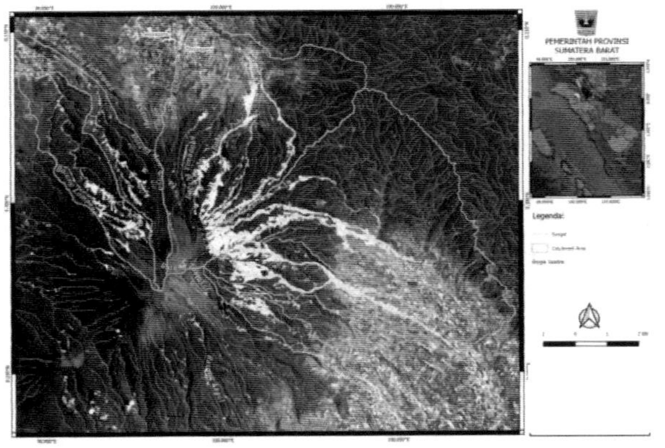

Figure 10. Map of potential avalanche debris flow of Mount Talamau.

Figure 11. Results of debris flow simulation in Bt. Kilangan River, Malampah subdistrict, Pasaman district.

4.2.2. Debris flow mitigation efforts. The simulation results on the pattern of debris flow that threatens residential areas, roads and bridges and agricultural areas when the rainfall intensity has been more than 10 mm/hour hastened the evacuation of settlements, because the potential debris flow velocity is 28 m/sec, to reach a distance of 7 km takes 4.3 minutes destroying the bridge on Bt. Kilangan River, as shown in figure 11. This is a reference that the characteristics of the tributaries of Mount Talamau which have steep slopes and the density of the river network is prone to sediment disasters.

Structural efforts require sandpocket buildings, checkdams/sabodams around Mount Talamau, but a combination with efforts to restore open land after the earthquake is needed to reduce water runoff in watersheds around Mount Talamau.

5. Conclusions and recommendations

5.1. *Conclusions*

Based on the analysis and discussion above, it can be concluded as follows:

a. The cause of the debris flow was triggered by the 6.1 magnitude earthquake and landslides on the slopes of Mount Talamau.
b. The volume of debris in one Bt Kilangan watershed reaches 5,000,000 m^3 in the deposit zone from the production of debris avalanches covering an area of 140 hectares.
c. For 10 watersheds around Mount Talamau, the total area of the erosion outcrop is 1,200 hectares, the average 2-meter soil collapse is estimated to be 24 million m^3 in the sediment zone.

5.2. Recommendations

Structural efforts to develop sandpockets, checkdams/sabodams in 10 tributaries that have the potential for debris and erosion flows, this is the zone of high ground movement and the Semangko fault can again trigger *galodo*/debris flow.

Efforts to restore open land erosion outcrops in the transportation zone and sediment zone in order to reduce the flow velocity of 28 m/sec with a steep slope in the watershed area around Mount Talamau.

Acknowledgments
We would like to thank the Universitas Andalas for the Indexed Publication Research Grant 2022.

References

[1] *The earthquake that shook West Pasaman occurred 15 aftershocks, National media Tempo*, https://nasional.tempo.co/read/1564607/gempa-guncang-pasaman-barat-terjadi-15-kali-gempa-susulan
[2] *The West Pasaman Earthquake triggered by the Sumatran Fault National, Meteorology and Geophysics Agency*, Media Liputan6.com (2022-02-25). *BMKG* Sebut Gempa Pasaman Barat Dipicu Sesar Besar Sumatera. liputan6.com
[3] Sieh K 2007 The Sunda megathrust - Past, present and future *J. of Earthq. and Tsu.* **1** 1-19
[4] *Pasaman Info, Pasaman Regent Conveys Condolences, 14 People Declared Dead,* (in Bahasa: Bupati Pasaman Sampaikan Dukacita, 14 Orang Dinyatakan Meninggal. www.pasamankab.go.id) Opened 2022-03-08
[5] Takahashi T 2006 *Mechanisms of sediment runoff and countermeasures for sediment hazards*, Kinmirai Sha: 420p (in Japanese)
[6] Rapp A 1960 Recent Development of Mountain Slopes in Kärkevagge and Surroundings, Northern Scandinavia *Geografiska Ann* **42** 65-200
[7] Jibson R W, Harp E L and Michael J A 2000 A method for producing digital probabilistic seismic landslide hazard maps *Eng. Geo.* **58** 271-289

Analyzing the impact of community-based approaches on disaster preparedness to the risk of tropical cyclone induced flooding in Chimanimani District, Zimbabwe

W Rangarirai[1], C Marnani[1] and A Rahmat[2]

[1]Disaster Management Study Program, Faculty of National Security, Republic of Indonesia Defense University, Jl. Salemba Raya No. 14, Jakarta Pusat, Jakarta 10430, Indonesia
[2]Indonesian Institute of Sciences, Republic of Indonesia

E-mail: rangawaini@gmail.com

Abstract. The rate of recurrence and intensity of hydro-meteorological disasters affecting Zimbabwe is increasing, leading to severe butterfly impacts on socio-economic development. This paper aims at reporting the findings of the case study on the impact of Community-based approaches to disaster preparedness to the risk of tropical cyclone induced flooding in Chimanimani district, Zimbabwe. The findings might help bridge gaps in community preparedness and future response to the risk of tropical cyclone flooding, thus reducing vulnerability and exposure. Primary data for research was obtained through online semi-structured interviews. Seventy-eight (78) research participants were engaged for a deeper understanding of the impact of CBA on disaster preparedness. Secondary data was gathered through document analysis by perusing, evaluating and reading summaries and conclusions from previous research work, books and journals. The Delphi model was used to measure the impact of community based approaches to disaster preparedness. The major study finding was that community-oriented training has a significantly high impact on individual level of disaster preparedness. The study recommended policy makers to propagate the implementation of community oriented disaster management training programs as the most prudent strategy to reduce the impacts of tropical cyclone induced floods in Chimanimani district, in Zimbabwe.

1. Introduction

Since 2000, the frequency of tropical cyclones originating from the Mozambique Channel Island affecting Zimbabwe has been increasing, often leading to severe flooding, droughts and other forms of hydro-meteorological disasters. The intensity and frequency of climate change related disasters in the Southern African region, Zimbabwe included, is projected to increase in the future, attributable to global warming induced extreme weather events for example, cyclonic activity in the Indian Ocean [1, 2]. Again, due to climate change, several places are anticipated to experience significant variations in rainfall frequency and patterns, prompting need for greater preparedness in-order to mitigate the impacts [3]. This is also true for Zimbabwe where the trend of cyclone induced floods has continued to increase until the present day with most recently, Cyclone Idai hitting Manicaland Province in 2019 leading to 347 deaths, 17 608 displacements and extensive damage to infrastructure valued at about $1,1 billion. Therefore, the need for community based hazard mitigation measures and reducing vulnerability among communities in Zimbabwe becomes a priority issue to minimize casualties and losses caused by tropical cyclone induced flood disasters.

The huge after-effects and scale of floods in Zimbabwe can partly be attributable to lack of disaster preparedness measures and limited capacity to effectively counter increasing community vulnerability, [4]. More than often, inadequate resources, policy gaps and a reactive as opposed to a responsive approach to disaster response have been coined as some of the major factors leading to the severity of disaster impacts in Zimbabwe. Additionally, communities living in disaster prone areas lack up-to-date disaster risk information and their response has often been very slow and uncoordinated, yet scientific predictions conclude that climate change induced hydro-meteorological hazards are increasing in intensity and frequency. This is also true of the risk of tropical cyclone induced flooding in Chimanimani district. The district's proximity to the Mozambican Island Channel where tropical storms are formed and its location along the cyclone path renders it highly prone to the risk of future flooding. More than often and in the past, little or no tangible corresponding disaster mitigation measures have been proffered, added to limited government capacity and other systemic shortfalls. To this end, implementing community oriented disaster response mechanisms become one solid option.

Many studies show that community oriented approaches have been effective in enhancing preparedness and mitigating vulnerability. This is so because in most instances, affected communities are the first to respond to disasters, and when effective and efficient, have a higher potential to reduce impacts [5]. Embracing community-based approaches has been proven effective in equipping communities prone to disasters with adequate information on disaster preparedness, particularly triggering swift response. In Zimbabwe, some district emergency disaster response teams might be located about 20-40km away from some villages. Moreover, coordinated disaster management from government and external support services usually reach disaster points later, and experiences witnessed when Cyclone Idai occurred, they were able to reach only a few of the affected villages approximately 4 days later. Helicopters could not fly into affected areas owing to bad weather while roads were inaccessible as bridges and road banks were damaged.

Prior to Tropical Cyclone Idai in 2019, communication and early warning by the Meteorological Services Department of the possibility of the tropical cyclone reaching the eastern parts of Zimbabwe was rather effective. This shows that available weather related forecasting equipment is reliable and valid. However, because information dissemination is still centrally controlled, it did not reach all prone communities or regions in appropriate time. Therefore, at risk communities did not take swift action to evacuate citing reasons that past tropical cyclone predictions turned out less destructive. Minimum awareness to conscientise the public was given, thus no evacuation or emergency response plans were triggered or effectively enforced. The experiences and attendant lessons learnt from the impacts of Tropical Cyclone Idai motivated this research, whose objectives are to analyse the impact of: a) community oriented training on disaster preparedness, b) appropriate land use on disaster preparedness and c) structural designs on disaster preparedness in facing tropical cyclone induced flooding in Chimanimani district, Zimbabwe. Improving the effectiveness of these approaches has proven to be appropriate in reducing vulnerability and exposure to disasters. Moreover, many studies concur that the degree of community disaster preparedness is significantly related to resilience and disaster risk reduction [6]. Since these initiatives are developed at grassroots levels, they are deemed relevant in generating appropriate responses to the risk of tropical cyclone induced flooding in Chimanimani district, Manicaland Province, Zimbabwe.

2. Conceptual framework

2.1. Community based disaster risk management

Community-based disaster risk management is indeed a broad concept that focuses on all elements of a disaster cycle, from prevention and mitigation to relief, response, and recovery. Additionally, community-based disaster risk management (CBDRM) is a strategy that lessens vulnerabilities and improves people's ability to deal with hazards. It aims to: (1) lessen vulnerabilities and improve vulnerable groups' and communities' capacities to deal with, prevent, or minimize loss and damage to life, property, and the environment, (2) lessen human suffering, and (3) speed up recovery [7]. It serves as "an umbrella word for programming approaches that include beneficiaries in their identification, design, or management, including the collection of approaches, employed inside community-level projects or as a component of national programs [8].

The concept of CBDRM has also been quiet familiar in Indonesia where the country's Disaster Management Unit (Badan Nasional Pennggulangan Bencana) (BNPB) has made great strides in empowering communities with knowledge on disaster risk management from grassroots level. Communities in disaster prone regions actively participate in "Desa Tangguh Bencana" which basically taps on local wisdom to enhance disaster preparedness. Communities are trained on emergency disaster response and disaster information sharing through *InaRisk* platform as capacity building initiatives geared towards reducing vulnerability and exposure [9].

2.2. Disaster preparedness

Disaster preparedness or readiness is strongly linked to capability of an individual, community, institution or government to apply disaster knowledge and developed capacities so as to effectively respond and reduce impacts of disasters. Furthermore, preparedness entails ability to effectively manage emerging threats or disaster risks in future [10]. Disaster preparedness is a process consisting of priority plans and actions geared towards mitigating negative impacts of disasters. It generally refers to being on top of the disaster situation at the most opportune time, including taking pragmatic actions associated with perceiving disaster risks as opportunities, and encompass capacity building measures and learning how to anticipate, predict and, where feasible, mitigate or prevent disasters, lessen disaster impact on vulnerable communities and, effectively respond to and deal with their after-effects at the local, national, regional, and global levels [11].

Inculcating effective disaster preparedness for vulnerable communities facing increased risks prevalent in specific regions has been an important component of disaster risk management [12]. In addition, disaster capacity and preparedness is reflected by the application of commensurate proactive rather than reactive action for disaster risk reduction and moreover, must be the basis for disaster risk reduction and building resilience both at the local, national, regional, and international levels [13]. The main objectives of disaster preparedness are to minimize and/or prevent property damage, loss of life, and damage to human and ecological resources, as well as to secure the economic means of survival for the vulnerable populations and social institutions exposed to hazard events. Many academics concur that preparedness is positively correlated to factors including capacity, capability, and prior disaster experience, among others. Soliciting for the active involvement, commitment, cooperation, collaboration and full participation of all stakeholders in disaster risk reduction are the key ingredients towards implementing an effective disaster preparedness plan. Undoubtedly, reduced disaster impact is the goal of rigorous investment in disaster preparedness.

3. Research methodology

This research uses a case study approach to understand individual level of disaster preparedness in Chimanimani district. In determining and weighing the impact of community-based approaches to individual disaster preparedness, this study employed a collaborative process namely the Delphi method. The collaborative process enabled researchers to rate the degree of impact as "significantly high, moderate or low". This method allows researchers to purposively select eight (08) experts in disaster risk management because their opinions are regarded credible and valid. Semi-structured interviews were used to obtain data from seventy-eight (78) informants in Chimanimani district via zoom meeting. Research participants were engaged in open-ended conservations for a deeper understanding of the problem under study. A thorough analysis of previous related literature sources including research journals, research articles, and books were used to get deeper insights on CBA and disaster preparedness. Researchers extensively read, examined and reviewed literature sources related to community preparedness towards flooding, disaster resilience, knowledge and vulnerability to disasters, and disaster risk mitigation among others. Analysis of published knowledge enable researchers to gain more insights on the effectiveness of community-oriented approaches towards preparedness to tropical cyclone induced flood risks and how these could suit the socio-cultural perspective of Chimanimani district in Zimbabwe.

4. Results and discussion

The study results are discussed in two sections, based on the main themes that arose from the data analysis process. First, issues of perceptions towards community-based approaches to preparedness towards tropical Cyclone Idai flood risks are presented. The next section focuses on indicators of community preparedness to tropical cyclone flood risks in Chimanimani district.

4.1. Community-based approaches to disaster preparedness

This study analysed the following community-based approaches to disaster preparedness, in particular to the context of Chimanimani district in Zimbabwe, in the aftermath of the devastating impacts of Tropical Cyclone Idai.

4.1.1. Community training in disaster risk and preparedness. The majority (94%) participants opined that there is a significantly high impact of community training in disaster risk towards individual level of disaster preparedness. 6% others were of the opinion that training must be complemented by adequate resources and other salient factors towards disaster risk management. Participants concurred that lack of disaster management training militate against early response, both for slow and rapid onset disasters. The findings are similar to a study conducted in 2020, Latur district, Maharashtra to determine the effectiveness of training in reducing the risk of exposure and community vulnerability to disasters. 74 % of the informants agreed that local level training in disaster management was a very effective strategy in reducing the risk of exposure and vulnerability. The primary objective of disaster management training was for community capacity building for high and medium risk hazards. The objectives of the training were to educate and build the capacity of stakeholders to improve their preparedness, and response at all levels and ensures active participation in all phases of disaster management among others (14). Trainings were organized according to onset of disaster with possible effects (intensity) of each of the disaster event on a scale of High (H) /Medium (M) /Low (L) on the basis of past history, probability of disaster occurrence and vulnerable population.

A United Nations Development Programme capacity assessment model (shown below) was used to identify the needs of stakeholders including communities and conduct appropriate training [15].

Figure 1. Capacity development process
(Source: UNDP, 2007)

In another study conducted in Chimanimani district, Zimbabwe in 2021, it was found out that institutional disaster management capacity plays a very crucial role in reducing community vulnerability to disasters as well as enhancing disaster resilience and overall sustainability. The study also found that equipping communities with disaster awareness through education and increasing their knowledge of climate change were considered key towards increasing community capacity to fully prepare for and effectively respond to the risk of hydro-metrological induced flooding [4]. In situations where communities do not possess adequate and relevant disaster information, their level of vulnerability and risk exposure to disasters and their attendant impacts was found to be significantly high. In Chimanimani district, Agricultural extension (Agritex) officers had conducted training on some communities (39.5%) on climate change and extreme weather conditions. It was however noted that despite having undergone training, communities in Chimanimani possessed extremely low knowledge levels on climate change induced hazards. Very few respondents represented by only 0.9% were rated highly knowledgeable about climate change induced extreme weather events while 46.3% had little knowledge. 17.7% others had no knowledge of climate change induced risks. The findings from this study question the effectiveness of the training program in achieving its intended goals. Further research may need to be undertaken regarding the attitude of participants and the capability of Agritex trainers to deliver knowledge on climate change and its butterfly impacts. This is significant because low knowledge levels on climate change induced extreme weather conditions is inextricably linked to high community vulnerability and exposure to disasters. When no tangible risk management measures are employed as a matter of urgency, the problem would negatively impact on the achievement of sustainable development goals as well as national economic sustainability [4].

Based on the above findings, it can be inferred that capacity building through training programmes targeting local communities in disaster risk management is very important and can equally be applied to Chimanimani district to enhance disaster preparedness. The involvement of communities in the design and formulation of training objectives are critical elements in promoting community ownership of the training programme, [16, 17].

4.1.2. Appropriate land use. Appropriate land use planning and ecosystem-based environmental management policies have been touted as effective priority actions for building disaster resilience and sustainability [13]. Poor land utilization methods increase losses from the increasing frequency of natural hazards, apart from also negatively impacting on the ecological ecosystem and the socio-economic sphere. Flood-plains and river banks are potential hazards for exposure and vulnerability [18]. Furthermore, those industries, settlements and societies located in costal and river flood plains are deemed most vulnerable and exposed to tropical cyclones, thunderstorms, hailstorms and other flood risks [19].

In a survey conducted in Chimanimani in the year 2021, it was revealed that the region is very prone to flood-related risks due to several factors including its physical or geographical conditions. After tropical cyclone Idai, some houses were found to be located along waterways and flood plains, thus putting inhabitants at high risk and exposure given the growing climate change induced risk of flooding [20]. The report also noted the practice of river and stream bank cultivation as one of the reason why some crops were washed away. Therefore, community exposure to the risk of Cyclone Idai was partly due to the physical location of their homesteads. In Ngangu and Kopa villages, Chimanimani district, some houses were swept or completely engulfed by floodwaters as they were particularly in the cyclone's path. As such, these houses were more susceptible to flooding due to their increased exposure. Most housing units, schools, medical facilities, townships, and crop-fields were at risk due to their closeness to and location alongside river basins and flood-plains.

Many scholars also agree that due to overpopulation, habitable land is diminishing at a fast rate, forcing some communities to settle on dangerous and risky areas where they remain prone to climate change induced risks. Poor farming methods and overgrazing have been touted as some of the common practices leading to soil erosion and flooding, also forcing people to opt residing in areas in which they may be more susceptible to the risk of flooding, mountain rock falls and landslides among others. Some regional town planning and local authorities also create settlements in very risky and vulnerable to disasters owing to shortage of habitable space. For example, it is now commonplace to find residential and infrastructure development projects in flood plains and water-logged areas as part of town expansion even though such places might be prone to landslides. The rampant cut down of trees for fire energy, deliberate starting veld fires in preparation for subsistence farming land clearing and other various reasons has left mountains without significant tree cover in many African countries. Arguably, trees and the natural vegetation have been found to reduce soil erosion and flooding. In the same vein, research indicates that inadequate watershed management lead to butterfly impacts on biodiversity. Siltation in rivers has often led to declining water volumes, while depletion of forest cover lead to more impacts of global warming such as rising temperatures and heat waves in different parts of the world. In addition, land degradation causes soil erosion whose net effect is increased vulnerability and exposure to the risk of flooding and landslides [21]. Stream bank cultivation especially around rivers has also been a major problem facing many governments, and water-proof strategies must be devised in-order to nip this problem in the bud.

25% of the participants agreed that planting trees, reducing over-grazing and conservation farming were good practices to reduce flooding. This shows that appropriate land was moderately rated, thus participants were of the view that there is minimal relationship between tropical cyclone induced flooding and individual level of disaster preparedness. Instead, they believe that even though these initiatives must be undertaken, tropical cyclone flooding was likely to affect communities as it originates from oceanic activity, thus more effort should be channeled towards scientific based research and the use of early warning technology in risk or hazard prediction. However, it is a fact that more education and awareness is needed for communities to comprehend the relationship between the environment and climate change. Therefore, the lessons and findings from these and other previous studies, this study therefore, concurs that inappropriate land use in Chimanimani district, Zimbabwe significantly contribute to risk exposure and vulnerability.

4.1.3 Structural designs. Some modern structures are designed to withstand and absorb the shocks from possible hazards. Structural measures include physical buildings intended to lessen the overall impact of hazards. In Japan, for example, modern engineering techniques are being incorporated into building structures to adequately achieve hazard resistance and resilience [22]. The construction of sea-walls in tsunami prone regions has also been very helpful in reducing their attendant impacts on people and property. Non-structural measures are those measures in which knowledge, science and practice or agreement is applied to reduce the impact of disaster risks. This includes the use of legal frameworks (insurance cover), increasing community awareness, community driven training and educational programs among others. In the same vein, to reduce vulnerability to disasters, Japan is currently implementing earthquake resistant buildings which are basically reinforced using two steel beams, added with concrete walls with steel bars as well as adding shock absorbers [23].

Owing to incapacity, most structures in the study area do not meet modern standards for absorbing shocks emanating from violent tropical cyclone induced floods. Although respondents agreed that structural designs have a significant impact on individual level of disaster preparedness, the implementation of this measure in the study area was cited as one hurdle. This strategy is not common among majority (90%) of the people who depend on subsistence farming and animal rearing and thus have limited capacity to install modern engineering techniques to achieve hazard resistance. However, the region receives high rainfall and is mountainous with steep slopes in which rivers and streams are ever flowing throughout the year, thus structures built on steep slopes risk being washed away by violent storms. The area is famous for banana and timber plantations, and commercial tea production. This might explain why some structures are located in water logged areas while others are built on steep slopes or hill tops where they can hardly withstand destructive and strong winds associated with tropical cyclones [24].

4.2. Individual level of disaster preparedness

To analyse an individual's level of disaster preparedness, it is important to first identify individual disaster resilience indicators. As already highlighted above, disaster preparedness encapsulates knowledge and capacities developed over time, either through local wisdom, experience or formal or informal training by an institution, community or individual himself to form some professional response and recovery when faced with a disaster. The overall end result of disaster preparedness is manifested through exhibiting positive virtues, actions or behaviours for example, effectively anticipating, responding to, and recovering stronger (build-back-better) after impacts of probable disaster risks [25]. Individual disaster preparedness can be described as an amalgamation of disaster knowledge and skills, disaster emergency plan, wealthy of disaster experience and, valuable and reliable disaster information critical for the formulation of response and mitigation decisions [26].

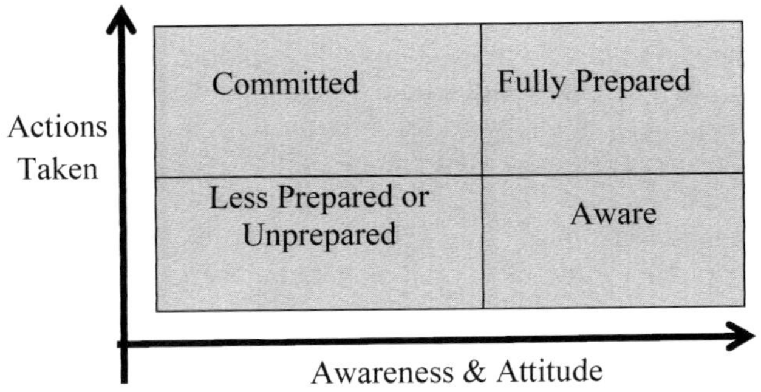

Figure 2. Indicators for analyzing an individual's level of disaster preparedness.

Individuals committed towards disaster preparedness actively participates and are greatly involved in disaster risk reduction processes including in the application and formulation of training and early warning information sharing. Those fully prepared are very aware of and demonstrate positive attitudes towards disaster risk management. It therefore follows that less prepared individuals take little or no action and do not demonstrate positive attitudes towards disaster response and in most instances are worst affected by disasters.

4.2.1. Disaster knowledge. The knowledge of communities living in areas at risk for an impending disaster is the first aspect of concern for disaster management stakeholders, if their level of disaster preparedness is to be improved. Inadequate preparedness increases community vulnerability to disasters [14]. For communities to have an understanding and awareness of disasters, prior knowledge of impending or future risks common in their locality is crucial. This disaster prior knowledge can be measured using some of the following indicators among others: knowledge of disaster early warning systems, available evacuation options such as high level shelters, evacuation maps and routes, ability to carry out disaster emergency response, disaster experience, and disaster simulation/drill exercises.

In a study conducted in Chimanimani district, 68% of the respondents confirmed that they received early warning information about tropical cyclone Idai and that it would make windfall into Zimbabwe [4]. Of those, the majority (85%) spread the information to fellow members within their localities. Surprisingly, however, only a few of those informed (represented by 36.7%) took prior action by evacuating early. About 12,6% of the respondents confirmed moving to safer places in anticipation of tropical cyclone Idai. Due to reasons such as lack of knowledge, perception and lack of disaster experience, a considerable number (52.6%) stayed indoors while a few (2.1%) decided to monitor the progression of the cyclone by listening to incoming news bulletins on national radio and television channels on the best safety procedures. Some even doubted the accuracy of early warning information given by authorities, arguing that just like previous warnings, tropical cyclone Idai might not reach their region. The majority of people thought the floods would just come to pass without notable impacts, an indication that perception and disaster inexperience were at play leading to inaction for early emergence response [4]. As such, this study concurs that strengthening disaster knowledge among communities through training is a very useful concept for disaster risk reduction and promoting resilience, more specifically because it enhances disaster preparedness.

4.2.2. Disaster emergency plan. Possessing knowledge about an impending disaster or risk, without a detailed plan of action may not lead to a reduction in the overall aftermath impacts of disasters. Communities need to be encouraged to collaborate, cooperate and co-exist in the formulation of a compact emergency preparedness and response plan. Team response and the capability to swiftly activate the plan at the opportune time of an emergency situation cannot be overemphasized. The degree to which vulnerable people understand hazards and risks they might face and the significance of preparedness is critical in saving them during and after disasters. Furthermore, willingness and zeal to adequately prepare for impending disasters by taking pragmatic actions is necessary. It follows therefore that, formulating an emergency plan or strategy and religiously putting the plan into practice through targeted disaster preparedness simulation or training rehearsals becomes a top priority. This also further entails encouraging citizens to protect assets by taking insurance property and life cover, evacuation routes and disaster information communication lines among others.

4.2.3. Information and communication. During an emergency situation, access to accurate and reliable information is key. Usually, disaster victims always need to know the latest information regarding the safety of their family members and how they can get help. When victims are short of accurate and relevant disaster information, anxiety and depression creeps in. this shows that lack of information or a delay to convey it creates a barrier that heightens vulnerability [27].

5. Conclusion

The butterfly aftermath effects of climate change induced extreme weather events are projected to continue affecting societies in several negative ways. Globally, the frequency of storms and tropical cyclones leading to severe flooding continue to be felt more than before, affecting one country or region and the other. In this respect, implementing community-based approaches towards disaster preparedness has been found to be very effective in reducing vulnerability and exposure. The study concluded that community-oriented training was significantly related to individual level of disaster preparedness. Readiness to disaster overall reduces its aftermath impacts, particularly saving lives and damage to critical infrastructure. Thus, developing home-grown disaster preparedness approaches become a key concept in capacitating communities, who in the majority cases are the first to experience impacts of disasters. Structural designs were rated low in terms of impact to disaster preparedness in Chimanimani district, while appropriate land use need to be considered as a key priority after training in disaster. Taking a que from the experiences of the 2019 Cyclone Idai and its socio-economic impacts on Zimbabwe, policy makers need to consider enhancing community oriented training in order to increase individual preparedness for emerging and unpredictable risk of tropical cyclone flooding facing communities in Chimanimani district. It is also paramount that rural agricultural extension workers increase community participation in land conservation practices as these are interwoven to climate change whose impacts include tropical cyclone flooding and severe droughts among others. Arguably, training impart upon individuals, the required knowledge and attitudes critical for swift response to disasters, and this initial action taken before a disaster strikes is strongly related to disaster impacts. Individuals exposed to effective training in disaster emergency response have been shown to be committed and fully prepared to face disasters. Ultimately, lower disaster aftermath impacts help build resilient communities.

Acknowledgements

We are greatly indebted to the Republic of Indonesia Defence University for funding this research. Moreover, we deeply express our sincere appreciation to all participants who thoughtfully contributed in this study and the several researchers whose insightful previous researches made it easier for us to read thoroughly and come up with recommendations applicable to Chimanimani district in Zimbabwe. To all anonymous reviewers, we are grateful for your inspirational comments which helped us think critically and improving the quality of the article.

References

1. Brown D, et al 2012 *Climate change impacts, vulnerability and adaptation in Zimbabwe* (London: International Institute for Environment and Development Publishers) p 80-86
2. Phillips M, Cinderich A, Burrell J, Ruper J, Will R. and Sheridan S 2015 The effect of climate change on natural disasters: A college student perspective *Weather, Climate, and Society* **7** 60–68
3. Dibyosaputo S, Cahyadi A, Nugraha H and Suprayogi S 2016 Estimation of the impact of climate change on lava flood insecurity in Magelang, Central Java *Proceedings of the National Seminar on Geography UMS* Vol. 37 No. 2
4. Chanza N, Siyongwana P, Williams-Bruinders L, Jakarasi V, Mudavanhu C, Sithole V and Manyani A 2020 Closing the gaps in disaster management and response: drawing on local experiences with cyclone idai in Chimanimani, Zimbabwe *Int. J. Disaster Risk Sci.* **11** 655–666
5. Appleby-Arnold S, Brockdorff N, Jakovljev I and Zdravković S 2018 Applying cultural values to encourage disaster preparedness: lessons from a low-hazard country *Int. J. of Disaster Risk Reduction* **31** 37– 44
6. Winderl T 2014 *Disaster Resilience Measurements: Stocktaking Of Ongoing Efforts In Developing Systems For* Measuring *Resilience:* United Nations Development Programme (UNDP). Available via https://www.preventionweb.net/go/37916. Accessed 26 May 2022
7. Mutsau S 2010 Disaster Management. (Harare: Zimbabwe Open University) p 76-78
8. Slaymaker T and Christiansen K 2005 *Community-based Approaches and Service* Delivery*: Issues and Options in Difficult Environments and Partnerships*
9. Badan Nasional Penanggulangan Bencana 2019 *Risiko Bencana Indonesia* (Jakarta: BNPB)
10. United Nations Office for Disaster Risk Reduction (UNISDR) 2013 Roles, Mandates and results of Key UN entities (Geneva)
11. Lavell A, Oppenheimer M, Diop C, Hess J, Lempert R, Li J, Muir-Wood R, Myeong S 2012 *Climate change: new dimensions in disaster risk, exposure, vulnerability, and resilience.* Accessed *online on 23 June 2022*
12. Cavallo A 2014 *Integrating disaster preparedness and resilience: a complex approach using system* of *systems.* Aust J Emerg Manag 29(3):46–51. ISSN: 1324 – 1540
13. United Nations Office for Disaster Risk Reduction (UNISDR) Sendai Framework for Disaster Risk Reduction (2015–2030) adopted at the Third UN Conference in Sendai, Japan on 18 March 2015. The successor instrument to the Hyogo Framework for Action (2005–2015).
14. Creswell J W 2018 Research Design: Qualitative, Quantitative, and Mixed Methods Approaches (4th ed.) (California: SAGE Publication, Inc.) p 41
15. United Nations Development Programme 2013 Community Based Resilience Analysis (CoBRA) Conceptual Framework and Methodology. Disaster Risk Reduction Action
16. Nadler Z and Nadler L 1994 *Designing Training Programs*, 2nd Edition Routledge, New York.
17. Boyle P G 1981 *Planning Better Programs.* (New York: McGraw-Hill)
18. Garrido J and Saunders W S A 2018 *Conference paper on Disaster Risk Reduction and* Land *Use Planning:* Opportunities to Improve Practice, accessed online 15 June 2022.
19. IPCC 2014 *Climate Change 2014: Impacts, Adaptation and Vulnerability. Part B: Regional* Aspects. *Contribution of Working Group II to the Fifth Assessment Report of the Intergovernmental Panel on Climate Change.* (Cambridge: Cambridge University Press)

20. Munsaka E, Mudavanhu C, Sakala L, Manjeru P and Matsvange D 2021 When *Disaster Risk Management Systems Fail: The Case of Cyclone Idai in Chimanimani District, Zimbabwe*, accessed online on 10 May 2022.
21. Government of Zimbabwe 2019 Ministry of Environment, Water and Climate, Environmental Management Act, Chapter (20:27) Harare.
22. United Nations Office for Disaster Risk Reduction 2008 *Developing early warning systems: A checklist. United Nations International Strategy for Disaster Reduction (UNISDR).*
23. Mizan B 2022 Proceeding on International Conference on Disaster Management. *Japan-Indonesia Collaboration in the Transfer of Knowledge, Technology, and Expertise on Disaster Risk Management*, Jakarta.
24. Government of Zimbabwe 2019 Ministry of Local Government and Public Works: Tropical Cyclone Idai International Humanitarian Assistance Appeal, April 2019.
25. website https://www.lawinsider.com. Accessed on 26 June 2020
26. Patrisina R, Emetia F, Sirivongpaisal N, Suthummano S, Alfadhlanir A and Fatrias D 2018 International Conference on Disaster Management: Key performance indicators of disaster preparedness: a case study of a tsunami disaster. (Padang: Universitas Andalas) **229**
27. Chapungu L 2020 Policy Briefing: *Mitigating the impact of cyclone disasters: Lessons from Cyclone Idai.* Johannesburg: South African Institute of International Affairs. https://media.africaportal.org/ documents/Chapungu_-_Final.pdf. Accessed 22 May 2022.

Identification of Lean Implementation in the Post-Disaster Housing Reconstruction

D M Octavia[1,2], T Ophiyandri[1*], A Suraji[1] and B Hidayat[1]

[1] Department of Civil Engineering, Universitas Andalas, Padang,Indonesia
[2] Department of Civil Engineering, Padang Institute of Technology, Padang, Indonesia

*ophiyandri@eng.unand.ac.id

Abstract. Housing reconstruction is a significant determinant of the post-disaster recovery process and one of the most complex challenges, especially in developing countries. Post-disaster housing reconstruction sometimes results in poor environmental quality, delayed construction completion, low performance, and wasteful materials. Previous research has shown the success of the lean application in construction projects as well as in reconstruction projects. Its means that post-disaster housing reconstruction can also apply lean principles and techniques. This study aims to identify and analyze lean theories and techniques that can be implemented to overcome, minimize or resolve problems in Indonesia's post-disaster housing reconstruction stage to obtain a lean model (Lean Housing Reconstruction). This study is the initial stage of developing a lean implementation model in post-disaster housing reconstruction. A systematic literature review (SLR) was conducted to obtain lean principles and techniques for post-disaster housing reconstruction. The results of the SLR obtained five lean construction techniques that can be used for housing reconstruction, including Lean Six Sigma, Last Planner System, 5S (Sort, Straight, Shine, Standardize, and Sustain), Value Stream Mapping, and Pull Scheduling.
Keywords: post-disaster, housing reconstruction, lean implementation, lean reconstruction

1. Introduction

Injuries, loss of life, disease, and other negative effects on physical, mental, and social well-being, as well as damage to assets and property, are consequences of disasters that occur in a country [1]. Disaster management has always been a significant concern in many countries, including Indonesia[2]. In post-disaster reconstruction situations, substantial obstacles or challenges arise, especially in large-scale disasters [3]. The most crucial stage of disaster management is reconstruction. The most significant difficulties in the post-disaster reconstruction phase are ad hoc decision-making, poor planning, haphazard recovery, waste of materials, environmental degradation, poor quality of the built environment, excess costs, and long reconstruction life cycles[4].

According to research conducted in Kobe, Japan, the post-disaster recovery process is influenced by seven factors: housing, community reconstruction, readiness, social ties, physical and mental health, relationship with the government, and financial and economic condition. [5]. One of the main factors affecting the post-disaster recovery process is housing or housing reconstruction. Housing reconstruction initiatives have often failed due to problems with community involvement, communication, resettlement, and cultural conformity [3].

In Indonesia, in the last five years, quite a large natural disaster occurred in 2018. Based on BNPB data, throughout 2018, 2 significant disasters occurred in Indonesia, namely the Lombok earthquake in July and August 2018, with damaged houses reaching 80,031 units. Furthermore, the earthquake and tsunami in Palu and Donggala in September 2018 damaged 99,726 units of houses. Post-disaster housing reconstruction is one of the most challenging aspects of post-disaster recovery operations, particularly in developing nations. [6]. Conditions following a disaster provide tremendous obstacles for reconstruction and call for government funding. While post-disaster home rehabilitation necessitates a sizeable portion of global property investment.[7].

The mechanism of development initiatives must include awareness. Given that it has a substantial impact on stakeholder views, one of the crucial factors in the decision to invest in safer housing, it must

IOP Conf. Series: Earth and Environmental Science 1173 (2023) 012071 doi:10.1088/1755-1315/1173/1/012071

be considered carefully [2]. Issues of communication, community participation, cultural conformity, and resettlement in the implementation of housing reconstruction programs have been challenges that continue to cause the failure of housing reconstruction projects, where many beneficiaries reject the provided housing or modify the housing; some even dismantle the house and sell its components. These challenges must be adequately managed for the reconstruction program to be successful [8; 9].

With all these challenges and problems, it is necessary to have an approach that can be used to overcome, minimize or solve some or all of the challenges and issues of post-disaster housing reconstruction. Lean construction is one strategy that can be used for post-disaster housing reconstruction. Lean first appeared in the manufacturing industry, specifically in the Toyota company's lean production method, which enables high-quality, diversified, and low-cost production to suit client needs. [10]. Lean thinking is a term to define the intended use of resources effectively to minimize wastage and maximize value [11]. The Toyota Company's success evidences lean's success in the manufacturing industry. This condition creates a massive demand for knowledge of lean thinking and adapting it beyond manufacturing industries such as services, logistics and distribution, retail, construction, maintenance, healthcare, and even government [12].

In the post-disaster housing reconstruction and rehabilitation process in Aceh, Indonesia, there are 4 (four) essential indicators: the process of providing housing, government involvement and policies, procurement of building materials and skilled workers, and community participation in housing design and supervision [13]. This indicator can be a value that must be achieved by implementing post-disaster housing reconstruction. This value can be realized optimally and adequately with lean principles.

In order to boost construction productivity and decrease construction waste and be able to provide affordable housing, the application of the concept of lean construction emerged [14]. In general, lean construction projects are cheaper, have better quality, are easier to manage, can be completed faster, and are safer. Thus, the implementation of lean construction in the housing reconstruction process can improve housing reconstruction performance, speed up the reconstruction process, minimize adverse environmental impacts due to the reconstruction process, and reduce material wastage.

Lean has also been shown to assist improvements in project performance as evaluated by a variety of KPIs. The adoption of lean construction in construction projects, particularly housing projects, can increase project performance, according to a growing body of literature. This shows that lean can help post-disaster housing restoration projects function better. The application of lean construction to post-disaster restoration has been the subject of various research in recent years. Started Mojtahedi and Oo [4] stated several lean construction approaches are feasible for post-disaster reconstruction, such approaches reduce waste, streamline workflows, improve environmental quality, and enhance post-disaster reconstruction performance. These approaches include Six Sigma, Quick mobilization, Just-in-Time, and Pull Scheduling.

In addition, El jazzar et al. [15] analyze the adoption of lean construction through workflow optimization by investigating the causes of workflow disruptions and providing ways that can enhance project performance. Its primary purpose is to improve the benefit delivered to project beneficiaries. Implementing the Last Planner System (LPS) to optimize project workflow in 10 shelters from the United Nations' "self-help" rehabilitation project is the next step [16]. This study is conducted by identifying the reasons of delays and adding value by removing construction workflow impediments so that projects can be completed more quickly and for less money.

Several studies related to lean construction or post-disaster reconstruction mentioned above show that lean can be implemented in post-disaster reconstruction projects, and this means that lean can certainly also be applied to post-disaster housing reconstruction. The purpose of this study is to identify possible applications of lean in post-disaster housing reconstruction. The method used in this research is to review several related papers through a systematic literature review (SLR) to determine the potential benefits of implementing lean and studying lean techniques that can be implemented to post-disaster housing reconstruction.

2. Disaster

UNISDR [17] defines a disaster as a significant disturbance to the operation of a society or community, resulting in losses and impacts on the human, economic, material, and broader environment, which cannot be mitigated by the afflicted community's resources. A disaster can also be defined as an abnormal or infrequent occurrence of a hazard that impacts a vulnerable community or geographic area, producing major damage, the chance of loss of life, disruption, and necessitating aid from outside sources because people cannot live normally [18].

Disasters are a big global issue and pose a serious danger to sustainable development. In addition to the loss of life, destruction, disease, injury and loss of assets and property, the impact of disasters can also cause economic and social disruption, infrastructure and environmental damage and other services damage [19]. According to the Sendai Framework for Disaster Risk Reduction [20], between 2005 and 2015, more than 1.4 million people were injured, more than 700 thousand individuals lost their lives, and over 23 million were rendered homeless as a result of disasters.. Figure 2.1 shows an illustration of the phases of disaster management with relevant activities and approaches. Activities carried out before a disaster occurs to effectively minimize the disaster's impact can be said to be a proactive approach. Meanwhile, response and recovery activities are reactive approaches [21].

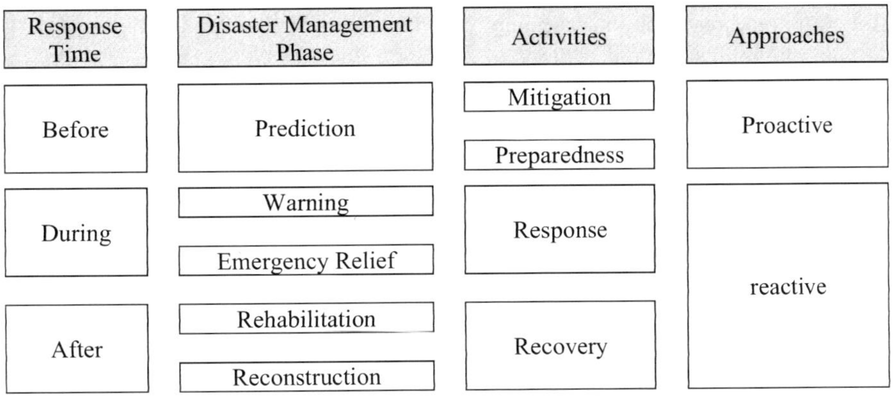

Figure.1 Disaster Management Phases, Activities, and Approach [21]

3. Post-disaster Housing Reconstruction

Reconstruction is a long-term activity [21] that starts immediately after the emergency phase ends. Pre-existing strategies and policies must form the basis of reconstruction, facilitating institutional responsibility for recovery actions and enabling public participation. Post-disaster reconstruction requires multi-sectoral involvement, significant resources, and considerable skills [22]. Institutional bureaucracy, corruption, poor coordination, lack of experience in construction management, and pressure from the government and humanitarian organizations for quick project completion can further hinder reconstruction efforts. [23; 24].

Based on Sakamoto & Yamori [5], the key determinant of how well the post-disaster recovery process goes is housing. For the affected communities to relocate, housing rehabilitation is a top priority in post-disaster recovery initiatives. For the majority of individuals, housing is crucial because it affects their welfare. Housing contributes to the security, health, economy, social, and educational well-being of families. This brings about pride and social connection [9].

Post-disaster housing reconstruction discussions since the 2004 Indian Ocean tsunami have emphasized "build back better," emphasizing creating more resilient structures than before the disaster and promoting disaster risk reduction. [25]. As the UN Secretary General's Special Envoy for Tsunami Recovery, former US President William Clinton suggested a "build back better" approach to disaster recovery in 2006. This concept has subsequently become a standard in contemporary post-disaster rehabilitation efforts [26]. Housing reconstruction can be divided into three general phases [3]:

1. Strategic and empowerment phase, institutional and financial reconstruction framework.
2. The phase of reconstruction planning in which particular plans and requirements are drafted and decisions are made.
3. In the phase of implementation, the actual reconstruction work is performed.

Problems of communication, community participation, cultural appropriateness and resettlement were recurring challenges in implementing housing reconstruction programs that failed. Many beneficiaries immediately rejected the housing provided or modified the house, and in some cases, even dismantled the house and sold its components. These challenges must be managed appropriately for the reconstruction program to be successful [8; 9]. In addition, limited capacity in the technical aspect is only one of the causes of housing reconstruction problems. Other issues also cause problems such as poor planning and coordination of the parties involved, scarcity of resources and the amount of aid distributed simultaneously, lack of knowledge and experience in the provision and construction of post-disaster settlements, sociocultural considerations, and not following the expectations and needs of beneficiaries[9; 27].

Bilau [3] identified problems affecting the implementation of the reconstruction from various literature, which was categorized into eight categories (logistics and supply chain; monitoring and control; human resource; risk management; quality; health and safety management; financial management and coordination problems). However, this categorization seems not optimal because there is still overlap between categories; other categories may also appear.

The main issue for house restoration in the devastated districts, according to Rouhanizadeh & Kermanshachi [28], was that the majority of homeowners did not want to transfer because they were afraid of the unknown and uncertain threats they may encounter. This held up the recovery process. In Aceh, Indonesia, the provision of housing, community participation in housing design and supervision, procurement of building materials and skilled labor, and government involvement and policy are all indicators that strengthen and support the post-disaster housing rehabilitation and reconstruction process [13].

Accuracy of beneficiary data in the housing provision process is essential to avoid overlapping housing deliveries. Early in the process of rebuilding homes, government capability must be strengthened. Depending on the situational circumstances, community involvement in housing reconstruction needs to be carefully explored, evaluated, and managed. Additionally, preparing communities for house construction through training and capacity building will greatly increase their ability to contribute to jobs. An uninformed community can trigger disputes with contractors and builders, resulting in construction delays [13].

Early community participation, especially in decision-making at the planning stage, is essential in shaping subsequent participation. Participation shouldn't be viewed as a collection of distinct jobs because it is a connected process that takes place throughout various project phases. These findings drive project plans that incorporate community involvement throughout the project life cycle, which in turn shape disaster recovery procedures. [29].

4. Lean Construction

The phrase lean first arose in the manufacturing sector, specifically in reference to the Toyota Company's lean production strategy, which provides high-quality, varied, and affordable output to satisfy client needs. [10]. Lean thinking is a phrase that describes the efficient use of resources to reduce waste and increase value for users [11]. The Toyota Company's success evidences lean's success in the manufacturing industry to date. This increases the need for lean thinking expertise and its application to non-manufacturing sectors like government, logistics and distribution, services, retail, healthcare, and the healthcare industry [12].

In lean production, the causes of failure to complete the product are identified, and action is taken to prevent the recurrence of the loss. In addition, reducing variation in workflows allows for reduced time and costs. Time is reduced due to the precise matching of jobs to human resources and resources, and costs are reduced because predictable workflows enable the timely execution of jobs and supplies [30].

Since 1992, researchers have persuasively supported adopting lean production principles in the construction industry [31; 32]. Lean production, which is based on a complex set of principles including continuous improvement, a fluid organizational structure, teamwork, the elimination of waste, effective resource utilization, and cooperative supply chain management, gave rise to the idea of lean building [33].

Koskela [31] revealed 11 principles of Lean construction, 1) including lowering the amount of activities that do not add value, 2) raising the value of product by methodically taking into account customer wants, 3) decreasing variability; 4) shortening cycle times; 5) streamlining by reducing the number of steps, sections, and relationships; 6) enhancing output flexibility; 7) enhancing process transparency; 8) concentrating control on the entire process; 9) incorporating continuous improvement; 10) increasing conversions to improve flow balance; and 11) benchmarking.

Lean construction is an approach to increase productivity and continuity. It is a systematic method for creating buildings with optimal engineering, design, and environmental responsiveness. Lean construction seeks to address the needs of clients in the construction sector by reducing waste and increasing productivity, health, and safety [14].

5. Lean Construction and Post-disaster Housing Reconstruction

Given that lean construction is a method to minimizing waste time and effort in order to maximize value the philosophy of lean construction and its individual components may be able to improve post-disaster reconstruction performance [4], this suggests that it may also be able to enhance the effectiveness of post-disaster house reconstruction. Experts have employed a variety of lean methods and tools in production and construction, including Lean Six Sigma, the Last Planner System, the 5S (Sort, Straight, Shine, Standardize, and Sustain) method, Value Stream Mapping, Quick Mobilization, Just in Time, and Pull Scheduling. On the basis of the five fundamental lean concepts of value, value stream, flow, pull, and perfection, the lean tools applicable to construction projects can be classified. Additionally, they believe this technique is relevant to lean principles. Even though it was originally designed for a manufacturing setting, it is also applicable to the construction industry because the emphasis is on process optimization. Consequently, it is considered that some of these strategies can be applied to housing restoration following a natural disaster. The results of identifying each lean technique applicable to post-disaster housing reconstruction are provided below.

5.1. Lean Six Sigma

Since their introduction the quality management techniques Lean and Six Sigma have been incredibly popular. They are also frequently applied in tandem and known as Lean Six Sigma. Lean generally focuses on eliminating waste and tasks that don't improve a particular product. Lean includes a plan to improve quality and cut costs in addition to its primary objective of reducing waste [34]. While Six Sigma focuses on the critical quality features of goods that are vital to consumers by identifying and eliminating errors flaws, or failures that can affect operations. Lean and Six Sigma are very congruent. Six Sigma is a statistically-based strategy for minimizing user-defined defect rates to minimize impacts and faults in each process [4]. Six Sigma can be regarded a helpful strategy for firms to implement in order to improve quality and operations while adhering to environmental requirements [35]. Consequently, applying both can reduce defects, increase delivery time, and save costs. Housing restoration following a natural disaster is a task that must be expedited, resulting in shoddy workmanship. Therefore, using Lean Six Sigma is very suitable for overcoming this.

5.2. Last Planner System

One of the most well-known lean methods is the Last Planner System (LPS). Glenn Ballard developed LPS in 2000. The Last Planner™ Production Control System has entered the Lean Construction theory literature [36]. Last Planner reduces wastage by preventing workers from making do starting work before all the items needed to complete the job are available [37]. The last Planner is critically responsible for reducing Work-in-Progress (a form of waste) reported in case studies where it has been implemented [38]. LPS has also proven to be a handy tool for managing the construction process and for continuous monitoring of planning efficiency, helping to develop forecasts, reducing workflow variation, and reducing/eliminating uncertainty in the construction process [34].

In residential house construction, the application of LPS has also been shown to improve the performance of the production system [39]. Thus, the use of the LPS technique in post-disaster housing reconstruction, especially at the reconstruction implementation stage, is also believed to increase the efficiency of reconstruction planning, reduce workflow variations to accelerate reconstruction implementation, and reduce uncertainty in housing reconstruction implementation. In addition, LPS is also an instrument for minimizing security breaches and scheduling problems, which encourages collaborative work. Collaborative planning is crucial in post-disaster management [11] so that LPS may be utilized effectively in post-disaster housing reconstruction where collaborative work is necessary and a necessity.

5.3. 5S (Sort, Straight, Shine, Standardize, and Sustain).

5S is a system suitable for use at the construction stage because this system creates and maintains a good, efficient and successful work environment. 5S comes from Japanese, which means [40]:

a. Seiri/Sort means isolating everything from the work environment that is needed and what is not needed at the moment. By implementing this process, improvement occurs by reducing costs and stock and preventing tool loss.
b. Selton/Set in order, which means arranging materials for easy use and labelling them so they are easy to find. This can increase effectiveness and efficiency, shorten the time, and improve safety.
c. Seiso/Shine, meaning cleaning and grading, to understand working conditions. This can create a clean construction environment, easy to inspect and remove the reasons for work accidents.
d. Seiketsu/Standardize, meaning creating a standard strategy for consistency when all three of the first 5Ss have been actualized. With this, it can improve safety, reduce pollution, and the construction process can run according to procedures.
e. Shitsuke/Sustain means measuring progress and planning for continuous improvement. This can increase awareness and morale, reduce errors due to inattention, and improve internal communication processes.

In the post-housing reconstruction process, some problems commonly occur in ordinary construction processes, such as wasted time due to searching for materials, poor management of the workplace, disasters, material storage areas that are not well defined, and unnecessary materials that are in the workplace can cause problems, affect worker morale. Thus, the use of 5S in the process of implementing post-disaster housing reconstruction will be able to minimize or even eliminate these problems.

5.4. Value Stream Mapping

VSM has been widely utilized in the construction industry including in the residential housing sector where it is used to identify productive flows and waste [39]. VSM is a lean tool that helps represent existing processes and improvements that need to be made in processes [41]. Three strategies - process modification the elimination of pointless operations and enhanced activity - are used by VSM to study the state of existing processes and increase productivity [42], so that VSM can be a management tool to help evaluate waste in workflows [43]. By identifying all operations that add and do not bring value to construction execution through VSM, the workflow can be enhanced. Adopting a continuous improvement strategy eliminates or reduces activities that do not add value.

This technique begins with the documentation of process stages and the flow of work item values and then employs a systematic strategy to analyze the process and generate an improvement plan. In its implementation, it is essential to create a VSM for existing conditions, then create a VSM with improvements and finally create a VSM for future projects based on the implementation of the improved VSM. Obviously, VSM can be utilized in post-disaster housing reconstruction as well. It starts by identifying the steps of the reconstruction process and the flow of values and creating a VSM based on the current state of implementation of post-disaster housing reconstruction. Next, identify processes that add value to those that do not. Processes that do not add value are reduced even if possible can be eliminated.

5.5. Pull Scheduling
As a lean construction tool, Pull Scheduling has been introduced to aid project managers in the timely completion of construction projects. Pull Scheduling emphasizes resource selection and assignment for work activities in order to reduce resource waiting time. The selection of resources must not only suit the requirements of earlier efforts, but also those of ongoing and subsequent endeavors [44]. During reconstruction, supplies are usually not delivered on time, resulting in delays in completing work [4]; the same is true for house reconstruction following a natural disaster. Pull Scheduling is therefore one of the lean methods that can be utilized to enhance the performance of post-disaster housing rebuilding implementation.

6. Conclusion
Delays in reconstruction planning, reconstruction steps, and delays in completion are problems that often occur in the implementation of post-disaster housing reconstruction. In addition, the issue of community participation and communication of cultural incompatibility causes housing reconstruction projects to fail where beneficiaries modify houses or refuse housing that has been provided. This paper identifies lean building strategies applicable to post-disaster housing reconstruction. The proposed lean construction techniques include Lean Six Sigma, Last Planner System, 5S (Sort, Straight, Shine, Standardize, and Sustain), Value Stream Mapping, and Pull Scheduling. The five techniques need to be explored further with the current and previous post-disaster housing reconstruction implementation conditions to propose a lean reconstruction model for post-disaster housing reconstruction.

References

[1] M K Lindell And C S Prater 2003 *Assessing Community Impacts Of Natural Disasters Nat Hazards Rev* Vol 4 No 4 Pp 176–185
[2] K S Pribadi D Kusumastuti S A H Sagala And R Wimbardana 2014 *Post-Disaster Housing Reconstruction In Indonesia: Review And Lessons From Aceh Yogyakarta West Java And West Sumatera Earthquakes* Pp 197–223
[3] A A Bilau E Witt And I Lill 2015 *A Framework For Managing Post-Disaster Housing Reconstruction* Procedia Econ Financ Vol 21 Pp 313–320
[4] Mojtahedi And B L Oo 2012 *Possibility Of Applying Lean In Post-Disaster Reconstruction: An Evaluation Study*
[5] A Sakamoto And K Yamori 2009 *A Study Of Life Recovery A Study Of Life Recovery And Social Capital Regarding Disaster Victims* J Nat Disaster Sci Vol 31 No 2 Pp 13–20
[6] I Ahmed 2011 *An Overview Of Post-Disaster Permanent Housing Reconstruction In Developing Countries* Int J Disaster Resil Built Environ Vol 2 No 2 Pp 148–164
[7] M Vahdatmanesh *Real Options Analysis Of Revenue Risk Sharing In Post-Disaster Housing Reconstruction* 2021
[8] P L Delaney And E Shrader 2000 *Gender And Post-Disaster Reconstruction: The Case Of Hurricane Mitch In Honduras And Nicaragua Decision Review Draft*
[9] S Barakat 2003 *Housing Reconstruction After Conflict And Disaster* Overseas Development Institute
[10] Lean Construction Institute 2019 *What Is Lean?* https://www.leanconstruction.org (accessed Sep 08, 2022)
[11] S Demirkesen 2020 *Investigating The Synergy Between Lean Construction Practices And Post Disaster Management Processes* Chall J Struct Mech Vol 6 No 1 P 23
[12] C Protzman et al 2018 *Brief History Of Lean* In The Lean Practitioner's Field Book Boca Raton Fl : Crc Press 2016 : Productivity Press Pp 2-1-2–36
[13] Azweri C Mutiawati N Al-Huda And H Mufiaty 2017 *Disaster Recovery Indicators Of Housing Reconstruction: The Story Of Post Tsunami Aceh Indonesia* Int J Disaster Manag Vol 1 No 1 P 35

[14] A Shastri H S Chauhan S B Chidapareddi V A Chopade And R S Kahane 2022 *Implementation Of Lean Construction Method To Ensure Effectiveness Of A Residential Construction Project* Int J Res Eng Sci Manag Vol 5 No 5 Pp 66–71 2022
[15] M S Eljazzar A A Beydoun And F R Hamzeh 2013 *Production System Design Optimizing Workflow For Shelter Rehabilitation Projects In Refugee Camps* In: Formoso C T & Tzortzopoulos P 21th Annual Conference Of The International Group For Lean Construction Fortaleza Brazil 31-2 Aug 2013 Pp 491–501
[16] M S El Jazzar And F Hamzeh 2015 *Post Measuring The Last Planner Metrics In Shelter Rehabiliation Projects* In: Proc 23rd Ann Conf Of The Int'l Group For Lean Construction Perth Australia July 29-31 Pp 783–792
[17] UNISDR 2009 *UNISDR Terminology On Disaster Risk Reduction* Geneva Switzerland
[18] C Benson And J Twigg 2004 *Measuring Mitigation' Methodologies For Assessing Natural Hazard Risks And The Net Benefits Of Mitigation-A Scoping Study*
[19] J Twigg 2015 *Disaster Risk Reduction* Overseas Development Institute United Kingdom
[20] UNISDR 2015 *Sendai Framework For Disaster Risk Reduction 2015 - 2030* Geneva
[21] T L Moe And P Pathranarakul 2006 *An Integrated Approach To Natural Disaster Management: Public Project Management And Its Critical Success Factors* Disaster Prev Manag An Int J Vol 15 No 3 Pp 396–413
[22] J Da Silva 2010 *Lessons From Aceh : Key Considerations In Post-Disaster Reconstruction* Practical Action Publishing Ltd
[23] J D Baranstein 2006 *Housing Reconstruction In Post-Earthquake Gujarat : A Comparative Analysis* Overseas Development Institute
[24] T Ophiyandri R Amaratunga And C Pathirage 2010 *Community Based Post Disaster Housing Reconstruction: Indonesian Perspective*
[25] J Twigg et al 2017 *Key Messages Self-Recovery From Disasters An Interdisciplinary Perspective* Overseas Development Institute United Kingdom
[26] G Fernandez And I Ahmed 2019 '*Build Back Better' Approach To Disaster Recovery: Research Trends Since 2006* Progress In Disaster Science Vol 1 Elsevier Ltd May 01
[27] C H Davidson C Johnson G Lizarralde N Dikmen And A Sliwinski 2007 *Truths And Myths About Community Participation In Post-Disaster Housing Projects* Habitat Int Vol 31 No 1 Pp 100–115
[28] B Rouhanizadeh And S Kermanshachi 2019 *Investigating The Relationships Of Socioeconomic Factors Delaying Post-Disaster Reconstruction* ASCE No 2010 Pp 105–113
[29] C Venable A Opdyke A Javernick-Will And A Liel 2018 *Community Participation In Post-Disaster Shelter Programs: Examining The Evolution Of Participation In Planning Design And Construction* In Construction Research Congress 2018: Infrastructure And Facility Management - Selected Papers From The Construction Research Congress Vol 2018-April Pp 222–232
[30] G A Howell And L Koskela 2000 *Reforming Project Management: The Role Of Lean Construction*
[31] L Koskela 1992 *Application Of The New Production Philosophy To Construction*
[32] G A Howell And G Ballard 1999 *Bringing Light To The Dark Side Of Lean Construction: A Response To Stuart Green*
[33] S D Green 1999 *The Dark Side Of Lean Construction: Exploitation And Ideology*
[34] J Tamošaitienė And T Starta 2019 *Integration Of Lean Last Planner System In Construction Processes*
[35] N Chugani V Kumar J A Garza-Reyes L Rocha-Lona And A Upadhyay 2015 I*nvestigating The Green Impact Of Lean Six Sigma And Lean Six Sigma: A Systematic Literature Review* International Journal Of Lean Six Sigma Vol 8 No 1 Emerald Group Publishing Ltd Pp 7–32
[36] Z K Rybkowski 2010 *Last Planner And Its Role As Conceptual Kanban*
[37] L Koskela 2004 *Moving On-Beyond Lean Thinking Title Moving On-Beyond Lean Thinking Moving-On-Beyond Lean Thinking* Lean Constr J
[38] H G Ballard 2000 *The Last Planner System Of Production Control Acknowledgements*

[39] D Murguia And X Brioso 2016 *Applying Lean Techniques To Improve Performance In The Finishing Phase Of A Residential Building Process Integration Framework For The Design Phase Of A Residential Building View Project A Model Of Systemic BIM Adoption View Project*

[40] A Hiwale A Wagh V Waghmare D Khairnar S Champanerkar And P Mane 2018 *Effectiveness Of 5s Implementation In Lean Construction (Commercial Building Construction Project* International Journal for Research in Applied Science & Engineering Technology Vol 887

[41] P V Ramani And L K L Ksd 2019 *Application Of Lean In Construction Using Value Stream Mapping* Eng Constr Archit Manag

[42] A Kuiper R Van De Hoef M Wesseling B A Lameijer And R J M M Does 2016 *Quality Quandaries: Improving A Customer Value Stream At A Financial Service Provider* In Quality Engineering Vol 28 No 1 Pp 155–163

[43] L R Espinoza R F Herrera And X Brioso 2021 *Use Of Value Stream Mapping In A Case Study In Basement Construction* In IGLC 2021 - 29th Annual Conference Of The International Group For Lean Construction - Lean Construction In Crisis Times: Responding To The Post-Pandemic AEC Industry Challenges Pp 995–1004

[44] I Tommelein *Pull-Driven Scheduling For Pipe-Spool Installation : Simulation Of Lean P Ull - Driven Scheduling For Pipe - Spool Installation* J Constr Eng Manag Vol 9364

Underground-caves connectedness preview using the geoelectrical profiling method as an effort to revive natural tourism in "Gua Pandan," East Lampung Regency

R M Antosia[1], H H Akbar[1], N A Santoso[1], I A Putri[1], B A Farishi[2] and H C Natalia[2]

[1]Department of Geophysical Engineering, Sumatera Technological Institute, Indonesia
[2]Department of Geological Engineering, Sumatera Technological Institute, Indonesia

E-mail: martin.antosia@tg.itera.ac.id

Abstract. The Covid-19 pandemic has made one of the tourists in the East Lampung regency inactive, namely "Gua Pandan." Due to its unique location geologically, efforts must be made to revive the area. One of the approaches is applying an indirect method, namely the geophysical method. One of them is the geoelectrical profiling method. Six lines are spread in the study area with a length of 70-105 m. The target is a cave filled with air because it is suitable as a tourist spot. Based on the measurement results, a very high resistivity value of more than 5,000 Ωm is suspected to be an air-filled cave. The cave area that is probably safe is about 120 m long with 1-3 m depth below the ground surface, 2-8 m width, and 2-6 m height. However, a safety assessment is necessary for feasibility so that the cave is worthy of being re-activated as a tourist spot.

1. Introduction

"Gua Pandan" is one of the geodiversity in East Lampung regency, Lampung-Indonesia [1,2]. In addition, the natural potential has eco-cultural value for the East Lampung regency [3]. Before the Covid-19 pandemic, this cave was one of the most visited tourist objects by the public [2-4]. Based on available information, this cave has a length of about 200-250 m [1,4]. This cave is also unique because it is located in the Sukadana Lava lithology unit [1], which usually caves are found in karst areas.

To revive this natural potential, conducting a feasibility study as a tourist place is necessary. One of these studies needs to re-checked the continuity direction or the connectedness between the cave mouths because, in some locations, there are cave mouths that stone ruins have covered. According to information from residents, there is also reasonably high mud in the cave, which makes it difficult to pass through the cave. Therefore, one of the studies that can be employed and environmentally friendly in the study area is to use the geoelectrical profiling method.

A profiling method is a 2D method that applies an active injected electrical field to the ground surface to get the resistivity distribution in the subsurface area [5,6]. The technique has many utilities, such as aquifer prospecting [7,8], subsurface structure [9,10], mining materials [11,12], and also for caves [13-18] or underground cavities [19-21]. In the caves/ cavities studies, most are in the karst area [14-16], even though there is an artificial cave [13]. The resistivity profiles show a high-value anomaly because of the air-filled cave/ chamber [13,14,18,21]. The technique of [21] worked to detect a shallow cave (a depth of fewer than 4 m with dimensions of about 2 m in height and 12 m in width) with an electrode spacing of 3 m. The study area has a similar target depth to [21], so the authors will implement an electrode interval of 2-3 m with the approaching Wenner-Schlumberger array because it has an acceptable horizontal resolution [22].

2. Method

When the electrical current through a pair of electrodes is injected into the earth's surface, the response will produce the potential difference between them [5,6,23]. Figure 1 shows the illustration of the electrode positions. According to [5,6], the observed electrical parameter is called apparent resistivity (ρ_a). It is calculated from the ratio of an injected current (I) and a potential difference (V), then multiplied by a geometrical factor K (based on the electrode positions, like in Figure 1). The statement can be simpled by the following equation [5,6],

$$\rho_a = K\frac{V}{I} \tag{1}$$

All variables are in The International System of Units (SI). The observed apparent value assumes that the earth is homogeneous. The reality is heterogeneous, so the pseudo-value will be inverted to get the estimated actual resistivity value using the inversion tool RES2DINV software from Geotomo Software [22].

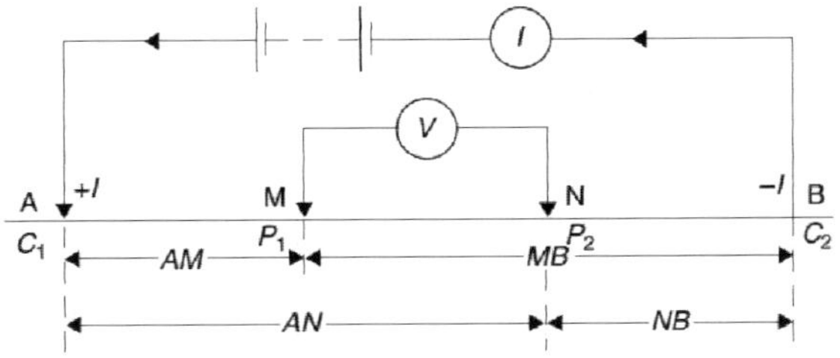

Figure 1. Electrode positions between the current electrode (AB or C_1C_2) and the potential electrode (MN or P_1P_2) [6].

This study will apply the Wenner-Schlumberger array of the electrode configuration in the field measurement process. The illustration of the 2D data acquisition can be viewed in Figure 2. Because the spacing of the potential electrodes is relatively constant, it will provide a suitable resolution in the lateral direction along the cross-section [22]. It can be seen in the illustration that with every change in the data levels, the potential electrodes are always steady. Each data level represents the deeper subsurface information. According to [22], the following equations are the Wenner-Schlumberger array's geometrical factor (K) and the average ratio of an investigation depth (Z_m) and the array's total length (L),

$$K = \pi n(n+1)a \tag{2}$$

$$\frac{Z_m}{L} = 0.19 \tag{3}$$

with n as a data level ($n = 1,2,3$, etc.), and a is a fixed electrode interval on the ground surface. The shallow target will employ an interval of 2-3 m in this study. In addition, 36 electrodes are used, so the total length is between 70 m and 105 m. By referring to (3), the maximum penetration depth is about 13.3 m and 19.95 m. For more detail about the survey design, Figure 3 presents the field data acquisition schemes in the area with six lines. There are several mouth caves (red dot), and some subsidences or rocks fall that cover the cavity (green zone on the map).

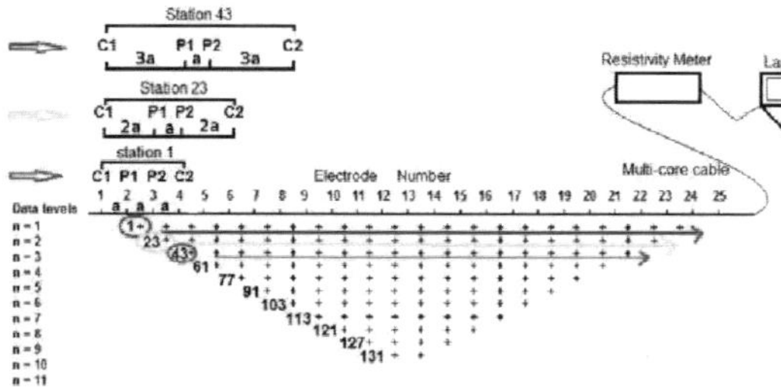

Figure 2. The 2D field measurement illustration for creating a resistivity profile [9].

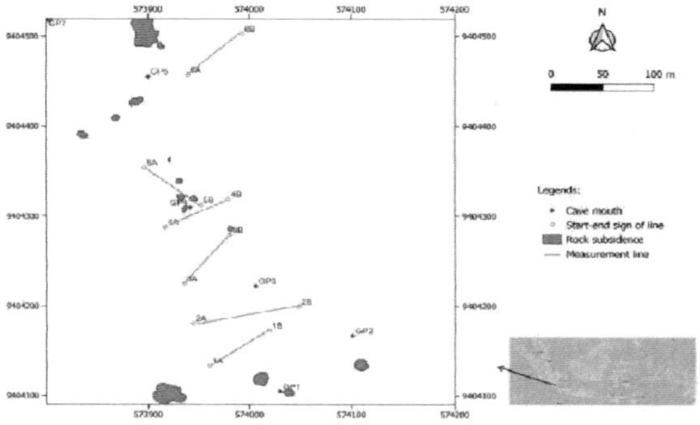

Figure 3. The field measurement map in the study area.

3. Result and Discussion

The location is at a Sukadana Basaltic Formation [1]; most are basaltic lava, one of the igneous rocks. So theoretically [5], the rock in the formation will have a high resistivity value. It is shown in Figures 4-9 that the matter is more than 400 Ωm. The target is an air-filled cave; hence the grade will be focused on the enormous resistivity anomaly. The distribution value over the depth gradually changes exponentially [22], then the profile display defaulted in logarithmic contour (upper profile of Figures 4 to 9). But sometimes, it leads to ambiguous interpretations. It needs to display a linear shape (bottom profile of Figures 4 to 9) to see the distribution result in a more contrasting value. After comparing each other cross-sections of each line, the authors categorize three cave anomalies. A resistivity value of more than 5,000 Ωm is expected for an air-filled cave. The value of about 4,000-5,000 Ωm is a cave filled with mud or stone materials. The last is the value between 1,000-2000 Ωm fathomed as a fully mud-filled cave, only found in line 6 (Figure 4). The estimation of air-filled and mud/rock-filled caves occurs in all profiles of Figures 5-9. Based on all resistivity cross-sections, the cave depth is about 1-3 m below the ground surface with a width of 2-8 m and height of 2-6 m.

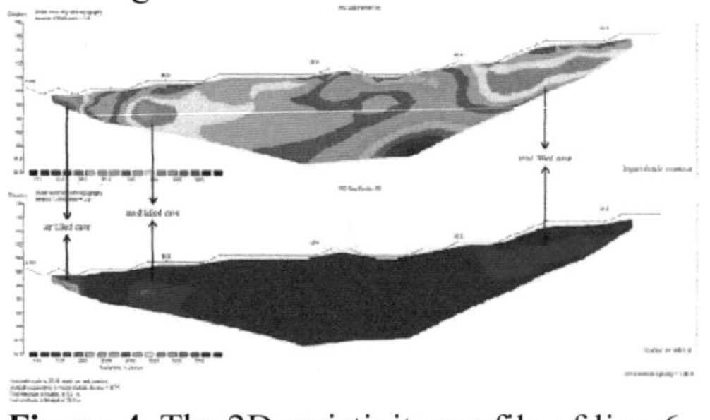

Figure 4. The 2D resistivity profile of line 6.

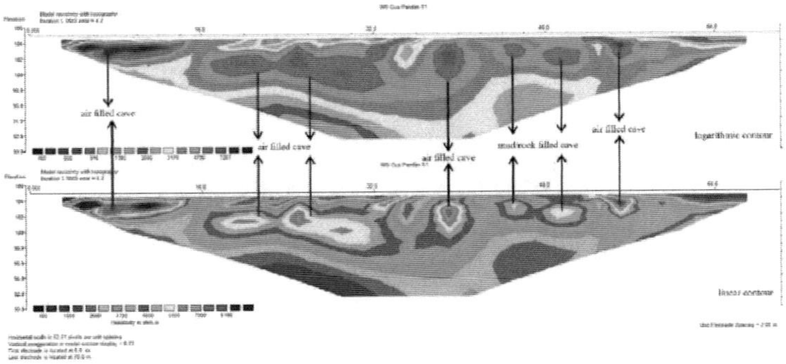

Figure 5. The 2D resistivity profile of line 1.

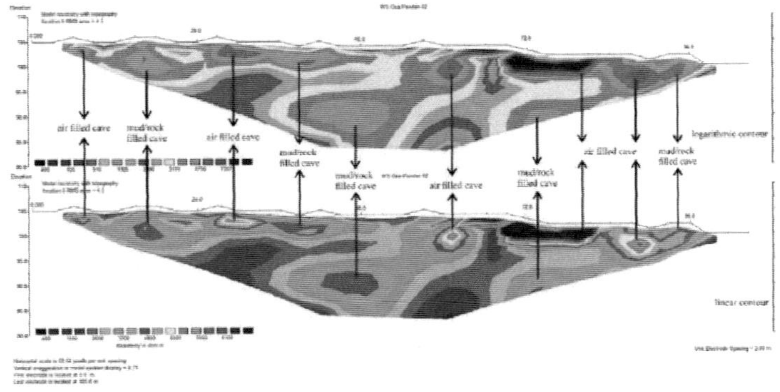

Figure 6. The 2D resistivity profile of line 2.

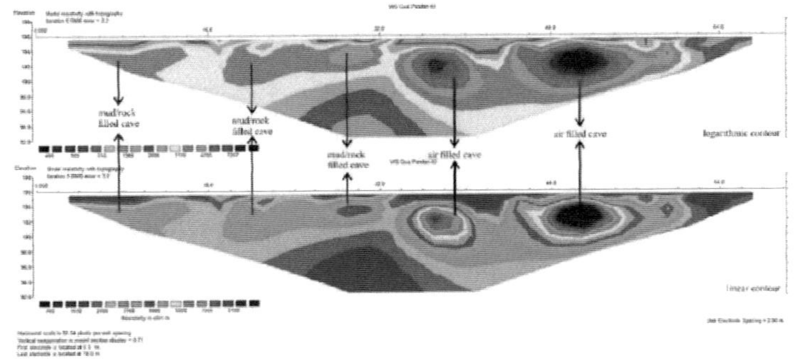

Figure 7. The 2D resistivity profile of line 3.

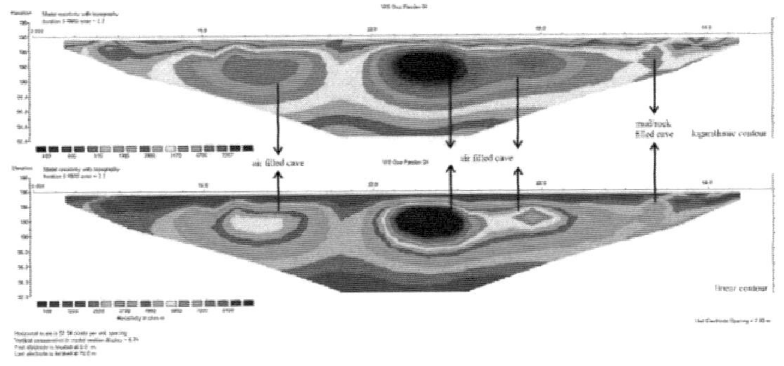

Figure 8. The 2D resistivity profile of line 4.

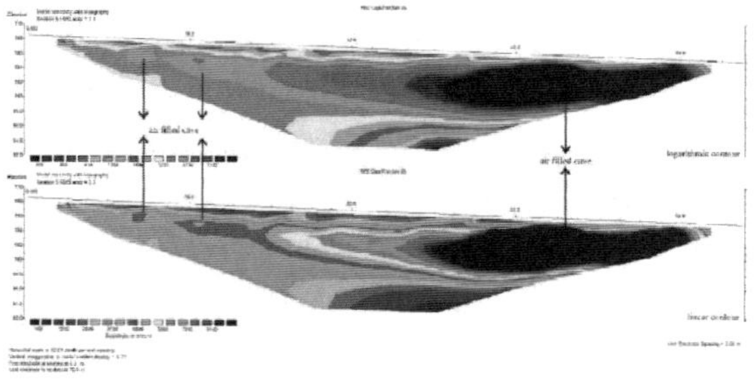

Figure 9. The 2D resistivity profile of line 5.

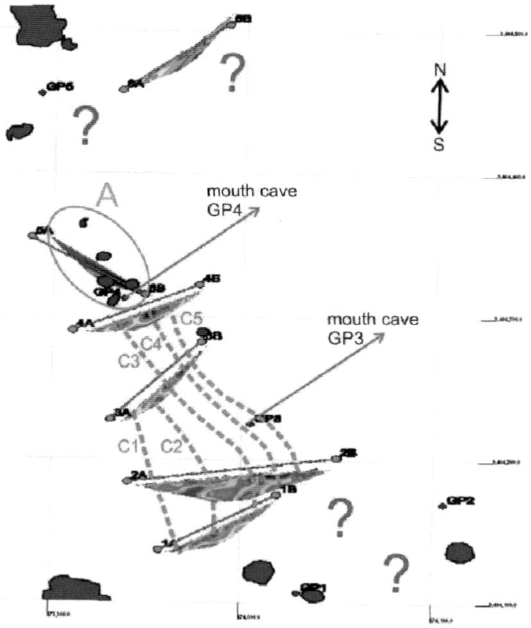

Figure 10. The cave connectedness probability is based on the resistivity profiles.

Each 2D profile is placed according to the position of the line map in Figure 3 (see the result in Figure 10) to provide a representative result. Then, find the connection between profiles. In Figure 10, there are several connectedness C1-C5 (dashed-blue line). Lines C3-C5 are the possible connection between the shape of lines 1-4 with a length of about 120 m. It has also confirmed that on the surface, there is a cave mouth GP3 between lines 2-3 and GP4 near line 4.

On the other hand, the mouth of GP4 is in the presence of a lot of rock subsidence area (area A in Figure 10). The highest resistivity distribution in line 5 also confirms that the air-filled cave might be mixed with the subsidence because the cave width is unreasonable (approximately 40 m based on the profile). In the next step, it is necessary to conduct a safety assessment on the site if GP3-GP4 wants to be activated as part of tourism. There are also other cave mouths, GP1 and GP2, in the south to the southeast region of the map (Figure 10). In the subsequent study, other geoelectrical lines should be added near the cave mouths in question to confirm the possible connection to the lines C1-C5 or perhaps a connection between GP1 and GP2 (question mark "?" in the southeast area of Figure 10). Meanwhile, for the GP5 cave mouth is not necessary to add the data because, according to line 6 (Figure 4), there is a possibility that the cave is filled with mud (question mark "?" located in the north/ northwest of Figure 10). Also, it is not easy to carry out other geoelectrical surveys because of an active agricultural area, so it is not feasible to become a tourist object.

4. Conclusion

This study shows that the area has a chance to be re-activated as tourism. A very high resistivity distribution (more than 5,000 Ωm) presents an air-filled cave, the investigation target. One concern note is that a lot of stones are declining or collapsing in the vicinity close to sites that have the potential to become a tourist attractions again. The site has a possible connectedness estimation length of 120 m with 1-3 m depth below the ground surface, 2-8 m depth, and 2-6 m height. For feasibility, it is recommended to carry out a safety inspection in the area.

Some of the caves filled with mud are not a problem because they are in certain spots. Then the cave, which is thought to be covered with dirt, is far away from the potential area (located north of the study area). The next question is on the measurement map's south and southeast sides. Because the connection is not yet known, adding geoelectrical data in the vicinity is necessary.

Acknowledgment

This study was funded through a 2022 research grant program by the Research and Community Service Institution (LPPM) Institut Teknologi Sumatera. Also, it was supported by the Watershed and Protected Forest Management Agency (BPDASHL) and Giri Mulyo-East Lampung Government. The authors thank the Department of Geophysical Engineering and Geological Engineering students for assisting with the data acquisition.

References

[1] Natalia H C, Harbowo D G and Ikhram R 2021 Potensi Geodiversity di Sekitar Kawasan Anak Krakatau-Way Kambas, Provinsi Lampung, Sebagai Kandidat Geopark Indonesia *Journal of Science and Applicative Technology* **5** 47–57

[2] Admin 2019 Potensi Alam Lampung Timur *TRANS7* https://www.trans7.co.id/seven-updates/potensi-alam-lampung-timur. [Accessed: 10-Aug-2022].

[3] Muklasin and Sitanggang H 2018 Daya Tarik Ekobudaya Lampung Timur *Antara News Lampung* https://lampung.antaranews.com/berita/307383/daya-tarik-ekobudaya-lampung-timur. [Accessed: 10-Aug-2022].

[4] Ato 2019 Pesona Goa Pandan di Lampung Timur *Bintangpost.com* http://bintangpost.com/read/2400/pesona-goa-pandan-di-lampung-timur. [Accessed: 10-Aug-2022].

[5] Telford W M, Geldart L P, Sheriff R E and Keys D A 1990 *Applied Geophysics*, Second Edition (Cambridge: Cambridge University Press)

[6] Reynolds J M 2011 *An introduction to applied and Environmental Geophysics*, Second Edition (Chichester: Wiley-Blackwell)

[7] Minarto E, Christy N N and Ruchimat A 2021 Identification of groundwater potential using Wenner configuration 2D resistivity method (Kupang, Nusa Tenggara Timur) *Journal of Physics: Conference Series* **1805** 012034

[8] Sikah J N, Aning A A, Danuor S K, Manu E and Okrah C 2016 Groundwater Exploration using 1D and 2D Electrical Resistivity Methods *Journal of Environment and Earth Science* **6** 55–63

[9] Wijaya A S 2015 Aplikasi Metode Geolistrik Resistivitas Konfigurasi Wenner Untuk Menentukan Struktur Tanah di Halaman Belakang SCC ITS Surabaya *Jurnal Fisika Indonesia* **19** 1–5

[10] Rizka, Fadhilah R, Piskora B A, Putraloka A, Alansyah A and David K H 2021 The identification of Krakatoa tsunami deposits based on comparison of geological and electrical resistivity tomography method, in Kunjir, South Lampung *IOP Conference Series: Earth and Environmental Science* **830** 012026

[11] Farduwin A, Lumbatoruan P G, Karyanto and Triyanto D 2021 Identification of zeolite using electrical resistivity tomography in Campang Tiga, South Lampung Regency *IOP Conference Series: Earth and Environmental Science* **882** 012046

[12] Antosia R M, Mustika, Putri I A, Rasimeng S and Dinata O 2021 Andesite prospect at West Sungkai of North Lampung: Its distribution based on electrical resistivity tomography *IOP Conference Series: Earth and Environmental Science* **882** 012086
[13] Ardi N D and Aryanti M 2015 Profil Resistivitas 2D pada Gua Bawah Tanah dengan metode Geolistrik Konfigurasi Wenner-Schlumberger (Studi Kasus Gua Dago Pakar, Bandung) *Jurnal Pengajaran Matematika dan Ilmu Pengetahuan Alam* **14** 79–86
[14] Pranantya P A and Sadikin N 2016 Aplikasi Metode Tahanan Jenis dalam Studi Geologi Karst Gua Seropan di Gunung Kidul, Yogyakarta *Jurnal Sumber Daya Air* **12** 105–116
[15] Ciputra R C and Muhammad A G 2019 Deteksi Gua Bawah Tanah dengan Metode Tahanan Jenis: Studi Kasus Gua Bribin, Kabupaten Gunung Kidul *Prosiding Seminar Geologi Nuklir dan Sumber Daya Tambang* **2019** 167–174
[16] Soma A Z, Arsyad M and Tiwow V A 2021 Analisis Karakteristik Kawasan Wisata Gua Leang Lonrong Taman Nasional Bantimurung Bulu Saraung *Jurnal Sains dan Pendidikan Fisika* **17** 93–103
[17] Zeid N A, Bignardi S, Santarato G and Peresani M 2017 Exploring the paleolithic cave of Fumane (Italy): Geophysical methods as planning tool for archaeology *SEG International Exposition and 87th Annual Meeting* **2017** 5125–5129
[18] da Gama M, Braga M A, Barbosa M R, de Paula R and Brandi I 2019 Geophysics applied to litho-structural mapping of Iron Caves, carajás, Brazil *Proceedings of the 16th International Congress of the Brazilian Geophysical Society* **2019** 1–5
[19] Hassan A A 2017 Numerical Modelling of Subsurface Cavities Using 2D Electrical Resistivity Tomography Technique *Diyala Journal For Pure Science* **13** 197–216
[20] Abbas S A, Saeed M, Ghani M and Ahmad T 2020 Subsurface Cavity Detection Using Electrical Resistivity Tomography (ERT); A Case Study from Southern Quetta, Pakistan *Pakistan Journal of Geology* **4** 101–105
[21] Thabit J M and Abed A M 2014 Detection of Subsurface Cavities by Using Pole-Dipole Array (Bristow's Method)/Hit Area-Western Iraq *Iraqi Journal of Science* **55** 444–453
[22] Loke M H 2022 Tutorial : 2-D and 3-D electrical imaging surveys *Geotomo software – download* https://www.geotomosoft.com/downloads.php. [Accessed: 10-Aug-2022].
[23] Antosia R M 2020 Voltmeter design based on ADS1115 and Arduino Uno for DC resistivity measurement *JTERA (Jurnal Teknologi Rekayasa)* **5** 73-80

The capacity of Polyurethane Sponge to Adsorb Nitrate, Ammonium, and Phosphate

T Edwin[1], M Mera[2], P S Komala[3] and Zukarnaini[4]
[1]Doctor Student in Civil Engineering at University of Andalas, Indonesia
[2]Lecturer in Civil Engineering at University of Andalas, Indonesia
[3]Lecturer in Environmental Engineering at University of Andalas, Indonesia
[4]Lecturer in Environmental Engineering at University of Andalas, Indonesia

Email: mas_mera@eng.unand.ac.id

Abstract. Nitrate, ammonium, and phosphate are the main contaminants in agricultural runoff that can harm aquatic organisms if in high concentrations reaches water bodies, for that contamination must be controlled. Polyurethane sponge is a durable material and can be used for filtering those contaminants from wastewater, but studies on the mechanism of adsorption by this material are still rare. Therefore, the purpose of this study is to predict the adsorption capacity, removal efficiency, isotherm, and adsorption kinetics of polyurethane sponge on ammonium, nitrate, and phosphate removal. The results of the study show that polyurethane sponge can remove nitrate, ammonium, and phosphate by adsorption mechanism. The highest maximum adsorption capacity is nitrate in 5-hour mixing time, followed by ammonium in 2-hour mixing time, and the last one is phosphate in 3-hour mixing time. The isotherm that is suitable for presenting the removal of nitrate and ammonium is the Freundlich isotherm, while the Langmuir isotherm is more suitable for the removal of phosphate. The adsorption kinetics which fit to describe the mechanism of nitrate and ammonium adsorption is the intra-particle diffusion model, while the phosphate adsorption kinetics is more suitable to be described by the Pseudo first-order model.

1. Introduction

In general, farmers use synthetic fertilizers and compost containing nutrients in the form of nitrogen and phosphate to increase agricultural production [1]. Fertilizers that are not absorbed by plants can be washed away and enter rivers or lakes through sewers. Fertilizer drift is an accumulation of nitrogen and phosphate. Most of the nitrogen in water is in the form of nitrate and ammonium. The presence of excess nitrate, ammonium, and phosphate can cause eutrophication, resulting in the growth of abundant aquatic plants. This can reduce oxygen levels in water bodies and harm the biota in these water bodies. The presence of nitrogen (nitrate and ammonium) alone in a water body will not cause eutrophication, but the presence of accumulated nitrogen and phosphate (nitrate and phosphate, or ammonium and phosphate) can cause eutrophication [2].

Polyurethane sponge is formed by chemical reaction of a monomer containing at least two isocyanate functional groups with a monomer containing at least two alcohol functional groups. This material has high porosity and a developed surface. The use of this material as a filter medium and as an attachment medium for microorganisms in charge of removing contaminants from industrial and domestic wastewater, is also applied to a fish pond filter [3,4]. Polyurethane sponges are proven to be harmless to the environment. The advantages compared to other adsorption materials, polyurethane sponge is resistant to weather conditions because it is not easy to decompose, odourless and harmless to humans

[5]. The ability of polyurethane to coat biochar is effective as an adsorbent in removing phosphate [6]. Modified polyurethane sponge can effectively remove ammonium [7]. Apart from being a medium for attaching microorganisms, polyurethane sponges have the ability to remove contaminants, one of which is through the adsorption process, namely the phenomenon of the transfer of contaminants to the solid surface. Therefore, this study is aimed to predict the adsorption capacity and efficiency of ammonium, nitrate, and phosphate removal using polyurethane sponge alone as adsorption material. The isotherm analysis and adsorption kinetics are used to estimate the adsorption capacity of polyurethane sponges.

2. Material and methods

The adsorption material used is polyurethane sponge with a black color and a density of 15 to 50 kg/m^3. The sponge is then cut into cubes with a volume of 1 cm^3 each. The artificial solution for each parameter used KNO_3 for nitrate, NH_4Cl for ammonium, and KH_2PO_4 for phosphate. The level of biochar used is 0.5 g/100 mL with a contaminant concentration of 20 mg/L [8]. The adsorption ability of the polyurethane sponge is observed at various mixing times of 1, 2, 3, 4, 5, and 6 hours [9]. These data are also used for the analysis of adsorption kinetics. Stirring is carried out with an *orbital shaker* at a speed of 180 rpm after the solution is filtered through a paper filter before analysis. Experiments with variations in the initial concentration of contaminants are 0.5; 10; 20; 40 and 80 mg/L. They are performed for the analysis of adsorption isotherms [8]. All experiments were carried out in triplicate.

Measurement of nitrate, ammonium, and phosphate concentrations is measured using a UV-VIS Spectrophotometer (Shimadzu UV-2600) with a procedure based on the Indonesian National Standard [10]. The removal efficiency (*%R*) is calculated using equation (1), and the adsorption capacity (q) is calculated using equation (2) [11].

$$\%R = \frac{C_0 - C_t}{C_0} \times 100\% \quad (1)$$

$$q = \frac{(C_0 - C_t)V}{m} \quad (2)$$

%R=Removal efficiency, C_o = initial concentration of solution, C_t = concentration up to time t (mg/L), V = solution volume (L), and m = weight of polyurethane sponge (g).

The adsorption isotherm plays an important role because it provides certain useful information such as understanding the adsorption mechanism, surface characteristics and adsorbate-adsorbent affinity. The Freundlich and Langmuir isotherm equations respectively are shown in equations (3) and (4).

$$x/m = K_f(C_e)^{1/n} \quad (3)$$

K_f= Freundlich constant (L/g) and n = adsorption capacity and adsorption intensity, x/m = amount of contaminant adsorbed to 1 g adsorbent (mg/g), C_e= contaminant concentration at equilibrium (mg/L).

$$x/m = \frac{abC_e}{1+bC_e} \quad (4)$$

x/m = amount of phosphate adsorbed to 1 g of adsorbent in equilibrium condition (mg/g), C_e = contaminant concentration at equilibrium (mg/L), a, b = empirical constant.

The adsorption kinetic model approach used in this study is Pseudo first-order kinetics (equation 5), Pseudo second-order kinetics (equation 6), and also the intra-particle diffusion model (equation 7). Adsorption kinetic approach is carried out to determine the rate of contaminant adsorption by adsorption media in a certain period of time [12].

$$q_t = q_e[1 - \exp(-k_{1p}t] \quad (5)$$

$$q_t = \frac{K_{2p}{q_e}^2 t}{1+K_{2p}{q_e}^2 t} \tag{6}$$

$$q_t = K_p t^{0,5} + C \tag{7}$$

q_t = adsorption capacity along mixing time t (mg/g), q_e = adsorption capacity at equilibrium (mg/g), k_{1p} = Pseudo first-order constant (/h), and t = mixing time (h). K_{2p} = Pseudo second-order constant (g/mg/h), K_p = constant of intra-particle diffusion (mg/g/h$_{0,5}$), C = related to the thickness of the boundary layer (mg/g).

3. Results and discussion

The results of adsorption capacity and removal efficiency of nitrate, ammonium, and phosphate are presented in Figure 1. For the range of pH conditions observed were 5.9-6.1 for nitrate removal experiment, 5.2-6.6 for the removal of ammonium, and 5.6-6.0 for the removal of phosphate. From the literature study conducted by Ambaye (2020), the pH of the solution and the ionic strength affect the efficiency of the adsorption process [13]. From this study, polyurethane sponge can remove nitrate, ammonium, and phosphate in water by adsorption process. The trend of phosphate removal was inversely proportional to nitrate removal based on the experimental time even though both ions were negatively charged. This can be influenced by the higher pH of the solution in the nitrate removal so that the surface of the adsorption media is more positively charged and causes a higher removal of nitrate than phosphate [14,15]. Figure 1 shows the adsorption capacity and removal efficiency of nitrate and ammonium which stabilized within 1 hour of mixing. The maximum adsorption capacity for the removal of nitrate and ammonium is 3.23 mg/g at a mixing time up to 5 hours and 2.83 mg/g at a mixing time up to 2 hours, respectively. Meanwhile, the phosphate adsorption capacity began to stabilize at a mixing time of 1 hour to 2 hours, then decreases at a mixing time of 3 hours and increases then stabilized again from a mixing time of 4 hours. The maximum adsorption capacity for phosphate removal is 2.55 mg/g at a mixing time of 3 hours. The results show that the adsorption capacity of nitrate is higher than that of ammonium and phosphate using a polyurethane sponge.

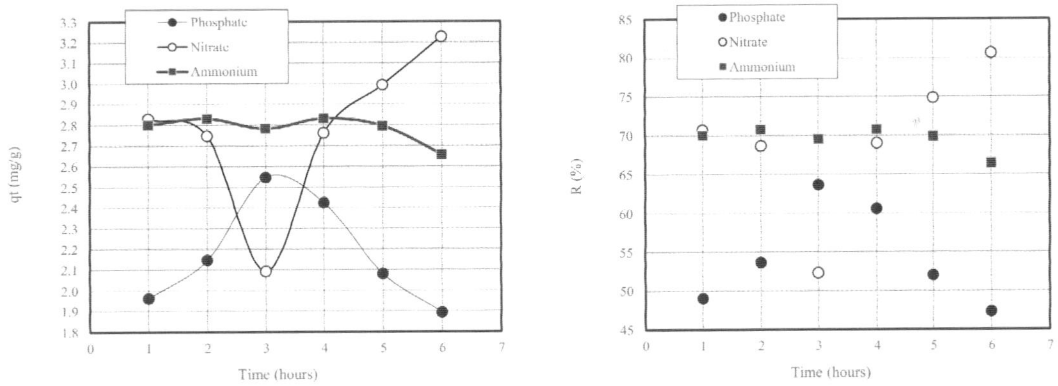

Figure 1. Adsorption capacity (left) and removal efficiency (right)

Figure 2 shows that the adsorption of nitrate, ammonium, and phosphate using a polyurethane sponge show agreement with the Freundlich isotherm. The Langmuir model describes monolayer adsorption, assuming that adsorption occurs at a certain number of adsorption location in the surface media, where each site is occupied by one adsorbate or contaminant molecule, and between the adsorbed molecules assumed no interaction [16]. While the Freundlich model represents adsorption that is not ideal, with many adsorption sites on the sorption media, which assumed as heterogeneous surfaces. This is based on the assumption that the active binding site is occupied first by adsorbate, and the binding of adsorbate decreases as the site occupancy increases. For illustrations of the respective nitrate, ammonium, and phosphate removal isotherms with polyurethane sponge materials (see Figure 2).

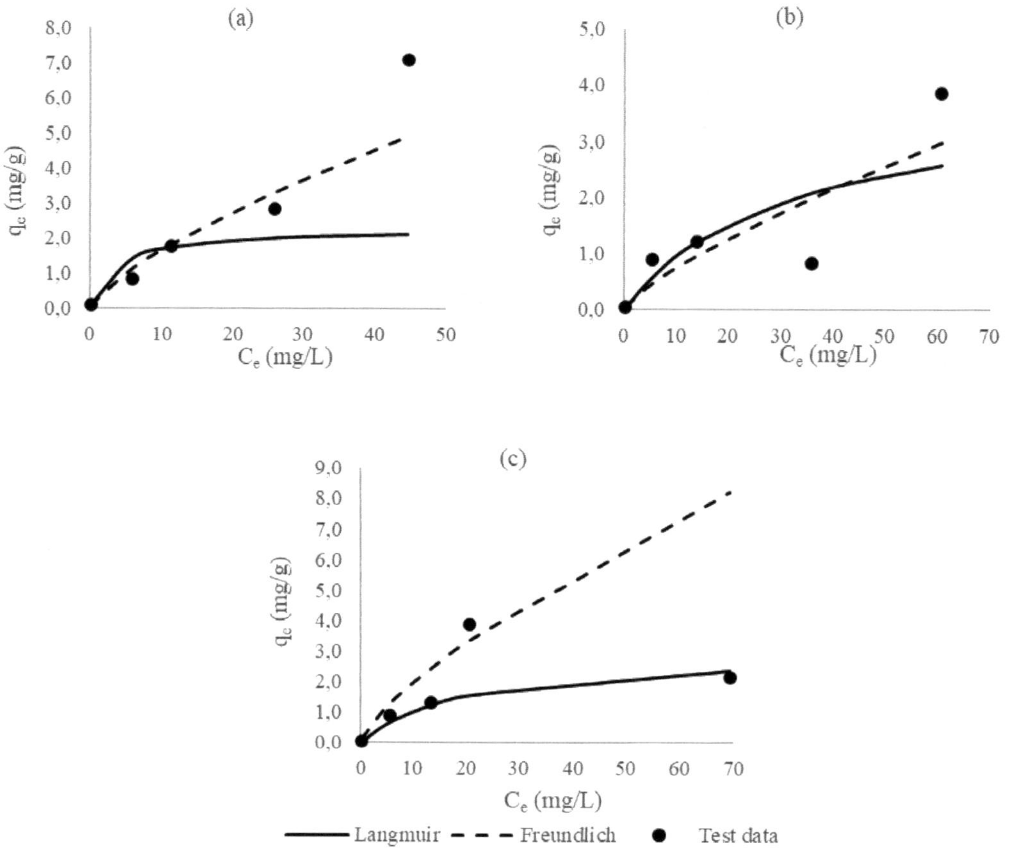

Figure 2. Adsorption isotherm for the removal of (a) nitrate (b) ammonium, dan (c) phosphate

Table 1. Adsorption isotherm

No	Contaminant	Langmuir			Freundlich		
		Q_{max}	K_L	R^2	n	K_F	R^2
1	Nitrate	0.28	0.07	0.70	1.29	0.12	0.98
2	Ammonium	0.10	0.03	0.23	1.37	0.30	0.88
3	Phosphate	0.17	0.07	1.00	1.35	0.36	0.97

Table 2. Adsorption kinetics

No	Contaminant	Pseudo 1st order			Pseudo 2nd order			Intra-particle diffusion		
		K (/h)	q_e (mg/g)	R^2	K (/h)	q_e (mg/g)	R^2	K (/h)	C (mg/g)	R^2
1	Nitrate	0.021	0.706	0.002	0.098	2.794	0.490	1.107	0.665	0.710
2	Ammonium	0.108	0.505	0.011	0.053	3.517	0.340	0.992	0.850	0.627
3	Phosphate	2.387	3.438	0.800	0.041	4.520	0.312	0.790	0.642	0.602

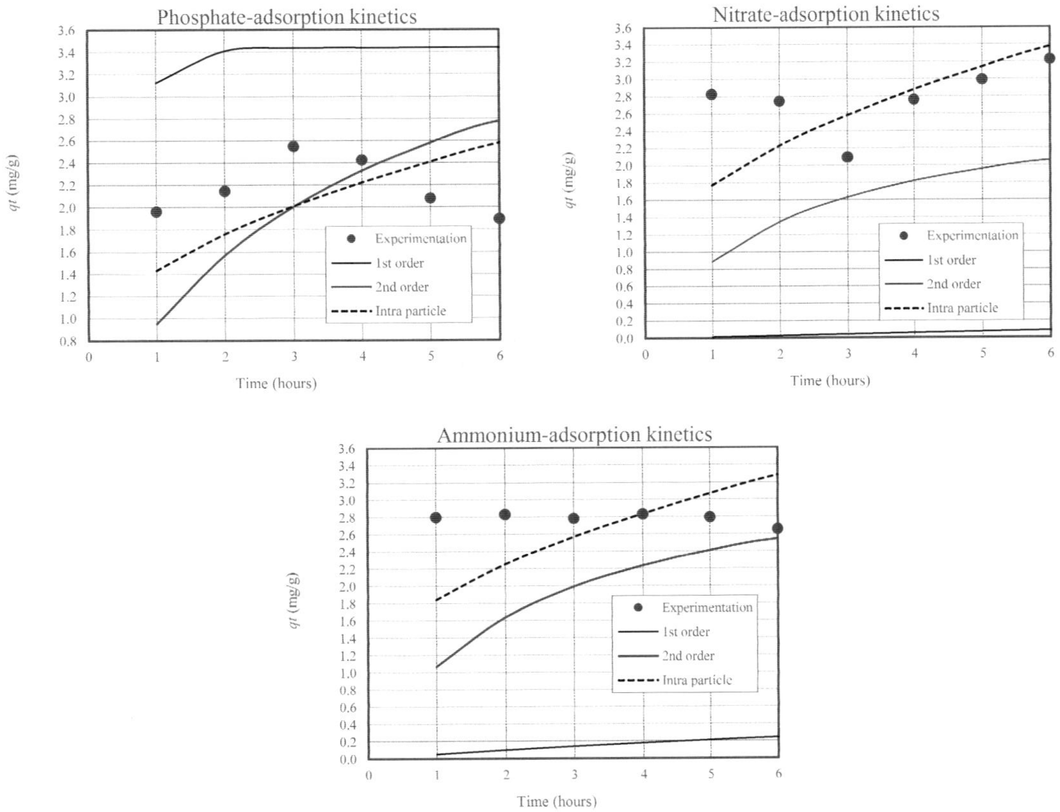

Figure 3. Adsorption kinetics for nitrate removal (left above) ammonium (right above), and phosphate (below)

The results of the calculation of adsorption kinetics are presented in Table 2. The R^2 value closest to 1 is considered more suitable to explain the adsorption that occurs. It appears that all materials for adsorption for nitrate and ammonium removal are more suitable to be described by the intra-particle diffusion model (IPD), while the phosphate removal is more suitable to be described by the first-order pseudo model. the mechanism with the IPD model is adsorption on the internal solids by diffusion into the interior of the adsorbent [16]. While the Pseudo First-Order model is fit to describe adsorption on the outside of the adsorption media [17].

4. Conclusions

From the results of the study, it is found that polyurethane sponge is capable to remove nitrate, ammonium, and phosphate by adsorption process. The highest adsorption capacity is nitrate, followed by ammonium and phosphate. In the removal of nitrate and ammonium, the Freundlich isotherm is the most suitable, while in the removal of phosphate the most suitable isotherm is the Langmuir isotherm. The most suitable adsorption kinetics for the removal of nitrate and ammonium is the intra-particle diffusion model, while the Pseudo First-Order model is the most fit for phosphate removal.

5. References

[1] Mateo-Sagasta J, Zadeh S M and Turral H 2017 *Exec. Summ.* 35
[2] Zeng Q, Qin L, Bao L, Li Y and Li X 2016 *Environ. Sci. Pollut. Res.* **23** 21008–19
[3] Bouabidi Z B, El-Naas M H and Zhang Z 2019 *Environ. Chem. Lett.* **17** 241–57
[4] Wei D, Li B, Huang H, Luo L, Zhang J, Yang Y, Guo J, Tang L, Zeng G and Zhou Y 2018 *Chemosphere* **197** 165–80

[5] Dacewicz E 2021 *Materials.* **14**1–19
[6] Meng Y, Wang Y and Wang C 2021 P *Polymers (Basel).* **13** 1–17
[7] Ahmed Z, Kim K P and Shin J 2016 *Desalin. Water Treat.* **57** 14849–57
[8] Zhou L, Xu D, Li Y, Pan Q, Wang J, Xue L and Howard A 2019 *Water (Switzerland)* **11** 1–16
[9] An T and Nguyen H 2015 Ph.D Thesis, University of Technology Sydney, 2015
[10] Rodger Baird, Eugene W. Rice Andrew D E 2017 *Standard Methods For The Examination OF Water and Wastewater, 23rd Edition* ed A D E E.W. Rice, R.B. Baird (USA: American Water Works Association)
[11] George Tchobanoglous F L B 1991 *Wastewater Engineering: Treatment Disposal Reuse-3rd edition* ed B J C and J M Morries (United State: McGraw-Hill, Inc)
[12] Wang J and Guo X 2020 *J. Hazard. Mater.* **390** 122156
[13] Ambaye T G, Vaccari M, van Hullebusch E D, Amrane A and Rtimi S 2020 *Int. J. Environ. Sci. Technol.*
[14] Ahmad M, Rajapaksha A U, Lim J E, Zhang M, Bolan N, Mohan D, Vithanage M, Lee S S and Ok Y S 2014 *Chemosphere* **99** 19–33
[15] Mukherjee A, Zimmerman A R and Harris W 2011 *Geoderma* **163** 247–55
[16] Kumar P, Sudha S, Chand S and Srivastava V C 2010 *Sep. Sci. Technol.* **45** 1463–70
[17] Revellame E D, Fortela D L, Sharp W, Hernandez R and Zappi M E 2020 *Clean. Eng. Technol.* **1** 100032

Junior high school students preparedness toward earthquake in urban and rural areas in Pesisir Selatan

R Salsabila[1], A R Alfian[1] and A Rahman[1]

[1]Faculty of Public Health, Andalas University, Padang, 25163

Email: azyyatiridhaalfian@ph.unand.ac.id

Abstract. The earthquake on September 30th, 2009, damaged 35 schools in Pesisir Selatan, so student preparedness really needs to be improved to reduce fatalities and the impact of earthquake. This study aims to determine preparedness differences of junior high school students in urban and rural areas against earthquake in Pesisir. This study is quantitative with a cross-sectional design. The research was conducted in junior high school located in urban and rural areas in Pesisir Selatan in July-August 2021 with 84 students selected using the proportional systematic random sampling technique. Data collection was done through interview with questionnaire. The parameter employed was LIPI's which consists of multiple facets such as knowledge, emergency response, early warning, and resource mobilization. Data were then analyzed using the Mann-Whitney test with $\alpha = 0.05$ and CI = 95%. The results showed there were differences preparedness of junior high school students in urban and rural areas with p-value = 0.000. Therefore, the Government through the Health Office and Education Office, and regional disaster management agency can cooperate regularly and periodically regarding the provision of simulations/training and monitoring regarding disasters to students.

1. Introduction

An incident that can be categorized as a disaster is when an event has damaged or interfered with human activities, either causing casualties or damages to existing facilities and infrastructure. In the latest data from the EM-DAT (Emergency Events Database) in 2020, there have been 389 natural disasters in the world. In 2020, 41% of the total disaster in the world happened in Asia. Indonesia has a very high prevalence of natural disasters, data from the 2018 world risk report, Indonesia is at number 36 with a total index of 10.36 out of 172 countries prone to earthquakes and tsunami. The reason is due to the country's geographical location. Indonesian territory is tectonically 3 (three) plates, namely the Indo-Australian plate, the Pacific plate, and the Eurasian plate, and Indonesia also has nine small plates that form complex meeting paths [1-5]

Earthquakes are natural disasters that have the potential to be the biggest threat that can cause damages in Indonesia, the impact of earthquakes can be fatal, and can occur in dense urban and rural areas. In Indonesia, earthquakes occur almost every year, which is approximately 3 earthquakes measuring 7 on the Richter Scale each year which causes loss of life and damage to infrastructure and the environment [6]. According to data from the National Disaster Management Agency (BNPB) it is stated that in a period of 15 years (2004-2018) Indonesia has experienced 240 earthquakes with a large magnitude and 14 tsunamis which caused an impact on more than 4 million people's lives. West Sumatra Province is one of the provinces in Indonesia which is located on the west coast of the central part of the island of Sumatra which stretches from northwest to southeast [7,8]

Districts and cities on the coast of West Sumatra that have a high potential for earthquake disasters, including West Pasaman Regency, Pasaman City, Agam Regency, Padang Pariaman Regency, Pariaman City, Padang City, and Pesisir Selatan Regency, and this area are also feared to be affected become the next place to be hit by a tsunami [8]. Pesisir Selatan Regency is a district that has a high disaster risk score of 168 and is ranked 79th out of 496 districts and cities in Indonesia [9]. Pesisir Selatan Regency is an area that is often the epicenter of earthquakes, the latest 3 earthquakes centered in Pesisir Selatan Regency, namely the 4.1 SR earthquake on June 16, 2020, the epicenter was in the sea, at a depth of 50 Km and located 6 Kilometers (KM) southwest of Painan, The next South Coast earthquake with a magnitude of 3.5 SR on June 29, 2020, the epicenter was at 22 Km Southwest of Painan, Pesisir Selatan. The most recent earthquake was an earthquake measuring 5.2 on the Richter scale on August 5, 2020, centered 56 Km Southwest of Pesisir Selatan [10]

According to LIPI-UNESCO/ISDR in 2016, disaster preparedness had to be be carried out because disaster risk reduction activities were the most important element in the pre-disaster stage so that the impact of disasters is reduced. Because the impact of a disaster is so large that it can cause physical and psychological impacts, as well as property losses to people who live in urban and rural areas. Urban and rural areas are areas that have equal opportunities for obtaining information and facilities related to disaster preparedness and other matters, but at this time there are still differences between areas located in urban and rural areas [11]. Research conducted by Irma Widiastar in 2016 on Community Preparedness in the face of public health emergencies (study on the Makasar Urban Village-East Jakarta Community and Campaka Village-Cianjur Regency in 2016) states that there are differences in the preparedness value of the people who live in rural and urban. For people who live in urban areas, the percentage of preparedness is 62.3%, while for people who live in rural areas it is 41.3% [12]. Another research result, which was found in the research of Ling Chai, et al about the differences in disaster preparedness between urban and rural communities in China in 2020, results are that there are urban and rural differences in disaster preparedness, mainly explained by socio-economic factors, such as education, income, and social status. The data obtained showed that the material for preparedness in urban areas was 37.48%, while in rural areas it was 34.86%. The level of knowledge in preparedness in urban areas is 77.52%, while in rural areas it is 75.97%. Disaster preparedness measures in urban areas are 24.50%, while in rural areas are 18.11%. In overall disaster preparedness, it was found that urban area preparedness was 49.43% while in rural areas it was 45.55% [13]. Because of the differences that have been described, it is very interesting to discuss the differences in preparedness that exist in urban and rural areas further.

The Indonesian Institute of Sciences (LIPI) and UNESCO conducted a research in 2006 on the level of preparedness using the parameters of knowledge and attitudes, early warning systems, emergency response plans, policies, and resource mobilization. The results can be concluded that schools are public spaces that have a high level of vulnerability, and the level of disaster preparedness in schools is lower than the level of preparedness of the general public and officials. The experience of the West Sumatra earthquake shows the magnitude of the impact of damage, especially on classrooms, another impact of the earthquake is that teaching and learning activities in schools do not run normally. For this reason, school preparedness is very important to pay special attention to. Adolescents are one of the age groups most at risk of being affected by disasters. Vulnerability in adolescents to disasters can be triggered by a lack of knowledge in understanding the level of risk of disaster threats around them which results in no preparedness in dealing with disasters carried out by adolescents [14]. Therefore, this study is needed to determine the difference in the level of preparedness for earthquake disasters in junior high school students in urban and rural areas in the Pesisir Selatan Regency.

2. Method

This study is a quantitative study using a cross-sectional design conducted in July-August 2021. The population in this study were all public junior high school students in urban and rural areas where the selection of urban and rural areas was carried out based on the regulation of the head of the statistical center number 37 of 2010 on the Classification of Urban and Rural Areas in Indonesia [15]. Junior High

School 2 Painan was chosen as the representative from the urban area and Junior High School 1 Tarusan was chosen as the representative from the selected rural area because it is located in the red zone area with the closest distance from the beach, which is 200 m from the whole school. The research sample was 84 students who were selected by proportional random sampling technique. The study used primary data collected through filling out standardized questionnaires by LIPI/UNESCO/ISDR in 2006 regarding guidelines for measuring community and school community preparedness which was measured using 4 preparedness parameters, namely: knowledge, early warning systems, emergency response plans, and mobilization resource. The index of each parameter is measured by the formula (1) and (2) [16]:

$$\text{Parameter Index} = \frac{\text{Skor } Rill \text{ Parameter}}{\text{Skor Parameter Max}} \times 100 \quad (1)$$

Then the preparedness index was measured using the formula:

$$\text{Preparedness Value} = (0.52 \times \text{Indeks K}) + (0.16 \times \text{Indeks EP}) + (0.22 \times \text{Indeks WS}) + (0.06 \times \text{Indeks RMC}) \quad (2)$$

While the overall preparedness value can be obtained by finding the average value of the overall index value that has been obtained [16]. Data analysis to see the difference in the level of earthquake preparedness for junior high school students in urban and rural areas was carried out using the Mann-Whitney test using an alpha of 0.005 and 95% CI.

3. Result and discussion

The results of the study show that the level of earthquake preparedness of junior high school students in urban and rural areas in the southern coastal district is shown in Table 1 as follows:

Table 1. Level of earthquake disaster preparedness for middle school students in urban and rural.

No.	Region	Mean
1	Urban	59.05
2	Rural	25.95

Based on the observations and research conducted on the preparedness of students of Junior High School 2 Painan or schools located in urban areas, this school already has a fairly good preparedness, the reason is that students not only have gained knowledge related to disasters through social studies lessons but students have also been exposed education about the disaster regularly by teacher. Teacher explains how students act before a disaster occurs, when a disaster occurs and after a disaster occurs. An example of providing information by this teacher is telling students about the map of the evacuation route and where the gathering point is in the event of a disaster at school. Students are also told to remember the telephone number of the family who can be contacted and ask the related family. The tsunami disaster gathering point, the teacher has also told students about how to avoid the danger of an earthquake if they are in the room and the teacher has also provided knowledge regarding siren information that will sound if there is a tsunami warning.

Another cause is that extracurricular activities such as PMR and scouting have been helpful at Junior High School 2 Painan and the high interest of students to carry out this extracurricular. Scouting is extracurricular that is required by schools for all students, this has resulted in the number of students who have been trained in capacity building such as saving themselves and saving other students during a disaster than students who have not been trained.

Meanwhile, at Junior High School 1 Tarusan which is a rural area, there has never been any provision of information related to disasters either by the school or related agencies, namely the Education Office and Regional Disaster Management Agency of Pesisir Selatan Regency. Other causes in the school environment are the insufficient number of posters and leaflets related to disasters with the number of students, as well as at schools there are no evacuation route signs, and there is no sign of a gathering point in the event of a disaster.

The results of the bivariate analysis that has been carried out, show that there are differences in the preparedness of public junior high school students in urban and rural areas in Pesisir Selatan Regency for earthquake disasters. The result of the Mann-Whitney test, the asymp sig (2-tailed) value is 0.000, meaning that there is a significant difference between the preparedness of public junior high school students in urban and rural areas in Pesisir Selatan Regency, as shown in Table 2.

Table 2. Differences in earthquake preparedness levels for middle school students in urban and rural areas in Pesisir Selatan district

No.	Variable	p-value
1	Urban	0,000
2	Rural	

The results of this study are in line with the results of Irma Widiastar's research (2016) on community preparedness in the face of public health emergencies (study on the Makasar Urban Village Community-East Jakarta City and Campaka Village-Cianjur Regency in 2016). Live in rural and urban areas. For people who live in urban areas, the percentage of preparedness is 62.3%, while for people who live in rural areas is 41.3% [12]. Other similar research results are researched by Sunarhadi and Khoirunisa (2015) on knowledge of disaster preparation for urban and rural students in the Solo city. The average level of student preparedness in rural areas is 52.02% in the less prepared category [17]. In another study, which was found in the research of Ling Chai, et al about the difference in disaster preparedness between urban and rural communities in China (2020), the result is that urban area preparedness is 49.43% while in rural areas it is 45.55% [13].

Preparedness is an anticipatory action taken by individuals or groups in an effort to reduce the number of casualties and property losses caused by a disaster. Preparedness is also a form of effective and proactive response carried out to reduce risks arising from the impact of disasters, or it can be concluded that the key to safety in disaster events is preparedness [18]. To inculcate a culture of disaster preparedness for students, schools have a large enough role, namely to create feelings in the context of planning and self-control against tsunami disasters, earthquakes, and other disasters, namely with students knowing how to protect and survive disasters, knowing what actions to take. What should be done during and after a disaster occurs?

Consistency and active participation of the school community are needed to maintain and improve preparedness for earthquake and tsunami disasters. Students and teachers must continue to play an active role and collaborate to maintain and improve preparedness facilitated by schools. The form of the teacher's role in maintaining and improving student preparedness, namely the teacher always provides information to students related to disasters and increases knowledge related to existing disasters because disaster management information will always increase according to technological developments, for students must always listen to directions from teachers related to this disaster. And always present and active in participating in PMR extracurriculars and scouting because these extracurriculars are very useful for disaster management. The follow-up of maintaining and improving school preparedness requires the participation of related parties, for example, the Education Office, Health Office, and Regional Disaster Management Agency of Pesisir Selatan Regency in supporting the implementation of student preparedness activities, which can provide simulations related to disasters, especially earthquake disasters.

IOP Conf. Series: Earth and Environmental Science 1173 (2023) 012074 doi:10.1088/1755-1315/1173/1/012074

4. Conclusions

There is a difference between the preparedness of public junior high school students in urban and rural areas for the earthquake disaster in Pesisir Selatan Regency, namely the p-value = 0.000. Schools in rural areas should be more active in collaborating with Public Health Center, Education Offices, Health Offices, and Regional Disaster Management Agency in providing information to students through training activities, simulations, and completing all facilities related to disaster preparedness, especially earthquakes.

References

[1] Mustofa N 2010 Gempa Bumi, Tsunami Dan Mitigasinya *J. georafi* **vol 7** no 1
[2] Centre for research on the epidemiology of disasters CRED 2019 *Natural disaster* Indonesia
[3] Centre for research on the epidemiology of disasters CRED 2020 Disaster Year in Review 2020 *Global Trends and Perspectives*
[4] Hadi H Agustina S Subhani A 2019 Penguatan Kesiapsiagaan Stakeholder Dalam Pengurangan Risiko Bencana Gempa Bumi *J Kajian Ilmu dan Pendidikan Geografi* **vol 3** no 1 pp 30–40
[5] Idrus M 2012 Mutu Pendidikan Dan Pemerataan Pendidikan di Daerah **vol 1** no 2
[6] PMI 2016 *Lokakarya Disaster Management oleh Palang Merah Indonesia Cabang (Bengkulu)*
[7] Badan Nasional Penanggulangan Bencana (BNPB) 2019 *Bencana Alam di Indonesia Tahun 2018 s/d 2019* (Jakarta: BNPB)
[8] Gersanandi and Subardjo P Anugroho A 2013 Analisa Spasial Kerentanan Bencana Tsunami Di Kabupaten Dan Kota Pesisir Provinsi Sumatera Barat *J Oseanografi* **vol 2** no 3 pp 232–7
[9] Badan Nasional Penanggulangan Bencana (BNPB) 2013 *Indeks Rawan Bencana Indonesia* (Jakarta: BNPB)
[10] Badan Meteorologi Klimatologi dan Geofisika 2020 *Gempabumi Dirasakan*, Indonesia.
[11] LIPI-UNESCO/ISDR 2006 *Kajian Kesiapsiagaan Masyarakat dalam Mengantisipasi Bencana Gempa Bumi dan Tsunami* (Jakarta: Deputi Pengetahuan Kebumian LIPI)
[12] Irma W 2016 *Kesiapsiagaan masyarakat perkotaan dan pedesaan dalam menghadapi darurat kesehatan masyarakat (studi pada Masyarakat Wilayah Kelurahan Makasar-Kota Jakarta Timur dan Desa Campaka-Kabupaten Cianjur tahun 2016)* (Depok: Universitas Indonesia)
[13] Tjan H Lintong F Wenny S 2013 Efek Bising Mesin Elektronika Terhadap Gangguan Fungsi Pendengaran Pada Pekerja Di Kecamatan Sario Kota Manado, Sulawesi Utara *J Biomedik* **vol 1** pp 34–39
[14] Hermon D 2009 *Dinamika Pemukiman dan Arahan Kebijakan Pengembangan Pemukiman pada Kawasan Rawan Longsor di Kota Padang Sumatera Barat* (Bogor: IPB)
[15] Kepala Badan Pusat Statistik 2010 *Peraturan Kepala Badan Pusat Statistik Nomor 37 Tahun 2010 Tentang Klasifikasi Wilayah Perkotaan Dan Pedesaan Di Indonesia* (Jakarta: Badan Pusat Statistik)
[16] Triyono A K 2011 *Panduan Monitoring dan Evaluasi Sekolah Siaga Bencana* (Jakarta: LIPI Press)
[17] Muh R Amin S Nanda K 2015 Pengetahuan Persiapan Bencana Siswa Perkotaan Dan Perdesaan Di Wilayah Solo *J Tata Kota Dan Drh* **vol 7** no 1
[18] Sudiartha G 2017 *Jangan Panik! Beragam Praktik Baik Kebencanaan* (Jakarta: Badan Nasional Penanggulangan Bencana)

Vulnerability Assessment of non-engineered houses related to earthquake in West Pasaman District

Rahmat Fadli Zakir[1], Abdul Hakam[2*], Bambang Istijono[1], Melinda Noer[1], [3]Syafira Rahmadilla Hape[3]

[1] Post Graduate School, Andalas University, West Sumatra, Indonesia,
[2] Department of Civil Engineering, Andalas University,West Sumatra, Indonesia
[3] Research Center for Geological Disaster, National Research and Innovation Agency, Indonesia

E-mail: ahakam@eng.unand.ac.id, abdulhakam2008@gmail.com

Abstract. The earthquake on February 25, 2022, in West Pasaman resulted in damage to people's houses as well as fatalities. The quake happened just under the Mount of Talamau, 18 km northeast of the city of Simpang Empat City, with a magnitude of 5.2 M. This earthquake was followed by small earthquakes that occurred until February 27, 2022. Damage to houses in the earthquake-affected area became the focus of attention in this research to assess the housing vulnerability related to the earthquake hazard. The vulnerability assessment of house buildings is critical in earthquake-prone areas because it can provide information about the impacts on the buildings. The vulnerability of house buildings is closely related to the potential for damage due to earthquake hazards. The benefit of the vulnerability assessment of house buildings is to arrange the mitigation action plan to reduce the level of damage to building structures. The method, in turn, can reduce the number of fatalities as well. Buildings with a high level of vulnerability to earthquakes are non-engineered or residential buildings that are built without considering essential factors in planning, such as the necessary technical requirements.

1. Introduction

The earthquake on February 25, 2022, in West Pasaman resulted in damage to 4,359 residential units with 902 minor damage, 2,906 units moderate damage, 551 units heavily damaged, and 10 people were killed, 45 seriously injured, 336 minor injuries [1]. The foreshock earthquake occurred at 08:35:51 WIT, with the epicentre on land at coordinates 0.14° North Latitude and 99.99° East Longitude, with a magnitude of 5.2 at a depth of 10 km, 18 km east of the Indian Ocean. The United State Geological Survey released that the earthquake occurred at coordinates 99,984° East Longitude and 0.222° North Latitude with a magnitude (M 5.0) at a depth of 10 km. The main earthquake occurred on Friday, February 25, 2022, 08:39:29 WIT, with the epicentre on land at coordinates 99.98° East Longitude and 0.15° North Latitude, about 17 km northeast of West Pasaman Regency, with a magnitude of 6.2 M at a depth of 10 km. After the main earthquake, several aftershocks of lower magnitude [2].

Most of the victims were caused by falling buildings or house materials. The damage to the residential/house building was caused by several factors, such as the way of building, the shape of the tread of the house, the poor quality of the housing material resulting in the construction of the house not meeting building standards so that the building was very vulnerable to earthquake shocks.

The amount of damage that occurred on the February 25, 2022, earthquake shows the magnitude of the impact caused by the disaster. Efforts are needed to reduce the impact of disasters that threaten the future. One way to reduce the impact of disasters is to strengthen the vulnerability aspect. The importance of the vulnerability assessment of house buildings in earthquake-prone areas is that it can provide information about the impacts that occur on the building due to the potential for earthquake hazards and reduce the level of damage to building structures prone to earthquakes and reduce casualties. One of the buildings that have a high level of risk for earthquakes is non-engineered buildings or residential buildings that were built without planning or application requirements.

2. Method

2.1 Study area

The research location chosen was West Pasaman Regency. This location was chosen because many people's houses were damaged during the February 25, 2022 earthquake. The damage ranged from light, moderate to severe damage. With the same earthquake strength, the difference in the level of damage is certainly caused by different vulnerabilities.

West Pasaman Regency has an administrative area with the border of Mandailing Natal Regency, Pasaman Regency, and Agam Regency, which can be seen in Figure 2.1. The location that is the focus of this research is the West Pasaman area on Mount Talamau. The epicentre of the 2022 pasaman earthquake is right under the mountain.

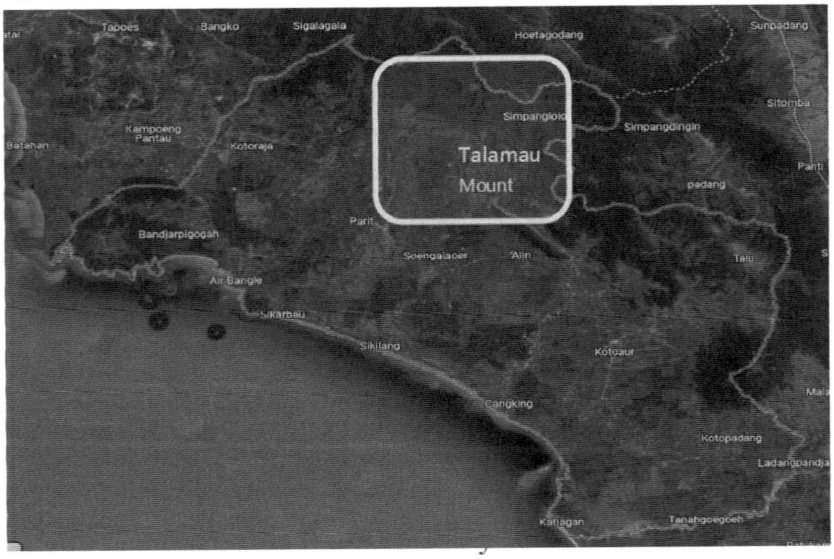

FIGURE 1. West Pasaman map for the research area

2.2 Data collection
This study uses a research method as a survey that uses primary data, including observations, recording and taking sample points in the field using rapid visual screening. In preliminary data collection, the reference guideline uses the steps guided by the modified FEMA 154 2015 Rapid Visual Screening. Secondary data was obtained from the Geological Agency, the Ministry of Mineral Resources obtained data on house damage from the West Pasaman Regency Government and the National Disaster Management Agency Team. At the same time, the determination of the level of vulnerability is obtained from the variables contained in each building indicator.

2.3 Literature Study
An earthquake occurred in West Pasaman on February 25, 2022. The foreshock earthquake occurred at 08:35:51 WIT with a magnitude of 5.2 at a depth of 10 km, 18 km east of the Indian Ocean. The main earthquake occurred on Friday, February 25, 2022, 08:39:29 WIT, with the epicentre on land at coordinates 99.98° East Longitude and 0.15° North Latitude, about 17 km northeast of West Pasaman Regency, with a magnitude of 6.2 M at a depth of 10 km. After the main earthquake, several aftershocks of lower magnitude [2]. Figure 2 shows the distribution of the West Pasaman earthquake. The earthquake impacted the number of damaged residential buildings, as seen in Figure 3—distribution of damage to infrastructure.

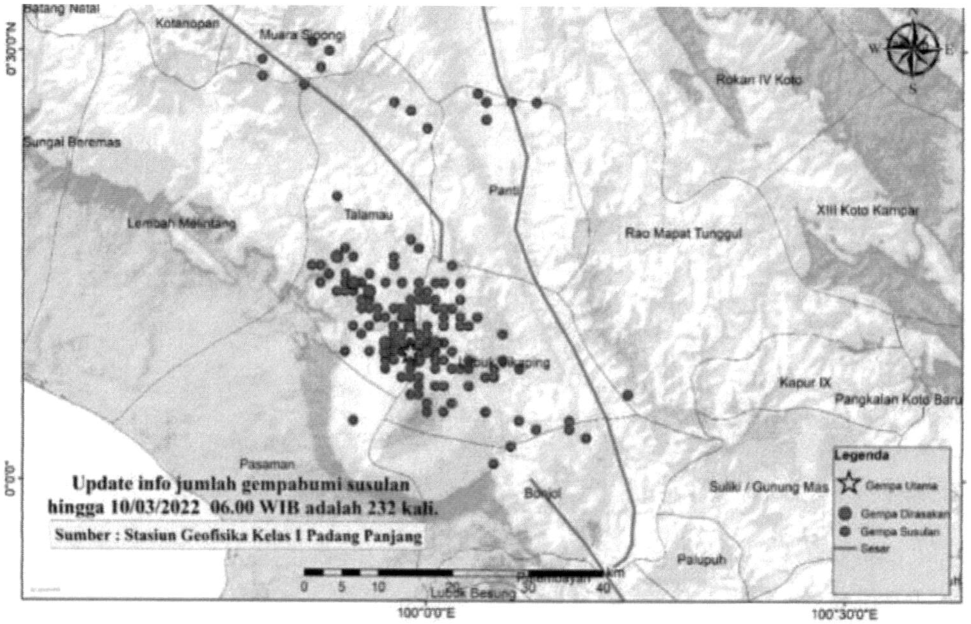

FIGURE 2. West Pasaman shocks in 2022 [2]

Earthquake-resistant buildings are buildings that can reduce earthquake energy that occurs through a combination of forces generated from the structural and non-structural parts of the building. In the event of an earthquake, the building can protect its occupants so that they have the opportunity to save themselves [3].

The philosophy of earthquake-resistant buildings is divided into 3 [4]:

a. In light/minor earthquakes, the main structure of the building cannot be damaged and must function correctly. However, minor damage to non-structural elements is still allowed.
b. In moderate/medium earthquakes, the main structure of the building may be damaged/lightly cracked but can still be repaired. Likewise, non-structural elements can be damaged but can still be replaced with new ones.
c. In a large/strong earthquake, the main structure of the building may be damaged but may not collapse completely, which aims to protect humans/building occupants to the maximum or there is still enough time for the building occupants to get out.

Anti-earthquake construction techniques aim to avoid casualties caused by building collapse during an earthquake. (a design earthquake or an ultimate limit state earthquake), reducing injuries and building damage (including its contents) caused by a moderate quake (serviceability limit state) [5].

In general, building vulnerability is the factor that can cause a building to be damaged or unable to meet the expected performance in the event of an earthquake. The expected performance is the performance of the building structure that keeps the building from collapsing in the event of an earthquake. For this reason, efforts to reduce earthquake risk (earthquake risk reduction) are needed by realizing earthquake-resistant buildings in earthquake-prone areas. By knowing the level of vulnerability of a building to earthquakes, it is possible to take mitigation steps when an earthquake occurs. This is a form of preparedness in dealing with earthquake disasters that may occur in the future [6].

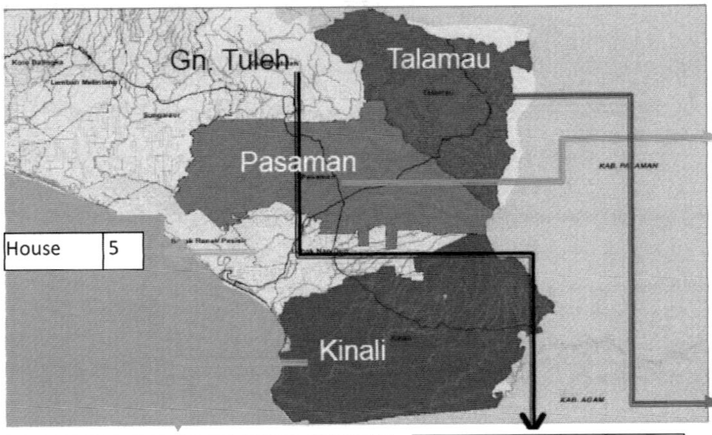

Houses	483
Education Facilities	25
Health facilities	6
Religious buildings	13
Infrastructure	2
Office buildings	23

houses	3.636
Education buildings	26
Healt facilities	10
Religious buildings	26
infrastructure	15
Office buildings	4

Houses	233
Education facilities	24
Health facilities	1
Religious buildings	1
Infrastucture	9

houses	2
Education buildings	1
Office building	1

FIGURE 3. Damage and Lost in Pasaman [1]

2.4 Result And Discussion

Based on the results of this study, the vulnerability of residential houses can be assessed by identifying the factors that influence the construction of houses. These factors have been summarized in the form of table 1 below. Furthermore, each of these factors is assigned a specific value which is part of the overall vulnerability of the building. These values are determined based on expert decisions through intensive focused discussions. Furthermore, the final sum of these values is the vulnerability of the building.
The level of vulnerability of house buildings to earthquakes is seen from the higher the final total value indicating the higher the strength of the building and the smaller the value of the house building vulnerability in the event of an earthquake. And conversely, the lower the final total value, the lower the strength of the building and the higher the vulnerable value of the building in the event of an earthquake. Furthermore, the vulnerability of buildings can be classified into 3 parts, :

a. Low vulnerability, for the value range 100 – 70
b. Medium vulnerability, for value range 70 – 40
c. High vulnerability, for values less than 40

TABLE 1. Vulnerability assessment form

Name :	Latitude :
Address :	Longitude:
Districk :	Ss Score :
	S_1 Score :

Num	Indicator	Weight and parameter	category	Class score	score	Description
1	location 30	Seismicity 15	Very hight	1		
			hight	2		
			Moderate hight	3		
			moderate	4		
			low	5		
		Soil Types 15	Soft soil	1		
			Medium soil	2		
			Hard soil, soft rock	3		
			rock	4		
2	Construction 70	Development based on planning 2,5	No	1		
			Yes	2		
		Symmetrical Plan 2,5	No	1		
			Yes	2		
		foundation 7.5	No	1		
			Deficient	2		
			Yes	3		
		Plinth Beam 7.5	No	1		
			Deficient	2		
			Yes	3		
		Steel binder fondation with plinth beam 5	No	1		
			Deficient	2		
			Yes	3		
		Column 12,5	No	1		
			Deficient	2		

	Yes	3
Walls **5**	No	1
	Deficient	2
	Yes	3
Steel binder walls with column **5**	No	1
	Deficient	2
	Yes	3
Top beam **7.5**	No	1
	Deficient	2
	Yes	3
Reinforced concrete frame **5**	No	1
	Deficient	2
	Yes	3
Roof truss construction **5**	No	1
	Deficient	2
	Yes	3
Steel binder top beam with plinth beam **5**	No	1
	Deficient	2
	Yes	3
Total score		

2.5 Conclusion

The earthquake-resistant house buildings are buildings that are able to reduce earthquake energy to the building. However the real application many building have vulnerability due to the earthquake. This paper proposes the method to assess the house vulnerability based on several important factors that influence the construction of houses. These factors have been assigned a certain value which is part of the overall vulnerability of the building. These values are determined based on expert decisions carried out through intensive focused discussions. Furthermore, the final sum of these values is the vulnerability of the building.

Acknowledgements

Authors would like to thank the head of the *Post Graduate School of Andalas University* for the support for this research. The author also thanks to any parties who helped finishing this research.

References

[1] Data Center. 2022. Overview of Refugee Data, Patients and Damage Due to the February 25 Earthquake in West Pasaman District. https://pasamanbaratkab.go.id (Opened 28 Agust 2022).

[2] Geological Agency, Ministry of Energy and Mineral Resources, Republic of Indonesia, https://vsi.esdm.go.id/index.php/gempabumi-a-tsunami/laporan-singkat-dan-rekomendasi-teknis/3969-laporan-singkat-tanggap-darurat-bencana-gempa-bumi-tanggal-25-februari-2022-di-daerah-pasaman-barat. (Opened 28 Agust 2022).

[3] SNI 1726. 2012. Earthquake Resistance Planning Procedures For Building And Non-Building Structures. Jakarta, National Standardization Agency of Indonesia.

[4] Satyarno, I. 2009. Chapter: Some Practical Aspects in the Post Yogyakarta Earthquake Reconstruction of Brick Masonry Houses, The Yogyakarta Earthquake of May 27, 2006, Edited by Karnawati D., Pramumijoyo, S., Anderson, R., Husein, S. Star Publishing Company, INC, Belmont, CA.

[5] Sulendra, I Ketut. 2010. Evaluation And Building Hazardous Reduction Action Based 2010 Earthquake Zone Map. Palu, Tadulako University. Infractructure Vol. 1 No. 2 December 2011:71-78.

[6] Perdana, Intan Putra. 2017. Evaluation Of The Vulnerability Of Community House Buildings To Earthquakes In the Bugisan Tourism Village, Prambanan District, Klaten Regency. Thesis. Yogyakarta : Gadjah Mada University.

AUTHOR INDEX